Materials and Manufacturing

About This Book

This book is meant to provide an introduction to materials, their properties, and common methods to make useful items from them. Materials and manufacturing are very deep topics, and a brief introduction is somewhat like introducing someone to the ocean by going for a boat ride. It is informative and meaningful, but ultimately you only experience the surface. There is so much more waiting just beneath the surface. The trouble is that if you dive in too enthusiastically, it is easy to get in too deep. The goal with this text is to find a balance between understanding the basics while appreciating the complexities. This textbook is aimed at readers without significant prior knowledge, or those who may not have understood the topics in previous studies. To keep the length manageable, some topics could not be fully explored, but hopefully it inspires you to go further in areas you find interesting.

You might find the writing in this book is different than in most others. It is meant to be read, so it is presented in a conversational tone, much like it would be verbally taught. The intention is to make the reader feel engaged and to explain and clarify the expected points of confusion. My hope is that it feels like you are being taught the material, not simply having facts and equations listed for your consideration. The scholar will find the descriptions too imprecise, but this text is not designed to be fully quantitative. The goal is to be accessible. If your career requires you to get more quantitative or to develop your knowledge in a specific field, I suggest using this book to set a foundation and another, more focused book to build upon that foundation. What is presented here is an attempt to combine my passion for manufacturing and materials with straightforward and engaging discussion and to transfer both the knowledge and excitement to the reader. For both of us, I hope I am successful!

Materials and Manufacturing

An Introduction to How They Work and Why It Matters

Mark A. Atwater

Mc Graw Hill Education

New York Chicago San Francisco Athens
London Madrid Mexico City Milan
New Delhi Singapore Sydney Toronto

Library of Congress Control Number: 2018936320

Materials and Manufacturing: An Introduction to How They Work and Why It Matters

1 2 3 4 5 6 7 8 9 QVS 23 22 21 20 19 18

ISBN 978-1-260-12231-2
MHID 1-260-12231-X

Sponsoring Editor
Summers Scholl

Editorial Supervisor
Donna M. Martone

Acquisitions Coordinator
Elizabeth Houde

Project Manager
Jyotsna Ojha,
Cenveo® Publisher Services

Copy Editor
Surendra Shivam

Proofreader
Raghu Narayan

Production Supervisor
Lynn M. Messina

Composition
Cenveo Publisher Services

Art Director, Cover
Jeff Weeks

Contents

About the Author

Mark A. Atwater, Ph.D., is an associate professor at Millersville University, where his areas of expertise include materials and manufacturing theory and practice. He has an undergraduate degree in manufacturing engineering technology and graduate degrees in mechanical engineering and materials science and engineering. He merges these fields in research on the development and production of nanostructured materials.

Acknowledgments

This book is the outcome of years of preparation, learning, and effort, and I am grateful to many who have aided that journey along the way. I am fortunate to have the opportunity to give my take on these important topics, and I am equally fortunate to have the students who inspired it, the facilities and equipment that are featured, and the colleagues who supported me at Millersville University. I believe one of the defining characteristics of this work is the originality of the figures, of which more than 500 were created to complement the text. As someone not specifically skilled in art, it was a massive undertaking to create the numerous drawings and other images. That is why I am particularly grateful for the work of Christopher Smith, an undergraduate student in Art and Design, who assisted with capturing many of the photographs. The best ones are undoubtedly his.

I am also grateful to the individuals and companies who assisted with, created, and/or granted permission to use other images and information for this text. These contributors include:

- Cindy English, MFA
- Lance Nelson
- Brian Monteith and Blaine Bracelin (American Keg)
- Rick Valverde and Henry Trabal (AMES)
- Matt Dabbo (Atlas Cutting Tools)
- Jeff Ostman (IHI Ionbond AG)
- Judy Bass (Industrial Press)
- Bill Young (Kent USA)
- Lisa Johansson (Koenigsegg Automotive AB)
- James Deisley and Jeffery Heger (Millersville University)
- Chief Pete Anders and Sgt. Jason Flood (Millersville University Police)
- Elena Howlin (Oerlikon Metco)
- Robert Ehrmann (Penn State Center for Nanotechnology Education and Utilization)
- Rick Hendricks, Keith English, and Tom Livingstone (Pennsylvania College of Technology)

- Nikki Stokes (Sandvik Coromant)
- Dennis Sharples (Sharples Machinery)
- Evan Syverson (Sodick)
- Jonathan Lim (Structo)
- Beau Kibbey (Superior Tire and Rubber)
- Andrey Rudenko (Total Kustom)
- David Birunas and Michael McGarry (TRAK Machine Tools)
- Bruce and Andrea Van Sant (Van San Distributing/MetalAce)

Finally, I am most grateful to my wife, Cammie. This book is just one of many crazy endeavors she has supported me through, and I am deeply grateful to have her as a companion.

Materials and Manufacturing

CHAPTER 1

Considerations in Materials and Manufacturing

1.1 Introduction

In Daniel Defoe's novel, *Robinson Crusoe*, the title character states, "I had never handled a tool in my life, and yet in time, by labour, application, and contrivance, I found at last that I wanted nothing but I could have made, especially if I had the tools ..." Crusoe was stranded on an uninhabited island with limited resources and in need of basic shelter and provisions. He spent a great amount of time and "infinite" labor concocting methods to make rudimentary items such as boards, wicker baskets, and ceramic vessels. He was able to live comfortably, even to the point of being pleased with his lot in life, because he could manufacture the goods he desired. Often the processes he employed were extremely slow and labor-intensive, but one thing he had plenty of was time.

In our modern society, it is no less important that we are capable of providing products to make life more comfortable, convenient, and prosperous. The challenge we are faced with is much different than Crusoe's. Whereas he had limitations regarding tools and skill, he had an abundance of time. We have highly advanced tooling and skilled labor but must reduce the time investment as much as possible. To maintain quality and improve efficiency is the goal of nearly all manufacturing processes. Consumers are not inclined to pay more for something that does not add value to the final product. The challenge facing manufacturing now is to decrease cost and increase value. This is not a trivial problem.

1.2 How It Works

How is this balance of lower cost and higher quality achieved? There are many methods to reach this goal. One way is through executive changes, such as lean manufacturing, continual improvement (Kaizen), statistical control (Six Sigma), and design engineering (improve manufacturability, use lower cost materials, redesign products to use less material, etc.). We will look primarily at process-level methods in manufacturing. In order to make intelligent decisions about what techniques and materials to use in producing a product, it is important to understand the materials and processes themselves. That is where we will build our foundation.

By starting from the basic principles of materials and manufacturing processes, we can make informed decisions regarding any future changes in the industry. Specifically, if we understand the way materials respond to temperature, pressure, and environment, we can judiciously choose the process or processes most appropriate for creating a specific product. Conversely, if we know what our manufacturing capabilities are, we can determine what materials we can readily work with and what order of processing to use.

This knowledge is highly transferable. By knowing the underpinnings of materials and processes, you are able to apply that in any field whether it involves casting steel pipe fittings, injection molding plastic cups, or machining titanium implants. This text is not written so as to make you an expert in any one material or process. Rather, it is designed to provide enough depth for you to be immediately effective in any manufacturing field. Of course, learning does not happen all at once and end there. It is likely that you will be or are even currently required to increase the depth and breadth of your knowledge further. By the end of each chapter, you should be able to explain *how it works*, how the material is useful, how the process is accomplished and/or how the parameters of processing (temperature, force, etc.) affect the interaction of the material and equipment. Based on knowing how it works, you should then be able to determine and describe why you would or would not apply it in a given situation. That is, *why it matters*. Interactions between the process and the product must be understood before they can be controlled and exploited. We will start by looking at this from a historical view point.

1.3 Historical Perspective

To illustrate the importance of materials and processes, we will take a moment to consider milestones in manufacturing and materials advancements. Perhaps you already know that the periods of civilization are defined by the materials in use during that time, such as the Stone Age, Bronze Age, etc. In absence of any other evidence, the fact that materials have continually defined the progress of civilization should be convincing proof of their importance. Materials and processing have developed together and have enabled one another. Advances in one area allowed advances in both areas. This relationship has been critical in allowing societies to thrive in the past, but they are still relevant to our progress today. We will start our investigation into this importance with the oldest of the "materials" ages*: the Stone Age.

1.3.1 Stone Age

The Stone Age (approximately 3300 B.C. and earlier) is typically divided into Paleolithic, Mesolithic, and Neolithic eras, but taken as a whole, we can consider the reason it predates others. Why is stone, the first of the historical materials† periods? When we talk of materials, there are two major divisions that can be made: naturally occurring and those that must be processed before being available. Stone is naturally occurring, so there is

*Materials ages overlap and are generally reported to start/end at different times. The years identified are for illustrative purposes, but they are generally accepted.

†"Materials" is often plural even if it seems like it should be singular. For instance, the study of materials is known as *materials science*, not material science.

no processing to form the material and little processing to create tools from it. To form tools, two rocks could be hit together, a very simple manufacturing procedure. By choosing different types of stone, one would be preferentially broken. This could be applied to materials such as flint, which will cleave to form a sharp edge so that arrow heads and cutting tools could be made.

In addition to stone, other "natural" materials such as wood, bone, and animal skins can be used to form tools and products. Because these materials are readily available does not mean that production in the Stone Age was trivial. Very much the opposite, without the advantage of precision machinery to produce items, there was a great need for craftsmanship in producing reliable products. Some examples are shown in Fig. 1.1.

1.3.2 Bronze Age

The Stone Age was characterized by using naturally occurring materials, but a transition to more advanced processing occurred in the Bronze Age (approximately 3300–1200 B.C.). During this period, the ability to process raw material into different, more useful forms was exploited. Copper ore, a rock and ceramic, was converted to a more useful, metallic form through a high-temperature process called **smelting**. Pure copper is rather soft, and to increase the strength, it may be alloyed (mixed, see Sec. 3.2.4 for more) with other metals such as tin. The hot metal can be formed readily at high temperature and this allowed for creating tools and weapons which were not brittle like stone, but still strong enough to survive repeated use (Fig. 1.2).

Here we see the change from using native materials to material design where specific elements were combined to create new, more useful structures. This material conversion requires much more sophisticated processing and measurement and created a precedent for materials research that is still ongoing.

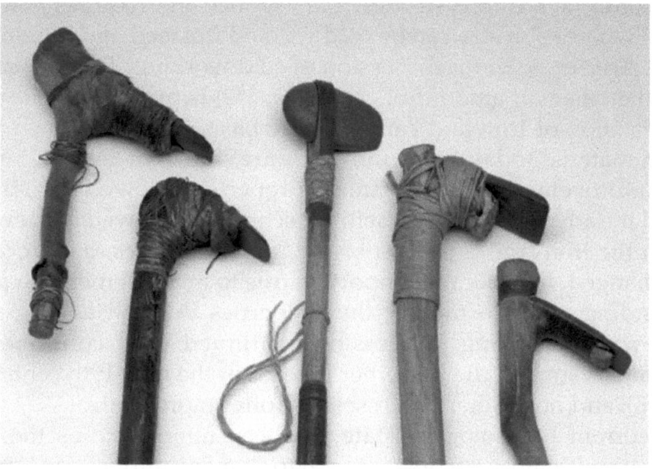

Figure 1.1 Tools were originally made from natural materials such as stone, wood, bone, and skins. These primitive tools were essential but very limited in their use.

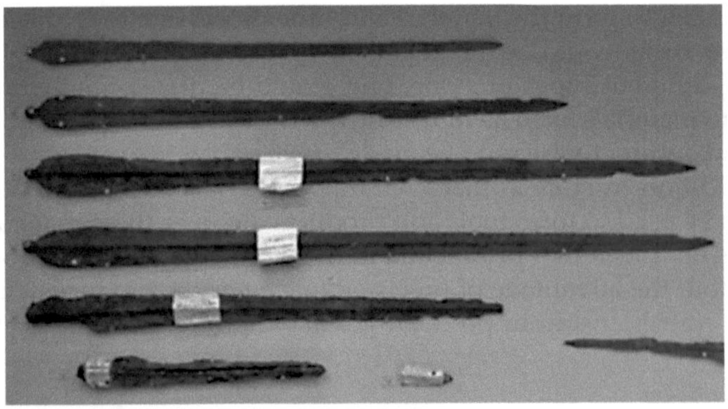

Figure 1.2 Advances in the Bronze Age allowed for durable tools to be made. Bronze is less brittle than stone (its toolmaking predecessor) and can be more easily shaped, allowing for intricate tools to be developed.

The invention of bronze and the utilization of smelting allowed for tailoring tools with specific properties. The amount of tin added to copper will change its strength and therefore durability. To create uniformity in the tools and weaponry, the process would have to be repeatable. No doubt much experimentation and testing would be required to perfect the process and products, but that design approach continues today.

1.3.3 Iron Age and Beyond

The next stage of materials development comes in the Iron Age (approximately 1200 B.C. to 50 A.D.), where the practices of the Bronze Age were carried over to new materials. Specifically, iron and steel (a mixture of iron and carbon) were brought into prevalent use. The smelting conditions required for iron ore are more exacting than for copper-based materials such as bronze. The temperature must be higher and the atmosphere must be controlled. Iron is stronger than bronze and requires hot working (e.g., blacksmithing) whereas bronze can be cold worked (formed and hammered near room temperature). See Sec. 9.2.1 for discussion of cold working. The major advantage with iron is that, when a small amount of carbon (< 1%) is present, the iron is greatly strengthened. This alloy of iron and carbon is the basis for steel, which is the most common structural material today. Some iron tools are shown in Fig. 1.3.

The improvements in mechanical properties allowed for stronger weapons and tools, and the advancement of smelting techniques allowed for new alloys to be created. Although the Iron Age is the last of the "material" ages, the process for improvement has not changed. Iron became important due to improvements in processing and better understanding ways to modify the properties through adding secondary elements, now known as alloying. Processing continued (and continues) to improve with advancements such as the Bessemer process in the mid-19th century (Henry Bessemer, 1813–1898), and now other processes that offer more control.

Our current technological state has been referred to as the Silicon Age. That is because computing has become integral to so many applications, and much of the hardware that makes it possible is based on the element silicon. There is little doubt silicon will be replaced eventually and a new age will reign. Even if silicon still plays an

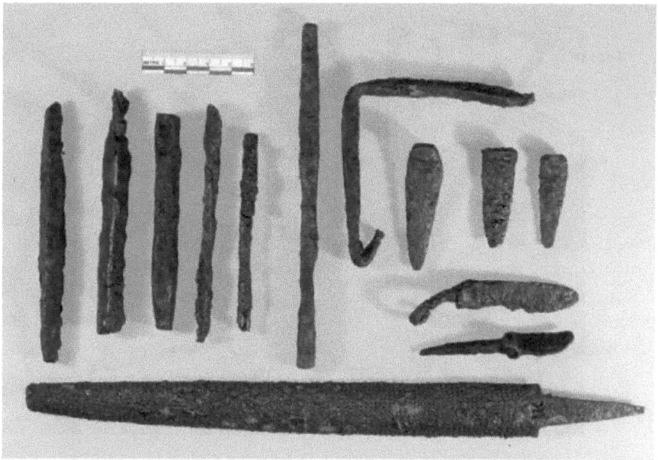

FIGURE 1.3 The development of iron tools allowed high-strength weapons and tools to be made. These held a sharp edge better than bronze and were more durable.

important role, it will be in support of new materials which enable the boundaries of current technology to be surpassed.

Transitions from age to age are not immediate and complete. The progress of processing and materials went hand in hand and occurred incrementally. Stone was used even after bronze was invented. Bronze continued being used even when iron gained popularity. Indeed, materials are developed and modified and can be perpetuated along in some capacity forever. There are a limited number of elements (about 92 naturally occurring*) that make up every single material we have and will use. Even fewer (closer to 10) are common structural materials. By combining materials in a thoughtful way, we can achieve "ideal" balances between strength and weight or other properties. Even advanced composites, which have allowed for space exploration and exceptional performance under extreme conditions, are not new. The ancients understood that mixing clay and straw together resulted in a superior material, which we now classify as a composite (see Sec. 6.4 for other examples). They constructed magnificent structures from this "advanced" composite long before any materials were studied scientifically, at least by our modern standards.

New research is revealing that the periodic table (see Fig. 2.1) is three dimensional, so to speak, and the properties of materials are size dependent. This extends the possibilities well beyond our historic applications. The nanoscale version† of a material can respond differently to heat, light, electricity, etc. The melting point, color, reactivity, and strength can change drastically at the nanoscale. For example, stained glass (Fig. 1.4) produced in antiquity was colored by mixing in metals, which formed nanoparticles that appear as different colors (e.g., gold can appear red). Although they did not understand the exact reasons, craftsman knew how to control the outcome. We now know much more about these size-dependent properties. They present new tools for the engineer, but they also present new challenges in controlling

*This number is debated depending on how "naturally occurring" is defined.

†"Nanoscale" refers to 1 billionth of a meter, or about 1000 times smaller than the thickness of a hair.

FIGURE 1.4 The properties of materials are size dependent. For instance, the color of gold changes based on size, and gold and other metals have been used to dye stained glass for centuries thanks to this property.

and manufacturing these materials. As society advances, wood and stone are still popular and appropriate in many applications and bronze and iron are now better understood and applied than ever before, but the importance in continuing to advance materials and manufacturing lies in the benefit that can *still* be realized today and tomorrow.

1.4 Present Challenges

As mentioned in the introduction, there is a grand challenge in manufacturing of maximizing quality while minimizing cost. We briefly looked at the progress of materials and the basic concepts of the processing required for their production, but there is much left to be said about the current circumstances of production technology. Historically, advancements have significantly shifted the way products are made and what is possible. This includes technology and methods such as interchangeable parts, mass production, and automated manufacturing. These took manufacturing from being a slow, custom, skill-intensive business, to be a rapid and universal process. This forever changed the way products are made, but is not applicable to all sectors of manufacturing. That is to say, there is no one-size-fits-all solution to achieving the cost/quality balance.

Manufacturing is highly competitive. This competition exists locally, regionally, nationally, and globally. The decision to use one company over another is a complex mix of considerations about costs for labor and shipping, production rate and capacity, quality constraints, etc. For large orders where high volume and **continuous manufacturing** will be used, the equipment tends to be a principal cost. Highly specialized, equipment designed for high-rate production is often used. The specialization of the equipment reduces the need for **skilled labor**, but also reduces flexibility (i.e., the equipment is dedicated to one specific job and has little use other than that).

For **medium-** to **low-volume manufacturing**, the equipment becomes more generalized and the labor must be more highly skilled. The equipment can often be used to produce many different types of products, but must be carefully setup and qualified when the process is changed. This is where the skill is required to ensure that quality is maintained. The trade-off here is that there is higher cost associated with labor and the down time while the machine is being converted to run another product. In this type of processing, scheduling becomes critical and a host of considerations must be made to ensure efficient process flow is maintained.

In any type of manufacturing, material cost is another important aspect. Following customer design specification such as strength, weight, dimensional tolerance, etc. is required, but the way in which the specifications are met can vary. For example, assume a structural member can be made from hollow steel or solid aluminum and still meet weight and strength requirements. Which one will you use? Where possible, cheaper material and easier processing will always win out. This is assuming there is no explicit design requirement on the material. For example, if the component cannot be magnetic, aluminum would be required since steel *is* magnetic. If there are few constraints, it is more a matter of whether it can be processed using the equipment currently available or that can be purchased for a reasonable cost.

Typically, the steel component will be cheaper from a material perspective, but what other considerations remain? Processing steel and aluminum will proceed quite differently. Take machining for example (see Chap. 13 for more on machining). Aluminum is a softer material and generally can be machined at a faster rate and cause less wear on tooling. The steel equivalent will end up costing more in supplies (tooling) and time (machine cycle time and operator wages). The difference between material and processing cost must be established to determine the more cost-effective choice.

There are many process-oriented variables that must be weighed, but other aspects of service or design requirements must be considered as well. Is there a specified service temperature (hot or cold)? Is this component likely to be loaded and unloaded repeatedly? How much will that load be? Is the service environment corrosive? What is the expected lifetime of the product this component will go into?

The steel can withstand elevated temperature more readily, but the aluminum will tend to corrode less in most environments. Steel performs better under cyclic (repeated) loading, known as fatigue (see Sec. 7.6.5), but aluminum may be considered more aesthetically pleasing (do not underestimate this aspect). Either material can be treated to perform better in any circumstance or a different grade of the material can be selected.

This is only the tip of the iceberg when considering a material and process! Machining was mentioned, but similarly, forging, casting, welding, or any other process can run through a similar set of questions to determine which material is the best to process by which technique. The designer's job (possibly *your* job) is to know how to make these decisions. This book is intended to help you get started.

The challenge in processing and material selection is to meet both the desired specification (customer requirements) and the company's interest (minimizing cost and maximizing profit). There is global competition for manufacturing, and to win in that game, you must find the route by which to maximize value. Choices, such as automating processes, ultimately come down to whether it will eventually improve the company's competitiveness and profit margin. A company's most valuable investment is in skilled employees capable of identifying methods to achieve that end. The goal of this text is to increase your knowledge and versatility in this competitive environment and to help you identify the best practice in processing.

1.5 Future Considerations

After considering where materials and processing originated from, the question arises, "Where do we go from here?" Advances in manufacturing commonly take place by improving rate and repeatability. This is often accomplished by automation of processes. Faster computers, more accurate controllers, and improvements in tooling have all allowed for processes such as high-speed machining, robotically controlled work cells, computer-integrated manufacturing (CIM), rapid prototyping, advanced casting techniques such as directional solidification, etc. (see Chap. 22 for more on advanced manufacturing). Any process can benefit from being both faster and more accurate.

With regard to materials, the push for new breakthroughs is equally alive. The demand is for stronger, lighter, and cheaper. Advancement in composite materials has allowed a whole host of new, lightweight structures to be designed. Biologically inert or active materials are allowing for quicker healing, better performance, and longer life from medical devices such as artificial heart valves and joint prostheses. New alloys enable new engine technologies and more efficient operation. Add to this the profound changes that have arisen due to advances in semiconductor processing. Advanced electronics has driven the technological advancement in just about every field.

Research to improve current processes and materials is largely in the details. By using computer simulation, processes can be modeled and studied without ever touching a production machine. The vast amount of data that can be collected from simulations can be cost effective in determining the feasibility of new processes. This simulation process can be extended to materials also. New alloys are being developed at the atomic scale, and new processes are being employed to surpass traditional "limits" in alloying, strength, and other properties.

Nanotechnology, the study and manipulation of materials at the atomic or molecular level, is being employed to create new materials with unique properties and holds promise to advance our society to a level unfathomable at present. The difficulty in making these "nano" materials and devices is that nanofabrication or nanomanufacturing techniques must be used to make them, and these techniques, broadly speaking, are in their infancy.

To put nanoscale processes and materials into perspective, we will compare them to common objects. For example, a human hair is approximately 80,000 nanometers (nm) across, and a red blood cell is about 5000 nm across. There are 1 million nanometers in 1 millimeter, and 25,400 nm in one *thousandth* of 1 inch (in) (0.001 in). Nanoscale materials and technologies must be controlled at scales significantly smaller than cells. The challenge with these advanced techniques and materials is to control the quality and repeatability and to bring them to production rates of traditional manufacturing. Though many exciting technologies and materials can be demonstrated on a small, laboratory scale, transferring the techniques to a level that would make them feasible for consumer-level distribution is not trivial. Imagine machining gears smaller than a dust mite (Fig. 1.5). This presents an entirely different manufacturing challenge where nanoscale devices and phenomena must have production scaled up reliably and cost-effectively. This is the reality of manufacturing. Challenges do not tend to get easier with time. New techniques, not commonly associated with production manufacturing, are used to meet the challenge. Solving unconventional problems requires unconventional thinking. Innovative approaches must be applied to take these cutting-edge technologies from science to society.

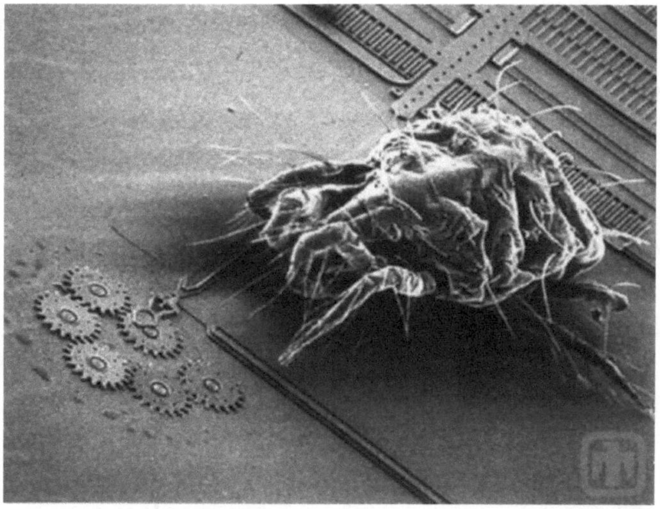

FIGURE 1.5 Dust mite approaching micrometer-sized gears (1 μm = 0.00003937 in). The current stage of society's development is largely due to silicon processing which allows for microprocessors and miniaturized technologies to exist, but other developments in materials and processing are being made concurrently. (*Photograph from Sandia National Laboratories, SUMMiT Technologies.*)

1.6 Why It Matters

Technology tends to benefit from itself. As manufacturing advances, materials can advance, and as materials advance, equipment for manufacturing can also advance. It is a self-perpetuating cycle. Technology is not simply marching forward. The pace of technological development is gaining speed. The last century has seen more scientific advancement than all of the preceding centuries combined.

Often people do not see the science behind technology; they simply see the result. This can be thought of as a growing tree (Fig. 1.6) where the root system is scientific understanding and technology (the visible result) is the trunk, branches, leaves, etc. As the understanding of fundamental scientific laws has advanced, the roots have grown deeper. This has allowed for technologies to be developed, first as the foundation (trunk), but later as more advanced off-shoots (branches) which have spurred a number of improved devices and techniques (the leaves). As the roots grow deeper, they can support the above-ground development, which gives shade (ease of living) to the societies which harness them. This, of course, is assuming that we can responsibly apply the technology we create, which is not always the case.

What is important to realize from this is the interconnectedness of the fundamental understanding of materials and processes and the products they facilitate. Discoveries and improvements in either materials or processing work to advance both. Processing is consistently happening at higher speeds, stresses, temperatures, etc. More extreme conditions demand more extreme materials. Materials and manufacturing are truly inseparable. It is in that disposition that we can begin unraveling the exciting facets of each.

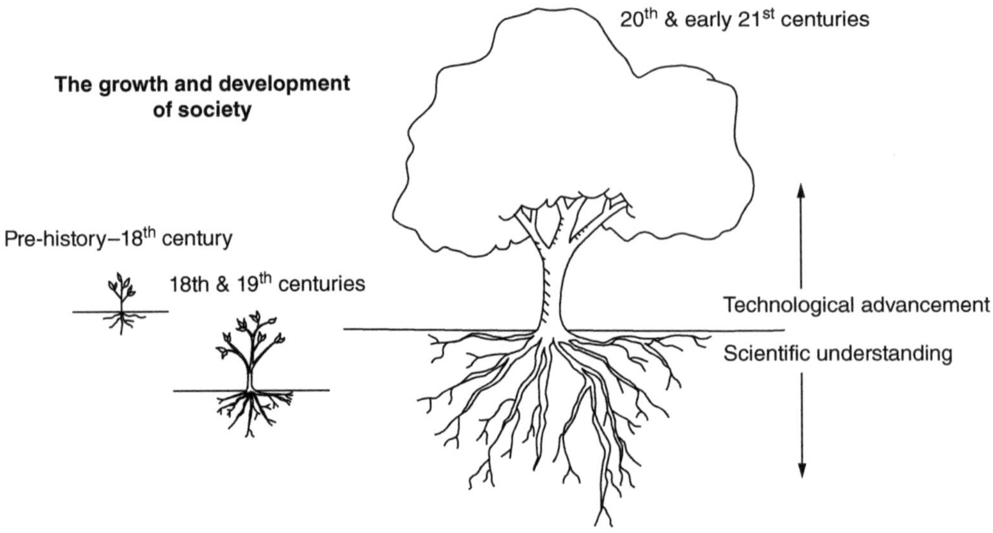

FIGURE 1.6 Society, in general, has benefited from advanced technology. That technology is generated through scientific study. Although important advancements were made in the early centuries, the 20th and 21st centuries have greatly outpaced previous ones. (*Illustrations courtesy of Cammie Atwater.*)

1.7 The How and Why of Considerations in Materials and Manufacturing

1.7.1 Chapter Review

1. Robin Crusoe, much like the earliest societies, had no formal training in manufacturing. He developed processing methods by trial and error. What limits does this put on designing new products and processes? What negative consequences might arise?
2. Natural materials were the foundation for technological advancement in early civilization. What are some natural materials in use today?
3. Stone is a common material, still in wide use today. Name three applications it can be found in.
4. What are two benefits and two disadvantages of creating stone tools by hand?
5. Bronze is made primarily of which element?
6. What are the three advantages of bronze over stone? What is the disadvantage?
7. How was the development of iron and steel tools an advantage over bronze?
8. Steel is a mixture of which two elements?
9. Steel is the most common structural material. Name five applications of steel.
10. What is the "grand challenge" in manufacturing?

11. As the number of products decreases, how does the manufacturing skill required change? What about the equipment used?
12. When making design decisions, how can you cut material cost?
13. Name three differences between processing steel and aluminum.
14. Name two active areas of material advancement.
15. Name two active areas of manufacturing development.
16. Give a general definition of nanomanufacturing. What is a fundamental problem?
17. Science and technology go hand in hand. Which has more of an impact on society? Which is more important?
18. Name an important technology. How has this technology been affected by advances in manufacturing and materials?

1.7.2 Concept Connections and Extensions

1. Is there value to knowing how things are made? Should the average consumer be concerned about it? Section 22.5 might help generate some ideas.
2. Is newer always better? Consider the phrase, "They don't build them like they used to." Is that a good thing or not?
3. Why is silicon considered the current age of material development by some? What properties make it useful? Sections 6.2.4.1 and 22.4.2.1 will be helpful.
4. Why is education in manufacturing becoming more important? Consider the changes in how we produce goods (e.g., Sec. 22.3). What roles do skilled and unskilled labor play?

CHAPTER 2

Concepts of Materials

Everything we touch, taste, smell, feel, and see (and much we do not see) is made of matter. **Matter** can be roughly described as anything that has mass and volume. That means this book, your brain, and your neighbor's cat all have something in common. By understanding the connections and shared properties of matter, we can make informed decision about processing and material selection. Even if you are not familiar with a specific material, if you know the properties of similar materials, or materials within the same class, you can make reasonable assumptions about other materials. Of course, there are limitations, but we can take advantage of the many shared characteristics of materials.

2.1 How It Works

Matter is generally described by reducing it into classifications. The most common classification is the state of the matter. The state of matter is typically broken into three subsets: solids, liquids, and gases. Most engineering materials are solids, so we will spend our time discussing the properties of these but do not neglect the other states. Processing commonly employs hydraulic (liquid) and pneumatic (gaseous) components, and more about the way these function in processing can be found in texts on fluid power. The form of matter depends on the element or combination of elements.

Elemental materials have the simplest structure because they are made from one type of atom. Examples of this include aluminum and iron, which are both common engineering materials and elements. Even though aluminum and iron are common materials, they are rarely used in pure form. As will be discussed in Sec. 3.2.4, they are more commonly mixed with other elements to create new materials, such as high-strength low-alloy (HSLA) steel, which we will build up to understanding.

Within the solid state, we find many more subdivisions. The periodic table, which arranges the elements into useful categories, is commonly divided into **metals, nonmetals**, and **metalloids** (an intermediate between metals and nonmetals). If you look at the distribution between nonmetals, metals, and metalloids in Fig. 2.1, you will see that metals dominate the periodic table. Except for astatine (At), which is a metalloid, all of the unstable and synthetic elements which have been created or observed are metals as well. There are a very large number of metallic elements and relatively few metalloids and nonmetals. How, then, do we explain that so many products are nonmetals (e.g., plastics, glass, wood, etc.)? The answer is that nearly all the materials we use are not in elemental form. Instead, two or more elements are combined to make them.

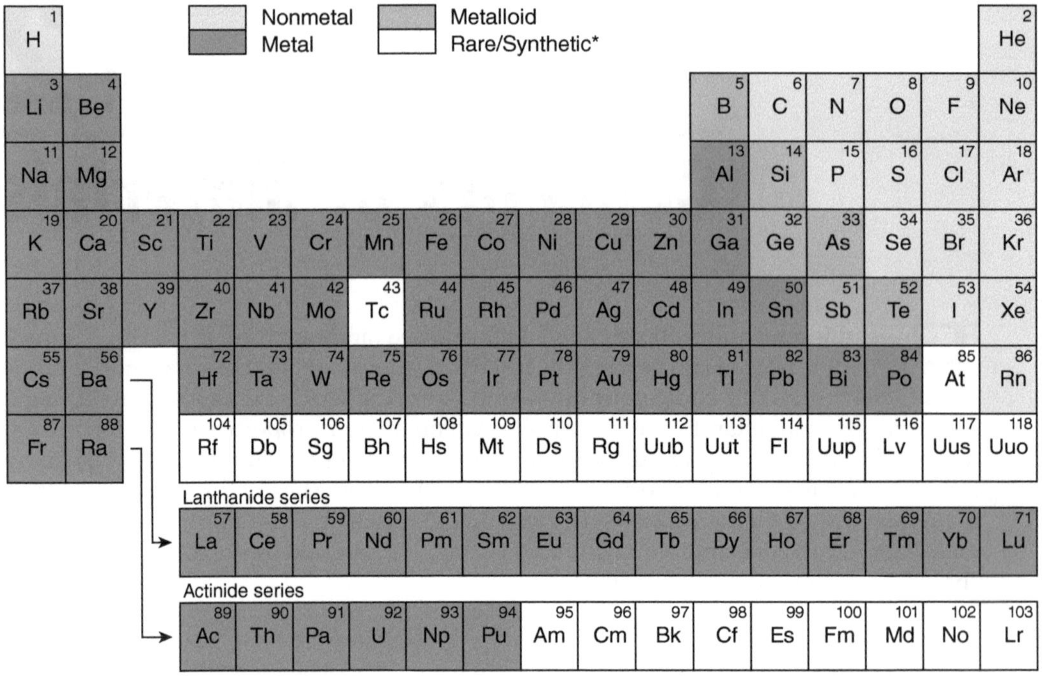

FIGURE 2.1 Periodic table of the elements showing those which are metals, nonmetals, and metalloids. It is apparent that most elements are metallic, but not all behave similarly.

In many instances, the combination of elements may involve a metal, but the final product is a nonmetal. These nonmetallic compounds or molecules will be described more in Chap. 5. The relative abundance of an element is also very important. Just because there are many types of metals does not mean there is a large supply. Out of the 92* or so naturally occurring elements, just 3 elements, carbon, oxygen, and hydrogen, which are all nonmetals, make up the majority of compounds critical to life. Likewise, a small number of metallic elements (about 10) are the basis for most engineering materials. Although there are many elements, we can focus our attention on a much smaller group.

Identifying nonmetals and metals does not provide us with much knowledge of the actual materials. It is much more useful to know *why* a material is a metal or nonmetal. What makes the difference? Referencing chemistry texts or materials science texts (a good idea!) will reinforce that the properties of a material are directly related to the electron configuration of the atoms, compounds, or molecules that make up the material. Although we will not concern ourselves with the great depth of information on atomic theory, it is important to get a basic understanding so we can apply it to the many materials we will be discussing in subsequent chapters.

---

*92 is a number commonly cited, but some would argue for 88 and others for 94. It depends on what is considered naturally occurring, either by whether it exists in nature in measurable quantities, or whether it *can* exist in nature, even for very brief periods.

2.2 Atomic Structure

Atoms are extremely small. For instance, if you were to stack copper atoms on each other until they were an inch high, it would take about 100 million atoms. In 1 g of copper (the weight of a dollar bill), there are about 13.5-million billion atoms. Even though atoms are so small, they are made up of even smaller parts. The structure of an atom is often depicted like the solar system, where the sun is circled by planets. In atoms, the nucleus is circled by electrons. This simple depiction is known as the Bohr model (after Niels Bohr, 1885–1962), and though it is not strictly accurate, it will suffice for helping us understand materials. This model is shown in Fig. 2.2.

As shown in Fig. 2.2, atoms are made of electrons, protons, and neutrons. The nucleus, around which the electrons revolve, is made of the protons and neutrons. The number of protons determines the type of atom. For instance, iron atoms have 26 protons, oxygen has 16 protons, lead has 82 protons, and so on. All iron atoms have 26 protons, but may have different numbers of electrons or neutrons. A difference in the number of electrons or neutrons is the basis for **ions** and **isotopes**, respectively, but for the purpose of this text, we will assume that the number of protons, neutrons, and electrons are the same (a fairly good assumption). Specifically, iron would have 26 protons, 26 neutrons, and 26 electrons. The number of protons in the atoms of an element is also called the **atomic number**, so the atomic number of iron is 26.

The mass of a material comes primarily from the protons and neutrons in the nucleus. Lead atoms, which have 82 protons, weigh much more than iron atoms. Lead is a very heavy material, but it is not only because the atoms are heavy. The way the atoms orient themselves will also affect the density (weight per volume) of a material, and this will be discussed in Chap. 3. Although the density tends to increase with increasing atomic number, it does not strictly follow that pattern. For example, copper

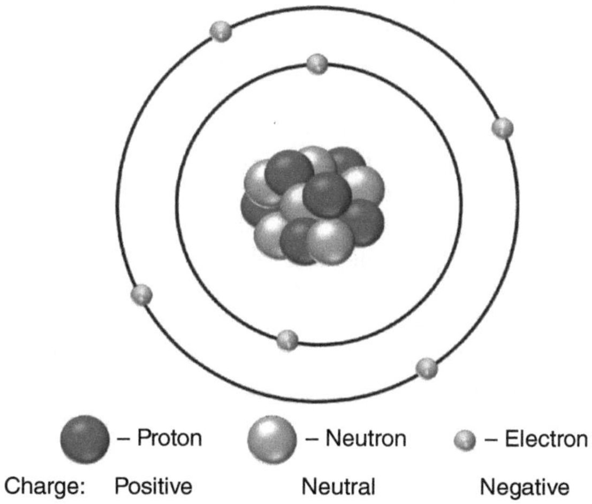

FIGURE 2.2 Simplified model of a carbon atom: atomic number 6 and 6 each of protons, neutrons, and electrons. Negatively charged electrons encircle a nucleus of positively charged protons and neutral (no charge) neutrons. In reality, the shape of an atom is complex and three dimensional.

(atomic number 29) has a higher density than tin (atomic number 50) because the copper atoms are packed closer together.

Another important aspect of an atom's structure is the distribution of charges. Protons are positively charged, neutrons have no charge, and electrons are negatively charged. While this has many important effects on the positions of electrons and the properties of atomic interactions, the major consideration in our case is the way it affects material properties. In the stable state, there are equal numbers of electrons and protons so that each negative charge (electron) has an equal positive charge (proton). This results in the atom having zero charge overall.

Where the charge of these atomic particles is seen most clearly is in electricity. Electrical current is simply the flow of electrons from one point to another. That indicates that the electrons have some ability to travel. In some materials they can leave the atom they are currently attached to. When an electron leaves the atom, that atom then has one more proton than electron, and the positive charges outnumber the negative charges. The atom has an overall positive charge, and this attracts electrons. The behavior of electrons is the basis for how materials are held together. The way a material responds to applied force, heat, electric field, or any other external influence is ultimately determined by **atomic bonds**, or the way electrons are distributed between its atoms. Understanding these relations is important because this distribution of electrons determines the bonding of atoms which we will consider next.

2.3 Atomic Bonding

Atomic bonding is the reason materials exist. If it were not for bonds, all atoms would exist in a sort of gaseous soup, and no meaningful process or part could be created. As discussed in the previous section, the electrical charge balance between electrons and protons is very important, but the way electrons are organized around the nucleus is also very important. Atoms are happiest (most stable) when they have a complete **valence shell**. Recall that there are an equal number of protons and electrons in a stable atom, but the electrons are not equally distributed around the nucleus.

If we return to the Bohr model of an atom (see Fig. 2.2), which has us imagining the atom as something like a solar system, we can add more detail. The electrons are not in unique orbits like the planets. **Electron shells** describe the most probable distance of electrons from the nucleus. Continuing with the solar system theme, they can be pictured as multiple earths circling the sun. They can be at the same distance from the sun, but not necessarily circling in the same direction. Electrons can move in any direction along the orbital shell as long they stay at the appropriate distance from the nucleus where the protons are located.

While this is a simplified approximation, it helps introduce us to the concept that electrons are really present as a sort of "cloud" and can move in all directions, not only in rings like the planets. Like layers of an onion, there are multiple clouds or shells when there are many electrons. Figure 2.2 will be valuable for illustrating the way electrons are distributed. Most bonding conditions are achieved to reach 8 electrons in the outermost shell (the valence shell). This represents the most stable state an atom can have. When two (or more) atoms bond, there is a strong attraction between the atoms. How that attraction is formed and the way electrons are distributed between the atoms determines the type of bond.

2.3.1 Primary Bonds

Primary bonds are the strongest bonds and directly determine the properties of a material. Whether the material is strong, brittle, has a high melting point, etc. all comes down to bonding. Which type of bond occurs is based on the combination of atoms. One atom can be involved in any of the three primary bonds, depending on the type of atom it is bonded with. For instance, when two metal atoms are bonded, it results in a metallic bond. When two nonmetals are bonded, the bond is typically covalent. A metal bonded to a nonmetal can be covalent, ionic, or some of both. We will look at each type and how they form.

2.3.1.1 Metallic Bonds

Metallic bonds are characteristic of metals (go figure!). Metals have many similar properties that are related to this type of bonding. These properties include electrical and thermal conductivity, ductility (bending without breaking), reflectivity (shiny appearance), and so on. Not all metals exhibit these properties equally, but they are common to the class of materials. So what makes metals, metals? What is it about the bonds that results in these properties?

In a metal, atoms are arranged in a periodic fashion. This can be thought of as a box of Ping-Pong balls. If you shake the box, the balls pack together fairly closely. Though not all metals have atoms which stack so nicely, they *can* be thought of as hard spheres that are in contact. Unlike the Ping-Pong balls, there is an attractive force between the atoms of a metal. The attractive force between these atoms allows them to move in relation to one another, but they do not separate. This attractive force is related to the electron and proton charges.

The electrons in a metal are **delocalized**. That means they are able to leave the atom they are originally associated with. There are trillions of atoms that contribute these "free" electrons, and this creates a "sea" of free electrons which surround the positively charged atoms. The positively charged atoms are known as ions (i.e., cations). The ion cores are "glued" together by the electrons. This is what makes metals able to undergo significant bending and stretching without breaking. The ion cores can move around significantly without exceeding the force holding them together. A schematic of this bonding configuration is shown in Fig. 2.3.

Delocalized electrons are the source of metallic properties. Metals appear shiny because light interacts with the free electrons and reflects it. Metals conduct electricity because the free electrons can *flow* relatively easily. The mechanical properties of metals are attributed to the *nondirectional* character of metallic bonds. Nondirectional bonding means that the atoms can move away from one another without breaking a rigid bond. Instead the bond is transferred to another atom, so the material can *flow*. This is critical for forming operations that require the material to deform without breaking.

2.3.1.2 Covalent Bonds

Covalent bonds are formed between atoms that share electrons, but this is different than metallic bonding. Metallic bonding has many *free* electrons, which are not associated with any one or two atoms. In covalent bonding electrons still travel between atoms, but only two. This restricted movement of electrons means that a covalent bond is **directional**. The bond exists in a very specific configuration. Some factors that arise because of this are bonds of longer length (distance between atoms is greater), that are rigid, and that are strong.

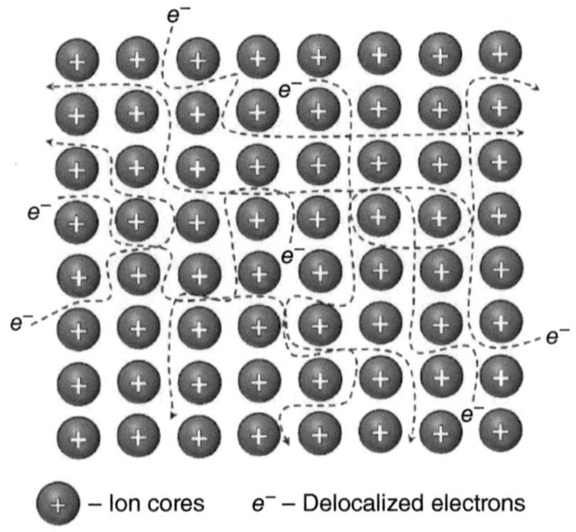

FIGURE 2.3 Illustration of metallic bonding. Electrons (e–) can move freely around the atomic cores. Because the electrons are not positioned on the atomic cores, there is a net positive charge present. This "free" electron behavior allows metallic atoms to move relative to one another without breaking apart. This is why metals tend to be very ductile, or able to be bent without breaking.

An example of this type of bond appears in carbon (think of graphite, for instance). Carbon atoms each have four electrons of their own, and they want four more electrons to reach a stable configuration (eight electrons in the valence shell). In order for two carbon atoms to form a stable bond, they would have to share four electrons with each other. Although atoms can share more than one electron, it is more likely that they will share electrons with multiple atoms. That means it is more likely that each carbon atom will bond with four other carbon atoms so that each pair shares one electron.

A practical example of this type of bonding occurs between carbon and hydrogen. In the simplest example, one carbon atom bonds with four hydrogen atoms. The carbon shares it electrons with the hydrogen atoms and vice versa so that all are stable. When a molecule like this is stable, it will not bond with other atoms or molecules (though it may *exchange* bonds in favorable conditions). The molecule that contains four hydrogen atoms and one carbon atom is called methane. It has a chemical composition of CH_4, where the "C" is the atomic symbol for carbon and "H" is the atomic symbol for hydrogen. Methane is a flammable gas that is the main component in natural gas which is commonly used for household heating and gas-fired stoves. The structure of this molecule and the distribution of electrons are shown in Fig. 2.4.

In methane, carbon shares one electron with each hydrogen atom. Because one electron from each atom is shared in the bond, it is called a **single bond**. When two electrons are shared, it is termed a **double bond** and three shared electrons comprise a **triple bond**. Molecules comprised of carbon and hydrogen are known as hydrocarbons, and the number of atoms in these are very important. The difference between gases, oils, and plastics are all related to their molecular structures, even though they contain similar ingredients. More on the importance of hydrocarbons to polymers is discussed in Sec. 5.3.

Hydrocarbons are not the only materials to exhibit covalent bonds. Ceramics, such as glass (mostly silicon dioxide, SiO_2), alumina (aluminum oxide, Al_2O_3) and rust (iron oxide, Fe_2O_3), have similar properties such as high hardness, brittleness and are not shiny

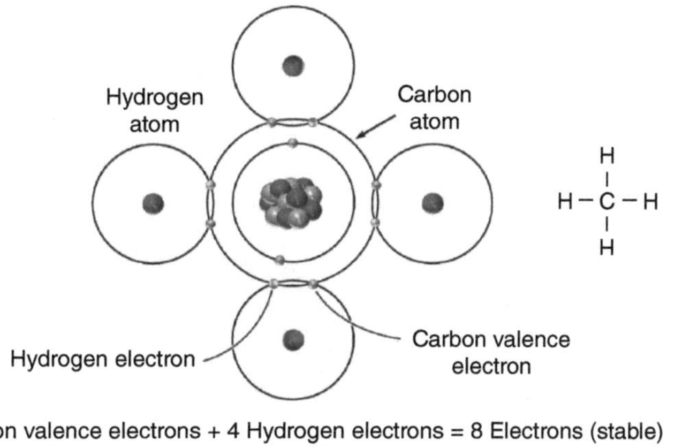

Figure 2.4 Methane, a simple hydrocarbon, has a central carbon atom which is covalently bonded to four hydrogen atoms. Because the bond is covalent, the electrons are shared between the hydrogen atoms and the carbon.

like metals. Covalent bonds bring these properties about by restricting the motion of atoms and electrons. The electrons are not free to "wander" about and the bonds are very rigid and strong. This makes forming operations and machining untenable in most cases.

Many covalently bonded materials also have high melting points, and this makes covalently bonded materials difficult to process by casting, too. Polymers, like ceramics, have portions of their structure that are covalently bonded. The major difference is that they are not entirely bonded in this fashion. Secondary bonds, to be discussed in Sec. 2.3.2, are also a main structural feature which gives them a low melting temperature and high formability. Do not let this be a point of confusion! Polymers and ceramics both have covalent bonds. It is the long-range three-dimensional structure that determines the overall properties. We will discuss the specific differences between polymers and ceramics in Chap. 5.

2.3.1.3 Ionic Bonds

Ionic bonds occur between atoms with very different valence states. The valence (outermost) electron shell of one atom is nearly full and the other is nearly empty. The major identifier in ionic bonding is the electronegativity difference between the atoms. **Electronegativity** is the attractive force an atom has on electrons. In general, the electronegativity of atoms decreases as you move from the top-right to the bottom-left of the periodic table. This trend is shown in Fig. 2.5. If you look to the top-right corner, you will see that fluorine (atomic symbol, F) and chlorine (atomic symbol, Cl) both have high electronegativity values. They both attract electrons. This indicates both will want to take electrons, but neither wants to give any away. The really important point of this trend is that atoms near to each other in the periodic table (i.e., with similar electronegativities) will not form ionic bonds, or the bonds will be only weakly ionic.

The difference between ionic bonds and other types is that in order to form an ionic bond there must be an *exchange* of electrons. Metallic bonds have free electrons that are not associated with one or two atoms in particular. Covalent bonds share electrons between atoms and are confined to that bond. In ionic bonding one atom donates an electron (or more than one) to another atom. The force that bonds the atoms together is coulombic. **Coulomb force** (named after Charles-Augustin de Coulomb, 1736–1806) is generated

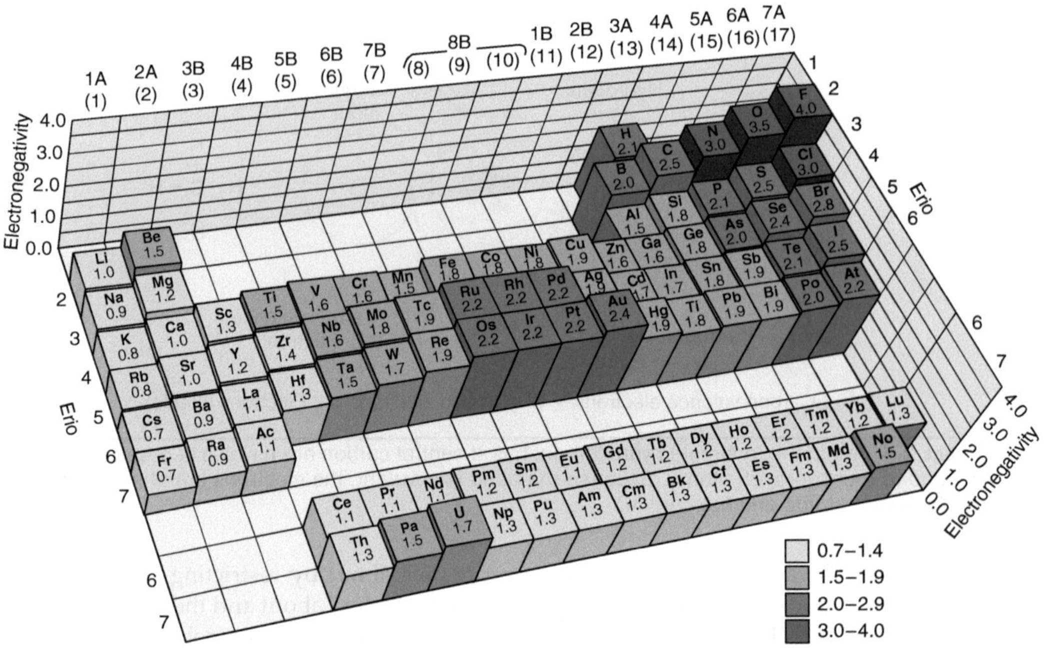

FIGURE 2.5 Periodic table of the elements with the electronegativities listed. The general trend is that the electronegativities decrease as you move down and to the left [from fluorine (F) to francium (Fr)]. Elements far away from each other (significantly different numbers) will tend to form ionic bonds. Those close to each other (similar numbers) will form covalent bonds.

between any positively and negatively charged items, not just electrons. If you want a practical example of this, rub a balloon on your head. When you go to remove the balloon, your hair (if you have any!) will *stick* to the balloon. This is an electrostatic (Coulomb) force.

To illustrate this with a common material, consider table salt, which is sodium chloride (NaCl). Refer again to Fig. 2.5, and note that sodium (atomic symbol, Na) and chlorine (atomic symbol, Cl) are on opposite sides of the periodic table. Chlorine has an electronegativity of 2.9, and sodium has an electronegativity of 1.0. The difference between them is 1.9, which creates a strong, ionic bond. Chlorine strongly attracts electrons but sodium does not. For this reason, both atoms can benefit by sodium giving an electron to chlorine. This same process occurs for other salts such as magnesium chloride ($MgCl_2$) and calcium chloride ($CaCl_2$). The number of atoms involved in the bonding is determined by the number of electrons in the valence shell of either atom.

2.3.1.4 Calculating Ionic Character

So what do we make of hydrocarbons? Hydrogen and carbon have different electronegativities, so they really form an ionic bond, right? Carbon (atomic symbol, C) and hydrogen (atomic symbol, H) have electronegativity values of 2.5 and 2.1, respectively. The difference is fairly small, so the bond is only weakly ionic. This means that a bond can be partially ionic and partially covalent. To determine how ionic a bond is, a fairly simple equation can be applied to determine the **percent ionic character** (%IC).

$$\%IC = [1 - \exp(-0.25(X_A - X_B)^2)] \times 100 \tag{2.1}$$

In the equation, "exp" is the exponential function (i.e., e^x), and X_A and X_B are the electronegativities of atoms A and B, respectively. If we apply this equation to sodium chloride (NaCl), where X_A is Cl and X_B is Na, we get

$$\%\text{IC (Na} - \text{Cl}) = [1 - \exp(-0.25(2.9 - 1.0)^2)] \times 100 = 59.4\% \tag{2.2}$$

This shows that the bond is primarily ionic, but there is also a covalent character to it as well (i.e., the remaining 40.6%). Now let us consider the carbon-hydrogen bond. We will replace X_A, this time with carbon, and X_B will be hydrogen. Using these values, we get

$$\%\text{IC (C} - \text{H}) = [1 - \exp(-0.25(2.5 - 2.1)^2)] \times 100 = 3.9\% \tag{2.3}$$

Even though the bond is considered covalent, it does have a small amount of ionic character.

Ionic bonds tend to be strong and rigid. Even though they are nondirectional, like metallic bonds, ionic materials cannot be easily deformed. This occurs because the atoms are specifically arranged so that there are negatively charged atoms surrounding positively charged atoms. If you think of magnets, they only attract one another from a single end. If you point similar poles of a magnet together, they repel each other. To deform an ionic material, such as bending, this requires atoms to slide past each other. When atoms with the same charge move past one another they repel each other and it causes the material to break.

For this reason, strongly ionic materials, such as table salt, are brittle. Because the bonds are very strong, these same materials tend to be very hard. Ionic bonds tend to be formed between a metal and nonmetal. The nonmetal is most likely to take an electron (or more) from the metal. The positive metal ion is called a **cation** and the negatively charged nonmetal is called an **anion**. The structure of an ionic material is shown in Fig. 2.6.

2.3.2 Secondary Bonds

Secondary bonds are weaker than primary bonds, but they play an important role in material properties. Perhaps the best example of secondary bonding we will examine

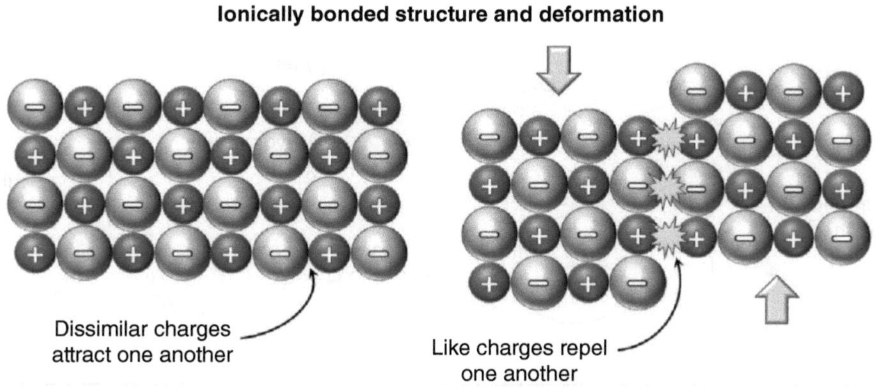

FIGURE 2.6 Ionic bonds are formed by the attraction between positively and negatively charged ions. These ions must be arranged so that like charges are not positioned adjacent to each other. Ionic materials are brittle (fracture readily) because like charges repel each other when the material is deformed (see right image) causing the material to cleave.

occurs in polymers. As will be discussed in more depth in Sec. 5.3, polymers are held together by a combination of primary and secondary bonding. Another common example of secondary bonding is in water, which produces the unique properties making it suitable for supporting life among other important functions.

To understand secondary bonds, it is beneficial to understand the concept of polarity. Without belaboring the details, a **dipole** is an uneven distribution of electrons within a *molecule*. The positively charged end is denoted with δ^+ (Greek letter delta), and the negatively charged end is denoted by δ^-. A molecule is simply a collection of primary-bonded atoms, and electrons will tend to spend more time around certain atoms in a molecule than others. This is analogous to people in a house. Although they may go anywhere in the house, there are certain rooms in the house where they will spend most of their time. With electrons, this leads to an uneven distribution of charge, and, therefore, the molecule will have a positively charged region and a negatively charged region.

If you recall the principle of ionic bonding, you will discover that it too was developed by charge distribution. The atom that gained an extra electron was negatively charged and the one that lost that electron was then positively charged. It is still the Coulomb force that attracts these unlike charges, but in secondary bonding it is not as strong. There is no permanent exchange of electrons so the charge difference is not equivalent to a discrete number of electrons. Rather, the strength of the dipole is determined by the electronegativity differences in the molecule.

Water, a wonderful example of many things, can be used to illustrate this property of polarity. Water is comprised of molecules of oxygen and hydrogen. The shape is something like a "V" where the two hydrogen atoms are positioned opposite the oxygen atom (Fig. 2.7). The oxygen and hydrogen atoms are bonded covalently, which means that they *share* the electrons. As in many cases, the sharing is not equivalent and the electrons spend more time near the oxygen atom than near the hydrogen atoms. The charge distribution (positive and negative sides) leads to an attraction between water molecules. The hydrogen atoms are attracted to oxygen atoms in *other* molecules. This attraction is known as hydrogen bonding and is described in the next section.

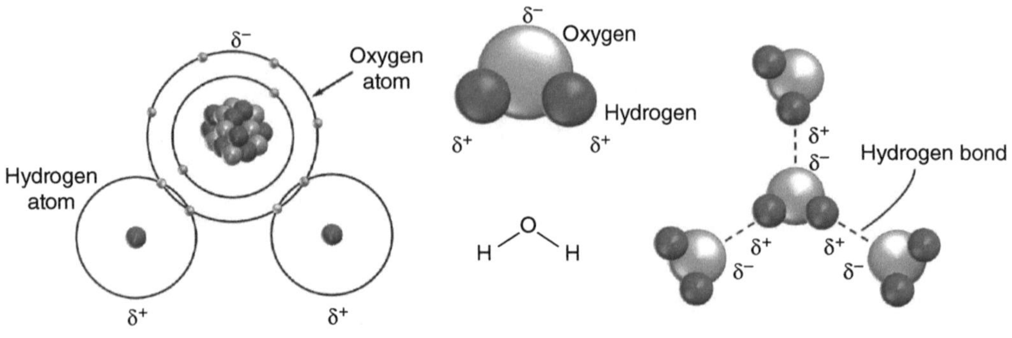

FIGURE 2.7 Water molecules (see left) have shared electrons, but the electrons tend to spend more time with the oxygen atom giving the molecule a net negative charge on that side. The Greek letter δ (delta) signifies the end of a dipole with the charge present (positive or negative). Due to the distribution of charge, the hydrogen atoms (positive) are attracted to the oxygen atoms (negative) in different molecules.

2.3.2.1 Hydrogen Bonds

Hydrogen bonds are formed between a positively charged (electropositive) hydrogen atom and a negatively charged (electronegative) atom on another molecule or on the same molecule. These bonds are not the same as those where hydrogen is bonded within a molecule. If a hydrogen atom is strongly bound to another atom in a molecule such as in water, the bond is covalent. Instead of having this strong, directional connection, the atoms are simply attracted to one another by their electrical charge difference.

The hydrogen bond that occurs between molecules of water is responsible for its properties. Although hydrogen bonds are weak in comparison to primary bonds, the effect can still be quite significant. For instance, water has a boiling point of 212°F (100°C). Why is this important? Consider hydrogen sulfide (H_2S), which has a similar molecular structure to water, but does not exhibit strong hydrogen bonding because hydrogen and sulfur have similar electronegativities (refer back to Fig. 2.5). Despite the similarities to water, hydrogen sulfide has a boiling point of –116°F (–82°C). Think if water behaved similarly! Water would not condense to liquid form even in earth's polar regions, even at temperatures where carbon dioxide is a solid (i.e., dry ice). Because of hydrogen bonding, water has the fortunate chemical properties that enable it to go from freezing to boiling over a relatively small range of temperature. Hydrogen bonding is also responsible for the expansion of water upon freezing while the majority of substances contract upon freezing. If ice did not float, we would have a whole host of issues.

Although water is a valuable example for understanding the importance of hydrogen bonding, it is not a practical engineering material. The place we encounter this type of secondary bonding most significantly in the materials we will explore in this text is in regard to polymers. The effect hydrogen bonding has on the properties of polymers is similar to that of water, changing the physical, chemical, and mechanical properties. We will consider the structure of polymers and the distribution of bonding in Sec. 5.3.

2.3.2.2 van der Waals Forces

Although **van der Waals bonding** (named after Johannes Diderik van der Waals, 1837–1923) is sometimes used to describe the total sum of intermolecular forces, it is really limited to those not created by hydrogen (previous section) or electrostatic interactions (Coulomb force). Like primary bonds, electron distribution is central. Recall that the electrons in an atom or molecule are not static (stationary). Quite the opposite, electrons move around quickly and randomly (there is a statistical probability of finding an electron at any one location, but we will leave that to the quantum physicists for now). As a consequence of their "uncoordinated" movement, they may not be evenly distributed. This uneven electron distribution causes an uneven charge distribution.

If all of the electrons are momentarily located on one side of an atom or molecule, that side will have a net negative charge. By default, the other side will have a net positive charge. We saw that this uneven charge was the mechanism for ionic bonding and hydrogen bonding. Unlike those prior two cases, van der Waals forces are a consequence of temporary, fluctuating dipoles. The way these arise is by the *chance occurrence* of an uneven charge distribution. This produces a dipole, and nearby atoms (or molecules) will then respond to the charge. The electrons of the responding atom(s) will move toward the positive side of the first atom and now both are dipoles. As this occurs throughout the material, between many molecules or atoms, there is an attractive force that develops. Figure 2.8 demonstrates how dipoles can be created, even in single atoms.

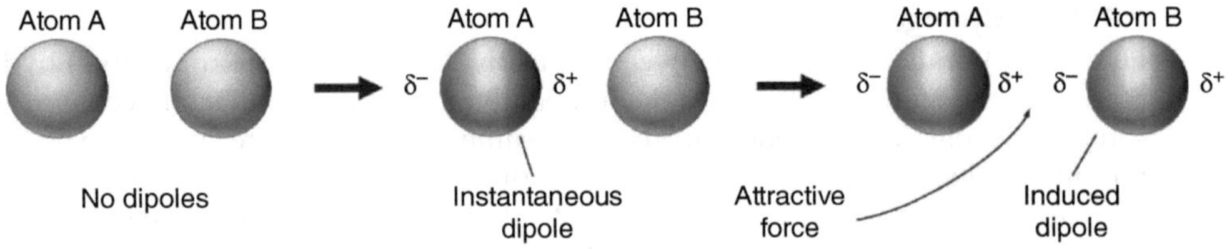

FIGURE 2.8 Illustration of the creation and induction of dipoles. In this figure, the Greek letter δ (delta) signifies the end of a dipole with the charge present. Although these dipoles create an attractive van der Waals force, the magnitude is relatively small compared to primary bonds.

Because these fluctuating dipoles are present in all atoms or molecules to some degree, van der Waals forces (not bonds) are present in all materials. Usually these forces are very weak and are therefore of little structural consequence, but they should not be neglected. Indeed, emerging technologies such as "gecko tape" exploit this force for adhesive purposes. Ever wonder how a lizard can climb a wall? An extremely large amount of contact area is created on a gecko's feet by an abundance of hair-like features call setae, and these maximize the interaction area and van der Waals forces. This force can be quite significant in certain situations. Figure 2.9 shows progressive magnification of the setae on a gecko's foot. These structures are copied in the synthetic setae used in man-made adhesives exploiting this concept. This reproduction of natural phenomena by artificial means is known as biomimicry or as the field of biomimetics.

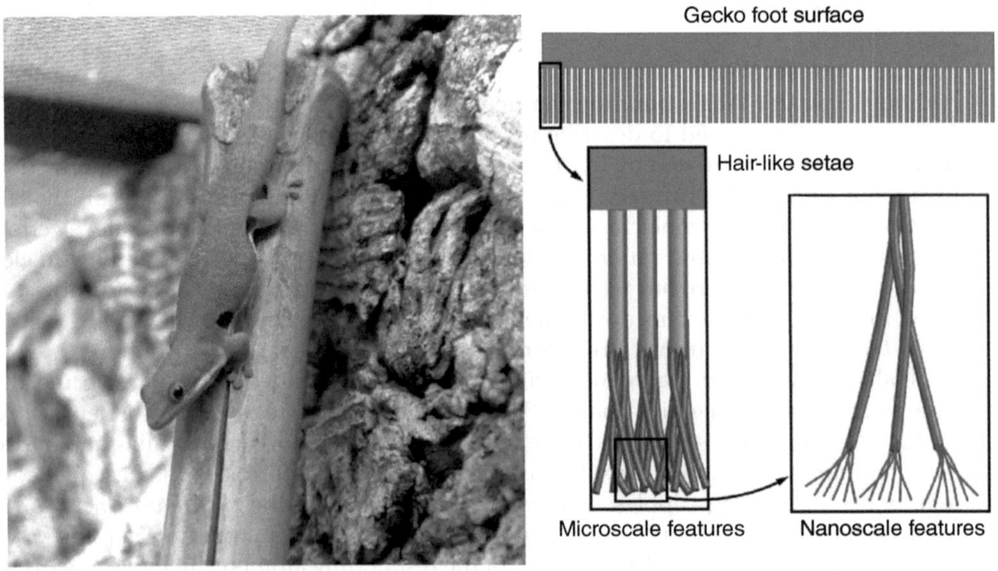

FIGURE 2.9 A gecko can scale walls (left) thanks to the van der Waals forces generated between its feet and the surface. Close examination of a gecko's feet (right) reveals that many hair-like features called setae give it the ability to cling to flat surfaces.

2.4 Material Phases

We have considered the structure of atoms and the types of bonds they form with each other. These are foundational to understanding the properties of materials in bulk form, but they do not tell the whole story. As with any good story, there is more to it; a twist or two yet to be revealed. Let us first consider the term "bulk." When we talk about atoms, we generally consider them as isolated or with a limited number of other atoms such as in molecules. When we build things we do not generally talk about what kind of atoms we would like to make the product from. Instead, we talk about materials. Materials, that we see and work with, are massive collections of atoms (millions and billion and trillions...). They are "bulk" collections of atoms. Unlike buying candy or nuts in bulk at your local grocer, atoms are not loosely arranged and separate in solid materials. This is where bonding begins to become quite important on a grander scale.

As mentioned in the introduction to this chapter, materials are present as solids, liquids, or gases (for completeness, substances can also be in a plasma state). Materials can be useful in all phases (Table 2.1). In gases, the atoms are separated, able to move freely past on another. There are no strong bonds between the atoms or molecules (do not forget, van der Waals forces are always acting). Gases have indefinite shape and volume. They can be easily compressed into a container of any size and shape, and therefore work well for inflating tires. Gases can be a component of structures, but they are not the entire structure. For instance, a "bouncy house" for children relies on air for proper operation, but there must be a material that contains the air. We will see the various ways gas pressure can be used to shape materials (e.g., the vacuum forming of polymers in Sec. 12.3.1.2).

In a liquid, the cohesive forces between atoms and/or molecules are larger, and great enough to restrict their motion. They can still move (flow) relatively easily, but they have a definite volume. That means they can fill a container of any shape, but they cannot be compressed (for practical purposes, they are considered "incompressible"). Liquids are used in some processes to apply force in shaping materials (e.g., isostatic pressing), and they can be used to cut materials as well, but they too must have a solid container for operation. The transition of materials from solid to liquid and the properties of that liquid are critical for casting and molding operations (see Chap. 11).

Solids have very strong interatomic and/or intermolecular bonding compared to liquids and gases. Solids hold their shape, and therefore will not fill a container of arbitrary dimensions and are incompressible. That is, they have a definite volume and definite shape. Solids are used for structural materials, but not all of them can be used in all situations. They behave in vastly different ways. Understanding the properties in

Phase	Volume	Shape	Uses in Manufacturing
Gas	Indefinite	Indefinite	Pneumatics, forming weak materials, cooling
Liquid	Definite	Indefinite	Hydraulics, forming strong materials, cooling, cutting, casting/molding of materials
Solid	Definite	Definite	Forming, punching, cutting, engineered components

Note: Even though solids are the primary structural materials, gases and liquids are very important in processing as well.

TABLE 2.1 The Phase of a Material Determines Whether It Has a Fixed (Definite) Volume and Shape relation to their structure is where we will devote our time. We will begin this in Chap. 3 with a closer look at the structure of metals.

2.5 Why It Matters

We began this chapter considering what an atom is and what it consists of. The basic structure and components are going to be a theme in this text, please review these concepts often if they are unfamiliar. Protons, neutrons, and electrons are the major parts we considered. Know the differences. This is especially true for electrons. Electrons are the "ambassadors" for the atom. Most interaction between atoms is dictated by the electrons, and these interactions are commonly defined as bonds. Both primary and secondary bonds determine the overall behavior of the material. Neither type should be disregarded, especially in polymers.

The bonding determines the bulk behavior of the material. Most processes will be concerned with bulk properties, so the bonding is very important, even at the industrial scale. You can think of it as a brick building (Fig. 2.10). No matter how large, it is made up bricks (atoms) and mortar (bonds). The type of brick is very important. The weight, strength, etc. will all determine the suitability of the building. Collectively, the bricks must fit together as well.

The mortar is equally important. If you have suitable bricks but no mortar or weak, crumbling mortar, that is a building you do not want to enter. The small pieces and how they are connected is critical. The pattern the bricks are laid in will determine how it performs, and this is also true in materials, as atomic arrangement is responsible for bulk properties, just like the type of bonds. The overall structure of the building, the thousands of bricks and all their connections are used to create a larger, overall shape. This is like a bulk product or component that is created from these smaller units. To quote the ancient philosopher, Aristotle, "The whole is greater than the sum of its parts." There is more than meets the eye.

This chapter should give you a basic understanding of how materials are constructed at an atomic level. Why do we care? Does it really matter? To answer both, consider the interplay between scientific understanding and practical application. This was discussed in Chap. 1 which culminated with a tree-based concept of how society benefits from knowledge and technology. Without a basic understanding of the atomic and structural phenomena in materials, we cannot fully appreciate the way they behave. We can only guess why

Figure 2.10 Like a brick building, materials are made of small pieces (bricks or atoms) connected together (mortar or bonds) that form a large-scale, useful product (building or bulk material).

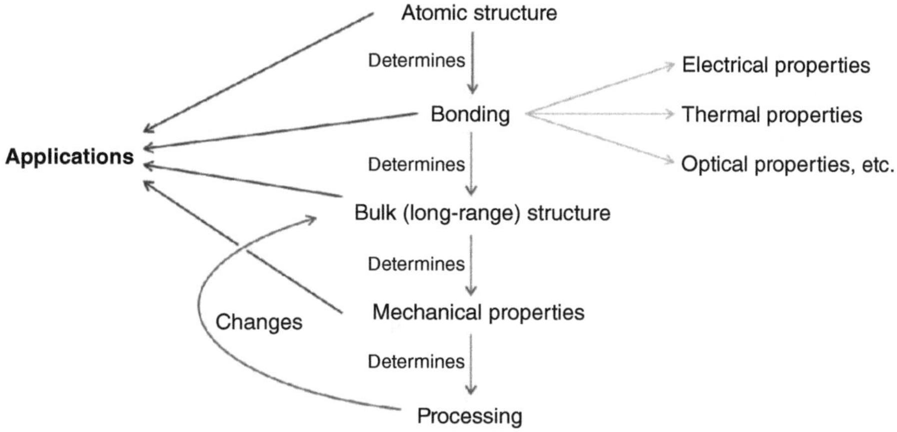

FIGURE 2.11 The structure of materials, from the atomic level up to bulk structure, will determine the way it is used and the way it is processed. Processing will affect the structure too. There is an understanding that must be reached in order to achieve efficient manufacturing and sufficient material properties.

salt breaks apart so easily, and we cannot imagine how lizards are running up and down walls! Rather than resort to trial and error and confusion, we can use the fundamental concepts in materials to make informed decisions about manufacturing. Without seeing the material or the equipment or ever trying the first process, we can make educated decisions that save time, money, and increase quality. A diverse background is valuable. Do not neglect to make material structure a tool in your problem-solving arsenal.

To emphasize the importance of these concepts, consider the diagram in Fig. 2.11. It shows us that we can understand the properties of a material by understanding its structure. The atomic structure determines bonding, which determines the long-range structure (many atoms or molecules), which determines the mechanical properties. The way you perceive materials as hard, brittle, soft, formable, shiny, etc. can always be traced back to the atomic structure. The way a material is used and processed is often controlled by its mechanical properties. Those properties are a function of the way it has been processed. Some materials become stronger during processing. This means materials need to be assessed with regard to their composition and their prior processing. There is an understanding that must be reached in order to achieve efficient manufacturing and sufficient material properties.

2.6 The How and Why of Concepts of Materials

2.6.1 Chapter Review

1. What is matter?
2. Figure 2.1 shows three classes of elements. What are they?
3. Even though the vast majority of elements are metals, most materials are not metals. Why?

4. What are the three types of primary bonds?
5. What are three major types of subatomic particles?
6. The atomic number of an element signifies the number of ___________.
7. An ion has a different number of ___________ than protons.
8. An isotope has a different number of ___________ than protons.
9. What type of charge does a proton have? What about an electron? What does this mean for the number of each to create a stable atom?
10. What determines the type of bond two atoms form?
11. The electrons in a metal are capable of moving freely. What is the term for this?
12. What about the bonding allows many metals undergo significant bending and deforming without breaking?
13. Name three characteristic properties of metals and how the bonding allows those properties.
14. Covalent bonds are *directional*. What does this mean?
15. The number of shared bonds atoms will make is determined by what?
16. Covalent bonds are rigid and strong. Name three characteristic properties this creates in ceramics.
17. What is electronegativity? What general trend does it have in across the period table?
18. Covalent bonds share electrons. What happens to electrons in an ionic bond?
19. What force holds ionic bonds together? Describe how that occurs.
20. Calculate the percent ionic character for CuCl, HF, HCl, and FeO. Which is most ionic?
21. Calculate the percent ionic character for diamond (carbon bonded to carbon). What does this mean?
22. Ionic bonds are nondirectional. What does this mean?
23. What are ionic materials brittle?
24. A cation has what type of charge? How about an anion?
25. What are secondary bonds?
26. What is polarity?
27. What are hydrogen bonds? How are these formed?
28. Name two beneficial properties hydrogen bonds give to water.
29. Hydrogen bonds are common in what engineering material?
30. van der Waals bonding is due to random movement of what?
31. What is the term for the unequal positive/negative charge creating van der Waals forces?
32. The increase of surface area does what to the total van der Waals force?
33. Name the three states of matter. Describe how they behave with regard to their shape and volume.

2.6.2 Concept Connections and Extensions

1. The Bohr model of an atom is inaccurate, but it gets the concepts across. Find out what electron orbitals are actually like by searching the internet or other resource and describe them.
2. Using the example of any engine, describe how solid, liquid, and gaseous phases are important to that application.
3. Refer to Sec. 5.3 and describe how primary and secondary bonds are important to the structure.
4. Describe Coulomb's law, a situation where it is experienced in everyday life, and a practical application of it.

CHAPTER 3

Metallic Materials: Structure

Metals have the simplest structures. While they may not be considered *simple* by many, the way we will look at them is actually pretty basic. We will consider atoms as spheres and look at the way they pack together to form a long-range structure, known as the crystal structure. Most of the metals used in common materials consist of one of three types of structures. There are many others, but limiting our focus to those three will still give us enough understanding to appreciate others. We will also find out that the structure of metals is not perfect. That means there are defects, and those defects are actually very important to manufacturing protocol. They can be controlled and exploited to improve material properties, but they can be detrimental if not controlled. For these reasons, it is quite important to spend time in understanding the principles of metallic structure.

3.1 How They Work

There are few topics more important to understanding materials than that of structure. Unfortunately, there are few topics less understood than structure. There are a couple reasons for this. First, the study of structure is as deep as you wish to go. Structure can be defined many ways. It can be considered *macro*scopically such as the cross-section of an I-beam or the linkage of members in a truss. This level of structure is dealt with in texts on component and structural design, though some the concepts will be visited in Chap. 7.

At the other end of the spectrum is *micro*scopic structure,* usually called the **microstructure**, which concerns collections of atoms, their long-range orientation and compositional differences. At this limit we are primarily concerned with small-scale features, not visible by the unaided eye. This is the realm of atoms, grains, imperfections, alloying effects, etc. This type of structure is nearly limitless in detail because the information and difficulty in attaining that information increases as the length scale decreases. That is, when you begin to study phenomena at an atomic scale, the amount of detail you need to accurately describe behavior increases drastically. When you study atoms, to get the structural information you need to describe bonding of each atom to each its neighbors, which requires in-depth calculations of thermodynamic energies and electron

*If you want to be strictly accurate, the limit of small structures is even below atoms and concerns subatomic particles.

interactions that require even more detail to accurately quantify. All of this only gets us a span of a few atoms, or if using massive computing power, may get us a few million atoms (a couple of micrometers if we are lucky).

So why bother? Well, properties are determined first at the atomic level and then extend cumulatively to the macroscale which we can easily see and work with. The microstructure and macrostructure are directly connected. The connection between atomic and macroscopic structures is shown in Fig. 3.1. By understanding atomic phenomena we can understand the behavior of large workpieces during processing and during use. There are useful approximations that make life easier for describing atomic interactions, and in most cases we can make good, informed decisions with that level of detail.

One area that requires a detailed, atomic-level understanding is alloy development for future materials. So do not believe highly detailed calculations that are only useful in theory. Some of the most important materials for our future are likely being "number-crunched" in such a manner now.

We have established one reason that structure is not well-understood is that there is a nearly unlimited depth to which it can be studied. This is not really problematic since there are a limited number of people who actually need this level of understanding. Another, more concerning reason that structure is poorly understood is that microscale structure is perceived as too abstract and unimportant to processing and applications. The average person engaged in manufacturing sees it as being too complicated or not worth the time. The purpose of this chapter is to scratch the surface of structural considerations in metals and to provide insight into macroscale behaviors that are entirely controlled by microscale phenomena. We cannot begin to appreciate the importance of selecting the material large components like I-beams are made of without building a foundation on structural concepts.

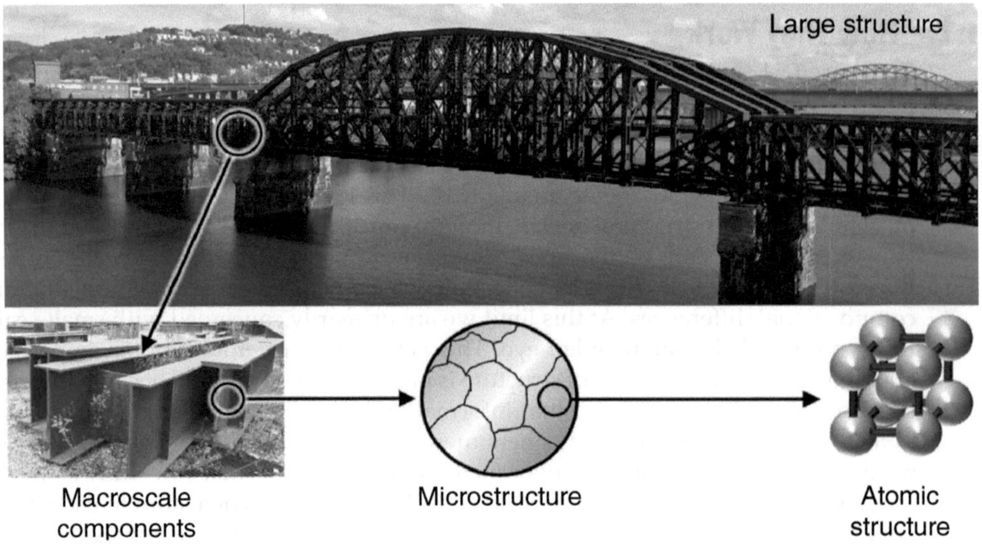

Figure 3.1 Structure can be defined in many ways. Very large structures are made up of macroscale components. Those components possess a microstructure and atomic structure determine their properties. Structure at the smallest scale directly impacts all larger scales, but this does not apply in the reverse direction.

3.2 Structure of Metals

Metals have the simplest structures and are among the most common structural materials. The conditions employed during manufacturing must be appropriate for each material. For instance, aluminum and steel are formed, machined, cast, etc. under considerably different conditions. Why? What makes them so different?

Many of the differences between metals, such as aluminum and steel, can be attributed to their bulk structure, or the way their atoms are arranged over "long" distances. When talking about atoms, long can be very short! One thousandth of 1 in (0.001 in) is well beyond "long" in regard to atoms even though it is quite small by many other standards.

Atoms are not usually haphazardly piled together in a metal. That would make them amorphous, a structure more common in polymers and ceramics (see Sec. 5.2.2 for ceramics, for example). Instead, they are arranged in a very orderly manner. Atoms can be thought of as hard spheres, and the way they pack together is roughly governed by this concept. Please do not misunderstand, atoms are *not* hard spheres! This way of approximating their shape and interaction is only to simplify the discussion.

As mentioned in Chap. 2, if you take a box full of Ping-Pong balls and shake it, the balls will fit together more tightly and occupy less space. It is the force of gravity that helps this process. At the atomic level, the forces of interatomic bonding are much stronger than gravity, but atoms, like Ping-Pong balls, also have an ideal way of fitting together. This self-arrangement is dictated by the lowest energy configuration. The Ping-Pong balls lower their height (potential energy) and the metal atoms lower their bond length (bond energy). Think of each bond as a spring. It can be compressed (pushed together) or stretched (pulled apart) but when released, it goes back to its original (ideal) size.

The ideal bond length and configuration is determined by the electronic* interaction of the atoms. If atoms get too close or too far apart, the energy increases. The three-dimensional, grid-like arrangement of atoms relative to one another is known as the crystal **lattice**. We think of atoms as being positioned at lattice points, which represent ideal locations, but as we will find out in Sec. 3.2.3, ideal and actual are very different concepts.

The majority of engineering metals have one of three arrangements: body-centered cubic (bcc), face-centered cubic (fcc), or hexagonal close-packed (hcp). These structures are shown in Fig. 3.2. There are other less common types of structures and the interested reader should look up Bravais lattice designations in a dedicated materials science text.

We will consider most materials as being of one crystal structure. For instance, iron is bcc and aluminum is fcc. That is accurate at room temperature and at atmospheric pressure, but many elements will undergo a change in structure at high or low temperature or at extreme pressure. The ability for a single material to exhibit multiple structures is known as **polymorphism** or **allotropy**. This is critical to the heat treatment of steel as iron changes from bcc to fcc at high temperature. These concepts will be elaborated on in Chap. 9. Some metals, like plutonium, can go through many structural changes. It has six different allotropes occurring from room temperature to around

*Related to electrons, not tech gadgets.

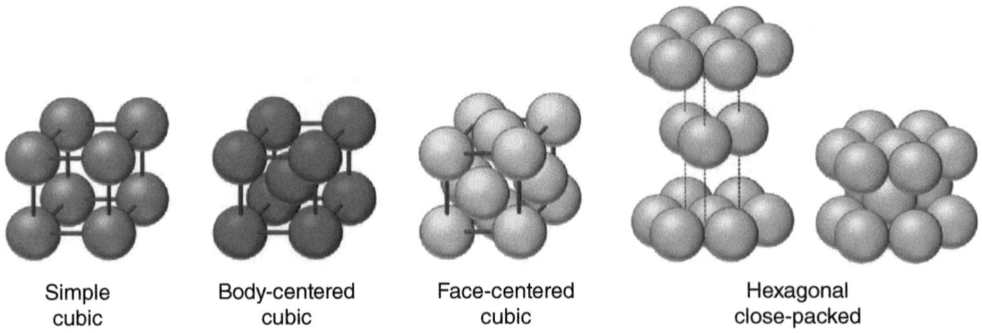

Figure 3.2 Metallic materials are commonly categorized by their structure. There are three primary types of atomic ordering in metals: body-centered cubic (bcc), face-centered cubic (fcc), and hexagonal close-packed (hcp). Many of a metal's properties can be attributed to its atomic structure. Note: Face and corner atoms are shared between adjacent unit cells.

850°F (454°C). The temperatures of those changes are within normal processing conditions. Imagine trying to work with a material that changes properties while you are working with it! Allotropy is an important property to account for, and it can have devastating consequences, as mentioned about tin in Sec. 4.4.2.

3.2.1 The Unit Cell

So what are we looking at in Fig. 3.2? Yes, there are atoms (spheres) stacked together, but what does this tell us about the material, and by extension, about the way it can be processed? To begin unveiling the importance of atomic arrangement we will consider aluminum, iron, and zinc. Aluminum is relatively soft and formable (ductile). Iron is stronger than aluminum but is not as ductile. Zinc is weaker than iron, stronger than aluminum, and less ductile than either steel or aluminum. What does this tell us? At least from this simple comparison, it appears that strength is not tied to ductility across these materials. So what *does* determine these properties? It is an important question because it is foundational to the processing of these materials and those like them. To better understand it, we will expand on Fig. 3.2.

In Fig. 3.2, the three most common arrangements of atoms—fcc, bcc, and hcp—are not isolated in small clusters as pictured. Rather, these clusters are called the unit cells. What that means is that if you want to know the pattern of atoms over long distances (millions of atoms) all you need to do is look at the unit cell. If we apply this to something more tangible, like a brick patio, we can get a feel for the concept. If you visit a friend to see his new patio (and get free food!) you could make an educated guess of what the brick pattern looks like by only seeing a few bricks. This is because bricks only fit together tightly in a few configurations.

So, unless your friend is a Picasso of sorts, you can safely assume the pattern that exists over a few bricks is the one that will extend across the entire patio. It is similar with metals. The way atoms pack together over just a few atoms can be repeated to form a bulk material (now in three dimensions instead of two). The smallest piece of the pattern that contains the whole pattern is the unit cell as pictured in Fig. 3.2.

The concept of building larger structures with smaller units is provided visually in Fig. 3.3. The "unit cell" (also known as a motif) of the bricks and the unit cell of the atoms in the fcc arrangement share the same relationship to the larger structure. To get

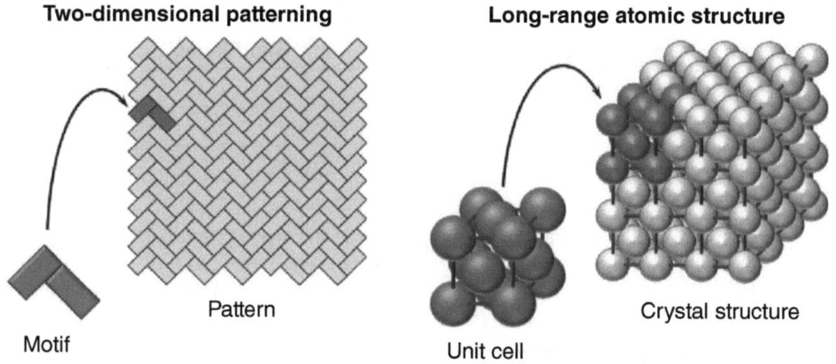

Figure 3.3 With a brick patio, you do not need to see the whole patio to know the pattern. Similarly, you do not need to see all of the atoms in a material to know its structure. A unit cell, which is repeated many times, can be used to describe bulk material structure.

something as large as an I-beam, in theory we just need to stack a whole bunch of unit cells together. You can think of this as building with LEGO® blocks or a cinderblock wall. When it comes to patterns like this, the whole really is just a sum of its parts.

3.2.2 Crystal Structure

So, we have established that metals consist of repeated pattern of atoms, and the simplest representation of this pattern is the unit cell. The long-range pattern, called the **crystal structure**, is the foundation of understanding where many material properties originate. The most common atomic arrangements are face-centered cubic (fcc), body-centered cubic (bcc), and hexagonal close-packed (hcp) are shown in Fig. 3.2.

In fcc crystals, the atoms are arranged in a very closely packed order (**packing factor** of 0.74 or 26% open space). In a bcc crystal, the structure is somewhat more open (0.68 packing factor or 32% open space). The way the atoms are oriented in an fcc structure means that all of the atoms are touching 12 other atoms. This is known as the **coordination number**. In a bcc structure the atoms are touching only eight other atoms. This actually plays a pivotal role in the way a material deforms (strength, ductility, etc.).

The way that it shows up is in the ability of atoms to slide past one another. When you permanently change the shape of a material, atoms have changed positions. The ease with which they can move past one another is directly responsible for the mechanical properties of a material. This has a major influence on the property differences between iron, aluminum, and zinc. The values for fcc, bcc, and hcp materials are summarized in Table 3.1.

Take a look at Fig. 3.4, which illustrates the concept of atomic motion. In the bulk material, sheets (planes) of atoms move together during deformation (bending, stretching, hammering, etc.). In order for the atoms to move past one another, they must "slip" over each other. This is like rumble strips on a road, and the atoms must bump over each other to move. The deeper the atoms sit between one another, the harder it is for the planes to move.

The difficulty of motion is known as the **lattice friction** or **Peierls stress** (after Sir Rudolf Ernst Peierls (1907–1995)). Metals with a bcc structure have atoms that are spaced farther apart and that sit further into open spaces. Therefore, bcc metals, which

Structure	Packing Factor	Coordination Number	Slip Planes	Slip Systems
bcc	0.68	8	3	48
fcc	0.74	12	1	12
hcp	0.74	12	3	12

Note: The structure of a metal will determine the way it performs. Some basic considerations in how atoms pack together in a crystal are given above. These numbers indicate how tightly atoms are arranged and how easily they can move relative to one another.

Table 3.1 Crystal Structure and Properties

have a high degree of atomic interference, will usually have higher strength and lower ductility than fcc metals. Metals with an fcc structure are tightly packed and there is less space between atoms. The planes do not interfere as much, so they can slide past one another more easily.

The cumulative effect of slip is realized as a change in shape. Figure 3.4 shows two lines of atoms, but real materials are three-dimensional and involve billions of atoms. The process is the same, however, but slip occurs at multiple locations simultaneously. Any time a metal is bent, stretched, rolled, etc. there are billions of atoms being relocated, and they will follow the path of least resistance, known as the slip planes. A two-dimensional illustration of the concept is shown in Fig. 3.5.

Where does this leave hcp metals? Although hcp metals have the same packing density as fcc metals (26% open space), they do not have as many favorable slip planes. **Slip planes** are "sheets" of atoms which slide together. You when you look at crystals from multiple directions, you can see that atoms form different patterns. The patterns which have atoms packed closely together form slip planes. Atoms can slide past one another in any direction, but there are a limited number of directions that are realistic. The atoms will take the path of least resistance, and in an hcp crystal, there are not many of those *active* paths when compared to fcc.

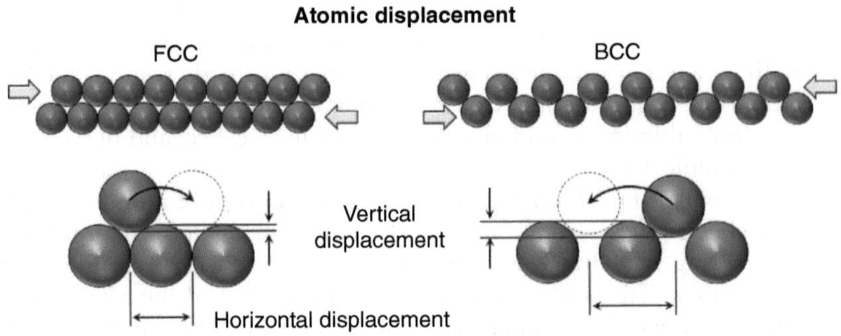

Figure 3.4 The deformation of materials results from atoms moving (slipping) relative to one another. The crystal structure determines how easily they can move. Large movements with greater atomic interference, such as in the bcc crystal shown above, will increase strength and reduce ductility.

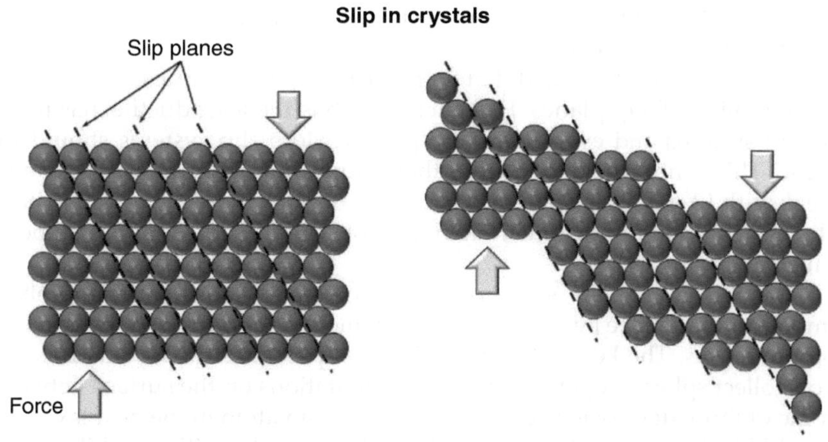

Figure 3.5 In bulk materials, multiple sets of atoms will slide past one another during deformation, and the way they deform is controlled by the crystal structure. The orientation of the atoms in motion follows basic rules for each crystal structure where the least amount of force is needed. These "favorable" orientations are called slip planes.

Table 3.1 lists the number of slip planes and the number of slip systems for the three most common structures. A **slip system** is a combination of a slip plane and **slip directions** within that plane. If you "rack" billiard balls in a triangle, this actually has the same structure as the slip plane in an fcc metal. If you put the cue ball on top of the racked balls and try to move it around, you will notice that the ball will follow the crevices between the racked balls. The racked balls are atoms in the slip plane, the movement of the cue ball indicates slip directions. The plane and direction together are a slip system, and the total number of slip systems for a crystal structure is the number of slip planes multiplied by the number of slip directions. These concepts are shown in Fig. 3.6. In real materials, movement will not be limited to one atom, and collections of atoms

Figure 3.6 When materials deform, their atoms move relative to one another. The movement occurs on certain slip planes in certain slip directions. Combined, these represent a slip system. This process can be thought of as the movement of hard spheres, such as billiard balls, but unlike moving a single cue ball, in real materials large collections of atoms would move together.

move together. This can be more accurately represented by two sets of billiard balls moving over one another.

We have just established that an fcc structure is the easiest to deform, but it has the least number of slip planes. Even hcp, which is the least ductile, has the same number of slip systems and even more slip planes. More slip systems should make it easier, right? This is not true; it is quality that counts, not quantity. This is like trying to cut a steak with 13 spoons instead of one knife. It is much more important to have a "useable" slip system than to have many that are not. The fcc slip systems can be "activated" at much lower forces.

Despite having the same packing factor and coordination number, there is an important difference between fcc and hcp metals. It is subtle, but has a dramatic impact on properties. The key difference is in what positions the layers of atoms take. When you collect spheres together, there are indentations on the surface between them. Only some of those depressions can be occupied by an atom in the next layer since that atom would block another atom from taking the nearest position. While hcp metals repeat the same two layer positions throughout the structure, fcc metals assume three different positions, and this makes them easier to deform three-dimensionally. The differences in these structures are shown in Fig. 3.7. Always keep in mind that these are really three-dimensional problems and representing them two dimensionally is never perfect, but it is often easier to understand.

Slip systems are not the only important feature to consider. Even when correctly accounting for the slip systems and crystal structure, the amount of force required to

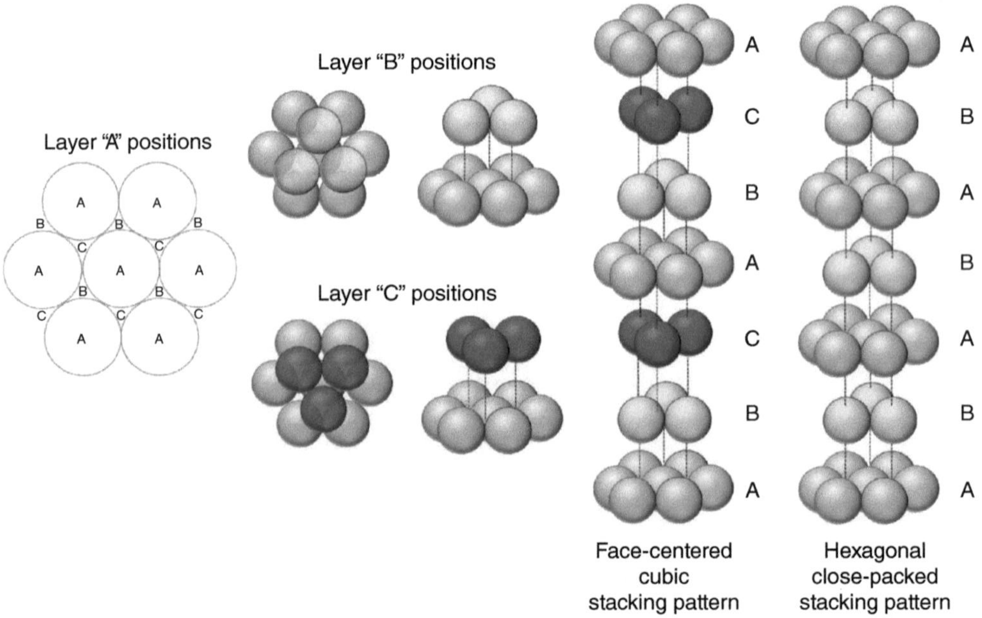

FIGURE 3.7 Although they have the same packing factor, the difference in the way atoms stack in fcc and hcp metals makes a very big difference in their mechanical properties. fcc metals are ductile and hcp metals are more brittle because the three-dimensional packing of atoms creates more favorable slip planes in the fcc configuration.

deform a material is much lower than would be predicted by the force required to move entire planes of atoms past one another. This indicates that we cannot accurately describe the deformation of a material as the coordinated movement of many atoms. There must be a difference from theory and reality, and that difference is imperfections, which we will consider next.

3.2.3 Crystal Imperfections

Nobody is perfect ... materials included! We have just discussed crystal structure; the long-range pattern of atoms in a bulk material. We now come to the point where we start finding exceptions. Like most exceptions to the rules, they can be confusing and frustrating, but they can also be very informative and useful. The topic of **imperfections** (or **defects**) in crystals is very complex when taken in detail. For our purposes, we can highlight some of the most important types and effects of these imperfections. Defects are categorized as point, line and planar types. A **point defect** involves only one atom (or atomic position), a **line defect** involves a row of atoms and a **planar defect** will affect a plane (sheet) of atoms. To simplify our analysis, we will not concern ourselves with planar defects, though these are important. Rather, we will focus our time on point defects and touch upon line defects.

3.2.3.1 Point Defects

A point defect is an imperfection involving one atom. There are two ways that this can manifest: an atom is present that should not be there or an atom is missing that should be there. To start with the first, consider an "extra" atom being stuck between the normal atoms in a crystal. This is kind of like packing for vacation with a suitcase that is much too small. As you force more into the suitcase the contents get tighter and tighter. The next item forces the current items to move from their position and puts stress on the suitcase and everything inside.

A similar thing happens when an extra atom is placed in a normally perfect crystal lattice. The atoms cannot be where they normally would be, and that puts stress on the bonds. This "prestress" on the bonds strengthens the material and is one of the benefits of alloying (mixing metals). The extra atom can replace a normal atom position (**substitution**) or occur between normal atom positions (**interstitial**). It can be of the same material (**self-interstitial**) or a different material (substitutional or interstitial **impurity**). If there is an atom missing from a normal position, it is called a **vacancy**. These various configurations are shown in Fig. 3.8. We will discuss more about point defects in the section on alloying (Sec. 3.2.4.1).

3.2.3.2 Line Defects

Line defects can be thought of as a collection of vacancies in a line. Instead of one lattice site being affected, a *half plane* of atoms is missing (or conversely, there is an *extra* half plane). What is a half plane? Fig. 3.9 shows a schematic of an edge dislocation, the simplest type of line defect. It can be seen that the bonds (lines between the circles) are stretched near the extra half plane. Like point defects, line defects can increase the strength of metals, but there is more to them than this.

Line defects, or dislocations, have a number of important impacts on metals. If we consider the way materials deform by atomic motion, as discussed in Sec. 3.2.2, we would expect that all of the atoms move together and they must all shift their bonds together. This type of motion would require a great amount of force and make the metal

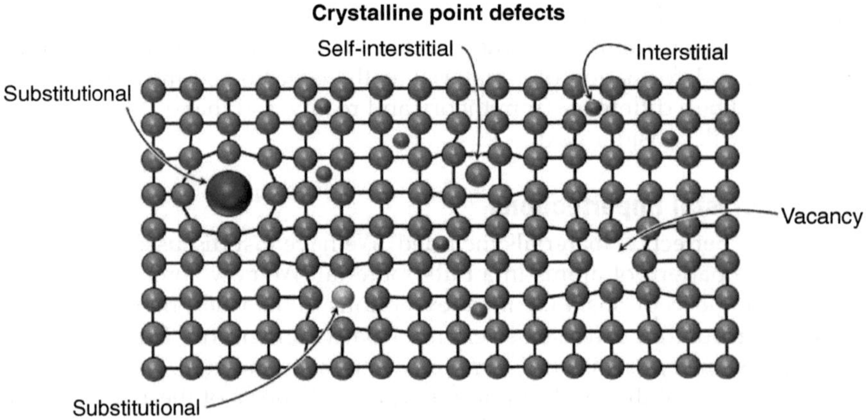

Figure 3.8 Point defects can occur in variety of ways. The defect can be caused by the host material or an impurity as shown above.

extraordinarily strong. Because of dislocations, the actual strength of metals is much lower. Dislocations can move with much less energy because they only exchange one bond (or row of bonds) at a time.

When the dislocation finally reaches a boundary, such as the surface of the material or a grain boundary, the material is now permanently shifted (Fig. 3.10). In Fig. 3.10, the two rows of atoms represent those most closely associated with the dislocation. It can be seen that atom "B" in the top row starts without a bond to the bottom row. As the dislocation moves, the bonds shift by one position. At the surface of the crystal, the dislocation results in an atomic "step" being formed. This is a very small change in shape, and bulk deformation occurs by multiplying this effect over millions of dislocations. This is analogous to a deck of cards. By sliding the cards (atomic planes) over each other, more distance can be traveled than with one card alone (assuming they must stay in contact with one another).

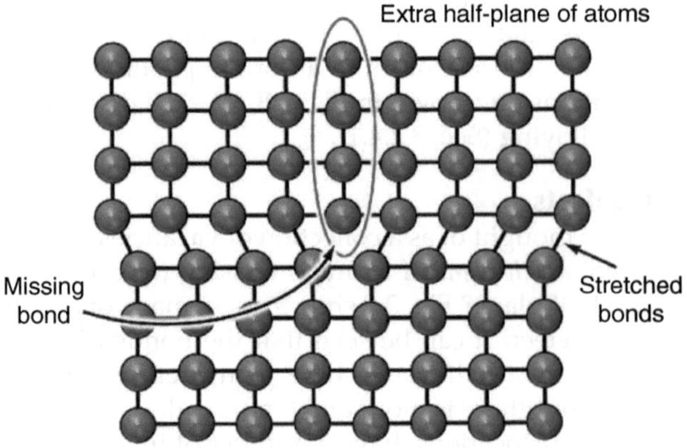

Figure 3.9 A line defect known as an edge dislocation is shown. This type of imperfection is made by an extra half plane of atoms. The additional atom at the interface creates an unsatisfied (missing) bond which, in turn, weakens the material by allowing easier movement at that location.

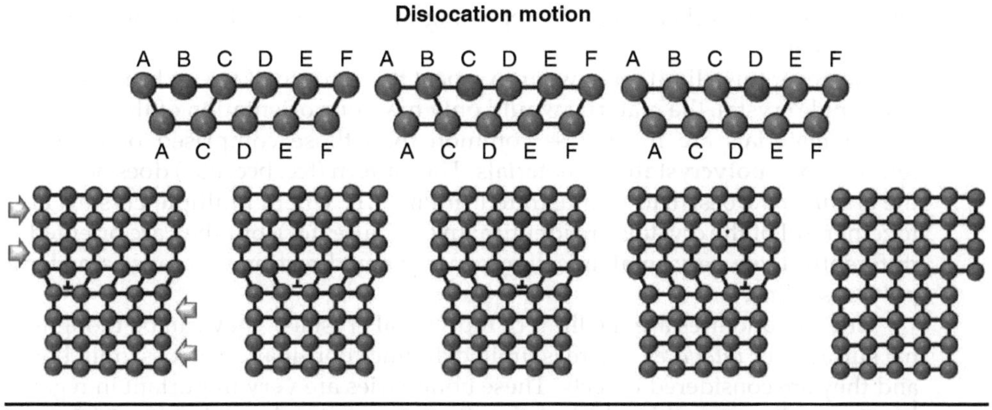

FIGURE 3.10 Dislocations travel through the material, exchanging bonds as they go. During deformation, the dislocation continues through the material until it reaches a surface or other boundary.

Dislocations allow a material to deform under less stress than expected from an "all-bonds-at-once" model, and the material is more ductile as a result, to a point. Dislocations weaken a material by allowing atomic movement through localized (single) bond exchange. When many of these dislocations exist, they can actually entangle* with one another, and this spaghetti-like collection of dislocations cannot move easily and becomes locked. When dislocations become stuck, they no longer contribute to ductility, but rather they work against it. Under significant strain (deformation) dislocations multiply. Although the mechanisms for their creation are beyond the scope of this discussion, it is valuable to know that cold working will increase the *number* of dislocations and heating (annealing) will reduce their number and the **defect density**, or the number of imperfections within a given volume of material.

Metals deform by the motion of dislocations, so they are very important for processing and strengthening. Grain boundaries, precipitates, impurities, and other "imperfections" hinder the movement of dislocations and thereby strengthen a metal. Dislocations weaken a material with respect to the hypothetical, all-bonds-at-once strength, but they can act to strengthen a material as well. Though they do not bring the material to the hypothetical strength, even when many are present, they can be utilized to improve the mechanical properties of metals (see Fig. 7.18 and related discussion for more).

3.2.3.3 Grain Boundaries

Like dislocations, grain boundaries have strength in numbers. That is, the more there are, the higher the strength. So what exactly is a grain boundary? How do we control the number? Well, to start, a grain is simply a single crystal. We have been developing an understanding of atomic order throughout Sec. 3.2, and grains are what determine the extent of the order. Recall from Sec. 3.2.1 that metals have a set pattern of atoms. Typical arrangements include face-centered cubic (fcc), body-centered cubic (bcc), and

*You may rightly wonder how a line of vacancies can possibly entangle. The restriction they place on one another is related to the lattice stress (stretched bonds) they create, not a physical entanglement such as a collection of ropes.

hexagonal close-packed (hcp). By repeating the unit cell in three dimensions you get a bulk crystal.

We never put a limit on how far to repeat this pattern. If we had a material that was one, single crystal, the material would only have one orientation of the unit cell. Single crystal materials are much less common than those comprised of many crystals, referred to as **polycrystalline** materials. The pattern (fcc, bcc, etc.) does not change, but the *orientation* does. This is illustrated in Fig. 3.11 where multiple crystals are joined together. All of the crystals are identical in structure (fcc), but they are oriented slightly differently. Even very small misalignments in any direction will cause a material to be polycrystalline.

Grain boundaries are the limit of individual crystals. They can be thought of internal surfaces, or *interfaces*. There is limited atomic coordination across grain boundaries, and they are considered defects. These boundaries are very important in regard to dislocations. The ease with which dislocations can move through a material determines how strong and ductile a material is. Therefore, anything that would inhibit a dislocation from moving will strengthen a material.

As discussed in the previous section, dislocations can inhibit each other when many of them are present and this increases the strength of a metal. Grain boundaries also inhibit dislocation motion. The dislocations cannot easily move from one grain to another, and therefore they get "stuck" at grain boundaries until the force to move them exceeds a critical amount. We can extrapolate this effect and surmise that the more barriers to motion (boundaries) the stronger a material gets. This is like getting randomly selected at the airport for additional screening. The extra barrier inhibits your progress relative to others.

In order to get more boundaries we have to have more grains. To have more grains in any given volume, we must make them smaller. The volume of the metal does not change, so the pieces have to add up to the same volume, but they can be different sizes. This is like creating a jigsaw puzzle. The same size picture can be made into a 250-piece puzzle or a 1000-piece puzzle. It does not change the picture or the size. There are just more units in that area, or in the case of metals, there are more units in that volume. This is represented in Fig. 3.12.

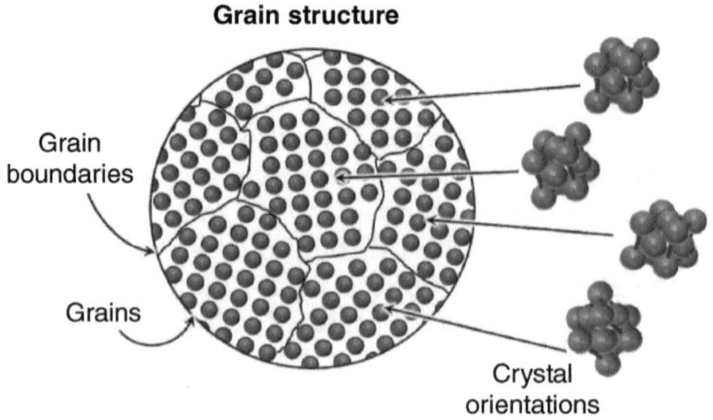

Figure 3.11 Grains boundaries are formed where crystals of different orientation meet. The grains all have the same structure (fcc, bcc, hcp, etc.), but they are tilted relative to one another. Materials made of multiple grains such as shown above, are called polycrystalline.

FIGURE 3.12 A smaller grain size results in more grains over the same area. The concept is shown here using a space-filling pattern similar to that of grain structure. The dark outline indicates the size of an average polygon, or what would be a single crystal, or grain, in a polycrystalline material.

So, being the ambitious folks we are, we want to make the grains as small as possible and the material as strong as possible. Does it work? Is this a good idea? To understand the effect of grain boundaries we can use the Hall-Petch equation (E.O. Hall and N.J. Petch):

$$\sigma_y = \sigma_o + \frac{k_y}{\sqrt{d}} \tag{3.1}$$

In this equation, σ_y (Greek lowercase letter sigma, subscript y) is the yield stress (i.e., the strength of the metal), σ_o is a material-dependent constant related to the difficulty in moving dislocations, k_y is another constant related to how much a material strain hardens (gets stronger as it is deformed), and d is the grain size. What is most important for our consideration is that the strength of a metal, σ_y is related to grain size, d, by an inverse square root function. If you plot 1 divided by the square root of a range of numbers, you will see a very important trend, especially as the numbers get small. This is shown in Fig. 3.13, which illustrates how strength rises drastically as grain size decreases.

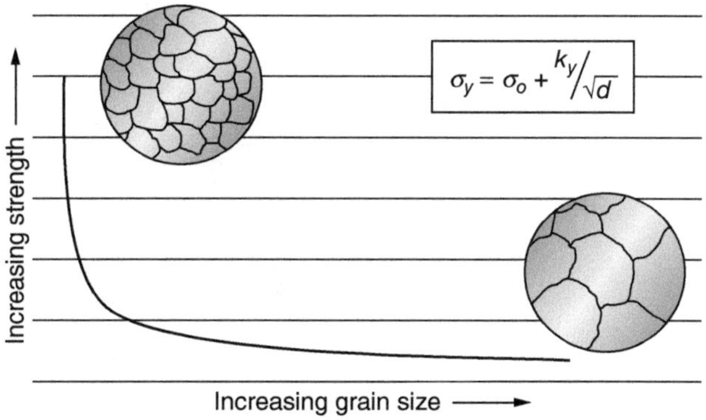

FIGURE 3.13 As grain decreases, the strength increases. This is displayed using the Hall-Petch relation. This indicates that at very small grains the strength will become infinite, which is an unreasonable prediction.

As the grain size is reduced the strength of metals increases, but unlike the Hall-Petch equation predicts, the strength does not become infinite at infinitely small grain sizes. Although the grain size of metals can be reduced to miniscule dimensions (nanoscale), and the strength does increase drastically, there is actually a softening that occurs at the smallest sizes.

For traditional processing, this limit is not reached and is of little practical consequence in current manufacturing, but metals with nanoscale (1 billionth of a meter) grains, called nanocrystalline metals, may be very important in the future. For the purposes of this text, we will consider grain size reduction to strictly strengthen metals. Nanoscale grains are shown in Fig. 3.14.

So yes, it does work, but is it a good idea? There are always trade-offs in life and materials are not exempt. Typically, as strength is increased by reducing grain size the ductility decreases. Therefore, it makes the metal harder to deform, but it will not deform as much without breaking. As with all material choices, the answer to a question of appropriateness is determined by the intended application. The physical factors in use and processing are discussed in Chap. 7. In general, you must know your application before you can say what properties are right for a given metal or alloy. Like the structural considerations we have been discussing, changing the *composition* by adding other elements can also achieve increases in strength or ductility. This will be our next topic.

3.2.4 Alloys

It is rare that metals are used in pure, elemental form. By far, it is more common to mix metals to improve the properties of a metallic material. A metal that consists of two or more components is known as an **alloy**. Alloying is conducted to balance or enhance

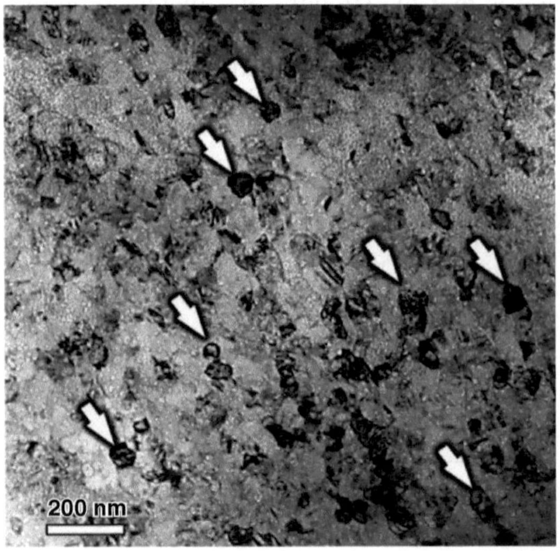

Figure 3.14 This transmission electron microscope (TEM) image shows nanoscale grains in brass. The alternating light and dark areas are the grains. It can be seen that most are 100 nm or below.

the properties of the constituent elements. For instance, aluminum, a soft, relatively weak metal, is alloyed with other elements such as magnesium or copper to greatly enhance the strength. Even when added in small quantities (a few percent) the properties can change greatly.

The degree of change may also be determined by thermal processing (i.e., heat treatment, see Sec. 9.3). If you recall, in the opening chapter we saw a distinct link between materials and society. The Bronze Age (Sec. 1.3.2) is a good example of how alloying can benefit technology, but it also emphasizes that alloying is not a new technique. Designing metals by mixing elements together is a highly valuable science, and many modern improvements to metallic materials are achieved through precision alloying.

Alloys can be as varied as solder (created to melt at low temperature) and superalloys (designed to work under high stress at very high temperature). We will look at some general categories of alloys, but the breadth and depth of commercial alloys is too vast for us to consider in detail. Knowing the general properties of alloy families is a good starting point for selection criteria, but it is no guarantee that all alloys in a family share properties or that a distant cousin in another subset might not serve the purpose better. For this reason, alloy design and selection is a fully developed topic itself, and many, good texts exist to aid the interested reader. The classes and designation of alloys are discussed in Chap. 4. Here we will focus on structural aspects.

The major alloying element (often > 50%) is considered the base material, and the minor alloying elements are technically considered impurities. This reveals that not all impurities are unwanted. Impurities may be accumulated during processing (e.g., hydrogen enters liquid metal during casting), but these are not planned. Alloying impurities are purposely added to the material. Both wanted and unwanted impurities will affect the performance of a metal, and process constraints typically attempt to reduce or account for the unwanted ones. Some common, unwanted impurities are carbon, sulfur, hydrogen, and oxygen. Typically, environmental impurities are common elements that occur in the work environment. Although alloying elements are rightfully categorized as impurities, we will not consider them that way in order to reduce confusion.

3.2.4.1 Alloy Structure

Alloys typically have a structure quite similar to pure metals. For instance, if you add copper (Cu) to nickel (Ni) the structure is fcc. Both Cu and Ni have fcc crystal structures, so this should not be too surprising. The Cu and Ni both occupy regular lattice sites, so it is a **substitutional alloy** (refer to Sec. 3.2.3.1 as a reminder). This indicates that alloys are created through point defects. Recall, crystal imperfections (defects) tend to strengthen a material. Therefore, alloys tend to be stronger than elemental materials. This is one of the main benefits of alloying, and it is commonly referred to as **solid solution strengthening**.

Because two elements do not have the exact same size (atomic radii), they cannot fit together as easily as atoms of a single element will. That means the crystal lattice will be distorted, and the bonds between the atoms will be stretched. This happens to a greater extent if one element is significantly larger (or smaller) than the other.

If the minor alloying element (the solute) is very small in comparison to the host material (the solvent), the element typically occupies a position *between* normal lattice positions. This type of alloy is called an **interstitial alloy**. A common interstitial alloy is steel, where carbon atoms (small) assume positions in the gaps between iron atoms (large). Even though the atoms take positions between regular lattice sites, there is still

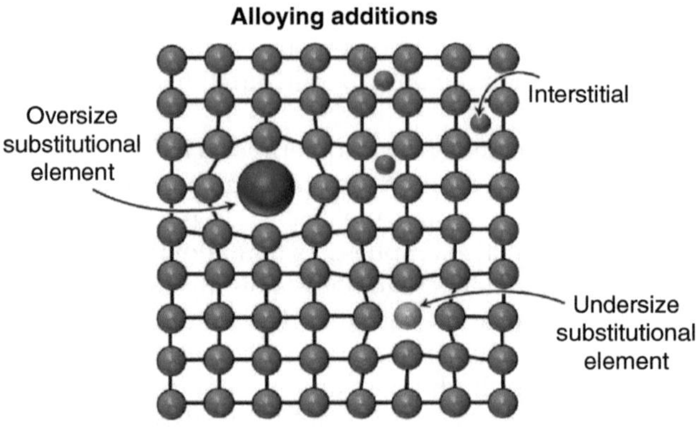

FIGURE 3.15 Alloying can be achieved by substitution of another element in normal atomic positions or by placement between those positions. The first case creates a substitutional alloy and the second creates an interstitial alloy. Comparing this figure to Fig. 3.8 reveals that alloying elements are actually point defects, although they are desirable ones.

not enough room for them and they will still strengthen the material by stretching the atomic bonds and hindering dislocation motion. The three basic configurations of atomic mixing are shown in Fig. 3.15.

3.2.4.2 Atomic Solubility

Cu-Ni alloys are fairly easy to envision since they are quite close in size and both elements have the same crystal structure in pure form. What happens when there are drastic differences between the elements in an alloy? How do we know if two elements can be mixed to form a stable alloy? Like oil and water, some elements will not readily mix, and instead they form separate areas of each, individual material. The separation of phases like this is known as **phase segregation** and can be useful in alloy development.

There is a general set of rules to determine the liklihood two elements will mix, known as the Hume-Rothery rules (after William Hume-Rothery, 1899–1968). The four rules for elements to mix well are

1. The atomic radii should differ by no more than 15%.
2. The crystal structures of the two should match.
3. The valency* of the two should match.
4. The two should have a similar electronegativity.†

If all of these criteria are very nearly satisfied, the two metals will be fully soluble (fully mix) in one another. If they are not met, the two may exhibit limited solubility. If there is a large difference in atomic size, the minor alloying element will tend to segregate to interfaces such as grain boundaries and surfaces. If they have different crystal structures, they will tend to phase separate into distinct regions of elemental composition.

*The valency of an atom is determined by the electrons in its outermost shell. This was discussed in Sec. 2.3.

†Electronegativity was discussed in Sec. 2.3.1.3.

If the valencies are different, the metal with the lower valency will have a higher solubility in the higher valency metal than vice versa. For example, more zinc (Zn) will dissolve in Cu than Cu will dissolve in Zn. Cu has a higher valency than Zn. The two elements cannot mix completely in a 50:50 ratio. Beyond their solubility limits, the elements separate and/or form different structures.

If the two metals have different electronegativities they tend to form intermetallic compounds that have a high strength and melting point and low ductility. These compounds are usually ordered structures, consisting of two or more metals and may also include nonmetals. Their properties are somewhat of a compromise between the good ductility of metals and the high stiffness of ceramics.

We have only considered two elements mixing together so far, but commercial alloys typically contain many elements, even if the alloy itself is considered to be made of two (binary alloy) or three (ternary alloy) *major* alloying additions (Table 4.3 is a good example of this). The way metals mix will have a direct impact on the microstructure. If they do not mix well, there will be distinct features in the structure involving many atoms. Remember, structure can be defined at many different length scales.

3.2.4.3 Phase Diagrams

The solubility of two metals is determined by the Hume-Rothery factors listed above as well as temperature. The separation of phases, the creation of intermetallic compounds, and the composition of the phases is a complex function of *composition* and *temperature*. To map this dependence out, **phase diagrams** are used.

Have you ever made rock candy? This is a fairly simple (and tasty) way to illustrate solubility and phase separation (Fig. 3.16). Rock candy is made by adding sugar to water and heating the mixture to create "simple syrup." At room temperature the sugar dissolves until there is too much to continue dissolving and the rest is present as solid granules in the sugar-water mixture (two phases: solid sugar + liquid syrup). As the mixture is heated and stirred, the extra solid sugar granules will now dissolve in the

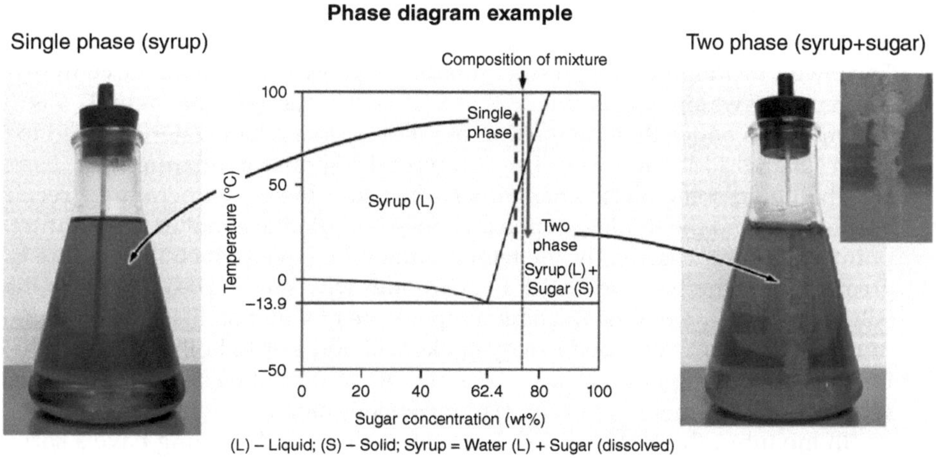

FIGURE 3.16 Phase diagrams are used to determine the phase(s) present for any temperature and composition. The sugar-water phase diagram shown here illustrates the concept. Rock candy is made by utilizing the fact that sugar will dissolve to a greater extent in hot water than in cold.

water (single phase: liquid syrup). For rock candy, a string or other item is then placed in the syrup and the mixture is allowed to cool.

Upon cooling the sugar separates out and forms crystals on the string (two phases: sugar + syrup). So why did the mixture go from two phases to one phase and back to two phases by changing the temperature? The simple answer is that materials tend to exhibit higher solubility at elevated temperature. The same thing happens in metals, but the number and complexity of the phases increase. Also, metals are used in a solid state, and so the rock candy could be also. If the water was cooled until it froze, the sugar would continually settle out until the water was solid. In metals, and materials in general, just because they are solid does not mean things are not happening. Even at room temperature, atoms can move around in a material, and although it may be slow, it can significantly impact a material.

Before we move to metals, let us glean what we can from this example. First, the lines on a phase diagram are very important. They represent **phase boundaries**. When the line is crossed, a new phase or phases are present. This is most intuitively experienced in the rock candy example during the cooling of the water. A change in temperature is equivalent to moving vertically on the phase diagram. We add a certain amount of sugar, represented in weight percent (wt%) along the horizontal axis. The arrows in the diagram are around 75 wt% sugar (chemical formula $C_{12}H_{22}O_{11}$). When we have 75 wt% sugar, we have 25 wt% water. The combination will always equal 100%. The balance is what varies. When we move vertically along these lines we change from a solid solution, single-phase mixture at the top to a two-phase, sugar-syrup mixture at the bottom. This phase change occurs at approximately 50°C, and the temperature is listed on the vertical axis (Fig. 3.17). In the same way, we can choose one temperature and move horizontally to determine how the phases change as a function of composition. For instance, at 50°C we could move right, increasing the amount of sugar, and we would have a single phase mixture until we reached approximately 75 wt% sugar. Phase boundaries can be between solid phases of different compositions as well, but in the sugar-water example, this is not encountered.

Additional information available in phase diagrams is the melting/freezing point of the pure materials and the composition which results in the minimum melting point. The lowest melting point for a solid mixture is known as the **eutectic composition**, and for the sugar water system it occurs at 62.4 wt% sugar (alternatively, 37.6 wt% water). You may also notice that the melting point decreases when water is added to sugar, or when sugar is added to water. This is a typical response in mixtures.

The liquid-solid phase change is shifted to a lower temperature (**freezing point depression**) and the liquid-gas phase change is shifted to a higher temperature (**boiling point elevation**). Essentially, the temperature range over which the mixture is liquid is greater than either pure material. Ever wonder why salt is spread on ice in the winter? Salt, like sugar, lowers the freezing temperature of water so that it stays liquid instead of turning to ice. Ever wonder why cooks will add salt to boiling water? The salt will increase the boiling point of the water and act to raise the cooking temperature (there are also, perhaps more important flavor considerations).

In mixtures, the temperatures between melting and freezing have a sort of slushy region where there is some solid present as well as some liquid. You can find the balance by choosing one temperature and composition. From that point on the diagram, draw lines directly left and right (Fig. 3.18). The phases present at that specific

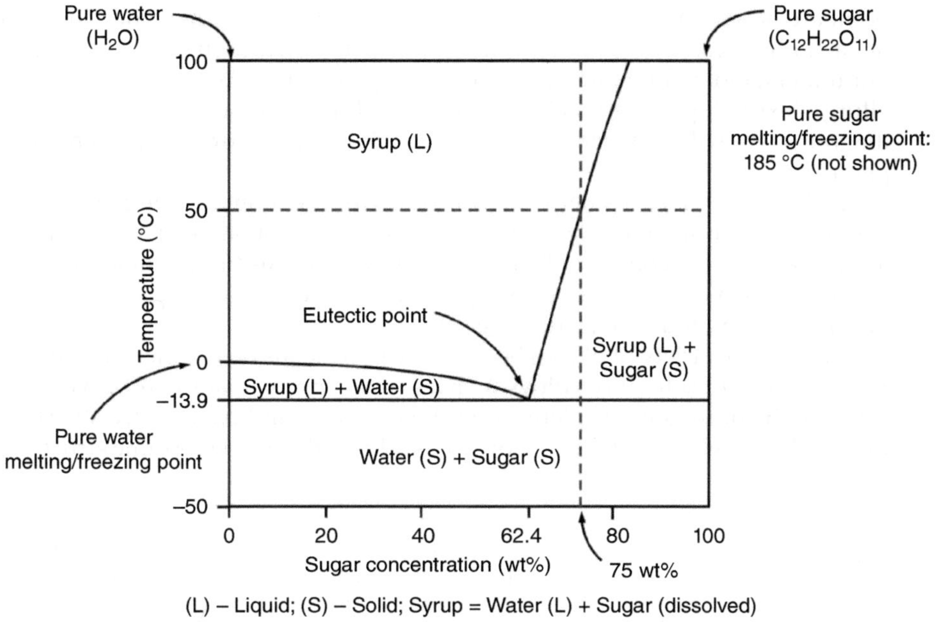

FIGURE 3.17 Phase diagrams, such as the sugar and water diagram above provide information on phase boundaries (the lines) where the mixture changes phases. This can be determined by moving vertically (temperature) or horizontally (composition).

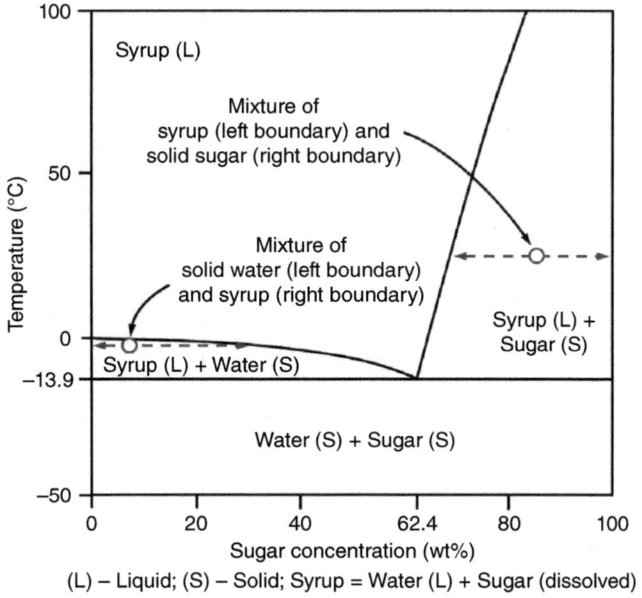

FIGURE 3.18 The phases present and the effect on melting/solidification temperatures are shown in phase diagrams. At any temperature and composition the phases present can be identified. Examples are presented at different location in the phase diagram, but the process is the same regardless of the composition and temperature.

point are determined by whatever phases are located on either side. The composition of each phase can also be determined by the "lever rule," but this is not pertinent to our understanding of materials in context to processing. Any materials science or metallurgy text will present this information (and much, much more) in regard to phase diagrams. We will only concern ourselves with what phases are present at what temperatures.

Although the sugar-water system is useful for exemplifying the concepts of alloying, it is not very informative of how metals will behave. Most phase diagrams involving metals are much more complex. Take, for example, the Cu-Zn system shown in Fig. 3.19. These two elements are the major components of brass, a common material used for bushings, knobs, musical instruments, etc.

There are a variety of phases because the two metals do not meet the Hume-Rothery requirements for solubility. Alloys with less than approximately 30% Zn (by weight) are called alpha brasses (Greek letter α in the diagram) and have an fcc structure like the major alloying element, Cu. As more Zn is added, the bcc beta-prime structure (symbol

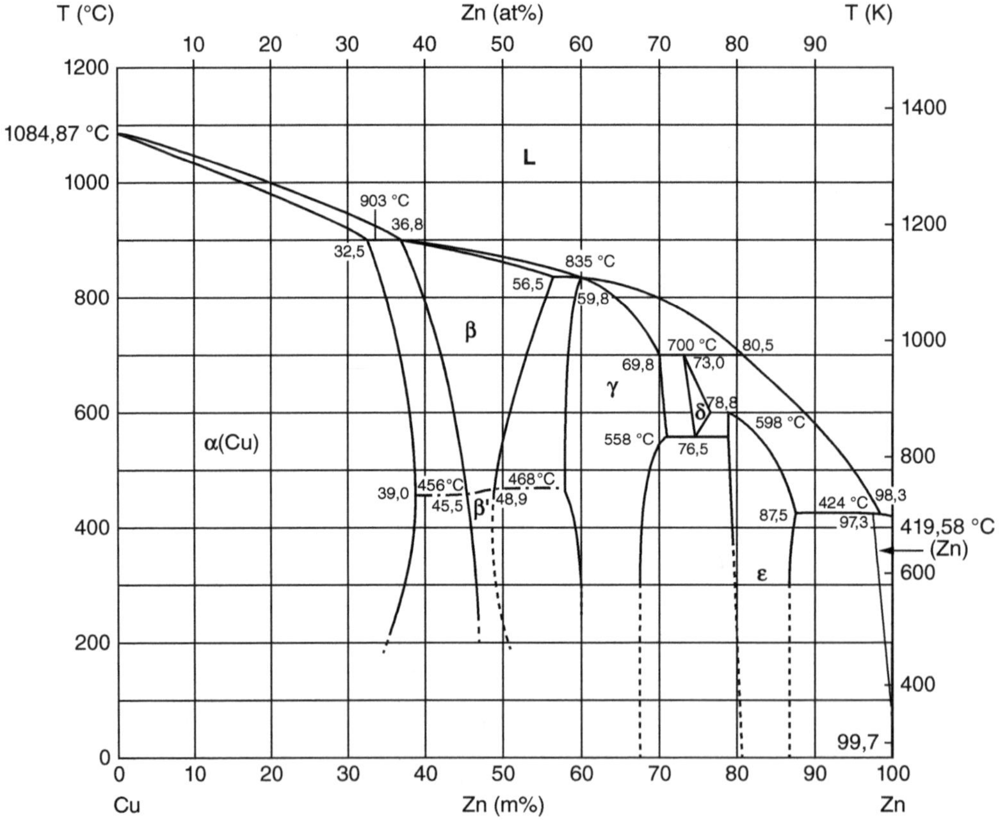

FIGURE 3.19 The Cu-Zn binary phase diagram is much more complicated than the sugar-water diagram considered above. Regardless of the number of phases present, the basic information and utilization remains unchanged.

β' in the diagram) starts to appear. Finally, pure Zn is hcp, but high zinc brasses are brittle and not common commercially. You can see on the diagram that within any particular area, the phases present are a mixture of those present to the left and right as was the case with the sugar water.

For the purpose of this text, phase diagrams will be called upon primarily in heat treating to help us understand how it works and why it is done that way. A great depth of information can be had by fully understanding phase diagrams, and it is a must for alloy development and the advancement of materials. We have been discussing binary phase diagrams (two materials), but other, more complex phase diagrams involving three (ternary), four (quaternary), or more elements also exist. To accurately determine the way materials will react when mixed is typically analyzed through computer simulation. This allows materials scientists to generate and interpret higher-order phase diagrams without excessive experimental validation.

3.3 Why It Matters

In this chapter, we have considered some basic structural concepts in metals. So, why is structure important? How does it inform those of us who are primarily interested in processing? Consider the following: When we want to describe a material, we should first consider what length scale we are interested in. Materials can be described by atomic properties, and this is vital for developing new materials, but it is rarely useful to someone working on the shop floor. If we increase the distance over which we describe our material, we get the crystal structure next. This contains more direct information about the way we can process the material, but is typically not specific enough.

When we move out further still, we will see that the crystal structure has variations throughout the material. Outside of the way atoms interact through bonding (Sec. 2.3), their arrangement is the next most influential aspect of structure. Small-scale structure affects large-scale structure. For that reason, structural information on atomic packing is very useful in choosing processing. For example, bcc materials tend to be less ductile than fcc materials. If we know our material's crystal structure, it can inform us of the amount of deformation or temperature of processing we can or should apply. When we begin looking at imperfections in the structure, we can appreciate why a material fails under excessive deformation (strain). Accumulation of dislocations and grain boundaries will strengthen a material and we can use this fact to our advantage.

Alloying is perhaps the most obvious change we can apply to a material. The choice of the composition will change a material from being weak to strong, corrosion prone to corrosion resistant, etc. Small variations in composition (i.e., a few percent) can have dramatic effect on the properties of a metal. How these compositional changes occur is also important.

The alloying elements may be incorporated between the host atoms (interstitial) or in place of the host atoms (substitutional). The size difference between the atoms is important, and a large difference will strengthen the material, but also limits the solubility of the alloying element. There are commonly many different alloying elements added to metals to help control behavior during processing, so the interactions tend to be complex. We looked at phase diagrams which provide a wealth of information on the way atoms interact. Phase diagrams serve as a map of material structure. We can look at composition or temperature and identify what structure is present.

Alloying, which is technically an imperfection, has important effects on the properties of the bulk material, but the atoms added are considered a permanent feature (i.e., thermal or physical treatment will not significantly alter chemical composition). Knowing a metal is a specific alloy will help identify the best way to process it since some metals are more sensitive to heat or cutting conditions than others. Since most metals are alloys, we need to know how they react to temperature which most production processes either produce as a by-product (e.g., machining) or use as a primary mechanism for processing (e.g., welding).

Phase diagrams describe the alloy's structural behavior at elevated temperature and can be used to determine processing temperature limits or to identify appropriate heat treatment parameters. Each structural level has valuable applications to material understanding. The purpose for that understanding is to improve product quality and capability. Well-informed manufacturing and design personnel are an asset to a company. Material structure is anything but abstract and unrelated. It is at the very core of processing and a working knowledge of the interplay between microscale features and macroscale properties can save a tremendous amount of time and money when choosing manufacturing methods. The way these concepts are related is shown in Fig. 3.20.

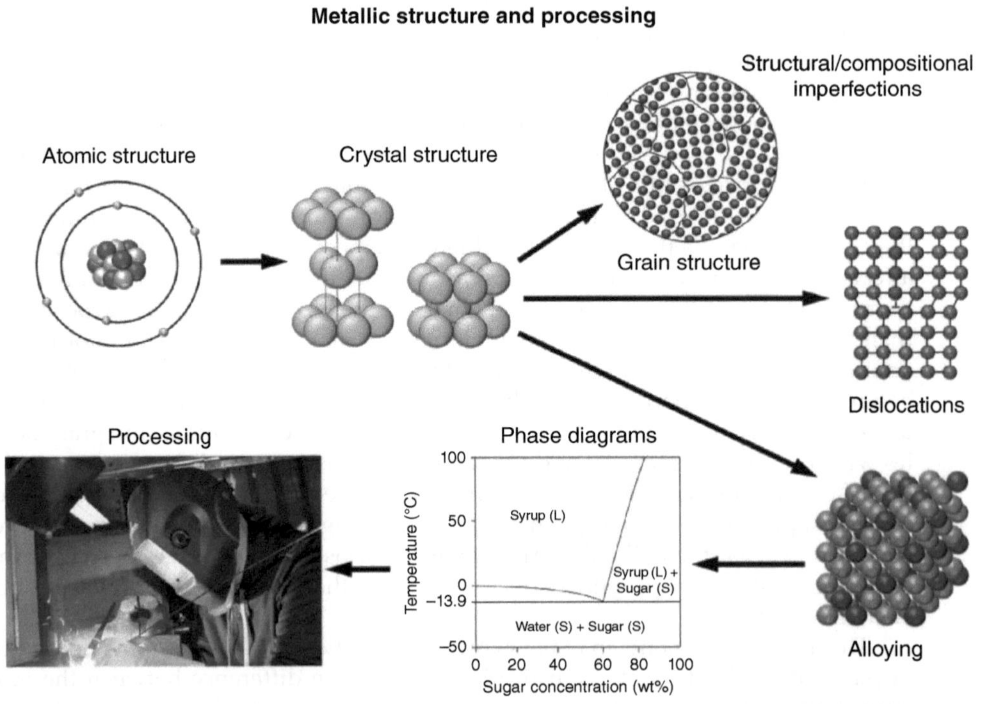

FIGURE 3.20 Material parameters used during processing are informed by structure. Atomic structure determines the crystal structure. Crystal structure is modified by imperfections but also determines the type and prevalence of defects. Alloying, a specific type of imperfection, is where phase diagrams are derived based on the mixing behavior as a function of temperature and composition. The phase diagram gives valuable information on how to process materials. For instance, welding must consider composition and temperature so the materials form a suitable, cohesive bond.

We have approached structure in a very limited way, but these concepts are all important and useful to processing. There are many textbooks dedicated just to structure, and there more books on each aspect of structure such as grains, dislocations, alloying, etc. For our purposes it is important that we see the big picture and to understand the importance of what we do not see. That is, the very active world of metallic structure.

3.4 The How and Why of Metallic Materials: Structure

3.4.1 Chapter Review

1. What comes to mind when you think of structure? How might this vary between people?
2. Why is studying microscale structure important for a large-scale process like forging?
3. Why does small-scale structure affect large structure, but not the opposite?
4. What is a "long range" when discussing atoms?
5. What are three most common arrangements of atoms?
6. What information does a unit cell provide?
7. Write out the full name for fcc, bcc, and hcp and name a metal for each that has that structure.
8. What is the crystal structure of a metal?
9. Why is the spacing between atoms important for processing?
10. What is "slip" when discussing structure?
11. What are the three types of crystal defects?
12. What is a point defect? Name three types of point defects.
13. Describe the difference between an interstitial and substitutional defect. How does this relate to alloying?
14. Do point defects strengthen a metal? Why or why not?
15. How do dislocations weaken metals? How do they strengthen metals?
16. Draw a graph with the general trend of strength versus the number of dislocations in a metal.
17. How do dislocations allow materials to deform?
18. What will deforming a metal do to the number of dislocations in a metal? What about heating?
19. What is the term used for a metal made from many crystals?
20. The grains of a metal all have the same structure. What difference *do* they have?
21. How do grain boundaries strengthen a metal?
22. What happens to the strength of a metal as you reduce the grain size?
23. When you reduce the grain size by 10%, the strength will not change as much as when you reduce it another 10%. Why?

24. Is working a material to increase the number of dislocations and decrease the grain size always a good idea?
25. What is an alloy?
26. Why is alloying used?
27. How were alloys important in the past? How are they important today and in the future?
28. How do impurities relate to alloying? How is it that impurities can be both good and bad?
29. Why are alloys stronger than pure materials?
30. What is the difference between a substitutional alloy and an interstitial alloy?
31. What basic rules determine whether two elements will mix? Write these rules in your own words.
32. What happens if two metals do not mix well? Give at least three possibilities.
33. What information does a phase diagram give?
34. Why do phases mix more easily at high temperature?
35. What value changes in a phase diagram as we move left or right (horizontally)? What about up and down (vertically)?
36. What do solid lines within a phase diagram represent?
37. How do we know what phases exist at any point in a phase diagram?
38. How does metallic structure inform processing?

3.4.2 Concept Connections and Extensions

1. How does the structure of metals resemble the way we make structures? Give some examples.
2. What might happen when two dislocations with opposite orientations (one pointed up, the other down) line up? Draw a diagram to demonstrate.
3. What other disciplines (nothing to do with atoms) might consider the concept of packing factor? Is it more or less difficult with nonspherical units? Explain.
4. Phase diagrams are used extensively to map the interaction of elements in alloys. How are they useful in processing and applications? Consider discussion in Sec. 8.2, for example.
5. Defect density is related to strength in metals. Develop an analogy to describe the concept of how defects can weaken a material as well as strengthen it (i.e., use a too little/too much scenario).
6. Describe the increase in strength in the stress-strain curve (Sec. 7.4.1) based on defect density. How does the material change structurally during testing?

CHAPTER 4

Metallic Materials: Identification and Application

Metallic materials are ubiquitous in everyday life. Some applications are obvious such as handrails, chain-link fence, automotive components, cookware, etc. Others are less obvious, such as some cell phone and computer cases, reinforcement of concrete structures, tire construction, and electrical wiring. Metals can vary so greatly that it is difficult to lump them into any one application or set of applications. Some of the principal functions metals occupy are covered in this chapter, but there could be (and are) volumes of books filled with those applications not covered. Again, because metals vary so much, there must be multiple classifications in order to divide them into categories with similar composition and/or properties. We will begin with compositional classifications and then move to property-based classifications. Some metals can fall into multiple categories, and they will be discussed where most appropriate.

4.1 How They Work

In the past couple of chapters, we have been developing a fundamental understanding of materials from the ground up. There are important designations given to metals that are useful for identifying their underlying characteristics. We have looked at the importance of atomic structure (Sec. 2.2), bonding (Sec. 2.3), and microstructure (Sec. 3.2), but there are more general ways to delineate metallic materials. These classification methods are shortcuts for materials selection and identification. For example, 316L stainless steel is a common metal, but what do the numbers, letters, and naming indicate?

Most obviously, the material is steel and stainless (rust resistant), but have you ever wondered, "What is steel?" There is no element named "steel," so it has to be a mixture of elements, but which ones? Perhaps you are already familiar with it, but it is common to use and rely on materials every day that we do not know much about. We trust that an engineer has chosen the proper material and that it is safe for use, and the vast majority of the time, our trust is well-placed.

We might take for granted that materials have some intrinsic properties that make them suitable for their application, like gold being resistant to tarnishing and having a pleasing appearance, but the reality is that there is a great deal of research and development that goes into designing and choosing metals. Even gold is most often not pure

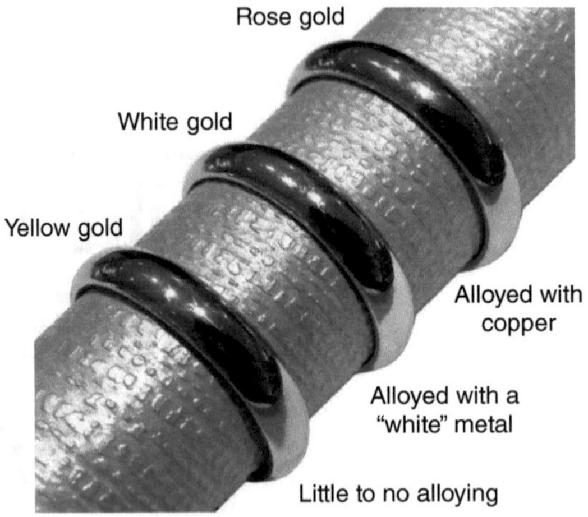

FIGURE 4.1 Gold is available in a variety of colors, and these color differences arise because of the alloying additions used. The appearance and major alloying compositions are listed.

(pure gold is referred to as 24 karat or 24k). It is mixed with other elements to increase its strength or change its appearance. For example, white gold is actually a mixture of gold and silver (or another "white" metal), and rose gold is a mixture of gold and copper, and the three can be mixed in varying proportions to achieve the desired balance of colors (Fig. 4.1). Even simple materials like gold are more complex than you might expect.

Materials are not haphazardly chosen in structural applications either. What works for the frame of a car may not work for a suspension spring. Both are often steel, but they have what might be considered "small" variations in composition. Again, knowing what steel really is comes up, and small variations can make big differences. These variations are where the numbering and lettering comes from (e.g., 316L). Although they have a lot in common, there are still important differences which we will consider.

We have raised some questions, and hopefully it gets you thinking about what metals should be used where. Even more importantly, we can begin unraveling the plethora of metallic material families. Metals are commonly distinguished by their composition or by their properties (of course, the properties are largely determined by the composition). There are thousands of different metallic materials, mostly alloys, and a standard system for identification is crucial to creating a common language. This language is essential for scientists, engineers, and designers to communicate.

4.2 Compositional Classes of Metals: Ferrous

Metals can be divided into two major classes: ferrous and nonferrous. Ferrous metals contain iron as the major alloying element (often greater than 50%). Being that iron is the basis for naming all metals in this way, you can imagine iron is a very important engineering metal. It is the basis for the numerous varieties of steels, but alloying elements also play a decisive role and further subdivide ferrous metals into even more categories. We will start with the common thread, iron.

Figure 4.2 Pure iron is rarely used in modern structural engineering. It still finds use in wrought form for decorative applications such as the balcony railing shown (left) and in magnetic applications such as electrical transformers (right).

4.2.1 Iron

"Pure" iron was a crucial material in the development of society through tools and weaponry. As mentioned in Sec. 1.3.3, the Iron Age was a defining period in history, and the last of the so-called "material ages." The material used then is what we now consider ***wrought* iron**, or iron with a small percentage of impurities in it. You may have seen wrought iron fencing, railings, or other decorative uses (Fig. 4.2). Iron is no less important to us today.

Although rarely used in pure form, it shows up in a majority of alloys. Iron is favored is due to its low cost, ease of processing, and versatility. Iron is the fourth most abundant element in the earth's crust (after oxygen, silicon, and aluminum), economical to refine from mined ore, and is extremely versatile. Iron alloys can be made to possess any range of properties from high hardness to high ductility, magnetic or nonmagnetic, corrosion resistance, etc. by relatively small additions of other elements.

Additionally, iron can be efficiently recycled without degradation and the rate of recycling is very high. According to United State Geological Survey (USGS),* 90% of steel or more was recycled from sources such as construction materials and appliances. Steel recycling from automobiles exceeds the requirement for new vehicles (domestically), so the recycling rate is actually greater than 100%.

In its pure form, iron finds limited use in modern society. Some products now made from steel used to be made from wrought iron. These include common items such as nails, chains, pipes, etc. Modern use of pure iron is mainly limited to nonstructural applications such as electrical transformers (Fig. 4.2), where its magnetic properties are most desirable. Even in that case, alloying additions are typically employed (i.e., silicon).

4.2.2 Plain Carbon Steel

Plain carbon steel is simply iron and carbon mixed together. There are often small quantities of other elements, but the two most important ingredients are iron and carbon. Despite the apparent simplicity, there is a large variety of steels based on these two elements. The amount of carbon is very important in dictating the resulting properties

*U.S. Geological Survey, Mineral Commodity Summaries, January 2018.

of the alloy. Although iron and carbon are the basis for many plain carbon steels, those two elements will also be the basis for many other alloy steels as well. It is worth our time, then, to start by discussing the naming conventions.

4.2.3 Naming Designations for Steels

Steel is not just steel. Steel is such as a broad category that entire industries, professional societies, and handbooks are dedicated to its understanding. To maintain a universal language, a naming convention was devised to provide information about a type of steel in a compact format. The composition and properties of standard steels must conform to set guidelines to ensure consistency. If you order SAE 1045 steel from one manufacturer then the next week from another, you should be able to use them interchangeably. There *should* be no appreciable difference between the two.

So what is SAE 1045 steel? Well the "SAE" lettering is for Society of Automotive Engineers, but you will also see steel designations for AISI (American Iron and Steel Institute), ASTM (American Society for Testing and Materials), and others. The problem with many standards is that they are developed after multiple naming conventions have coexisted for years. Although the Unified Numbering System (UNS) designation was developed more recently to bring these various identification systems together, old habits die hard, and you will still see the original designations in use. It is our burden to look at both and be diligent in our interpretation of the coding.

4.2.3.1 Alloy Designation under the Unified Numbering System

We will start by considering the UNS system, which is standardized for all metals and alloys. This system is based on broader categorical principles. A single letter is used to designate the metal and alloy class, and then five digits are used to indicate the specific alloy within that group. The letters used in the UNS numbering system are shown in Fig. 4.3. The UNS system, then, is used similarly to describe ferrous and nonferrous metals. Older conventions are specific to individual categories of alloys, as we will look at next.

4.2.3.2 Alloy Designation under the AISI-SAE System

The AISI-SAE numbering system is used for plain carbon steels, as well as alloy steels. Depending on the amount and type of alloying additions, the designation method will change. For plain carbon and low-alloy steels, this designation uses four numbers, and their meanings are shown in Fig. 4.4.

The AISI-SAE numbering system is nice for a couple of reasons. First, the number system is relatively simple. There are really only two factors to remember: the alloy class and the carbon content. Even better, the designation is quantitative. The first two digits give you alloying composition, and the last two allow you to quickly determine exactly how much carbon is present. Also, the numbering is grouped by alloying elements (most of the time). The 1000-series alloys are pretty much just carbon and iron with minor additions of other elements. The subclasses 10XX, 11XX, 12XX, etc. are then modifications to this central theme.

For instance, 1018 carbon steel is considered plain carbon steel, and 1118 is *resulfurized* steel, which has better machinability and heat treatment properties than the plain carbon steel. Note that both steels have 0.18% carbon content. When using the BBYY format, the percent carbon content can be determined by putting a decimal before the YY digits. For instance, 4140 steel is a chromium-molybdenum steel (BB = 41) and has 0.40% carbon content (YY = 40 $\rightarrow$ .YY = .40% carbon).

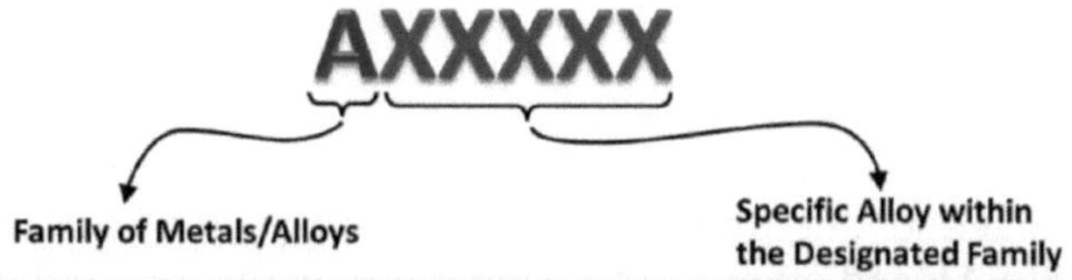

UNS Designation Letter	Family of Metals/Alloys
A	Aluminum and Aluminum Alloys
C	Copper and Copper Alloys
E	Rare Earth and Related Metals and Alloys
F	Cast Irons
G	AISI and SAE Carbon and Alloy Steels
J	Cast Steels
L	Low-melting Metals and Alloys
M	Magnesium and Magnesium Alloys
N	Nickel and Nickel Alloys
P	Precious Metals and Alloys
R	Reactive and Refractory Metals and Alloys
S	Heat and Corrosion Resistant (Stainless) Steels
T	Tool Steels (Wrought and Cast)
Z	Zinc and Zinc Alloys

Figure 4.3 The Unified Numbering System (UNS) combines the various designation protocols into one, all-encompassing format. There is some carry over from previous formats to aid in ease of use.

AISI-SAE Designation	Common Alloy Name(s)
10XX, 11XX, 12XX, 15XX	Carbon Steels
13XX	Manganese Steels
23XX, 25XX	Nickel Steels
31XX, 32XX, 33XX, 34XX	Nickel-Chromium Steels
40XX, 44XX	Molybdenum Steels
41XX	Chromium-Molybdenum Steels
43XX, 47XX, 81XX, 86XX, 87XX, 88XX, 93XX, 94XX, 97XX, 98XX	Nickel-Chromium-Molybdenum Steels
46XX, 48XX	Nickel-Molybdenum Steels
50XX, 51XX, 52XX	Chromium Steels
61XX	Chromium-Vanadium Steels
72XX	Tungsten-Chromium Steels
92XX	Silicon-Manganese Steels

Figure 4.4 The common numbering convention for steel alloys is based on SAE-AISI designations. The first two numbers simply denote the class of alloys. The second two numbers are directly related to the carbon content. To determine the percent carbon content, divide the last two numbers by 100.

The UNS system attempts to bring together all metals and alloys under a single naming convention. In order to ease the transition, the UNS designations often incorporate AISI-SAE and other alloy designations on the basis of familiarity. This adoption is made with many plain carbon and alloy steels. For instance, 1018 steel (AISI-SAE) is G10180 (UNS) and 4140 steel (AISI-SAE) is G41400 (UNS). There is a departure for some alloy classes such as tool steels. For instance, A2 tool steel and O1 tool steel (AISI-SAE) are T30102 and T31501 (UNS), respectively. Tool steels will be discussed more in Sec. 4.2.5.

The carbon content is decisive to resulting properties and is used to further delineate steel alloys. You can find a variety of alloys within each category, each with a different carbon content. Often the variation in carbon is quite small. For example, you can get alloys with SAE-AISI designations of 1017, 1018, 1019, 1020, etc. Each alloy has a specified composition of carbon as well as other elements. Small differences in composition make big differences in processing and performance. We will use the carbon content to divide steel into low, medium, and high carbon subcategories.

4.2.3.3 Low-Carbon Steel

Low-carbon steel has less than 0.30% carbon with a minimum of 0.05% (e.g., AISI-SAE 1005-1030). Steel needs carbon by definition, as 0% would be iron, hence the lower limit. Low-carbon steels are widely used in applications that are not highly demanding in regard to temperature, corrosion, wear, or other physical stresses. It is not very strong compared to other steels, but that becomes a benefit in processing. Low-carbon steel is relatively ductile and easily machined making processing cost-effective.

Modifications to the composition can improve machinability further. Sulfur and phosphorous, which are often considered *impurities* (unwanted elements), can be intentionally added in quantities greater than normal to improve the machinability (i.e., AISI-SAE 11XX and 12XX alloys, respectively), but this is at the expense of formability. Lead can also be added to these steels to improve the cutting characteristics, but with an associated weakening of the material. "Leaded" alloys carry an "L" in their AISI-SAE designation (e.g., 11L18 or 41L40). In general, low-carbon steels are vital in engineering design based on their low cost and acceptable properties. They are easily machined, welded, and formed and used in applications such as structural tubing, pins, screws, bolts, forgings, and sheet goods.

4.2.3.4 Medium-Carbon Steel

Medium-carbon steels have a carbon content above 0.30% and below or equal to 0.60% (e.g., AISI-SAE 1031-1060). Due to the higher carbon content, the strength of these alloys is higher. Applications are generally more demanding mechanically, but medium-carbon steel does not have notable improvements in corrosion or thermal performance. These steels can also be modified for free machining properties with similar benefits/drawbacks.

A benefit of higher carbon content, and one of its most important roles in many alloys, is the ability for the material to be significantly altered through heat treatment. By heating and quenching, the hardness and strength can be greatly improved. The *hardenability* of a steel is related to the carbon content (among other things), but discussion of this is saved for Sec. 9.3.6.3. It is sufficient to say that carbon content is directly linked to strength both is the "as-received" and the "as-processed" states.

Medium-carbon steels are often selectively hardened (e.g., a surface) or hardened through thin sections. Higher strength from the manufacturer means that machinability is slightly lower and forming operations require more force as well, but overall workability remains good. They may show up in similar applications as low-carbon steel, but those which require higher strength such as certain bolts, hand tools and structural members.

4.2.3.5 High-Carbon Steel

High-carbon steel has a carbon content greater than 0.60% but typically less than 1% (e.g., AISI-SAE 1064-1095) and follows the trend set by low- and medium-carbon steels. The higher carbon content means a higher strength and more ease of heat treatment to increase hardness further. High-carbon steels can be hardened through the full thickness more easily than medium-carbon steels. Although they are still relatively easy to process, they can be used in very demanding applications of wear and mechanical stress. Although processing is harder on tooling, the capabilities of these materials offset the additional cost and make them good choices in applications such as springs, wires, and cutting implements.

With all other factors being equal, the general effect of carbon addition in steel is shown in Fig. 4.5. The general trends shown are meant to be informative only, as performance is determined by many factors such as processing methods and operating temperature, which may alter the actual performance of the steel.

4.2.3.6 Processing Designations for Steel

Steels also carry other designations related to processing. The names are not intuitive and require some explanation. For instance, **killed steel** is deoxidized while it is in the

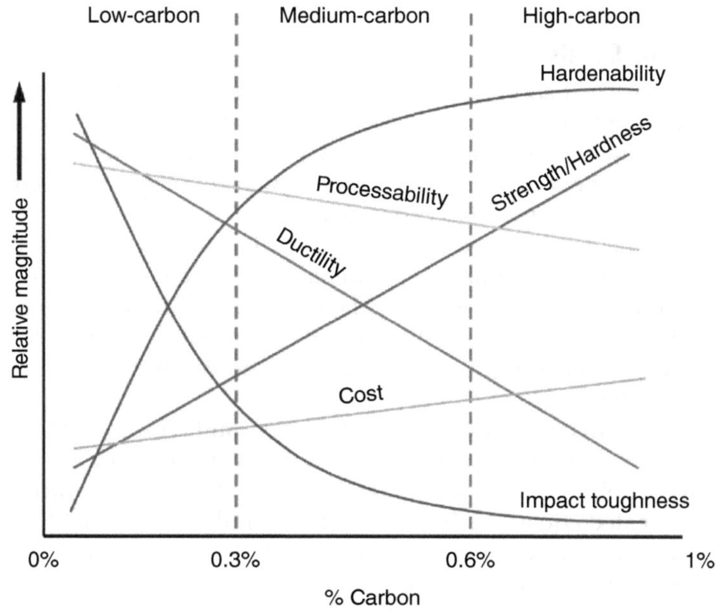

Figure 4.5 Carbon content is very important to determining the properties of steel. This chart shows some general trends in properties based on carbon content.

molten state so that no gas pockets form during solidification. This would occur because much more oxygen can dissolve in the molten steel than in solid steel. It must come out of solution (form bubbles) as the steel cools. This can result in porosity and inferior quality.

Deoxidation is accomplished by adding elements or compounds to the molten mixture. These bind to the oxygen and form slag and can be removed from the top. Common deoxidation agents include aluminum and manganese. **Semi-killed steel** is not fully deoxidized, but it gives higher yields (90% versus 80% in killed steel). Pore formation does occur during solidification, but this type of treatment is commonly applied to steels that are later rolled (see Sec. 12.2.3), which closes the porosity.

Rimmed steel uses little to no deoxidation agent. The lack of additives allows alloying elements to segregate and collect in the center of the ingot created. This leaves a "rim" of nearly pure iron around the ingot which has a very smooth finish. The nonuniform distribution of elements can be a disadvantage as well as the additional porosity. This type of material is usually cold worked through forming or drawing. A modification to rimmed steel is **capped steel**. A lid is put on the crucible of molten metal or a deoxidizing agent is added, which reduces the thickness of the iron exterior rim and reduces segregation of alloying elements.

Decarburized steel is similar to deoxidized steel in that it is designed to reduce a specific impurity. Instead of reducing oxygen, carbon is intentionally eliminated by bonding with oxygen or hydrogen. The temperature has to be relatively high to achieve this, and other alloying elements can also react. Low-carbon stainless steel uses only hydrogen to remove the carbon since chromium, a key alloying element, will react more readily with oxygen. Decarburizing can also occur unintentionally during processing, or it can be done in reverse to add carbon to a material. Carburizing will be discussed more in Sec. 9.3.8.

4.2.4 Alloy Steels

Alloy steels are just carbon steels with other elements added in significant quantities. The word "significant" does not give us a quantitative description, but the actual concentration of alloying elements may vary greatly, thereby making it hard to provide a single number. The concentration varies so much because the effect of an element on the characteristics of the steel also varies greatly.

Some elements have a profound effect on the alloy, even in small quantities. Boron, for example, is added to improve hardenability, but the concentrations it is used in are limited to only 0.0005% to 0.003%. Manganese, also added to improve hardenability (among other reasons), is typically present in concentrations of 0.30% to 1.00% in plain carbon steel. In this example, manganese has a lower limit that is 100 times larger than boron's *upper limit*.

It is evident that the type of element is important. This is due to a variety of reasons, including how alloying occurs (i.e., interstitial versus substitutional, whether the elements are soluble or insoluble, etc. See Sec. 3.2.4 on alloying). Since not all alloying elements are created equal, it is important to be familiar with what they do. The purposes of some common alloying elements are listed in Table 4.1. Most of these elements are intentionally added, but some, like phosphorus and sulfur, are often **trace elements** or **impurities**, which are present unintentionally and often have a stated *maximum* content rather than an acceptable range.

Element	Alloying Effect on Steel
Aluminum	Deoxidizer—negatively impacts processability.
Boron	Improves hardness and hardenability, no loss in processability at small amounts used.
Carbon	Improves strength, hardness, and hardenability.
Chromium	Increases low-temperature toughness, corrosion resistance, hardenability, and high temperature strength.
Cobalt	Improves strength and hardness at high temperature.
Lead	Increases machinability. Can cause embrittlement.
Manganese	Improves hot-working properties, strength, and hardness and used to deoxidize steel. Very effective at increasing hardenability.
Molybdenum	Increases toughness, corrosion resistance, hardenability, and high temperature strength, reduces embrittlement.
Nickel	Increases strength, toughness, corrosion resistance, and hardenability.
Niobium	Provides grain refinement, improves strength in low-alloy steels.
Phosphorus	Improves strength, corrosion resistance, and machinability. Too much embrittles steel.
Silicon	Increases strength, hardness, and electrical conductivity. Acts as a deoxidizer. Decreases workability.
Sulfur	Improves strength, corrosion resistance, and machinability when paired with manganese. Too much embrittles steel.
Titanium	Captures nitrogen and oxygen during processing.
Tungsten	Increases high-temperature hardness and strength and wear resistance.
Vanadium	Improves strength, toughness, and wear resistance.

Note: These elements are added in varying quantities to improve the processability or performance of steel alloys.

Table 4.1 The Effect of Alloying Elements on Steel

In **alloy steels** (or *low-alloy steels*), the principal alloying element(s) will be added in a concentration above 1%, but less than 4%. Higher concentrations typically result in special alloys with individual designations (e.g., tool steels or stainless steels). A couple of examples, using AISI-SAE designations, are 4140 steel (0.80–1.10% Cr, 0.15–0.25% Mo) and 5140 steel (0.70–0.90% Cr). Referencing Fig. 4.4, you will see that these are considered chromium-molybdenum and chromium steels, respectively. Both have other elements present, such as manganese and carbon, but these additional elements are also in plain carbon steels and are not definitive of these classes of alloy steels. Another "chromium steel" would be the AISI 400 series steels which have around 10.5% Cr or greater. This is an order of magnitude increase in chromium content, and these alloys get the special category of stainless steels (more on these in Sec. 4.2.6).

Low-alloy steels have an AISI-SAE designation that starts with two or greater (e.g., 2XXX, 3XXX, 4XXX, etc.). Manganese steels (AISI-SAE 13XX) are the only exception to this.

If you refer back to Fig. 4.4, you will see there are a variety of alloys containing nickel, chromium, molybdenum, and combinations of these and other elements.

Low-alloy steels offer a combination of moderate cost with good, general performance. Because there are so many low-alloy steels, listing the applications and properties of each would be prohibitively extensive. As a guide—and good overall practice—cross-reference the alloy number and class (e.g., 41XX is a chromium-molybdenum steel, see Fig. 4.4) with the effects of the alloying elements (i.e., chromium and molybdenum, see Table 4.1). In the case of a 41XX steel, chromium and molybdenum are the principal alloying elements. Chromium increases low-temperature toughness, corrosion resistance, hardenability, and high temperature strength, and molybdenum increases toughness, corrosion resistance, hardenability, and high-temperature strength while reducing embrittlement (see Sec. 7.4 for more on how these properties are defined).

For these reasons you might expect this class of alloys to be tough, strong, and suitable in impact or elevated temperature applications. These alloys, also known as **chromoly steels**, do indeed find applications in high-strength structural tubing (e.g., automotive roll cages and aerospace construction), gun barrels, and engine components.

It is important to note that elements used in alloying, while important, do not tell the whole story. There are various processing considerations which affect the final application. The alloy composition affects the way the steel responds to specific treatments such as carburization or nitriding, but it is always a *combination* of the composition and the processing parameters which ultimately determine the final properties. The effects of mechanical and thermal treatment (conditioning) will be discussed more in Chap. 9. The choice of an alloy is often based as much on how it will respond to processing as on the general alloy characteristics.

A list of automotive applications and suitable alloys is given in Fig. 4.6. The alloy lists are not exhaustive, but are representative of common alloys used. High-performance applications (e.g., race cars) will require different alloys, and many alloys used in these applications are nonferrous (see Sec. 4.3) to save weight or otherwise better perform under the given conditions.

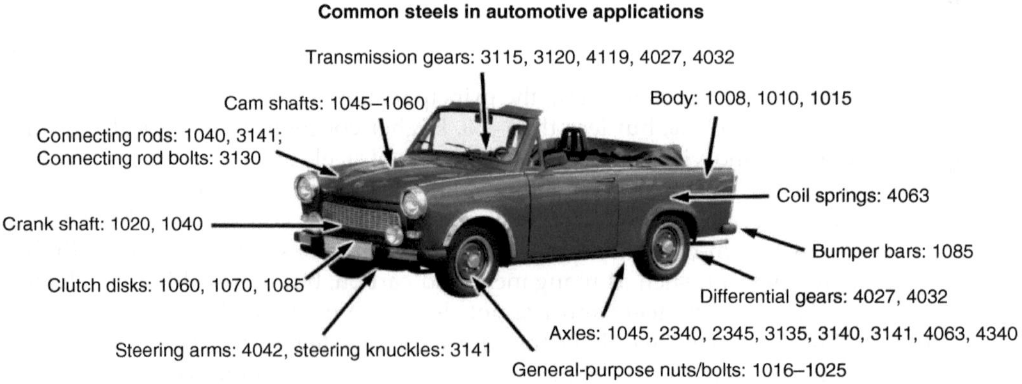

FIGURE 4.6 Steels are used in a variety of automotive applications. Some common alloys and where they are used are shown here. Many applications can be served by multiple types of steel, and the alloys lists here are not intended to be exhaustive, but are representative. The predicted service conditions will dictate the appropriate alloys.

4.2.5 Tool Steels

Tool steels are high-carbon and/or high-alloy steels. As the name implies, they are used in tooling applications. These applications often require exceptional strength, wear resistance, and performance at high temperature. Tool steels are most often hardened to improve the properties (see Sec. 9.3). Because hardening and application-specific performance are integral to these materials, they are often classified by these factors.

For instance, you might select a "shock-resisting" tool steel (S-grade tool steels such as S5) for impact intensive applications. The depth of hardening or hardenability of the alloy determines the appropriate processing steps used to manufacture the tooling and the conditions it can be used under. Although Chap. 9 will provide more perspective to these steels, we can preview it by saying that most ferrous metals will increase in hardness by heating to very high temperature then cooling the steel (quenching) to room temperature. The quenching step can be done in water (very fast), oil (fast), air (slow), or other mediums.

The method used to harden a tool steel, such as oil or water quenching, lends itself to designations of "O" or "W" grades. The more carbon and other alloying elements in the steel, the slower the part can be cooled and still become very hard. Tool steels often incorporate carbon in relatively high amounts (~1% or more) to assist in hardening and carbide formation with other alloying elements. Tool steel designations often emphasize the major alloying element, quenching medium or as-treated properties. Some of these designations are given in Table 4.2.

The way tool steels are identified gives some insight into their uses. For instance, hot-work tool steels are used for hot-working tool applications, where the material being formed or cut at a high temperature. Cold-work tool steels are used in where the material is processed nearer to room temperature. Each category has a variety of alloys

Tool Steel Category	AISI Letter Designation	Primary Alloying Elements
High-speed tool steels	M	Molybdenum with moderate chromium, vanadium, tungsten
	T	Tungsten with moderate chromium, vanadium, cobalt
Hot-work tool steels	H	Chromium (H1-H19), tungsten (H20-H39), molybdenum (H40-H59)
Cold-work tool steels	D	Carbon, chromium
	A	Carbon, chromium (low alloy) → "air-hardening" grade
	O	Carbon, manganese (low alloy)→ "oil-hardening" grade
Water-hardening tool steels	W	Carbon, other elements in relatively low quantities

Note: Tool steels are often referenced by their major alloying elements, method of hardening or applications. Each letter designation represents a group of alloys which are further specified by a number (e.g., D2, M42, T5, etc.).

Table 4.2 Partial Listing of AISI-SAE Tool Steel Designations associated with it. Cold-work tool steels have D, A, and O subcategories, and each of those subcategories has further divisions into designations such as D2, A6, or O2. You can get general insight into tool steels by their designations, but due diligence is required to ensure you select the best tool steel for the particular operation.

In general, high-speed tool steels are often used in cutting tools due to their exceptional wear resistance and retention of high-temperature strength (for steel, at least). "High speed" refers to the cutting speed (related to rpm, see Sec. 13.5), which allows for faster machining of materials and higher productivity. Hot-work tool steels are used in high-temperature processing because they do not easily soften after extended periods at elevated temperature [e.g., 1000°F (~540°C)]. This gives them a long service life, even in demanding conditions of wear and stress.

Cold-work tool steels are used at or near room temperature. They cover a range of properties and compositions, some being relatively low-alloy steels. The less demanding applications and reduction in alloying for some grades reduces cost significantly. This cost saving carries over to water-hardening tool steels. These alloys have relatively little alloy content, but have high carbon content. They are more difficult to through-harden, hence the need to quench rapidly in water. They are also more sensitive to defects and warping during quenching, but the cost often offsets the drawbacks for their intended applications. Other grades of tool steel include shock resisting, mold, and special-purpose.

4.2.6 Stainless Steels

Stainless steels are high-alloy steels, with chromium being the principal alloying element. Stainless steels contain at least 10.5% chromium, but many stainless alloys contain significantly more. Stainless steels are often applied in applications where corrosion must be minimized. Despite the name, these alloys are not entirely "stainless." They are most closely associated with chromium content because the effect of chromium is to create a passive oxide. This means they rust, but the layer is so thin it is invisible. It is also stable, so it stops rusting once formed. Stainless steels will still corrode in certain, harsh environments, but in many applications they provide substantially more corrosion resistance than plain or alloy steels. The concepts of corrosion are discussed more in Sec. 8.5.

Although chromium is the principal alloying element, there are often other alloying elements added, nickel being one of the most prominent. These additional elements often improve corrosion resistance further and increase both the strength and toughness of the alloy. Stainless steel designations are based on their structure and composition (which controls structure). While iron, plain carbon steels, and alloy steels all have a body-centered cubic (bcc) structure, high-alloy steels *may* have a face-centered cubic (fcc) structure (more these structures is discussed in Sec. 3.2).

The final structure is dependent on the amount of the alloying addition as well as *its* structure. Chromium, like iron, has a bcc crystal structure. Even when added in large quantities, the alloy is still bcc. Nickel has an fcc structure, and when enough of it is added, the alloy also assumes an fcc structure (although still mostly iron). The three most common categories of stainless steel are named for their structures: austenitic, ferritic, and martensitic. Precipitation-hardening and duplex stainless alloys are also available and serve in a number of applications. The AISI designations for stainless steels are given in Table 4.3. UNS designations begin with the letter "S" and typically incorporate the AISI designation in the numbering [e.g., 304 (AISI) is S30400 (UNS)].

Stainless Steel Category	AISI Letter Designation	Primary Alloying Elements
Austenitic	2XX	**Chromium** (16–19%), **Nickel** (3.5–6%), **Manganese** (5.5–15.5%), Carbon (0.12–0.25%)
	3XX	**Chromium** (15–28%), **Nickel** (2.5–37%, typically < 22%), Manganese (2%), Carbon (0.03–0.25%)
Ferritic	4XX	**Chromium** (10.5–27%), Nickel (< 0.75%), Manganese (1–1.5%), Carbon (0.08–0.2%)
Martensitic	4XX	**Chromium** (11.5–18%), Nickel (< 2.50%), Manganese (1–1.25%), **Carbon** (0.15–1.2%)
Precipitation Hardening	S17XXX[a]	**Chromium** (11.5–18%), Nickel (3–8%), Manganese (< 1.5%), **Copper** (3–5%), Carbon (< 0.07%)
Duplex	S3XXXX, S8XXXX[a]	**Chromium** (18–30%), Nickel (1–8%), Manganese (0.8–6%), **Molybdenum** (0.05–6%), Carbon (0.025–0.08%), **Nitrogen** (< 0.30%)

[a]UNS designations are given instead of AISI as naming convention varies for these categories.

Table 4.3 Designations for Stainless Steels and the Composition of Major Alloying Elements

4.2.6.1 Austenitic Stainless Steel

Austenitic stainless steels (AISI 2XX and 3XX) have an fcc structure which gives them unique properties such as high toughness and ductility, and unlike bcc stainless steels, they are nonmagnetic. Iron changes from bcc to fcc at high temperature, but the addition of nickel and other elements allows the retention of the fcc structure even at room temperature. Nickel is a relatively expensive metal, and 2XX austenitic stainless steels replace some of it with manganese to serve the same basic purposes at lower cost. There is a minor addition of nitrogen to help maintain the fcc structure without as much nickel (i.e., nitrogen stabilizes the fcc phase). The lower alloy content in 2XX steels means they are not as corrosion resistant as the 3XX steels, so they are applied in structural and decorative applications which are not particularly corrosive. Some common steels from this category include 201 and 202 which exhibit high work-hardening and serve general-purpose applications, respectively.

There is a wide variation of properties within 3XX stainless steels, so not all are equally suited for the same applications. The variations typically seek to address applications with especially stringent thermal, environmental, or mechanical requirements. Processing is also a major concern. The primary corrosion resistance is provided by chromium, but chromium is a fairly reactive element. Stainless steels often contain carbon in a large enough quantity to cause chemical reactions during thermal processing such as welding.

At high temperature the carbon will bond with chromium creating chromium carbide, a hard, brittle phase that does not prevent corrosion. If too much chromium is bonded to carbon, the steel will cease to be "stainless." For this reason, welded assemblies are made from low-carbon varieties, which are designated by an "L" after the AISI number designation (e.g., 317L).

FIGURE 4.7 Food service and storage containers are often made of austenitic stainless steel due to its corrosion resistance.

Unfortunately, stainless alloys that are easily welded are difficult to machine and vice versa. For example, 303 stainless steel machines easily but is not suited for welding. In welded assemblies, 304 is a better choice, but it requires lower machining speeds and can wear tools quickly if not properly set up. Where forming operations are used, the significant work-hardening tendency of many stainless steels should be considered. 316 stainless steel is a standard option for corrosion resistance, though it is not suited for all environments. 3XX stainless steels are often used in food handling (Fig. 4.7), chemical containers, turbine parts, and steam applications.

4.2.6.2 Ferritic and Martensitic Stainless Steel

Ferritic stainless steels (AISI 4XX) have a lower alloy content than austenitic steels, with chromium being the only major alloying element. Ferritic stainless steels are bcc at room temperature, are magnetic and have a relatively low strength and high ductility. They cannot be strengthened through heat treatment, but they do work-harden and can be strengthened through forming. These alloys have a high chromium content to balance the carbon content which can defeat the corrosion resistance. They perform well against high-temperature corrosion such as might occur in exhaust systems.

Martensitic stainless steels (AISI 4XX) are similar to ferritic in many regards. They differ most significantly in their ability to be hardened through heat treatment. They have a much higher carbon content which allows for the formation of the martensite phase, hence the name (see Sec. 9.3.4 for more about this phase). They respond readily to heat treatment and they can be quenched in oil or even air. Applications typically involve high-wear situations such as bearings (Fig. 4.8) and nozzles or where a sharp edge must be maintained such as cutlery and surgical instruments.

4.2.6.3 Other Stainless Steels

Precipitation-hardening stainless steels are fairly limited in variety. They are essentially 300 series stainless steels (e.g., AISI 302) that also include precipitate-forming elements such as copper and are often identified by their chromium and nickel compositions followed by "PH" for precipitation-hardening (see Sec. 9.3.7 for more on precipitation hardening). A common example is 17-4 PH, which has 17% chromium and 4% nickel

Figure 4.8 Hardenable grades of stainless steel, like 440C used in these bearings, can be applied in high wear environments and maintain good corrosion resistance.

(this is sometimes identified as AISI 630). What makes precipitation-hardening stainless steels unique is that they are not hardened during rapid cooling from high temperature, but rather during a subsequent heating at modest temperature. This allows tight control over the final properties, and the component can be machined in the annealed (soft) state and later hardened without loss of tolerance. These alloys find applications in jet engines, dies, and valves.

Duplex stainless steels consist of both fcc and bcc structures. These alloys have a relatively high chromium content and exchange higher nickel for more molybdenum and nitrogen to help maintain the fcc phase. Duplex stainless steels have a good combination of strength and corrosion resistance. They are designated under a variety of trade names such as Zeron or Ferallium, and many without an AISI designation. The UNS designations (S3XXXX and S8XXXX) are more uniform and are provided in Table 4.3. Note that S3XXX also covers some austenitic stainless steels. Further designation will sometimes include descriptors such as lean, standard, super, and hyper duplex stainless alloys. These indicate the alloying content and resistance to corrosion. With their lower cost and comparable performance to 3XX austenitic alloys and even some nickel alloys, duplex stainless steels are an attractive alternative in applications were high strength and/or excellent resistance to corrosion are needed, such as tubing, heat exchangers, and saltwater environments.

4.2.7 Cast Irons

Cast irons are basically *very* high-carbon steels. They have such a high carbon content (2–4%, typically between 2.5% and 3%), that they are characteristically hard and brittle, approaching ceramic-like behavior at times. Because of this, they are often difficult to process by machining or forming, so casting is the primary method used to manufacture components made from them (go figure!). It is not just because they are *not suited* for other processes, but also because they are *well suited* for casting. Their favorable properties include a lower melting point and higher fluidity than lower carbon steels (see Chap. 11 for more on casting).

Cast irons are designated by their structure. The structure is determined by the composition *and* the way they are processed. Although there are subcategories and structural details we will not cover in this discussion, the basic classes of cast irons and their differences are provided.

Cast irons tend to be hard and brittle because of the excessive amount of carbon that is available to form **cementite**, which is iron carbide (Fe_3C, a ceramic discussed in Sec. 6.2.2.2). As you might guess, a material that has "cement" in the name is going to be hard and brittle. Cementite is not the only form carbon takes. If the carbon does not bond with iron, but instead bonds to other carbon, it will form graphite regions within the iron. Graphite is much weaker and less brittle than cementite. It can also act as a lubricant for wear surfaces. The formation of graphite does not lend as much strength to the cast iron as cementite does, but it makes the cast material more manageable to work after casting in processes such as machining.

Whether the carbon forms cementite or graphite is determined by the composition (carbon as well as other alloying elements, most often silicon) and the rate at which the material is cooled. Commercial cast irons include gray iron, ductile (or nodular) iron, white iron, and malleable iron. These names (and their properties) are closely related to the way the graphite forms. Common applications for cast irons include engine blocks and engine components, heavy-duty machine bases, industrial piping, and many other fittings and structural components (Fig. 4.9). The type of cast iron is dependent on the

Figure 4.9 Cast iron is used in a variety of structural applications. The grade of cast iron determines how it is used.

structure, which in turn is dependent on the process used to make it. We will look at the common grades of cast iron and some important differences.

Cast irons are often classified according to their strength. Cast irons have varying designations from ANSI/ASTM and SAE, which include descriptive numbers. The number contains information on the mechanical properties such as 30A (a gray cast iron with 30,000 psi minimum tensile strength) or 60-40-18 (a ductile cast iron with a minimum tensile strength of 60,000 psi, a 0.2% yield strength of 40,000 psi, and a minimum 18% elongation over a 2-in length). The UNS designation for cast irons is FXXXXX (see Fig. 4.3). We will consider these alloys by their common names, as the properties within each category are similar.

4.2.7.1 Gray Cast Iron

Gray irons have a significant addition of silicon (1–3%), which improves graphite formation. Gray irons are named that way because when fractured, the surfaces appear gray. The graphite in gray iron forms in a "flake" fashion, similar to bran flakes. Gray irons are well-suited to complex castings as they have a high fluidity and low shrinkage. These alloys also have a relatively high machinability thanks, in part, to their graphite content. They are not particularly strong under tension and quite brittle, so applications must limit any tensile forces (for more on forces, see Sec. 7.2). They dampen vibration very effectively and are low cost. This makes them a nice fit for large components on heavy equipment and engine blocks.

4.2.7.2 Ductile (Nodular) Cast Iron

Ductile (nodular) irons are similar in composition to gray iron, but a small addition of inoculant is added to change the shape of the graphite phase. An **inoculant** is an element, such as magnesium, that causes the graphite to form into spherical inclusions, rather than flakes. The flakes act as stress risers that weaken gray iron. This is similar to packaging with a "tear here" label near a notch, such as packets of ketchup. That sharp point makes breaking the material easier, and the tips of the graphite flakes produce the same effect. By converting the graphite to a spherical (nodular) shape, the material performs much better under tension. Although not as ductile as steel, it can be used in comparable applications such as crankshafts, machine components, and gears.

4.2.7.3 White Cast Iron

White irons contain a large amount of cementite instead of graphite. This is due to low silicon content. Like gray iron, white irons are named for the appearance of their fracture surfaces that are silvery white. The abundance of cementite makes white irons especially hard and brittle. Although the castings are exceptionally strong, they require special care in design as they can crack during cooling if sharp corners or thin sections are present. Applications are restricted to those that demand high wear resistance and little deformation. Because white irons are extremely difficult to work after casting and have a limited set of applications, they are typically used as a precursor for creating malleable iron.

4.2.7.4 Malleable Cast Iron and Compacted Graphite Iron

Malleable irons are made by slowly heating and cooling white iron. The iron is heated to temperatures from 1470 to 1650°F (800–900°C) over a time of about 30 to 50 h, and cooling is conducted in about the same time or slightly longer. The heat treatment parameters will affect the final microstructure. The significant time investment pays off, as heat treatment converts cementite to compact, rosette-shaped aggregates of graphite, similar to nodular iron. Pipe fittings and applications similar to nodular iron and elevated temperature service conditions are common.

Compacted graphite iron (CGI) is the last type of cast iron we will look at. CGI is a newer class of cast irons (since the 1970s) where the graphite forms in **vermicular** (worm-like) shapes. Often there is an appreciable presence of nodules as well (< 20%). This produces a material with properties between ductile (nodular) and gray cast irons. This yields a material that has good strength, machinability, and damping. The microstructure is controlled through manipulation of composition and cooling, so quality control must be more rigorous and cost is somewhat higher. Applications are similar to gray and ductile iron, but CGI is particularly beneficial where machining is extensive.

4.3 Compositional Classes of Metals: Nonferrous

Nonferrous metals and alloys are based on elements other than iron. Although they may contain iron, it is not the primary element. Iron-based, ferrous alloys can possess a wide range of mechanical and physical properties, but there are even more applications that can benefit from nonferrous alloys. Nonferrous alloys are commonly used for extreme conditions. Where service conditions require very low temperatures, very high temperatures, a highly corrosive environment, low density, high density, low electrical conductivity, high electrical conductivity ... you get the idea. Nonferrous materials are often used in specialty applications, but some, like aluminum and its alloys, are used widely in structural applications.

Many nonferrous alloys are often rare or difficult to refine, so their cost is typically higher than that of iron-based alloys. Their use is driven by necessity. Sometimes nonferrous alloys are easier to process because they are weaker or more ductile than steels, but in other cases, they can be very difficult to work with. Looking at the periodic table (Fig. 2.1, for instance), you will see the majority of elements are metals but only one is iron. Sweeping generalizations, therefore, are difficult to make when discussing nonferrous metals. Some major alloy categories will be discussed in the following sections to further delineate their properties and applications. The universal truth of these materials is that our society and technology would not be possible without them.

4.3.1 Aluminum and Aluminum Alloys

Aluminum is only surpassed by iron in regard to its extent of use. Some of the properties that allow for its popularity are its light weight, high ductility, good electrical conductivity, corrosion resistance, and ease of manufacture. Some of its limiting characteristics are its lower strength, poor high-temperature performance, and higher cost. Even with these drawbacks, aluminum is a good fit for many applications.

4.3.1.1 Pure Aluminum

Pure aluminum is very weak, and its use is limited to applications where a high ductility and/or high conductivity are preferable. Unlike iron's bcc structure, aluminum has an fcc crystal structure that allows for extensive deformation without breaking. Because pure aluminum is so weak when pure, most structural applications are unsuitable. A common place you will find near-pure aluminum is in aluminum foil. Think about how you can take a sheet of foil, compress it into a ball and then flatten it out again without it breaking into pieces. If the foil was steel, it would take more force to compress it and flatten it and it would be more likely to crack. Low strength and high ductility as well as corrosion resistance (you do not want a metallic taste in your food, right?) make aluminum a good fit for the application.

Although pure aluminum is not suited for elevated temperature service, it does well at low temperature by maintaining ductility. Aluminum's high electrical and thermal conductivities also allow for applications in wiring. Special junctions should be used where aluminum is to be connected to copper wiring to prevent arcing as aluminum has a different coefficient of thermal expansion and will corrode galvanically when connected to copper wiring (these concepts are discussed in Chap. 8).

4.3.1.2 Aluminum Alloys

Aluminum alloys overcome many of the drawbacks of pure aluminum, but they also diminish the benefits. For instance, the strength of aluminum alloys can be as much as 30 times higher, but the ductility will decrease substantially. There are a number of aluminum alloys, most of which incorporate copper, magnesium, silicon, manganese, or zinc. These elements are usually added in minor quantities (< 10%) and they increase strength in two ways: solid solution strengthening or precipitation hardening. Solid solution strengthening is achieved when two elements with different-sized atoms are mixed. The movement of atoms (how the material deforms) is inhibited by the mismatch (see Sec. 3.2.4.1). Precipitation hardening is a different process where elements in the alloy form discrete regions of a different composition, thereby creating a "pinning" effect on deformation (see Sec. 9.3.7).

Because aluminum alloys can be hardened through cold work or heat treatment, the designation system they use includes numbers for identifying the alloy *and* its condition. The system for identifying aluminum alloys was originally developed by the Aluminum Association and includes four digits followed by the **temper designation**. The first of four digits identifies the class of alloys and the last three digits specify an alloy within that class.* There is also a temper or condition designation which indicates the processing involved in creating the alloy. This consists of a letter that designates the conditioning process category and one or more digits after that which gives the specific treatment within that category.

Sound complicated? Here is an example. A common aluminum alloy is 6061-T6. By referencing Fig. 4.10, you find that this alloy contains principal additions of magnesium and silicon and has good all-around performance mechanically and good corrosion resistance. Furthermore, you can determine that it has been thermally conditioned (strengthened) through heat treatment.

Aluminum alloys are popular because of their specific strength. **Specific strength** is simply the amount of force a material can support for its own weight. Aluminum has a density of 2.70 g/cc. That is about one-third of iron (7.87 g/cc), which is a good approximation for most steels. As mentioned above, aluminum alloys can be very strong, approaching that of plain carbon steel. For illustration, let us assume they have the same strength. A steel component of the same size will weigh three times more, so its

*Only for 1000 series alloys do the last *two* digits indicate the purity of the aluminum above 99.00%. For example, 1060 aluminum would have a 99.60% purity. This does not apply to other aluminum alloy series.

Aluminum Association Designation	Major Alloying Element(s)	Properties/Applications
1XXX	None	> 99.00% pure, good workability, used in electrical or thermal conductors.
2XXX	Copper	Heat treatable, can meet or exceed mild steel properties, lower ductility and corrosion resistance, used in sheet form, and usually clad for corrosion.
3XXX	Manganese	Good workability, used in applications requiring moderate strength.
4XXX	Silicon	Significant silicon content lowers melting point without embrittlement, used in welding wire and brazing applications.
5XXX	Magnesium	Can also incorporate manganese to create moderate to high-strength alloy used in applications requiring welding or corrosion resistance.
6XXX	Silicon and Magnesium	Silicon and magnesium in equal parts to form magnesium silicide during heat treatment. Moderate strength with good corrosion resistance.
7XXX	Zinc	Also includes some magnesium for heat treatment. Very high strength used in air frames and other high-stress parts.

Condition Designation	State of Alloy
F	As-Fabricated – no specific control on work hardening or thermal treatment during processing.
O	Annealed, recrystallized – forms the softest condition of wrought alloys.
HXYZ	Strain-Hardened – cold-worked alloys which have no annealing or tempering applied, the hardest condition of wrought alloys. Digits associated which the "H" designation indicate the specific condition. **X = 1** – strain-hardened only; **X = 2** – partially annealed; **X = 3** – stabilized **Y = 1 through 9**, higher numbers indicate higher strain-hardening **Z = 1 through 9**, arbitrary number for variations on the HXY condition.
W	Solution Heat-Treated – an unstable condition preceding precipitation hardening, used only when a natural aging period is indicated.
TXX	Thermally Conditioned – creates a condition other than already listed, includes strain hardening and thermal conditioning. XX = 2 through 10. Other numbers may be added after the two digit designation to indicate specific modifications on the condition.

Figure 4.10 Aluminum alloy designations based on composition and processing condition.

specific strength will be one-third that of the aluminum. This only speaks of strength, and it does not indicate other properties such as ductility. In weight-critical applications such as aerospace, this provides a good starting point for material selection.

4.3.2 Copper and Copper Alloys

Copper and its alloys find a wide variety of uses due to their good corrosion resistance, high electrical and thermal conductivity, and ease of processing. As discussed in Sec. 1.3.2, copper is a metal that has been in use for millennia. The Bronze Age was a period which centered, technologically, around that specific copper alloy. Some of factors which made copper and its alloys popular then, as well as now, are related their availability and diversity of properties.

4.3.2.1 Pure Copper

Pure copper is very soft and weak. These properties make it exceptional for forming operations where a high ductility is beneficial. Copper is also slow to work harden, so it can be extensively formed without breaking. Copper is so ductile that can actually be almost "gummy" to machine, tending to "flow" around cutting tools rather than forming chips. The formability of pure copper is a perfect match for creating electrical wiring, which is its most common application.

The more pure copper is, the better it is at conducting electricity and heat. Highly pure copper for these purposes is often referred to as "oxygen-free, high-conductivity" (OFHC). Copper has a density greater than iron (8.94 g/cc and 7.87 g/cc, respectively), so it does not provide a lightweight alternative for structural or other applications. Copper's mass makes it unsuitable for long wire spans like high-voltage lines as it can break under its own weight. Aluminum can be used over longer spans, but it too needs reinforcement for reliability.

Number designations for copper and it alloys are well-represented in the UNS system (see Fig. 4.3 for more on the UNS system), as their previous designations are included in their UNS designations (see Fig. 4.11 for copper classifications). Copper and its alloys are also referred to by a variety of characteristics including composition, appearance, and application. The two major subcategories are brasses and bronzes with many further divisions within each of those. Wrought alloys (C1XXXX-C7XXXX) and cast alloys (C8XXXX-C9XXXX) also have different classifications. The wrought alloys will be the focus of this section, but casting alloys are widely used as well.

4.3.2.2 Brass Alloys

Brass alloys are made by alloying copper and zinc. Depending on the amount of zinc added, the crystal structure of brass can be fcc or a mixture of fcc and bcc phases. Up to about 35% zinc, the alloy is α (lowercase Greek letter, alpha) brass, which is fcc like pure copper. Alpha brasses also have good ductility and are easily worked, but they machine more easily than pure copper. For even better machinability lead is added (leaded brass) to create a "free-machining" brass.

Higher zinc content increases strength and hardness, and brasses work-harden well. They are often sold with a temper designation such as "half-hard" or "quarter-hard," which indicates the amount of cold work that has been applied (Table 4.4). Subcategories within the brass alloys include cartridge brass (30% zinc), yellow brass (35% zinc), and naval brass (39.25% zinc, 0.75% tin). These, as well as other names, are based on appearance or applications. Although they have UNS designations, these common

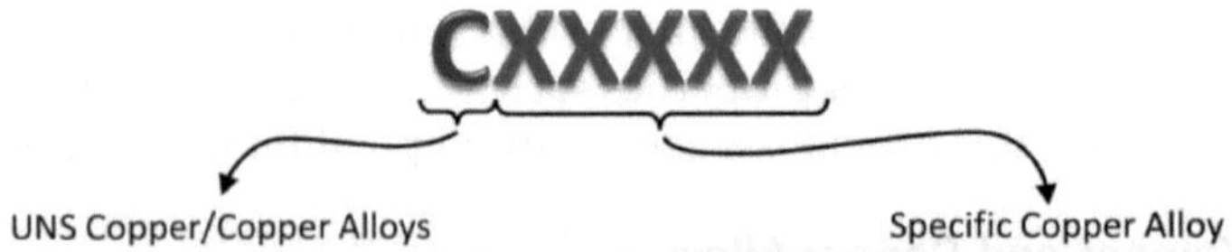

UNS Designation(s)	Major Alloying Element(s)	Alloy Family
C10XXX – C19XXX	none	High-Purity Coppers (> 97%)
C20XXX – C28XXX	Zinc	Brasses
C31XXX – C38XXX	Zinc, Lead	Leaded Brasses
C40XXX – C49XXX	Zinc, Tin	Tin Brasses
C50XXX – C52XXX	Tin	Phosphor Bronzes
C53XXX – C54XXX	Tin, Lead	Leaded Phosphor Bronzes
C60XXX – C644XX	Aluminum	Aluminum Bronzes
C647XX – C661XX	Silicon	Silicon Bronzes
C70XXX – C72XXX	Nickel	Copper Nickels
C73XXX – C79XXX	Nickel, Zinc	Nickel Silvers
C80XXX – C99XXX	Varies	Casting Alloys

Figure 4.11 UNS designations for copper and copper alloys.

Condition Designation	% Cold Work
Annealed	0
Quarter-hard	10.9
Half-hard	20.7
Three-quarter hard	29.4
Hard	37.1
Extra hard	50
Spring	60.5
Extra spring	68.7

Table 4.4 Condition Designations for Copper and Copper Alloys

names are often used. Applications for brasses include ammunition casings, radiator cores, instruments, such as horns, trumpets, etc., fasteners, low-cost jewelry, plumbing components, and many more.

4.3.2.3 Bronze Alloys

Bronze alloys are made using a major alloying element other than zinc. When examining Fig. 4.11, you will see there are many bronzes, and they may contain aluminum, tin, or silicon (they often contain other elements as well). So when differentiating between brass and bronze, the key difference is that brass alloys are based on zinc, and bronze

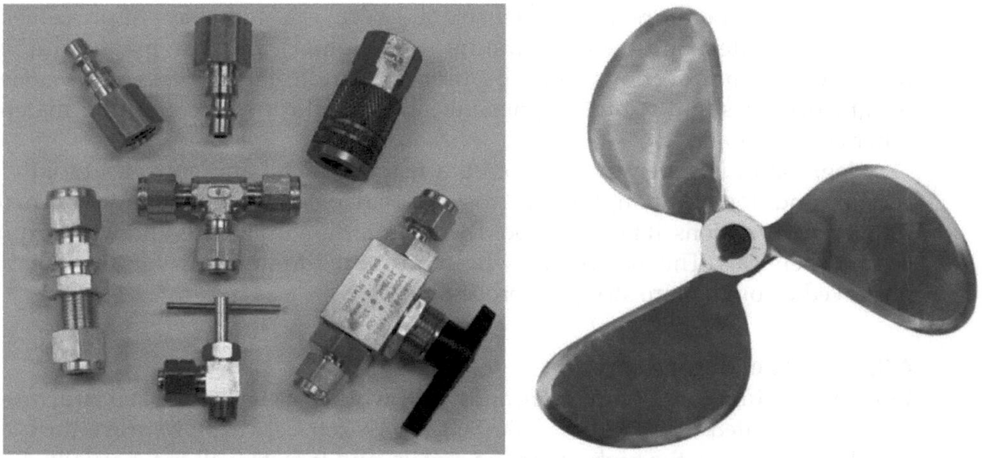

Figure 4.12 Copper and copper alloys find uses in electrical, wear, and small structural items. Brass is used extensively in fittings for liquid and gas delivery (left) and bronze alloys are often used where strength and corrosion and wear resistance are needed, such as the boat propeller pictured on the right.

alloys are based on something else.* Bronze alloys are typically stronger than brasses (though not as strong as steel) and are also corrosion resistant. Many of their applications are similar to those of brasses and include springs, bearings, bushings, pump components, valve stems, and more (Fig. 4.12).

4.3.2.4 Other Copper Alloys

Other copper alloys include nickel and beryllium alloys. Nickel and copper are completely miscible in one another, meaning they can be mixed in any quantity without separating (unlike oil and water). These alloys also go by the name **cupronickel** and perhaps the most common application is in monetary coins such as the U.S. nickel, dime, and quarter. This alloy contains about 25% nickel, but despite being mostly copper, it appears silvery. Copper-nickel alloys are exceptionally well-suited for saltwater applications and are employed in many maritime applications such as propellers and radiators and piping that handles saltwater.

Beryllium copper alloys are some of the few heat-treatable copper alloys, and they can be strengthened significantly through precipitation hardening, even exceeding many steels. These alloys are expensive because of the beryllium addition, and they are also a health concern because of the toxicity of beryllium. They are used in highly wear-resistant electrical contacts, nonsparking and nonmagnetic tools, cryogenic applications, and in some medical tools.

4.3.3 Magnesium and Magnesium Alloys

Magnesium is the lightest structural metal, and most of its applications are based on weight-critical requirements. Magnesium is not very strong and does not perform

*There are exceptions to this rule such as nickel and beryllium-based alloys which get their own designations, but in the majority of cases, if zinc is not the primary alloying element, it is considered bronze.

well at high temperature. Its low density (1.7 g/cc) is a significant improvement over steel (80% lighter) and even aluminum (40% lighter). Because magnesium is relatively weak, thick sections must be used to handle large forces. This offsets the weight benefit slightly, but magnesium and its alloys still cannot be matched by other classes of alloys.

Magnesium is generally considered a reactive metal and corrodes rapidly under certain conditions. Although the corrosion behavior is acceptable under normal environmental conditions, it performs poorly in marine environments and when exposed to certain chemicals. This has partially been attributed to impurities in the metal and is improved through more stringent processing.

4.3.3.1 Pure Magnesium

Pure magnesium cannot be significantly cold worked as its hcp crystal structure limits ductility. Elevated temperatures [300–500°F (149–260°C)] vastly improve the formability. Machinability is the highest among metals, and it casts well too, partially due to its low melting point [1203°F (651°C)], which works well in die casting (see Sec. 11.2.2.3). Magnesium has actually replaced plastics in certain applications such as electronics casings (Fig. 4.13) as it is easier to recycle and cheaper to produce.

4.3.3.2 Magnesium Alloys

Magnesium alloys have higher strength than pure magnesium, but the improvement is not as dramatic as what is seen in aluminum alloys. Some alloys can be precipitation hardened, and common alloying elements include aluminum (≤ 10%), zinc (≤ 6%), manganese (≤ 1.2%), and zirconium (< 1%). A popular magnesium alloy for die casting is AZ91. The designation system for magnesium alloy (Fig. 4.14) follows the pattern where the first two letters identify the major alloying elements and the last two digits are the percent composition of those two elements (rounded to the nearest whole number). For example, AZ91 would have 9% aluminum and 1% zinc. There is often a letter following the designation which further specifies variation on the composition. Designations in addition to that indicate processing and follow the basic system used for aluminum (see Fig. 4.10).

Figure 4.13 Magnesium is the lightest structural element, and many of its applications are based on that property. It is relatively weak, however, so thicker sections or lower demand applications are used. Examples include electronics casings (on the left) and high-performance automotive wheels (on the right). (*Right photo by Rayukk, CC BY-SA 2.0.*)

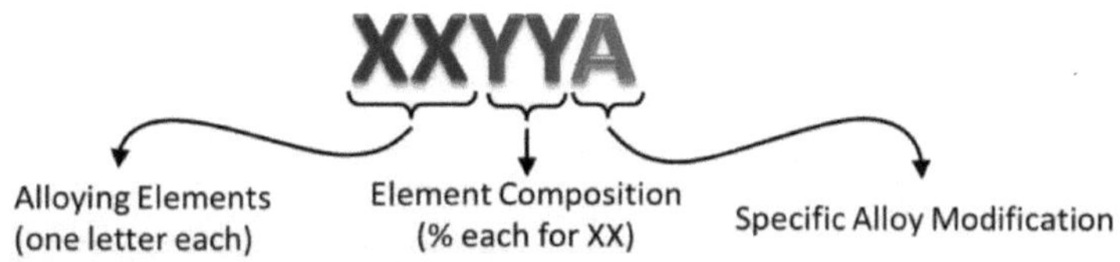

Elemental Alloying Designations (XX)			
A	aluminum	L	beryllium
B	bismuth	M	manganese
C	copper	N	nickel
D	cadmium	O	silver
E	rare earth	R	chromium
F	iron	S	silicon
H	thorium	T	tin
K	zirconium	Z	zinc
Alloying Composition (YY)		**Alloy Modification**	
1-9: indicates % of elements 1 & 2		A, B, C, etc. : for specific alloy variation	

Figure 4.14 Magnesium alloy designations follow ASTM standard nomenclature and include direct information on the alloying elements and their respective amounts. UNS designations for magnesium alloys are in the form MXXXX (see Fig. 4.3).

4.3.4 Nickel and Nickel Alloys

Nickel finds extensive use as an alloying element, but it is also used as a base material. Nickel alloys are typically employed in applications requiring high temperature and/or corrosion resistance. Nickel is one of three elements that are magnetic* (iron and cobalt are the others). Nickel has an fcc crystal structure and is ductile even at cryogenic temperature. This is coupled with high strength that makes nickel-based components very tough. In general, these characteristics do not severely limit processing, but nickel-based "superalloys" can be quite challenging to process.

Nickel and nickel alloys can be broken into categories based on composition. The first is pure or low-alloy nickel (> 94% nickel), referred to as Ni 200. The others are based on alloys containing chromium, copper, cobalt, iron, molybdenum, or some combination of these. Previous sections on stainless steel and copper-based alloys have included nickel as an alloying element, and depending on the composition, the nickel-based alloys may be in the same family of materials.

For instance, the UNS alloy N08800 actually has more iron than nickel (46% and 32%, respectively), but such high nickel content is rare in what are considered iron alloys, so it gets labeled as a nickel alloy. Another possible source of confusion is the naming convention for nickel alloys. Although the UNS system covers these alloys (as NXXXXX), they often go by trade names such as Inconel, Monel, and Hastelloy, which have further subcategories under them (Fig. 4.15).

*Technically *ferromagnetic*, or permanently magnetic, is more accurate, and room temperature is assumed. All elements respond in some way to a magnetic field, but others tend to have a weak, temporary response under what might be considered "normal" conditions. Other elements are paramagnetic or diamagnetic.

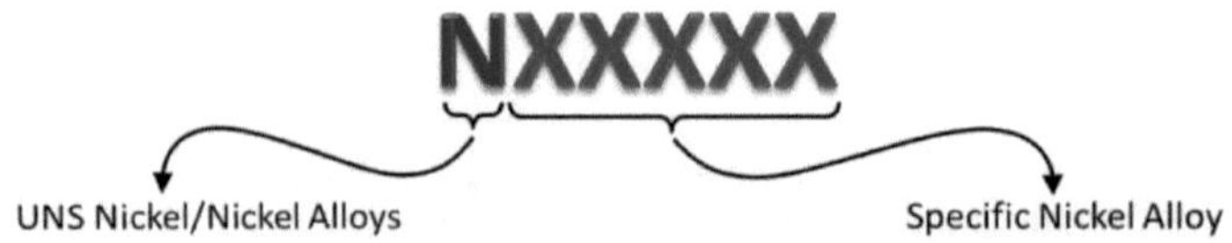

Major Alloying Element(s)	Common Trade Name(s)[a]	Properties/Applications
None (N02XXX)	Commercially Pure	High ductility, good corrosion resistance. Food and chemical processing equipment.
Copper (N04XXX, N05XXX)	Monel	High to moderate strength and ductility, excellent general corrosion resistance. Marine applications, valves pumps, impellers, petroleum refinement.
Chromium (N06XXX)	Inconel, Nichrome	Excellent mechanical properties at cryogenic and high temperature, excellent corrosion resistance. Oil and gas production, turbine components, marine components, heat treating equipment, heating elements, electronic components.
Chromium, Cobalt (N07XXX)	Waspalloy, Udimet	Excellent strength, high temperature strength and corrosion resistance, especially oxidation. Heat treatable. Turbine components, nuclear reactor components, some forming tools.
Iron (K9XXXX)[b]	Invar	Moderate strength and ductility with very low thermal expansion. Used in high precision measuring and manufacturing equipment, glass/metal interfaces such as in optical devices, satellite components.
Iron, Chromium (N08XXX, N09XXX)	Incoloy	Good strength, moderate ductility depending on composition, very good at high temperature and corrosion resistant in a variety of atmospheres. Heat treatable. Valves, fittings, furnace parts, chemical piping, heat exchangers in high temperature, corrosive environments.
Molybdenum (N10XXX)	Hastelloy B	High strength and very good ductility. Good in high temperature and nonoxidizing applications. Select chemical handling.
Molybdenum, Chromium (N10XXX)	Hastelloy W, Hastelloy C226	High strength and ductility, excellent corrosion resistance, heat treatable. Filler rod for welding, chemical handling.

Figure 4.15 Nickel and its alloys often go by trade names, but their UNS designations start with "N." [a]The naming conventions are not rigidly connected to composition, and a variety of alloys exist under each name. Most alloys contain appreciable amounts of other elements than those listed as the primary alloying element(s). [b]Invar alloys are correctly considered iron alloys as they typically contain 36% Ni. The UNS "K" category is designated for miscellaneous ferrous alloys.

As can be seen in Fig. 4.15, most applications for nickel alloys involve highly corrosive environments and/or high temperatures. The dimensional stability of many nickel alloys under load at high temperature (known as *creep* resistance, see Sec. 8.2.6.2) makes them particularly popular for turbine applications.

Although all of the alloys show good corrosion resistance, Monel (also known as Ni 400), which consists primarily of nickel (~68%) and copper (~30%), is suitable to a very wide range of environments, such as marine, acid, steam, petroleum, etc. One limiting factor to the use of nickel alloys is the density of nickel (8.91 g/cc), which is similar to copper, though it varies somewhat with composition. Another factor which controls the use of nickel even more tightly is the cost, which is three to four times that of copper and nearly 100 times more than iron (depending on economic conditions). You can imagine, then, that the use of nickel alloys must be thoroughly justified by demanding operational conditions.

4.3.5 Titanium and Titanium Alloys

Titanium and its alloys have been increasing in popularity due to their unique combination of properties. They are relatively light (Ti has a density of 4.51 g/cc, about half that of iron), can be used at elevated temperatures [1022°F (550°C) continuously], are corrosion resistant, biocompatible, and quite strong, especially for their weight. This wide range of properties means they can serve in a variety of applications (Fig. 4.16), but they do have drawbacks.

Titanium is an abundant element on earth, but it is difficult to refine from ore. The refined metal is difficult to machine so tooling costs are high and cycle times are relatively long. This has been a very active area of manufacturing development and the processes are becoming much more efficient. Titanium is reactive and must be processed carefully to avoid the formation of oxides, nitrides, and carbides during high-temperature processing. For instance, proper shielding during welding, both on the front and back sides, is critical to forming a strong bond, and casting is often done under vacuum or inert atmospheres to avoid contamination of the material.

All of that means titanium is expensive, around twice the price of copper and 10 times the cost of plain carbon steel. Nickel, the topic of the previous section, costs about twice that of titanium, but titanium is more commonly used in large-volume, structural applications where cost differences are magnified.

Titanium and titanium alloys are often classified by their *grade*. For example, you might see "Grade 5" titanium sold. Titanium products are also classified by composition,

FIGURE 4.16 Titanium is used where lightweight, high strength, and resistance to heat and corrosion are needed. Examples in the bicycle frame on the left and the automotive exhaust system on the right. (*Left photo by Richard Child, CC BY 2.0. Right photo by Frank Derks, CC BY 2.0.*)

such as Ti6Al4V (usually pronounced "tie-six-four"), which contains 6% aluminum and 4% vanadium as major alloying elements. Another categorical separation is based on microstructure. Normally, titanium has a hexagonal close-packed (hcp) crystal structure at room temperature, but at elevated temperature [~1634°F (890°C)] the structure changes to body-centered cubic (bcc), which is more ductile. By alloying the titanium, the bcc phase can be "stabilized" at room temperature.

The hcp structure is called the alpha (α) phase and the bcc structure is called the beta (β) phase. Titanium alloys can be alpha, beta, or a mixture of the two, and they are classified as alpha (or near alpha) alloys, beta alloys, and alpha-beta alloys, respectively. This is similar to stainless steel alloys which can be bcc or fcc depending on the type and quantity of alloying elements present. Their UNS designation starts with an R (for reactive and refractory alloys) as they are suitable for elevated temperature service.

Designations generally follow a major alloying element or elements, but other alloying elements may be present in appreciable quantities (>1%). Figure 4.17 lists

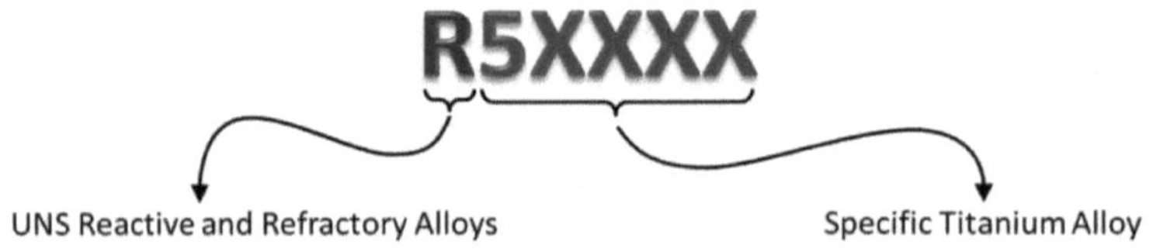

Major Alloying Element(s)	ASTM Grade(s)[a]
None (R50XXX)	$1\text{-}4^{CP}$
Palladium (≤ 0.25%) (R52XXX)	7^{CP}, 11^{CP}, 16^{CP}, 17^{CP}
Ruthenium (≤ 0.14%) (R52XXX)	26^{CP}, 27^{CP}
Nickel, Molybdenum (R53XXX)	$12^{n\alpha}$
Aluminum, Tin (R54XXX)	6^{α}
Aluminum, Vanadium (R56XXX)	$5^{\alpha/\beta}$, 9^{α}, $18^{n\alpha}$, $23^{\alpha/\beta}$, $28^{n\alpha}$, $29^{\alpha/\beta}$
Chromium, Vanadium, Molybdenum (R58XXX, some with no UNS number)	19^{β}, 20^{β}, 29^{β} (some with no grade)
Alloying Element	**Effect**
Aluminum	Alpha stabilizer
Chromium	Beta stabilizer
Molybdenum	Corrosion resistance, beta stabilizer
Nickel	Corrosion resistance
Palladium	Corrosion resistance
Ruthenium	Strong beta stabilizer, grain refiner
Tin	Alpha phase strengthener
Vanadium	Beta stabilizer

FIGURE 4.17 Some common titanium and titanium alloys and the effect of alloying elements. There are many other titanium alloys available, but some do not have a corresponding grade or UNS number associated with them. Others may be sold under the manufacturer's trade name. [a]CP—commercially pure, α—alpha alloys, nα—near-alpha alloys, α/β—alpha-beta alloys, and β—beta alloys.

some common titanium alloys and the effect of their alloying elements. There are many other titanium alloys available, but some do not have a corresponding grade associated with them.

4.3.5.1 Pure Titanium

Pure titanium is used for aircraft ducting and tubing, chemical and food processing, medical instruments, and naval equipment. Due to the lower strength of "unalloyed" or commercially pure titanium, it is not used in critical load-bearing applications. Do not be confused by commercially pure grades of titanium. They incorporate very small amounts of alloying elements such as palladium and ruthenium which make a significant difference, but the quantity does not warrant the designation of being "alloyed" in a traditional sense.

4.3.5.2 Titanium Alloys

Titanium alloys are subdivided by their structures: alpha, beta, or a combination. **Alpha and near-alpha alloys** have good processability characteristics (for titanium) and often contain aluminum which is an alpha stabilizer. They have very good oxidation resistance but cannot be strengthened through heat treatment since there is only the single, alpha phase. They find use in applications such as space vehicles, airframe construction, missile components, recreational equipment, automotive components, hydrocarbon processing, and brine processing.

Alpha-beta alloys contain a significant fraction of the bcc beta phase (higher than those of the near alpha alloys). This is accomplished by high concentrations of beta-stabilizing alloys such as vanadium. These two-phase alloys are heat treatable and can be made very strong. Grade 5, Ti6Al4V is the most common titanium alloy and accounts for nearly half of the market for titanium. Alpha-beta alloys are used in many of the same applications already mentioned, but also can be used in ballistic armor, gas turbine components, and landing gear.

Finally, **beta alloys** consist primarily of the bcc phase and contain large amounts of beta stabilizers, typically chromium, molybdenum, and/or vanadium. These tend to be some of the strongest titanium alloys and can be through-hardened. They fit similar applications to the alpha-beta alloys, but are more often applied for their structural characteristics than for corrosion resistance (although it is an important factor in many structural applications as well).

4.4 Property-Based Classifications of Metals

There are a variety of metals that do not have their own category. For instance, although tungsten is a common metal, it does not see a broad enough range of alloys and uses to warrant a specific UNS category or other extensive classification system. We will classify these less common and specialty metals by their properties, not their composition.

4.4.1 Refractory Metals

Refractory metals have the common characteristic of a high melting point. These metals include the elements of tungsten, molybdenum, niobium, rhenium, and tantalum. **Tungsten** is commonly used in light bulb filaments (incandescent lights, Fig. 4.18) and welding electrodes. While pure **molybdenum** has modest thermal performance, it is

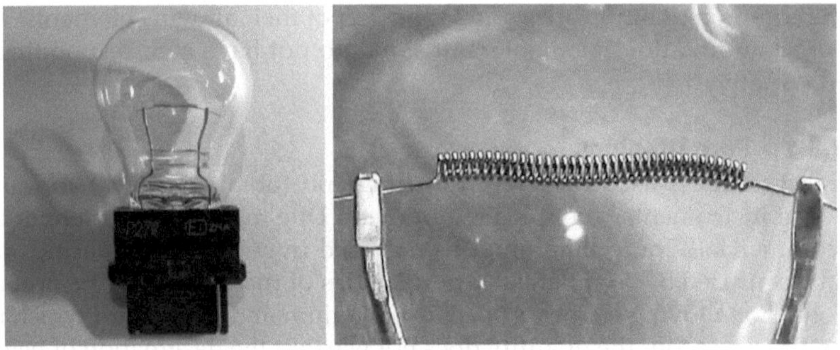

FIGURE 4.18 Refractory metals, like tungsten, are capable of withstanding very high temperatures. That has made tungsten especially useful for incandescent light bulb filaments, as pictured.

often alloyed with small amounts of zirconium and/or titanium to increase its high-temperature strength. These alloys, such as TZM (for titanium-zirconium-molybdenum), possess excellent high temperature properties [up to 2000°F (1093°C)] and are used extensively as die casting molds. Molybdenum is also used in some jet engine components and heating elements.

Niobium (also known as columbium) has good ductility compared to other refractory metals, which makes manufacturing processes easier. Niobium is used in missile, nuclear, superalloy, and superconductor applications, but most of its applications come as an alloying contribution. **Rhenium** is a very dense metal (21 g/cc, about twice that of lead), and it is used in limited quantities, mostly as an alloying element or in catalytic processes. This rare element is the only refractory metal to possess a hexagonal close-packed (hcp) structure [all others are body-centered cubic (bcc)]. **Tantalum** is nearly impervious to chemical attack at temperatures below 150°C and has good strength and ductility. It is used in chemical applications as well as in electronic components.

Although a very high melting point is the primary commonality within the refractory group, they also tend to be strong, wear resistant, and corrosion resistant. These metals tend to be difficult to process because of their high strength and melting temperatures. Powder metallurgy is one of the most common methods for forming parts from them as complex shapes can be created with limited postprocessing (see Sec. 12.2.5 for more). Although five elements are commonly considered refractory, an extended definition is sometimes used with a lower melting temperature requirement. This also includes metals such as rhodium, iridium, osmium, zirconium, titanium, vanadium, cobalt, etc. The top 10 elements, by melting temperature, are given in Table 4.5. Traditionally defined refractory metals are shown in bold.

4.4.2 Low-Melting Metals

Having just looked at the metals with the highest melting points, we now look at metals with low melting points. In general, these metals are used primarily for their ease of processing as they can be readily cast, formed, and machined. Metals within this classification include lead, tin, and zinc. Die casting, discussed in Sec. 11.2.2.3, and other liquid-state processes benefit from lower temperatures as it reduces the cost of equipment and operation. Some common metallic elements with low melting temperatures are given in Table 4.6.

Element (Atomic Number)	Melting Temperature (°C (°F))	Crystal Structure[a]
Carbon (6)	3500 (6332)[b]	simp. hex.[c]
Tungsten (74)	3410 (6170)	bcc
Rhenium (75)	3180 (5756)	hcp
Osmium (76)	3045 (5513)	hcp
Tantalum (73)	2996 (5425)	bcc
Molybdenum (42)	2617 (4743)	bcc
Niobium (41)	2468(4474)	bcc
Iridium (77)	2410 (4370)	fcc
Boron (5)	2300 (4172)[d]	rhomb.
Ruthenium (44)	2250 (4082)	hcp

[a]simp. hex.—simple hexagonal, bcc—body-centered cubic, hcp—hexagonal close-packed, fcc—face-centered cubic, rhomb.—rhombohedral.
[b]Carbon sublimes, it does not melt.
[c]Simple hexagonal structure assumes graphite allotrope, the most stable form of carbon.
[d]The melting temperature of boron is sometimes cited 2067°C.
Note: Bolded elements are refractory metals.

Table 4.5 Ten Elements with the Highest Melting Temperatures

Element (Atomic Number)	Melting Temperature (°C (°F))	Crystal Structure[a,b]
Mercury (80)	–39 (–38)	rhomb.[c]
Gallium (31)	30 (86)	ortho.
Sodium (11)	98 (208)	bcc
Lithium (3)	180 (356)	bcc
Tin (50)	232 (450)	ortho.
Bismuth (83)	271 (520)	rhomb.
Lead (82)	327 (621)	fcc
Zinc (30)	420 (788)	hcp
Magnesium (12)	639 (1182)	hcp
Aluminum (13)	660 (1220)	fcc

[a]rhomb.—rhombohedral, ortho.—orthorhombic, bcc—body-centered cubic, fcc—face-centered cubic, hcp—hexagonal close-packed.
[b]Crystal structure assumes room temperature.
[c]Mercury crystal structure assumed below –39°C, no structure at room temperature (liquid).
Note: Bolded elements are commercially dominant low melting point metals and alloys.

Table 4.6 Common Low-Melting Point Elements

4.4.2.1 Lead

Lead is a soft, dense metal. These characteristics are due in part to its fcc crystal structure and its low position on the periodic table (larger atomic size). It is very weak and easily worked through forming. It is so soft that it is used as a solid lubricant in hot forming operations and for engine components when leaded gasoline was used. It is relatively corrosion resistant and was used in antiquity for plumbing applications. More recent understanding of its toxicity has reduced its appeal (and legality) for many applications. That has prompted research into alternative materials and alloys with similar properties. Lead is widely applied in car batteries (lead-acid batteries), solders with low melting points and bearing alloys (Babbitt alloys) that are solid and can be poured in place. It is also used as a soft, sacrificial surface for clamping, and in nonmarring hammer heads.

4.4.2.2 Tin

Tin is a weak metal with a relatively low density (7.37 g/cc) compared to its high atomic mass. Tin is commonly used as an alloying element (e.g., bronze, which heralded the Bronze Age, discussed in Sec. 1.3.2) and in solders (e.g., lead-tin solders). It is used as a corrosion-resistant coating for steel in can liners ("tin" cans). Tin is often alloyed with antimony, copper, and/or lead to create "white" metals used for coating applications such as dinnerware and ornamental applications.

Tin undergoes a phase transformation at 55.8°F (13.2°C). As the temperature is lowered, tin transforms from a ductile tetragonal crystal structure (the beta phase or "white" tin) to a diamond cubic structure which is very brittle (the alpha phase or "gray" tin) and nonmetallic in character. Gray tin has a much lower density (5.77 g/cc), which means the volume must increase by about 27%. Once some of the component has changed it makes it easier for the rest to transform, making it autocatalytic. This phenomenon, known as *tin pest*, will cause the object to simply disintegrate into powder. Tin pest has been branded as a factor in Napoleon's men freezing during their Russian campaign as the tin buttons on their uniforms crumbled in the cold winter weather (though this is disputed). Alloying (and unwanted impurities) can reduce or eliminate the problem with tin pest.

4.4.2.3 Zinc

Zinc is a popular option for die casting, and it is used in a variety of applications such as alkaline batteries, appliance components, automotive trim, low-stress machinery components, and plumbing fixtures. Zinc is the fourth most common engineering metal, but most of its use is not as zinc components. Instead, it is used extensively in conjunction with other metals. More than half of all zinc is used for galvanizing. Zinc is very reactive under environmental conditions and readily corrodes to form a stable oxide, which makes it useful in galvanization (see more in Sec. 8.5.2.3). It is also used in wide array of alloys such as brass.

4.4.3 Noble (Precious) Metals

A handful of metals are especially resistant to corrosion or chemical attack. These are known as precious metals or noble metals. Because they do not corrode easily, they are found in pure form in the earth's crust. Whereas most metals are refined from ore, elements such as gold can be found as nuggets or in veins running through surrounding rock. Although these elements are easy to extract in certain situations, their rarity makes them difficult (and expensive!) to extract in significant quantities.

Noble metals typically include silver, gold, platinum, palladium, rhodium, ruthenium, iridium, and osmium. Probably the three most familiar are silver, gold, and platinum due their extensive use in jewelry, but these metals have many other applications and more important properties than their good looks. Most of these metals are good catalysts, meaning they can be used to convert one chemical to another without being changed themselves. The catalytic converter in your car, for instance, converts harmful exhaust gases into more benign gases. Platinum is the most effective catalyst for this application, but due to its high cost, palladium and rhodium are often used, but they too are expensive. Fortunately, when appropriately applied (high surface area with good distribution), the catalyst can be used in relatively small quantities.

Looking back to Table 4.5, you will find many catalytic metals also have high melting temperatures. This is another beneficial property, as the efficiency of catalytic processes (catalysis) is typically higher at elevated temperature. Catalysis is not their only application; however, gold is used in high conductivity electrical contacts, silver is the best electrical conductor available, and platinum and iridium have found a market in automotive spark plugs for longer life and more efficient combustion.

The applications of noble metals must be thoroughly justified due their high cost, and they are often applied in as small a quantity as possible by using coatings or distributions of small particles. The prices for some common noble metals are given in Table 4.7. Pricing of rare metals is highly volatile and dependent on international social and economic conditions. To get perspective on their cost, pricing for some commodity base metals are also given. For instance, iridium is 10,000 times more expensive than aluminum, so its use needs to be carefully thought out. Prices for metals are not always directly linked to their relative abundance. Although a strong factor, the difficulty in extracting and refining them is not *necessarily* related to their availability (e.g., titanium, Sec. 4.3.5).

4.4.4 Superalloys

Superalloys are a special class of materials which are designed for extreme conditions of heat and stress. A classic example where these alloys are applied is in turbine design. Turbines rotate at very high speed which puts constant centrifugal force on the blades. Jet turbines often operate at high temperature for extended periods of time, and the blades tend to slowly stretch after long service periods. This time and temperature dependent deformation is called *creep* (more in Sec. 8.2.6.2).

The tolerances required to achieve efficient turbine operation will not allow for any appreciable creep, and superalloys are specifically suited to resist this deformation. Superalloys also resist environmental degradation during their demanding operation. For an alloy to be strong, temperature resistant, and corrosion resistant over long periods of harsh service, "super" is an apt description.

We have seen in earlier sections that a variety of elements possess either corrosion resistance, high melting temperature, or other valuable properties. One of the common themes is that the more you incorporate "specialty" metals, the more you pay. Therefore, superalloys are not cheap, and their use is limited to high-demand application such as jet engines and nuclear reactors. Another hallmark of metals which have outstanding properties of strength and temperature resistance is that they tend to be difficult and expensive to process.

Also, superalloys often operate near their melting temperature, so single crystal components are desirable but have to be carefully made. Polycrystalline structures will

Commodity Precious Metals		
Element (Atomic Number)	Cost (USD/ozt)[a]	Relative Abundance (out of 78 elements)[b]
Platinum (78)	1476	71st
Gold (79)	1307	72nd
Rhodium (45)	1240	77th
Palladium (46)	881	74th
Iridium (77)	620	76th
Silver (47)	201	65th
Ruthenium (44)	68	73rd
Commodity Base Metals		
Element (Atomic Number)	**Cost (USD/ozt)[c]**	**Relative Abundance (out of 78 elements)[b]**
Aluminum (13)	0.06	3rd
Copper (29)	0.22	26th
Lead (82)	0.07	37th
Nickel (28)	0.60	24th
Tin (50)	0.70	49th
Zinc (30)	0.08	25th

[a]Prices (U.S. dollar per troy ounce) are based on midyear market conditions, 2014.
[b]Only elements which are found in appreciable quantities are considered. For example, noble gases such as argon are not present as part of the crust, but oxygen, which bonds to other elements, is. Highly unstable radioactive elements are also not included (e.g., astatine).
[c]Prices are converted from USD/metric ton and based on midyear market conditions, 2014.

Table 4.7 The Cost of Noble Metals Compared to Common Base Metals

creep more quickly and columnar (aligned) grains can also be created to improve performance, but single crystal turbine blades will maximize lifetime (see examples in Fig. 4.19). The most common subdivisions of superalloys are according to their composition. These are iron-based, cobalt-based, and nickel-based superalloys. These alloys achieve high strength through a combination of solid solution (see Sec. 3.2.4) and precipitate strengthening (see Sec. 9.3.7).

4.4.5 Amorphous Alloys (Metallic Glass)

Amorphous alloys are not technically classified by their properties. Rather, they are classified by their atomic structure. As usual, however, properties are determined by structure. The unique and defining characteristic of amorphous alloys is that they *do not* have a regular crystal structure. While every metallic element has a crystal structure in solid form, when certain elements are mixed, they do not fit into any distinct pattern. This disorganized structure makes them a glass, like window panes, but the properties of **metallic glasses** are much different in most regards. Like ceramic glasses (and amorphous polymers), they do

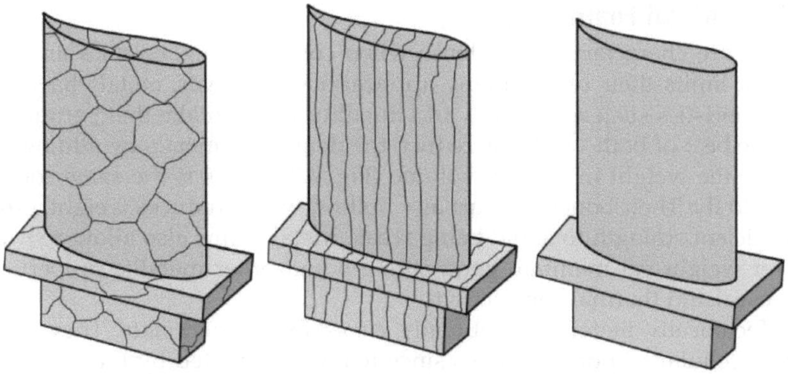

Figure 4.19 Superalloys use a combination of composition and structure to achieve excellent thermal, corrosion, and creep resistance. These high-performance alloys are often applied in turbine operation, such as the turbine blades illustrated here.

not have a distinct melting point and they are often cast or molded in a viscous state. Unlike ceramic glasses, amorphous alloys are not as brittle, they are electrically conductive (though not as much as crystalline metals) and they tend to have higher yield strength than comparable crystalline alloys but lower ductility.

There are a couple of important factors that determine whether an alloy will be amorphous. The elements being mixed together will have large atomic size differences making them more viscous in the liquid state and making it harder for them to organize as they cool. Imagine trying to create an orderly stack of tennis balls. Then try to do the same thing with tennis balls, baseballs, golf balls, beach balls ... you get the idea. Large variations in size make it more difficult to organize (or *crystallize*).

The rate of cooling is another important factor. Early glass-forming alloys could only be made in very thin sections which were cooled from liquid very rapidly (cooling rates of 1,000,000°C per second). This rapid cooling can be achieved through melt spinning, which is essentially a process of pouring molten metal onto a very cold (cryogenic), rotating drum. This creates thin ribbons, but thicker sections are useful in more applications. By carefully research on atomic chemistry and its effect on amorphization, new alloys have been developed that may have millimeter or centimeter scale cross sections.* These alloys and the thicker parts they create are known as **bulk metallic glasses**. Amorphous alloys are often made from several elements in specific ratios. Examples of these include $Ti_{40}Cu_{36}Pd_{14}Zr_{10}$ and $Zr_{41.2}Ti_{13.8}Cu_{12.5}Ni_{10}Be_{22.5}$ (subscripts indicate atomic percent in the alloy).

Amorphous alloys find applications in high-efficiency electrical transformers (as ribbons), sporting goods, medical devices, and cell phone cases. These alloys go by trade names such as Vitreloy and Liquidmetal. Sometimes metallic glasses are called *liquid metal*, but do not be confused, these alloys are always solid in commercial applications.

*Remember, the center of a part is the last to cool, and it too must be cooled at the critical rate to achieve a fully amorphous component.

4.4.6 Metal Foams

Metals are characteristically dense when compared to ceramics and polymers. Their weight limits their use in many applications; however, metals have many favorable characteristics such as strength and ductility which make them attractive. One way to get the best of both worlds is reduce the density of metals by adding voids (pores) to reduce the weight in a given volume (Fig. 4.20). This is the same concept that allows birds to fly. Their bones are partially hollow, which reduces weight, but they maintain a sufficient strength to avoid being weak. Metal foams also allow a structure to have a lower weight yet maintain good strength and other metallic properties such as high electrical and thermal conductivity.

Technically, metals are not really *foamed* in the solid state. They are more properly called cellular or porous metals since foams are created by the mixing of a gas and a liquid. The loose use of "foam" is common, however, and it gets the general idea across. Just be aware there is a difference.

Metal foams are created by two general methods: liquid-state or solid-state foaming processing. In liquid-state methods, the molten metal either has gas bubbled through it (direct foaming) or a foaming agent added to the melt (indirect foaming). This **foaming agent** (or **blowing agent**) gives off gas as it decomposes, and titanium hydride (TiH_2) is a common example. This creates a sort of froth on the metal (like blowing bubbles into milk), and that froth is solidified and collected. Aluminum is the most common metallic foam because it has a low melting temperature and is relatively stable in air. Not all metals are well suited to liquid-state processing because they have a high melting temperature and/or they are very reactive with air in the molten state.

There are two main distinctions in metal foams based on the pore structure. In **open cell foams** the pores are connected to each other and the surface. **Closed cell foams have** pores that are separated from one another. Closed cell foams are stronger and better at absorbing impact energy and blocking sound and heat. They are more common for structural applications. Open cell foams allow the passage of fluids and have a high surface area.

The shape and size of pores are also important. Metals that are created by solid-state processes can be tailored to possess very specific pore properties, but liquid-state methods are less precise. For instance, if you blow bubbles in milk, you can only indirectly control how those bubbles form. While the liquid-state foaming process is much more scientific in practice, the general concept still fits.

Metal foams have a number of applications. This is attributed to their low density as well as their unique properties. Just like polymer foams (e.g., seat cushions), metal foams can efficiency absorb energy. They are much stiffer, however, and that makes them suitable for applications such as automotive crash barriers or crumple zones in the vehicle itself. As foam is crushed, it absorbs energy at a constant rate until all of the pores are collapsed. Once that occurs, it then behaves like a solid metal.

Metal foams are often incorporated into other structures as well. They may be contained between sheets, known as sandwich panels (see Sec. 6.4.3), or within tubing. This enables a convenient package to use the foams, as they are inherently difficult to build structures out of alone, since there is little area to secure fasteners or weld components together. Metal foams are not just interesting for their mechanical properties. They are used in "functional" applications such as filters, in thermal management (either as heat exchangers or insulation), or as a support for catalysts. The applications are dependent on the size and amount of porosity, whether fluids can pass through the foam structure

Figure 4.20 Metal foams share many characteristics with polymer foams, but they are stronger and more resistant to temperature. When they deform they crush permanently, properties are useful in energy absorption.

(open versus closed porosity) and the metal which makes up the structure. Again, aluminum dominates the foam market, but foams of titanium, steel, copper, and most any other metal can be made. Their extent of use varies widely. In solid-state foaming titanium is one of the most studied as it has applications in biomedical implants and advanced aerospace structures.

4.5 Why It Matters

Metals heralded the earliest progress in society and are still essential to modern engineering and technology. There are thousands of alloys which are carefully crafted to satisfy demanding applications. That is among the key aspects of metallic materials; they are versatile. Light, strong, dense, soft, and anywhere between metals can be applied. The development of metals continues to be a scientific effort, and new alloys with exceptional properties are rapidly being developed. This is partly due to unique compositions and multi element alloying, but it is equally driven by clever processing and by controlling the microstructure of the materials.

Even with the wide array of alloys and new developments, there are a variety of "stalwarts" in popular use. These are materials that are common, but that does not make them outdated. We saw that ferrous metals account for the majority of structural metals, and that steel can have an enormous range of properties depending on alloying contributions. Knowing the classes and uses is important to identifying what type of steel will perform the necessary task for the least amount of money. This should also alert you to the fact that "steel" is a very generic term, and to be precise is important.

Even broader than ferrous metals is the nonferrous category. That is because they can have any element (except iron) as the primary metal. This leaves the category so diverse that it is necessary to further subdivide it into groups based on composition or properties. Some of the most common nonferrous metals and alloys are based on copper, aluminum, nickel, zinc, and other relatively abundant metals. Nonferrous alloys are also grouped into classes such as refractories and noble metals that consist of elements of similar properties.

Nonferrous metals tend to be costlier than iron-based alloys, and it is the designer's job to make informed decisions on alloy choice. Some applications cannot avoid the cost. For instance, medical implants must be biocompatible, and there are relatively few metals that meet that requirement. You cannot save money by choosing cheaper materials. Although the conditions of use will often limit the material choice, there is typically some range of options available. Also, the raw material cost is only part of the equation, as the ease of processing can quickly offset the cost of stock, either good or bad.

While the numbering systems are "unified" in the UNS system, older conventions are still prevalent. Being familiar with all of them is the only way to really cover all bases, but it requires memorization and time. Handbooks provide an excellent reference for quickly determining the properties, composition, and related alloys. The diversity of metals gives us the many possibilities we have in designing products, but it puts the burden on us to understand their proper use and processing. Fortunately, with time and effort, the burden is not so great, and the complexity of metallic materials becomes a great asset.

4.6 The How and Why of Metallic Materials: Identification and Application

4.6.1 Chapter Review

1. Think of an application that is most often satisfied by a metallic material. How would you replace that material with a nonmetallic material? Could you? What makes a metal the right choice?
2. Gold comes in many colors. How is that possible?
3. Why is it necessary to have designations for metallic materials?
4. Name three reasons that metals are alloyed.
5. Ferrous alloys have what element as the primary ingredient?
6. What are some advantages of ferrous alloys?
7. What benefits does plain carbon steel have over pure iron?
8. Plain carbon steel contains which two alloying elements?
9. Why are there multiple designations for the same metal? What is the primary goal of the UNS system?

10. Using the SAE/AISI steel alloy designation, what do the first two digits represent? The last two?
11. Why is it important to subdivide plain carbon steels into low, medium, and high carbon varieties?
12. Referring to Fig. 4.5, why can we not have the best of everything in one alloy?
13. What is the benefit of killed steel?
14. What is the difference in composition between plain carbon steel and low-alloy steel?
15. What benefits can be achieved using alloy steels?
16. Why are tool steels used for tooling? Relate their properties to their applications.
17. What process do tool steel designations like the "A" in A2 refer to? Why is it important?
18. Stainless steels contain what element that gives them corrosion protection? What is the minimum quantity?
19. How can alloying elements change the crystal structure of stainless steel?
20. Is stainless steel magnetic? Explain your answer.
21. What applications do stainless steels most often show up in?
22. How are processes such as machining and forming affected by the properties of stainless steel?
23. Why must low carbon stainless steel alloys be used in welded assemblies?
24. What is the meaning of "duplex" in duplex stainless steels?
25. Do cast irons have more or less carbon than plain carbon steels?
26. What properties are characteristic of cast irons?
27. Why are cast irons mostly processed by casting? Name a reason they are well suited for casting and why they are poorly suited for other operations.
28. The designations of cast iron are primarily in regard to the form what element takes?
29. How can white cast iron be the starting point for multiple cast irons?
30. What are some common applications of cast irons?
31. Nonferrous metals and alloys are defined as having a minority of what element?
32. What are some reasons that nonferrous metals are more expensive than ferrous metals?
33. Why are the properties of nonferrous alloys so varied?
34. Why is aluminum alloyed so often instead of being used in pure form?
35. Aluminum has an additional temper designation as part of its numbering system. Why is this needed?
36. What are some of the most important benefits of aluminum over steel?
37. Copper is used extensively in wiring. Why is this, and why is not used over long spans?

38. Pure copper is very soft but can be difficult to machine. Why?
39. Aside from copper, what is the main alloying element in brass?
40. How do the processing characteristics differ between copper and brass?
41. Why is bronze an ambiguous term for copper alloys?
42. What is the main benefit of copper-beryllium alloys? What is the main disadvantage?
43. What is the most useful property of magnesium? What are some drawbacks?
44. Magnesium alloy designations give information about what?
45. What applications do nickel alloys often serve in?
46. How do additions of nickel influence the properties of steel?
47. What is a drawback to using nickel alloys?
48. Titanium is one of the most common elements on earth. Why is it so expensive?
49. What crystal structure does pure titanium have? What is mean by an alpha-beta alloy?
50. Refractory metals are known for what common characteristic?
51. What are some applications of refractory metals?
52. In what applications are low-melting point metals beneficial?
53. What applications does lead serve now, and what uses have been phased out?
54. What is a "white metal?"
55. Noble metals are known for what characteristics?
56. Why are noble metals used in small quantities where possible?
57. Superalloys are used under what conditions?
58. What are the benefits/disadvantages of polycrystalline, columnar, and single crystal superalloy components?
59. Amorphous alloys are also called metallic glasses. How are they similar to window glass, and how are they different?
60. What are some applications for metallic glasses?
61. Metal foams maintain some metallic properties but reduce which property?
62. What are the two methods for creating metallic foams/porous metals?
63. Why is "foam" a misnomer when referring to a solid metal?
64. Where are metal foams applied?

4.6.2 Concept Connections and Extensions

1. Identify a product that markets its value by incorporating steel. What type of steel is it and what properties does it have?
2. Consider a product that used to be made from metal but is now plastic. What is the application of it, and why do you think the change was made?

3. Name an application that the properties of metal are critical to but cannot be used without incorporating a nonmetal as well?
4. Describe how refractory metals are required for incandescent lighting applications. What is the disadvantage of this method of lighting in comparison to LEDs (Sec. 6.2.3.2)?
5. Propose an alternative designation system for metals. What challenges are there in doing this?
6. How can lower carbon be combined with high strength? Refer to Sec. 9.3.6 for more information.
7. How are low-melting metals and alloys useful in products and processes? Consider the cases of sprinkler systems and soldering/brazing (Sec. 19.3.1).
8. Contrast the properties of metals foams with solid metals and polymer foams (e.g., Fig. 6.31). How does this allow it to serve in unique applications?
9. Describe the structure of metallic glass and provide a sketch. Why is "liquid metal" a good marketing tool but a bad description?
10. Research the applications for toxic metals such as lead, beryllium, cadmium, and others. Why are they used, and how are they being replaced?
11. Noble metals are known for their high cost. Discuss how that may be driven by their functional properties and how their role as jewelry and in aesthetic applications may contribute.
12. Titanium is one of the few biomaterials. What challenges occur in an orthopedic implant application? Use discussion from Chaps. 7 and 8 to support your answers.

CHAPTER 5

Nonmetallic Materials: Structure

Nonmetallic materials are actually nature's default state in most cases. Most metals are not found as metals. They are found in ores, which consist of metallic and nonmetallic elements bonded together. Nonmetallic materials, then, may contain metallic elements, but they do not possess metallic *properties*. This very broad classification means that nonmetallics exhibit a range of characteristics. From cement to silly putty, the general description of being nonmetallic does not provide much insight into the properties, and we need to know more about these materials.

The biggest differences, as you might guess, are based on structure. Nonmetals are typically separated into the subcategories of polymers and ceramics. Polymers are characteristically weak and flexible while ceramics are typified by high hardness and brittle behavior. Because of the wide variation within nonmetallics, we need to be careful about generalizing them. As we will explore in this chapter, the atomic and molecular structures of ceramics and polymers are very different from one another, and they are both very different from metals. Their unique structures give them unique properties and help us satisfy a great number of common applications.

5.1 How They Work

One of the hallmarks of metallic bonding is the presence of delocalized electrons, but in nonmetals, the electrons are restricted to certain atoms, and these materials are dominated by covalent and ionic bonding (for the basics, see Sec. 2.3.1). These are the basis for nonmetallic structure and properties. While there is much more to the structure, we will make some broad assumptions to simplify our discussion. The reason of simplification is important because nonmetals tend to have complicated structures.

Metals were relatively simple, and their structures could be represented by hard spheres in contact with one another. Bond distances are longer in ceramics, and their unit cells can contain multiple elements. Polymers consist of molecular chains, and their bulk properties are determined not just by composition, but also by chain length, orientation of atoms and groups of atoms along the molecule and the bonding between chains. While most commercially important metals have one of the three crystal structures, common ceramics and polymers vary widely. That is not to say there are not general rules of structure and common types. We will look a variety of examples and determine some defining characteristics.

5.2 Ceramic Structure

Ceramics share some basic structural features with metals, but there are very important differences. We will continue with the concepts of unit cells (Sec. 3.2.1) and crystal structure (Sec. 3.2.2), but ceramics typically contain more than one type of atom in their unit cell. Having multiple types of atoms in a unit cell can increase the size of the unit cell. That means the complexity of the unit cell increases, and more information is needed to describe the material's structure, even when considered in the most basic form. Although ceramic structure is a rich and deep subject, we will only consider some basic structures and the importance of structure on processing.

5.2.1 Ceramic Crystal Structure

Like their metallic counterparts, there are some general structures that ceramics assume. Remember, ceramics are characterized by covalent or ionic bonding. This tends to render the materials rigid and brittle. Long bond distances and open structures usually provide a corresponding decrease in density compared to metals. Another unique characteristic of ceramics is their chemistry. That is, they have a specific composition that is often assumed. For instance, alumina, a common abrasive material, has a chemical composition where there are three oxygen atoms for every two aluminum atoms (Al_2O_3).

The structure of the material assumes that this is true. When it is, and the chemistry is exactly correct, the material is a **stoichiometric** compound. A stoichiometric compound has no excess or deficit of an element (e.g., there are not two aluminum atoms and two oxygen atoms). For multielement ceramic structures, this chemical balance is typically used for determining structure, but it is not always true. Deviation from it can be considered a crystal imperfection, such as discussed for metals, and the variation is typically slight. For instance, one out of every 100,000 aluminum atoms may be missing. With this in mind, we will consider some simple, two-element ceramic structures.

5.2.1.1 Sodium Chloride (Rock-Salt) Crystal Structure

Ceramics with only two elements typically distribute those elements evenly throughout the structure. We can start by returning to a familiar friend, table salt. In the discussion on ionic bonds (see Sec. 2.3.1.3), we established that there is an exchange of electrons between atoms and that they will naturally arrange into an ordered structure so that like-charged atoms (positive or negative) are not adjacent to each other (see Fig. 2.6 for a reminder). What we did not establish is how they are arranged in three-dimensional space, though it can be ascertained from the two-dimensional diagram.

The 3D structure is shown in Fig. 5.1. This **sodium chloride structure** is face-centered cubic (fcc) in nature; however, there are now two atoms that comprise the unit cell as shown on the left. If you consider only the corner and center atoms, you can see that the fcc structure has additional atoms within it. The same can be done with the other atoms, but it is less apparent from that particular image. Therefore, this rock salt structure can be thought of as two "interpenetrating" fcc cells, and this creates a much larger unit cell.

5.2.1.2 Cesium Chloride Crystal Structure

The rock-salt structure has an fcc pattern consisting of two elements. This same principle applies to the bcc configuration. The bcc analogue is referred to as the **cesium**

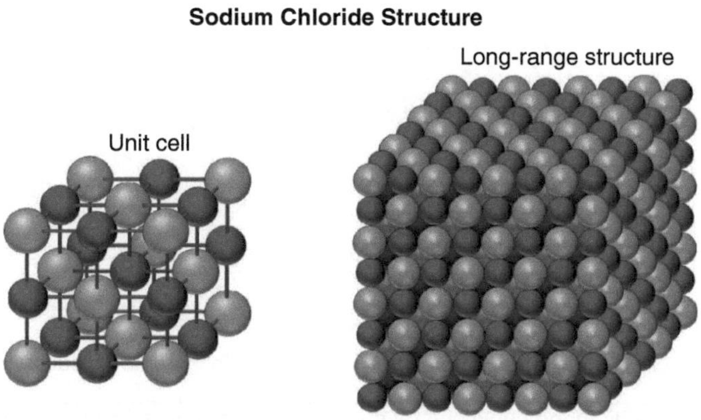

Figure 5.1 Ionically bonded ceramics often assume a "rock salt" configuration as shown in the image. This structure is analogous to an fcc lattice as described for metallic bonding.

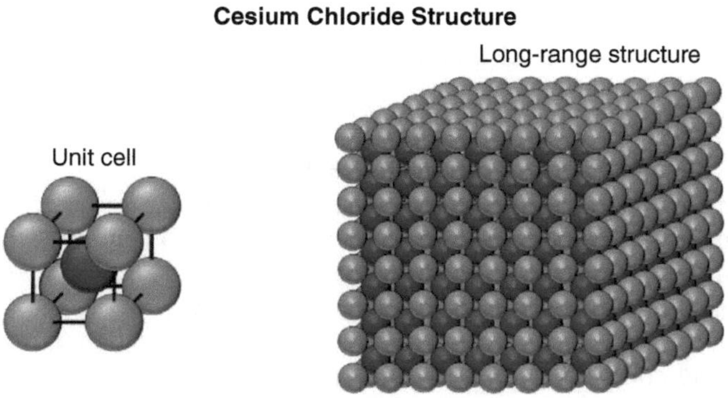

Figure 5.2 The cesium chloride structure has a similar arrangement to the metallic bcc structure. In the ceramic version, a second element replaces the body-centered position. Either element in the pair can be thought of as the centered atom.

chloride structure (Fig. 5.2). The name is based on the compound, cesium chloride, which has this structure, but it is also a generic description for any two elements organizing in the same fashion. The body-centered atom is one type of element while the corner atoms are of the other type. In this case, the unit cell is the same size (nine atoms) as in metals, but one or the other element must be chosen for the center position.

5.2.1.3 Diamond Cubic/Zincblende Crystal Structure

Another relatively common structure for ceramics is **diamond cubic structure** (Fig. 5.3). This is the structure of diamond (no surprise!), silicon, and some semiconductor compounds. When two atoms are involved, it is called a **zincblende** structure. It can be hard to picture, but the diamond cubic structure is actually in the same family as the face-centered cubic structure, but it has two atoms at each location where there would normally be only one. Since the atoms do not fit as tightly together, the packing factor (a measure of the filled space in a unit cell) is only 0.34, whereas metals are typically

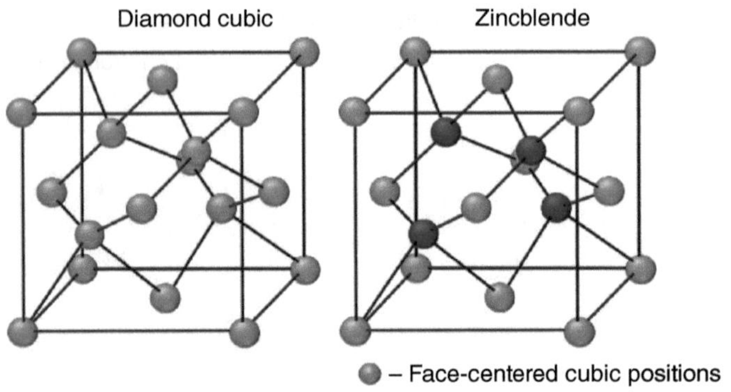

Figure 5.3 The diamond cubic crystal structure is common among ceramics and is called zincblende when multiple elements are present.

0.68 (bcc) or 0.74 (fcc and hcp). This is part of the reason some ceramics have a lower density than metals in many cases. The concepts behind crystal structure are discussed in more detail in Secs. 3.2.1 and 3.2.2 for metals, but ceramics follow many of the same principles.

5.2.1.4 Other Crystal Structures in Ceramics

There are a variety of different structures that ceramics assume. The common theme to take away is that they are more complex than metals (due to the ordering of multiple elements), have larger unit cells in many cases, and require a stoichiometric or near stoichiometric composition. We have considered three of the most basic structures, but other, more complex structures are quite common. For instance, it was noted earlier that the chemical composition for alumina is Al_2O_3. The unit cell for alumina is shown in Fig. 5.4. The larger distances between atoms, rigid bonding, and lack of slip planes cause ceramics to be characteristically brittle.

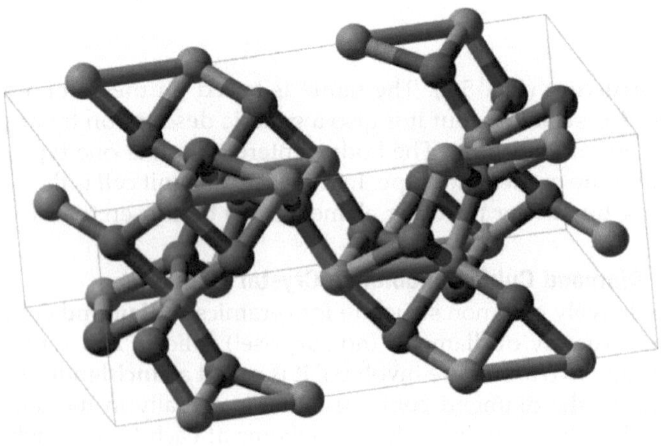

Figure 5.4 Ceramic structures can be complex in shape and distribution of elements. Even a two-element structure like alumina (Al_2O_3) is not as easily described as most metals.

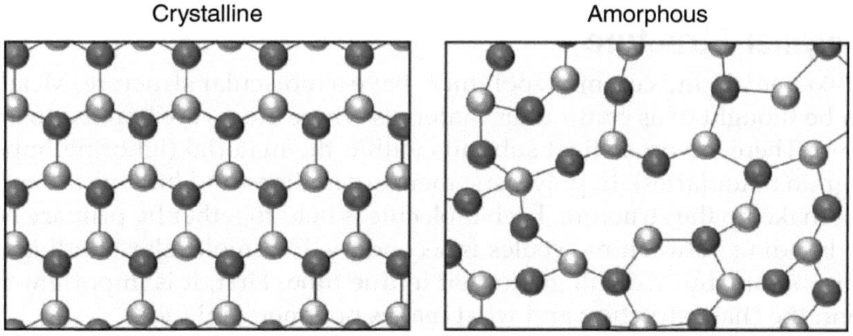

FIGURE 5.5 Ceramics can have a well-defined structure, but they can also be formed with little to no long-range pattern. Materials which do not possess long-range order are considered amorphous.

5.2.2 Amorphous Ceramics

We have been considering the way atoms arrange into crystals in ceramics, but there are instances where no defined crystal structure exists. Ceramics with a disordered structure are known as being *amorphous*. What this means is that there is no long-range order (crystallinity) even though all the same ingredients may be present as a crystalline version.

A common example of an amorphous ceramic is glass.* Glass is made by melting silicon dioxide (SiO_2) and other minerals, but we will focus on the major ingredient, SiO_2. Those two elements, when cooled from a highly viscous, "glassy" state can freeze into a disorganized pattern. Although the atoms are rigidly bonded together, there is no definite pattern to their structure.

See Fig. 5.5 for instance. The same two elements (silicon and oxygen) can be arranged in a well-ordered pattern (left) or they can be arranged in a disordered state (right). Even though all of the same ingredients are present, the properties will vary because of the difference in structure. One of the most telling properties is that the amorphous structure will be optically transparent while a polycrystalline structure will appear hazy or even opaque.

With enough time and at high temperature, an amorphous structure can rearrange into a crystalline pattern. The opposite is also true. A fast cooling rate from the molten state facilitates amorphization as the atoms have less time to arrange in an ordered configuration. Amorphous materials have many interesting and useful properties, including optical, mechanical, and chemical, so their production is important, especially in nonmetallic production.

Because of their simpler structure, metals can organize into a crystal structure much faster than ceramics, and it is very difficult to form amorphous metallic alloys like in ceramics. Metallic glasses *can* be formed (see Sec. 4.4.5), but they are not as common as amorphous ceramics. Polymers are difficult to create crystallinity in, and they are always at least partially amorphous (see Sec. 5.3.2.7). To understand why polymers behave differently, we will consider their structure next.

*Although "glass" as used here is intended to identify the material used in windows, any amorphous (noncrystalline) material can be considered a glass, including polymers and metals.

5.3 Polymer Structure

Unlike metals and ceramics, polymers have a molecular structure. Metals and ceramics can be thought of as continuous materials where atoms are bonded to atoms again and again. There are no distinct subunits within the material (ignoring imperfections such as grain boundaries). In polymers, there are millions and billions of separate molecules that make up the structure. Each molecule is held together by primary bonds, but often the bonding between molecules is secondary. Intermolecular bonding can be of a primary nature, but we will get to that in due time. First, it is important to better understand the chain structure and what makes polymers unique.

5.3.1 Polymer Chain Structure

Polymer literally means many units, and each molecule is made from the same repeating structure. This repeated structure determines the properties of the molecule, and the molecules of polymers directly affect their bulk properties. A polymer molecule is essentially a string of carbon atoms* with other atoms or groups of atoms attached along the length. Similar to a charm bracelet, the carbon atoms act as the support (bracelet) for the other atoms or groups of atoms (the charms).

Polymer chemistry and structure can be (and are) engineered at the atomic level. This is accomplished by controlling how the molecules form, what elements are present, and how long each molecule, or chain, is. The number and location of chemical units control the properties of the material, so we will first look at how polymers are created.

5.3.1.1 Formation of Polymers

Polymers are formed through polymerization. **Polymerization** is the process of combining chemical units together to create a larger molecule. Each unit is caller a **monomer**, or simply a **mer**. The mechanisms used to cause the formation of polymers are rich in chemical theory, but we will focus on the end products without much attention to reaction process. That said, there are two basic types of polymerization: addition and condensation.

Addition polymerization is accomplished when monomers rearrange bonds to generate a single, larger molecule. This type of polymerization is also known as *chain-growth*. There is no waste product, and the final product is simply the polymer itself. This can occur because the starting product typically contains a double bond. One of the bonds can be reallocated to join with another molecule and thereby create the polymer chain.

Often the reaction uses and **initiator**, sometimes called a **catalyst**, to start and/or accelerate the reaction. An example of this would be a peroxide molecule which reacts with the monomers to begin the process of polymerization. The peroxide is used up during the reaction, and it is therefore not truly a catalyst. Catalysts, which are *not* used up during the reaction, are also used to help facilitate the chain growth process or control the output. These may be metallic or organometallic† in nature, such as a

---

*There are some inorganic polymers that do not include carbon, but these are much less common. One commercially important inorganic polymer is based on silicon and oxygen is polydimethylsiloxane (PDMS), a rubbery material used for silicone products (see Sec. 6.3.3.5).

†Organometallics consist of a metal atom (or atoms) bonded to an organic compound by a carbon.

Ziegler-Natta catalyst (after Karl Ziegler and Giulio Natta), which often pair titanium chloride and an aluminum organometallic such as triethylaluminum ($Al(C_2H_5)_3$).

Condensation polymerization, also known as step-growth polymerization, creates by-products during the polymerization reaction. This is because the monomers have reactive ends on them, and they do not need the addition of peroxide or other free-radicals to assist in reorienting bonds. The monomers undergo direct reactions with each other; they also yield additional compounds (**by-products**), which are not included in the polymer (e.g., water).

The reaction is often started using elevated temperature and requires two different chemicals. This is unlike addition polymerization that could combine one chemical in longer and longer sections. The process can only continue if both chemicals reacting have two functional ends each. That way, after the first reaction occurs, there are still ends available to continue the reaction. This is kind of like LEGO® blocks. You can connect them from both ends to create as large of a component as you like since there are always free spaces to add more.

Polymerization can be driven by other factors as well. For instance, photopolymers will polymerize in the presence of light, usually a specific wavelength of light (e.g., UV light). The type of polymer will determine the required polymerization process. Some of the more common polymers are listed in Table 5.1. The final structure of a polymer is always determined by the monomer(s) used.

5.3.1.2 Mer (Unit) Structure

To illustrate the structure of a polymer, we will look at polyethylene, which has one of the simplest structures. Polyethylene is used in applications as varied as children's toys and surgical implants. Like many materials, we can glean information just from its name. Polymers, in general, start with "poly." As with many other uses, the "poly" in polymer indicates "many" or "multiple." So a polymer is many "mers." So what is a mer? In this case, a mer is a chemical base unit (mer means "part"). So when we speak of polyethylene, we are essential stating that it is many "ethylenes" or ethylene units.

The appropriate terminology for hydrocarbons and other side groups is a chemistry topic that we will not delve into very deeply, but there are both rhyme and reason for

Addition Polymerization	Condensation Polymerization
Polyethylene	Polyester
Polypropylene	Polycarbonate
Polyvinylchloride	Polyamide (nylon)
Polytetrafluoroethylene	Polysiloxanes
Polyacrylonitrile	Phenolics
Polystyrene	Polyurethanes[a]

[a]The formation of polyurethane occurs by the reaction of two different monomers, but it does not result in the condensation of any by-product.

Note: The polymerization methods for polymers can differ. The two most common methods, addition and condensation polymerization are listed with polymers created using those methods.

Table 5.1 Polymerization Method Used for Common Polymers

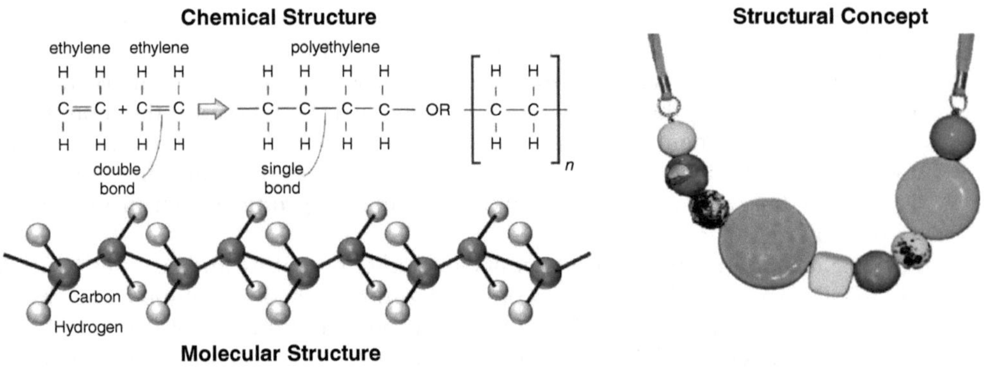

Figure 5.6 Polyethylene is a common polymer with a simple repetition of ethylene molecules. Through polymerization, the base units are locked together in a chain. Some polymers may have various units combined together, similar to a necklace shown on the right.

the names plastics are given, and you can see the correlation between polymer naming and structure throughout Sec. 6.2.4. So, as with metals and ceramics, we have a base unit or unit cell, so to speak. In polyethylene the base unit is an ethylene molecule. Ethylene has the composition C_2H_4. These molecules combine together (through polymerization) to form a longer molecule. The many connected units form the polyethylene molecule. Like a beaded bracelet or necklace, as shown in Fig. 5.6, the many units connect together to form a long chain that can vary in structure.

The polyethylene (PE) molecule has only hydrogen atoms connected to the **carbon backbone**, or the central carbon atoms along its length (Fig. 5.7). Other atoms or groups of atoms can fill these positions as well, and this is where the compositional differences in polymers arise. A very simple difference can be achieved through replacing the hydrogen atoms with fluorine atoms. This results in polytetrafluoroethylene (PTFE), also known as Teflon™. There is a vast difference in the properties of PE and PTFE even though the physical structure is similar.

Figure 5.7 Small differences in chemical composition create huge differences in properties. Polyethylene is used in low-cost container applications while PTFE is used in nonstick situations. The only chemical difference is that hydrogen atoms are replaced by fluorine in PTFE.

You may imagine that when we exchange small hydrogen atoms with a group of atoms, known as a **side group**, we will achieve even greater differences. This is true, but do not underestimate the importance of chemistry. The way electrons are shared in the material makes important differences as well. Another example of this importance can be seen in polyvinylchloride (PVC). This common piping material only differs from PE by exchanging one in four hydrogen atoms for a chlorine atom, but it has a higher strength and lower ductility and very different applications.

To emphasize the importance of side groups we will look at polystyrene (PS). The mer unit for polystyrene is shown in Fig. 5.8 on the right. The side group is called an **aromatic ring**. The ring consists of carbon atoms bonded together in hexagon with hydrogen atoms connected to each carbon atom (see the right side of Fig. 5.8). This group of atoms is bulky and interferes with the polymer chain movement. Polyvinylchloride, which has a single chlorine atom instead of an aromatic, is also bulky because the chlorine atom is very large relative to the hydrogen atoms.

To permanently deform a polymer is to cause chains to move relative to one another (i.e., slide past each other). Side groups interfere with chain motion because they can physically interlock with one another. Weak polymers can be easily shaped into new configurations. Stiff, brittle polymers do not take bending or stretching well at all. There are a variety of reasons for their behavior, but side groups are a basic and important factor.

Think of untangling a pile of loose extension cords. Probably not too difficult, but it does present some trouble. This represents "smooth" molecules that do not have side groups protruding from them (Fig. 5.9). Now imagine a pile of Christmas lights. Untangling those can prove *very* difficult. The difference is that the light bulbs (side groups) extend out from the central wire (carbon backbone). They get stuck on each other and on the wires when you try to move them. It takes more force to move an individual strand and the side groups get stuck repeatedly. A polymer with this type of structure requires more force to deform.

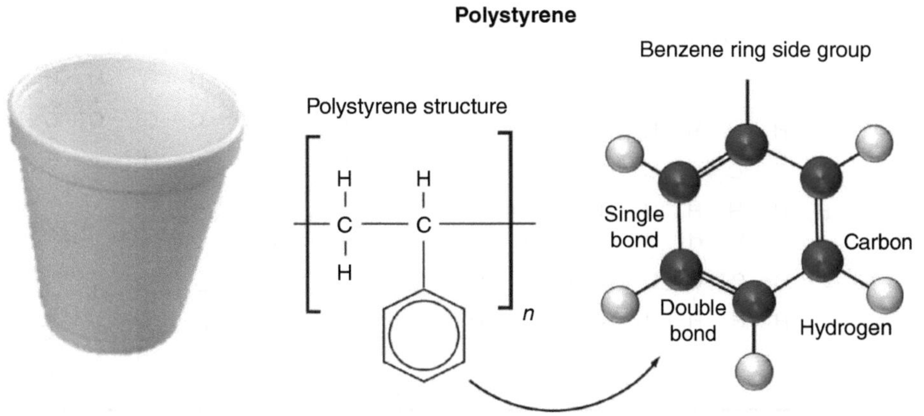

FIGURE 5.8 Polystyrene is commonly used in commercial products and is often associated with Styrofoam™ containers. Polystyrene has a large benzene side group repeated throughout its structure (right), which influences it properties.

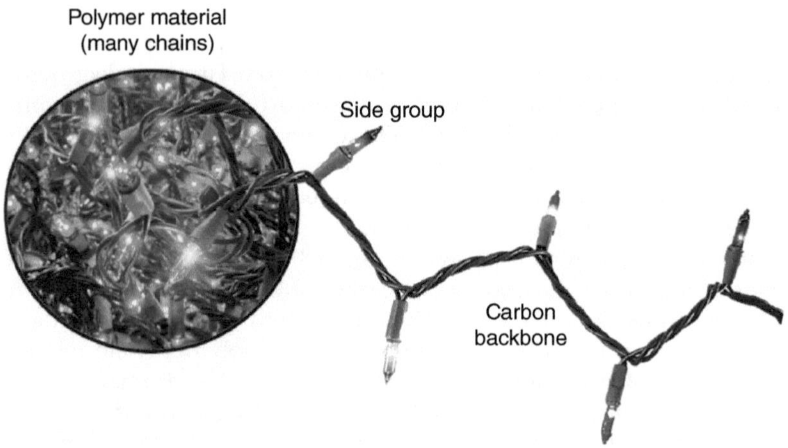

Figure 5.9 Side groups on polymer chains, like lights on a strand of wire, make motion more difficult. Untangling Christmas lights emphasizes the difference protrusions from a central chain can make. The size, orientation, and spacing of side groups will affect the polymer's properties.

To better visualize the effect of how side groups affect chain movement, see Fig. 5.10, which shows polyethylene and polystyrene chains. The polyethylene chains will be less likely to interact mechanically, and therefore, less force is required for forming the material.

The two-dimensional representations of the polymer chains in Fig. 5.10 are not realistic. From Fig. 5.6, the three-dimensional representation shows there is more to consider with physical interactions. Also, there are two important assumptions that were made in the example of Fig. 5.10. First, the polystyrene chains were assumed to strictly interact between side groups. Second, all of the side groups were assumed to be on one side of the molecule, but neither assumption is very good. Polymers are not that

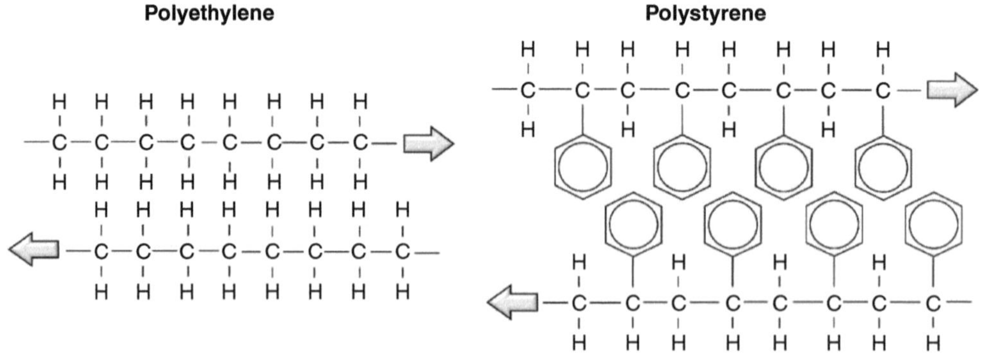

Figure 5.10 Polymer side groups and large pendant atoms can hinder motion due their size and interlocking effect. The type of side group is very important in determining the magnitude of this affect. For purposes of illustration, this image compares polyethylene (no bulky side groups) and polystyrene which has an aromatic group as part of the styrene monomer.

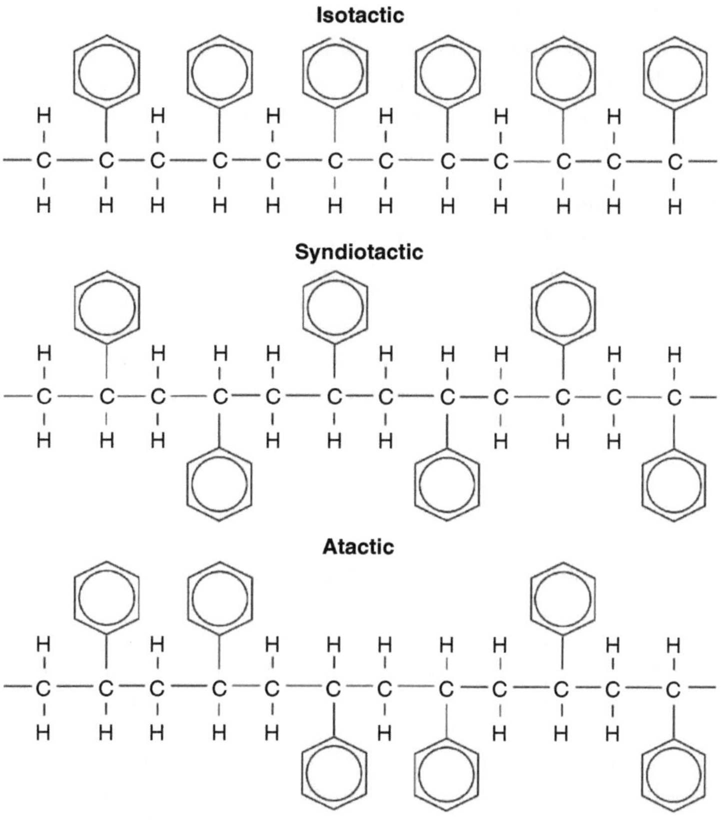

FIGURE 5.11 The way side groups are distributed along a polymer chain is broken into three categories, depending on the distribution and regularity.

uniform and predictable over billions of atoms and molecular chains. The interactions tend to be much more complex. For instance, side groups can be distributed off of a molecule in regular or irregular patterns. The way the molecular side groups are arranged is known as the **tacticity** of the molecule. These can be broken down into three primary types, as shown in Fig. 5.11.

The configuration of the example in Fig. 5.10 is isotactic. An **isotactic** polymer has all side groups positioned on the same side of the molecule. A **syndiotactic** molecule has side groups that alternate from one side to the next in an orderly pattern. An **atactic** molecule has side groups that occur randomly along the length of the molecule.

You may wonder, "Why should I care? I don't see the molecules!" Indeed, you do not see individual molecules and atoms, but you see their effects. The tacticity of a polymer will affect many important properties. Among these are the crystallinity (Sec. 5.3.2.7) and related mechanical and optical properties, as well as chemical and thermal properties that we will discuss later. What you do not see, then, is the most important. The molecular variations do not end with tacticity, and there are conformational features that extend beyond one molecule or side group.

5.3.2 Multichain Effects in Polymers

We will consider the following structural features as being the interaction of multiple molecules. In reality, if these structures are connected through primary bonds, they can be considered single molecules which are longer and more complex in nature. The first category of multichain structures we will look at is copolymers.

5.3.2.1 Copolymers

A polymer comprised of a single mer type is considered a **homopolymer**, and this was our assumption in the previous discussion. The combination of multiple polymer types makes it a **heteropolymer**, which is another name for a **copolymer**. Copolymers are created through mixing monomers of different types during the formation of the polymer chains. For instance, nitrile, a common polymer used in nonlatex disposable gloves, is made from the combination of acrylonitrile and butadiene mers. These mer units are more complex than those we have considered to this point, but the basic concepts remain the same (Fig. 5.12).

The monomers used to create the copolymer are important, but that is not all that can be said about the structure. The way the two polymers combine is also a fundamental piece of information. Each mer unit can combine to form a "pure" polymer like polyethylene or polystyrene for a short distance, but then these chain *segments* are combined with other, different segments to form the copolymer.

The way the combination occurs will dictate the terminology used to describe it. When the mers combine in small yet distinct segments of one type or the other, they are called **block copolymers**. When the mers are evenly dispersed in repeating sequence of one type after the other, it is called an **alternating copolymer**. When the mers are sporadically positioned along the length of the chain, the material is a **random copolymer**. In the final occurrence, a homopolymer has another polymer attached along the length in a branching fashion. Like grafting branches of one tree to another, this type is called a **graft copolymer**. These types of polymers are shown schematically in Fig. 5.13.

Copolymers allow for the tailoring of properties. This is analogous to alloying metals (Sec. 3.2.4). By combining dissimilar materials, the new material is expected to have different properties than either constituent, but the properties are likely to lie somewhere between the two. This is used to create polymers with greater toughness so that

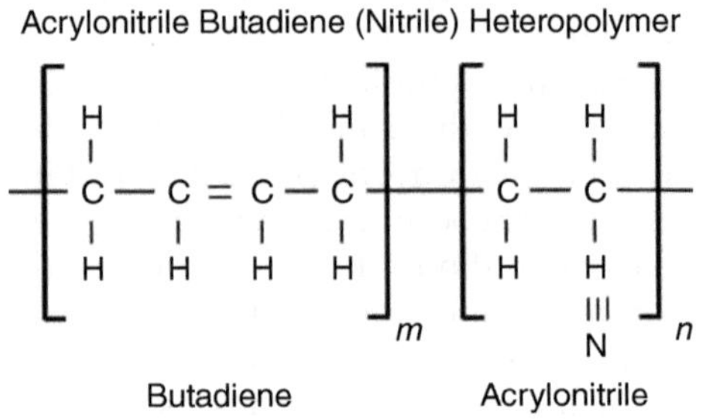

FIGURE 5.12 The combination of two mer units is used to create nitrile, a common copolymer. The location and number of each monomer (*m* for butadiene and *n* for acrylonitrile) determines the class of copolymer.

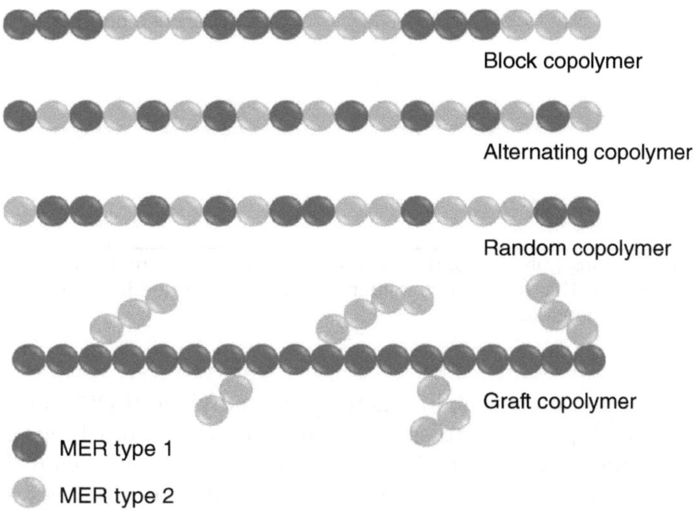

FIGURE 5.13 Copolymers can be subdivided by their structure. How the different mers combine will determine the terminology used to describe them and the properties of the resulting material.

they can resist cracking, to modify the softening temperature, reduce crystallinity for improved forming properties, etc. The mers and processing parameters must be carefully controlled to create the desired product, but the effort is well worth it as it allows for entirely new compositions with new properties to be created.

For example, ABS plastic, a common material in household goods, is a copolymer. ABS is an abbreviation for acrylonitrile butadiene styrene, a ter-polymer containing three mers (see Sec. 6.3.1.6). This yields a tough, easily processed polymer capable of a host of different applications. As just indicated, copolymers are not limited at two mers. The creation of new polymers is a matter of carefully choosing the type and number of mers, their relative quantities, and then polymerizing them to create the desired structure. It is not an easy feat, but it has made many products cheaper and better performing.

5.3.2.2 Polymer Chain Branching

We considered the formation of copolymers, which have a branched structure. These branched copolymers consisted of a homopolymer chain with another type of polymer attached along the length. This same branching structure can be achieved in a single polymer as well. This happens when a polymer forms another molecule at a bond that would normally be taken by a single atom (e.g., hydrogen) or a side group. The branched structure is shown in Fig. 5.14.

Just like side groups, branching increases the interaction volume of a molecule. The branches interact with other molecules and increase the spacing between molecules. This creates a more open structure which reduces the crystallinity (see Sec. 5.3.2.7, for more on crystallinity). Also, the melting temperature tends to decrease as do strength and hardness. The large molecular spacing increases the transparency, but allows oxidative and chemical permeability to increase. The structure is vitally important to just about all properties (see Table 5.2).

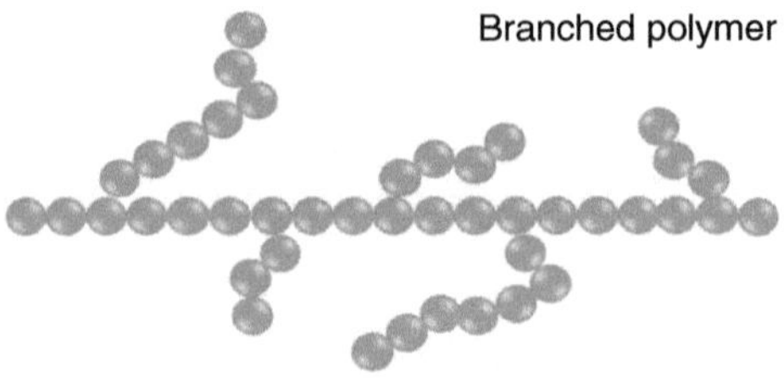

FIGURE 5.14 During polymerization, the formation of polymer molecules which branch from another can occur. This occurrence influences many properties including mechanical, thermal, and chemical responses.

Two common materials, low-density polyethylene (LDPE) and high-density polyethylene (HDPE), are examples of the importance of branching. Extensive branching "opens up" the structure of LDPE, and among the properties mentioned above, also decreases the density (hence, the name). HDPE does not possess extensive branching, and therefore, it has a higher density (~0.95 g/cc versus the 0.92 g/cc density of LDPE). Some of the properties of these polymers are discussed in Sec. 6.3.1.1. There you will also find another type of polyethylene, ultrahigh molecular weight polyethylene (UHMWPE). As the name implies, the molecular weight is extremely high. We will now look at what that means and why it is important.

5.3.2.3 Polymer Chain Length (Molecular Weight)

We have looked at the chemical composition of the mers that make up a polymer, the combination of different mers to create a copolymer, and the effects of chain branching. There is still another, basic chain parameter that greatly influences the properties and applications of polymers. That parameter is the chain length, more often referred to as the **molecular weight**.

Recall, all polymer chains are really just molecules. The length of the molecule will determine the weight of the molecule. Just as a steel chain will get heavier when more

As Polymer Branching Increases...	
Density	Decreases
Melting point	Decreases
Strength	Decreases
Hardness	Decreases
Toughness	Increases
Weather resistance	Decreases
Transparency	Increases
Shrinkage during molding	Decreases

Source: Adapted from Table 6.1. Brent, Strong A. "Plastics materials and processing." Prentice Hall (2000).

TABLE 5.2 Effects of Polymer Branching

links are present, the mass of a polymer molecule increases as well. Because polymer molecules are comparatively light, we are not so much concerned about the actual mass of the polymer molecule, but instead we can use that as an indirect measure of its length. The length is quite important as the interactions between chains will be determined by the length. If you consider the polymer molecules as entangled steel chains, the longer each chain is, the harder it will be to pull it out of the pile. The molecular weight (more appropriately the **molecular mass**) is calculated by summing the atomic weights of each atom in the molecule:

$$M = N_1A_1 + N_2A_2 + N_3A_3 + \cdots + N_nA_n = \sum_{i=1}^{n} N_iA_i \tag{5.1}$$

In Eq. (5.1), N_1 is the number of atoms of element 1 in the molecule and A_1 is the atomic mass of element 1. The *total* number of elements, n, and their respective masses *in each molecule* will determine the molecular mass. The condensed form of this summation is given with Σ (capital Greek letter sigma). Sound confusing? We will work through some examples. Polyethylene has two carbon atoms (atomic mass = 12.011 g/mol) and four hydrogen atoms (atomic mass = 1.008 g/mol) in each mer (i.e., the mer unit C_2H_4 shown in Fig. 5.6). So, each *mer* can be found to have a molecular mass of

$$M = \sum_{i=1}^{n} N_iA_i = 2\left(12.011\frac{\text{g}}{\text{mol}}\right) + 4\left(1.008\frac{\text{g}}{\text{mol}}\right) = 28.054\frac{\text{g}}{\text{mol}} \tag{5.2}$$

In the above example, $n = 2$ for a single mer, since there are two elements, but polymers, by definition, are made up of many mers. *The molecular mass of the chain is simply the number of mers multiplied by the molecular mass of each mer*. In a chain of 1000 mer units, the molecular mass would be 28,054 g/mol [i.e., Eq. (5.2) multiplied by 1000]. That means 1 mol* of those polymer *chains* would weigh 28,054 g. Another way to look at it is that fewer chains would be needed to form the same mass as many individual mers (i.e., 1 chain instead of 1000 mers). This general description gives us a basis for using the molecular mass as an indicator for other properties such as strength.

We can begin to understand the term "ultrahigh molecular weight" used to describe certain types of polyethylene (UHMWPE) and why it is categorized separately. UHMWPE has chain lengths of around 100,000 to 250,000 monomers per molecule. This is about a thousand times longer than HDPE, which only has 700 to 1800 monomers in each molecule. The chain length makes movement more difficult, and that allows for some interesting properties and applications.

One of the processing parameters that chain length directly impacts is the ability to extrude or flow the plastic. This can be measured using the **melt index**, which is a simple method that allows for the rate of flow through a small orifice to be measured at a specified temperature. The shorter the chain length (molecular mass), the faster the polymer will extrude. Conversely, the longer the chains, the slower the polymer extrudes. UHMWPE, then, is much more difficult to extrude than a lower molecular

*A mole (abbreviated *mol*) is a measurement of the amount of a substance. There are 6.022×10^{-23} atoms/mol (or molecules, or anything else), also known as Avogadro's constant.

mass polymer like HDPE, so processing that involves this type of flow is more difficult or even unrealistic. As with all materials, the mechanical, thermal, and chemical properties influence processing just as much as the end applications. Properties and applications of polymers will be discussed in more depth in Chap. 6.

5.3.2.4 Variations in Molecular Mass

Polymers never have all their molecules exactly the same length. There will be some variation, and the amount of variation is an important material property. The distribution of molecular weight will influence strength, toughness, melting properties, etc. If all the molecules have very similar molecular masses, the polymer has a narrow distribution. If a large range of sizes exists, the polymer has a wide distribution. The difference is important.

For processing and performance, this has a major influence on thermal behavior. If a polymer has a narrow distribution, all the molecules will behave similarly. This results in a melting behavior where all molecules start to move freely at about the same temperature. This case is good for injection molding where a high fluidity and narrow solidification window are desirable. This is not good for processes such as extrusion where the polymer is too weak in its molten state. A wide distribution results in smaller molecules becoming "fluid" at lower temperature and acting somewhat like a lubricant for larger molecules. The larger molecules maintain a good working strength at elevated temperature. This high-temperature property is known as **melt strength**.

Due to a high molecular mass and narrow size distribution, UHMWPE requires a very high processing temperature and the narrow size distribution means that extrusions are not favorable. UHMWPE is actually processed quite differently than many polymers because of its "difficult" thermal behavior, and it is often sintered. Lower molecular mass molecules can be added to increase the melt strength at lower temperature, and this effectively lowers the average molecular mass.

The average molecular mass of a polymer can be calculated by the number average, $\bar{M}_n$, or the weight average, $\bar{M}_w$. The **number average molecular mass** considers the relative occurrence of molecules, *i*, with a given size in comparison to the overall number of molecules, N_i. The number average is calculated by

$$\bar{M}_n = \frac{\sum N_i M_i}{\sum N_i} \tag{5.3}$$

The numerator in Eq. (5.1) is simply the total number of molecules, N_i, which have molecular mass, M_i. For example, if you want to find the number average molecular mass over 1000 molecules, you first need to subdivide your molecular mass. You could look at increments as small as one monomer, but this would likely be more trouble than it would be worth. Instead, you might look at groups in increments of 10 or 100 monomer units. The value for M_i will represent a *range*, not a single value. Let us say you found that over 1000 molecules the distribution in Table 5.3 existed.

To calculate the number average molecular mass, you would plug in values from Table 5.3. Equation (5.3) requires a single number, and here we have ranges. For Eq. (5.3) to be strictly accurate, we would need to count every single molecule and its length. A more reasonable, though less accurate, method would be to take the midpoint of the

i	N_i	M_i (g/mol)
1	50	100–200
2	150	201–300
3	300	301–400
4	250	401–500
5	150	501–600
6	100	600–701
N_{total}	**1000**	

TABLE 5.3 Fictional Distribution of Molecular Masses in a 1000-Molecule Sample

range. For instance, the midpoint of 201 and 300 would be about 250. The wider the range, the more information we lose, so be careful with your choices. By taking the midpoints, we get

$$\bar{M}_n = \frac{\sum N_i M_i}{\sum N_i}$$

$$= \frac{[(50)(150)+(150)(250)+(300)(350)+(250)(450)+(150)(550)+(100)(650)]\left(\frac{\text{g}}{\text{mol}}\right)}{50+150+300+250+150+100}$$

$$\bar{M}_n = \frac{410{,}000\frac{\text{g}}{\text{mol}}}{1000} = 410\frac{\text{g}}{\text{mol}} \tag{5.4}$$

The number average molecular mass of this hypothetical polymer is 410 g/mol. This would be a very short chain length [recall from Eq. (5.2) that a polyethylene monomer is 28.054 g/mol], but the math is identical for any chain length. The number average will report a value that is skewed toward the greatest population of molecular masses. For example, if a polymer sample has 1000 chains with a molecular mass of 410 g/mol and one molecule with a mass of 410,000 g/mol, the value will be greatly skewed toward the lower end ($\bar{M}_n = 819$ g/mol). The troubling fact is that 1000 of the smaller molecules are equal in mass to one large one. They contain the same number of atoms and *roughly* the same volume but are counted much differently.

This, of course, is an exaggerated example, but there is a need to determine the molecular mass by a means that incorporates chain length more directly. This is done using the **weight average molecular mass**, and the formula is similar to the number average:

$$\bar{M}_w = \frac{\sum N_i M_i^2}{\sum N_i M_i} \tag{5.5}$$

All of the variables were previously described in Eq. (5.3). In fact, $\sum N_i M_i$ is already calculated in Eq. (5.3). If we return to our example from Table 5.3, we can plug in values as follows:

$$\bar{M}_w = \frac{\sum N_i M_i^2}{\sum N_i M_i}$$

$$= \frac{[(50)(150)^2 + (150)(250)^2 + (300)(350)^2 + (250)(450)^2 + (150)(550)^2 + (100)(650)^2]\left(\frac{\text{g}}{\text{mol}}\right)^2}{[(50)(150) + (150)(250) + (300)(350) + (250)(450) + (150)(550) + (100)(650)]\left(\frac{\text{g}}{\text{mol}}\right)}$$

$$\bar{M}_w = \frac{1.855 \times 10^8 \left(\frac{\text{g}}{\text{mol}}\right)^2}{410{,}000 \left(\frac{\text{g}}{\text{mol}}\right)} = 452.4 \frac{\text{g}}{\text{mol}} \tag{5.6}$$

This calculation gives us a slightly higher average molecular weight than the number average. The distribution of sizes in this case was relatively even or what is termed *normal*. In the case of a normal distribution, the number average and the weight average will be comparable. If there is a skewed distribution with a preference toward one size or another, the weight average will be different than the number average.

In the second polymer described above (equation not shown), there were 1000 small chains and 1 longer chain of equivalent mass to the other thousand. The number average molecular mass was 819 g/mol, but if recalculated using the weight average formula, the molecular mass is 205,205 g/mol. It should be evident that in cases where there is a skewed distribution, the method of calculation becomes very important.

5.3.2.5 Thermoset versus Thermoplastic Polymers

There are two major groups of polymers: thermosets and thermoplastics. **Thermoset polymers** form cross-links at elevated temperature and become harder with increasing heat. **Thermoplastic polymers** soften with heat and are not characterized by an increasing number of cross-links between chains. These results are depicted in Fig. 5.15. For this reason, thermosets cannot be recycled by heating, but thermoplastics can be remelted and reformed into new products during recycling. Rubber, such as in tires, cannot be recycled, but beverage containers, such as milk jugs, can be recycled.

Much of plastic manufacturing is based on thermal processing, such as injection molding (Sec. 11.3.3), which is not suitable for thermosets. Thermosets are used extensively as well, but usually they will be processed at lower temperatures or through chemical reaction such as two-part epoxies (Sec. 6.3.2.3). Heating may be used to cure the thermoset, thereby making it more rigid. The magnitude and duration of heating controls the number of cross-links formed and the resulting stiffness.

5.3.2.6 Polymer Chain Cross-Linking

UHMWPE, discussed above, can be further modified to tailor its properties. In fact, nearly any polymer can modified through **cross-linking**. Cross-linking involves the

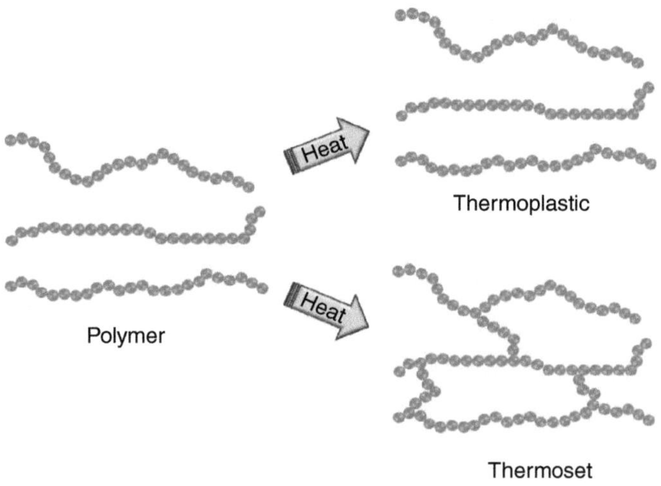

Figure 5.15 Thermoset and thermoplastic polymers are defined by their reaction to heat. Thermoset form rigid bonds between polymer molecules at elevated temperature while thermoplastic do not. This has a number of important consequences when it comes to thermal processing.

formation of primary bonds *between* polymer molecules (Fig. 5.16). From Sec. 5.3.1, we have established that primary bonds occur within the polymer "back bone." That gives strength within the chain but allows for movement between chains. Side groups and branching restrict movement through increased frictional forces, but there is still no rigid connection.

Cross-linking creates a *rigid* connection between molecules, effectively creating a larger molecule with a complex shape. Cross-linked molecules cannot permanently slide by one another without breaking a strong, primary bond. Cross-linking effectively increases molecular mass. As multiple chains combine rigidly, the properties follow the

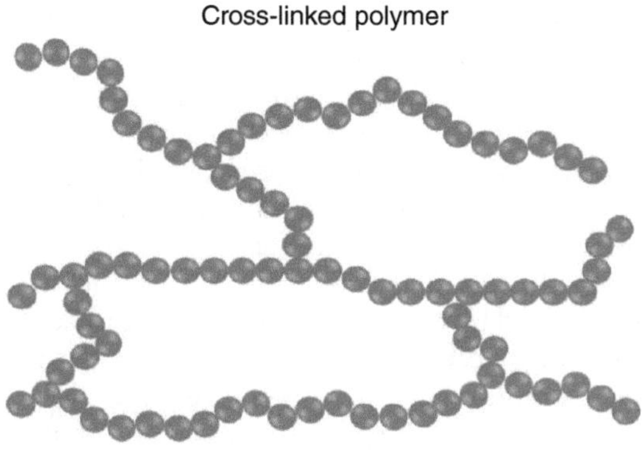

Figure 5.16 Cross-linked polymers have primary bonds between chains. This can make the polymer very rigid or create other properties such as elasticity. The amount of cross-linking directly determines the resulting properties.

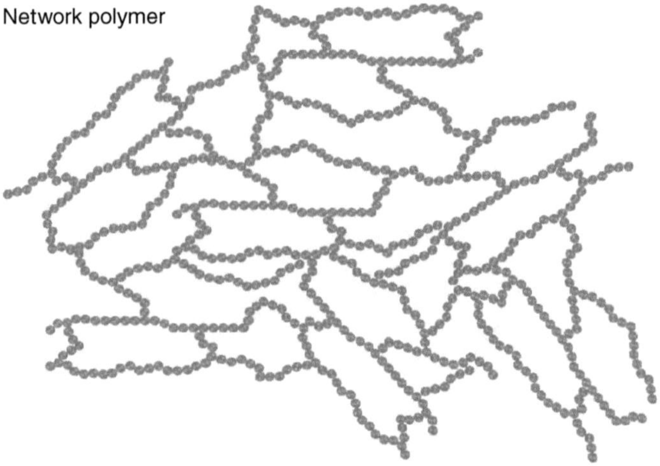

FIGURE 5.17 Extensive cross-linking results in a strong, rigid, and brittle material. Because the molecules are so thoroughly connected, there is not much movement allowed before primary bonds within the carbon back bones are being stretched.

same basic trend as discussed in Sec. 5.3.2.3. The major difference is that there is no softening point with temperature. The material becomes increasingly linked forming a rigid, **network polymer** (Fig. 5.17). A network polymer is simply one where the molecules are so interconnected that it is more like a net or chain-link fence than entangled strands.

Cross-linking can occur by means such as temperature, chemical additive, or irradiation. UHMWPE can be made even more rigid through cross-linking. In the case of UHMWPE, the polymer is irradiated (e.g., gamma radiation) to create the cross-links. Two-part epoxies, on the other hand, use a chemical reaction to harden through polymerization and cross-linking. In that case, a **hardener** must be added to the epoxy resin in the correct ratio or the cross-linking will be ineffective at strengthening the polymer.

Heat is also used to increase cross-linking in *thermoset* polymers. The degree of cross-linking determines the degree of rigidity. Polymers with only a few cross-links may still be fairly ductile or elastic (such as rubber bands). Extensively cross-linked polymers form networks of cross-links that result in a very strong but brittle material. They are rigid due to the locking together of the polymer molecules. The amount of cross-linking and variation on polymer chemistry can modify the final properties. An example of a network polymer structure is shown in Fig. 5.17.

Cross-linking can be intentional or unintentional. Environmental factors such as UV radiation from sunlight and weathering (moisture and temperature) can cause polymers to become brittle over time (see Sec. 8.5.3 for more on polymer corrosion). Cross-linking is achieved by connecting individual molecules into a network. In epoxy, where the cross-linking occurs during the formation of the molecules, the cross-link bonds are created simultaneously with the regular molecular bonds. In a polymer, which is already fully formed, the primary bonds within molecules must be broken and reformed into new bonds which connect molecules that were previously independent of one another. Radiation, moisture, heat, and chemicals can all cause the breaking of bonds in that manner.

5.3.2.7 Crystallinity in Polymers

We have already considered a variety of structural features in polymers and seen at least a little about the effects of the structure has on properties. The composition of two polymers can be identical, but one might be wax and the other rigid plastic. So, what is the difference between grease, wax, and a plastic like HDPE? You might not expect all of them to have the same composition, but polyethylene can be used to make all three. Composition alone does not guarantee any particular set of mechanical properties.

We talked about molecular weight in Sec. 5.3.2.3 and that indeed affects the bulk properties of the material. The length of the polymer chains is not the only important attribute, however, and the way the chains align, or the **crystallinity**, is also important. The concept of crystallinity has been discussed for metals and ceramics so far, and polymers have a version of it as well. Recall that crystallinity is simply long-range periodicity (repeated pattern) that involves multiple atoms, sets of atoms or in the case of polymers, molecules. The amount of crystallinity and the length of the molecules (molecular mass) will determine the strength and rigidity of a polymer. This concept is shown in Fig. 5.18 for polyethylene.

Not all polymers are crystalline, and no polymer is completely crystalline. At best they are *semicrystalline*. Thinking back to metals and ceramics, it is much easier to orient millions of individual atoms than millions of molecules. Imagine trying to pack a box of Ping Pong balls so that all are oriented in a regular pattern. Now try it with ropes. Long, linear objects, like molecules, do not lend themselves to being periodically oriented. Even when they are highly oriented, the folds and bends in the molecules are disordered areas.

The structure of polymers makes them less likely form an ordered structure. They do form *regions* of crystallinity where the molecules are highly aligned, but these are always combined with amorphous regions. The relative amount of crystallinity is determined by factors in the polymer molecule that include branching and the presence of side groups. Polymers that are not branched and that are isotactic with small side groups tend to crystallize more readily. Polystyrene, for example, does not crystallize well due to the large benzene side group (see Fig. 5.9). Ideally, the molecule will also be flexible in order to coil over on itself to form these ordered regions (Fig. 5.19).

Amorphous polymers are defined as having very little or no crystallinity. These are common, but they tend to be weaker than crystalline polymers. Crystalline regions are stronger than amorphous ones, which act more "rubbery." The higher the crystallinity,

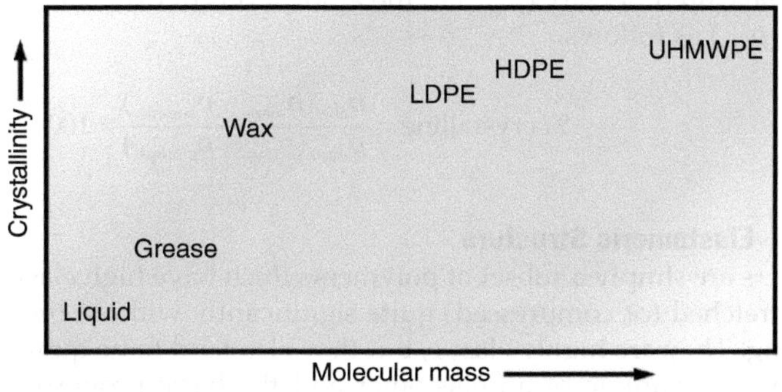

FIGURE 5.18 The properties of a polymer change drastically as the molecular weight and crystallinity are increased. In polyethylene, shown here, the same monomer structure can go from grease to hard plastic by increasing both.

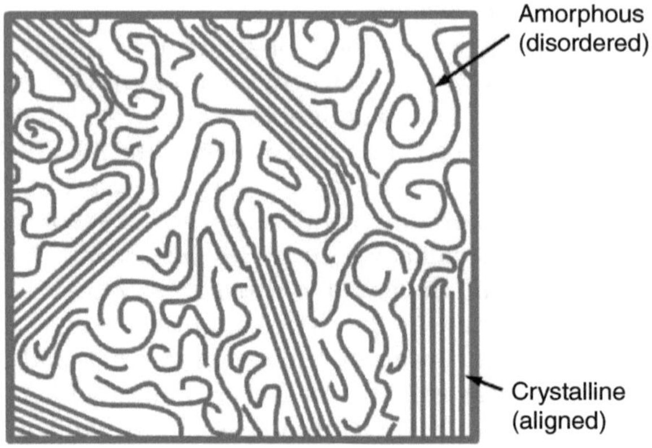

FIGURE 5.19 Polymers can show a high degree of order in regions, but these are always accompanied by some amorphous regions as well.

the tougher the polymer, but amorphous polymers have other benefits such as optical clarity, a larger workable temperature range, and lower shrinkage upon cooling. These properties will be discussed more in Sec. 6.3.

Some polymers exhibit a wide range of crystallinity depending on processing. Polyethylene and polytetrafluoroethylene (PTFE or Teflon™) are among these. Because of their simple, linear chain configuration, they readily form ordered regions. The extent of the crystallinity can be controlled by the rate of cooling during processing. Even highly crystalline polymers become amorphous at high temperature. The slower the cooling rate, the more time is available for chains to move into alignment (crystallize). Fast cooling will reduce the crystallinity, but may not entirely eliminate it, especially for polymers that readily crystallize. Likewise, polymers that do not form ordered regions easily will not benefit as much from slow cooling.

Although optical clarity is a method of assessing crystallinity, the actual fraction of crystallinity can be more precisely determined through a measurement of density. Where chain segments are well aligned, the atoms per volume will be higher and the density will be higher as well. Amorphous regions will have a lower density due to the larger molecular spacing. The percent crystallinity can be determined by comparing the measured density (ρ_{meas}) to the amorphous density (ρ_{amorph}) and the crystalline density (ρ_{crys}) as follows:

$$\%\ \text{crystalline} = \frac{\rho_{crys}(\rho_{meas} - \rho_{amorph})}{\rho_{meas}(\rho_{crys} - \rho_{amorph})} \times 100 \tag{5.7}$$

5.3.2.8 Elastomeric Structure

Elastomers are simply a subset of polymers which have high elasticity. This means they can be stretched (or compressed) quite significantly without breaking or permanently deforming. They are highly elastic, but they also tend to be quite weak. A rubber band is a common example of an elastomer and the basic properties they have. You can stretch a rubber band without much effort, at least compared to something the same size made of rigid plastic. When you deform a rubber band, it comes back to its original shape. You cannot bend rigid plastics very far without permanently changing

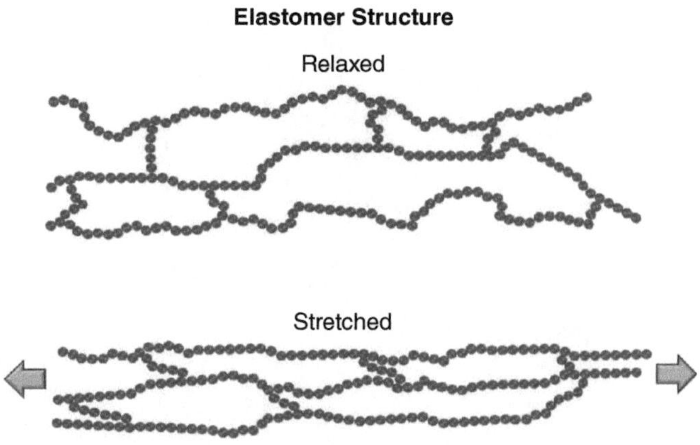

Figure 5.20 Elastomers are highly elastic because their structure allows for reversible movement between chains. There are few cross-links compared to network polymers, and the molecules can slide relative to one another until the existing cross-links restrict motion.

their shape. Not all rubber is the same, though, and some versions can be quite rigid, even though it is still elastic.

Elastomers are discussed in more depth in Sec. 6.3.3, but we can look more at what makes them unique here. Of course, it is based on their structure. Elastomers can be thermoplastics as well as thermosets, but we will consider elastomers to derive their properties from limited amounts of cross-linking, a structural feature most often identified with thermosets. Elastomers are highly disordered, or amorphous, and they only contain a small fraction of the cross-links a rigid plastic would have (Fig. 5.20). Because the cross-links are few and far between, the chains can move quite a bit before their linkages will restrict their movement. This is kind of like a dog on a leash, movement is easy until the full length of the leash is used up, then it is strong and rigid and prevents further motion. In elastomers, the amount of force required to stretch them will increase as their length increases, a property that is discussed more in Sec. 7.5.3.

5.4 Composite Structure

Composites are made from two or more distinct materials combined together to make a single material. These constituent materials come from the metal, ceramic, or polymer categories, each of which has been described previously. The structure of composites, then, will be primarily discussed from the macroscopic viewpoint since the atomic and microscopic structure of materials has already been covered (see Chap. 3 and earlier discussion in this chapter).

Composites are extremely versatile. Composite properties depend on what components are used and how they are combined. Among the key factors in composite structure are the type of materials (composition), the relative concentration of each material (rule-of-mixtures effect), and the size, shape, and orientation of the materials in combination (morphological or geometric considerations).

Composites encompass a great number of materials, both natural and synthetic. For instance, many biological materials are composites as they are made from multiple **phases**, or distinct regions of one material or another. Bone is rigid without being too

FIGURE 5.21 Bone, a natural composite, can be replicated for function using synthetic composites such as carbon fiber. Although the design and composition are vastly different, the purpose can replicated by careful composite design decisions.

brittle. This is because it consists of crystalline and amorphous mineral phases (hydroxyapatite), organic phases (collagen), and liquids.* Bones must be strong and rigid enough to support our bodyweight, but they must also resist breakage due to impact.

The hydroxyapatite mineral phase provides high rigidity while the organic, collagen phase improves toughness. The ratio of these phases determines the exact properties of the bone. In contrast, materials that must replicate this behavior can be designed synthetically. An example of this is with prosthetics, as shown in Fig. 5.21, which can be tailored to give a range of properties. We will now look at various structural factors that influence composite performance.

5.4.1 General Composite Design

Composites can be made up of many materials, but for the sake of clarity, we will focus on two-phase composites. The two phases are a **matrix phase** and a **reinforcement phase** or **dispersed phase**. Reinforcement and dispersed terminology are considered interchangeable for our purposes, but the dispersed phase does not always reinforce (add strength to) the matrix, and you will find both terms in discussions on composites. The matrix phase is the material that acts as the binder and often completely encases the dispersed phase.

You can think of this like raisin bread where there is a phase occurring in multiple, small segments (raisins, the dispersed phase) and a unifying material that holds them together (bread, the matrix). The dispersed phase is commonly present in subunits distributed throughout the composite. These dispersed materials are often in the form of particles or fibers, each of which will be discussed below.

Some composites do not have this matrix/dispersion structure, and they can be in layered form known as laminates. **Laminates** are simply materials which consist of

*If you are interested in this biological composite, you can read more about it in "Analysis of Bone as a Composite Material," by K. Piekarski (1973) among other places.

Figure 5.22 Examples of composite structures are all around us. Raisin bread (left) is a two-phase, particle-reinforced composite and layer cake (right) is a laminate composite. Although many advanced technologies are enabled through synthetic composites, simple versions are part of everyday life as well.

multiple layers permanently bonded together to create a solid material. Continuing the food theme, this is along the lines of a layer cake where distinct regions are combined to form a single, composite structure (Fig. 5.22). Laminates will be discussed in Sec. 5.4.4, and may consist of multiple *composite* layers. Obviously, composite structure can become very complex, but we will limit our concentration to major themes.

5.4.1.1 Compositional Classification of Composites

Although composites are largely defined by macroscopic structure, the atomic-level composition is still the basis from which the properties are determined. Composites are often classified by the material category that comprises the matrix phase. If you mix metal powder into a polymer, the composite is a **polymer matrix composite** (PMC). You could also mix in a ceramic powder, but the naming will not change. The polymer matrix is the important factor for naming.

Likewise, a metal with polymer or ceramic mixed into it is a **metal matrix composite (MMC)** and a ceramic with a polymer or metal dispersed within it is a **ceramic matrix composite (CMC)**. The composite does not require a material from a different class, but this is often the case. Two polymers, metals, or ceramics can be mixed together to form a composite, so long as they maintain distinct regions.

Common material combinations include a polymer and a ceramic, such as with fiberglass and carbon fiber composites, both of which are PMCs. Metallic composites, such as oxide-dispersion strengthened (ODS) superalloys, are an example of MMCs that have ceramic particles distributed throughout. CMCs are often two ceramic materials combined to decrease the brittle behavior inherent to many ceramics.

An increasingly common CMC is transformation-toughened zirconia (ZrO_2), which relies on a secondary ceramic such as yttria (Y_2O_3) to maintain a tetragonal crystal structure at room temperature even though it prefers a monoclinic structure. The benefit comes from the presence of different *crystal structures* within the parent (matrix) phase in response to crack formation and is not simply a compositional matter. This transformation behavior of zirconia is also used to toughen alumina (Al_2O_3), where small particles of ZrO_2 are distributed throughout the Al_2O_3 matrix (see Sec. 7.5.2.1 for more on the ZrO_2 toughening mechanism). Some basic examples of composites are shown in Table 5.4.

For all classes of composites, the ratio of the materials used is one the most important factors in determining the resulting properties. Take fiberglass, for example.

	Matrix	Reinforcement	Example
Polymer matrix composite (PMC)	Polymer	Any material	Fiberglass
Metal matrix composite (MMC)	Metal	Any material	ODS superalloys
Ceramic matrix composite (CMC)	Ceramic	Any material	Zirconia toughened alumina

Note: Composites can be used in widely varying applications since they can be made from any material.

TABLE 5.4 Naming Convention and Applications of Composites

The polymer matrix phase is *relatively* flexible and weak, and the glass is *relatively* strong and rigid. The reason "relatively" is italicized is because composites often combine materials with complementary properties. The two phases have different and distinct properties. Although neither phase may be the most rigid, most ductile nor lightest of the available materials, the two are significantly different from each another.

One phase (reinforcement) will typically be very rigid but characteristically brittle. The other phase (matrix) will be flexible but weaker. When the two are combined, the properties of the composite will land somewhere between the two constituent materials. The ratio of the phases will affect how closely the composite represents one or the other phase. For instance, if you add a few glass fibers to a large sheet of polymer, the fibers will change the properties of the sheet very little. As you add more fibers, they begin to contribute more.

In the extreme case, you would have nearly all glass fibers with a small amount of polymer holding them together, and it would act more like the fibers than the polymer. The properties of the composite are represented by the **rule of mixtures,** which predicts that the properties of a composite are proportional to the volume fraction of the reinforcement (or matrix) phase. A composite with 99% matrix phase will have properties about 99% similar to the matrix and 1% similar to the reinforcement phase. A 50% mix will have properties directly between the two phases. The general rule-of-mixtures behavior is shown in Fig. 5.23.

We have now covered some important concepts of how the materials chosen for composite and the relative ratio of phases affect performance, but there is still more to consider about the size and shape of the reinforcing (dispersed) phase. We will look at the basic configurations of composite reinforcement next.

5.4.2 Fiber-Reinforced Composites

One of the most common ways that a reinforcement phase is combined with a matrix phase is in the form of fibers. These fibers may be woven like cloth or they may be separate. They may be long, short, aligned, random, etc. There is much more to composites than simply adding fibers. The characteristics of the fibers and their orientation within the matrix are both very important.

5.4.2.1 Fiber Properties

The fibers that are used in composites comprise an entire area of study alone. We are most interested in general concepts, but much work has and is being done to optimize fiber properties. Size is one factor that influences the fibers' ability to reinforce the matrix. This is taken into account by the diameter and length.

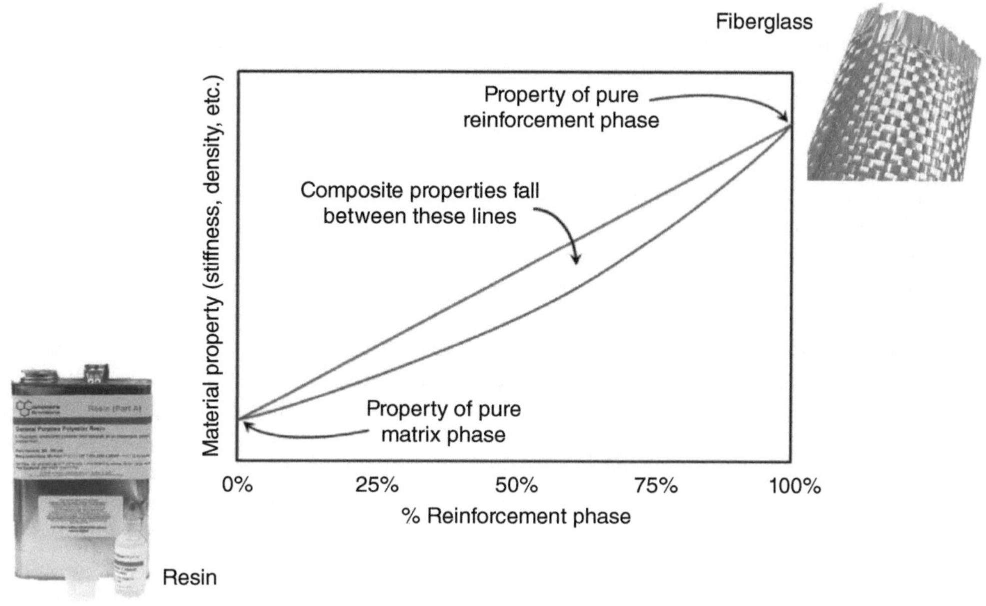

Figure 5.23 The rule of mixtures will predict physical and mechanical properties of the composite material. In this case, glass fibers are represented on the right axis and the epoxy resin matrix is represented on the left axis. Determining the upper and lower property boundaries (graph lines) is discussed in Sec. 7.5.4.

Assuming the volume fraction of the reinforcement phase is constant, when the diameter of the fibers decreases, the strength increases. Very small fibers tend to be more difficult to produce and work with, but the overall performance of the composite can be improved. The size dependence of strength, based on diameter, is shown in Fig. 5.24.

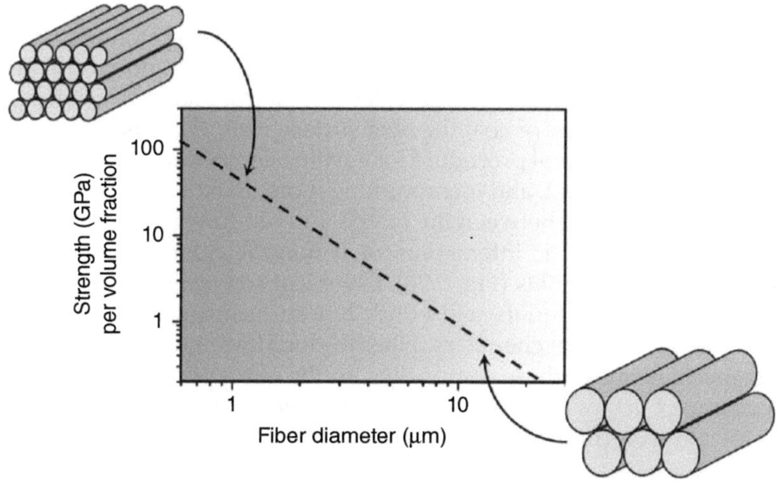

Figure 5.24 Fiber diameter is one aspect that influences the strength of a composite. Smaller fibers result in a higher strength when the volume fraction is constant.

Note that axes of the graph in Fig. 5.24 are log-log,* meaning that a straight line (as shown) would be exponential on linear axes. This behavior is commonly represented on log-log plots to condense the information. The general appearance of the curve on linear axes is shown in Fig. 3.13 for strength based on grain size reduction, which has a similar trend as is found with fiber size and strength.

Fiber diameter is not the only size factor. The length of fibers is also important. Longer fibers will be able to transfer load more efficiently since the fiber is continuous. The fiber material is pulled directly when the composite is pulled instead of the matrix pulling on the reinforcement. To understand the concept, consider a chain and a bungee cord. The chain is very strong and rigid when pulled, and the bungee cord is relatively weak and elastic. This property combination is common in composites, but the way the two are combined is important. If you have bungee cords (matrix) and chains (reinforcement) running parallel to one another along the length of the section, the chain will take the majority of the force. If the chain is cut into small sections that are *connected* by bungee cords, the resulting combination will be much weaker. The bungee cord will stretch first, and then the chain will begin to contribute to the strength.

This is the same for fiber composites. Short fibers will be connected *end-to-end* by the matrix material. While this is a greatly simplified description, it emphasizes the importance of **continuous fibers** versus **discontinuous fibers**. The reinforcement phase is always connected by the matrix in some fashion, but for discontinuous (often called *chopped*) fibers, the matrix must physically bridge the gap along the axial fiber separation. This situation is shown in Fig. 5.25. The amount of matrix material will determine the efficiency of transferring load and can be controlled by modifying the volume fraction of the dispersed phase. More fibers will result in smaller gaps. Figure 5.25 shows fibers that are well aligned in a single direction, which is not always the case.

Another fiber characteristic to consider is the fiber morphology. **Morphology** is simply the structure or form of the fibers. Most importantly, the surface condition of the fibers affects how well the fibers bond to the matrix and, in turn, how well the composite responds to applied load. If the fibers can slide within the matrix, they do not help support the load. The **interfacial bonding** is the means and degree of connection between the matrix and the reinforcement.

The interface is the boundary between the two phases, and the only way that the two are truly combined is by locking this boundary in place. Materials do not always have a chemical compatibility. Like oil and water, the matrix phase, which is typically added in liquid form, may not *wet* or coat the fiber surface well. These issues have been addressed and are part of the general procedure for creating composites, but there are multiple ways the two phases can bond, and incorporating more than one is good practice.

Interfacial bonding between the matrix and reinforcement can occur by mechanical, chemical, or electrostatic interaction, or it may be achieved by reaction or diffusion between the two materials (Fig. 5.26). Chemical compatibility is a major part of determining if the two materials will bond, but surface roughness is another that is less sensitive to the material chemistry. Fiber pullout from the matrix is one of the principal failure mechanisms in these composites, so the matter of bonding is quite important to performance. Surface roughness creates an interlock between the two phases that must be physically overcome before the fiber can slide.

---

*The term "log" is short for logarithmic. Notice that same distance can represent 1 or 100 units, depending on where it is located along the axis.

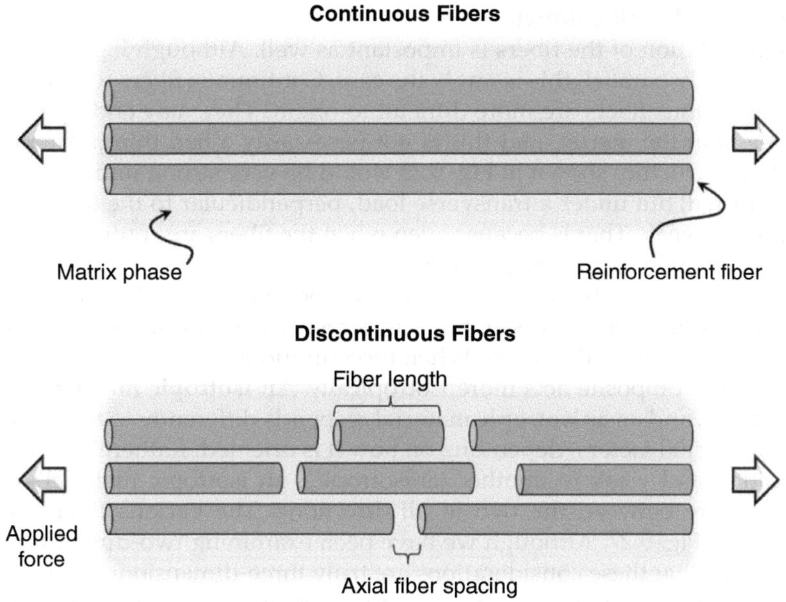

FIGURE 5.25 The length and continuity of reinforcement fibers is important to composite strength. Spacing between fibers, in the longitudinal (axial) direction, requires the matrix phase to support more of an applied force. This may or may not be desirable depending on the properties of the matrix and the intended application.

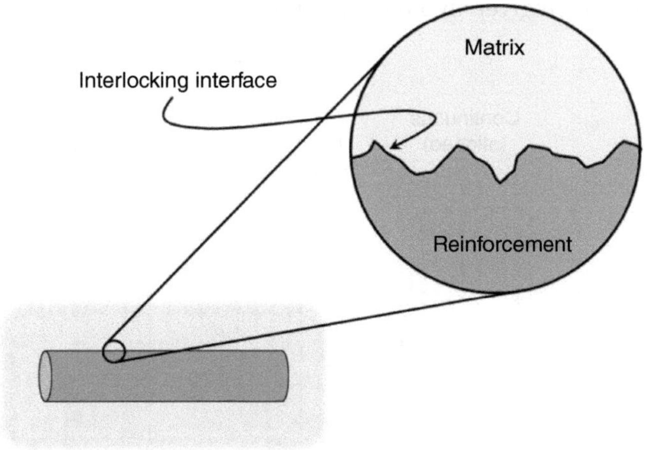

FIGURE 5.26 The bond between the matrix and reinforcement phases is critical to efficient load transfer and fracture resistance. The bond strength may be attributed to multiple factors, including mechanical interlock (shown).

5.4.2.2 Fiber Orientation

The orientation of the fibers is important as well. Although Fig. 5.25 shows fibers that are perfectly parallel, this is rarely the case. Continuous fibers will often be aligned, but discontinuous fibers are more difficult to orient. They may end up randomly oriented throughout the matrix, and this is not necessarily a bad thing. An aligned composite with the structure shown in Fig. 5.25 would be very strong under longitudinal loading as depicted, but under a transverse load, perpendicular to the one shown, it would be relatively weak. That is because even when the fibers are continuous, the matrix material is crucial in supporting the force.

The weaker matrix is the principal load-bearing material in the transverse direction. The fibers are actually discontinuous across the width of the composite, which goes across the width of the fibers. When discontinuous (chopped) fibers are randomly oriented, the composite acts more isotropically. An **isotropic material** acts equally in all directions, and an **anisotropic material** responds differently to force, electricity, heat, or other external factors, depending on how it is oriented. Rather than being strong in one direction and weak in another (anisotropic), an isotropic material's strength will be somewhere between the two in all directions. The various fiber configurations are shown in Fig. 5.27. Although we have been examining two-dimensional examples, do not forget that these considerations are truly three-dimensional in nature.

The orientation of fibers is not only considered when they are separate from one another. Often fiber reinforcements such as fiberglass and carbon fiber are woven into a fabric before being combined with the matrix. There are a number of weave patterns and multiple fiber materials can be woven together. Some examples of these woven reinforcements are shown in Fig. 5.30. The fabric itself can be heavier or lighter, too. This is comparable to the difference between burlap and silk, and some weaves are stiffer while others are more flexible, even before any matrix phase is added. The weight of a fabric is often given in ounces (or grams) and indicates the weight per square yard

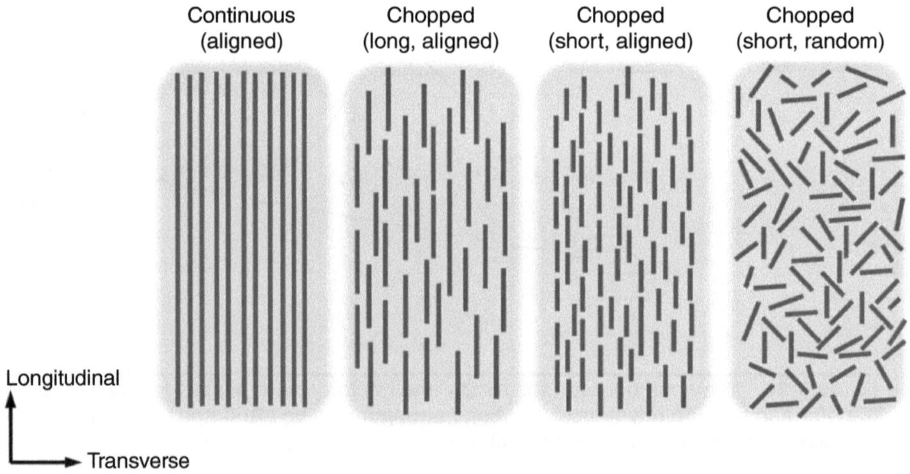

FIGURE 5.27 Fiber orientation is an important parameter in composite strength. Oriented fibers are very strong in the longitudinal direction but weak in the transverse direction. Randomly oriented fibers will have strength between the two in all directions of loading.

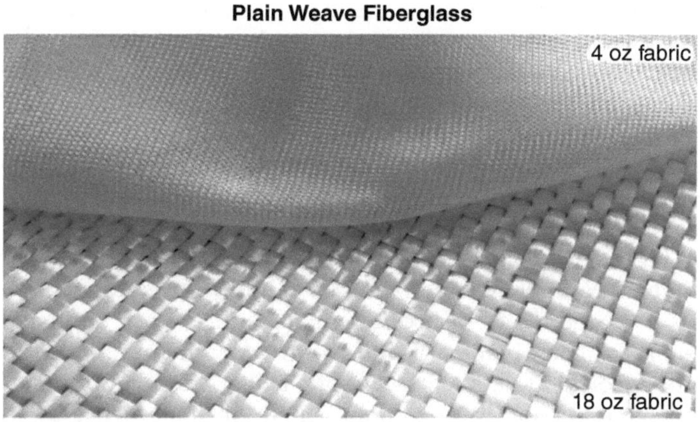

Figure 5.28 The weight of woven composite reinforcement determines how flexible it is and how many layers will be needed to build strength. Lighter fabrics require more layers for strength, but can replicate mold details much better than heavier fabrics.

(or square meter) of the fabric. The differences between fabric weights of fiberglass are shown in Fig. 5.28.

Thicker fabrics will be stronger with a single layer, but they do not replicate geometric detail as well when forming the fabric around a mold. Multiple layers of thinner fabrics can be used where fine detail and high strength are required. To create components, the fabric is layered on a mold surface, which has already been coated with resin (the **gel coat**). A perforated plastic sheet called the **peel ply** is then applied which allows resin to seep through when the liquid resin is pressed into the fabric (often using vacuum or positive pressure). The **bleeder cloth** absorbs excess resin, but without the peel ply it would stick to the fabric after the resin hardens. The final layer is the vacuum bag that holds the layers together and is sealed to itself or the mold surface using **gum tape**, which accommodates surface irregularities to create an airtight seal. The resin may harden under ambient conditions or be heated under pressure in an autoclave. The basic materials are shown in Fig. 5.29.

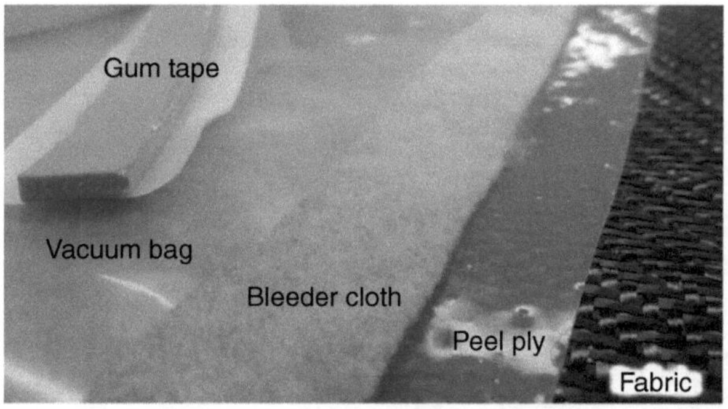

Figure 5.29 The process for creating composite structures often utilizes multiple layers which control resin flow and allow the resin to be forced into the reinforcing cloth using vacuum pressure.

One of the important parameters that determines the flexibility and weight of the fabric is the **tow size**. The tow size is the number of smaller fibers that combine to make one strand. You may see a weave listed with 3k or 12k as part of the description. These numbers indicate the number of fibers in each strand (3000 and 12,000, respectively). Some weaves are denoted as "wide tow." These have extremely wide strands, closer to the weave of a basket, and not as much like fabrics. The reduction of crossover points means less fiber crimping and associated resin trapping, which results in a stronger and lighter composite.

From Fig. 5.30 you can see that multiple fiber types can be combined to for what is called a **hybrid composite** which consists of multiple reinforcement materials. A simple extension is to use fibers of a different material with a similar structure (diameter, length, etc.). The combination of different reinforcement can also allow for other structures to be incorporated, such as chopped fibers and continuous fibers to be paired together. Cost reduction is a potential benefit of this practice by mixing in lower cost fibers with more expensive ones.

The weave orients the fibers in a specific pattern. Fiber orientation, as discussed above, is important to the mechanical properties of the composite. When force is applied to a woven composite, the orientation of force relative to the orientation of the weave has a similar effect as with linear fibers. Take, for example, the plain weave. It has fiber orientations aligned perpendicular to each other. The composite will be strongest in the direction of the fibers, so it will be strongest in the transverse and longitudinal directions.

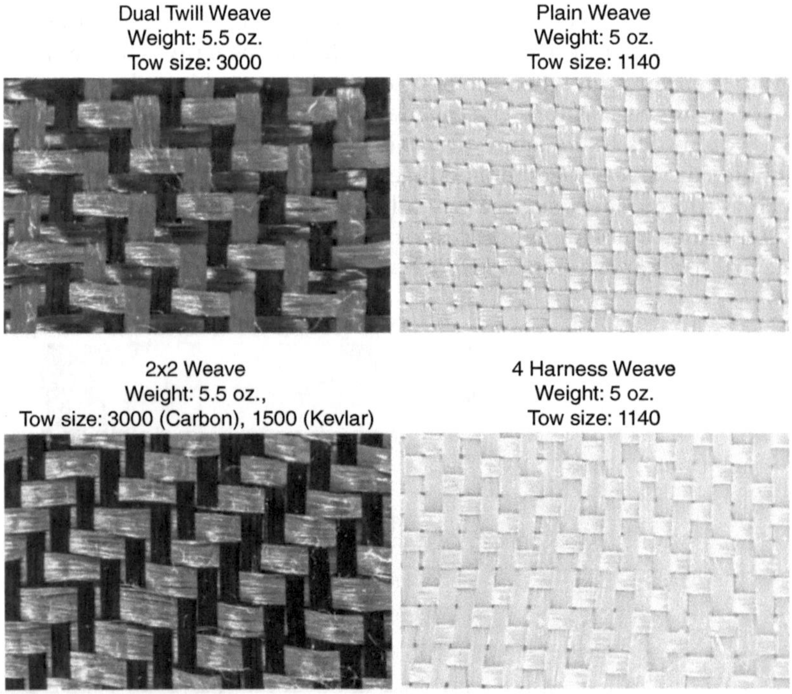

FIGURE 5.30 Fiber reinforcements may be woven in a variety of patterns and multiple materials of varying colors can be incorporated for visual impact. The weave, weight, and tow size are important to the drape of the fabric and strength of the final product.

Although this may seem as though it solves all of our problems from Fig. 5.27, there is still a weakness. When force is applied at 45° to the fiber directions, it will be at its weakest. Complex forces that are not linear and well aligned to the fibers have to be supported as well, and one way to improve the mechanical response is to layer the composite sheets with their fiber orientations at 45°. While one sheet will be at its weakest orientation, the next will be at its strongest. This improves the isotropy of the component. This intentional alignment of anisotropic materials is often employed in laminate composites (see Sec. 5.4.4). The general behavior of a single, plain-woven sheet is shown in Fig. 5.31.

5.4.3 Particle-Reinforced Composites

Particle-reinforced composites have many of the same considerations as fiber reinforcement. The size, bonding, concentration, etc. will determine the performance of the composite. In the case of particles, the shape may be regular [e.g., spherical, cubic, lenticular (lens-shaped), etc.] or irregular, having no consistent morphology. The particle size may be monodisperse (single size) or polydisperse (wide range of sizes). The particles and matrix can begin as two distinct phases, or a second phase can be formed *within* the matrix, such as happens in precipitation hardening of metals (see Sec. 9.3.7). When a second phase is formed in situ (during processing), the particle size can be controlled by processing parameters such as time and temperature, and the overall particle size tends to be very small (tens of nanometers). These materials are rightly termed composites since there are distinct, intermixed phases, but for our purposes, we will concentrate on composites that start with distinct phases.

In our discussion of fiber reinforcement we established the importance of size and bonding with the matrix. Those same concepts transfer directly to particle reinforcement. Maximum strength is achieved by incorporating the smallest possible particles

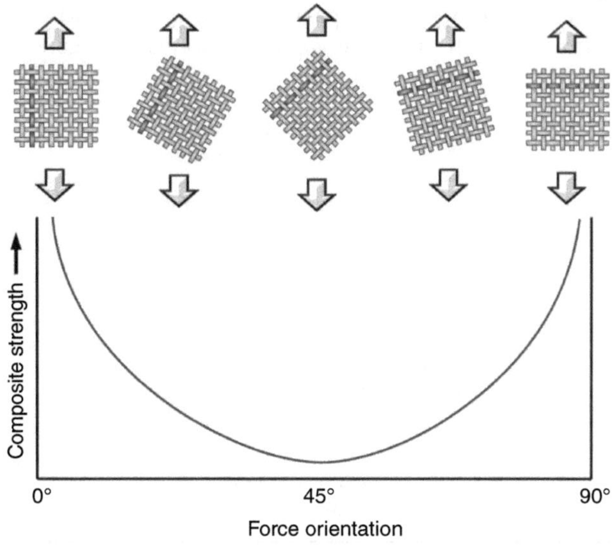

Figure 5.31 Woven fiber reinforcement, like loose fibers, shows anisotropy when loaded from different angles. In the plain woven example shown here, the minimum strength occurs 45° to the longitudinal fiber axes.

with a uniform distribution. The rule-of-mixtures behavior applies as well. The more reinforcement, the more the composite acts like the reinforcement phase. For instance, a common MMC is tungsten carbide (WC) mixed with cobalt (Co). The Co acts as the matrix that holds WC particles together. Co coats (wets) the surface of the WC efficiently so that the mixture is mechanically robust and load transfer is efficient.

As you may know, and as will be discussed in more detail in Sec. 6.2.2, carbides are used in applications where wear resistance and thermal performance are important. Carbides, being a ceramic, are very hard and maintain good strength at high temperature. The problem is that they are also brittle and chip easily. When they are mixed with Co, the Co acts as a buffer between particles and imparts better toughness while maintaining good high-temperature performance. Considering our rule-of-mixtures concepts, we can balance toughness (Co) and wear-resistance (WC) by judiciously selecting a ratio of Co (matrix) to WC (reinforcement).

In this case, the fraction of WC is very high (about 90%). The Co, then, acts primarily as a binder to keep the WC particles together and aids in producing the tools and devices that incorporate this material. The reinforcement phase does not need to be present in a greater quantity than the reinforcement phase. Particle bonds without Co would be very brittle and unsuitable for most cutting applications where this material is used extensively. The structure of this composite is shown in Fig. 5.32.

Carbides are not the only materials that benefit from a composite structure. A classic and quite familiar example is concrete (Sec. 6.4.2.1), which is the most widely used construction material in the world. Although the terms concrete and cement are used interchangeably by many, concrete is really cement mixed with aggregate. The aggregate (gravel and sand) fills volume and adds strength to the cement matrix. Portland cement, most commonly used in construction, is a mix of calcium silicates and other minor constituents (see Sec. 6.2.1.2). The matrix is relatively expensive, so the addition of aggregate also serves to reduce the overall cost.

For the aggregate to be most effective, there should be a range of sizes, which helps lock the particles together. Large particles (stones) should be combined with smaller particles (pebbles) which can fit between the stones. This can then be supplemented by

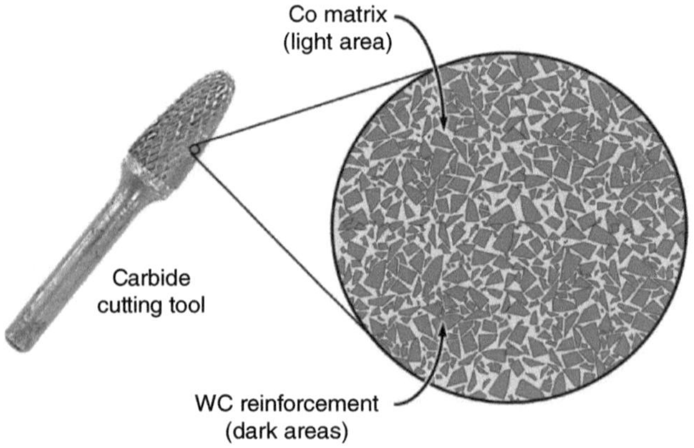

Figure 5.32 Carbide cutting tools are metal matrix composites with a very high concentration of the reinforcement phase. Cobalt (light areas) acts as the matrix in which the tungsten carbide (darker areas) is distributed. The high carbide content ensures superior wear resistance and thermal performance.

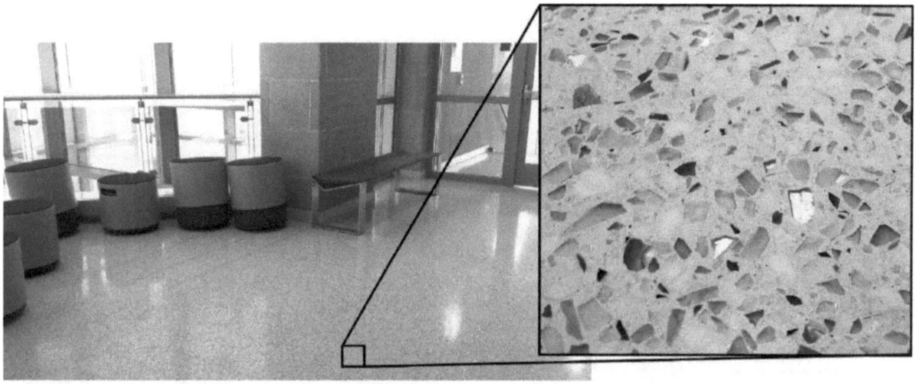

FIGURE 5.33 Concrete is a particle-reinforced composite consisting of cement (matrix) and aggregate (reinforcement). The composite structure is strikingly displayed in terrazzo flooring as shown.

adding even smaller particles (sand), which fit between the pebbles. The cement with a carefully measured volume of water, is mixed together with the reinforcement to ensure an adequate bond. A clear example of this sort of composite structure can be seen in terrazzo flooring, which is concrete that has been ground down to reveal the particle phase within the matrix. In this case, the choice of particle reinforcement is as much decorative as it is functional. This composite structure is shown in Fig. 5.33.

5.4.4 Laminate Composites

Laminate composites, also called laminar composites, are a significant departure from the particle and fiber-reinforced materials described above. Instead of having a reinforcement phase within a matrix, the composite material consists of layers of different materials, or materials with different orientations. This concept was mentioned in the discussion on fiber orientation in woven reinforcement (see Sec. 5.4.2.2), but the same basic principles apply to any anisotropic material. If a material is stronger under longitudinal loading than transverse, the material can be layered to orient the strong axis perpendicular to the previous layer.

A common example is plywood. Each layer (ply) serves a different purpose. While there may be decorative face veneers, the core and structural portion of the material has the grain of the wood (the strong axis) oriented at 90° angles in each layer. This gives it good stiffness in either direction and reasonable stiffness at 45° which is the weakest point. A third layer can be oriented at 45° to enhance the isotropy further. A general plywood structure with 90° layer orientation is shown in Fig. 5.34.

Although the benefits of laminar composites are often structural, they can also be decorative or functional in other ways. For instance, the outer layers in the plywood pictured in Fig. 5.34 are for decorative purposes. These are thin wood veneers chosen for an attractive grain pattern and do not contribute significantly to strength. Other layer functions include weight reduction at constant thickness, optical variation (e.g., color or transparency), or as a barrier to diffusion of gas or liquid.

The latter function can be seen when packaging liquids in "paper" containers such as juice boxes. There are multiple layers that contribute to strength, liquid containment, and artwork. This is shown for a juice box in Fig. 5.35, but similar practices are employed in producing soda bottles, soup cans, and other packaging applications. When a single

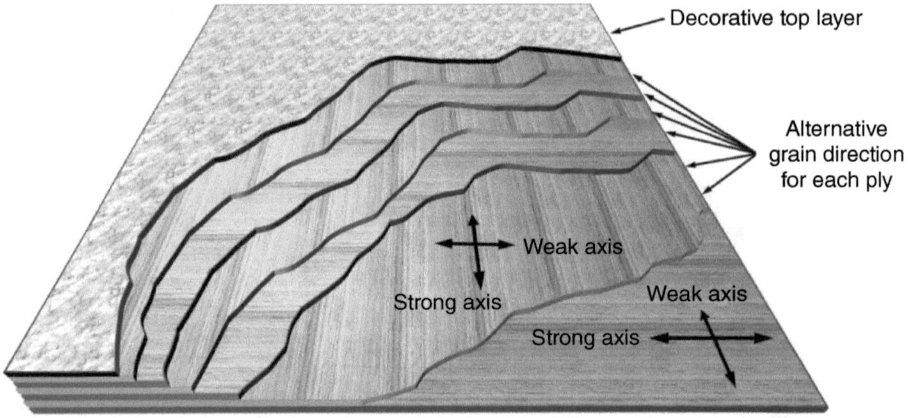

Figure 5.34 Laminar composites such as plywood (shown) utilize multiple layers in specific orientations to increase the isotropy of a composite. Wood is much stronger along the grain than across it, so plywood orients the grain at right angles to reinforce the weak axis of the layer above/below. In addition to these structural layers, decorative layers are also added to enhance appearance.

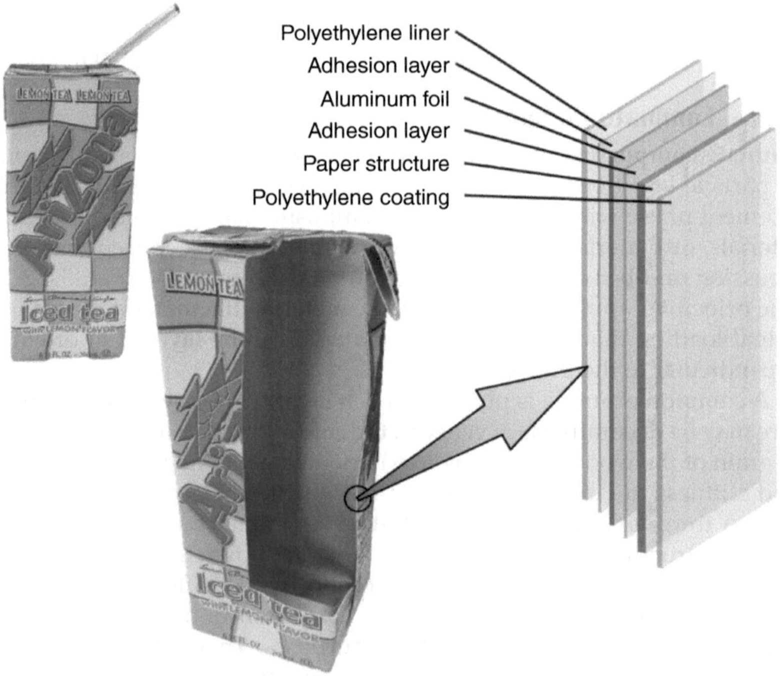

Figure 5.35 Laminar composites are used extensively in packaging such as the juice box shown. Each layer contributes to containment, strength or decoration. This same practice is used in other containers as well.

material is not sufficient to fit an application, multiple layers can be used to find a balance between function and cost.

5.5 Why It Matters

Nonmetallic materials are all around us. They make possible some of the most advanced materials for the most demanding applications, but they are also some of the most common and what might be considered mundane of materials. Although metals can serve in a wide array of applications, nonmetals are even more diverse. They consist of nonmetallic bonds such as ionic and/or covalent bonds.

This bonding gives nonmetals some general characteristics, such as having low electrical and thermal conductivity compared to metals, but it does not limit the range of properties nonmetals can have overall. Some are very weak, ductile, and melt at low temperature. Others are rigid, strong, and can serve in applications at very high temperatures. Though we looked at engineering materials, nonmetals are more common in all states of matter, being gas, liquid, or solid. In fact, one chemical (e.g., ethylene) can be any of the three, depending on the degree of polymerization. Understanding the structural factors that give nonmetals their properties is vital to correctly choosing, processing, and applying that material.

We saw that most ceramics are composed of more than one element. In fact, ceramics are the default (or natural) state of many materials, especially metals. We see this when metals are derived from ores, which are ceramics. Metals bonded to nonmetals, such as oxygen and carbon, yield nonmetals. Ceramics are excellent in high-temperature, high-wear applications; a perfect combination for high-speed tooling in manufacturing. Their limitations often result from being brittle.

Ceramics are perhaps the least common material in consumer products. Glass is one of the most ubiquitous applications, and obviously it is quite important, but many advanced technologies and processes rely on ceramics or coatings of ceramic to improve performance. These will be discussed in more depth in Sec. 6.2. Ceramics are used extensively in construction such as cement, concrete, and asphalt, though the latter two are actually ceramic composites. They also see extensive applications in insulation, both for heat and electricity. Ceramics may not be as visible in consumer products as polymers, the other major nonmetallic classification, but they are essential to our everyday lives.

Polymers share little in common with ceramics or metals. That is part of the reason they have become exceptionally popular in modern society. Think of packaging and disposable products. Look around you for a few seconds. Chances are that you will see a handful of polymers, more often just called plastics. Also, if you see anything painted or clear-coated, that is a common application for polymers as well. Plastics can be vastly different in appearance and behavior, so more needs to be known before choosing the right polymer for the job.

Polymers used for beverage containers will differ from those used in joint replacement, and there are some general structural considerations that inform us about the properties. The chain chemistry, length, degree of crystallinity, presence of cross-links, etc. all change the properties of a polymer. Polymers deform by molecules sliding by one another. The more difficult the sliding is, the higher the strength, and cross-links cause the polymer to be very rigid, but this also decreases the ductility.

Transparent polymers are amorphous, and they are usually amorphous because the chain structure does not allow them to pack closely together. This disordered structure also renders them better able to withstand impact and reduces shrinkage during molding. Two major classes of polymers were defined: thermoplastics and thermosets. These two differ in their response to heat. Thermoplastic soften with heat while thermosets harden. Since much of polymer processing centers around high-temperature processes, knowing the difference is essential. Their structure is very important. More about some common polymers and their applications are given in Sec. 6.3.

Composites are rapidly gaining market share in product design, but they have long been part of daily life. Their basic premise is that by combining dissimilar materials, a favorable balance of properties can be achieved. Composites are classified by the matrix material, which can be metal, polymer, or ceramic. Some of the most common composites are polymer matrix composites, such as fiberglass, but the other classes are used regularly as well. As mentioned above, concrete is a composite and used extensively in construction. It is often reinforced with steel bars (rebar) as well, making it a hybrid composite, since it contains multiple reinforcement phases. The reinforcement can be in the form of particles or fibers, or the composite can be layers of materials, each with a specific contribution to the overall application. The type and characteristics of the reinforcement are central to the final properties.

Tires are a common example of nonmetallics combined into a complex composite. The combination of materials is important, but the properties of each material are important as well. For instance, the outermost material is a carbon black, particle-reinforced rubber for durability and elasticity. Inner layers (plies) use polymer fiber reinforcement, such as polyester, and the innermost layer uses butyl rubber, which prevents air from leaking out. Additionally, there are steel belts and beads which stabilize the tire and help lock it onto the rim. In tires equipped with studs, the studs may be made out of a carbide ceramic, bringing together all the material classes in one, "simple" package. Modern tires are really quite an engineering undertaking, and they make travel more reliable and efficient by thoughtfully combining materials. It is rare that a tire fails dramatically, as shown in Fig. 5.36, but when it does it reveals the complex internal structure. Common composites are discussed more in Sec. 6.4.

Figure 5.36 Modern tires are complex products that incorporate all classes of materials to result in specific properties not achievable using any one material alone. Normally this complex structure is not visible unless the tire wears excessively or blows out, as shown here. (*Photo by Tire Zoo*, **CC BY 2.0**.)

5.6 The How and Why of Nonmetallic Materials: Structure

5.6.1 Chapter Review

1. What is "natural" about nonmetallics?
2. What makes a nonmetal, nonmetallic?
3. How can nonmetallic materials contain metals?
4. How are the crystal structures of ceramics similar to those of metals? How are they different?
5. What is meant by a stoichiometric composition? Why is this rarely achieved?
6. What metallic crystal structure is the sodium chloride structure related to?
7. What metallic crystal structure is the cesium chloride structure related to?
8. What is the difference between the diamond cubic and zincblende structures?
9. What differences are there in crystalline and amorphous ceramics?
10. Based on temperature, how can an amorphous structure be created? How can it be crystallized?
11. How is the structure of polymers fundamentally different than ceramics or metals?
12. What is the difference between addition and condensation polymerization?
13. What is a mer? A polymer?
14. What is the difference in atomic bonding within and between polymer chains.
15. Why are side groups important to the strength of polymers?
16. What is the difference between a homopolymer and a heteropolymer?
17. What purpose do copolymers serve?
18. How does branching affect the properties of a polymer?
19. What does molecular weight indicate about a polymer?
20. Recalculate Eq. (5.2) for polytetrafluoroethylene (fluorine has an atomic mass of 18.998). See Fig. 6.21 for the structure.
21. Recalculate Eq. (5.2) for polyvinylchloride (chlorine has an atomic mass of 35.453). See Fig. 6.15 for the structure.
22. What effects does an increasing atomic mass have on the properties of a polymer?
23. Determine the number and weight averages for the following polymer:

i	N_i	M_i (g/mol)
1	200	100–200
2	150	201–300
3	50	301–400
4	75	401–500
5	225	501–600
6	300	600–701
N_{total}	**1000**	

24. Determine the number and weight averages for the following polymer:

i	N_i	M_i (g/mol)
1	50	100–150
2	80	151–200
3	100	201–250
4	115	251–300
5	500	301–350
6	300	351–401
N_{total}	**1145**	

25. Why is heat so important when discussing polymers?
26. What is the primary difference between thermoplastic and thermoset polymers?
27. What is a cross-link, and what kind of bonding is involved?
28. A polymer with extensive cross-linking is known as what type of polymer?
29. How does crystallinity occur in polymers?
30. Why is no polymer fully crystalline?
31. What structural features prevent polymers from crystallizing?
32. Assume a polymer has a density of 1.5 g/cc when crystalline and 1.2 g/cc when amorphous. The density is measured as 1.43 g/cc. What is the percent crystallinity?
33. Assume a polymer has a density of 1.5 g/cc when crystalline and 1.3 g/cc when amorphous. The density is measured as 1.39 g/cc. What is the percent crystallinity?
34. What properties are different between amorphous and crystalline polymers?
35. What is the defining characteristic of elastomers?
36. Structurally, how can elastomers stretch so far and so easily? Why do they get harder to stretch the farther they have already been stretched?
37. List five applications where elastomers are used. Could the elastomer be replaced with another material?
38. What is a composite? Name a natural and a synthetic version.
39. What role does the matrix play in a composite structure? What is the role of the reinforcement?
40. Why is "dispersed" a more general term than "reinforcement" when talking about composite phases?
41. Why is the ratio of phases so important in composites?
42. How can you determine the final properties of a composite by choosing the type and amount of matrix and reinforcement phases?
43. In fiber-reinforced composites, name three design factors regarding the fibers.
44. Why is the interface of a fiber with the matrix so important?
45. What benefits and disadvantages anisotropic materials?

46. What is the reason that fiber reinforcement is woven?
47. How does the design of a weave determine the properties of a composite?
48. What is a hybrid composite? Why are they used?
49. Why are loading and reinforcement orientation important? How can strength be balanced in all directions?
50. Name three types of particle shapes that are used as reinforcement.
51. Can there be less matrix material than reinforcement in a composite?
52. What is the most common particle reinforced composite material?
53. What are the benefits of laminate composites over solid materials?
54. What nonstructural applications do laminates serve in?
55. What is a common product that incorporates many principles of nonmetallic materials and composite structure? Did you think of that product the same way before learning about these materials?

5.6.2 Concept Connections and Extensions

1. Consider the tow size in Fig. 5.32. What do you notice about the 2×2 twill that is different from the others? Why is their relationship important?
2. How can one material, like alumina, be opaque or transparent? Sketch the structure and how light passes or does not.
3. Why are ceramics lower density than most metals, or at least than the pure versions of the metals they contain (e.g., WC versus W)? Include the structural and compositional factors.
4. Why do metals bond with nonmetals to form stable ceramics? Use the discussion in Sec. 2.3 to justify your answer.
5. Which side group configuration would be most likely to interlock, atactic, syndiotactic, or isotactic? How does the side group affect it? Use an illustration to support your answer.
6. Strength increases in composites as fiber size decreases. Using the discussion in Sec. 22.4, describe the potential benefits of nanoscale fibers and the challenges to using them.
7. Identify a laminate product. Describe what you think the materials are and what the purpose of each layer is.
8. Given the angular dependence of weave on strength, devise a layering scheme to maximize the strength from any angle. How many layers will it require?
9. Some composites use more reinforcement phase than matrix phase materials. Why do you think this is true of carbide cutting tools. See the additional information in Sec. 13.4.5.

CHAPTER 6

Nonmetallic Materials: Identification and Application

Nonmetallic materials have nearly limitless variety of properties since they can be made from any combination of elements that result in a primarily covalent or ionic bonding structure. Strong, rigid materials like ceramics with exceptionally high melting points and weak, ductile materials such as polymers with low melting points are both nonmetallic materials. With such variation in nonmetallics, we must define their differences and naming conventions. As described in Chap. 3, metals have a universal numbering system, but not everyone uses it. Different professional societies might refer to the same metal differently. The same happens with nonmetals, especially in polymers. While there is some rhyme and reason, it is not always evident. We will look at some of the most common nonmetals, their naming conventions, and the uses they are employed in.

6.1 How They Work

Chapter 5 described the basic structures and properties of nonmetallic materials. It was emphasized that generalizing those materials can be misleading. The materials are classified as polymers, ceramics, or composites, but each category can vary widely in appearance, properties, and applications. To help refine our understanding of these materials, we will look at some more specific ways that they are categorized and applied. We will start with ceramics, which are fairly manageable. They tend to be named and referred to by their chemical composition.

Polymers will be described next, and they can be quite unruly. This is simply because there are so many trade names in use that their chemical names, which are based on composition, are much less known. Also, a variety of polymers can be referred to under a single name, so we will have to be diligent in defining them carefully.

Last, we will look at composites. Many of these are named for their reinforcement material, as discussed in Sec. 5.4.1.1, but it is not always that simple, so there too, we will give them due attention.

6.2 Common Ceramics

Ceramics have been an important part of society since prehistoric times and continue to enable new technologies that are fundamental to societal advancement. Although ceramics can include natural minerals such as those found in geologic formations such as granite, marble, obsidian, and others, our concern is primarily with engineering ceramics. That is not to say that the other materials are not important to engineering challenges, but they are not manufactured in the way other ceramics are. **Engineering ceramics** are those with a controlled composition and structure. This gives them predictable properties and allows their use in critical applications.

We will classify ceramics by their composition. Engineering ceramics are often created by combining a metal or metalloid (see Fig. 2.1 as a reminder) with one of three elements: oxygen, carbon, or nitrogen.* While these elements are not the only ones used, they represent the majority of structural ceramics. The ceramics that contain these elements are known as oxides, carbides, and nitrides, respectively. We will look at some common examples of each in the next few sections.

6.2.1 Oxides

Oxides, in their most basic form, are simply oxygen-containing compounds. Oxides are the most common ceramics because they easily form in nature. For instance, rust is a form of iron oxide, which is a ceramic. Corrosion (rusting), then, can be thought of as the process of forming ceramics by the breakdown of a metal (much more on this in Sec. 8.5). Of course, it is not quite that simple, but the natural tendency toward oxide formation can be a good thing as well as a destructive one. Here we will look at some manufactured oxides and their applications.

6.2.1.1 Aluminum Oxide

Aluminum oxide (Al_2O_3), also called **alumina** or garnet, occurs naturally and is used in natural form as well as being created synthetically. While there are many crystal structures alumina can have, the most common is α-Al_2O_3 (alpha-alumina), known as corundum. It can be very valuable or inexpensive, depending on the size of the crystal. Large, single-crystal alumina is used for jewelry, and goes by the names of ruby or sapphire. The color of the gems is determined by small amounts of metallic impurities such as chromium (ruby), iron (blue), titanium (yellow), or other metals. Corundum is also a good example of how useful ceramics can be for their high hardness. Alumina, which has a Mohs hardness of 9,† is often used in abrasive applications. Most "sand" paper will actually use alumina particles (see Fig. 6.1). Abrasives using alumina are relatively low cost, and that reflects the abundance of abrasive-grade alumina.

6.2.1.2 Calcium Oxide (Cement)

Cement is a ubiquitous construction material, and it has been in use for centuries. **Portland cement**, the most common type, is made from multiple ingredients, and calcium oxide (derived from limestone) is only the main ingredient. It also contains significant silica (~20%), alumina (~5%), and some ferric oxide and sulfate. The ingredients

*Other examples include ceramics containing sulfur (sulfides), phosphor (phosphides), and many more.
†See Sec. 7.6.1.1 for more on the Mohs hardness.

Figure 6.1 Aluminum oxide, also called corundum and alumina, is a common ceramic that can be used in applications such as jewelry or industrial abrasive (shown). It is very hard, and its color (as gem stones) varies based on impurities within the crystal.

are added as natural minerals containing one or more of the constituents, and the exact composition varies and is tailored for the application.

Cement is used in mortar, some grout, and as a binder in concrete (see Sec. 6.4.2.1). It comes in powdered form and is mixed with water to initiate the setting process and can be easily worked with until it hardens (see Fig. 6.2). **Non-hydraulic cement** cures by reaction with carbon dioxide in the air, and it will not set in high-moisture conditions.

Figure 6.2 Cement is a mixture of ceramics with CaO being the primary constituent. By varying the composition, a variety of properties can be achieved. Cement is often mixed with aggregate to create concrete. (*Photo by Army Corps of Engineers, CC BY 2.0.*)

Hydraulic cement forms a more chemical and water-resistant ceramic that can harden even if immersed in water.

There are many types of cements, most of which are specific blends of Portland cement. These are used for refractory applications, are colored, account for shrinkage during curing, etc. A drawback with Portland cement is the high energy consumption required to process the material [including heating to 2642°F (1450°C)] and the associated environmental effects, which are quite significant. Alternatives, such as "energetically modified cement," can reduce the overall impact by lowering energy input.

6.2.1.3 Silicon Oxide and Silicates

Silicon oxide, more commonly called silicon dioxide (SiO_2), is so common you probably do not even think about it as an engineering material. You have likely built castles from it, broken objects made of it, and look right past it every day. Sand is primarily SiO_2, or **silica**. Silica occurs naturally in huge quantities, and silicon (Si) is the second most abundant element in Earth's crust (the first is oxygen). Silica has two oxygen atoms and only one silicon atom per molecule, so the prevalence of silica necessitates oxygen is more prevalent.*

Perhaps the most common way that silica is used is in glass. Windows, spectacles, baking ware, lab ware, etc. all rely on a majority of silica, but the composition will vary widely between them. These variations make some less sensitive to changes in heat, give them specific optical properties, or can change their electrical and mechanical properties.

Quartz, which is highly pure SiO_2, is the base for semi-precious gems such as citrine and amethyst, and occurs naturally within geologic rock formations such as granite. Commercially, quartz finds distinct applications based on its crystallinity. When completely amorphous, it is known as "fused" quartz. The amorphous structure and high purity give it improved high-temperature strength, a very low coefficient of thermal expansion (CTE), and exemplary optical properties. It is used in optics, specialty laboratory implements, and for semiconductor processing, typically as an electrically insulating layer on top of silicon or circuitry in microprocessors (more in Sec. 22.4.2.1). It is also widely used to make tubing (see Fig. 6.3) and other labware exposed to high temperatures and rapid temperature changes.

When crystalline, silica has much different properties, one of which is piezoelectricity. The **piezoelectric effect** results in an electric charge being generated from a mechanical stress, or, alternatively, a mechanical stress can be generated by an applied charge. By squeezing or otherwise applying stress to a quartz crystal, you can generate electricity, or you can apply electricity to change the size of the crystal. The magnitudes of the changes are not large, but it is very useful in applications such as electrical-spark cigarette lighters, acoustic guitar pickups, strain gauges for mechanical testing, small speakers and buzzers, quartz clocks, and many more.

Silicates are essentially an *impure* form of SiO_2. Silicates can have varying ratios of silicon and oxygen, and additions of other elements such as hydrogen, sodium, boron, and calcium change the properties appreciably. When other elements are added to silica, the glass transition temperature lowers, allowing for easier processing, but it also changes the performance as well (see Sec. 8.2.6.1 for more on glass transition).

The "recipes" for silicate glasses can be quite complex. Soda-lime glass, for instance, contains sodium carbonate (Na_2CO_3), known as "soda," calcium oxide (CaO), known as

*Silicon rarely occurs in pure form. Rather, it occurs in more complex silicates where oxygen is not necessarily included (e.g., SiF_6), though it almost always is.

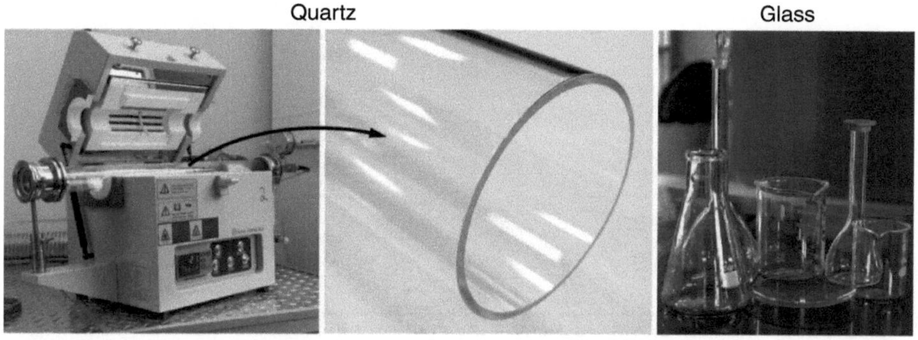

Figure 6.3 Silica can be used in pure form as quartz, which is often used in high temperature tube furnaces, as shown on the left, or mixed with other elements or compounds, such as borosilicate glass (right). The composition will affect the properties which affect the manufacturing methods.

"lime," and oxides of magnesium and aluminum. Because it is relatively weak and has poor thermal properties, soda-lime glass is common for window panes and other low stress, moderate-temperature applications.

Borosilicate glasses are primarily comprised of boron oxide and silica, and provide much better thermal performance due to a lower CTE. Pyrex is a trade name for this type of glass, and this type of glass is used in cookware and laboratory items, such as those shown in Fig. 6.3. There are other types of silicates as well, most of which are named for their composition. These include lead "crystal," which contains lead oxide but is not crystalline, and aluminosilicates (containing alumina), which are used for its strength.

6.2.1.4 Titanium Oxide

Titanium oxide, or titanium dioxide (TiO_2), is known as *rutile*, and is one of a variety of titanium oxides. It is used extensively in everyday items such as sunscreen, paint, and toothpaste. It is popular in those applications and others as a pigment, because it has a bright white appearance, is highly opaque, and is inert, so it will not fade and does not pose a health risk* in cosmetic applications or where ingestion may occur (like the toothpaste shown in Fig. 6.4). Rutile efficiently scatters visible light (hence the bright white appearance) and absorbs ultraviolet light, hence its use in sunscreen.

These optical properties also make it useful as a photocatalyst and in dye-sensitized solar cells. Photocatalysis is a process where a reaction is driven by light using a non-consumable material (the catalyst) to accelerate the reaction. Examples of this include the formation of hydrogen gas by splitting water molecules and self-cleaning glass, where TiO_2 creates free-radicals that oxidize and break down organic matter. TiO_2 is typically modified chemically to enhance its performance, and other structures of TiO_2 (anatase, for example) can be used as well. In solar cells, nanoparticles of TiO_2 are often used to support organic dye molecules in the electrolyte.

*There are other factors to consider for normally "safe" materials such as the particle size, route of ingestion, and quantity encountered.

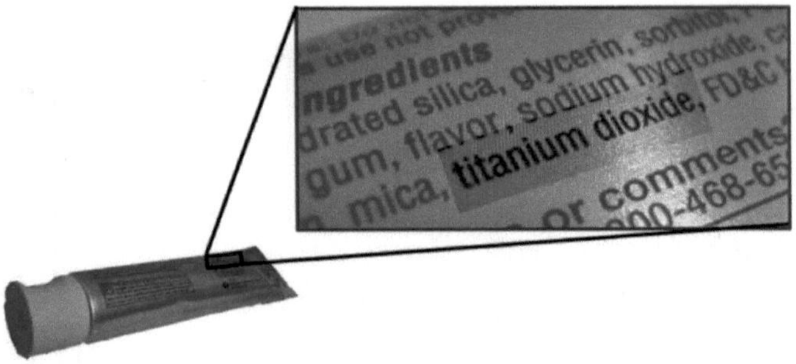

Figure 6.4 Titanium dioxide is a common ingredient in products with a bright white color. These include paint, sunscreen, and toothpaste (pictured above).

6.2.1.5 Zinc Oxide

Zinc oxide (ZnO) is a white ceramic with pigment applications similar to those of titanium dioxide and as an additive to polymers, lubricants, and batteries. ZnO finds significant use in rubber manufacture as it assists with vulcanization and helps protect the rubber from environmental damage. It is used extensively in "ceramics" (i.e., in decorative, pottery-type applications) for glazing, enameling, and structural performance, and is also a useful constituent of paint used on steel as it overcomes common compatibility issues. In an indirect application, ZnO forms on Zn used to galvanize steel and prolong the useful life of structures (see Sec. 8.5.2.3).

ZnO also appears in fortified foods as a source of zinc, which is an essential mineral for normal bodily function. Electronically, ZnO is a wide-bandgap semiconductor, which makes it useful in electronics applications such as thin-film transistors, LEDs, and transparent electrodes. ZnO is one of a few ceramics (also including titanium dioxide) that exhibits **thermochromism**, or a change in color with temperature. ZnO changes to a yellowish hue when heated to high temperature due to oxygen loss, but recovers its white appearance when cooled. Thermochromism is exhibited by a variety of organic compounds and is the concept behind mood rings.

6.2.1.6 Zirconium Oxide

Zirconium oxide (ZrO_2), also called **zirconia**, is a ceramic where polymorphism (or allotropy) plays a central role in its applications. Although many ceramics can assume multiple structures, zirconia's structural transitions can be controlled to benefit applications in unique ways. The three, temperature-dependent crystal structures that occur are monoclinic [below 2138°F (1170°C)], tetragonal [2138–4298°F (1170–2370°C)], and cubic [above 4298°F (2370°C)]. One of the more interesting aspects of these structures is that when zirconia changes from tetragonal to monoclinic, it undergoes a 3% to 4% volume increase. This leads to the phenomenon known as transformation toughening. Most ceramics are quite brittle, but in this case, as a crack develops it can be resisted because the tetragonal phase is converted to the larger, monoclinic phase when a crack forms. The transformation causes a compressive stress at the crack tip, which suppresses further crack growth. This is described in more detail in Sec. 7.5.2.1.

The monoclinic phase is the stable form at room temperature, so in order to have the tetragonal phase present at room temperature, additives such as yttrium oxide (Y_2O_3),

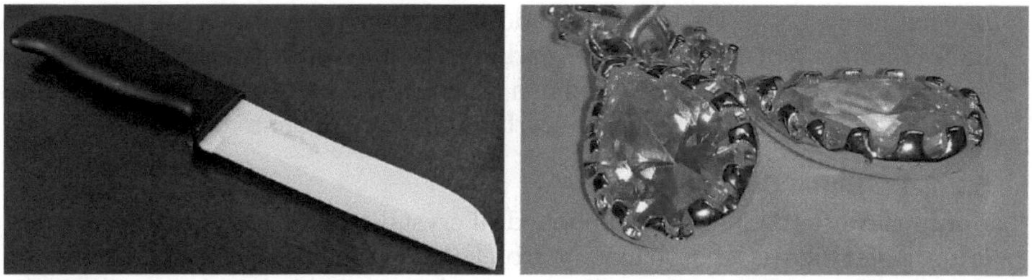

Figure 6.5 Zirconia can possess a range of properties depending on its structure. It can appear white and opaque and be relatively tough when tetragonal (on left) or it can be clear and transparent when cubic (on right).

magnesium oxide (MgO), or cerium oxide (CeO_2) are added.* These additions "stabilize" the tetragonal phase, and so the ceramic is typically referred to as **partially stabilized zirconia** (PSZ). Transformation toughening allows this ceramic to be used in applications with large stresses that are not always compressive. Have you ever seen or used a ceramic knife, such as the one in Fig. 6.5? The white blade is a stabilized zirconia. Did you read the package? You are not supposed to wash it in the dishwasher. Heat and moisture can cause the transformation to occur before a crack develops, thereby eliminating the ability to resist a crack when it forms. PSZ is also used in applications of high stress and wear such as high-end ball bearings.

The last form of zirconia, **cubic zirconia**, is used much differently than the other two. Cubic zirconia, or CZ, is very hard, transparent, and colorless. One of the most noticeable applications is as a diamond simulant (see Fig. 6.5). The cost for diamond varies widely based on the color and clarity, but CZ can be made flawless with perfect clarity, something that would cost about 500 times as much in a natural diamond.† Of course, there is a difference between the two materials. CZ has a Mohs hardness of about 8, while diamond is 10, the highest for a natural material. This means CZ will wear over time and lose its clarity. Diamond is also an excellent thermal conductor and CZ is not. You can actually tell if a stone is CZ by breathing on it. The condensed moisture (fog) will clear faster on the diamond than on the CZ because the diamond transfers the heat faster. CZ also has somewhat different optical properties and a higher density that can be used to confirm the difference. Diamond is discussed more in Sec. 6.2.4.2.

6.2.2 Carbides

Carbides are much like oxides in the sense that they are simply compounds containing carbon. Again, the compounds are typically formed with metals, but they also form with metalloids such as silicon. You may come across carbides being categorized with metals, but recall what makes metals, metals: metallic bonding. Bonding between carbon and a metal is covalent and ionic in nature [you can determine exactly how much

*In addition to compositional stabilization, the tetragonal phase can also be stabilized at very small, nanoscale grain sizes. Polycrystalline zirconia is often abbreviated TZP for tetragonal polycrystalline zirconia.

†A loose, 1.0 carat diamond of excellent color and clarity (D and VVS1, respectively) costs about $20,000. An "equivalent" CZ stone costs $30 to $40 *with a sterling silver ring included*!

of each using Eq. (2.1)]. Carbides tend to be very hard and wear resistant, because of their bonding. Not all metals form carbides, or the carbides are relatively unstable. Others form very stable carbides and can be used in widely varying applications. We will look at some of the most common carbides, but there are many others.

6.2.2.1 Boron Carbide

Boron carbide (B_4C) has a rhombohedral crystal structure, but instead of atoms at each lattice point there is a boron icosahedron (20-sided polyhedron). This is an example of the complexity ceramics can have. Boron carbide is probably best known for its low density and extraordinary hardness. Its low density (2.52 g/cc, a little less than aluminum) makes it ideal for reducing mass in structures, and its extreme hardness (~9.5 Mohs) make it suitable for high-strength, high-wear applications. Among its applications are nozzles for sand blasting and water jet equipment, armor plating for vehicles and in bulletproof jackets, and in cutting tools and abrasives. Boron carbide is also very good at absorbing neutrons, so it is used in nuclear control rods (in conjunction with enriched boron).

6.2.2.2 Iron Carbide

Iron carbide (Fe_3C), also known as **cementite**, has an orthorhombic structure and is most commonly formed within carbon-containing ferrous alloys. Iron is the most common structural metal, and it can be strengthened by cementite within it. The most practical way to do this is through heat treatments of a steel high in carbon (see Sec. 9.3.5). The higher the carbon content, the more cementite can form and the greater the strength of the steel. This is because iron carbide is a strong, hard ceramic which imparts those characteristics to the metal. White cast iron, which is highest in cementite, is extraordinarily hard for an iron alloy, but it is brittle and difficult to work. Cementite's applications outside of metallurgy are rather limited in comparison to its importance in alloy design. It does not see any widespread use as a pure material.

6.2.2.3 Silicon Carbide

Silicon carbide (SiC), also known as **carborundum**, most commonly has a hexagonal structure but also exists in over 200 variations, most of which are two-dimensional modifications, similar in concept to chain length in polymers. It is a common ceramic with many applications related to its high hardness. Although it does form naturally, it is rare and the vast majority of SiC is synthetic. Historically it has been used in cutting applications such as abrasives (see Fig. 6.6), but more recent uses include semiconductor applications and as a diamond simulant.

SiC cutting tools can be "whisker" reinforced by the addition of very fine alumina whiskers as reinforcement. SiC carbide was an early LED material and still plays an important role in LED technology, though gallium nitride (described in Sec. 6.2.3.2) has gained popularity because of its greater light output. SiC is very hard (9–9.5 Mohs) and has a thermal conductivity closer to diamond than cubic zirconium (see Sec. 6.2.1.6), and can be used in place of diamond for decorative and functional purposes, often under the name moissanite, or synthetic moissanite. SiC is also used in high performance vehicle components such as brakes and clutches. These are often called "carbon-ceramic," as the carbide is combined with carbon fiber to create a lightweight, durable composite.

6.2.2.4 Titanium Carbide

Titanium carbide (TiC) has a rock salt structure (see Fig. 5.1), is extremely hard (9–9.5 Mohs) and like other carbides, many of its applications center on that property. For instance, TiC is

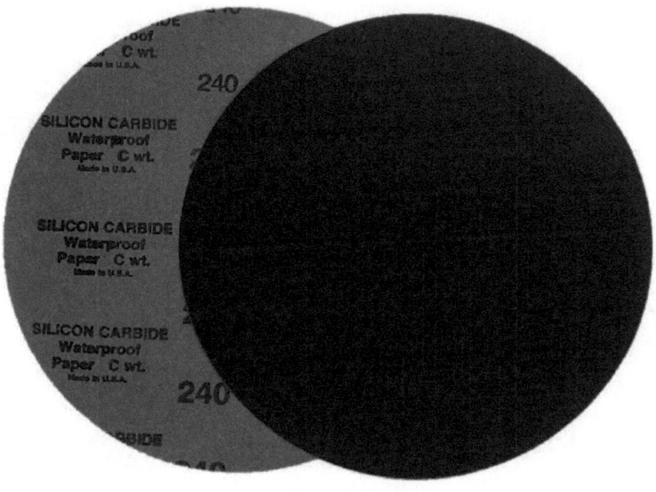

Figure 6.6 Silicon carbide is very hard and is used in high-wear and abrasive applications such as grinding wheels and sanding disks.

used in cutting tools for high-speed machining. TiC powder is bonded together using a nickel–cobalt binder, much like tungsten carbide (discussed in Sec. 6.2.2.5), and is one of a variety of cermet cutting materials (see Sec. 13.4.3). Its high-melting temperature (3160°C) allows for high-temperature applications as well. TiC is also used as a coating for cutting tools, but it is less common than other ceramics such as titanium nitride (see Sec. 6.2.3.4).

6.2.2.5 Tungsten Carbide

Tungsten carbide (WC) has a hexagonal structure at room temperature (α-WC) and is perhaps the most popular carbide. It is used in applications ranging from cutting tools to tire studs to jewelry. It is a relatively heavy material, since tungsten has a high atomic mass, and is quite hard (~9 Mohs). Its electrical resistivity is low for a ceramic, and is even comparable to some metals. Even so, most applications utilize WC for its hardness.

It is used extensively in production cutting tools (see Fig. 6.7) for increasing machining rate and decreasing tool wear as well as in surgical tools for improved performance. It is also used for armor-piercing ammunition (thanks in part to its high density) where the core of the projectile is W_2C. WC is also employed in wear point applications such as ball bearings, ski pole tips, snowmobile tracks, and ballpoint pen tips. Because of its high hardness, jewelry made from it will not readily lose its polish due to wear and tear, and the removal of WC rings in emergencies is facilitated by its more brittle nature (i.e., it can be cracked off without the use of ring cutters).

6.2.3 Nitrides

Nitrides are compounds involving nitrogen. The same elements that tend to form oxides and carbides, as discussed in preceding sections, also form nitrides. Although nitrogen is one of the most abundant elements in the universe and on Earth, it primarily occurs as a gas. This is because nitrogen prefers a diatomic configuration where two nitrogen atoms are connected by a triple bond, which makes it quite stable and unreactive. For this reason, nitrides are less common than either oxides or carbides, and are

Figure 6.7 Tungsten carbide is a popular cutting tool material. Although it is gray in appearance, it is often coated with other materials which may have a different color.

rare as naturally occurring minerals since oxygen is also highly available and more reactive. Most commercial nitrides have to be elaborately processed through high-temperature and/or chemical reactions. Nitrides are less common in commercial applications, though they do play important roles. The following sections will describe some of the more popular types.

6.2.3.1 Cubic Boron Nitride

Boron nitride (BN) can assume a variety of properties depending on the structure. For instance, hexagonal BN has a structure similar to graphite and can be used as a lubricant. **Cubic boron nitride** (abbreviated c-BN or CBN) is an extremely hard ceramic compound with a structure similar to that of diamond. CBN is second only to diamond in hardness (9.5–10 Mohs); it also shares characteristics such as a high thermal conductivity. Combined with better chemical and thermal performance, CBN has been an important substitute in high-wear, high-temperature applications. Abrasives and cutting tools are some of the most common examples, and the ease of sintering CBN makes it simpler to create large components than possible with diamond. Also, diamond reacts with certain metals during processing, and CBN is more effective for machining iron and nickel alloys. When used as cutting tool, it is most often as an insert only at the very tip of the tool (a similar configuration is shown in Fig. 13.15 for polycrystalline diamond).

6.2.3.2 Gallium Nitride and Related Compounds

Gallium nitride (GaN) has a hexagonal crystal structure and is somewhat different than other nitrides we will discuss. Although it too is very hard and has a high melting temperature, its most common applications involve semiconductor fabrication. GaN is a direct bandgap semiconductor, and it is used to produce high output LEDs of varying color (such as the LED strip light shown in Fig. 6.8). By modifying the GaN with additions of other elements such as aluminum and indium, or pairing gallium with arsenic (GaAs, an *arsenide*), and phosphorus a color range from infrared to ultraviolet can be created. Ga compounds like these are typically deposited as a thin film and are not used

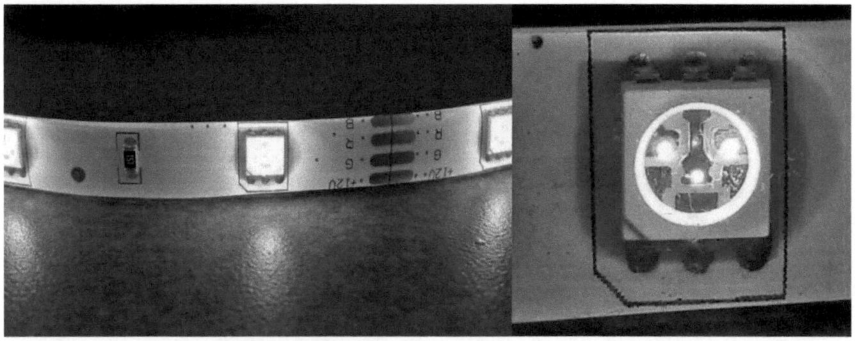

Figure 6.8 Gallium nitride (GaN) is an important semiconductor material used in LED lighting and other technology. Here it can be seen that white light (or any color) is created by the combination of red, blue, and green LEDs.

in bulk form. GaN-based laser technology is used in Blu-Ray video, which has a much higher information density than DVD.

6.2.3.3 Silicon Nitride

Silicon nitride (Si_3N_4) can exist in three crystal structures: trigonal, hexagonal, and cubic (α, β, and γ, respectively). The trigonal and hexagonal forms are much more common than cubic, but the hexagonal, β-phase is the most stable and what is typically considered for Si_3N_4. Like GaN, Si_3N_4 has semiconductor applications. Si_3N_4 is typically employed as an electrical insulation layer in microprocessor design. It is also an efficient passivation layer being very hard and resistant to diffusion of damaging elements into the circuitry.

Si_3N_4 also has bulk applications where its high hardness and excellent high-temperature performance are utilized. For example, it is used in various high-temperature and wear components within engines, including turbocharger rotors where is low mass improves response. Si_3N_4 can also be employed in high-end bearings due to good resistance to impact and lower frictional forces than steel bearings. Cost is significantly higher than the materials it replaces, and it is typically applied only where the specific properties of light weight, corrosion resistance, and wear resistance are required.

6.2.3.4 Titanium Nitride and Related Compounds

Titanium nitride (TiN) has a cubic, rock-salt structure, and is most often applied as a coating on other materials. It is very hard and possesses a golden hue when applied as a coating and can be identified on many coated cutting tools, such as drill bits like the one shown in Fig. 6.9. A TiN coating can double or triple tool life due to its low coefficient of friction and high hardness. It is often applied to lower-cost, high-speed steel (HSS) tools as an intermediate between cemented carbides and uncoated HSS. It is also used as a coating for wear surfaces in medical implants and medical instruments since it is biologically inert. Its golden appearance is also used for purely decorative purposes in household applications.

In addition to nitrogen, other elements can be added to create variations on TiN. Common examples of this include titanium carbon nitride (TiCN, also called titanium carbo-nitride) and titanium aluminum nitride (TiAlN). TiCN, which has a bluish-gray color, is about 30% harder than TiN and has an even lower coefficient of friction, but

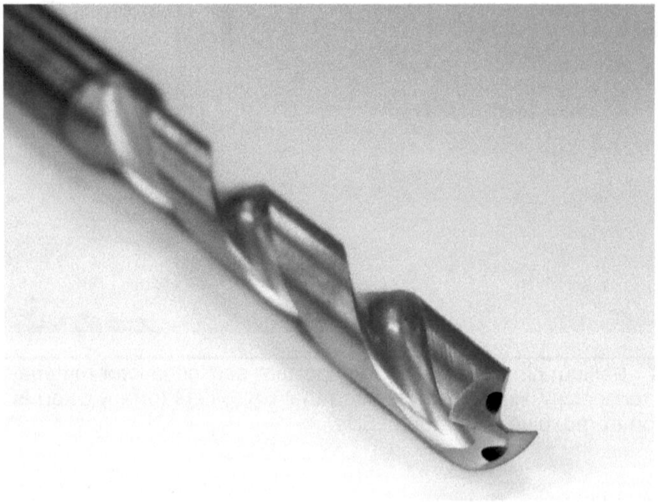

FIGURE 6.9 Titanium nitride is a common ceramic coating used on cutting tools. It can be identified by its distinctive golden color.

TiCN has a lower oxidation temperature and cannot be operated at as high of temperatures. TiAlN, which appears a purple-gray, is slightly softer than TiCN, but very resistant to high-temperature processing. When more aluminum than titanium is present, it is considered aluminum titanium nitride (AlTiN), which is almost twice as hard as TiN. This dark gray coating is also suitable at elevated temperature due to the aluminum content forming a passivating oxide layer. The high hardness does lead to increased chipping of the coating if not used properly.

6.2.4 Other Ceramics

The materials in this section are most closely identified as ceramics due to their structure and/or properties. We will look at silicon and carbon as examples of materials that do not fit nicely into our previous categories. Neither of these materials are normally considered a ceramic, because they are elements that possess properties that may differ from "traditional" ceramics. Nevertheless, these materials are profoundly important to modern technology, so it is beneficial to discuss them further.

6.2.4.1 Silicon

Silicon is a metalloid (or semiconductor), but it shares more in common with ceramics than with metals or polymers, so we will discuss it here. In Sec. 6.2.1.3, silicon dioxide, or silica, was described. It involved bonding with oxygen, which greatly changes its properties. For instance, silica is often colorless and transparent, but elemental silicon is gray and opaque. Silica is a good insulator, silicon is a semiconductor. Both are fairly rigid and hard, and both can be used as crystalline and amorphous structures, though silica is typically amorphous and silicon is typically crystalline.* The majority of silicon

*Crystalline quartz is used in piezoelectric applications and amorphous silicon is used as a thin film in some electronics.

used in industrial applications is not used as a pure material. Rather, it is used in the processing of other materials such as steel or in compounds such as silicates in cement.

Silicon has a diamond cubic structure (see Sec. 5.2.1), and it responds to processing differently in different orientations. It is not used as a bulk, structural material, but the structure is quite important for other reasons. Modern computer-based technologies are based on silicon microprocessors, and this integrated circuit manufacturing relies heavily on the production of single crystal silicon **boules** (large rods, essentially), which have a well-defined orientation. Because of the sensitivity of modern electronics, the silicon must be very pure to start (99.9999999% pure, known as *nine nines* pure). Further processing, such as ion implantation, introduces impurity elements to the silicon in defined locations to create a positive (p-type) or negative (n-type) regions required to form the circuitry by increasing electrical conductivity. Silicon processing is described in Sec. 22.4.2.1, and a basic electronic configuration is shown in Fig. 6.10.

6.2.4.2 Carbon

Carbon has a lot of tricks up its sleeve, so to speak, but we will only look at some here. If you reference a few different periodic tables, you will probably find carbon labeled as a metalloid and a nonmetal. Even though it is covalently bonded, carbon occupies a strange, middle-ground of properties that straddle those of metals and ceramics. For instance, it can be a very good electrical conductor, it can be a poor electrical conductor, or it may depend on the orientation.

The reason there can be such stark differences is because carbon can come in so many forms. It is polymorphic, meaning it can assume different structures. You probably know that coal and diamonds are made of the same, basic stuff, but

Figure 6.10 Silicon is used in many applications, but computer processors are perhaps the most important to our modern society. Without silicon and its unique properties, we would not have the cheap and abundant technologies we enjoy today.

understanding what makes them so different is important to their processing and applications. We will look at the most basic polymorphs of carbon, but materials research keeps unveiling new configurations, and nanoscale materials will be highlighted in Sec. 22.4. Carbon is critical for life, and it also plays a central role in polymer chemistry, but these topics are either beyond our scope (life) or discussed later (for polymers, see Sec. 6.2.4).

Amorphous carbon is a material where the carbon atoms do not form into any long-range, periodic pattern. Amorphous carbon commonly occurs as a constituent of coal and soot, but most often these are a mixture of graphite and amorphous carbon. There is no defined crystal structure, and though it can be in a glassy state, it is not typically thought of a glass.* Amorphous carbon is reactive, and it makes an effective filter media. Carbon filters utilize **activated carbon** that has been treated to contain microscopic pores that enhance its surface area and therefore its overall capacity for trapping contaminants in gas or liquid form.

A special form of amorphous carbon is **diamond-like carbon** (DLC). DLC is most often used as a surface coating to increase wear resistance and decrease friction. DLC can exist in a number of configurations determined by the bonding between carbon atoms, but the purely sp^3 tetragonal form (referred to as ta-C) is extremely hard, possibly more so than diamond, though it is difficult to measure.† The coatings can be applied to most any material and are most often used in demanding applications such as high-performance engines components, biomedical devices, and advanced machine tooling.

Diamond is the hardest natural form of carbon, and it has a diamond cubic crystal structure (for which it is the namesake). It can be found naturally and mined, but for many technical applications it is more cost effective to create it synthetically. It is by no means a simple process and requires extraordinary pressure and temperature [2732°F (1500°C) and 5 GPa (~49,000 atm), for instance]. It can also be "grown" by chemical vapor deposition (CVD) for greater control and through a variety of other methods, but the high-pressure, high-temperature (HPHT) process is still most popular for its relatively low cost. Diamond is prized for its optical and mechanical properties as well as its rarity.

Diamond is used in jewelry because of its durability and the way it refracts and disperses light. Diamonds are much more useful than simple adornment, however, and they are used extensively in high-demand applications. Diamond can handle high pressure, high wear, and high thermal conductivity applications. Because diamond is so hard, it is commonly used as an indenter to measure the hardness of other materials, including very hard ceramics. Its hardness makes it an excellent abrasive and one of the only materials that can grind advanced ceramics such as cubic boron nitride and tungsten carbide. The cost of diamonds soars with size, so most abrasive applications use very fine diamond powder.

Graphite, which has a hexagonal, planar structure, is the most stable form of carbon. This means that diamond is only metastable at room temperature. Given enough time, diamond will revert to graphite (at least theoretically). Under normal circumstances that will not happen, so do not put off buying that engagement ring or other piece of jewelry.

---

*Glassy carbon is not technically amorphous carbon do to its lack of "dangling" bonds.

†The sp^3 designation has to do with how the bond is "hybridized" between the s and p orbitals (the electron shells). Diamond is purely sp^3 bonded, but DLC can have a combination of sp^2 and sp^3 hybridization. These concepts are covered in introductory chemistry texts.

Figure 6.11 Carbon can possess a variety of structures including amorphous, graphite, and diamond. In addition to these common structures, it can also take on a number of nanostructures such as carbon nanotubes, fullerenes, and graphene. The applications of carbon are as varied as the allotropes.

Graphite's planar structure gives it unique properties. It is very much like stacked sheets of paper where each sheet is made from covalently bonded carbon atoms. That structure results in highly anisotropic properties. Each single sheet of carbon atoms (known as **graphene**) is very strong within its plane, but the sheets can be easily separated. Because the sheets can slide, graphite is good lubricant. Graphite also conducts electricity well along the planes but very poorly across the planes.

Graphite is used in varying applications including battery electrodes, high-temperature processing (e.g., as crucible material), and in fiber form as composite reinforcement. Even the humble pencil uses graphite as the "lead." Graphite can be highly ordered or somewhat disordered (not amorphous). The higher the order, the more perfectly aligned and spaced the sheets are. The in-plane properties improve with greater order. Graphite with misaligned planes is called *turbostratic*, and poor-quality graphite is sometimes called "amorphous" although it is not entirely disordered. The various forms of carbon are shown in Fig. 6.11.

6.3 Common Polymers

Polymers have escalated in importance in modern society. Polymers occur naturally in living organisms, so they are by no means new. Ancients utilized natural polymers including rubber and animal horns. Although natural polymers still play a role, modern polymers are highly controlled and often fully synthetic. The first man-made plastics (the term *plastics* is used synonymously with polymers here) date to the 19th century, but it was not until the early 20th century that they saw significant progress in understanding and diversity.

Once plastics hit the market, however, they revolutionized the design of everyday items. Their low cost, ease of manufacture, durability, and versatility allowed them to supplant materials such as wood, metal, and glass. Polymers can have a wide range of properties depending on composition and structure. We will only look at a very small cross section of polymers and their general properties and applications. They will be categorized according to the major classifications of thermosets and thermoplastics (discussed in Sec. 5.3.2.5). A section on elastomers, which can be either thermosets or thermoplastics, will be discussed separately due to their unique structures and applications.

6.3.1 Common Thermoplastics

Thermoplastics, by definition, soften with applied heat. This has two implications. They can be more easily shaped and processed at elevated temperature and they are limited to modest temperature during use. That broad statement will be refined for each polymer in this section as variation among thermoplastics is significant.

6.3.1.1 Polyethylene

Polyethylene (PE) is a simple and versatile polymer. It is comprised of only carbon and hydrogen atoms, but it can exhibit a wide range of properties depending on its crystallinity and molecular weight (these were discussed in Sec. 5.3.2). These factors determine the differences between oil, wax, and advanced plastics used in medical applications. They are all the same material, just different configurations. PE plastics are generally low cost because of their simple composition and ease of processing. It requires low temperatures to be formed, so it can be mass-produced economically and it is relatively easy to recycle. We will look at a few varieties of polyethylene, but there are others as well. The unit structure of polyethylene is shown in Fig. 6.12, as well as some variations on chain structure.

High-density polyethylene (HDPE), as the name suggests, is a form of polyethylene with a higher density than other types. The density is only few percent higher than low-density PE (LDPE), but the difference is important to performance. The reason for the higher density is that the polymer molecules are very "smooth." They do not exhibit appreciable branching, so they can pack together more closely, a feature that also contributes to its high crystallinity. HDPE has higher strength and stiffness than lower density PEs, so it can be used in thinner cross sections while still maintaining its shape. This lends itself well to containers such as milk jugs and trash cans (see Fig. 6.13), but it also resists chemical attack and can be used for chemical storage and piping. The linear structure of HDPE molecules allows them to align easily, and during processing this

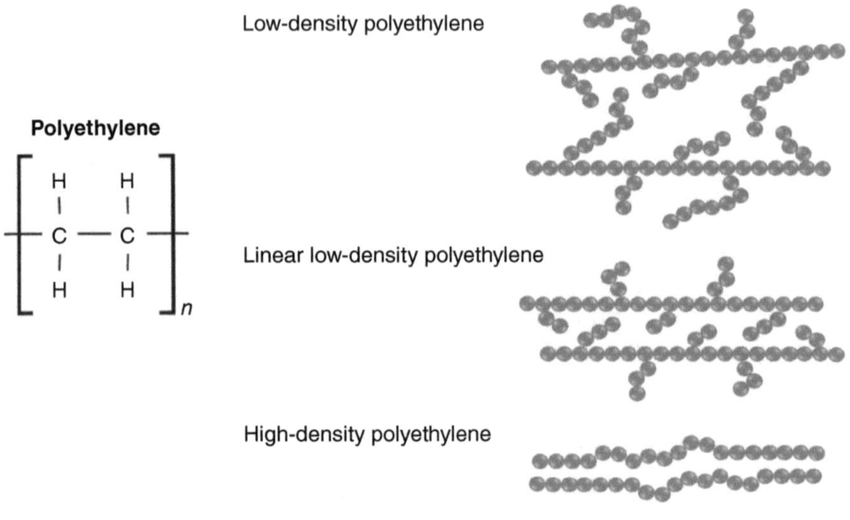

Figure 6.12 The mer unit or monomer of polyethylene is quite simple, but by modifying the branching and chain length, a variety of properties can be achieved.

Figure 6.13 Polyethylene is easy to process, recyclable, and durable. It is used in a multitude of applications, depending on the structure of the polymer chains.

strengthens the "melt" during molding. The molecular alignment also lends strength in the solid state, so it is a suitable polymer for applications such as trash bags, grocery bags, and hard hats. Due to its high crystallinity, HDPE is opaque in thick sections but can be colored through addition of pigments.

Low-density polyethylene (LDPE) is a highly branched PE structure. The branching prevents the PE molecules from packing together as closely as in HDPE, and the open structure gives LDPE a lower crystallinity and higher transparence. Its structure also allows it to be very flexible and resistant to cracking or tearing. These properties make it well suited for applications such as flexible tubing, clear packaging films, low-stress containers, and household cling wrap.* LDPE has a low processing temperature and is easily and cheaply processed.

Linear low-density polyethylene (LLDPE) could almost be considered a hybrid of HDPE and LDPE. It exhibits more branching than HDPE, but not as much as LDPE. It is still relatively linear, like HDPE, and has density between that of LDPE and HDPE. The properties of LLDPE also lie between those of HDPE and LDPE, but it is cheaper to process than LDPE since it requires lower temperature and pressure to make. LLDPE can serve in similar roles as LDPE and HDPE since it exhibits intermediate properties. Again, applications included plastic wraps, flexible lids, flexible tubing, and toys.

Ultrahigh molecular weight polyethylene (UHMWPE) is a special type of HDPE where the molecular weight (and therefore the chain length) is much greater than other PEs. UHMWPE has a molecular weight of 3 to 6 million g/mol, whereas HDPE has 50,000 to 300,000 g/mol. Because the chain length is so great, UHMWPE is tougher than HDPE, does not melt well, and can decompose before becoming workable. Like some metals and ceramics, UHMWPE is often sintered from powder, which limits the complexity of components made from it. Its applications often center on its high lubricity (near to that of Teflon) and resistance to wear. It is also biocompatible and has been a popular material in joint replacement, particularly as an acetabular liner.

*Cling wrap, under the trade name "Saran," was originally made from polyvinylidene chloride (PVDC), but the toxicity potential caused it to be removed from food contact applications.

6.3.1.2 Polypropylene

Compared to polyethylene, **polypropylene** (PP) has one hydrogen atom per monomer replaced with a methyl side group (CH_3). This increases the space each molecule occupies, but the orientation of the side groups has an even more profound effect. If you refer back to Sec. 5.3.1, you will see that polymer structure has important implications for properties, and that the orientation of side groups can take three forms, isotactic (all on one side), syndiotactic (orientation varies in a specific pattern), and atactic (random orientation). The bulky side groups in syndiotactic and atactic configurations prevent crystallization, and the resulting properties include a low stiffness and rubbery behavior. Most PP of commercial importance is isotactic, and has appreciable crystallinity, similar to HDPE (see Sec. 6.3.1.1).

Many of PPs applications are related to HDPEs, but there are some differences. For instance, polypropylene is stiffer and more wear resistant than PE, and has a higher service temperature than PE. PP can be bent repeatedly without degrading, making it a popular option for an **integrated hinge** or **living hinge**, which are molded as part of sections they connect, typically with a thinner section in the hinged area (e.g., flip-top lids on disposable and low-cost containers as shown in Fig. 6.14). PP provides a favorable balance between strength and cost, and it is important structural polymer. For structural applications it is often modified through additives such as fiberglass reinforcement or fillers which reduce shrinkage and lower cost.

6.3.1.3 Polyvinyl Chloride

Polyvinyl chloride (PVC) is similar in structure to polyethylene, except one hydrogen atom per monomer is replaced by a chlorine atom. Although the chlorine atom is a direct replacement for hydrogen (atom for atom), it represents a large size difference. In fact, it actually occupies about the same space as the three-atom methyl group, CH_3, found in polypropylene. This results in loose packing of the polymer molecules and low crystallinity. In polyethylene, rigidity increased with crystallinity. PVC, on the other hand, is highly amorphous but still rigid because of the hindrance the chlorine atoms cause to chain sliding. The vinyl chloride structure is also more susceptible to chemical attack, and this allows PVC to be welded chemically rather than through heating. This can be done very simply by applying a solvent-based adhesive to both surfaces

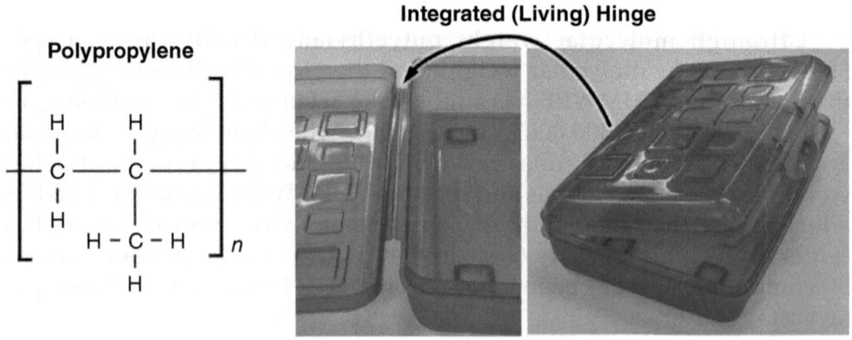

Figure 6.14 Polypropylene is a versatile plastic used in applications such as containers. It does not break easily under repeated bending, and it is used in flip top lids as a solid hinge, known as a living hinge.

Figure 6.15 PVC is normally a rigid polymer at room temperature, but it is often modified by mixing it with additives to make it more flexible. It can be used in applications from rigid plumbing to flexible products, like inflatables and wire insulation, depending on if and how it is modified.

and allowing them to set. This process is often employed in the joining of PVC piping, as shown in Fig. 6.15.

Chlorine not only makes a difference in mechanical properties, it can be both dangerous and useful chemically. The gaseous vinyl chloride monomer is a suspected carcinogen, so production of PVC must be highly controlled. PVC monomers are not supposed to be present in the final polymer, so the finished product is not considered carcinogenic. The PVC polymer can be dangerous when overheated, however, because it creates gaseous hydrochloric acid (HCl) during decomposition. Once thermal decomposition begins, it accelerates autocatalytically (self-propagating reaction). A thermally damaged PVC plastic will appear yellow to brown depending on the amount of decomposition. The thermal degradation and formation of HCl also causes PVC to be flame retardant since the HCl gas will actually block oxygen from the site since HCl is heavier than air, and it will extinguish the flame. Again, the harmful gases associated with high-temperature processing limit its use, but these can be mitigated through additives (discussed in Sec. 10.2.3). Additives are also crucial to the mechanical performance of PVC, and two categories are created based their use: rigid PVC and plasticized PVC, also referred to as vinyl.

Rigid PVC is stronger and stiffer than polyethylene or polypropylene, yet it is still a low cost, commodity polymer. The polarity of the chlorine atom in the vinyl chloride monomer causes significant intermolecular interaction which contributes to the mechanical properties as well as a rather high glass transition temperature [T_g = 140–180°F (60–82°C)]. The glass transition temperature is discussed in Sec. 8.2.6.1. Because most of rigid PVC's applications are below its T_g, it is in a brittle, glassy state. PVC is rarely used in pure form, and additives such as clay or talc can be employed to further reduce cost, and impact modifiers can be added to improve toughness or stability during processing. Rigid PVC can be processed through normal methods for thermoplastics, but temperature should be kept to a minimum and equipment should be carefully inspected for damage caused by the inevitable release of small amounts of HCl during processing. Rigid PVC is commonly applied in plumbing, conduit, house siding and trim and, to a lesser degree, in containers.

Plasticized PVC is a more flexible version of PVC achieved through the addition of additives known as *plasticizers* (see Sec 10.2.3.1 for more on additives). The plasticizers sufficiently reduce molecular interactions to make the polymer more ductile and flexible. The plasticizers effectively lower T_g below the application temperature. While this results in a weaker material, the usefulness in many applications outweighs any loss of strength. In most cases the plasticizers are added to PVC pellets before processing without unduly effecting processability. The plasticized PVC can be very flexible and is used in applications ranging from tubing (trade name, Tygon), to inflatable toys, rafts, and beds, cable coatings, and even as a leather simulant (polyurethane is also popular for making this poly-leather, or "pleather"). Because it can be made clear, flexible, kink-resistant, and sterilized, plasticized PVC is also popular for food contact and medical equipment use, though with some scrutiny. This is because some plasticizers, such as phthalates, are suspected of being harmful to health, and the leaching of these chemicals out of the PVC is a concern.

6.3.1.4 Polystyrene

Polystyrene (PS) has an "aromatic"* ring in place of a hydrogen atom when compared to polyethylene. This *phenyl group* is similar in structure to the chemical benzene and is comprised of six carbon atoms, and five of those atoms are bonded to hydrogen atoms. Polystyrene, then, contains only carbon and hydrogen. Like the chlorine atom in polyvinyl chloride, the size difference and location of the ring will greatly impact the performance of PS. The structure of this polymer is discussed more in Sec. 5.3. Such a bulky side group makes it difficult for the polymer chains to fit tightly together, so PS is almost entirely amorphous. It is a diverse polymer, and it finds applications in solid form and as a foam used in packaging and for insulation, as shown in Fig. 6.16.

The molecular structure of PS results in properties including high transparency, a low glass transition temperature (see Sec. 8.2.6.1), and chemical sensitivity to many organic solvents. It is resistant to water, however, and is used extensively in food service applications. PS is relatively sensitive to environmental degradation, so it is typically employed in short-term applications including packaging and disposable items.

It is difficult for PS chains to move relative to one another. This makes PS rather stiff and brittle, but it also has a low glass transition temperature, so it can be processed easily at modest temperature. The mechanical characteristics are determined by the molecular weight (chain length), and it is kept fairly low to maintain ease of processing. PS finds a broad range of applications including CD cases, disposable cutlery, and petri dishes.

Expanded polystyrene (EPS) is a foam often made by introducing gas into the plastic during polymerization. The plastic stock, typically beads, expands when heated as the

FIGURE 6.16 Polystyrene is a rigid polymer that is highly amorphous and used in solid and foamed structures as shown here.

*In chemistry, the term *aromatic* refers to flat, stable rings of atoms.

Figure 6.17 Expanded polystyrene can be made by heating pellets which contain the expandable gas. As shown, the beads expand to fill a mold and form a solid part.

gas trapped within the beads expands (see Fig. 6.17). The individual beads are contained within a mold that forces them fuse together during expansion. EPS is a very good thermal insulator, it absorbs impact well, it is highly buoyant, and it is inexpensive. These characteristics have made it extremely popular for a protective shipping material, in disposable coffee cups, in floatation devices, in "take-out" containers and in building insulation. The extensive use of EPS has attracted criticism since it does not readily break down in the environment and recycling is low compared to the amount used.

6.3.1.5 Thermoplastic Polyesters

Polyesters can be either thermoplastics or thermosets, but the discussion here is only pertinent to thermoplastics. Polyesters are defined by the presence of an ester group in the mer unit. Two common polyesters are **polyethylene terephthalate** (PET or PETE) and **polybutylene terephthalate** (PBT). The structure of PET is shown in Fig. 6.18, and PBT is nearly identical except for having two ethylene groups instead of one. The two ethylene groups connected in the backbone of PBT make it a butylene group.

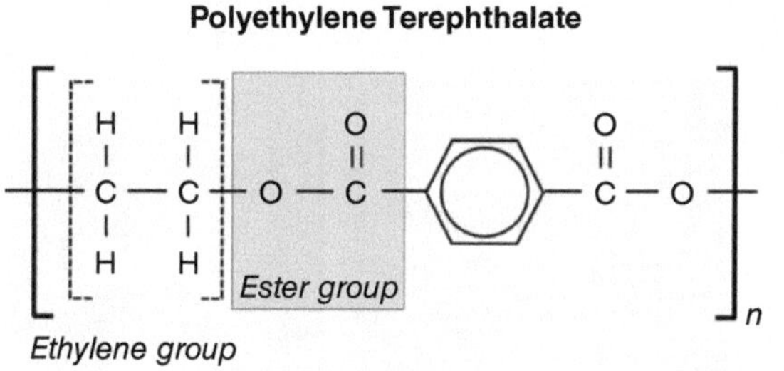

Figure 6.18 The mer unit for polyethylene terephthalate (PETE or PET) is shown. The ester group categorizes it as a polyester. It only differs from polybutylene terephthalate (PBT) in that PBT has two ethylene groups instead of one. One the most common applications for PET is in beverage containers.

Polyesters are popular in applications such as containers for carbonated beverages, in decorative films (trade name, Mylar) and light duty mechanical applications. Polyesters can have various levels of crystallinity and molecular orientation. The strength increases with both, but crystallization is slow and processing is more elaborate for achieving those conditions. Polyesters are mildly moisture sensitive and dimensional stability must be considered, since crystallization during and *after* processing can lead to shrinkage. PBT crystallizes somewhat faster than PET, which eases processing, but it is not as strong.

Polylactic acid (PLA) is naturally derived, biodegradable polyester. In each mer unit it has a methyl side group (CH_3), connected to the ester group. PLA is made from renewable organics such as cornstarch and sugar cane, and it is one of the most widely used bioplastics. It can be moderately crystalline (<40%) and has a relatively low glass transition temperature [140–149°F (60–65°C)]. See Sec. 8.2.6.1 for more on the glass transition temperature. PLA can be processed using most common thermoplastic manufacturing techniques such as extrusion and injection molding, but one of its more notable applications is as filament for 3D printing (see Fig. 6.19).

It is used extensively for this purpose as it is low cost, can be made in a variety of colors and effects (sparkle, translucence, glow-in-the-dark, etc.), and is easily extruded at modest temperature. Due to its biodegradation characteristics, PLA can also be used as a temporary surgical implant to provide support while the body performs natural repairs. The breakdown period can range from 6 months to 2 years. Packaging applications and disposable products can also benefit from the decomposition of PLA, but that also prevents its use in environmentally demanding applications where it may breakdown unwantedly.

6.3.1.6 Acrylonitrile Butadiene Styrene

Acrylonitrile butadiene styrene (ABS) is a mixture of the three monomers (i.e., acrylonitrile, butadiene, and styrene). Polymers can be combined during polymerization (*copolymerization*) or after polymerization as a homogeneous mixture (*alloy*) or where distinct phases are still present (*blend*). Copolymer structures were discussed in Sec. 5.3.2.1, and the other two are beyond the scope of our discussion, but can be found in texts devoted to polymers. This multi-component nature of ABS makes it different than other polymers we have discussed so far. There are a variety of polymers created through blending and alloying, but ABS will be the only one we consider. The principles apply to others as well.

Figure 6.19 Polylactic Acid (PLA) is a popular material for 3D printing filament as it has a low processing temperature, is inexpensive, and is biodegradable.

Figure 6.20 ABS is a durable plastic made up of three different monomers. The balance of those monomers determines the final properties of ABS. It very tough and able to withstand abuse, properties which make it a good fit for applications such as LEGO® blocks and wheels.

Instead of having properties dictated by the mer unit and molecular weight, the properties are also dependent on the ratio of the three monomers in ABS. Polybutadiene is a soft, rubbery polymer; polyacrylonitrile is very strong, heat and weather resistant; and polystyrene is transparent, rigid, and low cost. The ratio of these polymers controls the final properties of ABS. Styrene is the major component (about half to two-thirds of ABS) with acrylonitrile and butadiene making up the remainder.

This mixture allows ABS to have a favorable balance of properties including good toughness (even at low temperature), ease of processing, and chemical and weathering resistance. ABS costs about 1.5 times more than polystyrene, but its improved properties make it worthwhile. It can be processed through typical thermoplastic methods and is applied in a range of applications (see Fig. 6.20). ABS is found in wheels, pipes, appliance housings, helmets, LEGO® blocks, car trim, canoes, etc.

6.3.1.7 Polytetrafluoroethylene

Polytetrafluoroethylene (PTFE) is fluoropolymer with a structure reminiscent of polyethylene (PE), but instead of having hydrogen pendant atoms, it has fluorine atoms attached to the carbon backbone. PTFE is not the only fluoropolymer, but it is the most popular and representative of others as well. The fluorine atoms are highly electronegative and bond tightly while not interacting with external chemicals. This makes PTFE highly stable and resistant to chemical attack. PTFE is used as a lining in chemical tanks and as piping and valves for transporting chemicals. The unique bonding within the PTFE molecules also makes it an excellent electrical insulator. PTFE is best known as a nonstick material under the trade name, Teflon, where it has been used extensively in cookware (see Fig. 6.21). Its exceptional lubricity makes it suitable as a bearing surface, for non-lubricated assemblies, and as a thread sealant (i.e., plumber's tape, sometimes referred to as "Teflon" tape*).

Because PTFE has a structure similar to PE, it too can readily organize into a crystalline configuration. In fact, it can achieve very high crystallinity (>90%) and has an exceptionally high density (~2.2 g/cc)—greater than magnesium (1.74 g/cc) and

*DuPont states it has never licensed the Teflon brand for use as plumber's tape. It is made from fluoropolymer, just not under that brand name.

Figure 6.21 Fluoropolymers are often applied in situations where their high lubricity is exploited. They are widely applied in nonstick coatings (center), bearing surfaces, and thread sealing tape (right).

approaching that of aluminum (2.7 g/cc). The smooth nature of its molecular structure renders PTFE relatively weak, being in the range of polypropylene or HDPE. PTFE is also difficult to process since its decomposition temperature is similar to its melting temperature.

Like UHMWPE, it too is sintered in many cases. It can be made as simple extrusions, but complex shapes cannot be produced this way. To ease processing and allow other applications, modifications to the PTFE structure have been developed, much like deviations on the PE structure result in different properties. Other popular fluoropolymers include a copolymer, fluorinated ethylene propylene (FEP), and perfluoroalkoxy (PFA). Most fluoropolymers display the basic properties of PTFE with a balance on strength, processability, or chemical reactivity.

6.3.1.8 Acrylics

Acrylics, also referred to as acrylates, are generally clear, elastic polymers. They are used as additives in "latex" and acrylic paints as well as in clear plastic sheeting and adhesives. We will look at some of the more common acrylics below.

Polymethylmethacrylate (PMMA) is also known by the trade names Plexiglas or Lucite, or simply as "acrylic." It contains large side groups and is highly amorphous. PMMA has exceptional clarity, and UV and weather resistance making it a good choice for transparent outdoor applications. It competes with polycarbonates (see Sec. 6.3.1.11) in that regard and costs less, but it is not as impact resistant. It also is not scratch resistant, so coatings are required in abrasive applications such as automobile headlights and glass replacements. PMMA is relatively easy to process by common thermoplastic techniques such as injection molding and extrusion and has very good machinability as well. Sheets of PMMA can be extruded or cast, and casting eliminates the internal stresses and therefore optical distortions and makes them suitable for applications such as display cases (see Fig. 6.22). The precursor to the solidified sheet is an acrylic "syrup" of monomers, which are polymerized by peroxide or UV light. Casting of PMMA is not limited to sheet, and more complex structures with thick cross sections can also be created. Some of the applications in that realm are counter tops and sinks, which may be colored in variety of ways, tanning bed interior surfaces and a host of artistic works that leverage the glass-like appearance of PMMA.

Cyanoacrylates, better known as "superglue" or "krazy glue," are polymers that harden rapidly (through polymerization) when exposed to water. Often moisture in the air is sufficient to drive polymerization, but a "skin" forms quickly and drastically

Polymethylmethacrylate

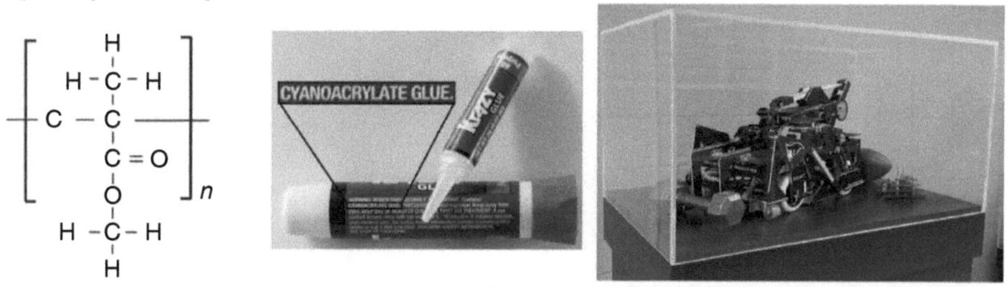

Figure 6.22 Acrylates are used for their high clarity and adhesive properties, depending on composition. Structural polymers, such as polymethylmethacrylate (PMMA), are often referred to as acrylic.

slows the continued hardening. As you might expect, it is primarily used as an adhesive. It does not have good shear strength, which makes it a good option for temporarily bonding some items. Additives are used to improve shear and impact strength, and many grades of these polymers exist. Medical versions, such as Dermabond, can even be used in place of stitches for some injuries. Cyanoacrylates can react exothermically with cotton which can lead to smoking or even ignition of the fabric, so care should be taken during use.

Polyacrylonitrile (PAN) is one of the three components of ABS plastics, and is often used in fibrous form. It is applied in high-efficiency filtration, textiles, and as a precursor to high-quality carbon fibers for lightweight composite reinforcement. For carbon fiber, the PAN fibers are oxidized at moderate temperature and then carbonized at high temperature. PAN has high strength and stiffness, but will vary with fiber size and surface treatment.

6.3.1.9 Nylons (Polyamides)

Nylon is a manufacturer name given to **polyamides**, and it is not based on the structure of the polymer. Polyamides, however, are now often referred to as nylons, regardless of the manufacturer. The polyamide name comes from the amide group in the mer unit, though there may be some chemical variation between different polymers in this class. The properties of these polymers come partly from nitrogen and oxygen in the amide group creating polarity that causes adjacent molecules to connect through hydrogen bonding. This allows for tight packing of molecules, and therefore high crystallinity as well as high strength. Nylons also tend to have good toughness, abrasion resistance, optical translucence and require high-temperature processing. Polarity within the structure also renders nylons susceptible to polar solvents, water being one of the most important. Nylons should not be used in applications where high temperature and moisture occur simultaneously, and processing should avoid these conditions as well. On the other hand, nylon is well suited for use with nonpolar chemicals such as oil and grease.

Nylon is used in a variety of applications including rope, bearings, zippers, and carpets. The specific chemistry of nylon is usually indicated by naming, such as nylon 6,6 (six-six), nylon 5,10 (five-ten), nylon 11, or nylon 6,12 (six-twelve). The numbers indicate what precursors are combined by the amide group, and how many carbon atoms are in each mer unit. In the structure shown in Fig. 6.23, R and R′ represent the precursors used in forming the particular nylon. In Nylon 6,6, R represents

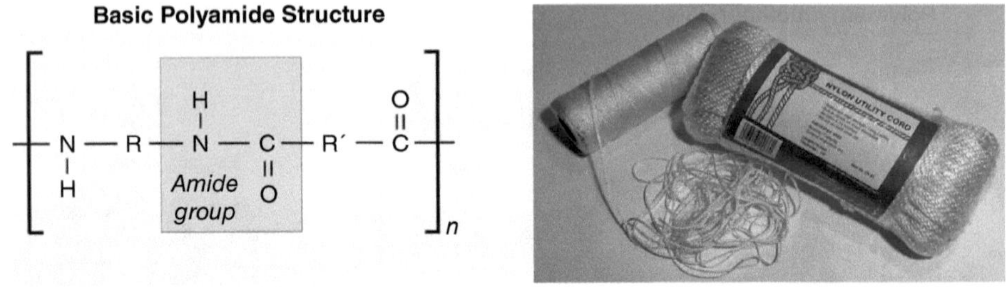

FIGURE 6.23 The general structure of polyamides includes the amide group and generic space holders for two carbon chains, R and R′. In Nylon 6,6, R and R′ are given by the chemicals hexamethylene diamine and adipic acid, respectively. Nylon is used in many applications, including rope (right).

hexamethylene diamine [$H_2N(CH_2)_6NH_2$] and R′ is adipic acid [$(CH_2)_4(COOH)_2$]. The nylon 6,6 mer unit consists of 12 carbon atoms where all are bonded to hydrogen atoms except those indicated in the general structure. The strength goes up and the flexibility goes down as the chain length is reduced, so the properties can be tailored through chemical variations.

Aramid is not nylon in the traditional sense, but it *is* a polyamide.* It is often referred to under the trade name of Kevlar, used for polymer fibers. Aramid has aromatic rings for R and R′, making it an aromatic polyamide. Its structure gives it the highest stiffness and strength of any nylon, partly due to the aromatic rings in the backbone, but also due to the stacking of these rings, which limits intermolecular movement. Depending on how the chain bonding occurs on the aromatic rings, the polymer is considered a para-aramid or a meta-aramid. Figure 6.24 shows the para-substitution. Meta-substitution would put a kink in the chain.

Kevlar (a para-aramid) has a strength-to-weight ratio about five times greater than steel, and is used over a wide range of temperature. It is suitable for cryogenic applications and can be used at elevated temperature [300–500°F (149–260°C)], but strength

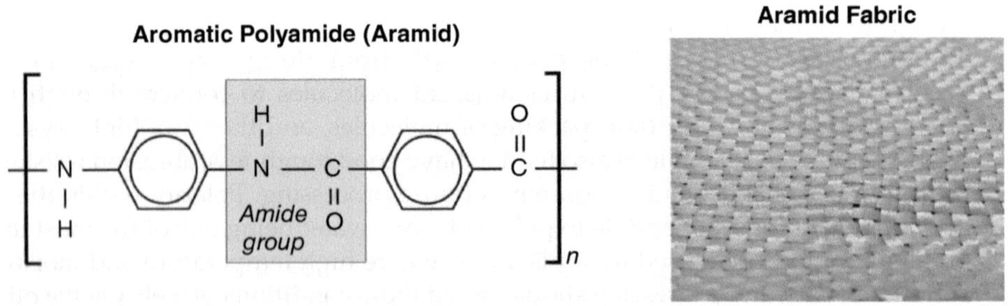

FIGURE 6.24 Aramid fibers, often referred to under the trade name Kevlar, are suitable for applications requiring high strength, heat protection, puncture protection, and cut resistance. Note that Aramid panels can be combined with steel plate to improve ballistic performance and reduce ricochet.

*Aramid is short for aromatic polyamide.

drops significantly at maximum temperatures. Aramid fiber is the most common form used, and appears in body armor, heat protection, in various vehicular armor applications, tire reinforcement, composite reinforcement, sporting goods, sails, etc. Anywhere woven or fibrous materials can be used, there is potential for this polymer to play a role. It is also used in "pulp" form for applications such as brake pads and in insulation. It is not easily processed, but a chemical modification to the structure, where the amide group bonds to a different carbon atom in the aromatic ring, yields Nomex (a meta-aramid) which is moldable and used as a flame retardant.

6.3.1.10 Acetals

Acetal polymers are actually polyoxymethylenes (POM). A common example is known by the trade name Delrin. These polymers serve in similar roles to nylons and have high strength and rigidity. They have a relatively simple mer unit, unlike nylons, and this allows them to reach very high levels of crystallinity. Acetals are much more resistant to moisture than nylons, and they serve in a number of applications involving prolonged water contact such as hose couplers and shower heads. The temperature must not be excessive, however, due a low maximum operating temperature. Acetals also have good lubricity and abrasion resistance making them suitable for use in gears (see Fig. 6.25), guitar picks and hinges.

The properties vary depending on the type of acetal. Homopolymers (including Delrin) are stiffer and stronger, whereas copolymers are more flexible. Both quickly degrade at elevated temperature releasing formaldehyde, but end capping and stabilization steps are taken to reduce the likelihood of this. Still, processing should be carefully conducted to avoid overheating and to provide plentiful ventilation. The high crystallinity of acetals makes molding and extrusion somewhat more challenging as the processing temperature window and solidification parameters need to be carefully controlled.

6.3.1.11 Polycarbonate

Polycarbonate (PC) is known by the trade name Lexan and is perhaps best known as "bulletproof glass." It is strong and tough but does not hold up well under abrasion, so it is often coated with a scratch-resistant material. Like aramid, the aromatic rings in the backbone of PC provide it with strength, and the methyl side groups between aromatic rings also increase interlocking of the polymer chains. The structure of PC makes it difficult for crystallization to occur, so PC is amorphous and transparent. Even so, its strength

Figure 6.25 Acetal polymers are rigid and wear resistant, which makes them suitable for applications such as gears in light-duty applications (e.g., printers, as shown here).

Figure 6.26 Polycarbonate is used in applications where clarity must be combined with high strength and toughness, such as the camera case and safety glasses shown here.

exceeds some crystalline polymers while maintaining a superior toughness. PC can be formed through tight bends, even at room temperature, but it also performs well at low and elevated temperatures [working temperature between –40 and 266°F (–40 and 130°C)]. This has made PC a capable replacement for some metal applications and simplifies processing, although PC has to be heated to higher temperatures than competitive materials such as polyesters or nylons for many types of thermoplastic processing.

Optical clarity and excellent toughness have made PC an attractive material for safety glasses, automobile headlights, CDs, DVDs and Blu-ray disks, "theft-proof" packaging (unable to be ripped open by hand), and other impact resistant enclosures (see Fig. 6.26). Again, PC is not scratch resistant and must be coated for certain applications. It is also only moderately UV-resistant, and must be protected from yellowing. PC has also been used in some "indestructible" containers. The containers, best known under the trade name Nalgene, leverage the clarity and toughness of PC, but there have been concerns over PC as a food-contact container.

PC is made by the reaction of bisphenol A (BPA) and phosgene. BPA has been identified as an endocrine (hormone) disruptor and phosgene is a toxic chemical, even being used as a weapon in World War I. The condensation reaction between these two chemicals releases hydrochloric acid as well. The final product will not contain these chemicals under *ideal* conditions. Leaching of chemicals from plastics has been a concern in a number of applications (for instance, see PVC in Sec. 6.3.1.3), but BPA has gained much attention recently and also appears in polymers other than PC, such as epoxy. The FDA has declared the levels of BPA leached into food or beverages are "safe." Interestingly, however, use of BPA-containing plastics has been "abandoned" by industry in infant product packing.* Now many containers, including Nalgene, are specifically labeled "BPA-free" to alleviate concerns.† It is up to the consumer, then, to decide if even low levels of harmful chemicals are acceptable, but the necessary information to make an informed decision is rarely available or easily accessed by the majority of the population.

6.3.2 Common Thermosets

Thermosets are used less extensively than thermoplastics, but they are commercially important nonetheless. Thermosets are often shaped and then cured, during which they form extensive networks of cross-links. This structural change creates something analogous to a highly interconnected cargo net. The bonds created during curing are permanent, and they

*See the FDA report, *Bisphenol A (BPA): Use in Food Contact Application*.

†The replacement for BPA, bisphenol S (BPS), has also raised safety concerns for similar reasons.

prevent the polymers from melting before degradation. Section 5.3.2.6, discusses more about cross-linking. What allows cross-linking to occur is the presence of "unsaturated" bonds, meaning the molecule is reactive enough to form new bonds between molecules. The majority of thermosets are stiff and strong, though they may have a variety of properties that determined by the reactants that form the polymer. Reactions and chemistries are more complicated than thermoplastics, and only the basic material properties will be discussed here.

6.3.2.1 Phenolics

Phenolics, or phenol formaldehyde resins, were some of the first polymers synthesized (under the trade name Bakelite) and had immediate applications in the replacement of ivory due to their high hardness (and the expense and controversy associated with ivory). They are still popular because of their low cost and ease of processing. Phenolics are cross-linked by heating (one-stage, **resole resins**) or by addition of a curing agent and heat (two-stage, **novolac resins**). Phenolics are highly adhesive, which gives them applications in that realm, but this can also produce problems during processing and mold release must be applied directly to the mold or as an additive to the molding powder.

Phenolics often contain a significant amount of filler, such as wood flour, which further reduces their cost. Phenolics have low electrical and thermal conductivity, and they are used extensively in printed circuit boards and other electrical components as well as handles and knobs on cookware and appliances where heat may be an issue (see Fig. 6.27). Phenolic resin is also used as the matrix in fiber-reinforced composites (mostly fiberglass), wood products (such as plywood), and cloth (used in speaker construction).

6.3.2.2 Thermoset Polyesters

Polyesters also appeared in the thermoplastics section (see Sec. 6.3.1.5), and that is because polyesters can be either thermoplastics or thermosets. **Thermoset polyesters** are referred to as being *unsaturated*. This means that there are carbon-carbon double bonds in the backbone of the molecule, which leaves the possibility for further bonding

Figure 6.27 Phenolics are strong and rigid, and they are used for knobs and handles, often to add thermal insulation. They also see extensive applications in composite lumber.

FIGURE 6.28 Thermoset polyester is a common polymer as a composite matrix and coating material. This includes aviation and marine structures such the boat shown here.

(cross-linking) to occur. The starting point for these polyesters is as short chain polymers that form a thick liquid or a solid with a low melting temperature. The cross-linking is initiated by a free radical addition, such as peroxide, and may incorporate heat as well. These thermosets are also sensitive to environmental influences and can cross-link unintentionally which limits their shelf-life. Cross-linking can be controlled by other additives which slow (inhibitors) or speed up (accelerators) the reaction.

Thermoset polyesters are the most popular thermosets and their resins can be used as a composite matrix, like other thermosets, but they are also used as coatings for other materials, including composites (e.g., the "gelcoat" on fiberglass boat hulls, see Fig. 6.28). The chemistry of the polyester determines the weathering and mechanical properties, and multiple types are often used in a single product. Molding resins, such as alkyds, are modified polyester resins which incorporate fillers to improve their performance. Molding compounds have low electrical conductivity and are used in electrical applications to encapsulate components. A special type of molding compound, **bulk molding compound**, has dough-like consistency and is extensively filled with additives and reinforcement (~70%). This allows for larger, more complex components to be produced, but molding must be carefully conducted to avoid separation of the ingredients. It often processed using injection or compression molding.

6.3.2.3 Epoxies

Epoxies are named for the presence of epoxide groups in the molecular structure. As shown in Fig. 6.29, the epoxide groups typically occur at the ends of the polymer molecules, though they can be within the molecule as well. R indicates the mer group comprising the majority of the epoxy chain structure. While the chemistry can vary, epoxy is commonly created by the polymerization reaction between bisphenol A and epichlorohydrin. Both of these chemicals present health concerns. Bisphenol A is described more in Sec. 6.3.1.11. Epichlorohydrin reacts with water to form a carcinogenic compound.*

*Epichlorohydrin hydrolyzes to form 3-monochloropropane-1,2-diol or 3-chloropropane-1,2-diol (3-MCPD).

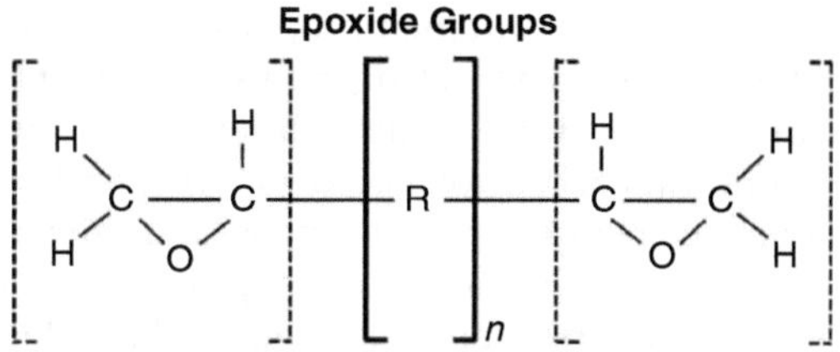

Figure 6.29 Epoxies often consist of a polymer resin molecule that has epoxide groups on both ends. In this case "R" represents the polymer molecule between the groups. The length and structure of the molecule will determine the final properties of the cured epoxy. The epoxide groups are sites where cross-linking occurs when a hardener is added to the resin.

Epoxies are perhaps best associated with two-part polymerization reactions. More specifically, they often come as two separate components known as the resin and the hardener (Fig. 6.30). The resin is often a highly viscous liquid, indicating chain length (molecular weight) is relatively low (high chain lengths would result in a solid). The polymer used for the resin controls the final properties of the epoxy, but the hardener also has a decisive effect. The hardener is responsible for creating cross-links between molecules. The hardener "opens" the epoxy ring and bonds it to the ring on another molecule, thereby linking them.

The resin molecules must have at least two epoxide groups to allow cross-linking to fully harden the mixture. If too little hardener is added, the cross-linking will not be as extensive and the resulting material will be soft, flexible, and thermoplastic in nature. If too much hardener is added, the same result will occur. The resin and hardener must be mixed in the proper ratio to achieve full curing and cross-linking. Some epoxies are aided in cross-linking by elevated temperature as well.

In general, epoxies are hard and brittle, but the properties depend on the resin and degree of cross-linking. A higher cross-link density will result in a higher hardness and lower toughness. Greater spacing between cross-links will decrease hardness and increase toughness. Due to the thickness of the resin, even when mixed with the hardener, it can be used to coat surfaces and encapsulate components or reinforcements such as glass or carbon fibers to create composites.

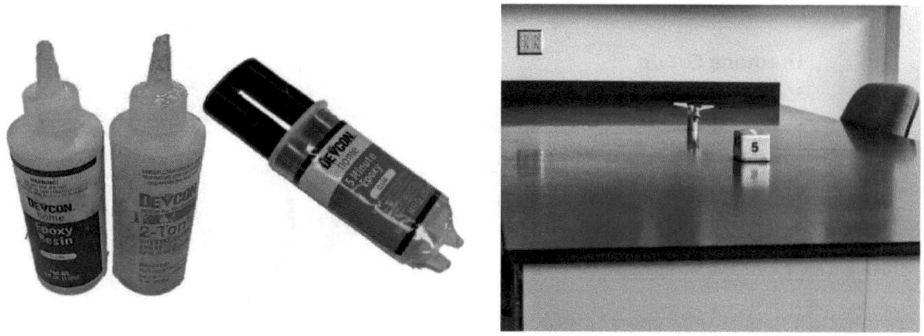

Figure 6.30 Epoxy is a durable polymer which is often mixed as two parts, resin and hardener (left), which can then be applied as an adhesive or used to create temperature and chemical-resistant surfaces applications where easy clean-up and long life are needed.

Epoxies are relatively costly, but their properties are superior to other thermosets such as unsaturated polyesters. They are stronger, more resistant to chemical attack, and can be used at higher temperatures. Epoxies also show less shrinkage during molding, which allows for tighter dimensional control. Epoxies can be used as a tough and durable coating material used to finish cement floors, truck beds, and other structures where chemical or abrasive damage may occur. They are also a popular choice for laboratory work surfaces where chemical and thermal resistances are required (see Fig. 6.30).

6.3.2.4 Polyurethanes

Polyurethanes are named for the urethane (or carbamate) groups in their structure. The precursor molecules can vary, but they are often a polyol and an isocyanate (more specifically, a di-isocyanate, which has two isocyanate groups). There are significant health concerns with isocyanates, and newer versions of polyurethane are being developed that do not use them. Polyurethanes can be thermoplastics or thermosets. Thermoset polyurethanes are more common and are the focus of this section. Polyurethanes may possess a wide range of properties. They can be soft or rigid, solid, or foamed, and they are generally resilient. This enables them to serve in diverse applications, some of which are given in Fig. 6.31.

Flexible foams are created by the evolution of gas during processing which is trapped in the polymer during solidification. Foaming agents, even as simple as water, are added to produce the gas required. The type and quantity of the foaming agents will determine the properties of the foam. These flexible, resilient foams are used in applications such as seat cushions and padding. Rigid foams often have similar densities but are highly cross-linked. They are used in packaging and spray-in, expanding foam insulation applications. Polyurethane will burn readily and additives are used to reduce their flammability. Burning polyurethanes will release hazardous gases, including hydrogen cyanide, which is an additional health concern.

Solid polyurethanes can exhibit strength and stiffness found in the acetals and nylons, but they can also act as a rubber substitute with high elasticity and pliability. These properties are often in opposition, but they depend on the precursor chemicals and amount of cross-linking. This wide variety of properties can be assessed by measuring the durometer of the polyurethane (see Sec. 7.6.1.4). High flexibility is

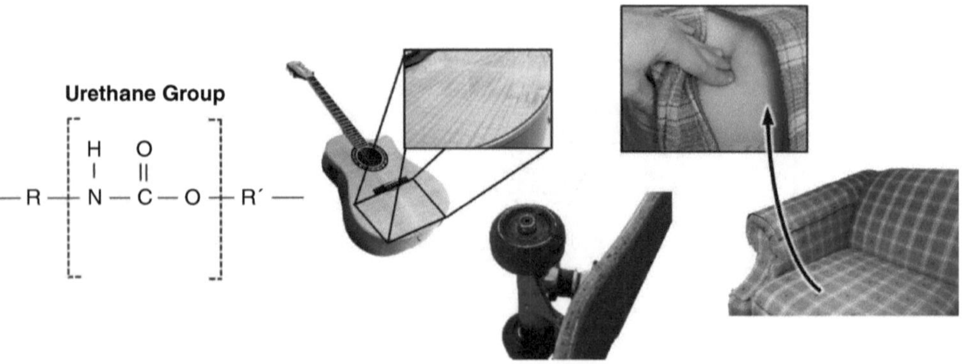

FIGURE 6.31 Polyurethane can vary widely in its properties. The composition and degree of cross-linking can determine whether it forms a transparent film used in furniture and instruments, skateboard wheels and other wheels and rollers, or is flexible for foam cushions.

exhibited by thermoplastic polyurethanes, and in fiber form is known as spandex (trade name Lycra). The molecular interactions allow a great deal of elasticity with little effort followed by a much stiffer stronger response after extensive stretching. Polyurethanes can be employed in a great breadth of applications depending on the properties they possess, including wire insulation, tubing, skateboard wheels, swim wear, and compressive clothing (as spandex) and many others.

6.3.3 Elastomers

Elastomers are grouped by properties, not by structure or thermal response. This is similar to the property-based classification of metals, where metals that exhibiting similar properties are put together, regardless of their composition or structure. Elastomers all possess exceptional elastic properties. Referring to Sec. 7.4, elasticity is a reversible change in shape, while plasticity is a permanent one. Plasticity does not refer to "plastics" specifically, and metals behave plastically as well. The ability to be reversibly deformed to a great extent, often to several times their original dimensions, is unique to elastomers.

Their elasticity most often arises from a weakly cross-linked structure, but elastomeric behavior can arise from other causes as well. In the case of cross-linking, the polymer molecules are connected at only a few locations, which allows the molecules to "unravel" before straightening out and behaving rigidly. This gives many elastomers the unique behavior of stretching extensively even though they are primarily brittle when they fail. More about the structure of elastomers is described in Sec. 5.3.2.8.

6.3.3.1 Natural Rubber

Natural rubber, known as **latex**, has been in use for centuries. It can come from a variety of plants, but the primary source is the rubber tree, *Hevea brasiliensis*. The liquid sap contains latex in colloidal form, which is processed to separate it out. Natural rubber is exceptionally water-resistant and can be used as waterproofing, but it is too soft and deforms permanently at modest temperature. This problem was solved by Charles Goodyear when he discovered **vulcanization**. The process involves heating rubber in the presence of sulfur (or other vulcanization addition), which introduces cross-linking. With extensive cross-linking, the rubber becomes quite rigid. Also, variations on the natural rubber structure, such as different plant species, have higher hardness and can be used in applications such as golf balls.

As was found upon later investigation, natural rubber is comprised of a mixture of **polyisoprenes**. Although they do not vary compositionally, the orientation of the atoms and side groups in the polymer molecules can differ. The two configurations are *cis*-isoprene and *trans*-isoprene. The prefix "*cis*" indicates the atoms are on the same side of the backbone and "*trans*" indicates they are on opposite sides (Fig. 6.32). *Trans*-polyisoprene

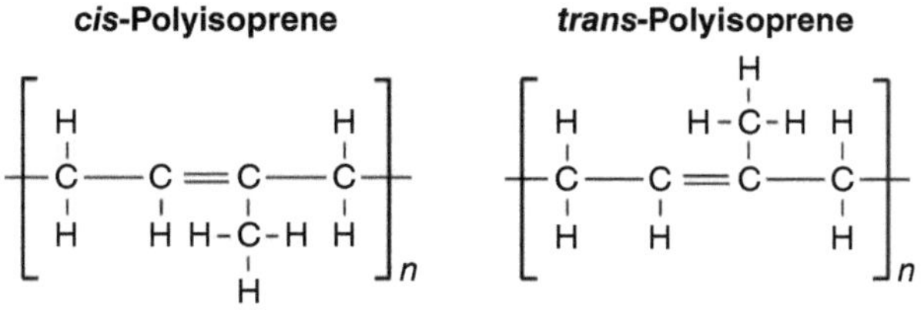

Figure 6.32 The configuration of elements in the mer unit will change the properties of rubber. The "-*cis*" configuration is softer and more flexible than the "-*trans*" configuration.

(or poly-*trans*-isoprene), known as **gutta-percha** (named after the tree *palaquium gutta* it comes from), results in a stiffer rubber as molecules can fit closer together. *Cis*-polyisoprene, known as **Hevea rubber** (for the tree *Hevea brasiliensis*), is more flexible and weaker. The variations in natural rubber and shortages during war efforts (WWI and WWII) led to a demand for synthetic rubbers that were not dependent on natural production and superior in terms of performance and uniformity.

Synthetic polyisoprenes were an early development in elastomer design. Though more expensive than natural rubber, the variations in *cis*-isoprene and *trans*-isoprene properties can be controlled in synthetic rubber. Polyisoprene of one type or the other can be produced during synthesis, so the final properties can be tailored for specific applications. Polyisoprenes are used in latex gloves, as shown in Fig. 6.33.

6.3.3.2 Butadiene Rubbers

Polybutadiene has a structure similar to polyisoprene, but there is no methyl group. The lack of a bulky side group in polybutadiene reduces intermolecular interaction, and for that reason it is weaker than polyisoprene. It does have some distinct advantages including low cost and good adhesion to metal. Polybutadiene, like polyisoprene, can come in "*trans-*" and "*cis-*" configurations. It is used extensively as an additive in other polymers. It is often used in combination with styrene to form styrene butadiene rubber (known as Buna-S). The balance of styrene and butadiene determines the stiffness (more styrene, higher stiffness) and resilience (more butadiene, more elasticity). This principle is also applied to ABS, where butadiene and styrene are joined by acrylonitrile (described in Sec. 6.3.1.6). ABS utilizes the **rubber toughening** properties of butadiene to achieve high strength and impact resistance.

One of the main uses for polybutadiene is in tire manufacturing (Fig. 6.34). Tire construction is quite complex, and multiple polymers are used, but butadiene is an important constituent. It is reinforced and modified by carbon black (see Sec. 6.4.1.5), which greatly enhances the wear properties. Polybutadiene is not particularly resistant to oils, oxidation, and UV light, so other additives are also incorporated. Other applications include bridge pads, "bouncy" balls, and some hose liners.

Nitrile butadiene rubber (NBR), or simply nitrile rubber, was created to overcome the poor performance of butadiene when in contact with oils. This type of rubber is also

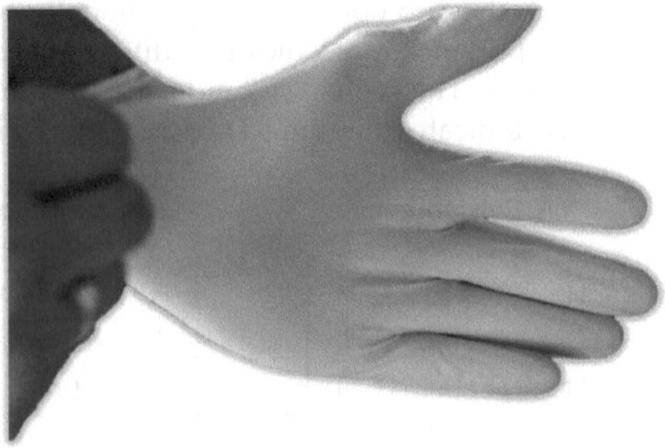

FIGURE 6.33 Natural rubber (latex) is often used in applications where resilience is needed, but where conditions of use are modest (chemical, thermal, stress, and other factors).

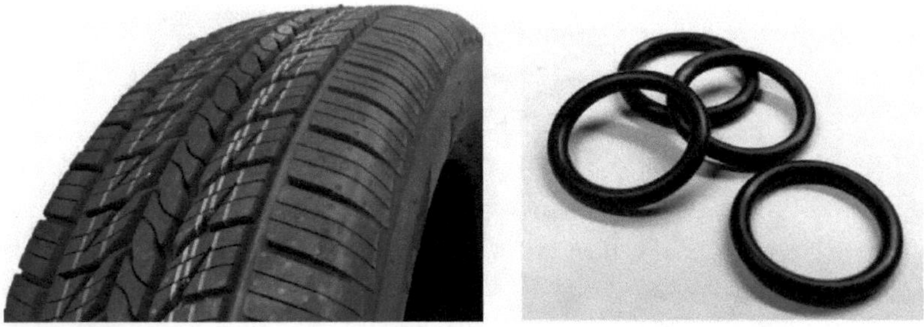

Figure 6.34 Polybutadiene is commonly used in tire construction (left), but it acts as a matrix for carbon black and is part of a more complex structure. It is also used in seals and some shock and vibration-dampening applications. For more oil-resistance, nitrile rubber (known as Buna-N) was developed and is used in O-rings and other seals.

known as Buna-N, and is a copolymer of acrylonitrile and butadiene. All of the ingredients in ABS, acrylonitrile, butadiene, and styrene can be used alone or in any combination with commercially viability. Nitrile rubber is more costly than butadiene, so it is used in applications where oil-resistance is important. Nitrile rubber is commonly used as an alternative to latex gloves used in medical or laboratory settings, and it is often blue (not always!). Its structure is shown in Fig. 5.13.

Neoprene is another oil-resistant elastomer, and it is also known as chloroprene. It is not directly related to butadiene, but it shares a similar structure and similar properties. It has one hydrogen atom replaced by a chlorine atom, and the result is nearly identical to that seen with polyethylene and polyvinyl chloride (PVC). The large chlorine atom restricts chain movement so the material is stiffer, and it also reduces flammability, and neoprene displays good performance at elevated temperatures. Neoprene is perhaps best known by name as a diving suit material, but it is used in many other situations as well. Its oil-resistance makes it well-suited for hoses carrying fuel or oil-based materials and gaskets for the same.

6.3.3.3 Olefin Elastomers

Olefins are purely hydrocarbons such as polyethylene and polypropylene. Thermoplastic elastomers can be created from copolymers of these, and they are categorized as ethylene propylene rubber (EPR). EPR shows good resistance to heat and weathering, is an electrical insulator, and is easier to process than many thermosets. It is not suitable for contact with solvents and is not as resilient as thermosets. EPR is used as automotive weather-stripping, wire insulation, gaskets, tubing, and belts. EPR can be modified by adding butadiene, thereby creating ethylene propylene diene monomer (EPDM) rubber. The addition of butadiene allows some cross-linking to take place (thanks to the double bond). EPDM is used in similar applications to EPR but with better weathering resistance. These applications include door seals, O-rings, rubber roofing, and others (Fig. 6.35).

6.3.3.4 Fluoroelastomers

Fluoroelastomers are useful for reasons similar to polytetrafluoroethylene and other highly fluorinated polymers, including superior resistance to chemical attack and nonstick properties. These polymers are often referred to under the trade name Viton and can be tailored by choosing monomers with the desired properties. Fluoroelastomers are exceptional at high-temperature service and in corrosive environments.

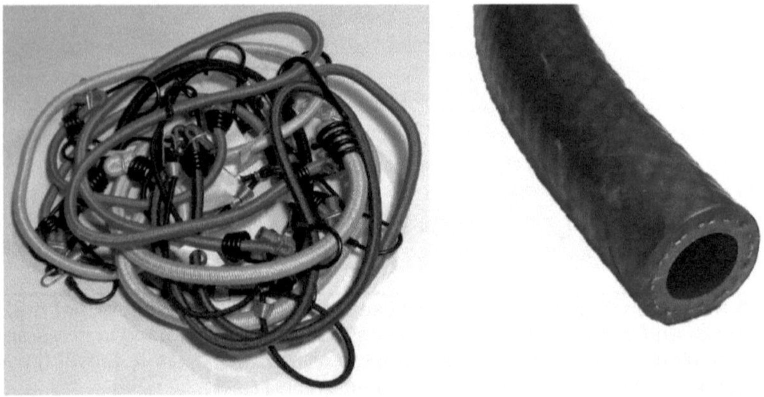

Figure 6.35 EPDM is used in a variety of applications, including bungee cords (left) and tubing (right).

Their excellent performance comes with an associated cost, but many applications require these materials regardless of the additional expenditure. These include chemical handling equipment, high-vacuum assemblies, high-radiation applications, gas seals, and cryogenic applications.

6.3.3.5 Silicones

Silicones, or *polysiloxanes*, are not based on carbon, but rather have silicon as part of the backbone. Silicon has some characteristics in common with carbon, but it differs in some important ways too. Silicon does not readily form double bonds, so that is not a viable method for cross-linking to occur. In fact, the backbone of silicone consists of alternating silicon and oxygen atoms that form a siloxane mer unit. Like carbon, silicon can form four bonds, so many other features of the molecular structure are similar. One of the most common silicone polymers is polydimethylsiloxane (PDMS), and its structure is shown in Fig. 6.36.

Silicones are excellent at repelling water, they have low thermal conductivity, low flammability, good strength from cryogenic temperature all the way to 400°F, they are highly resistant to weathering and UV degradation, inert and relatively nontoxic. Because they possess such a broad set of favorable properties, silicones are used for

Figure 6.36 Polysiloxanes share some characteristics with hydrocarbon molecules, but they differ greatly with their Si-O backbone chemistry.

everything from soft lithography and microfluidics to caulking, waterproofing, prosthetics, medical implants, and gaskets for engine components (RTV).*

6.3.4 Commodity versus Engineering Plastics

There are subcategories of polymers outside of the thermoplastic/thermoset designations. These categories are commodity plastics and engineering plastics. While these subdivisions are not precise, they are informative. *Commodity plastics* are "common" polymers used extensively. They are usually considered standardized in terms of composition and characteristics, but producers would likely argue that it is not so simple. *Engineering plastics* are polymers with advanced properties and typically used to a lesser extent.

Engineering plastics have superior strength, toughness, and stability, both during and after manufacture, when compared to commodity plastics. They often have a wider range of working temperatures and are more resistance to environmental degradation. They were (and still are) developed to replace metals, ceramics, or wood, so they must be strong and durable and able to be produced or finished with a pleasing appearance. The basic temperature and stress limits of some common polymers are shown for commodity plastics in Table 6.1. These same factors are given for engineering plastics in Table 6.2.

6.3.5 Identifying Polymers by Recycling Symbols

One of the unique identifiers for polymers is their recycling symbols. Unlike metals and ceramics, polymers are often branded with a number and abbreviation indicating their composition (Fig. 6.37). Not all plastics are labeled, and they may contain additives that are not accounted for. Because so many applications for plastics involve disposable containers, the rate of recycling is important to reducing overall waste. There are many difficulties associated with recycling plastics, including fillers and pigments, so they are often not reused for the same applications (see Sec. 22.5 for more on sustainability).

Commodity Plastics		
Polymer	**Maximum Use Temperature [°F (°C)]**	**Tensile Strength [psi (MPa)]**
High-density polyethylene	248 (120)	4640 (32)
Low-density polyethylene	176 (80)	1740 (12)
Polypropylene	275 (135)	3340 (23)
Polyvinyl chloride	158 (70)	8410 (58)
Polystyrene	194 (90)	7540 (52)

Commodity plastics are used in large quantities, but they are typically not as strong or resistant to environmental conditions such as temperature. Values for temperature and strength will vary for each polymer type, and the given values are not absolute but representative.

TABLE 6.1 Mechanical and Thermal Performance of Commodity Plastics

*RTV is a silicone that cross-links at room temperature, hence the name, room temperature vulcanized (RTV).

Engineering Plastics		
Polymer	Maximum Use Temperature [°F (°C)]	Tensile Strength [psi (MPa)]
Acetals	270 (110)	9570 (66)
Polycarbonate	275 (135)	8120 (56)
Polytetrafluoroethylene	500 (260)	4350 (30)
Polyethylene terephthalate	212 (100)	7400 (51)
Polymethylmethacrylate	194 (90)	5410 (58)
Polyamides (nylon)	250 (120)	10,590 (73)
Acrylonitrile butadiene styrene (ABS)	176 (80)	6100 (42)

Engineering plastics are used less extensively than commodity plastics, and their performance is typically better regarding strength and use temperature. Values for temperature and strength will vary for each polymer type, and the given values are not absolute but representative.

Table 6.2 Mechanical and Thermal Performance of Engineering Plastics

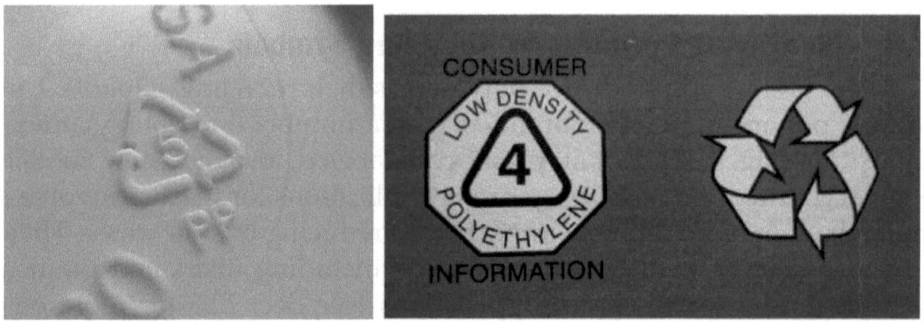

Figure 6.37 A unique feature of polymers among the materials classes is that they are often identified by recycling symbols. These may appear molded into the product, such as the travel mug show on the left, or on the packaging, as shown for the plastic wrap on the right.

Rather, they are used to produce other items where tight compositional and structural control is not as critical, such as composite lumber or lawn furniture.

There are seven recycling categories. The first six are dedicated to individual polymers, and the seventh is used to encompass all others. Five of the seven (#2–#6) are considered commodity plastics since they are used in large quantities while the other two (#1, #7) are for engineering plastics. Polyethylene terephthalate (PET) is symbol #1, and is widely used for beverage containers. The #7 designation is used for many copolymers, thermosets and blends, or polymers with high filler content. The designations are given in Table 6.3.

6.4 Common Composites

Composites appear in applications from the mundane to the most advanced. Any structure made up of two or more distinct materials that are intimately combined can be considered a composite. This leaves the door open for quite a few materials to be considered, but we will limit our discussion to the most prevalent and "quintessential" composites.

Polymer Recycling Symbols		
Symbol	**Polymer (Abbreviation)**	**Uses**
1	Polyethylene terephthalate (PET or PETE)	Polyester fibers, straps, beverage containers
2	High-density polyethylene (HDPE)	Milk jugs, bags, bins, hard hats, composite lumber
3	Polyvinyl chloride (PVC)	Pipe, fence and railing, "vinyl" siding, inflatables
4	Low-density polyethylene (LDPE)	Bags, bottles, containers, plastic wrap
5	Polypropylene (PP)	Containers, disposable cups, automotive trim, living hinges, laboratory ware
6	Polystyrene (PS)	Food containers, compact disk cases, disposable utensils, packing foam
7	Other (O)	Polymers include: acrylic, ABS, composites, nylon, PLA, PTFE, etc.

The recycling symbols are standardized for common plastics and are intended to make sorting an easier task. Not all polymers are equally recyclable, but most thermoplastics can be remade into different items. Thermosets can be repurposed in ways other than melting and remolding.

Table 6.3 Recycling Symbols and Common Uses for Polymers

6.4.1 Polymer Matrix Composites

Polymer matrix composites (PMC) have a polymeric material as a binder, or matrix. These are the most prevalent composites, partially due to the ease in creating large components from them. They are often lightweight, and can be very strong, a combination of properties that have made them popular for applications ranging from sporting goods to aerospace. The matrix polymer can vary quite a bit, but some of the most popular include thermosets such as epoxies and urethanes (see Sec. 6.3.2 for more on thermoset polymers).

6.4.1.1 Fiberglass

Fiberglass, or *glass-reinforced plastic* (GRP), can come as long, continuous fibers, short, chopped fibers, or as a woven mat. The glass reinforcement phase is often supported by a thermoset polymer matrix such as epoxy or polyester. The glass fibers, typically between 0.0002 and 0.001 inch (5 and 25 μm) in diameter, are produced by gravity-fed extrusion through small orifices with raw components being melted just prior to extrusion or by using premixed "marbles" as the feedstock (more on extrusion can be found in Sec. 12.3.4). After formed, the fibers are immediately "sized." The *sizing* process introduces a surface coating to the fibers that is useful for protecting it during shipment and use or to aid in adhesion of another material such as the matrix polymer.

A variety of glass chemistries can be used in creating composites. E-glass (E is for prior electrical applications) is one of the most common types, and is an alumino-borosilicate glass with a tensile strength of ~3.45 GPa. S-glass (S for "stiff"), is another common reinforcement popular for its high strength (~1/3 higher than E-glass). S-glass uses alumina

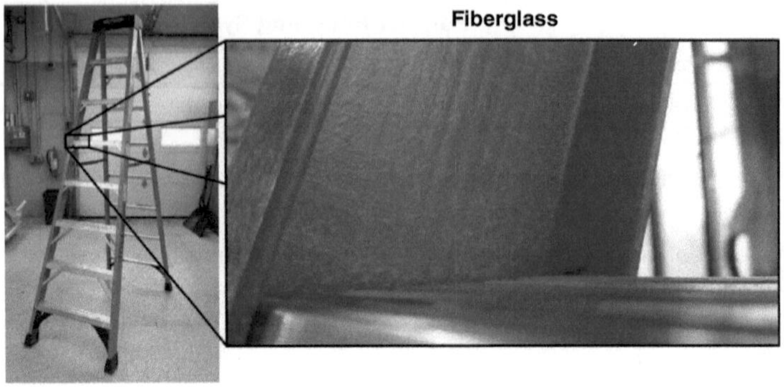

FIGURE 6.38 Fiberglass is popular composite due to its durability, light weight, versatility, and low cost. It is used in applications ranging from watercraft, aircraft, and pools and ladders.

and magnesia instead of boron, aluminum, and calcium oxides to achieve its high stiffness. Unfortunately, this comes with increased cost. Other types of glass fiber include C-glass (chemically resistant) and AR-glass (alkali resistant), which have better performance in the presence of corrosive media.

Glass fibers have a much higher strength than bulk glass components made from the same material. While this may seem counterintuitive, it is a common characteristic in fibrous materials. The reason is simply that the number of defects, which introduce weakness by concentrating stress, is reduced. Just like a chain is only as strong as its weakest link, materials are only as strong as their weakest point. Fewer defects allow the material to approach its "ideal" strength. The final strength of the fiberglass composite is determined by a mixture of the fiber and matrix properties (see Secs. 5.4 and 7.5.4.1 for more on composite strength).

Fiberglass composites are used in a variety of applications including boat hull construction and vehicle panels, antenna shrouding (it does not block signal), storage tanks, and lightweight pressure vessels, some roofing and outdoor structures, traffic lights, water slides, lightweight ladders (Fig. 6.38), and many others. Glass-filled polymers, such as nylon, are also composites, though they are used somewhat differently. Instead of being used in sheet form, they are often injection molded or extruded into bulk components.

6.4.1.2 Carbon Fiber

Carbon fiber composites, or *carbon fiber-reinforced polymers* (CFRP or CRP), are extremely lightweight and extremely strong. Carbon is an imprecise way to describe a specific material, as described in Sec. 6.2.4.2, because carbon has so many allotropes. Carbon fibers used as reinforcement are graphitic, though not entirely graphite. Carbon fibers are often created with a polymer fiber as a precursor. One of the precursors for high-quality carbon fibers is polyacrylonitrile (PAN). PAN is a component of ABS plastic and nitrile rubber, but must be carbonized to make carbon fibers. Generally speaking, the process involves creating polymer fibers and then heating those fibers to very high temperature in an inert atmosphere so that they do not burn, a technique called pyrolysis or *carbonization*.* Atoms other than carbon are driven off, but the remaining carbon

*Pyrolysis is the decomposition of organic matter in the absence of oxygen and carbonization is the conversion of a material to carbon only. Therefore, when taken to the limit, pyrolysis results in carbonization.

is not well-structured. To increase the atomic order, the carbon is heated to an even higher temperature (up to ~5400°F or 3000°C). The conversion of disordered carbon to graphite is known as *graphitization*.

One of the main benefits of carbon over glass for fiber reinforcement is its lower density, which reduces overall weight. Due to the more elaborate processing, carbon fiber costs significantly more, so uses are centered on high-demand applications where performance outweighs cost. The cost of carbon fiber has dropped significantly over the past few decades, however, and it is becoming more common in general, consumer applications. Additionally, carbon fibers can be treated in different ways or "sized" using materials for specific properties: bromine for electrical conductivity, titanium carbide for corrosion and abrasion resistance or polycarbonate as an adhesion promoter between the reinforcement and matrix.

Recent advancements in carbon materials have brought carbon *nanocomposites* to a greater prominence. Carbon nanotubes and nanofibers possess even higher strength than traditional fibers. That is because the strength of a composite is proportional to the fiber size, a concept discussed in Sec. 5.4.2. There is no doubt that nanocomposites will play an important role in advancing lightweight materials in the years to come. As with most new materials, cost reduction and manufacturing improvements will be critical in bringing them to market. Current applications served by CFRPs include aerospace (the Boeing 787 Dreamliner is approximately half composites), high-performance vehicles (Fig. 6.39), including lightweight electric vehicles, sporting goods, prosthetics, and many more.

6.4.1.3 Aramid Fiber

Aramid fibers, often referred to as Kevlar, are polymer fibers known best for their use in bulletproof vests (Fig. 6.40). These fibers consist of highly aligned polymer molecules, which provide exceptional strength at an even lower density than graphite fibers

FIGURE 6.39 Carbon fiber has an excellent combination of lightweight and high strength. It is relatively expensive, however, so most applications involve extreme conditions where the benefit outweighs the cost, such as the Koenigsegg Agera RS shown here, which has most of its structural components made from it, including the wheels. (*Courtesy of Koenigsegg Automotive.*)

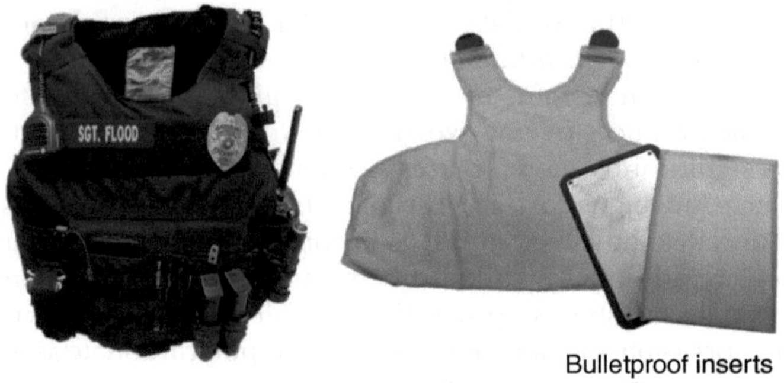

FIGURE 6.40 Aramid fibers, often referred to under the trade name Kevlar, are suitable for applications requiring high strength, heat protection, puncture protection, and cut resistance. Note that Aramid panels can be combined with steel plate to improve ballistic performance and reduce ricochet. When combined with a matrix polymer, these composites are very strong.

(the structure and properties of aramid are discussed more in Sec. 6.3.1.9). The Kevlar fibers come in different grades of varying strength and toughness. Kevlar composites find similar applications to carbon fiber but at even higher cost. Applications include high-performance vehicle components and high-end sporting goods such as tennis rackets, hockey sticks, and kayaks.

6.4.1.4 Hybrid Composites

Hybrid composites are simply mixtures of reinforcing phases and/or matrix phases. Just as composites can take many forms, hybrids can be in a wide variety of configurations. We will consider hybrids to share the most in common with polymer matrix composites, though that is not strictly required. For instance, with fiber reinforcement, two types of fiber can be used. The fibers can be within the same layer (intraply) or they can be stacked in distinct layers (interply). The materials do not have to be related either. Metal-matrix composites can be combined with polymer-matrix composites, fiber reinforcement can be used with particle reinforcement, and so on.

The rule-of-mixture performance (see Sec. 7.5.4.1) does not change. Just as varying the loading of reinforcement changes the balance between the matrix properties and the reinforcement properties, the balance between reinforcement phases will determine the final properties of the composite. For instance, carbon fibers are stiff but not good under impact, aramid fibers are good under impact but not as stiff. By combining the two into a single composite, the midpoint of those two materials can be achieved, or anywhere between based on the relative amounts used.

Glass fibers are relatively heavy, but they are low cost and good under compression. By mixing glass with aramid or carbon, the composite can have acceptable strength and remain affordable. Reinforcement can be metal, polymer or ceramic, fiber, particle, continuous, discontinuous, etc. There is unlimited potential for creating tailored materials by choosing matrix and reinforcement phases that represent important properties. The reality of cost and simplicity limit what combinations make sense. Hybrid composites are used in applications similar to other fiber-reinforced polymers, and some fiber-fiber hybrids are shown in Fig. 6.41.

Figure 6.41 Hybrid composites use multiple reinforcement phases and can combine a limitless variety of materials and configurations (e.g., fibers, particles, laminates, etc.). The fiber-fiber hybrids shown here combine Kevlar of varying color and carbon fibers, but the possible variations are endless.

6.4.1.5 Carbon Black

Carbon black is used as a particle reinforcing phase in polymers. Rubber is the major market for carbon black, specifically in the manufacture of tires. The carbon black particles help reduce wear and conduct heat, which helps the tire run cooler and last longer. The particles also strengthen the material, but the resulting properties are dependent on the type used and in what quantity. Carbon black is a by-product of petroleum production and in the form of nanoscale particles (20–350 nm). Smaller particles result in greater strength of the rubber composite. This reinforcement is used in belts (Fig. 6.42), hoses, engine mounts, and applications other than tires. The carbon, added in relatively large quantities (e.g., 30% by volume), gives the rubber composite a black coloration. For this reason, carbon black is also used as a pigment in paint, ink, and printer toner. For reinforced polymers that are not to be black in color, fumed silica can be used as an alternative. Carbon black can also be added to increase the electrical conductivity of polymers.

Figure 6.42 Carbon black is a low-cost powdered material (left) often employed as a pigment or reinforcement for polymers, such as in tires, hoses, and belts.

6.4.1.6 Engineered Wood Products

Perhaps the most common composites are those we do not normally associate with composites. Composites are often thought of as advanced materials with high-end, specialty applications. To be fair, that is partially true, but "simple" materials such as particle board are composites as well. Engineered woods are made from wood reinforcement with a polymer matrix. In most cases the polymer does not make up a majority of the composite. When you see composite wood, it looks more like wood than plastic.

Composite wood is typically created from the processing scrap of lumber. There are many varieties with standards governing their structure and properties. Sawdust, wood chips, and other wood debris are combined by a polymer adhesive to form a strong and stable product. The type of material and the way it is combined determines the category of engineered wood. Some commonalities include dimensional stability (they expand and contract much less than regular, solid lumber), and they offer mechanical isotropy so that orientation is not critical to performance. The high-binder content wears cutting tools faster than natural lumber and there are toxicity concerns associated with the binders (including formaldehyde).

Plywood is among the oldest and best known engineered wood products and consists of distinct layers of wood oriented at 45°, 90°, or some other angular variation so that when multiple layers are combined they act more uniformly in any, single direction. Plywood can be made from hardwoods or softwoods and can have veneer applied so that it is more aesthetically pleasing. This is especially important for applications such as furniture and cabinets.

Oriented strand board (OSB) is a variation on the concept behind plywood. Instead of using distinct, solid layers, or *plies*, OSB uses many small, rectangular wood chips that are oriented at varying angles to achieve greater uniformity. Unlike plywood, OSB is not as likely to have internal gaps and works well as a structural material for floors and wall sheathing. It typically costs a little less than plywood, but is not as stiff, is more sensitive to moisture (greater swelling), and does not hold fasteners quite as well.

Medium-density fiberboard (MDF) is a composite created by breaking down wood into fibers and then compressing those fibers at high pressure and temperature with the addition of a binder. Although MDF consists of fibers, like natural lumber, it performs like plywood. MDF has a higher density than other composite lumbers but works well and is stable. MDF, thanks to its uniformity, is a good substrate for decorative veneers and is relatively inexpensive. It also holds detail well with much less chipping or tear-out as can occur in other composite woods. It is used in cabinet construction, fabrication of woodworking jigs, and in general, paint-grade construction. It can be made moisture- and fire-resistant as well.

Particle board, or chipboard, is popular for low-cost furniture construction. It is a relative of MDF since it is a fine distribution of wood product, which results in higher uniformity than plywood or OSB. It is very low cost but performance is not especially good. Particle board is highly sensitive to moisture and will expand irreversibly if made wet. It is relatively weak, though it is suitable for indoor, light-duty construction. Particle board is nearly always finished with a waterproof veneer to improve durability and aesthetics.

Wood-plastic composites (composite lumber) are created by the combination of thermoplastics, wood flour, and other fillers such as stabilizers and colorants. These are most commonly seen in decking products such as Trex. These "wood" products

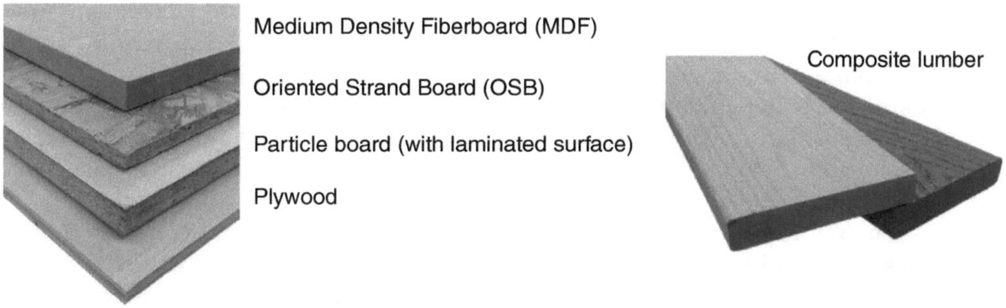

Figure 6.43 Engineered lumber can take many forms, all of which are polymer-based composites with wood. The size, orientation, and quantity of wood used will determine the applications the composite is used in. Each type shown here is described in the text.

are highly constituted by the polymer binder (~50%) so they display weathering characteristics such as rot and insect resistance similar to plastics. This gives them superior durability and longevity as well as color retention when compared to solid wood.

The composite boards are patterned to resemble natural wood and can be colored traditionally or any other appearance desired. They are highly formable and easily worked using traditional tooling. The composite boards are weaker, however, and not entirely maintenance free. They also raise concerns about fire hazards since they can burn hotter and melt. Wood-plastic composites can be environmentally friendly if recycled materials are used, but they themselves are more difficult to recycle unless being remade into composite boards. The various types of composite woor products are shown in Fig. 6.43.

6.4.2 Ceramic Matrix Composites

Ceramics are characteristically brittle, but they are often very strong and capable of operating at high temperature without loss of performance. Most ceramic matrix composites (CMCs) are designed to increase toughness, but depending on the additive, it might help reduce cost as well. Some CMCs are very common, such as concrete, but others are quite rare and expensive, and only applied under the most demanding of circumstances. Perhaps somewhat surprisingly, CMCs often employ other ceramics as reinforcement, but the composition and structure of the reinforcement can dramatically change the working properties of the composite.

6.4.2.1 Concrete

Concrete is one of the most ubiquitous materials, but it is rarely considered among advanced materials or as a composite. The reasons for this are partly due to its familiarity (which breeds contempt, or so the saying goes) and partly due to its appearance. Carbon fiber, for example, is often used without being painted over, just for the visual impact. In some applications, the only true selling point is the appearance (think of some cell phone cases, jewelry and accessories). Concrete does not get the attention because it is used in more mundane applications, but those applications, and the properties that allow them, are exceptionally important.

Concrete can vary in composition. The ceramic matrix material (cement) most often contains more than one reinforcement phase, making concrete a hybrid composite. It may contain any range of reinforcement phases, including gravel, sand, and steel bar. The recipe is dependent on final application, but each phase plays a role to improve strength or reduce cost. Cement is relatively expensive and energy-intensive to make (discussed in Sec. 6.2.1.2), so adding gravel and sand (aggregate) can reduce the cost per cubic foot.

The reinforcement also reduces the likelihood of catastrophic failure due to rapid crack propagation. Ceramics, and any brittle material, fracture readily because cracks are not resisted once they form. Particle reinforcement acts to block a crack from passing through the material as easily. Cement alone is poor under tension. To avoid cracks altogether, cement is used in compression, for which it is very good. When cement goes from being in compression to tension, such as in bending, the tensile force is likely to cause it to crack and fail. By adding steel reinforcement bars, or *rebar* (Fig. 6.44), the steel can help resist tension and stop the structure from failing suddenly. Even if the structure begins to fail, the rebar will help hold it together long enough to vacate or condemn the building, bridge, etc.

Concrete can contain multiple particle reinforcing phases and steel bar. The steel bar is similar to "fiber" reinforcement being that it is much longer than it is wide, and the orientation remains important. For concrete pads it is often placed in a grid pattern, and for other structures it may be placed asymmetrically with more rebar in areas prone to tensile force. The rebar can be prestressed when set in the concrete to add even more crack resistance. This process involves applying a tensile force to the rebar that is later removed, after the concrete is set. The steel then tries to shorten, thereby applying a compressive stress.

Not all concrete is created equal. Low strength concrete can be lightweight and inexpensive, but applications requiring large spans or high loads, as in skyscrapers, require high strength formulations. The strength of concrete is usually expressed in units of pressure it can withstand [psi (MPa)], with a range from about 2000 psi (14 MPa)

Figure 6.44 Concrete is a mixture of reinforcement phases. Aggregate particles and reinforcing bars help reduce cost and increase durability, respectively. (*Photo by U.S. Navy, CC BY 2.0.*)

to over 22,000 psi (150 MPa). Concrete strengthens with time due to the hydration of its constituents, which locks the material particles together. While curing can complete quickly (days), hardening may persist for years.

Corrosion can be a problem with steel reinforcement. As the metal corrodes, it expands and can crack the concrete. Other materials such as polymers, basalt, and composite reinforcement bars such as fiberglass can also be used which do not readily corrode and are particularly beneficial for high-moisture environments or those where corrosive chemical are more likely to be encountered.

6.4.2.2 Whisker-Reinforced Ceramics

Whisker-reinforced ceramics are often used in applications where high hardness and wear resistance need to be paired with toughness. Materials tend to lose their impact strength as their hardness increases, but a balance can be achieved by adding reinforcement. Whiskers are essentially fibers with small dimensions. They are thin (0.1–1 μm) in diameter and short (less than 100 μm long). Their small size reduces the likelihood of defects, so their strength is often very high. When reinforcing ceramics, whiskers act to prevent cracks from propagating by blocking their progression and bridging across the fracture.

A common example of a whisker-reinforced ceramic is alumina reinforced with silicon carbide (SiC) whiskers. The concentration of SiC is typically 25% or less, but improvements in strength and toughness of 50% are achieved in the composite. The thermal shock performance is improved as well. These properties make the alumina-SiC composites excellent for applications in cutting tools where temperature can fluctuate and high stresses and wear occur. The use of these composites can increase metal removal rates by as much as 10 times that is achievable by carbide on difficult metals such as superalloys. The reactivity of SiC and iron make this material unsuitable for machining ferrous alloys.

6.4.2.3 Carbon–Carbon

Carbon-carbon composites consist of the carbon fiber reinforcement of a carbon matrix. The production of such composites often starts with polymer matrix composites containing carbon fibers (Sec. 6.4.1.2). The polymer matrix is carbonized by pyrolyzing the polymer composite until any non-carbon elements are burned off. The resulting matrix may be amorphous and glassy, or graphitic if further heated to very high temperature [4532°F (2500°C) or above]. Depending on processing, the composite can be porous or dense. Obviously, there is large variation within this material type.

Porous carbon-carbon composites have entrained voids that make it an excellent thermal insulator. Very short fibers (a few millimeters at most) are used in the formation of these composites. It may have as high as 90% porosity, which weakens it mechanically. Temperatures above 752°F (400°C) can only be withstood in vacuum or inert atmospheres. The carbon will react with oxygen in the air to form carbon dioxide otherwise. *Dense carbon-carbon composites* can use short, chopped fibers or long continuous fibers. Typical thermoset polymer matrix materials such as phenolics can be used, as can any liquid which will infiltrate and coat the fibers and also yield sufficient carbon upon heating.

Multiple cycles of infiltration, polymerization, and carbonization are used to create a fully dense composite. To avoid damage to the material, gas must be released slowly during the production process which can take days, and even though they use relatively

FIGURE 6.45 Carbon-carbon composites can vary widely in their properties, but dense, graphitic versions are very strong, even at high temperature, which makes them suitable for extreme applications. A familiar example is in the construction of the leading edges of the Space Shuttle. (*Photo by NASA Goddard Space Flight Center, CC BY 2.0.*)

low-cost materials, carbon-carbon composites are very expensive. The dense composite can be employed in high-temperature and high-stress conditions such as missile components, heat shielding (Fig. 6.45), and brake disks for high-performance vehicles and airplanes.

6.4.2.4 Carbon-Ceramic

Carbon-ceramic composites are ambiguously titled, but most often this title is used to denote a SiC-reinforced carbon composite. This particular composite shares applications with carbon-carbon composites discussed in the previous section, but it is typically associated with somewhat higher abrasion resistance and does not suffer from a lowering of the friction coefficient in moist conditions like carbon-carbon composites, which makes carbon-ceramic better as a brake disk material (Fig. 6.46). Benefits of carbon-ceramic composites over metals also include lower mass, lower thermal expansion, and a higher rigidity at elevated temperature.

SiC-reinforced SiC is also used in the similar applications, but amorphous SiC fibers lose strength at lower temperature than carbon fibers. Like carbon-carbon composites,

Figure 6.46 Carbon-ceramic composites are used in extreme conditions of temperature and stress. While typical consumer applications are rare, brake discs are one of the most familiar. These composites are only used commercially on high-performance vehicles, race cars, and some aircraft. (*Courtesy of Koenigsegg Automotive.*)

SiC-carbon composites can be created from a fiber-reinforced, Si-based polymer as a precursor that is heated to drive off unwanted elements. Other methods tend to produce higher quality however.

6.4.3 Laminates

Laminates, or laminar composites, are created by layering different materials. The materials may be different in regard to composition and/or structure. The basic principles were covered in Sec. 5.4.4, and a common example, plywood, was discussed in Sec. 6.4.1.6. Plywood is a laminar composite where the composition is often the same for each layer, but the orientation (structure) is varied. Also, plywood is often finished on one or both faces with a decorative veneer with a different composition. Laminar composites, then, are created for aesthetics as well as performance. While there are many laminar composites, most seek to increase strength, reduce weight, create a protective barrier, or create some combination of these and other properties.

A more "exotic" example of a laminate is often referred to as a *sandwich structure* (or *sandwich panel*). Sandwich composites (or sandwich panels) have a core with significantly different properties than the face sheets. Often the structure is similar to a honeycomb where a series of hexagonal voids are oriented with the openings perpendicular to the face of the sheets (Fig. 6.47). This specific type is known as a *honeycomb sandwich structure*. The honeycomb structure is efficient at carrying load for its weight (high specific strength) and the face sheets stabilize the core while providing a barrier to wind, rain, and other outside influence.

Sandwich panels like this may be made out of any material or combination of materials, and the core does not have to have a honeycomb structure to be a sandwich composite. A common example is corrugated cardboard. The cardboard interior is wavy, and the result is a much stiffer material without excessive weight. The interior may be porous (common for insulation) or have any other functional or structural properties. Depending on the composition of the sandwich panel, glue, brazing, or other methods will be used to combine the face sheets and the core. Sandwich panels

FIGURE 6.47 Laminate composites can incorporate any material or combination of materials. Laminates can be solid, such as the familiar material, plywood (see left image), but a core material can be held between face sheets to reduce weight and maintain strength. The core may be patterned after a honeycomb (center) or other configuration which maximizes rigidity. This is demonstrated practically in corrugated cardboard (right).

are used in many applications from aerospace to packaging as they offer a customizable balance of strength, weight, and cost.

6.5 Why It Matters

Knowing the options available is a prerequisite to making the best decision in materials selection. Nonmetallic materials are so broad in their composition and structure that we cannot possibly be explicit in listing them, but some of the more common materials discussed here will get you started. Ceramics are sometimes too quickly discounted based on their low toughness and polymers for their low strength. Knowing there are exceptions to the rules makes it easier to find the right material for the job, even if you would not have thought it should be a good fit.

A couple examples of nonmetallic materials with surprising properties are partially stabilized zirconia (PSZ) and aramid fibers. You would not normally associate a ceramic with being tough, but thanks to clever materials design, specifically the transformation toughening phenomena, PSZ can be used in a wider range of applications than most ceramics. Also, you probably would not reach for a polymer as a first defense against bullets, but aramid fibers, better known as Kevlar, are the standard in personal protection for police and military. The diversity of nonmetallics allows us to balance cost, strength, weight, and many other performance factors.

Adding to our ability to customize materials for an application are composites. We can intentionally and exactly design complex materials by combining those with dissimilar properties. Strong, brittle fibers can be paired with a soft, flexible matrix to create a lightweight and strong material with properties between the two constituents. We can mix materials, but we can also layer them to create laminar composites and sandwich panels.

It matters because it is complicated. There are so many options when it comes to nonmetallics that we need to understand the small, but important, differences. Even within polymers, and even within a single composition of polymer, the properties can vary widely and the applications will too. Polyethylene is an excellent example of this. By changing the branching structure or chain length, the same two elements can combine to form a clear, flexible plastic (LDPE), or a very stiff one with a very low coefficient of friction (UHMWPE).

Likewise, "small" alterations to the monomer structure give us entirely different properties such as those displayed by PTFE or PVC. Depending on the application, you may be able to get acceptable properties at a much lower cost by knowing the differences. For instance, we saw that, in general, engineering polymers have a higher strength and service temperature than commodity polymers, but their cost and processing complexity is also higher.

By modifying the cross-linking of polymers, we can get a very rigid strong plastic or a very weak and stretchable elastomer. Elastomers are closely related to the polymers they are based on, and knowing the service environment is important. For conditions where oil is encountered, you would want to use nitrile rubber (Buna-N), not butadiene (Buna-S) as it does not degrade due to the exposure. That sort of decision can save a lot of money down the road.

That is why understanding nonmetallics is important, because seemingly similar materials can have profoundly different properties. Nonmetallic materials are among the most advanced available. When it comes to combining high strength and lightweight, they are largely unrivaled. This is especially true for high temperature applications. A small sampling of the nonmetallic materials in a supercar, where much of the engineering eventually trickles down to general consumer vehicles, is shown in Fig. 6.48. Keep in mind, just about every material was "advanced" at one point.

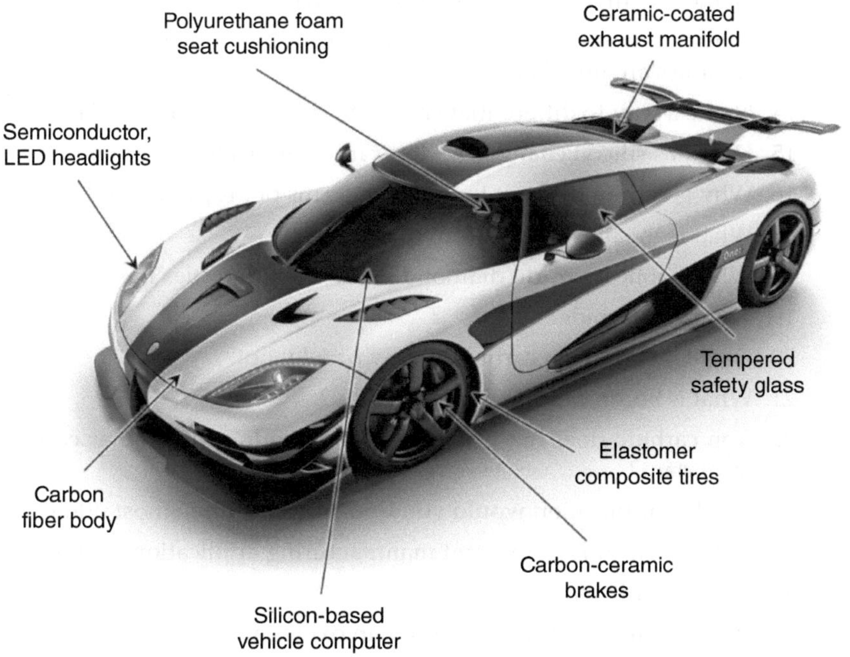

Figure 6.48 The advanced applications of nonmetallic materials, such as in this Koenigsegg One:1 supercar, are eventually incorporated into lower cost vehicles once the materials and manufacturing are optimized for large-scale production. Although it may take decades, advanced nonmetallics are now being employed in mainstream applications thanks to initial use and testing in more "exotic" circumstances. (*Courtesy of Koenigsegg Automotive.*)

6.6 The How and Why of Nonmetallic Materials: Identification and Application

6.6.1 Chapter Review

1. What types of bonding do nonmetallic materials possess?
2. How are engineering ceramics different than naturally occurring ceramics?
3. What are oxides?
4. In what manufacturing processes would aluminum oxide be found?
5. How is aluminum oxide used in decorative applications?
6. What is the most common application of calcium oxide?
7. Where might you find silicon oxide in nature? What about in engineered products?
8. What applications does quartz have? How does it depend on crystallinity?
9. What are silicates?
10. Why is titanium dioxide such a popular additive to products like paint?
11. Zinc oxide is thermochromic. What does that mean?
12. Name the three crystal structures of zirconium oxide.
13. Why can zirconium oxide be used as a cutting implement, like a knife, but other ceramics are unsuitable?
14. Name a way to differentiate between cubic zirconia and diamond.
15. How has silicon changed society? In what products can you find it?
16. Why is carbon difficult to define in terms of properties and structure?
17. What is the most stable form of carbon?
18. What applications do diamond's extreme properties make it suitable for?
19. What is a carbide?
20. Do all metals form carbides? Why or why not?
21. What is boron carbide used for? Why?
22. Iron carbide does not get used in end-products like other carbides. Why is it important?
23. In what application would you find silicon carbide most often?
24. What is the most important manufacturing application for tungsten carbide?
25. What is a nitride?
26. Why are nitrides less common than oxides and carbides?
27. What applications does cubic boron nitride appear in?
28. Gallium nitride is an important semiconductor used in what applications?
29. How is silicon nitride used in microprocessor design?
30. Why is titanium nitride used as a tool coating? How can the coating's properties be altered?

31. Why have polymers replaced metals, ceramics, and wood in many applications?
32. What is the primary difference between high-density, low-density, and linear low-density polyethylene?
33. Why is ultrahigh molecular weight polyethylene processed so differently from other polyethylenes?
34. What are some benefits of polypropylene over polyethylene?
35. Polyvinyl chloride is very rigid or very flexible. How is this achieved, and what applications fall under each type of material?
36. Why can polyvinyl chloride be dangerous if overheated?
37. Why is polystyrene highly amorphous?
38. What is expanded polystyrene, and what application is it used in?
39. The most commonly used polymer in consumer applications is polyester terephthalate. Why is it so common?
40. In what specialty application is polylactic acid popular?
41. How can the properties of acrylonitrile butadiene styrene be controlled?
42. What is most distinct characteristic of fluoropolymers like polytetrafluoroethylene?
43. Why is polymethylmethacrylate clear? How does this affect applications?
44. How do cyanoacrylates harden?
45. What are some applications for nylons (polyamides)?
46. Why should high moisture environments be avoided with polyamides?
47. What is another, more common name for aramid? What applications is this polymer used in?
48. What are some benefits of acetals over polyamides?
49. What is another, popular name for polycarbonate?
50. Why can the by-products and residual presence of monomers in polycarbonate be concerning?
51. What are some applications of phenolics?
52. What is the chemical difference between thermoplastic and thermoset polyesters?
53. Where are epoxies used?
54. Polyurethanes can be found in very rigid or very flexible form. How is this possible?
55. What is the structural difference between gutta-percha and Hevea rubber?
56. What is the primary advantage of synthetic rubber over natural rubber (polyisoprene)?
57. Where is polybutadiene often applied? How does it perform in contact with nonpolar solvents like oil?
58. What is benefit of nitrile butadiene rubber over regular butadiene rubber?
59. What is an olefin polymer? What elastomeric applications do they serve in?

60. Why are fluoroelastomers used even though they are costly?
61. What is fundamentally different about the structure of silicone?
62. In what applications are polysiloxanes used?
63. What benefits are there to engineering polymers over commodity polymers? What are the disadvantages?
64. How can polymer products often be identified?
65. Why is identifying the type of polymer relatively simple, but the exact ingredients may still not be known or difficult to recycle?
66. In what products is fiberglass used?
67. How do the glass fibers affect the properties of the fiberglass composite?
68. What is the main benefit of carbon fibers over fiberglass?
69. How are the carbon fibers treated for specific applications before being incorporated into the composite?
70. What are aramid composites used for?
71. What is a hybrid composite? Why is it beneficial?
72. What applications are common to find carbon black in?
73. Why are engineered wood products considered composites?
74. List differences in applications and properties of engineered wood products.
75. What is most common ceramic matrix composite? How is this often a hybrid composite?
76. What is main purpose of whiskers in ceramic matrix composites?
77. How can carbon and carbon combined together be a composite?
78. Where are carbon-carbon composites used?
79. What ceramic is most common in carbon-ceramic composites?
80. Why are laminates so versatile?
81. Laminates can be very basic and common or very high-tech. How is that possible?
82. How do engineered wood products and laminates overlap?
83. What is meant by a sandwich structure when referring to laminates?

6.6.2 Concept Connections and Extensions

1. Graphite is the most stable form of carbon. So, why do we not worry about diamonds transforming to graphite? Use external resources to identify the underlying principles.
2. Why do you think oxides are the most common version of ceramics? Consider the question from a nature perspective and processing perspective.
3. Cement is the most widely used material in construction but very energy intensive. Is there any alternative material? Explain your answer.
4. Look at the product ingredient label of various items you have at home. Is titanium dioxide included? Any other ceramic material?

5. You have probably seen packets of "silica gel" included in shoe boxes, purses or other bags, and in other packaging. What is the purpose of these packets and how do they achieve it?
6. Zirconium oxide is used in making knives. Describe how it is prevented from breaking based on the discussion in Sec. 7.5.2.1.
7. Why are so many plastics clear? Use the discussion in Sec. 5.3.2.7 to justify your answer.
8. Many polymers are modified using chemicals that are known or suspected of being harmful to humans. What challenges are there to eliminating the use of these chemicals, and how can manufacturers play a "shell game" by switching to other additives that are less known, but also are not guaranteed to be safe?
9. What barriers are there to recycling plastics? Include aspects of labeling and their use in products.
10. Describe how tires include various nonmetallic materials. Why are they so hard to recycle (consider both their properties and their composition).
11. Describe the difference between concrete and cement. Use the discussion in Sec. 10.3.1 to determine the beneficial role of the second phase. What other benefit(s) does that phase have to the final product?
12. Lightweight composites like sandwich panels and carbon fiber are limited to select, and often expensive, applications. What steps might be taken to make them more common?

CHAPTER 7

Mechanical Concepts in Processing

We intuitively understand much about how materials respond to force. Our personal experience and observations have given us a "feel" for what we will discuss in this chapter. The problem is that processing and applications are not so much "touchy feely" as they are quantitative. The general relationships between materials and forces can be understood from a qualitative (no numbers) point of view, and there is merit to knowing approximate behavior. Would you expect an engineer to build a bridge by "eyeballing" it? Do you want to drive a car that has engine components made from what is "probably" a suitable material? What about a plane that was assembled to specifications that "felt" right? Obviously these circumstances are unacceptable in most consumer applications. Engineers and manufacturers have precision measuring devices, analysis software, torque specs, certified compositions, and tolerances they utilize to ensure that each and every part, as well as their assembly, is suitable for its intended application. In this chapter, we will look at some of the most fundamental equations that describe forces and a material's response to those forces. This will allow us to develop baseline analysis tools that help ensure our design is right.

7.1 How It Works

Processing changes a material. The change may be physical or chemical. It may involve the shape, color, hardness, etc. There are many possibilities. That change may be achieved through a variety of means including chemical, thermal, or mechanical techniques, and many processes involve a combination of these. Mechanical forces are used extensively in processing, but they are also ubiquitous in product applications. Specifically, we use force to create products like a pry bar (Fig. 7.1), for example, but that pry bar is going to have force applied to it in its end use as well. Both aspects are important. Throughout this book we will look from three perspectives at this concept. These three are the material perspective (part being processed), the tooling perspective (the equipment doing the processing), and the application perspective (how the product is used after manufacturing). The mechanical aspects of materials and manufacturing are vital to understand and optimize processing.

We developed an understanding of materials through examining their structures in Chaps. 3 and 5. The applications of these materials were discussed in Chaps. 4 and 6. From that we get a fuller view of how structure affects properties and how that, in turn,

FIGURE 7.1 Most products will experience two distinct sets of forces: those which occur during production and those which occur during use. The material used determines the processing, but the intended application determines the material. This crow bar was bent and formed during production, and it will experience impact, bending, and torsion during use.

affects the suitable applications. The end use is not the only aspect we need to consider for materials. The way materials are processed will also vary greatly based on their properties as well. The processing will be described in detail starting in Chap. 9 and throughout the rest of the text, but it will benefit to have somewhat of a background in how processing applies force to materials and why that dictates appropriate choices in manufacturing.

7.2 Forces in Materials and Manufacturing

Most processes require some force be applied to the part being manufactured, but the object applying that force (the tooling) is also affected. Consider a hammer (Fig. 7.2). In its simplest form, a hammer will consist of a head and a handle. Often the handle is made from a different material than the head. While the head is a metal, the handle may be

FIGURE 7.2 Product design requires judicious use of materials. Even something simple, like the hammer pictured here, involves many considerations of force including the magnitude and rate of loading. The head must be able to withstand impact, while the handle must reduce vibration.

wood, plastic, or fiberglass. Why? Which is better? What is the purpose? Is it worth the extra cost? What metal is the head made of? Does it matter? Again, what about cost?

These are all very general questions, and many are intuitive. A hammer head is metal because it is used to hit hard objects and it must be durable to withstand that use. The handle does not require that level of abuse, so it can be lighter and less resistant to impact. There is much more behind even this simple example. The bigger picture we must consider includes the processing of these components. The material is also very important in processing. Just as material properties will determine applications, the material's properties will determine processing as well. We will use mechanical concepts to describe a material's response to applied force, and that will guide us in choosing our methods.

7.2.1 Terminology in Loading

Before we can go too far with describing what forces are present during processing, we need to establish a common vocabulary. This includes the way forces are applied externally and how they arise internally. There are five basic ways in which forces can be applied:

1. Tension
2. Compression
3. Bending
4. Torsion
5. Shear

Tension involves the application of forces that are collinear (in line with one another) and that are in opposite directions, away from one another. The effect on the material is that of stretching. For example, if you were to climb a rope, the rope would be under tension. **Compression** is essentially the same as tension, but forces are applied toward one another. The effect is to compress a material (as the name would imply). Perhaps you are sitting now. The chair/cushion you are on is under compression. Tension and compression can be applied to a material/component simultaneously. This arises during bending. Tension, compression and bending are shown in Fig. 7.3.

Bending occurs when forces are directed at each other, like compression, but they are offset, not collinear. If you have a pencil nearby, you can bend it by placing your index fingers at opposite ends and pushing with your thumb in the center. This is "three-point" bending, but other bending methods exist as well. The outer arc of the bend will be under tension while the inner arc is under compression. During bending, the part is essentially being lengthened on one side and shortened on the other. How are opposite sides under opposite load? There has to be a point where the stress changes from compression to tension, and at that point the stress is zero. This is referred to as the **neutral axis** and is a key concept in structural design and analysis. In symmetric cross sections it is in the middle. Calculating the location of the neutral axis is beyond our needs here, but any text on solid mechanics will discuss it in depth.

Torsion is related to rotation, but torsion requires no rotational motion. Instead, the *tendency* to rotate or twist is the source of torsion. If you still have that pencil around, you can easily spin it in your fingers. When you do that, there is effectively no torsion being applied to the pencil during that motion. Torsion can only develop if the object is

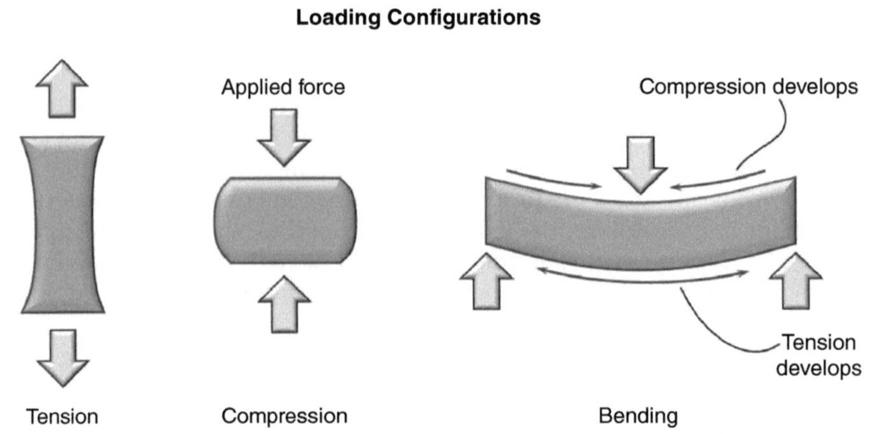

FIGURE 7.3 The result of loading a material in tension and compression are shown here. Bending, which occurs for noncollinear forces, results in a combination of tension and compression.

constrained. That is, the object must be held still under the applied force at some point. If you grab that pencil tightly in one hand and then try to twist it with the other, torsional force will develop. Your attempt to twist it (applied force) is resisted by the constraint (your other hand). This is similar to "wringing out" a wet rag.

The end does not necessarily have to be rigidly held, and the constraint could come from inertia, or the resistance to change in motion.* A car's wheels are not rigidly held during acceleration, but torsion develops in the drivetrain because the force being applied from the engine occurs faster than the forward velocity. Whenever there is a resistance to rotation, whether it comes from friction, inertia, or rigid constraint, torsional forces are present (Fig. 7.4).

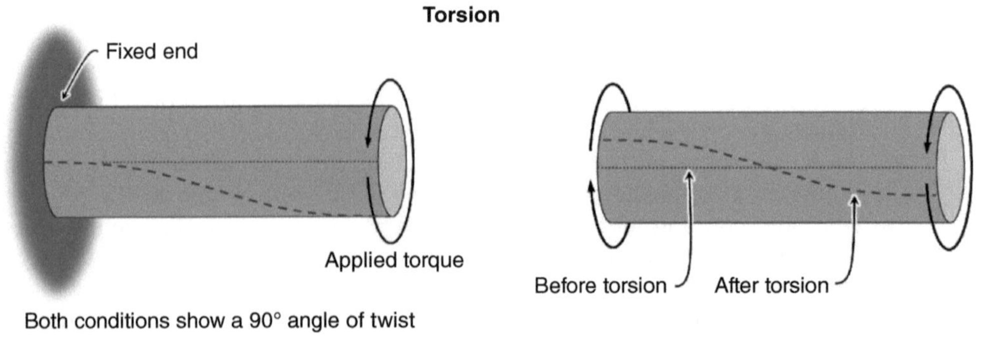

FIGURE 7.4 Torsional force develops when a rotational force (torque) is applied. The above illustration shows that the component twists under load whether it is fixed at one end or an opposite torque is applied simultaneously. The fixed end of the component on the left would develop an opposite torque due to the requirement of equilibrium, so the two situations are actually identical in theory.

*This is technically rotational inertia, usually called the *moment of inertia*.

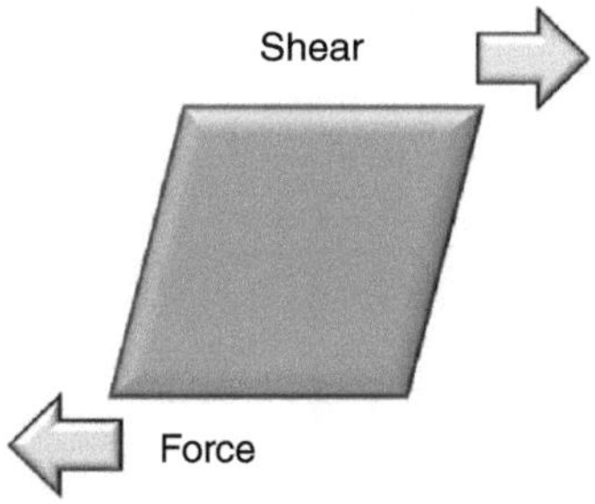

Figure 7.5 Shear loading is caused by forces which are offset but very close together, not like in bending. Shear develops as a result of other forces and is the primary mechanism by which materials deform.

Shear is similar in ways to to bending. Like bending, shear involves offset forces in opposition. The forces are so close together, however, that there is no bending (see Fig. 7.5). This causes the material to "slide" apart, something like a deck of cards being pushed at the top. The surface of each card would represent a *shear plane* and the motion of the card would indicate it is an active shear plane. Look back at Sec. 3.2.2, especially Fig. 3.4. The shear planes in crystals are called the slip planes and are the basis for all forming operations and many other processes that change the shape of a part. Polymers, which do not have a well-defined atomic crystal structure, also deform through shear, but it is related to the sliding of molecules, not planes of atoms. Shear forces arise within materials during nearly all loading and is the basis for plastic deformation. As it turns out, this is a very important concept.

7.3 Basic Mechanics

In the study of solid mechanics, there are two ways stationary structures and components are analyzed with regard to force: statics and strength of materials. Neither allows for motion in the analysis, but the findings can often be applied to understand performance in situations where motion may occur. The major assumptions for both are that forces are at equilibrium and no straight-line movement (translation) or rotation occurs. **Statics** describes force distribution throughout a structure or component assuming it is perfectly rigid, but **strength of materials** allows for the structure or component to deform (Fig. 7.6). All processes seek to change a material's size, shape, or other properties (i.e., deformation). Although statics is an important basis for strength of materials, it does not allow for material-specific parameters. Instead, it assumes structures are made from perfectly rigid materials that are infinitely strong. Unless a major loophole in the laws of Physics occurs after this book is in your hands, I will say with conviction that no material has those properties.

We will primarily concern ourselves with the strength of materials since our aim is to understand processing, which is material dependent. Our treatment will also be very fundamental. Basic quantitative and qualitative descriptions will be used to generalize processing effects. That is not to say the coverage will be unimportant. Quite the opposite, the basic concepts and analysis techniques are most universally applied and are essential to more complex treatments. There are a variety of excellent texts on statics and strength of materials for the reader interested in more advanced topics.

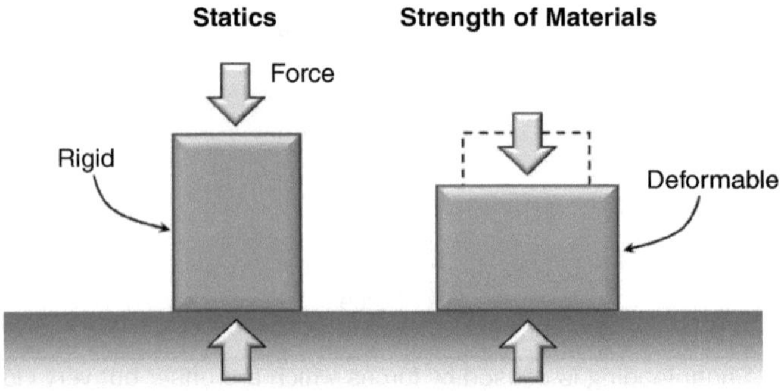

FIGURE 7.6 The study mechanics can be applied to perfectly rigid materials (statics), but the response of materials to force is essential to designing processes and predicting performance in applications. The deformation of materials is accounted for by the study of strength of materials.

7.3.1 Axial Stress and Strain

Stress and strain describe physical phenomena that act on a part. **Stress** is the force per unit area applied to an object. Stress develops regardless of size, shape, or composition of the object, and the force can be applied in any manner such as tension, compression, or torsion.* That means two different parts with different geometries can be under the same stress at the same applied force. The reason stress is not sensitive to shape or composition is because it simply relates force to the area the force is applied to. The equation for axial stress is

$$\sigma = \frac{\text{force}}{\text{area}} = \frac{F}{A} \tag{7.1}$$

where σ (lowercase Greek letter sigma) is the stress, F is the applied force (lb or N), and A is the cross-sectional area (in^2 or m^2). By assessing the units, you can see that stress is a measure of pressure [psi or Pa (Pascal)]. If you increase the value of force (numerator), the stress increases. Likewise, by decreasing the cross-sectional area (denominator), you will also increase the stress (Fig. 7.7). This is an important concept since the same amount of force can have a dramatically different effect on a material depending on the size of the object. Think of bolts, for example. If two bolts, one being 0.5 in. in diameter and the other 0.25 in, are tightened to the same tension, the smaller bolt will have much more stress on it. How much? To find out stress, we use Eq. (7.1). For a bolt, the cross-sectional area would be given as the area of a circle (i.e., πr^2). The radius for the two would be 0.25 and 0.125 in, respectively. The ratio of the stress can be given by keeping all other factors constant.

$$\sigma = \frac{F}{A}, \; A = \pi r^2$$

*Certain methods of applying stress have specific names, but the way stress develops follows the same principle (e.g., shear stress is caused by force that results in shear).

½ in bolt:

$$\sigma_1 = \frac{F}{\pi r_1^2} = \frac{F}{\pi(0.25\text{ in})^2} = \frac{1}{0.0625}\left(\frac{F}{\pi}\right) = 16\left(\frac{F}{\pi}\right) \tag{7.2}$$

¼ in bolt:

$$\sigma_2 = \frac{F}{\pi r_2^2} = \frac{F}{\pi(0.125\text{ in})^2} = \frac{1}{0.015625}\left(\frac{F}{\pi}\right) = 64\left(\frac{F}{\pi}\right) \tag{7.3}$$

$$\sigma_1 = \frac{1}{4}\sigma_2$$

You can see by the analysis above that reducing the diameter by half increases the stress by a factor of 4. Alternatively, if the smaller bolt is tensioned to one-quarter the value of the larger, they will be under equal stress. Stress is used to measure the strength of materials. The stress level at which they permanently deform is called the **yield stress**. This will be discussed in more detail in Sec. 7.4.1.

Strain is a measure of deformation. The amount of strain an object is under is measured by its change in dimension. If you stretch a rubber band, you are taking it from the unstrained (equilibrium) size to a different size under load. Strain is relative. Unlike stress, which can be calculated at any point at any time, strain measures a difference in two conditions, so at least two conditions must be known. Specifically, strain compares an initial length to a final length after a change in load. There must be two points of reference to determine strain. The equation takes the following form:

$$\varepsilon = \frac{L_f - L}{L} = \frac{\delta}{L} \tag{7.4}$$

where ε (lowercase Greek letter epsilon) is the strain, L is the initial length, L_f is the final length, and δ (lowercase Greek letter delta) is the change in length (i.e., $L_f - L$). Note that

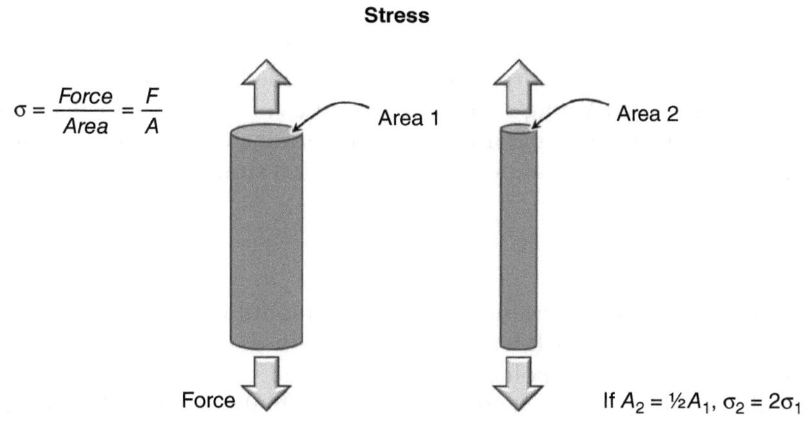

FIGURE 7.7 Stress only depends on the amount of force applied and the area over which that force acts. The response to stress is a material factor, and it is the basis for processing parameters and application requirements.

strain is unitless. The calculation involves a length (inches or meters) over a length (also inches or meters), so they cancel. As long as you are consistent with units, strain will simply be a number, and often a small number. When the decimal value is multiplied by 100, it represents the percent change in dimension. Therefore, a 0.002 strain value corresponds to a 0.2% strain and a 0.74 strain decimal is 74% strain. Although some polymers may exceed 100% strain, it is rare for metals to enter this "superplastic" regime. Most ceramics, being brittle, have very low values of strain at fracture and will not retain that strain permanently.

We will look more at how strain enters into the material performance in Sec. 7.4. For now, take the example of a polymer foam seat cushion. Let us say the cushion is 2 in thick and compresses by 1.625 in when someone sits on it. What is the strain?

$$\varepsilon = \frac{L_f - L}{L} = \frac{(2\text{ in} - 1.625\text{ in}) - 2\text{ in}}{2\text{ in}} = \frac{0.375\text{ in} - 2\text{ in}}{2\text{ in}}$$
$$= \frac{-1.625\text{ in}}{2\text{ in}} = -0.8125 \text{ or } -81.25\% \tag{7.5}$$

There are a couple of important notes to make about the findings of Eq. (7.5). First, we knew the amount the cushion was compressed from the beginning, and that *was* the change in length, δ. We did not need to find the final length, but doing so led us to another important point: strain is negative for compression, or where the length is reduced under load. Doing the math proved this out for us, but we could have gone directly to the final division of −1.625 in over 2 in if we knew that compression was negative. Finally, this gives us a decimal value of 0.8125, which when multiplied by 100, gives us the equivalent percentage of 81.25.

Is this a large value? Without more information, we cannot be sure. That is a large amount of strain, but a cushion is *designed* to compress, so it is not necessarily too high. We also do not know the amount of weight on the cushion. We cannot be sure if it is the appropriate stiffness. Both stress and strain are only understood by the effects they have on materials.

The way a material responds to stress determines the amount of strain. Compare the polymer foam cushion we just looked at to a solid wood seat. The wood will give very little, and for that reason it is considered much stiffer. That is, for the same stress, it exhibits less strain. This relationship is described by the **elastic modulus** (E) of the material, also called Young's modulus, modulus of elasticity, elasticity, or the stiffness. More about the material parameters that determine this is discussed in Sec. 7.4.2. The modulus of elasticity equation relating stress and strain is

$$E = \frac{\sigma}{\varepsilon} \tag{7.6}$$

To demonstrate the importance of the elastic modulus of a material, consider the ¼-in bolt example in Eq. (7.3). It was found that a ¼-in-diameter bolt would experience four times the stress of a ½-in bolt under the same load, and the stress would be 64 times the force divided by pi. If we set the force to 1500 lb, the stress will be 30,558 lb/in^2 (psi). That may seem like a large number, but we can only determine the effect it will have if we know the stiffness of the material. The elastic modulus is different for each material.

If we pick steel for the bolt material, its elastic modulus is 29×10^6 psi. If we rearrange Eq. (7.6), we can solve for the strain in this example.

$$\varepsilon = \frac{\sigma}{E} = \frac{30{,}558 \text{ psi}}{29 \times 10^6 \text{ psi}} = 0.00105 \text{ or} \sim 0.1\% \tag{7.7}$$

Even though the stress seems large in the example above, it only results in a 0.1% strain. If we put 1500 lb on that 1 in-long, ¼-in-diameter bolt, we stretch it by about 0.001 in. If the bolt was 2 in, it would stretch by about 0.002 in. If you want to know how much a component will stretch in inches, millimeters, etc., you can calculate it directly. Equation (7.8) comes from the elastic modulus [Eq. (7.6)] as shown below.

$$E = \frac{\sigma}{\varepsilon} = \frac{\left(F/A\right)}{\left(\delta/L\right)} = \frac{FL}{A\delta} \tag{7.8}$$

Equation (7.8) gives us a compact way to determine effects of stress and strain (see Fig. 7.8). Being able to directly relate a material to performance in this way is valuable. By rearranging Eq. (7.8), missing information can be determined. Stress and strain on a component are valuable to know in processing and end use, and they can be combined to determine the deflection, δ [a factor of strain and the change in length in Eq. (7.4)], as shown in Eq. (7.9).

$$\delta = \frac{FL}{AE} \tag{7.9}$$

Tungsten is much stiffer than copper, and the value of E will be much higher, which reduces the deflection. We can control E through material selection, but even simpler

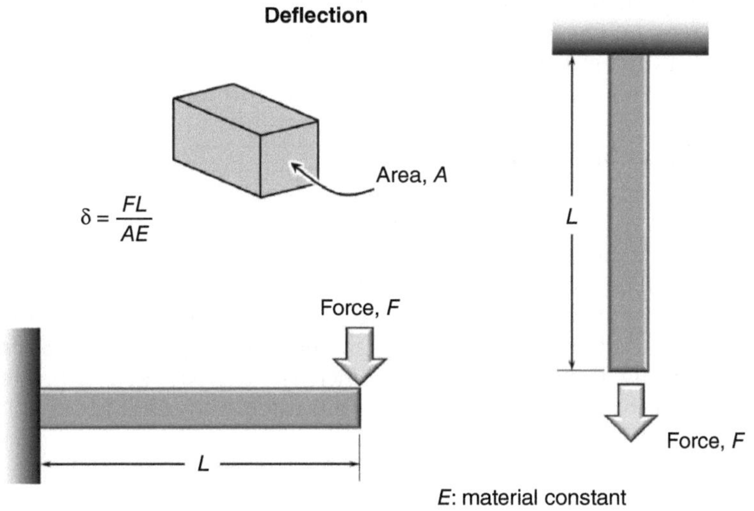

FIGURE 7.8 The amount a material deforms during loading is known as the deflection. Using this equation, the loading can be applied in line with the object or perpendicular to it, but it cannot deform plastically (permanently), only elastically.

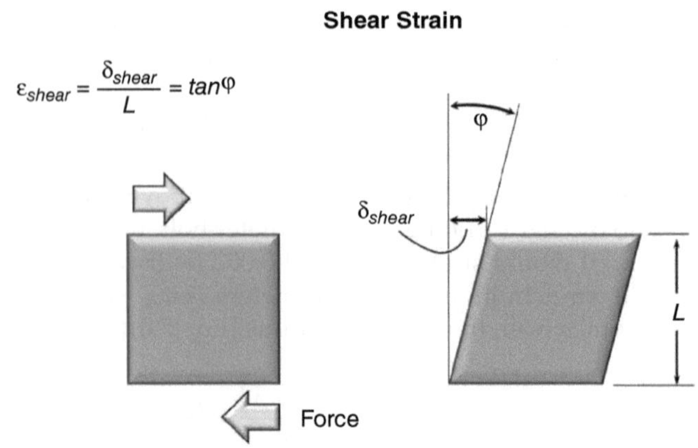

FIGURE 7.9 Shear strain develops when offset, opposing forces are applied. The deformation that occurs is angular.

ways to reduce deflection include reducing the force and/or unsupported length and increasing the cross-sectional area. It is important to note that the *effects* of strain can accumulate. After you stretch, bend, twist, or otherwise deform something far enough, it will not return to its original shape. The equations above are for reversible, or *elastic* deformation, and they do not describe permanent or *plastic* deformation (both will be discussed in Sec. 7.4).

For instance, if I stretch a metal bar slightly, then I hand it to you in the next room, you will not know that it was stretched. The metal bar, however, *will* know. When a material is permanently deformed (permanently strained), the atomic-scale structural changes remain. For that reason, when you try to stretch the bar more, there is already strain "built-up" in the material so you will find it harder to stretch it than I did. This principle of strain hardening will be described further in Sec. 9.2.1.

7.3.2 Shear Stress and Strain

We have looked at axial loading, but there is another, basic method of force application. Shear force is caused by opposing forces that are not collinear and cause a nonuniform deformation. If the starting cross-section is square, after shear force is applied, the resulting shape will be a parallelogram. This concept is shown in Fig. 7.9. **Shear stress** is calculated identically to axial stress [force per area, Eq. (7.1)], but the force is shear force, and the area used in the equation is parallel to the applied force. **Shear strain** is calculated differently, and instead of a linear increase/decrease in length, it results in an angular change in shape. The amount of strain is calculated using Eq. (7.10).*

$$\varepsilon_{\text{shear}} = \frac{\delta_{\text{shear}}}{L} = \tan\varphi \tag{7.10}$$

*According to small angle approximation, for angles less than 10° (0.176 radians), the tangent of the angle is approximately equal to the size of the angle *in radians*.

In a manner similar to the elastic modulus for axial loading, the **modulus of rigidity** is a measure of shear stress over shear strain. The equation for the modulus of rigidity is given in Eq. (7.11).

$$G = \frac{\sigma_{\text{shear}}}{\varepsilon_{\text{shear}}} \tag{7.11}$$

7.3.3 Torque and Torsion

It is very common to have rotating components involved in processing as well as in applications. For instance, in drilling and milling, the tool rotates. In turning, the part rotates. The rotational force on a tool or part often involves **torque** (the force applied), which results in **torsion** (the internal response). Let us consider drilling for a moment. The drill press has a motor that applies rotational force (torque) to the drill bit. That results in rotational motion. The action of drilling transfers the force from the motor, through the bit and to the piece. This basic principle applies to many common situations, and the torque developed, T, is often calculated by

$$T = Fd \tag{7.12}$$

where F is the force applied to the component, perpendicular to the moment arm, and d is the length of the moment arm. The **moment arm** is simply the distance between the applied force and the point to which the force is applied. This is illustrated through tightening a bolt. If you use a longer wrench, it *seems* easier to tighten (or loosen) a bolt. In reality, the same amount of torque is required, but the force can be smaller if the moment arm is larger. This is shown in Fig. 7.10, where d_1 is smaller than d_2. The force required to tighten or loosen the bolt would be greater and linearly proportional to the distance. That is, if you grasp the wrench twice as far from the nut, the torque will increase by twice as much for the same force.

Just the same, pulling a wrench perpendicularly will maximize the torque transferred, even for the same pulling force. So why not always pull perpendicularly? Nuts are often hexagonal and space is often limited. The position the wrench is in to start is not ideal. If

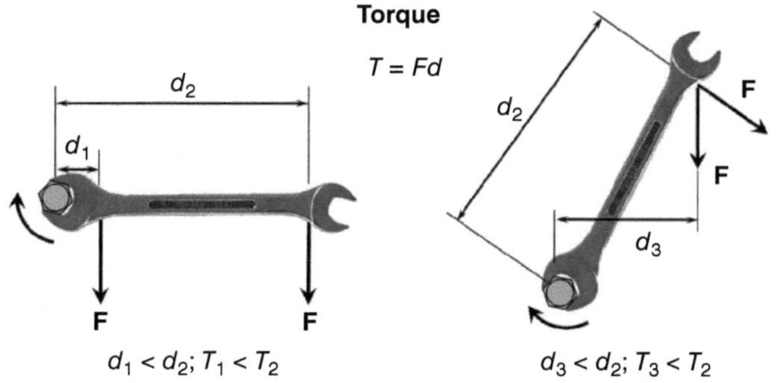

FIGURE 7.10 Much about torque is intuitive, such as the benefit of extra leverage at a greater distance from the point of application (the bolt). Also, applying force perpendicular to the moment arm (the wrench) is more effective. The same amount of torque is created with less force as the moment arm, d, increases.

you need maximum torque, position it such that it is perpendicular to a direction you can pull. The difference an angle makes in the moment arm, even when pulling from the same location, is illustrated in Fig. 7.10. Only the *component* (or amount) of force applied perpendicularly to the moment arm contributes to the torque. The actual (or *effective*) moment arm can be found by calculating the leg of a right triangle with one side (the hypotenuse) as the distance between the bolt center and the position the force is applied. The other sides can be found with the angle and/or magnitude of force, whichever is known more accurately. Common trigonometric formulas are provided inside the back cover.

Returning to drilling, there is also a force associated with removing the material. The material removal force opposes the motion of the drill and causes stress. Enough stress causes the drill to break. Many rotating components such as gears, shafts, and pulleys have to withstand a certain amount of stress during normal operation. Whereas most machine designs account for maximum acceptable load limits, processing does not necessarily have limits. There is no safeguard against exceeding the tool's capability.

For instance, if you are drilling a hole by hand, it is up to you how hard you push the drill into the workpiece. If you apply more pressure than the drill can physically handle, it breaks. There is no safeguard for the drill. Tool design and lifetime considerations will be discussed in more detail in Chap. 13, but there are some general concepts in torque and torsion which will aid in understanding other processing parameters as well.

First, the amount of stress is related to the applied force and the area it is applied to [see Eq. (7.1)]. The larger the force is and smaller the area is, the larger the stress will be. In this case, it is called the **torsional shear stress** (see Fig. 7.11). This is the stress that will break a component if it becomes too large. The torsional shear stress, τ (lowercase Greek letter *tau*), is given in Eq. (7.13):

$$\tau = \frac{T\left(\frac{D}{2}\right)}{J} = \frac{TD}{2J} \tag{7.13}$$

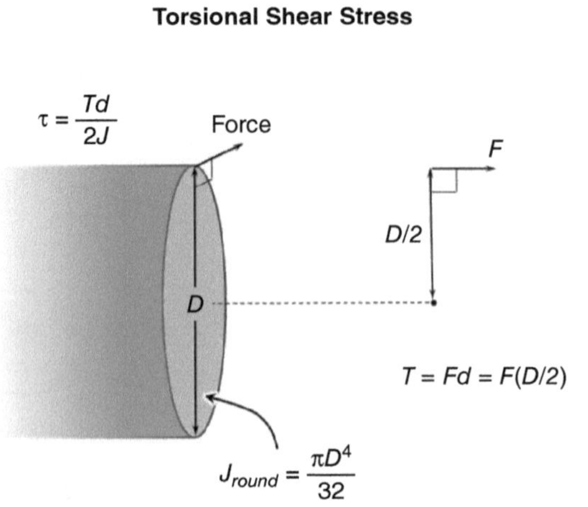

FIGURE 7.11 Torsional shear stress is dependent on component geometry and applied force. Although a circular cross section is shown here, the cross section can have any shape.

In Eq. (7.13), D is the diameter of the component, T is the applied torque (units of lb·ft, N·m or similar), and J is the **polar moment of inertia** (units of ft, in, m, etc. raised to the fourth power). The torque and inertia give us guidelines for how to avoid or utilize rotational force. Looking at Eq. (7.13), we see that increasing torque and diameter increase the stress, and an increase in the polar moment of inertia will decrease the stress. The polar moment of inertia is dependent on diameter, and the equation for a solid circular cross section is given by Eq. (7.14):

$$J_{\text{round}} = \frac{\pi D^4}{32} \tag{7.14}$$

This may seem contradictory since increasing the diameter increases the stress, but increasing the polar moment of inertia (which depends on diameter) decreases the stress. Equation (7.14) shows how the inertia is related to the cross-sectional area of the component (for a round cross section), and the two are not proportional. As the diameter increases, the inertia also increases, but its increase is *exponentially faster* than that of the diameter has (J is proportional to D^4).

Also, the geometry of the cross section (e.g., square, round, hollow, etc.) will determine the efficiency with which the cross section handles the applied load (more cross-sectional geometries are provided inside the back cover). Therefore, to reduce the stress in a rotating component, just like the case of axial stress, increase the area (cross section) or decrease the applied force (torque).

Often it is not only the amount of stress that is important, but also the allowable deformation. This is related to axial strain [Eq. (7.4)], but in the case of torsion, strain occurs in a twisting motion. Like wringing out a towel, one end will rotate relative to the other (Fig. 7.12). Because the movement is rotational, the change in dimension will be *angular*. The **shear strain,** γ, is related to the component diameter, D, the length of the component, L, and the **angle of twist,** θ, using Eq. (7.15).

$$\gamma = \frac{\left(\frac{D}{2}\right)\theta}{L} = \frac{D\theta}{2L} \tag{7.15}$$

If you want to minimize the shear strain on a component, you need to reduce the length of the component or increase the diameter. If you want to know how much a component will twist, the angle of twist can be calculated. This is the actual deflection under the given conditions of operation. The angle of twist (in radians*) is calculated using Eq. (7.16).

$$\theta = \frac{TL}{JG} \tag{7.16}$$

The only new variable is G, which is the **modulus of rigidity** or the **shear modulus.** This is a material constant, similar to the modulus of elasticity [see Eq. (7.11)]. Also, note the similarity to the equation for linear deflection, Eq. (7.9). The choice of material will determine the angle of twist under any given conditions. Steel, for instance, has a higher modulus of rigidity than aluminum, so when exposed to the same conditions of torque, it

*To convert to degrees, multiply by $180/\pi$.

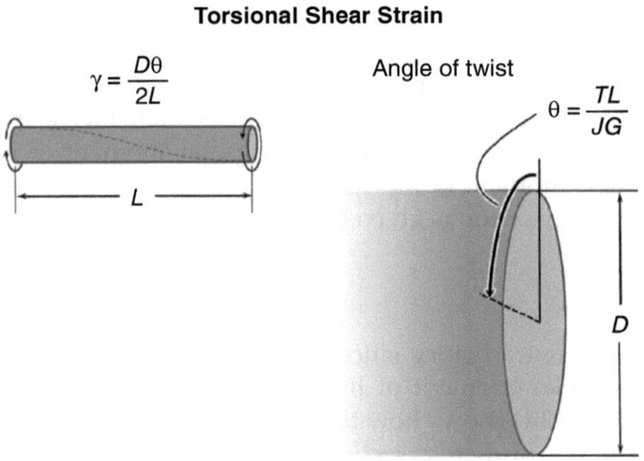

FIGURE 7.12 Torsional shear strain is rotational deformation and is proportional to the size and rigidity of the component.

will twist less. Values of G are tabulated for many materials and can be referenced when needed. A small selection of materials is provided in Appendix B.

7.4 Mechanical Properties

The goal of understanding forces is to apply those concepts at the material level. Knowing about forces does not give us information on how a material responds to them. Describing materials in context of their mechanical properties gives us the ability to quickly define sets of properties. That is, many properties such as hardness and ductility are often linked. Furthermore, classes of materials and subcategories also behave similarly. Metals tend to behave similarly to other metals and likewise for ceramics and polymers. Even though a category of material will have similar properties within it, classes of that material (e.g., copper-based alloys) will have even more in common. We will look at how materials respond to applied force and how we can use that information to determine appropriate processing parameters.

7.4.1 Stress-Strain Diagrams

One of the most compact ways to depict the mechanical response of a material is the stress-strain diagram. One curve provides information on strength, stiffness, ductility, and more. To clarify how useful these diagrams can be, we will look more closely at a typical curve in Fig. 7.13.

A stress-strain curve shows the strain (change in length) for a given applied stress. That stress is most often applied as tension due to the ease of completing the test. As we move vertically on the ordinate axis (y-axis), the force applied to the material increases. Since stress is simply force divided by area, the stress must increase as well. As the applied force increases, there is a tendency for the material to elongate. The elongation is the strain. You can see that the increase in stain is not uniformly related to the increase in stress. The stress and strain have a linear relation in many materials when the force is relatively small. This is the elastic region on the stress-strain curve.

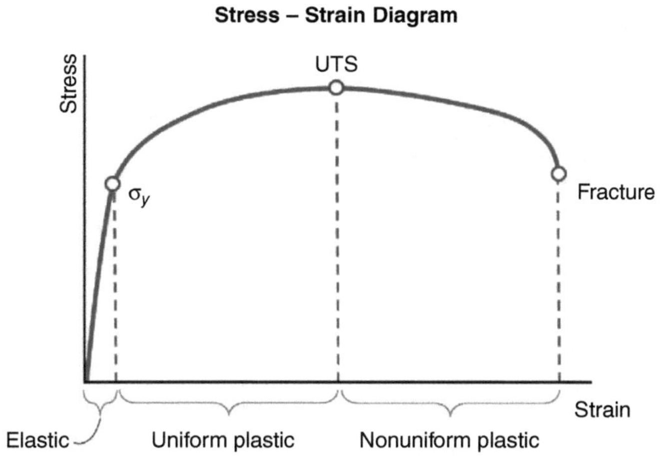

Figure 7.13 A typical stress-strain curve shows how a material responds to force. The curve contains information about strength, stiffness, ductility, work-hardening, toughness, and more. Main points include the yield stress, σ_y, ultimate tensile strength (UTS), and the fracture strength. The three types of deformation that occur during a stress-strain test include elastic, uniform plastic, and nonuniform plastic regimes.

The point at which strain increases more rapidly than stress is called the proportional limit or yield stress (given as σ_y). Strictly speaking, the **proportional limit** is where the stress and strain are no longer linearly related (i.e., Hooke's law is invalid, and E in is no longer constant). In many rigid materials, the proportional limit and yield stress are nearly identical, but in others, especially elastomers, the proportional limit comes before the yield stress and some of the elastic deformation is nonlinear (Fig. 7.22). The yield stress is where the material stops deforming elastically and begins plastically deforming, regardless of the behavior occurring before or after that point.

Plastic (permanent) deformation occurs two ways. It starts as uniform plastic deformation, which occurs when the entire component is stretching. In the final deformation mode the strain concentrates at a small region until failure occurs. The concentration of force and deformation at small region is called necking. Ductile materials will exhibit more extensive necking than brittle materials. To better understand these terms, we will spend a little time discussing them. The stages of deformation are shown in Fig. 7.14.

7.4.2 Elasticity

All material deformation is related to atomic motion. **Elasticity** is a material property that describes *reversible* changes in shape. You might associate this behavior with rubber bands, bungee cords, or elastic waistbands, which can be stretched and return to their original shape. This same behavior happens in metals and ceramics as well, but the amount they can stretch elastically is very small. Even though materials vary greatly in their elastic response, there are some commonalities as well. First we will look atomically.

The characteristic of elastic deformation is that atoms do not change nearest neighbors. After the material is pulled, bent, or otherwise deformed, the atoms have not changed position. They do not break or exchange bonds. The atomic movement in elastic deformation is merely the *stretching* of bonds. The bonds can be thought of as

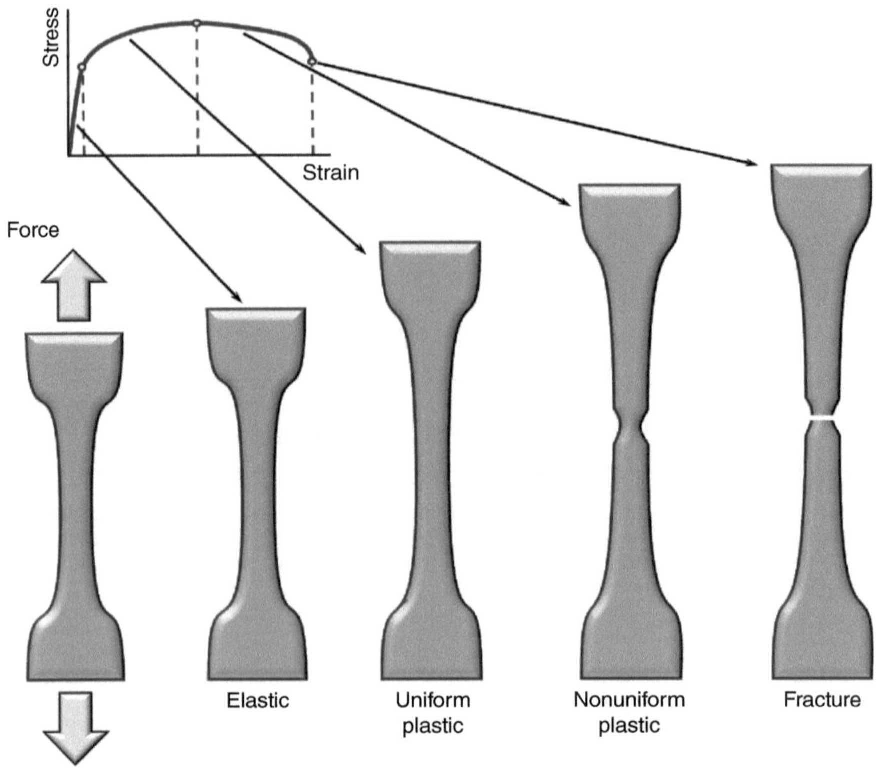

FIGURE 7.14 The stages of increasing strain are elastic, uniform plastic, nonuniform plastic, and eventually fracture. These are illustrated by the tensile specimen shown above.

springs (Fig. 7.15). Like springs, the farther you stretch them, the harder it gets to stretch them. Also like springs, they have a stiffness associated with them. The stiffness of a material is a consequence of the cumulative effect of the many atoms comprising it. Therefore, a material's bonds determine the stiffness, also called the elastic modulus, *E*, which is related to the stress and strain by Hooke's law (Robert Hooke, 1635–1703), which describes spring force. The elastic modulus is defined in Eq. (7.1).

The greater the elasticity of a material, the more stress that is required to deform it. All materials will exhibit some degree of elastic deformation as force is applied. The greater the elastic modulus the steeper the slope of the linear portion of the stress-strain curve (see Fig. 7.13). This is simply because more force must be applied to cause the same amount of deformation. This is shown below in Fig. 7.16. The initial portion of a stress-strain curve is linear (or close to it), and the stiffness of a material can be determined from the measuring the slope of the stress-strain curve in the elastic portion.

7.4.3 Plasticity

Plasticity is a much more complex description of material deformation. The major separation between elastic and plastic deformation is that plastic involves permanent atomic displacement, where atoms *change* nearest neighbors. Plastic deformation has nothing

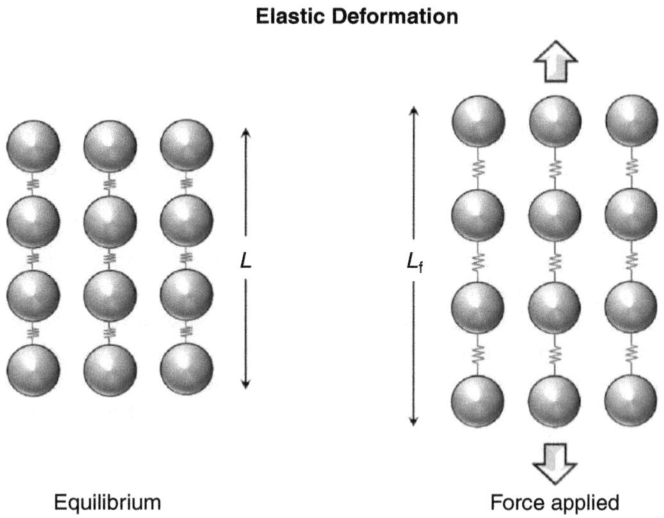

FIGURE 7.15 Elasticity is a measure of reversible change in dimension for an applied force. Atomically this means that atoms (or molecules) do not break or exchange primary bonds and return to their original positions after the load is removed.

to do with polymers. All material categories can deform plastically, but brittle materials will show little to no permanent (plastic) change in shape when they are loaded to failure. This is evidenced by the ability to recombine the separate pieces of a brittle fracture to recreate the part and maintain initial dimensions. If you break a window into six pieces, for instance, you can take those pieces and reassemble it like a puzzle. Once reassembled, the dimensions are identical to those before it was broken. That is an example of a perfectly brittle material. Most ceramics epitomize this behavior, but metals and polymers can be brittle too. They will often have a slightly higher ductility than the glass just mentioned, but they will still be relatively brittle compared to other metals and polymers.

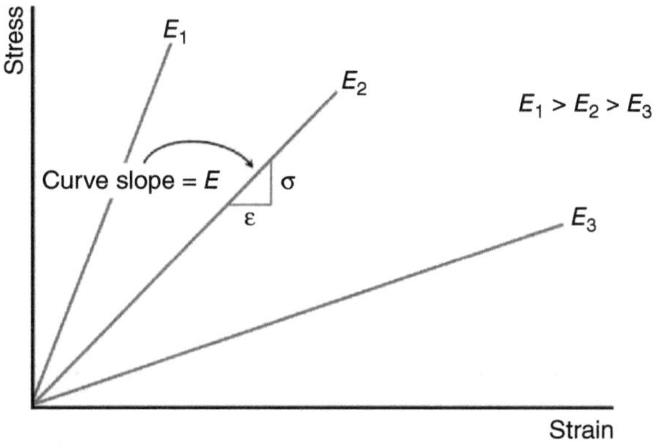

FIGURE 7.16 The elastic portion of the stress-strain curve is often linear, or can be approximated as so, and the slope of the line is the elastic modulus.

If a material is ductile, you cannot get the same dimensions before and after failure. It has permanently changed shape. Let us say you break a key off in the lock. The key, being metal, will clearly show a change in shape. If you were to extract the piece that was broken off, you would find that when the two pieces were reassembled, they do not appear like an unbroken key. This is also evidenced *before* breaking. Automotive panels will dent and crush without fracturing. Plastic bags and packaging will stretch and bend without cracking.

It is easy to see the failure of a component as the point where it breaks into multiple pieces, but in many applications a product or component is considered failed if it leaves the elastic regime (discussed above) because that means the product has changed dimensions and will be unsuitable or at least less effective for continued use. For instance, if a seal in garden hose deforms over time and begins to leak, it has failed because it no longer serves its original purpose or has exceeded the original design tolerance. If the leak is tolerable, you can still use it, so the failure is not catastrophic.

Even though plasticity is unwanted when products are in use, *it is required to produce products*. Think about it. Any material that is not permanently changed through processing will be identical to its raw, stock form. Manufacturing relies on controllable, permanent changes in shape. This is most clearly seen in operations such as forming (see Chap. 12), but it is also part of other processes as well. Because of how frequently plasticity is used in manufacturing, it is important to understand it a little better. We will only scratch the surface of it, however, as the basics will serve us well.

Plastic deformation is permanent deformation. That means that anytime you change the shape of something, such as balling up aluminum foil, bending a paperclip, stretching plastic wrap, etc., you are causing atoms or molecules to slide past one another and they do not return to their original position. Because the material is changing macroscopically, it must also be changing microscopically (Fig. 7.17). How this movement occurs is extremely complex and involves looking at parameters such as lattice friction, dislocation motion, bond exchange, etc. Metals are perhaps the simplest in their basic mechanisms.

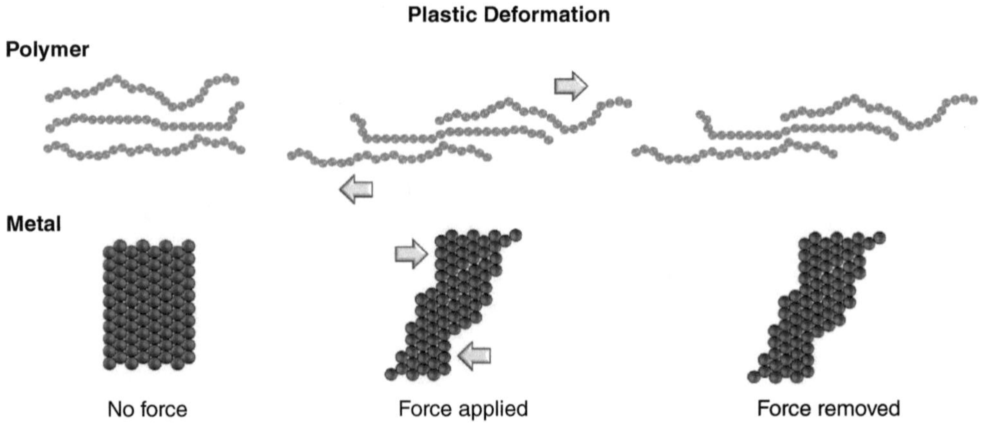

FIGURE 7.17 Plastic deformation involves the permanent change of atomic positions. After sufficient stress is reached to cause the material to yield, atoms change "nearest neighbors" and do not return to their original positions. This happens in all categories of materials, but the mechanisms and amount of deformation before fracture will differ. A simple polymer and metal are shown here.

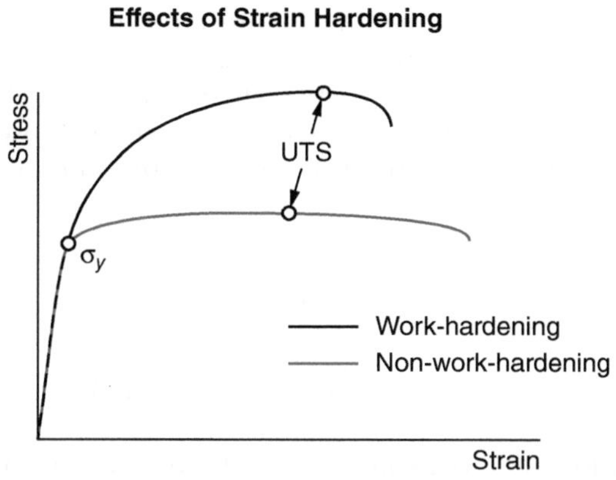

Figure 7.18 Materials that strengthen during deformation (i.e., work-harden) will continue increasing in strength until they reach their ultimate tensile strength (UTS) followed quickly by fracture. Nonwork-hardening materials will exhibit more uniform stress during deformation at a level near their yield stress (σ_y).

Dislocations (discussed in Sec. 3.2.3.2) are created and moved during plastic deformation, and they control the shape of the stress-strain curve. After leaving the elastic section of the curve (the yield point), the material begins behaving plastically (refer back to Fig. 7.13). The stress and strain are no longer linearly proportional. Once plastic deformation begins, it continues at a similar stress, but in many materials (including nonmetals) the stress will continue to increase slowly as strain increases. This indicates that the material is becoming stronger as it is deformed.

A material that strengthens with deformation is known as a **work-hardening** (or strain-hardening) material. This is the basis for cold working in metals, which is discussed in Sec. 9.2.1, and in polymers a similar strengthening can be achieved through increasing chain alignment during extrusion or drawing. Strengthening by plastic deformation can be applied to the industrial stock or to the product or component later in the processing. The difference in a stress-strain curve for a work-hardening material and a nonwork-hardening material is shown in Fig. 7.18.

Some important characteristics of materials that exhibit extensive strain hardening are that the strength increases steadily as deformation continues, and work hardening localizes the strain. More specifically, as a material deforms, the sections that have deformed will be stronger, so further deformation takes place at locations that are weaker. This causes the strain to spread out during deformation, but when the **ultimate tensile strength** (UTS) is reached, which is the highest stress on the curve, nearly all ductility has already been exhausted and fracture occurs soon after. Non-work-hardening materials deform uniformly at or very near their yield stress and tend to exhibit high ductility overall. Crystalline materials, which can reach very high strains, typically over 200%, are called **superplastic**, but this is usually only achieved by deformation at very high temperature.

7.4.3.1 Ductility versus Brittleness

What defines some materials as being brittle and others as ductile is that brittle materials fail in the elastic regime (or very near to it). Therefore, brittle materials show little to no plasticity. Perfectly brittle materials show none at all. Ductile materials will show significant plasticity under loading before failure. This does not mean they can be stretched like taffy though. For instance, a 2% elongation ($\varepsilon = 0.02$) at failure is not perfectly brittle, but it is not highly ductile either. Ductility includes a range of values, and it can be relative to other materials, similar materials, those with similar strength, etc. Quantitatively, it can be measured by the **percent elongation** or **percent reduction in area** at failure.

$$\%EL = \frac{L_f - L}{L} \times 100 \tag{7.17}$$

$$\%RA = \frac{A - A_f}{A} \times 100 \tag{7.18}$$

If the equation for $\%EL$ looks familiar, it should. Aside from multiplying by 100, it is identical to the equation for strain [Eq. (7.4)]. The maximum difference in the original dimensions (L or A) and the final dimensions (L_f or A_f) determines the amount of ductility. The more strain a material undergoes before breaking, the more ductile it is. Likewise, the more you can reduce the area, the more ductile it is. If the object is shortened/flattened, the values for elongation and reduction in area will be negative, but the amount of strain will be the same. Compressive forces cause this to occur, such as the process of "upsetting" a workpiece in forging (Sec. 12.2.1).

Material deformation is volume conservative. No material is added or subtracted during testing or forming, so when you stretch a bar, the cross-sectional area must decrease. The volume is a function of length and area simultaneously. Whenever one increases, the other must decrease (Fig. 7.19). If the bar only stretches a little before breaking, the material is not very ductile and the area does not need to decrease much. Both measures of ductility must be in agreement.

Remember, brittle materials are not necessarily weak. They are unable to stretch plastically, but that is it. Diamond, something you probably would not consider weak, is brittle. Do not get confused between strength and ductility. You can typically generalize that very strong materials tend to be more brittle than weak materials, but it is only a generalization. To further separate these properties, we will consider hardness and toughness next.

7.4.3.2 Toughness versus Hardness

Having just looked at ductility and how that affects material performance, we can extend the concepts to toughness and hardness. Like ductility, toughness and hardness describe

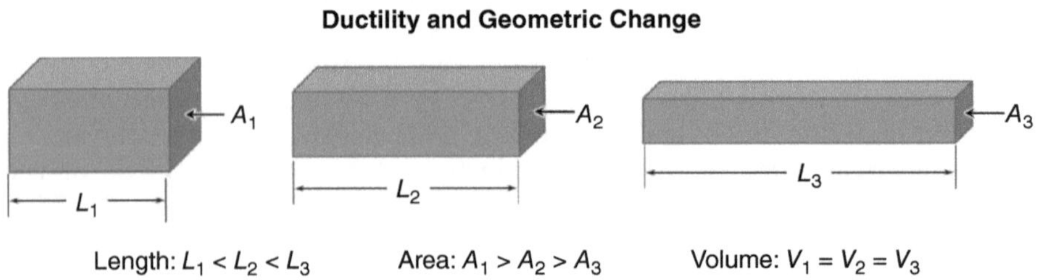

FIGURE 7.19 Ductility is often measured by the magnitude of permanent deformation a material can undergo before fracturing. It is measured by the change in length or reduction in area, which are directly linked since volume is conserved.

a material's response to applied force, but the application of force is much different. All material properties are related to some degree since all arise from the atomic characteristics, but there are two major categories of force application: static and dynamic.

Static force is constant. As you are sitting in your seat, you (through gravity) are applying a constant load. Your weight is not changing, so the force is static. Many standardized tests for assessing mechanical properties involve applying force in a slow and continuous manner. That is how hardness is often assessed, for example. The **hardness** of a material is the resistance to indentation or abrasion. A small indenter or point is brought to the surface of the material and a constant force is applied to it. Initially, the indenter will sink quickly into the material. Depending on the material, it may also stop quickly, or it may slowly "settle in" before ceasing to go further. The depth the indenter goes into the part is dependent on how hard the material is. Hard materials resist indentation, scratching, and abrasion.

Dynamic force is associated with a *rate* of application. If you hit a nail with a hammer, the force is applied by impact and carries momentum with it. This is the basis for dynamic loading. Think if you missed the nail you were hammering and hit your finger instead. That hurts! Now imagine if you put the hammer on your finger and gradually increased the pressure. It will still hurt eventually, but it will be quite different.

Materials respond to impact differently than they do to static loads. This leads to phenomena like strain-rate sensitivity, which causes normally ductile materials to behave in a brittle fashion if force is applied fast enough. Armor, which must withstand great impact, needs to be designed with dynamic loading in mind. Projectiles are traveling very fast with a great deal of force. The armor goes from an unloaded state to being highly stressed in a fraction of a section. The response cannot be accurately assessed by slowly applying a constant load during testing.

Toughness is a measure of how much energy a material can absorb before breaking. Technically, *toughness is the area under the stress-strain curve up to the rupture point*. The higher and longer the curve, the tougher the material (see Fig. 7.20). Brittle materials fail at low strain, and therefore are not considered tough. Ductile materials have much longer curves, which increases the associated area. The strength of brittle materials may be higher (look at the stress values), but the lack of any appreciable plastic deformation causes them to fail with less applied energy. The area under the *elastic* portion of the stress-strain curve is known as the **modulus of resilience** and is higher for strong materials, even if they are brittle.

A material with a high hardness does not necessarily have a high toughness. The two often conflict. In applications where a material is under large static or quasi-static* loading, a material with high hardness (high strength) is often preferable. If impact or other dynamic loading conditions exist, a material that has a high toughness is the better choice. These two concepts are often applied in processing. For example, forging (discussed in Sec. 12.2.1) can be done through continuous application of pressure (quasi-static) or through impact (dynamic) such as drop forging. The material being processed and dies used in processing must both comply with the given conditions to create parts reliably and economically.

---

*The term "quasi-static" is often used because you cannot apply force in a truly static manner or the force would not be able to change at all (including during application). Quasi-static is a better term because the force is being applied at a given rate (dynamic), but the rate is so low that it approximates a static (constant) load.

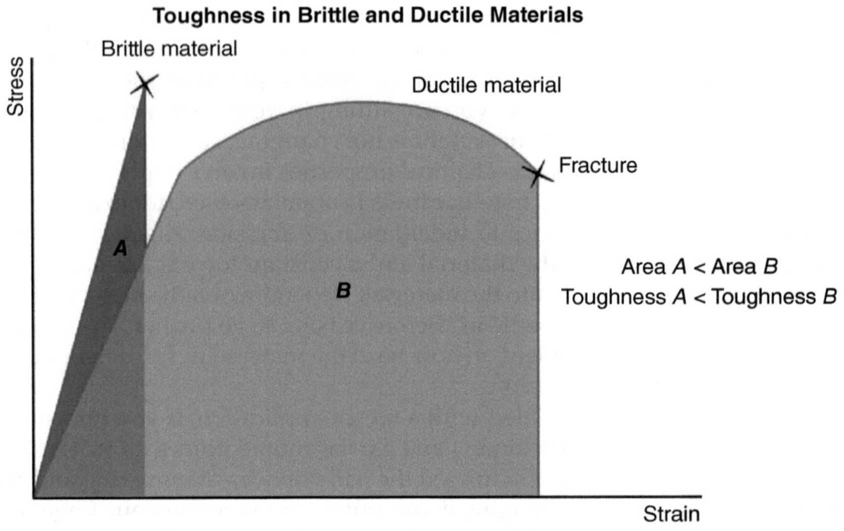

FIGURE 7.20 The toughness of a material can be determined by the area underneath the stress-strain curve up to the rupture point. The larger the area, the tougher the material is. The area is determined by the overall strength (stress values) and the ductility (strain values). Brittle materials break at low strain (curve A), which limits the area. Ductile materials (curve B) may not be as strong, but they can absorb more energy as evidenced by the increased area.

The **fracture toughness** is a measure of how much energy required for a crack to lengthen. It is the energy a material absorbs during breaking, and the sensitivity of a material to a preexisting crack. Loading can occur in three "modes," which are determined by the orientation of forces in relation to a preexisting crack. As is common, we will focus our attention on Mode I (Roman numeral one), which examines the effect of a crack that is being opened up by a tensile force. Let us say a material with fracture toughness, K_{IC}, contains a crack of length, a. The maximum amount of stress, or the critical stress, σ_c, for that material is given by Eq. (7.19). A schematic representation is given in Fig. 7.21.

$$\sigma_c \leq \frac{K_{IC}}{Y\sqrt{\pi a}} \tag{7.19}$$

The variable Y is a dimensionless constant (often 1.1) determined by the test sample's geometry when measuring the fracture toughness. If you work out the units in Eq. (7.19), you will see that the fracture toughness, K_{IC}, has units of stress multiplied by the square root of length (e.g., psi$\sqrt{\text{in}}$ or MPa$\sqrt{\text{m}}$ or alternatively psi·in$^{1/2}$ or MPa·m$^{1/2}$). While these units certainly seem like a strange representation of toughness, not all units are intuitive from a conceptual point of view, but they are mathematically accurate—think also of inertia [Eq. (7.14)] with its units of in^4 or m^4. By rearranging Eq. (7.19), we can solve for the maximum (critical) crack length at a given stress before failure will occur. This is shown in Eq. (7.20).

$$a_c = \frac{1}{\pi}\left(\frac{K_{IC}}{\sigma Y}\right)^2 \tag{7.20}$$

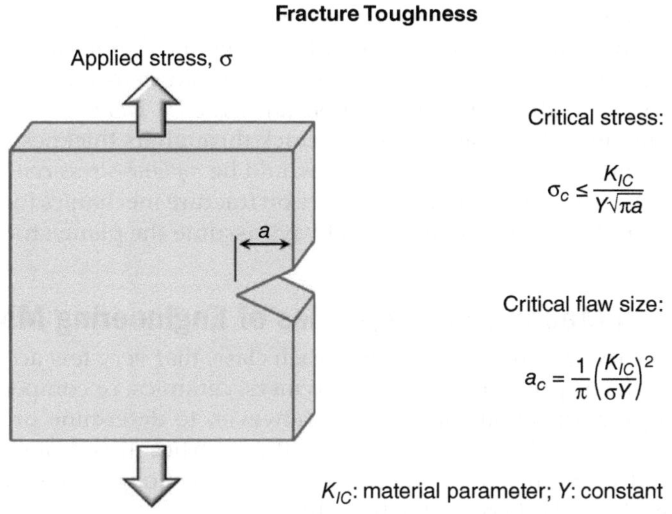

FIGURE 7.21 The fracture toughness is described by an equation relating a preexisting flaw (crack) to the stress a material can withstand which contains that flaw. The maximum stress for a fixed flaw size or the maximum flaw size at a fixed stress can both be assessed using this analysis.

So let us apply this to a couple of different materials and see what the effect of fracture toughness is. Although fracture toughness can be measured for an unknown material, we will use known values for common materials (see Appendix B for some common values). We will compare the critical crack lengths for aluminum (K_{IC} = 24 MPa·m$^{1/2}$), concrete (K_{IC} = 0.2 MPa·m$^{1/2}$), and polystyrene (K_{IC} = 1.2 MPa·m$^{1/2}$) at a stress of 35 MPa. The equations are shown in Eq. (7.21).

$$a_c = \frac{1}{\pi}\left(\frac{24\ \text{MPa·m}^{1/2}}{35\ \text{MPa}\ (1.1)}\right)^2 = 0.124\ \text{m or } 124\ \text{mm (aluminum)}$$

$$a_c = \frac{1}{\pi}\left(\frac{0.20\ \text{MPa·m}^{1/2}}{35\ \text{MPa}\ (1.1)}\right)^2 = 0.0000086\ \text{m or } 0.0086\ \text{mm (concrete)}$$

$$a_c = \frac{1}{\pi}\left(\frac{1.2\ \text{MPa·m}^{1/2}}{35\ \text{MPa}\ (1.1)}\right)^2 = 0.00031\ \text{m or } 0.31\ \text{mm (polystyrene)} \qquad (7.21)$$

You can see from the above comparison that the fracture toughness makes a tremendous difference. Of course, the K_{IC} for aluminum is much higher to start, but that number gets squared in the equation. When the denominator is the same, the numerator is critical. A couple of other notes are that we have said nothing about the yield stress of these materials and that the loading is tensile only. Both of these can be illustrated by looking at concrete in the example. Concrete is very strong, but it is brittle. Although it can handle a tremendous amount of compressive stress, if there is even a fraction of a millimeter crack, it will fail under modest tensile stress.

The polystyrene is not nearly so strong, but it can handle a larger flaw size under the given conditions, even though that size is also very small. Aluminum has a mid-range strength compared to the other two, but its superior toughness allows for a very noticeable crack to form before imminent failure. The calculations assume that the structure with the crack is sufficiently large (thick) to avoid deforming in a direction perpendicular to the loading axis, a situation known as a *plane-strain condition*. If the part being tested was sheet metal with the crack through its thickness, it would not be calculated using Eqs. (7.19) or (7.20) as it would be a *plane-stress condition*. Unfortunately, we do not have the space to discuss more on fracture mechanics (or *fortunately*, depending on your viewpoint!), so we will always assume the plane-strain condition.

7.5 Common Mechanical Properties of Engineering Materials

Materials vary so widely, even within each class, that very few accurate statements can be made which apply to all metals, polymers, ceramics, or composites. There are some common properties that can be used, however, to determine processing and predict performance. Recall from Chap. 2 that all properties arise from atomic structure and mechanical properties are a consequence of bulk (long-range) structure of atoms or molecules. Fig. 2.11 displays the basic hierarchy.

Since materials within each class share many structural features, they tend to behave at least somewhat similarly. Knowing the mechanical properties of materials, even in the broadest sense, can help us quickly reduce our possibilities and aid in materials selection. Much about the material properties within the classes has been discussed already (see Chaps. 3 and 4 for metals, and Chaps. 5 and 6 for nonmetals and composites). In this section we will look at properties that are specific to each class of material.

7.5.1 Metals

Our focus in the preceding sections of this chapter has been more centered on metals than other materials. For instance, Figs. 7.13 and 7.18 are common shapes for the stress-strain curves in metals. We will see shortly (Sec. 7.5.3) that polymers often have distinctly different behavior. Metals are characterized by a combination of good strength and ductility. This is related to their metallic bonds, which are nondirectional (review Sec. 2.3.1.1). Since plasticity is the basis for ductility, the way atoms move relative to one another is important. In metals, this atomic motion occurs by the movement of dislocations along slip planes (see Sec. 3.2.3.2). The majority of plasticity in metals is due to dislocation motion and the ease of dislocation motion determines how easy it is to deform the material.

In face-centered cubic (fcc) materials, where atoms can slide by one another without much resistance, the force required to form the material is relatively low and the amount of deformation the material can undergo without breaking is large. In hexagonal close-packed (hcp) materials, the ability to move atoms is lower because there are not as many favorable directions for the atoms to move. Recall that grains and grain boundaries also play a role in the motion of dislocations.

The Hall-Petch equation [Eq. (3.1)] demonstrates the importance of grain size to strength. When there are many grains, not all grains are aligned favorably for deformation to occur. Metals are nearly always polycrystalline, and these considerations have led to the development of equations to predict the minimum force required to initiate plastic deformation, or the so-called **critical resolved shear stress (CRSS)**, often referred to as the **Schmid factor** (Erich Schmid, 1896–1983).

The CRSS is a way to determine the force on a misaligned slip plane and requires knowledge of the slip plane and slip direction(s) orientations. The application of that level of analysis is beyond our scope, but what we should take away from these considerations is that the force required to cause plastic deformation is higher than what we might predict in the ideal case. Since metals have this ability to deform, they are processed by many forming operations (see Chap. 12), which apply pressure to change dimensions. Understanding the factors that control metallic deformation aids in correctly designing production protocol.

Most metals used in structural applications are alloys. A look at Chap. 4 quickly confirms that there are many options when using metals, and one of the reasons for this is because they possess different mechanical properties such as yield point, ductility, toughness, etc. A small amount of an alloying element can dramatically change the properties of a metal. This can be partially attributed to solid solution strengthening, second phase formation, or other structural variations (see Sec. 3.2.4).

For this reason, one category of alloys, such as aluminum alloys, can be very weak and very ductile or very strong with limited ductility. Although alloying is important mechanically, it is also used for other properties such as corrosion or thermal characteristics (Chap. 8). The way a metal is processed can also change the mechanical properties. Plastic deformation typically strengthens a metal. Not all metals are equally responsive, but the ability to create a stronger component during processing is often exploited through cold working (see Sec. 9.2.1).

A list of metals and their properties is given in Table 7.1. Notice the wide variation in strength and stiffness. Metals are tested by all variety of methods including tension, compression, and hardness. More on testing methods is presented in Sec. 7.6.

7.5.2 Ceramics

Ceramics tend to have very high strength and hardness (see Table 7.2), but they have little to no ductility. Forming operations involve plastic deformation, so ceramics are generally unsuitable for forming, at least at room temperature. Ceramics become less brittle at high temperature (like most materials), and because of their high melting points, they are often used in high-temperature applications. Because they lack ductility, ceramics are often processed "green," meaning they have been dispersed or "softened" with an additive (usually water) to enhance their formability (see Sec. 12.4).

Ceramics have covalent or ionic bonds (or a mix of these) that greatly hinder atomic motion. In metals the bonds are capable of being easily transferred due to the free electrons, but ceramics have rigid bonds that cannot be easily broken and reformed. Ionic materials have the additional consideration of like-charges aligning during deformation, which repel each other and contribute to failure. The bond distances tend to be larger (atomic spacing) and planes of atoms cannot slide as easily. Although dislocations exist in ceramics just as in metals, they are much more limited in how they enhance plastic deformation.

7.5.2.1 Transformation Toughening

Ceramics tend to have a high sensitivity to cracks and sharp corners. Stress concentrates at these locations, and since ceramics cannot resist crack propagation, they are likely locations for failure. The fracture toughness was described in Sec. 7.4.3.2, and concrete was shown to perform quite poorly in tension if any flaw is present. Ceramics have low fracture toughness values in general, but there are ways to improve the toughness of ceramics. One of the ways is transformation toughening. **Transformation toughening** is the process of a crack resulting in a structural change that makes it harder for the crack to continue. The progression of the crack causes the crack to slow, or at least

Mechanical Properties of Common Metals			
Metal	**Yield Stress ksi (MPa)**	**Ultimate Tensile Strength (UTS) ksi (MPa)**	**Modulus of Elasticity 10⁶ psi (GPa)**
Cast aluminum	8–27 (55–186)	19–46 (131–317)	10.3 (71)
Cast aluminum (heat treated)	8.5–43 (59–296)	20–48 (138–331)	10.3 (71)
Aluminum alloys	11–59 (76–407)	12–63 (83–434)	10–10.3 (69–71)
Aluminum bronze	62–66 (427–455)	71–110 (490–758)	16–19 (110–131)
Brass	10–65 (69–448)	34–99 (234–683)	15–17 (103–117)
Copper	10–53 (69–365)	32–57 (221–393)	17 (117)
Inconel	30–160 (207–1103)	80–170 (552–1172)	31 (214)
Cast magnesium	12–23 (83–159)	22–40 (152–276)	6.5 (45)
Magnesium	20–32 (138–221)	35–42 (241–290)	6.5 (45)
Monel	25–130 (172–896)	70–140 (483–965)	26 (179)
Nickel	15–115 (103–793)	55–130 (379–896)	30 (207)
Phosphor bronze	14–80 (97–552)	40–120 (276–827)	15–17 (103–117)
Silicon bronze	21–62 (145–427)	56–110 (386–758)	15 (103)
Titianium	40–120 (276–827)	50–135 (345–931)	15–16.5 (103–114)
Zinc	—	19.5–31 (134–214)	6.2–14 (43–97)
Gray cast iron (Class 20)	—	20 (138) minimum	11.6 (80)
Malleable iron	30–80 (207–552)	40–100 (276–689)	19.9 (137)
Nodular (ductile) iron	40–90 (276–621)	60–120 (414–827)	23 (159)
1025 steel (low carbon)	40–90 (276–621)	60–103 (414–710)	30 (207)
1045 steel (medium carbon)	50–162(345–1117)	80–182 (552–1255)	30 (207)
1095 steel (high carbon)	20–150(138–1034)	90–213 (621–1469)	30 (207)
Iron	23–32 (159–221)	34–54 (234–372)	28 (193)
3140 alloy steel	62–162(427–1117)	93–188 (641–1296)	30 (207)
4130 alloy steel	46–161(317–1110)	81–179 (558–1234)	30 (207)
4140 alloy steel	68–200 (469–1379)	109–220(752–1517)	30 (207)
5150 alloy steel	51–190 (352–1310)	98–210 (676–1448)	30 (207)
6150 alloy steel	59–210 (407–1448)	96–228 (662–1572)	30 (207)
9840 alloy steel	45–50 (310–345)	120–285 (827–1965)	30 (207)
303 stainless steel	35–95 (241–655)	85–125 (586–862)	28 (193)
304 stainless steel	35 (241)	85 (586)	28–29 (193–200)
316 stainless steel	35 (241)	85 (586)	28 (193)
440C stainless steel (annealed)	65 (448)	110 (758)	29 (200)
440C stainless steel (hardened, tempered)	27 (186)	285 (1965)	29 (200)

Data reproduced from Industrial Press, Inc., *Machinery's Handbook*, 29th edition, Oberg, Jones, Horton, Ryffel. Table 1. pg 513 and Table 13. pg. 432. ISBN 978-0-8311-2901-9.

Table 7.1 Mechanical Properties of Common Metals

Mechanical Properties of Common Ceramics			
Ceramic	Flexural Strength ksi (MPa)	Compressive Strength ksi (MPa)	Hardness (Mohs)
Aluminum silicate	4.5–10	12–25	1–6
Silica foam	0.4	1.4	—
Titania (TiO_2)	20	100	8
Zirconia (ZrO_2)	102	261	7.5
Zircon ($ZrO_2 \cdot SiO_2$)	16	90	7.5
Steatite ($MgO \cdot SiO_2$)	18–21	80–90	7.5
Alumina (Al_2O_3)	72	400	9
Machinable glass	14–15	32–50	5–5.5

Data reproduced from Industrial Press, Inc., *Machinery's Handbook*, 29th edition, Oberg, Jones, Horton, Ryffel. pg. 389. ISBN 978-0-8311-2901-9.

TABLE 7.2 Mechanical Properties of Common Ceramics

requires more force to progress. This effect is epitomized by zirconia (see Sec. 6.2.1.6) when it transforms from its tetragonal phase to the monoclinic phase. The phase transformation is accompanied by a volume expansion that acts to close the crack and resist the tensile force. Also, there is an energy associated with the phase change that must be overcome as well. This is shown schematically in Fig. 7.22. Ideally, the structure consists entirely of tetragonal grains to start, and only after a crack develops will the phase change occur. There are environmental factors that can cause premature phase transformation, and these are discussed in Sec. 8.5.3.

Other methods of toughening ceramics include the addition of "whiskers," but this creates a composite and is better described by composite behavior (Sec. 7.5.4). Because ceramics tend to be very strong in compression and weak in tension, they are often tested using bending. Bend tests are described in Sec. 7.6.4. Hardness testing of ceramics is also more limited and often uses Knoop hardness testing (see Sec. 7.6.1.5).

7.5.3 Polymers

Polymers are weaker than metals and ceramics, but they may be brittle or highly ductile. The strength and ductility of polymers tend to be inversely proportional, but good strength and ductility are combined in engineering polymers. The general aspects of polymer structure and strength effects were described throughout Sec. 5.3 and factors such as monomer structure, cross-linking, chain length, and crystallinity were all identified as controlling the mechanical characteristics of a polymer. Also, the unique molecular structure of polymers makes their deformation distinctly different.

Polymer molecules slide by one another during deformation. If cross-linking is present, then the amount of sliding is restricted, and in order to deform further the cross-links must be broken, which requires primary bonds within the chain to be broken. For that reason polymers tend to deform at low stress initially and then begin to resist further deformation. This gives a response similar to work-hardening in metals, but the phenomenon is completely different. The stress-strain curve for a "typical" polymer is shown in Fig. 7.23.

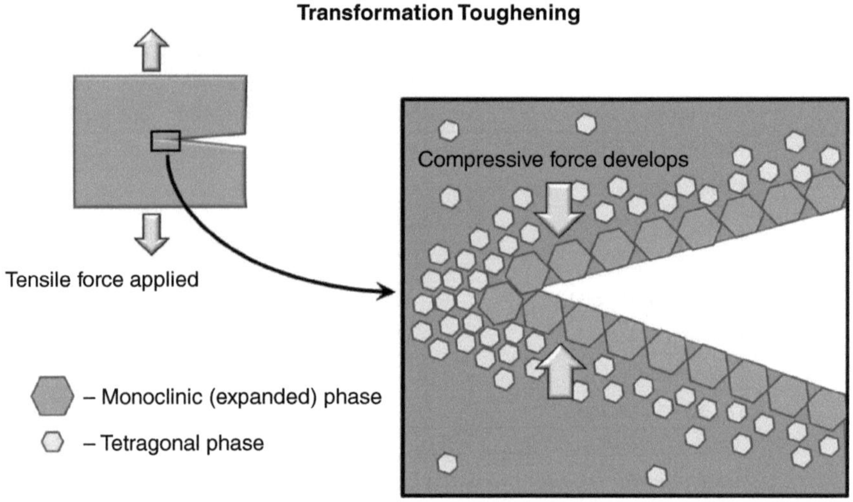

Figure 7.22 Transformation toughening is created by an increase in volume at the crack tip. The increased volume causes a compressive force to develop which helps balance tensile force which acts to extend the crack. Note that the entire structure is made of tetragonal grains before the crack develops.

When crystalline (or technically, semicrystalline) polymers are stretched, the amorphous regions are the first to elongate. As these regions extend, the molecules begin to align (forming more crystalline regions) and resist further stretching. The crystalline regions, which are somewhat randomly oriented depending on processing, begin to align to direction of force and eventually unravel. This process is dependent on the fraction of crystallinity and how easy it is for the chains to move and align. Because of this mechanism, semicrystalline polymers are often tougher than amorphous polymers, which cannot readily orient and crystallize.

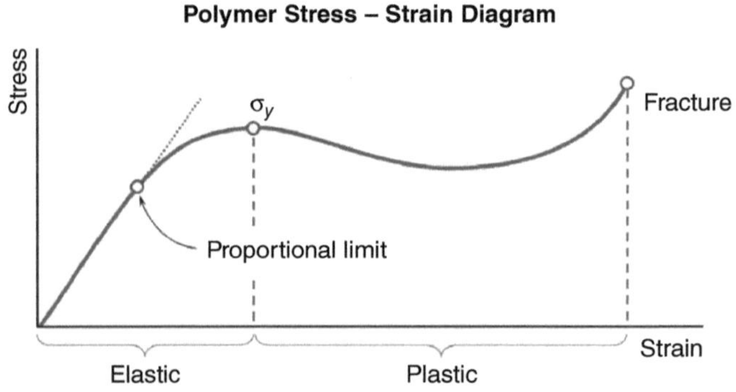

Figure 7.23 Polymers deform by different mechanisms than metals or ceramics. Polymers are dominated by chain sliding and eventually by the breaking of primary bonds within chains either due to cross-linking being present or the inability of chains to continue moving freely (interlocking). Polymers can behave nonlinearly, even in the elastic portion of the curve.

Another side-effect of chain alignment is a change in appearance. Have you ever bent a piece of plastic and seen it turn white? **Blushing** is the phenomenon where a change in appearance is caused by excessive strain. In clear polymers this is related to the increased chain density/alignment in the stretched region.

The elastomer subset of polymers has a response similar to other polymers but the elastic portion of the curve dominates the stress-strain diagram. What does it mean to have little plasticity? We have discussed in this chapter that low ductility and brittleness are the same thing, so that indicates elastomers are brittle. That might seem unlikely given that elastomers can stretch to double their original length (or more). Ceramics such as glass, which are the benchmark for brittleness, do not really stretch at all, or at least imperceptibly so.

One way you can test if something fails in a brittle manner is to take the pieces and attempt to reassemble them. If they fit together and look the same as before they broke, it is brittle. Try it with a rubber band. Stretch it until it breaks. Do the ends fit back together and appear as though it is not broken? A highly elastic material can be brittle. Although some elastomers do have a plastic region in their stress-strain curves, the cross-linking in many of them (thermoset elastomers) will prevent appreciable plastic deformation before failure. This behavior is shown in Fig. 7.24.

Due to their chain configurations, some polymers also exhibit a unique response to stress and strain where they may continue deforming at a constant load. This behavior may be best exemplified by memory foam as used in mattresses, pillows, insoles, and other applications. When a force is applied, the material yields slowly. So long as the force is not sufficient to cause plastic deformation or fracture, once the load is removed, it will return to its original dimensions slowly. **Viscoelasticity** is the deformation behavior where a material absorbs energy during loading and unloading and thereby does not respond instantaneously to applied force.

Viscoelastic materials have a stress-strain response between that of a viscous liquid and an elastic solid. As shown in Fig. 7.25, at constant stress the material continually increases in strain until it reaches equilibrium, but after the force is removed it does not rapidly return to its starting dimensions. This behavior is often marketed by showing

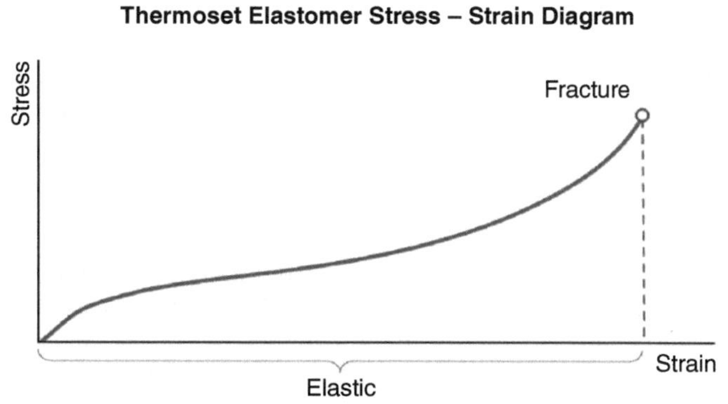

Figure 7.24 Elastomers have a stress-train curve dominated by elastic deformation. In thermoset elastomers, the plastic region is typically minor compared to the elastic portion, so they are characteristically brittle, even though it seems counterintuitive. Thermoplastic elastomers will exhibit higher plasticity, but they are still highly elastic.

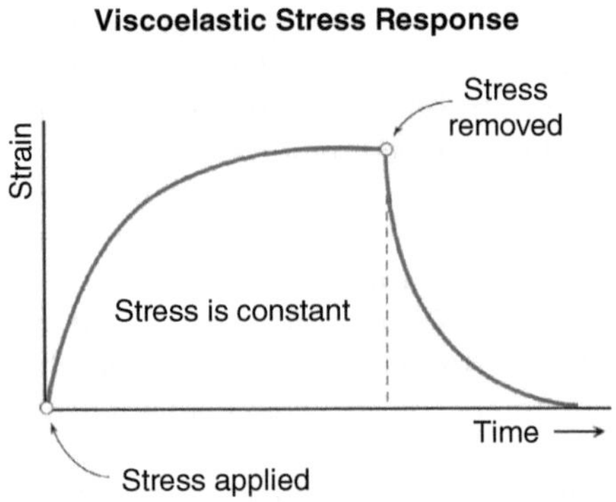

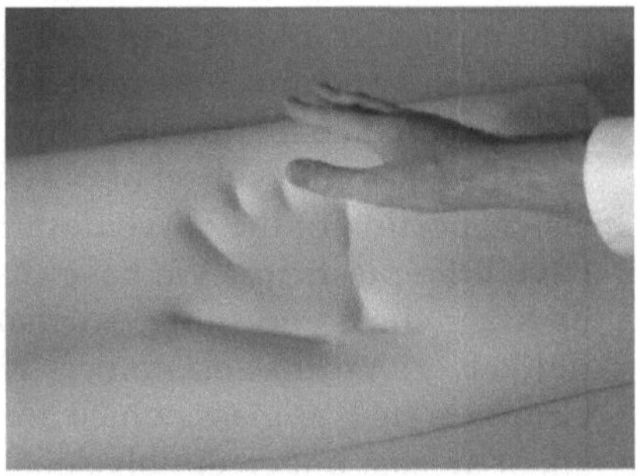

Figure 7.25 Viscoelastic materials have a delayed elastic strain response to stress. When stress is applied, the material deforms slowly until equilibrium is reached. Once the stress is removed, the material slowly regains its original dimensions. (Photo *on right by Johan*, CC BY-SA 3.0.)

how the foam will retain a handprint after the force is removed. Other materials will show little to no appreciable hesitation in *elastic* strain when stress is applied. When plastic deformation is reached, this behavior is not valid. All materials show some degree of **strain rate sensitivity**. The faster a material is stressed, the less ductile it will behave. Some materials are very sensitive while others are not. Viscoelastic materials will show a similar, muted response, but the initial portion of the strain curve will be even steeper and the final strain will be higher.

Polymers can be tested by most methods suitable for other materials. They often use durometer measurement for hardness and a variety of tests for determining the melt viscosity, which is not a mechanical property in the sense we are considering. Some basic properties of common polymers are given in Table 7.3.

7.5.4 Composites

Composites behave like the materials they are made from. They can combine contrasting materials to achieve a middle ground not found in either separate category. The final properties are determined by the ratio of materials and the type of composite it is, such as particle-reinforced, fiber-reinforced, laminate, etc. Much about the properties of composites was presented in Sec. 5.4, but Fig. 5.23 is particularly useful in presenting the balance of properties in a composite. In this section we will look at how we can quantify the final outcome.

7.5.4.1 Rule of Mixtures

The **rule of mixtures** is a mathematical concept that applies fractional values to composite properties based on the ratio of volumes. If a composite is 40% material A and 60% material B, we can expect that its final properties will be 40% like material A and 60% like material B. At least in theory we can predict it will be that way. While mixing can take a variety of forms, we will limit our assessment to continuous fiber-reinforced composites. The general form the rule of mixtures is presented is given in Eq. (7.22).

$$P_c = fP_f + (1-f)P_m \tag{7.22}$$

Mechanical Properties of Common Polymers			
Polymer	**Density g/cc**	**Modulus of Elasticity ksi (MPa)**	**Flexural Modulus ksi (MPa)**
ABS (extrusion grade)	1.05	275	300
Acetyl(homopolymer)	1.41	310	320
Acrylic	1.19	400	400
Nylon 6,6 (extruded)	1.13	390	—
PET/PETE	1.36	500	400
PTFE	2.19	225	80
PVC	1.38	550	400
Phenolics	1.38	—	1000
Polycarbonate	1.19	345	340
HDPE	0.97	156	160
UHMWPE	0.94	110	130
Polypropylene	0.91	155	200

TABLE 7.3 Mechanical Properties of Common Polymers and Their Densities

In Eq. (7.22), P represents any property for the matrix (P_m), fibers (P_f), and composite (P_c) and f is the volume fraction of the reinforcement phase. If we had 40% reinforcement, f would equal 0.4. The volume fraction can be calculated by Eq. (7.23) where V_f is the volume of the fibers and V_m is the volume of the matrix.

$$f = \frac{V_f}{V_f + V_m} \tag{7.23}$$

Let us consider the elastic modulus of a glass fiber-reinforced epoxy. The glass fibers have an elastic modulus of about 70 GPa and the epoxy is about 3 GPa. If we mix 35% glass fibers, the elastic modulus of the composite would be given by substituting values in Eq. (7.22) which gives us

$$E_c = 0.35(70\ \text{GPa}) + 0.65(3\ \text{GPa}) = 26.45\ \text{GPa} \tag{7.24}$$

By adding 35% glass fibers to the epoxy, we can increase its stiffness by more than eight times. Is there a downside? Of course the process is more difficult and the glass fibers are an added expense, but they are not especially costly. Aside from their mechanical properties, the biggest difference between the glass and epoxy is their densities. Glass has a density around 2.5 g/cc and epoxy is nearer to 1.2 g/cc. We can use the same rule of mixtures equation to determine the final density of the composite, ρ_c.

$$\rho_c = 0.35(2.5\ \text{g/cc}) + 0.65(1.2\ \text{g/cc}) = 1.66\ \text{g/cc} \tag{7.25}$$

You can see the glass causes a modest gain in density, but is it offset by the increase in strength? If we divide the strength by the density, we can get the **specific strength.** The strength is the force at which a material fails, but from Eq. (7.24) we are interested in the elasticity, which has the same units. The **specific stiffness** of the composite is about 16 GPa/(g/cc),* while the glass fibers have a specific stiffness of 28 GPa/(g/cc) and the epoxy is 2.5 GPa/(g/cc). So we lose stiffness from pure glass, but that is unavoidable. If we mix a stiff material with a weaker material, we will always end up somewhere between the two (i.e., stiffer than the weak material and weaker than the stiff material). The important thing is that we gain stiffness faster than we gain weight.

Considering pure epoxy as the basis, in the composite we increased density by about 38%, but stiffness increases by 882%. We more than make up for the difference in weight, as evidenced by the higher specific stiffness already calculated. We also gain flexibility. Glass fibers tend to be more brittle and are not much use unless supported by a matrix.

The farther apart the properties are of the components in a composite, the greater the difference a small addition of reinforcement will make. This is clearly seen by the comparison of density and stiffness change in the composite we have been discussing. The two materials have densities that are much closer than their elasticities. We can use this to determine the requirements of a composite by rearranging Eq. (7.22) to solve for a certain property or a set of properties. While the rearrangement can take a few different forms, we will look at a known matrix material with two constraints, density and stiffness.

We will use epoxy for the matrix and assume we need to achieve a stiffness of 30 GPa at a density no greater than 1.6 g/cc. This still leaves the volume fraction of reinforcement and the reinforcement properties unknown. These two properties are not independent of one another, so one cannot be solved without knowing the other and hand calculation is not feasible. Instead we can use basic guiding principles.

For instance, you can choose the properties of the fiber by narrowing to one type and determine the volume fraction required (likely), or you can determine the volume fraction you want and then determine the material properties (less likely). Let us assume we have the choices of glass, aramid, and carbon for reinforcement. The equations would require solving for the volume fraction, f, needed to satisfy either the mass or stiffness. Let us start with stiffness as the limiting factor.

$$
\begin{aligned}
f_{\text{glass}} &= \frac{E_c - E_m}{E_f - E_m} = \frac{(30-3)\text{ GPa}}{(70-3)\text{ GPa}} = \frac{27}{67} = 0.40 \text{ (glass)}\\
f_{\text{aramid}} &= \frac{E_c - E_m}{E_f - E_m} = \frac{(30-3)\text{ GPa}}{(130-3)\text{ GPa}} = \frac{27}{127} = 0.21 \text{ (aramid)}\\
f_{\text{carbon}} &= \frac{E_c - E_m}{E_f - E_m} = \frac{(30-3)\text{ GPa}}{(380-3)\text{ GPa}} = \frac{27}{377} = 0.07 \text{ (carbon)}
\end{aligned}
\tag{7.26}
$$

You can see that all reinforcements are capable of increasing the stiffness using 40% or less of the material, but there is also the density to consider. This is addressed next.

$$
\begin{aligned}
f_{\text{glass}} &= \frac{\rho_c - \rho_m}{\rho_f - \rho_m} = \frac{(1.6-1.2)\text{ g/cc}}{(2.5-1.2)\text{ g/cc}} = \frac{0.4}{1.3} = 0.31 \text{ (glass)}\\
f_{\text{aramid}} &= \frac{\rho_c - \rho_m}{\rho_f - \rho_m} = \frac{(1.6-1.2)\text{ g/cc}}{(1.45-1.2)\text{ g/cc}} = \frac{0.4}{0.25} = 1.6 \text{ (aramid)}\\
f_{\text{carbon}} &= \frac{\rho_c - \rho_m}{\rho_f - \rho_m} = \frac{(1.6-1.2)\text{ g/cc}}{(1.86-1.2)\text{ g/cc}} = \frac{0.4}{0.66} = 0.61 \text{ (carbon)}
\end{aligned}
\tag{7.27}
$$

*The proper units for specific strength are Pa/(kg/m^3), but it does not affect the analysis here.

This is where the field begins to narrow. For stiffness we were looking for the minimum amount of reinforcement, but in terms of density, we are looking at the maximum amount. Glass reinforcement is immediately disqualified because the maximum amount to comply with density is lower than the minimum amount for stiffness. The composite cannot be as stiff and light as required using glass. Next is aramid. It is quite peculiar since it is actually lighter than the required maximum density (1.6 g/cc). It appears you can add 160% aramid fiber, but that does not make sense.

What it does tell us is that stiffness is the only concern, and the 21% addition is our only limiting factor. For carbon fibers, the minimum strength addition is only 7% and the weight addition is limited to 61%. That means we can add fibers to any level between those two and we will satisfy both the strength and density requirements. You can pick any number of properties and matrix or reinforcement materials and determine their suitability this way. Other factors you may want to consider include cost and processing characteristics.

Composites can be tested using most common mechanical methods, but tests must be sure to include both (or all) phases. For instance, hardness is not as straight forward since the indenter is likely to encounter the two materials at different points or in different amounts. This is also important for tensile testing. As described in Sec. 5.4.2.2, the orientation of the reinforcement is important to the resulting properties. Composites are also tested for delamination by bend tests or Mode I fracture testing.

7.6 Methods of Testing Materials

The way forces are used in processing is vital to designing manufacturing protocol for specific materials. Some materials, like copper, may bend or form very well because they are soft, but they may actually be difficult to machine for that same reason. They can actually be too soft and gummy. Other materials, like cast iron, may be very brittle and nearly impossible to form, but they may machine well. Of course, it depends on the specific alloy and the processing used before the operation under consideration. Metals were used to illustrate this concept, but polymers can vary widely in processing characteristics as well.

With so much variation possible, it becomes imperative to accurately determine the properties of materials experimentally. Instead of making assumptions about how a material *should* respond, it can be measured. There is a vast array of testing procedures available, and each has its benefits and disadvantages. Most established testing methods have explicit standards, which are often determined and governed by ASTM International (formerly the American Society for Testing and Materials). There are more than 12,000 ASTM standards, which define proper methods for analysis and quality control. Standards are important to ensure that the same type of test is done the same way around the world so that results are consistent. We will look at some of the most common methods for testing and the information we get from them.

7.6.1 Hardness Testing

The **hardness** of a material indicates its resistance to indentation and/or scratching. A very hard material is resistant to scratches and indenting. Diamond is a common example of a very hard material and it is used to scratch other hard surfaces such as glass. In fact, diamond is so hard, it is often the material used for hardness testing by indentation. Because it is hard and rigid, it gives accurate and consistent results (more on diamond in Sec. 6.2.4.2).

While hardness testing procedures take many configurations, they all seek to determine a quantitative measure of how a material resists indentation or scratching. This means the sample being analyzed must be damaged to get the necessary information. That makes hardness testing a destructive test. Depending on the type of test used and the material/component being tested, the damage may be so minor that it is not really considered damaged. Also, the surface being indented should be very flat and smooth, which may not be possible with certain components.

7.6.1.1 Mohs

The Mohs hardness scale, after Friedrich Mohs (1773–1839), is not strictly a quantitative measure of hardness. Instead, it relies on comparing the hardness of one material to another. The harder of the two materials scratches the softer one. A comparative scale is then developed from 1 to 10, with diamond (being the hardest) as a 10 (Table 7.4). Most ceramics, especially engineering ceramics, are 6 and above and most metals are between 2 and 6. Very soft materials can actually fall below 1. The Mohs scale has limited appeal to engineers because it does not give an exact material property. It is quick and easy to see if one material scratches another in order to determine their relative hardness values, but it is typically not sufficient for making design decisions.

7.6.1.2 Brinell

Brinell hardness testing is based on the simple principle of indenting a sample with a spherical indenter at known load and then measuring the indentation to determine the hardness. The harder the material is, the smaller the indent will be for the same load. By applying Eq. (7.28), which was developed to relate hardness and indentation characteristics, the Brinell hardness number (BHN, or sometimes HB) can be determined.

$$\text{BHN} = \frac{2F}{\pi D(D - \sqrt{D^2 - d^2})} \tag{7.28}$$

Mohs Hardness Scale	
Material	**Mohs Hardness**
Talc	1
Gypsum	2
Calcite	3
Fluorite	4
Apatite	5
Orthoclase	6
Quartz	7
Topaz	8
Corundum	9
Diamond	10

TABLE 7.4 The Mohs Hardness Scale Differentiates Between Materials by a Scratch Test. If One Material Will Scratch Another, It Is Placed Higher on the Scale. The Traditional Number Designations Are Provided Here

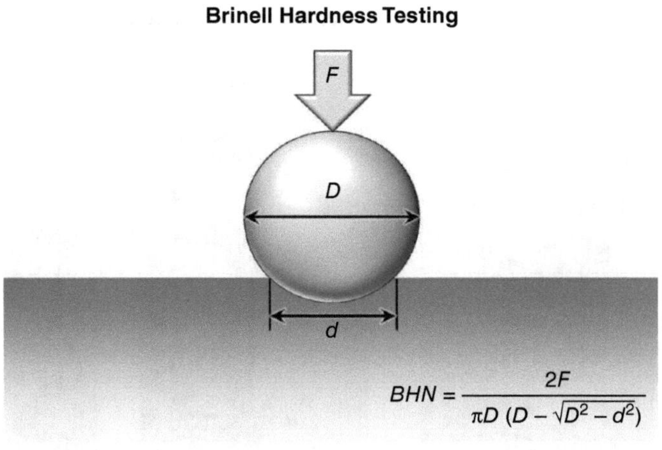

FIGURE 7.26 The Brinell hardness test involves indenting a material with a ball of known diameter, *D*, at a specific force, *F*. After the indentation is made, the indent diameter, *d*, is measured and the hardness is determined by the Brinell hardness equation.

The BHN requires a known force, *F*, the indenter ball size, *D*, and the diameter, *d*, of the resulting indent (Fig. 7.26). BHN numbers are determined using SI units of kilogram force (kgf*) and mm for force and diameter, respectively. The measurement of the indent becomes critical, and care should be taken to ensure its accuracy. Although the "BHN" of a material is commonly cited (e.g., BHN 150), it can be expressed as megapascals (MPa) by converting millimeters to meters and multiplying by gravity (9.81 m/s^2). The units of MPa are more easily compared to other mechanical tests and properties.

As an example, if you indented a material with a 20-mm diameter ball using 500 kgf and the indent was 7 mm in diameter, the BHN would be found as given in Eq. (7.29). The load can vary significantly (typically between 200 and 3000 kgf), and for softer materials a larger ball and/or a lower load are suggested to avoid error. A Brinell hardness tester is shown in Fig. 7.27.

$$\mathrm{BHN} = \frac{2(500)}{\pi 20(20 - \sqrt{20^2 - 2^2})} = 159 \qquad (7.29)$$

7.6.1.3 Rockwell

Rockwell hardness testing uses a variety of indenter materials and geometries to cover a wide range of hardness values. This makes it quite popular and allows for very soft materials and very hard materials to both be tested using the same equipment. Also, unlike the Brinell hardness test, it does not require independent measurement of the indentation. Rather, it determines and reads out hardness based on the *depth* of the

*The kilogram force (kgf) unit is equivalent to 1 kg mass multiplied by the acceleration of gravity (9.81 m/s^2).

Figure 7.27 There are several methods for testing hardness. Most common among these are the Brinell, Rockwell, Vickers, and Knoop tests. Each has benefits and disadvantages, but all are relatively simple. The Vickers and Knoop hardness values are often used in microindentation mode where the indent is so small it must be measured using a microscope. The chief advantage is that it can detect hardness variations over very small distances and do so nondestructively.

indentation. On newer models, digital readout and touchscreen control simplifies analysis, but models with dial readout and control are still very simple to use (Fig. 7.27).

There are only two principle settings to control: the indenter type to be used and the amount of force to be applied. By modifying these two factors, a variety of hardness scales are possible. That means there are multiple Rockwell Hardness scales, and each is indicated by a letter, such as Rockwell C (HR_C) or Rockwell B (HR_B). Each scale has specified testing conditions. The Rockwell C scale uses a diamond cone with a 120° included angle, which is also known as a "brale" indenter, and 150 kgf. The Rockwell B scale uses a 1/16-in ball and 100 kgf. A list of Rockwell hardness scales and their conditions are given in Table 7.5. The way these scales relate to one another is shown in Fig. 7.28.

Conducting the test requires that a preload is put on the indenter by bringing it into contact with the sample. This is also known as the minor load since it is much smaller than the final load. The major load is that specified for the scale (e.g., 60, 100, or 150 kgf). After the minor load is applied, a lever is released manually or automatically (in newer models) and slowly increases the load up to the desired level. After the major load is applied, time is given for the indenter to "settle" into the part. Once the indenter stops progressing into the part, the major load is removed (minor load still applied) and the hardness number is given on a dial or digital readout. That hardness number is really the difference in depth between the positions before the major load is applied and after it is removed. This concept is shown in Fig. 7.29. This method will indicate higher hardness values for lower indentation depths.

The Rockwell hardness test is perhaps the most popular because it is so easy and can be done on a wide range of materials. The sample is positioned on the "anvil," or

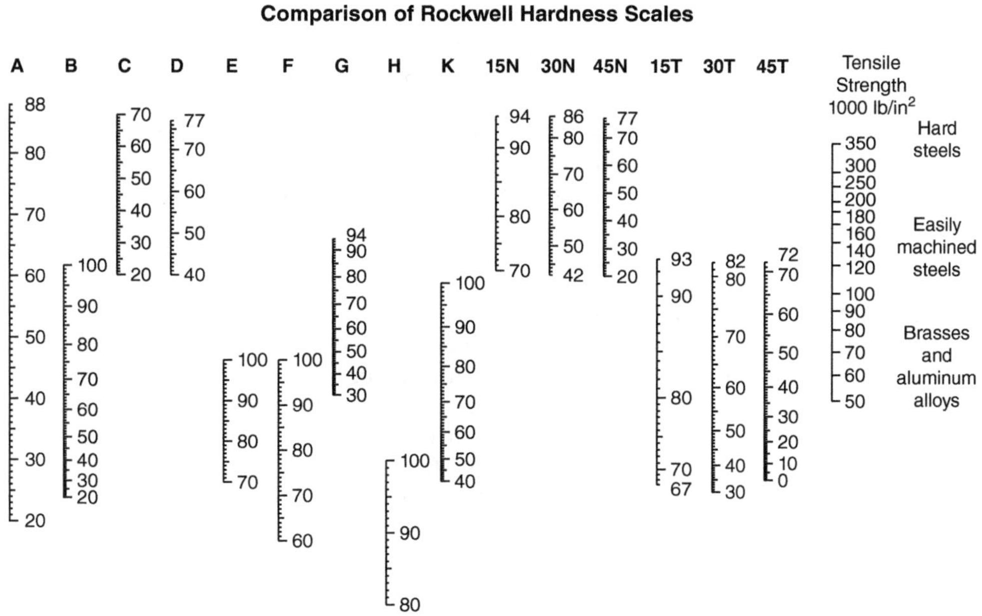

FIGURE 7.28 The various Rockwell scales are compared. Rockwell B and C scales are the most common. (Courtesy of Industrial Press, Inc., from *Machinery's Handbook*, 29th edition, Oberg, Jones, Horton, Ryffel. Table 3, pg. 512. ISBN 978-0-8311-2901-9.)

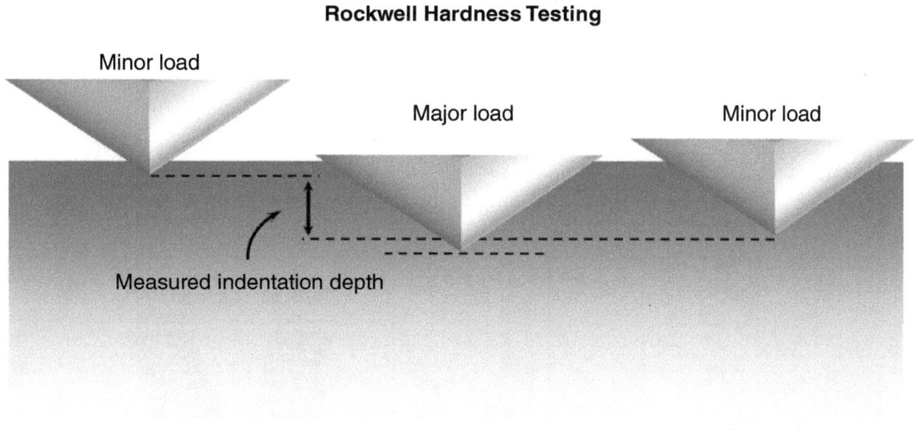

FIGURE 7. 29 The Rockwell hardness test involves applying a small preload, or minor load, and then following that with a major load. After the indentation is complete, the major load is removed and only minor load remains. The difference in the depth at minor load before and after the major load gives the hardness value. The force and indenter geometry determine the scale and resulting hardness value for a material.

the platform below the indenter, and multiple types of anvils are available for testing certain materials, such as cylindrical components. Depending on the load and indenter type, a disadvantage is that it is not surface sensitive. That is, if there is a very thin layer with different properties than the underlying material, the test will reveal more about the bulk properties than those of the surface. Also, when thin materials are indented, they are subject to an "anvil effect" where the underlying stage material skews the result. The Rockwell superficial scales are used for very soft and thin materials. For precisely correlating position and properties, such as surface effects, a microindentation technique is preferable (see Sec. 7.6.1.5).

7.6.1.4 Durometer (Shore)

Some materials are too elastic for Brinell or Rockwell hardness testing. Polymers, often elastomers, are best assessed using the durometer test. This hardness scale was developed by Albert Shore (1876–1936), and so the terms durometer and Shore hardness are often used interchangeably. Like other hardness tests, the durometer is measured using a known load and specific indenter geometry. Unlike other tests, the durometer is measured directly and instantaneously. The indentation depth is used to determine the durometer, which is a measure of the stiffness of the material. The major difference is that the durometer measurement need not leave any permanent indent. This allows for measuring the hardness of very soft and elastic materials, such as rubber.

The durometer test is extremely easy to conduct. A small, handheld device has an indenter protruding from the bottom. That spring-loaded indenter is pressed into a material, and the indenter is pushed upward. That motion results in the movement of a needle, which indicates the durometer number for that scale. When the device is removed, the indenter returns to the zero position and the next test can be conducted. Within seconds a test can be performed and the device is extremely portable. An example of this is shown in Fig. 7.30.

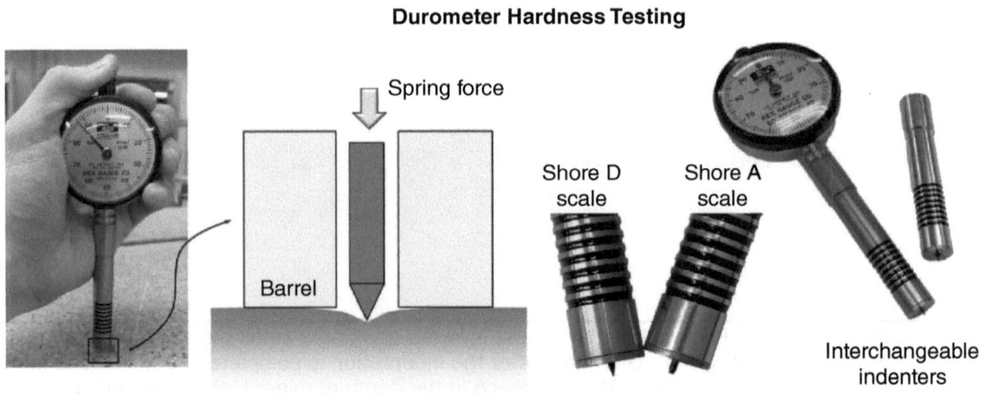

FIGURE 7.30 Durometer hardness testing is very simple and portable. It quickly gives a direct measurement of hardness using a specific load and indenter geometry for each scale. Unlike Rockwell and Brinell hardness tests, the durometer of a material can be determined in the elastic state and does not require that a permanent indent be left. Durometer is limited to softer materials such as plastics and rubber.

Some manufacturers sell individual durometers of a single scale, but others offer a single gauge (dial) that can be transferred from tester to tester, which reduces overall cost. The stiffness of the spring and the shape of the indenter determine what range of materials the scale is suitable for. Indenters may be sharp and needle-like (for hard materials) or they may be flat and broad (for soft materials). They can also be spherical or a truncated cone (like a traffic cone). Durometer hardness scales range from very hard plastics to those of chewing gum consistency. Each scale ranges from 0 to 100, and if the reading is very near the maximum or minimum of the range, the next range should also be tested to increase confidence in the reading. Note that the scale often includes "Shore" in the name, such as Shore A or Shore D, but this is not a technical requirement, and many manufacturers simply state it is "type A" or "type D." See Table 7.6 for details on the various durometer scales.

Rockwell Hardness Testing Configurations			
Scale	**Indenter Type**	**Load (kg)**	**Applications**
A	120° diamond cone (brale)	60	Cemented carbides, case hardended parts, sheet metal
B	Ball (1/16 in.)	100	Unhardened plain carbon steels, copper, and aluminum alloys
C	120° diamond cone (brale)	150	Hardened steel,alloy steels, cast iron, titanium
D	120° diamond cone (brale)	100	Thin steel, case hardened parts
E	Ball (1/8 in.)	60	Aluminum and magnesium alloys, bearing materials, cast iron
F	Ball (1/16 in.)	100	Thin, soft steel, annealed copper alloys
G	Ball (1/16 in.)	150	Malleable iron, bronze, beryllium copper
H	Ball (1/8 in.)	60	Aluminum, lead, and zinc
K	Ball (1/4 in.)	150	Lead alloys, graphite, Babbitt material, very soft, and thin materials
Rockwell Superficial Scales			
15N	120° diamond cone (brale)	15	Superficial scale used for very thin and/or brittle materials such as sheet metal and some ceramics
30N	120° diamond cone (brale)	30	
45N	120° diamond cone (brale)	45	
15T	Ball (1/16 in.)	15	
30T	Ball (1/16 in.)	30	
45T	Ball (1/16 in.)	45	

Note: The appropriate scale is dependent on the material being tested, its prior treatment (such as hardening) and the thickness of the sample.

Table 7.5 Each Rockwell Hardness Scale Has a Unique Condition of Load and Indenter Type

Durometer Scale	Indenter Geometry	Spring Force	Applications
A	Truncated cone 35° included angle	1.81 lbf (8.06 N)	Soft rubber, plastics,and elastomers, soft print rollers
C	Truncated cone 35° included angle	10 lbf (44.48 N)	Medium hard elastomers and plastics
D	Cone 30° included angle	10 lbf (44.48 N)	Hard rubber and plastics such as thermo plastics, flooring and bowling balls
B	Cone 30° included angle	1.81 lbf (8.06 N)	Harder elastomers and plastics. Paper and fibrous materials.Use above 93 Shore A.
M	Cone 30° included angle	0.17 lbf (0.76 N)	Thin materials (down to 0.050 in.)
E	Spherical 2.5 mm (3/32 in) radius	1.81 lbf (8.06 N)	Soft rubber, elastomer foams
O	Spherical 1.20 mm (3/64 in) radius	1.81 lbf (8.06 N)	Very soft elastomers, textile windings, soft granular materials. Use below 20 Shore A.
OO	Spherical 1.20 mm (3/64 in) radius	0.25 lbf (1.11 N)	Light foams, sponge rubber gels, animal tissue
DOO	Spherical 1.20 mm (3/64 in) radius	10 lbf (44.48 N)	Dense textile windings, medium hardness print rollers
OOO	Spherical 6.35 mm (¼ in) radius	0.25 lbf (1.11 N)	Ultra soft gels and sponge rubber

Note: This type of hardness test can be used on very soft materials such as gel as well as on hard materials such as polymers used in helmets. Data from Rex Gauge Company, Inc.

Table 7.6 The Test Conditions and Applications for Durometer Hardness Testing Are Given Here

7.6.1.5 Knoop and Vickers (Microindentation)

Microindentation is achieved using very small loads and precision diamond indenters to create microscopic indents that are measured by the aid of a microscope. There are two types of indenters used in this method: the Knoop diamond and Vickers diamond. The difference between the two indenters is the geometry of the diamond, with the Knoop indenter having different angles than the Vickers indenter.

The Knoop indenter has one diagonal that is seven times longer than the other. It is also very flat, with an included angle of 172° along the long diagonal and 130° across the shorter diagonal, compared to the 120° of the brale indenter used in Rockwell hardness testing. The Knoop hardness test is popular for brittle and thin materials and is among relatively limited resources for testing the hardness of ceramics. The Vickers indenter has equivalent diagonal lengths and a slightly sharper point, with an included angle of 136°. Vickers indenters are also used at higher loads (non-microindentation), but many times a Rockwell scheme is more common for the purpose. Loads for

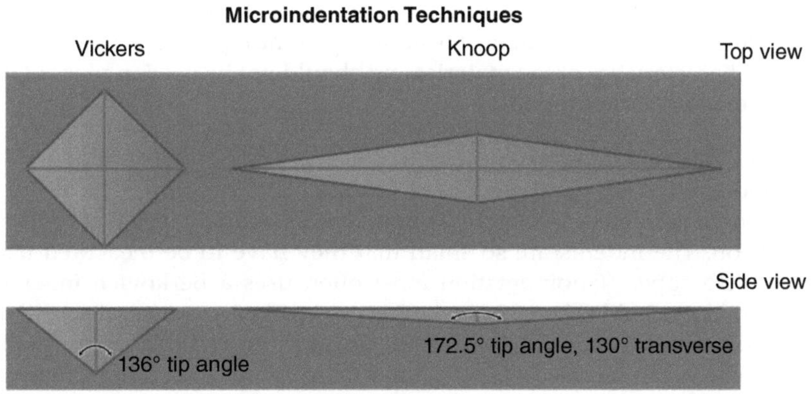

Figure 7.31 Microindentation is often done by using either a Vickers or a Knoop diamond indenter. The indent geometries are shown here. The measure of hardness is determined by the length on the sides of the indents.

microindentation are in grams of force (gf), often less than 1 kgf and typically down to 10 gf. The indent geometries are shown in Fig. 7.31.

The fact that microindentation testing requires measurement can introduce human error, but the equipment is designed to reduce the error by offering micrometer adjustment of the reference lines used to measure the diagonals. Modern version can even measure indents without human intervention through digital optics, but this can introduce other errors if the indents are not well-defined along the edges. Aside from surface sensitive testing and location specific information on hardness, microindentation can be used to determine mechanical properties in traditional units of MPa or GPa instead of arbitrary points on a continuum, such as HR_B 80. For instance, the Knoop hardness of a material can be determined by plugging in the measured length of the long axis, L, and the applied force, F, as shown in Eq. (7.30).

$$\text{Knoop hardness} = \text{HK} = \frac{\text{indenter force (kgf)}}{\text{area of indent (mm}^2\text{)}} = \frac{14.2290F}{L^2} \tag{7.30}$$

Likewise, the Vickers hardness can be calculated using Eq. (7.31).

$$\text{Vickers hardness} = \text{HV} = \frac{\text{indenter force (kgf)}}{\text{area of indent (mm}^2\text{)}} = \frac{1.8544F}{L^2} \tag{7.31}$$

To convert from kgf/mm^2 to GPa, multiply the hardness value by the force of gravity ($9.81\ kg{\cdot}m/s^2$) divided by 1000 (i.e., 0.00981).* This value can then be compared to other mechanical properties such as tensile strength, which are recorded in the same units of pressure. This is useful, as hardness can be used to estimate the tensile strength of a

*1 $kg{\cdot}m/s^2$ is equal to 1 N (newton). A Pa (pascal) is equal to 1 N/m^2. Since area is measured in mm^2, division by 1,000,000 (or 10^6) is needed for the conversion to m^2. This gives units of megapascal (MPa), and to get gigapascal (GPa) divide that value by 1000. Likewise, if area is measured in microns, conversion must account for the difference.

material using the Tabor relation, where the Vickers hardness (in GPa or MPa) is divided by 3. Although a quick estimate, the accuracy of the equation is dependent on the deformation characteristics of the material and should not be used in place of directly testing the material's tensile properties.

7.6.1.6 Nanoindentation

Nanoindentation is a relatively recent development in testing. The technique is capable of making nanometer-sized indents in materials, which gives it exceptional sensitivity to location. The indents are so small that they have to be measured using an atomic force microscope. Nanoindetation most often uses a Berkovich indenter, which is a three-sided pyramid. The technique uses extremely accurate and sensitive hardware to collect data on force and displacement during indentation. This actually results in a curve, which can be used to determine many properties not accounted for in other hardness methods. Other material properties, such as the elastic modulus, can be determined using the appropriate set of calculations (e.g., Oliver-Pharr analysis). Some machines can also perform other tests such as scratch tests to assess wear properties. This type of analysis is run using software.

Sample sizes are limited, the preparation is more rigorous and the equipment is extremely expensive and complicated compared to other tests such as Rockwell hardness testing. The reason nanoindentation continues to be an expanding technology is largely for research purposes where only small volumes of material are produced and multiple levels of analysis are desirable in a single test.

7.6.2 Tension and Compression Testing

The general mechanical properties of a material can be determined through tension and compression testing. These axial forces are described in Sec. 7.2.1 and are a basic component of mechanical assessment. Tensile and compressive forces develop in many applications, especially during bending in which both are present.

7.6.2.1 The Universal Testing Machine

To test a material's response to various forces, simple tests can be performed, often with the same equipment. This equipment is referred to as a **universal tester**, **mechanical tester** or a **load frame**, all of which describe hardware capable of controlled movement at specified loads. The machine has a moveable crosshead, which can have a variety of attachments for various test scenarios. The movement of the cross-head may be mechanical, using twin screws, or hydraulically powered.

The maximum force can be less than a Newton or pound or hundreds of thousands of Newtons or pounds (or more!). The resolution of the test is typically reduced at higher loads, but the variation is less critical. The force is measured by an electronic load cell. That load cell will have a maximum load, such as 1000 lb or 5 kN, but the machine frame may have a higher rating. This allows for a higher resolution load cell with a smaller capacity to be used so long as the material does not require a force beyond the load cell's limit.

Although the testing protocol often relies on the gradual application of force or increase in displacement in a single direction, cyclic tests are also possible. A sample may be cycled through varying amounts of tension or compression at a variety of rates, and hydraulic models are capable of faster loading cycles. The testing can also be done at high temperature by the addition of a furnace, and many other accessories are available

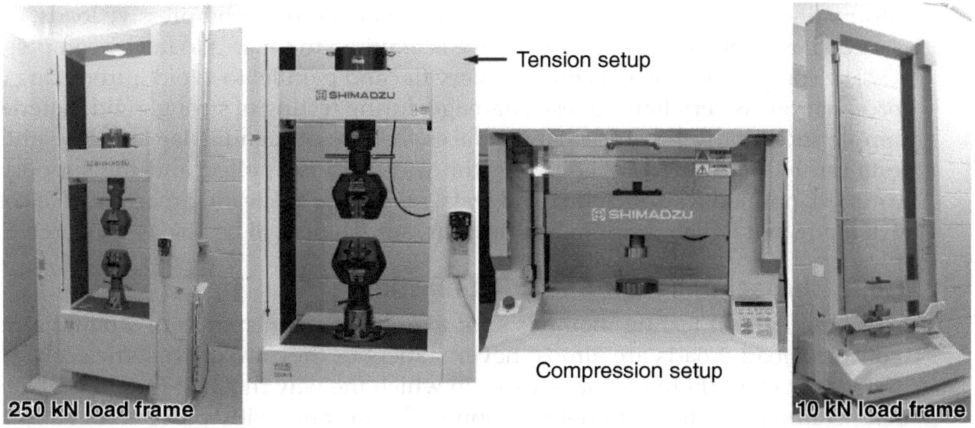

Figure 7.32 Universal testing machines like those shown here are very valuable for collecting data on a wide variety of materials in many different configurations. They can be set up to collect data on tensile (left), compressive (right), bending, and other properties. The rating of the load frame determines the maximum force that can be applied. Here, 250 kN (~56,200 lb) and 10 kN (~2250 lb) versions are shown.

for measurement and control of testing conditions. This versatility makes these machines very popular in industry for quality control, in research for assessing new materials and designs, and in education for demonstrating a variety of mechanical and materials concepts. Examples of these machines are shown in Fig. 7.32.

7.6.2.2 Tension Testing

Tension testing applies an increasing tensile force to a material at a controlled rate, often until it fails. The test sample is often clamped or screwed into fixtures, which hold it vertically in the machine. One end of the sample is fixed rigidly, and the other is slowly moved away, which causes the sample to elongate. The material is slowly pulled apart, or if brittle, will suddenly break once the stress has reached a sufficient level. Most tests rely on a sample "coupon," or standard specimen geometry, to ensure consistent test results. The material will determine what the standard size and shape is, but most often the sample has a reduced cross section in the middle to concentrate stress and control the area in which the final fracture of the material occurs.

The result of the test is a stress-strain curve, which was described in Sec. 7.4.1. The test is often done at a very slow rate, where elongation may be only a millimeter per minute or so. The rate of loading is important because some materials are sensitive to strain rate. At very slow loading they are ductile, but when they are loaded quickly, they act brittle. Low loading rates create a "quasi-static" condition where it approximates a constant load and deformation behavior can be studied in more detail as more data is collected during deformation.

7.6.2.3 Compression Testing

Compression testing is much like tension testing aside from the loading configuration. Standard test samples are used (usually cylindrical) and loading rates are slow. The compression behavior of a material is usually less informative than tension, as materials are commonly more susceptible to failure under tension. Very brittle samples may give better information, however, because they often excel at handling compressive loads. Concrete is a good example of this, and great force is required to cause it to fail by compression.

The compression platens must be very flat and parallel to avoid introducing off-axis forces as well as very hard to resist damage during testing of strong, rigid materials. The sample should also have flat parallel faces that are perpendicular to the height. Other tests can be performed by adjusting the way load is applied and the support of the sample. Common tests include bend tests, which use the testing machine in a compressive mode, but which result in bending stresses. These will be discussed in Sec. 7.6.4.

7.6.3 Impact and High-Strain-Rate Testing

Many mechanical testing methods seek to analyze the response of a material to static (constant) load. Loads are almost never static, but they are often sufficiently approximated that way. There are many cases in which the way the load is applied will have a distinct impact on the material's response. For instance, silly putty essentially "melts" in your hands when you hold it, but if you pull it apart quickly, it will snap into two pieces. What is the difference? The rate at which force is applied is different, and the strain rate, or the rate at which a material is deformed, determines the behavior under load. Although a compression test, as just described in Sec. 7.6.2.3, can give you information about a type of steel, it does not necessarily tell you whether that steel is suitable for a hammer head.

The hammer is meant for impact—situations where the strain rate will be very high, but the actual amount of strain will be very low. In the extreme case are ballistic applications, which encounter very high-impact forces, strain rates, and overall strain. Bullets and armor both need to be evaluated in a way that slow, quasi-static loading cannot capture. That is why impact and high-strain-rate testing are so valuable to many products where motion and force are closely linked.

By varying the temperature during an impact test, the conditions during service can be investigated. For instance, if a component will be used in a low-temperature, cryogenic application, a sample may be cooled in liquid nitrogen directly before testing. Also, if high temperature is needed, a sample may be tested while hot. Some testing machines even have provisions for heating/cooling samples directly in the tester, so no temperature change during loading occurs. This is particularly useful in identifying the ductile-to-brittle-transition temperature (DBTT) of a material (see Sec. 8.2.5 for more on the DBTT).

7.6.3.1 Charpy and Izod

Perhaps the simplest of the impact testing setups are Charpy and Izod tests. Both tests use a weighted pendulum that is released from height and allowed to swing freely until it impacts a sample held in position at the bottom of the swing. This is very similar in principle to a golfer's swing where the ball is now a material sample produced to a specific geometry* for the test. The concept behind the test is that the more energy the sample absorbs during breaking, the less height the pendulum will achieve after impact.

For brittle materials, the pendulum may swing nearly as high as it started, but for tough materials it may not swing far at all. Again, this can be very much like a golfer's

---

*The specimen dimensions are 10 mm^2 in cross section with a 2-mm-deep 45° notch having a tip radius of 0.25 mm. The length of specimen is determined by the test and machine, but 40 mm is the unsupported dimension for Charpy and 22 mm for Izod. The length should substantially exceed those values to avoid error, with 55 mm being the ASTM required dimension.

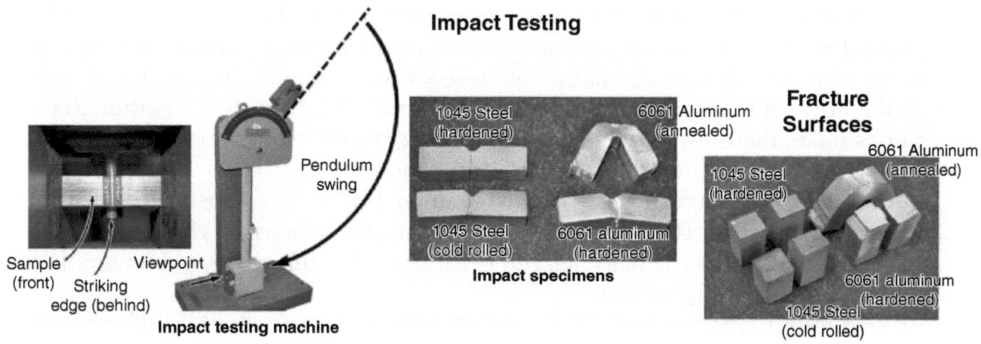

Figure 7.33 An impact tester, such as pictured here is used to determine the fracture properties of a material. Ductile samples deform greatly when impacted, but brittle samples fracture cleanly and with little deformation. You can see that the fractured 1045 steel samples can be fit back together with almost no change in shape. Aluminum, being more ductile under impact, shows more permanent deformation after testing, and the annealed sample is so soft that it does not fracture at all.

swing when teeing off. If the swing is good, he will "follow through" and complete the swing well beyond the point where the club contacts the ball. This situation is like a brittle fracture where very little energy is absorbed. On the other hand, if he is not very good and his swing contacts the ground, his "follow through" may be hindered or completely stopped as the ground absorbs the impact of the club much more than the golf ball will. This second situation is similar to the behavior when impact testing very tough materials. A common impact tester and some examples of brittle and ductile failure are shown in Fig. 7.33. The setup and basic principles of the tests are shown in Fig. 7.34.

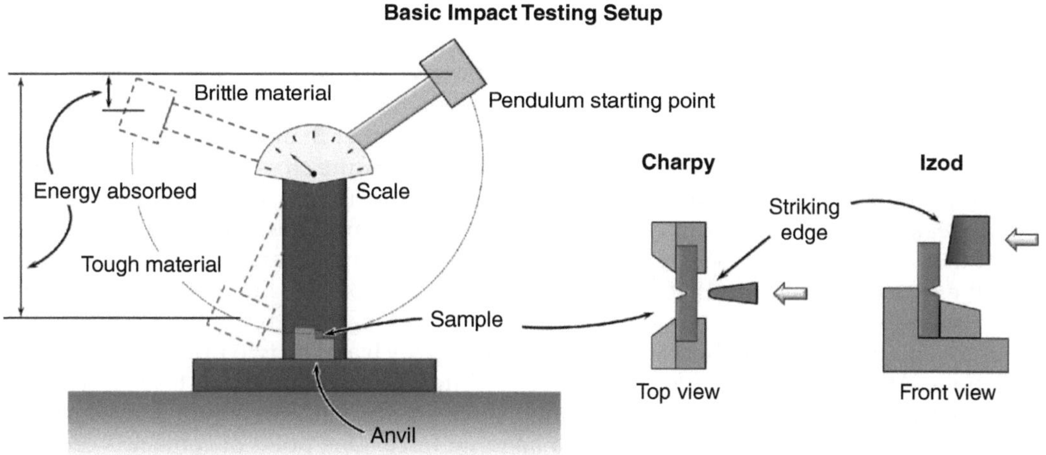

Figure 7.34 Impact testing using Charpy and Izod configurations is a simple and quick way to get basic material properties. A weighted pendulum impacts the sample, and the amount of energy absorbed by the sample (the toughness) is reflected in the height the pendulum swings to after the sample fractures. Brittle materials absorb very little energy and the pendulum swings to a high position. The opposite is true for materials which are very tough.

Most impact samples have a notch of some sort that acts as a stress concentration. When the pendulum strikes the sample, the notch is the location where the fracture will occur. Standard sample geometries include V-notch, U-notch, and keyhole notch samples that follow standards for the size, shape, and position of the notch. Although the two are very similar, there are differences worth noting, mostly related to the configuration of the tests. The Charpy impact test uses a notched samplethat is held horizontally and supported by the anvil. The striking edge impacts the side opposite the notch so it opens up during fracture. The Izod test holds the sample firmly in a vertical orientation. This time the notch faces the hammer, but because of the difference in orientation, the notch still opens up during fracture.

Although Charpy and Izod impact tests provide valuable information on the impact behavior of materials, the rate of loading during impact is not high enough to account for all applications. In situations where extremely high rates of impact are incurred, such as in ballistics, there can be very different behavior. Other tests, such as the **Split-Hopkinson bar**, sometimes referred to as a Kolsky bar, which is basically a gas-powered gun that causes an *incident bar* to impact a sample held against a *transmitted bar*. The main benefit of this method is that stress and strain can be measured at high rate, not just the energy absorbed. The test is not far from hitting your thumb (the specimen) with a hammer. The sample transmits plastic deformation and reflects elastic energy.

7.6.4 Bend Testing

Bend testing is a method of assessing strength in a variety of materials, but it is quite popular in ceramics. The tests reveal the flexural strength and the fracture toughness. Ceramics are particularly difficult to test otherwise since they are so good in compression, which makes them difficult to break, and poor under tension, which makes them difficult to get reliable data due to sensitivity to error. Bending, you may recall from Sec. 7.2.1, is a combination of tension and compression. By subjecting brittle materials to a combination of forces, more information can be gained.

During bending, the highest stress is at the "outer fibers," or what amounts to the upper and lower surfaces in rectangular sections. Because the force is concentrated at these surfaces, the strength of the material may be higher than when measured by tensile testing. This is because tensile testing subjects the entire cross section to equal stress, and any defect throughout the volume may initiate failure sooner. Defects are the "weakest links." These can be voids, compositional variations, such as impurities, or other features that cause stress to concentrate around that location. In bending, unless the surface is compromised, such as by scratches or notches, the defect concentration will be lower.

This brings up another facet to bend testing, which are the two principal configurations. Although bend testing is a simple modification to compression testing, it can be carried out between three or four points. Three-point bend tests are the most common and use three offset points to create a bending force. One point of contact is centered between two supports. In four-point bending there are two centrally located points of contact in addition to the two supports. Although they achieve the same basic result, the four-point bend test tends to report lower values of strength. The reason for this is that the stress is maximized over a greater distance and there is more area for a defect to cause failure. Calculation of the stress during loading will vary slightly and are given

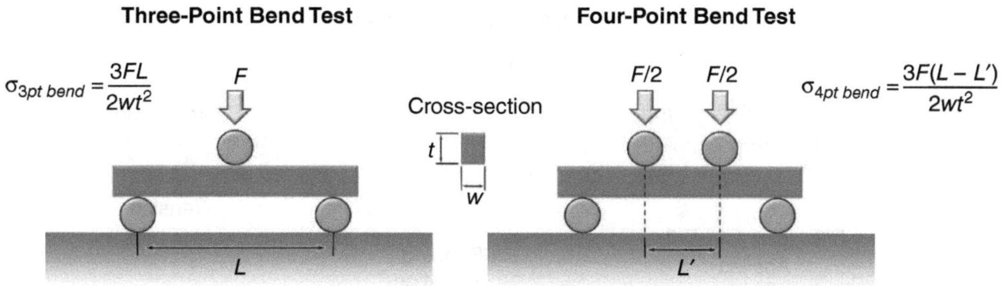

FIGURE 7.35 Bend tests are commonly applied to determine flexural strength of materials, but the configuration of the test determines how that value is calculated. This type of test is popular for ceramic and composite materials, but can be applied to any material.

for three- and four-point bending in Eqs. (7.32) and (7.33), respectively. The variables in the two equations are detailed in Fig. 7.35.

$$\sigma_{3\text{pt bend}} = \frac{3FL}{2wt^2} \tag{7.32}$$

$$\sigma_{4\text{pt bend}} = \frac{3F(L-L')}{2wt^2} \tag{7.33}$$

7.6.5 Fatigue and Creep Testing

Although the methods of testing we have been considering are relatively quick to conduct, they only tell part of the story. Many components are in service for long periods of time under varying environmental and stress conditions. Even something as simple as a spoke on a bicycle wheel goes through varying levels and types of stress thousands of times during its life. During one revolution of the wheel a spoke will be under tension when it is above the axle and compression when it is below the axle. During use the wheel undergoes many revolutions, so the spoke undergoes many cycles between tension and compression. This cyclic loading, even though well below the yield point, can eventually cause failure. This is known as **fatigue**.

Fatigue is not always a change between tension and compression. It may arise by a range of loading conditions and can simply be a variation in tension or compression or other force such as torsion. The suspension springs in a car are always under compression during normal use, but they are under varying amounts of compression during cornering, braking, bumps, etc. Not all materials and designs are capable of handling this type of cyclic loading well, and fatigue testing can reveal the sensitivity a material has to a given amount of cyclic stress. The number of cycles to failure is usually obtained as a function of stress magnitude and variation in loading.

Fatigue is something that typically happens over thousands or even millions of cycles. In order to be reasonable, testing must occur by rapid cycling. Although there are various configurations, one of the simplest and most common involves rotating a material under bending stress, as shown in Fig. 7.36.

As discussed in Sec. 7.2.1, bending causes tension and compression within a structure. If the sample is bent into a U shape, the top of it is under compression and the bottom is

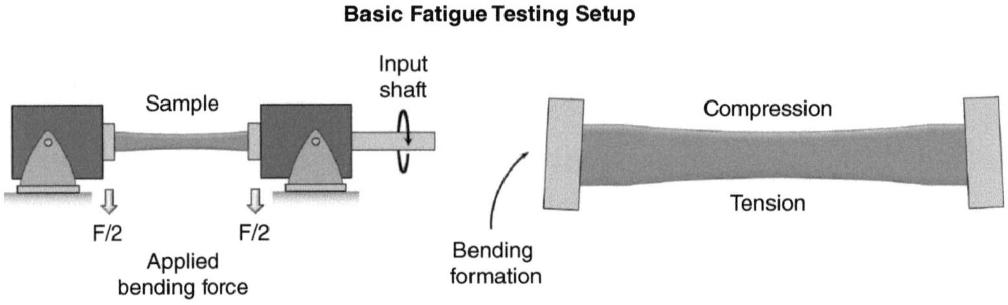

FIGURE 7.36 Fatigue testing often requires thousands to millions of cycles and can be done relatively quickly using a rotating sample holder that cycles the material through tension and compression rapidly by applying a bending force.

under tension. If we maintain the bending force, but rotate the sample by 180°, the side that was under compression (on top) is now under tension (on bottom). Rotation can be conducted at thousands of revolutions per minute, and likewise, cycling between tension and compression can be achieved at thousands of cycles per minute. This makes fatigue testing possible in relatively short time frames.

Fatigue properties are determined by structural, environmental, and thermal conditions. Any feature that causes a stress concentration, such as notches, scratches, or sharp corners, will cause a component to fail faster. Corrosive environments will also decrease the lifetime of a component. Even exposure to air and humidity can have a noticeable impact. Temperature also has a decisive effect. As temperature is raised, the fatigue life drops quickly.

All materials are affected by fatigue, but some are more sensitive than others. For instance, metals like steel and titanium exhibit what is known as an endurance limit. The **endurance limit** is the stress level at which the component can operate under fatigue indefinitely. Others, such as aluminum, show no endurance limit and will fail after a finite number of cycles even at very low stress. In many cases, if the stress is low enough, it is expected that other factors would cause the component to fail or the product would no longer be useful after millions of cycles.

Higher stress will cause all materials to fail by fatigue. In high-cycle fatigue failure, the stress is in the elastic region of a material's stress-strain curve (see Sec. 7.4 for more on this). In low-cycle fatigue, the material actually undergoes plastic deformation during cycling, which greatly reduces its lifetime. The stress level determines the lifetime of a component, and by comparing the stress to the number of cycles until failure, an S-N curve is created. The **S-N curve** shows the number of cycles to failure as a function of applied stress (Fig. 7.37).

Fatigue often does not occur as a regular fluctuation between tension and compression. It is much more common that stress variations happen at varying magnitude and sign (tension positive and compression negative). Think of a flag pole. The amount of force varies as does the direction depending on prevailing winds. There are some times of great magnitude (a hurricane) and other times with much smaller forces. This random application of force cycling is harder to account for, but fatigue testing does provide valuable insights.

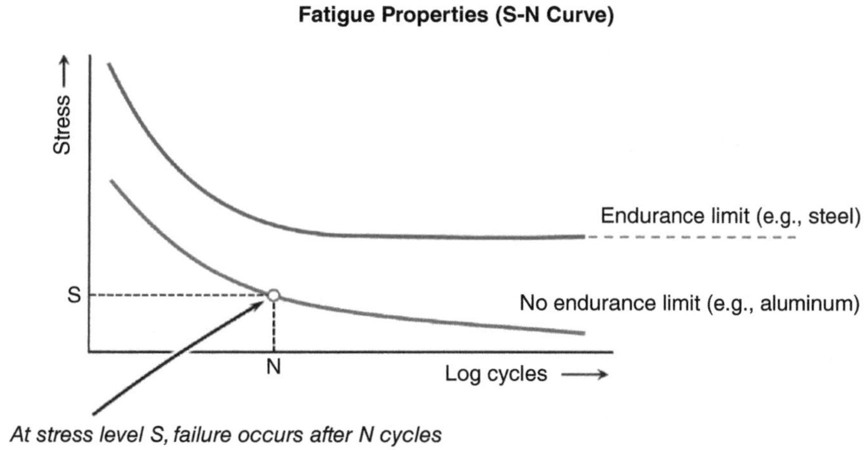

Figure 7.37 All materials are sensitive to fatigue, but some, like steel, can operate indefinitely at low stress. Others, like aluminum, will ultimately fail even at low stresses. All materials will fail sooner at higher stresses, but other factors such as environment and temperature will affect their performance. Note that "log cycles" is referencing the logarithmic scale of the axis.

Alternatively, some components are under constant stress, but they are exposed to varying amounts of heat. A bolt, for instance, is tightened and under constant stress even when the assembly it holds together is under none. This is similar for a tensioned cable (even guitar strings) and many other fasteners and components. At other times the component may have force intermittently applied during operation. What makes these situations unstable is the application of heat. Materials generally soften at elevated temperature and this can cause them to deform slowly over time under conditions that would not normally cause a change to occur. This is known as **creep** and is discussed in more detail in Sec. 8.2.6.2.

The way creep is tested experimentally can vary somewhat, but often it is done though compression. The material will be supported on a platform and a constant load is applied to a punch that contacts the top of the sample. A specific load is applied and the deformation/penetration of the sample will increase over time. The sample and testing rig is contained within a furnace, which is used to supply a constant temperature during the test.

Temperature and stress are very important to the rate of creep. At constant stress, the creep rate, having units of strain/time (e.g., percent per hour), will increase exponentially with temperature, but it also increases exponentially with stress at constant temperature. Therefore, both are very important to the final rate of creep. The reason creep is so important is that as components change size, they may fail to operate as needed and can even cause catastrophic failure.

Even at low stress, elevated temperature can cause the unwanted deformation of materials and it should be considered in design. By testing the behavior at a variety of temperatures, even if above the intended service temperature, a predicted lifetime can be extrapolated back to the actual service temperature. Although it is best to test a material at the exact temperature of interest, this can be time intensive as creep can occur over months and years of service.

7.6.6 A Note on Data Quality

The testing methods mentioned above have standards associated with the test conditions. This puts defined geometries (physical dimensions, surface roughness, etc.), environments (temperature, humidity, etc.) and equipment and testing parameters (rates, materials, accessories, etc.). This is very important because a single material tested under different conditions will give different results. When doing hardness testing, for instance, if the sample is not flat with parallel sides, has a smooth finish (usually mirror polish) and is firmly supported, the results will be skewed to lower values. The results may still be precise, but they will not be accurate.

Precision is the repeatability of a measurement, or the spread of values that result. **Accuracy** is the difference between the nominal result and the actual result of a measurement. If the equipment or sample is not ideally set up, the data will be inaccurate. If the error is consistent, it will still be precise. These concepts are illustrated in Fig. 7.38. This is perhaps the most insidious type of error, since the data appear to be "accurate" if they are consistent. When the results have a wide spread of values, it immediately causes concern about the test conditions. Inaccuracy, however, is easier to adjust for than imprecision. It may be that the samples need to be prepared more carefully or the machine needs recalibrated.

There is an even more troublesome issue with getting good data. It is YOU! Operator bias can affect measurements. Some tests remove this, but others are very susceptible. For instance, when collecting a stress-strain curve using a tension test, the machine records the data. You do not get to decide. When measuring or reading out on analog dials or gauges, if the indicator is between values, it is up to you to decide what the actual value is. It is common practice to add one digit beyond the readout of the machine or other device. If the arrow falls between a 5 and a 6, but is very close to the 6, you may *assume* the value is really 5.9, but not 5.95.

In most cases, the subdivisions will be sufficient that you do not have to do such guesswork. Brinell hardness tests are good example of where this might be a problem. You are required to measure the indent, and when measuring by "eye," you may make an error in either direction. If you have collected the indent diameters on ten indents, and your 11th is slightly larger, you may accidentally or otherwise skew your

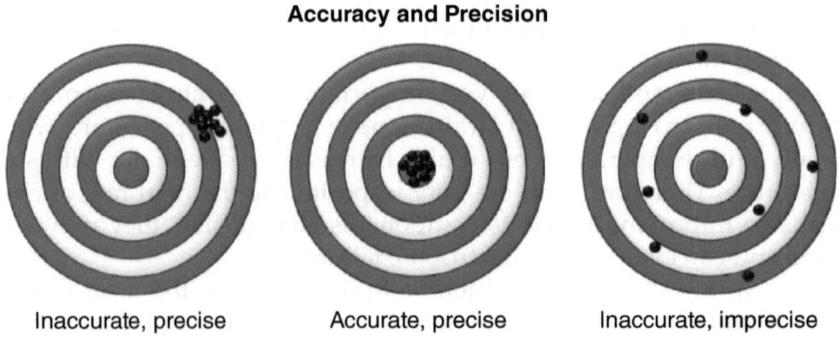

FIGURE 7.38 Data collection depends on many variables to ensure quality. The variables include human, machine, and material inconsistencies. Getting the same result many times is precise, but it does not mean it is accurate. Accuracy depends on getting the correct result. Sometimes it is easier to identify a test that is inaccurate when it is also imprecise.

measurement to the smaller side to keep it in line with what you have already collected. Do not do that! Truth is much more important than consistency. Hopefully you get both.

Not only operator bias, but the consistent placement of samples and reproduction of testing setups can affect data. How you place the sample in the machine. How you initiate or conduct the test, especially where manual input is needed, will greatly affect the outcome. For instance, if you load a tensile specimen at angle to the loading axis, it will fail sooner. If you are doing a manual Rockwell hardness test and you "drop" the indenter into the sample rather than lower it slowly, you will get different results. Be careful to do it correctly, and make sure to do it the same way every time.

There is safety in numbers. It is possible to have accuracy with one test, but it is not possible to have precision. No material, machine, or operator will be perfectly uniform or repeatable. Collect as much data as possible to help identify and remove outliers. Outliers are data points that are significantly different than the others in a set. How significant the difference is can be determined by a variety of methods. There are testing standards associated with identifying and dealing with "outlying observations."* While there is standard statistical analysis that can be applied to determine the error and whether a data point is an outlier and can be dropped, we will leave that for textbooks on statistics.†

7.7 Why It Matters

Again, all material properties are determined by atomic bonding and all processing is limited by the material properties. Much of processing involves mechanical force. Sometimes this force is quite large and beneficial, other times the force may be lower and/or detrimental. It is universally important to understand and control forces in processing as well as during use. We started this process by considering basic force configurations and equations. It is not arbitrary how force is applied. Some materials are isotropic and respond equally well to forces in all directions, but others do not. Ceramics are a common example of this as they tend to be very strong in compression and weak in tension.

Force is not our only concern. It is equally important to know the area over which the force is acting. A small force can cause a large stress. We saw that stress causes deformation, which we called strain. Strain can be temporary (elastic) or permanent (plastic). When you are using a product or producing one, it is exceptionally important to know what will cause the material to transition from elastic to plastic strain. This is displayed compactly in the stress-strain curve, which also gives us a good deal of other information.

The stress-strain curve informs us of the yield point, ductility, strain hardening behavior, and more. Materials have different stress-strain curves, but metals tend to look similar and polymers are distinct from metals. In polymers the stress-strain curves are different for elastomers than other polymers as well. All of it gives us insight on

*The interested read should see *ASTM E178—Standard Practice for Dealing with Outlying Observations* for more on this topic.

†Common practices include the use of graphs, models, or hybrids. Box plots, normal probability plots and other methods, such as Q tests, are all used for the purpose of determining outliers.

what to expect from the material. Some polymers also have a unique viscoelastic response in which they respond slowly to changes in stress. Most materials will react nearly instantaneously to stress.

Composites have properties that are dictated by their constituents and the relative quantities of each. This behavior is quantified by the rule of mixtures calculations. We can choose any property known for the matrix and reinforcement and determine the final property of the composite, at least within a reasonable approximation. Many composites have multiple layers of alternating composition or orientation, and at that point it becomes a very complex analysis. The simple rule of mixtures approach gives us a useful tool in determining basic properties. We can also use it to determine the required loading of a certain reinforcement to achieve a desired result.

Understanding the basics of force helps us understand how materials are characterized by mechanical testing methods. It is not enough to know what a material *should* be like. In many cases, quality control requires us to confirm that a material or component is able to withstand the loading conditions for which it is designed. That can be very simple, such as taking hardness measurements or it can be more involved as in bend tests or tensile tests, which are still relatively simple.

If you do not change the size and/or shape of a material, it is unlikely ready for sale. Most all manufacturing requires that permanent change be introduced, and often this is achieved through force. This chapter has highlighted major themes in force, stress and strain that are common in processing and in machinery, tooling, products, etc. It is intended to give you a quantitative basis for describing and predicting the suitability of a design.

Forces occur at all stages of a product's lifetime as shown in Fig. 7.39. Although we have considered a variety of experimental methods for assessing properties, material "testing" can now be done by simulation using computer-aided, three-dimensional modeling packages. This allows for very complex geometries to be analyzed efficiently

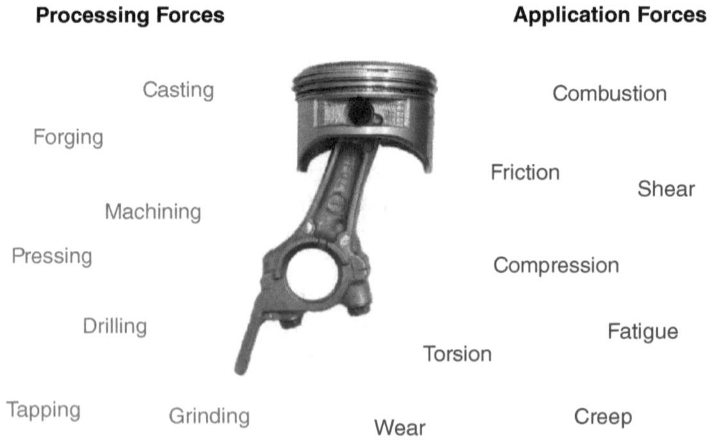

FIGURE 7.39 Knowing about forces is important from initial design to final application. Each step in the processing of components has force involved, and some of them depend on it directly. Forces are developed in a variety of ways during application as well, and knowing about the basic types can make you more effective in designing processes and products.

and accurately using expected conditions during processing or use. Although the computer does the number crunching, if you do not understand what it is doing, how will you know if the result is reasonable or not? You cannot blame a computer for the mistake, because you are controlling the computer. Do not trust "black boxes" and gut feelings. Learn to calculate it.

7.8 The How and Why of Mechanical Concepts in Processing

7.8.1 Chapter Review

1. Why is it important to quantify material properties?
2. How are processing and material properties interrelated?
3. Why is force ubiquitous in processing?
4. What are the five types of loading? Draw an illustration showing each.
5. How does bending result in tension and compression? Where does each force develop?
6. What is the neutral axis?
7. What is torsion? Under what conditions does it develop?
8. How are shear and bending related? How are they different?
9. Why shear so important for material failure?
10. What are statics and strength of materials? How are they different?
11. What is stress? What two ways can it be reduced?
12. Calculate the stress in the following objects:
    a. 500 lb applied to a 0.5-in-diameter rod
    b. 50 lb applied to a square rod 0.03125 in on each side
    c. 4500 lb applied to a strip 1.375 in wide and 0.125 in thick
    d. 12,000 lb applied to a tube with 0.5 in outside and 0.125 in inside diameter
13. Calculate the stress in the following objects:
    a. 600 N applied to a 6-mm-diameter rod
    b. 45 kN applied to a square rod 13 mm on each side
    c. 450 MN applied to a strip 10 mm wide and 3 mm thick
    d. 1200 N applied to a tube with 20 mm outside and 18 mm inside diameter
14. What is strain? How does the original length of component affect it?
15. Calculate the decimal and percent strain in the following situations:
    a. A 12-in-long bar is stretched by 0.04 in
    b. A 35-in bar was originally 36 in
    c. A 100-ft wire is stretched to 104 ft

16. Calculate the decimal and percent strain in the following situations:
    a. A 150-mm long bar is stretched by 2 mm.
    b. A 6-m column is now 5995 mm.
    c. A 20-m bungee cord is stretched to 55.6 m.
17. What does it mean for a material to have a high stiffness? Give an example using a spring.
18. Calculate the elastic modulus for each of the situations below.
    a. Stress of 40 ksi at a strain of 0.02
    b. Stress of 250 MPa at a strain of 0.015
    c. Stress of 110 ksi at a strain of 0.04
    d. Stress of 79 MPa at a strain of 0.005
    e. Stress of 12,000 psi at a strain of 0.032
    f. Stress of 65 MPa at a strain of 0.008
19. Using the stress and strain calculated, find the elastic modulus if you combine the results of
    a. Questions 12.a and 15.a
    b. Questions 12.b and 15.b
    c. Questions 12.c and 15.c
20. Using the stress and strain calculated, find the elastic modulus if you combine the results of
    a. Questions 13.a and 16.a
    b. Questions 13.b and 16.b
    c. Questions 13.c and 16.c
21. Using the upper values (ksi) in Table 7.1, calculate the strain at the yield point for each material below.
    a. Cast aluminum
    b. Aluminum bronze
    c. Monel
    d. Titanium
    e. 6150 steel
    f. 316 stainless steel
    g. 440C stainless steel (hardened)

22. Using the average values (MPa) in Table 7.1, calculate the strain at the yield point for each material below.
    a. Cast aluminum (hardened)
    b. Brass
    c. Copper
    d. Phosphor bronze
    e. Zinc
    f. 1095 steel
    g. 303 stainless steel
23. What is deflection and how does it change with
    a. Increasing force
    b. Decreasing length
    c. Decreasing area
    d. Increasing elasticity
24. How does the orientation between force and area differ between axial stress and shear stress?
25. How does the modulus of rigidity differ from the modulus of elasticity?
26. What is the difference between torque and torsion?
27. What is a moment arm in regard to torque?
28. What is torsional shear stress?
29. How does the torsional shear stress depend on the diameter of a component? Does an increase in diameter increase or decrease the stress for a constant torque?
30. What is the polar moment of inertia of a 1-in-diameter rod?
31. What is the polar moment of inertia of a 19-mm-diameter rod?
32. What is the polar moment of inertia of a tube with 2 in outside and 1.75 in inside diameters? See back inside cover.
33. What is the polar moment of inertia of a tube with 30 mm outside and 10 mm inside diameters? See back inside cover.
34. If a 0.625-in diameter bar is subjected to a torque of 200 in·lb, what is the torsional shear stress in psi?
35. If a 1.25-in-diameter bar is subjected to a torque of 400 ft·lb, what is the torsional shear stress in psi?
36. If a 40-mm-diameter bar is subjected to a torque of 600 N·m, what is the torsional shear stress in MPa?
37. What diameter, to the nearest 1/16th of an inch, is necessary to keep the torsional shear stress below 60 ksi if a torque of 350 in·lb is applied?

38. What diameter, to the nearest millimeter, is necessary to keep the torsional shear stress below 120 MPa if a torque of 800 N·m is applied?
39. What kind of deformation does torsional shear strain result in?
40. Is the angle of twist a direct indication of the strain? Why or why not?
41. How can shear strain be minimized through design?
42. What is the shear strain on a component 0.375 in. in diameter and 2 in long if it is subjected to an angle of twist of 0.01 rad?
43. What is the shear strain on a component 0.5 in. in diameter and 24 in long if it is subjected to an angle of twist of 2°?
44. What is the shear strain on a component 5 mm in diameter and 60 mm long if it is subjected to an angle of twist of 0.025 rad?
45. What is the shear strain on a component 30 mm in diameter and 2 m long if it is subjected to an angle of twist of 6°?
46. Calculate the angle of twist for Question 42 if 500 in·lb torque is applied to a material with rigidity of 3×10^6 psi.
47. Calculate the angle of twist for Question 43 if 80 ft·lb torque is applied to a material with rigidity of 7.2×10^6 psi.
48. Calculate the angle of twist for Question 44 if 70 N·m torque is applied to a material with rigidity of 79 GPa.
49. Calculate the angle of twist for Question 45 if 820 N·m torque is applied to a material with rigidity of 115 GPa.
50. Why are stress-strain diagrams so useful?
51. How are stress-strain diagrams most often generated experimentally (what type of loading)?
52. What is the proportional limit? How does it differ from the yield stress?
53. Why does the stress-strain diagram drop after reaching a maximum?
54. What is atomic basis for elasticity? That is, how does it occur?
55. How does the slope of the elastic portion of the stress-strain curve change with increasing elasticity?
56. What is plasticity? How does it used to define material failure?
57. Why is plasticity so important for processing?
58. What is work-hardening, and why does it occur?
59. How can ductility be quantified? How does it relate to stain?
60. Assume a workpiece 20 in long of square cross section starts 0.75 in on a side is subjected to a rolling operation where it ends with a 0.625-in square cross section. What is the final length? What is the percent reduction in area?
61. Assume a workpiece 60 in long starts with a 0.5-in diameter and is subjected to a rolling operation where it ends with a 0.125-in diameter. What is the final length? What is the percent reduction in area?

62. Assume a workpiece 3 m long of square cross section starts at 35 mm on a side is subjected to a rolling operation where it ends with a 25-mm square cross section. What is the final length? What is the percent reduction in area?
63. Assume a workpiece 87 in long starts with a thickness of 1.5 in and width of 3 in and is subjected to a rolling operation where it ends with a 1.125-in thickness and 3.5-in width. What is the final length? What is the percent reduction in area?
64. Assume a workpiece 150 mm long with a 10-mm diameter is subjected to a rolling operation where it ends with a 6-mm diameter. What is the final length? What is the percent reduction in area?
65. Assume a workpiece 600 mm long starts with a thickness of 12 mm and width of 50 mm and is subjected to a rolling operation where it ends with a 8-mm thickness and 45 mm width. What is the final length? What is the percent reduction in area?
66. How do hardness and toughness differ? How do they often depend on one another?
67. How do the applications differ for hard materials and tough materials?
68. How can toughness be measured?
69. Using Appendix B, calculate the critical flaw size for the following conditions:
    a. Steel subjected to a tensile stress of 40 ksi
    b. Titanium subjected to a tensile stress of 65 ksi
    c. Alumina subjected to a tensile stress of 200 MPa
    d. Polycarbonate subjected to a tensile stress of 40 MPa
70. Using Appendix B, calculate the critical stress for following conditions:
    a. Brass containing a 0.06125-in flaw
    b. Zinc containing a 0.005-in flaw
    c. Silicon carbide containing a 0.01-mm flaw
    d. High-density polyethylene containing a 2-mm flaw
71. What does the fracture toughness of a material tell you about its resistance to failure if a flaw is present? How does it affect the ability to detect a flaw before failure occurs?
72. What are three mechanical property generalities that can be stated about:
    a. Metals
    b. Ceramics
    c. Polymers
    d. Composites
73. How does transformation toughening reduce the likelihood of fracture toughness in a ceramic material?
74. How does the stress-strain curve for polymers differ from that of metals?

75. Why are semicrystalline polymers tougher than amorphous polymers?
76. Elastomers are dominated by what type of deformation?
77. If you break a rubber band, the ends fit back together without showing signs of stretching. What does this indicate, and why does it break in this manner?
78. What is viscoelasticity, and why does it occur?
79. What is strain-rate sensitivity, and how does it affect deformation?
80. Describe what the rule of mixtures indicates about the properties of a composite.
81. Determine the mass of a composite containing 30% reinforcement where the matrix has a density of 1.2 g/cc and the reinforcement has a density of 2.73 g/cc.
82. If you want to achieve a final density of 1.8 g/cc, what percentage of reinforcement (density of 2.3 g/cc) would you add to a matrix with a density of 1.2 g/cc?
83. What volume fraction of reinforcement (elasticity of 120 GPa) would you need to add to a matrix (elasticity of 72 GPa) to achieve an elasticity of 80 GPa?
84. What volume fraction of reinforcement (elasticity of 50 GPa) would you need to add to a matrix (elasticity of 18 GPa) to achieve an elasticity of 45 GPa?
85. Why is it important to verify material properties experimentally?
86. What is the purpose of standards in mechanical testing?
87. How does the Mohs scale determine the hardness of materials? What is the drawback to it?
88. How do Brinell and Rockwell hardness testing differ?
89. A 3-mm diameter indent is put into material using a 10-mm ball under 500 kgf. What is the Brinell hardness of the material?
90. A 4-mm diameter indent is put into material using a 20-mm ball under 2000 kgf. What is the Brinell hardness of the material?
91. What size would the indent be in a 200-BHN material after being indented with a 12-mm ball at 1000 kgf?
92. What Rockwell scale would you use for hardened steel? What are the indenter and load settings?
93. What is the purpose of the Rockwell superficial scales?
94. What are the steps involved in Rockwell hardness testing?
95. What is the anvil effect?
96. Why are hardness methods like Brinell and Rockwell not suitable for many polymers?
97. How does the indent differ in Durometer harness testing from that created by Rockwell or Brinell?
98. What applications is microindentation used for?
99. What are the differences between Vickers and Knoop indenters?
100. How is nanoindentation fundamentally different in the data it collects about a material?

101. What is a universal testing machine, and what kind of testing it can be used to conduct?
102. What is meant by the term "quasi-static?"
103. Why are high-strain-rate tests important to determine material properties?
104. What is the primary difference between Charpy and Izod tests?
105. What material class is often bend tested rather than tension tested?
106. Why do four point bend tests often result in failure at lower stress than three point bending?
107. Describe what fatigue is in relation to material performance.
108. Why is fatigue testing a time consuming process, and how can fatigue be tested in a timely fashion?
109. What is meant by an endurance limit? What does this mean for materials that do not exhibit it?
110. What is creep? How can it be accelerated?
111. In what kind of applications is creep most important?
112. What is the difference between accuracy and precision?
113. How can *you* be the biggest problem to collecting accurate data?

7.8.2 Concept Connections and Extensions

1. A 40-lb bag of cat litter is packaged in a polymer bag. On this bag there is a handle labeled as, "stretchy handle." Why do you think the manufacturer thought it was important to label it this way? Describe your answer in terms of typical mechanical properties of polymers.
2. Consider the discussion on fracture toughness and whiskers described in Sec. 6.4.2.2. Is this a practical way to improve the fracture properties of materials?
3. Consider a spoked wheel. Are the spokes under tension or compression? Explain your answer.
4. Show the steps required to go from Eq. (7.22) to the form used in Eqs. (7.26) and (7.27).
5. Torque is often listed as an important feature of automobiles, especially trucks. What does it mean practically?
6. How could hardness testing be difficult on a viscoelastic material? What can be done to make it more accurate?
7. If bending includes tension on one side and compression on the other, what does this mean for the stress in the middle? Can bending replace separate tension and compression testing?
8. How are fatigue and creep similar? What is different about them? How practical is it to study them under actual use conditions and why?

9. Identify three applications that are served by ceramic materials and specify the properties that make them suitable for those applications. Are the properties mechanical? Optical? Electrical?
10. Come up with your own description of the rule of mixtures and provide an analogy to explain it to someone with no background in materials.
11. Use the basic definitions of mechanical loading to describe three exercises (like at the gym). What types of loads are present?
12. Imagine you see someone doing a burnout in their car (because it would *never* be you). Plot what the required torque would be from the start to the finish. Considering static and dynamic friction will help. How can the threshold to getting the tires to spin be lowered?

CHAPTER 8

Thermal and Environmental Concepts in Processing

Temperature is something we can never dismiss. Whether it is high, low, or somewhere in between, it directly affects performance of large structures, tiny electronics, moving components ... everything! Likewise, the ambient conditions (humidity, salinity, oxygen, *temperature*, and so on) will also directly affect performance. What we find, however, is that these considerations are not just limiting to applications. Quite to the contrary, we rely on them extensively in processing. They are neither good nor bad, but if we understand the underlying effects they produce, we can be judicious about their use.

The basic considerations we will discuss in this chapter are employed in most all design. Whether it is an engine component in a lawn mower or a nuclear containment vessel, knowing and controlling the conditions that materials will be exposed to is essential. We can understand why some materials are good for insulation, why some are good as cookware, why some rust noticeably and some do not and when to use them. Components and products are *always* in some environment, it is our goal to consider the effects.

8.1 How It Works

Temperature affects materials by changing the energy of the atoms they are made from. Environmental factors, which we will broadly consider as corrosion, affect materials by changing their composition or molecular structure. Temperature is something that is always present. Atoms always have some amount of energy, which is manifest through motion. This means that at room temperature materials are quite active, though we cannot see it visually most times. Atoms can move around within a material and actually change the properties over time. By elevating temperature, we increase the activity of the atoms and molecules. When we lower the temperature, we slow things down.

This principle ties into corrosion as well. Corrosion is a chemical change in a material, often resulting in deterioration of that material, and it occurs more rapidly at

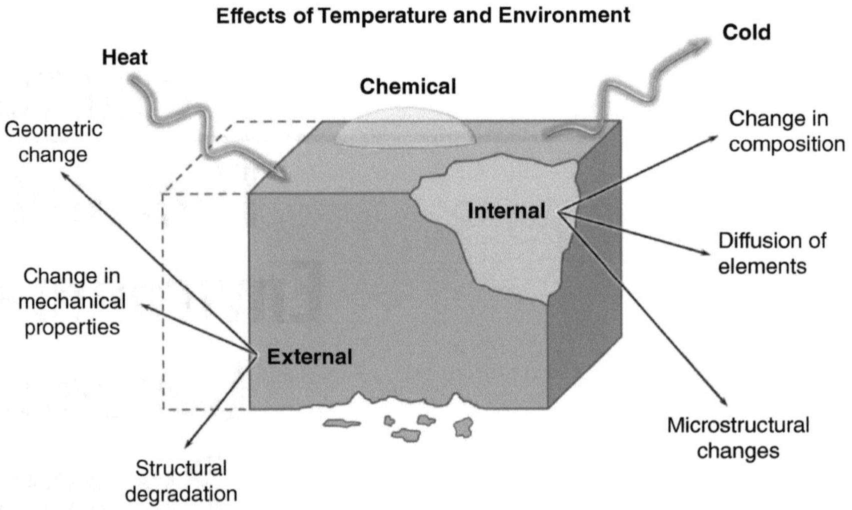

Figure 8.1 Temperature and environmental conditions have a broad impact on materials. The use of heat and chemicals during processing is very useful, but during application variations in ambient conditions can cause severe loss of integrity in materials.

elevated temperature and more slowly at lower temperature. Corrosion also depends on the humidity, any chemicals present, oxygen concentration, and much more, but mostly it depends on the material being corroded. Although corrosion is universal, the extent varies significantly. Some materials form stable corrosion products, and corrosion can be a good thing if it helps protect the base material. Unfortunately, it does not always do that, and it ends up causing a great deal of damage to products. We will look at internal and external considerations of temperature and environment regarding a variety of phenomena, which are summarized in Fig. 8.1. Hopefully it helps you understand some of the common experiences you may have had and aids you in controlling processing and use conditions in the future.

8.2 The Basics of Temperature Effects

First, we will look at how temperature influences materials. No material (none!) is unfazed by a change in temperature. Some have a rather muted response, but they all respond. This basic principle is founded on the fact that heat and energy are related. In fact, heat is simply a form of energy. To remove all heat would be the removal of all energy, which is physically impractical, if not impossible, under most circumstances.* The relationship between heat and energy is best captured by **absolute temperatures**, such as the **Kelvin** and **Rankine** scales, which do not have negative values for temperature. Both scales are

*Using lasers and magnetic fields, scientists have cooled very small collections of atoms to within a billionth of a degree above absolute zero and even slightly below zero (statistically speaking). This does not have any appreciable effect on our discussion, but if you find it interesting, look into Bose-Einstein condensates.

zero at **absolute zero**, which is the theoretical limit of low temperature, where matter is at its lowest energy.

Temperature measured in Kelvin (K) increases in increments equal to degrees **Celsius** (°C), and an increase in degrees Rankine (°R) is equivalent to degrees **Fahrenheit** (°F). For instance, increasing the temperature of something by 1°F is equivalent to increasing it by 1°R, but the actual temperature of the object would be different on the Fahrenheit and Rankine scales. The Kelvin scale does not use the term *degrees.* It is simply Kelvin (i.e., 400 K is four-hundred Kelvin, not four-hundred degrees Kelvin).

To convert from Kelvin to degrees Celsius subtract 273.15 (i.e., 1 K = –272.15°C). To go from degrees Rankine to degrees Fahrenheit, subtract 459.67 (i.e., 1°R = –458.67°F). To see how these four temperature scales compare, see Appendix A. This basic concept is important because many equations and concepts dealing with temperature are best understood thinking in terms of absolute temperature. We will now consider the foundational laws of how temperature affects materials.

8.2.1 Thermodynamic Laws

Carrying forward the heat-energy relationship, we will also use some general thermodynamics to characterize processing and material properties. As the name implies, **thermodynamics** considers how temperature/energy (thermo-) creates changes (-dynamics). While the discipline of thermodynamics is a deep topic within itself and has various uses in chemistry and physics, we will scratch the top-most surface to get a sense of how it impacts processing and materials applications. The foundational laws of thermodynamics are shown in Table 8.1.

The **zeroth*** **law** simply states that at equilibrium (no change is occurring) objects in contact will not exchange heat. When you put a bottle of water in the freezer, the air in the freezer cools the plastic bottle which cools the water inside. Once the water in the bottle is completely frozen and the same temperature as the freezer, all three are in equilibrium. Even though the air and the water are not in direct contact, they are in equilibrium.

Thermodynamic Laws	
Law	**Description**
Zeroth	When two, separate objects are at equilibrium with a third object, they are also at equilibrium with each other.
First	The total energy of an isolated system is constant. Although it can be converted from one form to another, it can be neither created nor destroyed.
Second	The entropy of an isolated system will increase as thermodynamic equilibrium is approached.
Third	The entropy of a system will approach a constant, minimum value as the temperature approaches absolute zero.

Table 8.1 The Four Laws of Thermodynamics Give Us Insight into the Effect Temperature Has on Materials and How We Can Mitigate or Exacerbate Them Depending on Our Final Goal

*Zeroth (0th) is like fourth or sixteenth, just with zero as the number. Have not heard of a zeroth anything before? The reason this terminology is used is because the zeroth law is foundational to thermodynamics, but it was not formalized until after the first law was already established.

The **first law** is related to the conservation of energy. Heat is a form of energy, and when you use heat to do work (like power an engine) that heat energy is converted to motion (kinetic energy) but none of it is lost. Although no energy is lost, the conversion of energy is never perfect, and some of the energy is wasted.

That brings us to the **second law**. There is always some amount of energy inefficiency. That is why perpetual motion machines just will not work. You always have to keep adding energy to keep a process going. **Entropy** is often associated with disorder or the tendency toward equilibrium. It takes energy to maintain differences. For instance, if you weld two parts together, the heated zone is much larger than the weld area. It is entropy that prevents the very hot weld from staying hot while the rest of the part stays cool. The same concept occurs if you add dye to water. The concentrated drop eventually disperses throughout the container to create an even distribution. The highest degree of randomization has the lowest energy (highest entropy). Just like cleaning your home, it takes more energy to keep things organized than to allow them to get disorganized.

The **third law** is related to the second in that entropy and temperature both affect the energy associated with a thermodynamic system. At absolute zero the entropy is zero in a perfect crystal. Real materials are *imperfect*, and at absolute zero there will still be some structural disorder even though they have no thermal energy. The entropy remaining is known as the residual entropy. The thermodynamic laws are *laws*, and that means they govern the response of materials to heat and other energy. The rest of the temperature-related discussion fits into this framework, and keeping in mind the basics of these laws will be useful as we go along.

8.2.2 Thermal Conductivity

One of the basic parameters a material possesses is its thermal conductivity. A material's **thermal conductivity** determines how quickly it absorbs and releases heat. Thinking of the concepts of equilibrium and entropy discussed above, it follows that a difference in temperature can only be eliminated by the transfer of energy. The material between the high- and low-temperature zones will determine how quickly equilibrium is reached. Let us look at a practical decision that involves thermal conductivity. On a cold day, if you have to wait for a bus, what bench would you rather sit on, wooden or metal? Why? Is the metal bench colder? Assuming the temperature is fairly constant, both benches will be in equilibrium with the air. If we apply the first law, then both benches are in equilibrium with each other. They are the same temperature. Why does the wooden bench make more sense?

This is a simple but illustrative example of thermal conductivity. Metals *feel** colder in the winter because they can transfer heat faster. Energy travels from high concentration (you) to lower concentrations (cold objects). You perceive the object as cold by the rate at which you warm it up. Metals have characteristically high thermal conductivities. Whether it is hotter than you or colder, the heat can move into and out of a metal object more quickly than most ceramics, polymers and their composites.

For our purposes, we will limit our consideration of heat transfer to conduction through a solid material. Although heat can be transferred other ways, we will ignore

---

*If two materials are the same temperature but feel differently, that indicates we do not actually feel *temperature*. Instead, what we perceive as hot and cold are rates of heat *exchange*.

the transfer of *matter* such as would occur by pouring water on a hot bar of steel, for instance. Structurally speaking, the metallic bonding in metals as well as their crystal structure will determine thermal conductivity (see Secs. 2.3.1.1 and 3.2 for more on structure). The majority of heat transferred in metals is accomplished by free electrons. The atoms will also contribute, however, and this is important in nonmetals. Heat is a measure of energy and that energy is contained within the material by movement of atoms. The atoms will not move much through the material, but they vibrate, rotate, etc. *in place*.

In their crystal positions, the atoms transfer energy to one another by jiggling about until they transfer some of that motion to their neighbors. Energy can travel as waves through a crystalline material (called a phonon). The spacing of the atoms and perfection of ordering will affect how well the waves travel. Just like a rocky beach breaks up waves, imperfections in the structure will do the same. This is minor compared to the electron contribution to heat transfer. Ceramics and polymers, which do not have an abundance of free electrons, do not conduct heat well since they rely on transfer of energy to occur between atoms. Polymers and ceramics can be modified by adding conductive materials, but in general they are not used for heat transfer purposes.

This means applications such as radiators, soldering irons, and teapots are best served by metals. In applications where heat needs to be blocked from traveling, nonmetals are ideal. Ceramics are an important insulation material as they tend to have low thermal conductivities and are able to withstand high temperature. To put some numbers behind this concept, the thermal conductivity of some common materials are compared in Table 8.2. Thermal conductivity is often denoted by κ (Greek lower case letter, kappa). It is measured as energy per unit time per temperature change. SI units are W/m·K and Imperial units are BTU/h·ft·°F.

8.2.3 Coefficient of Thermal Expansion

One of the most common properties associated with temperature is the **coefficient of thermal expansion** (CTE). The CTE describes the relationship between temperature change and dimensional change. This is a material property that is not readily affected by processing,* but which has important effects on process control. The amount of energy (heat) a material has determines the way the constituent atoms interact. If we consider atoms to be localized, they will not travel through the material, but they are not motionless. They vibrate about an average position.

This is kind of like your movements throughout your home. Even if you do not leave it, you move about within that space. You are in a single location, broadly speaking, but you still move around some. Atoms vibrate more and more at higher temperature and they require more space to do so. This concept is shown in Fig. 8.2. The curve shown in Fig. 8.2 describes the behavior of atoms in a material. If the atoms get too close they repel one another (left of point A). As they get farther away (right of point A), they want to move closer (absence *does* make the heart grow fonder!). This distance-energy relationship is known as the Lennard-Jones potential (after Sir John Lennard-Jones, 1894–1954). Under normal conditions, the atoms will be spaced at the most favorable distance, which is at the lowest energy in the "trough" of the curve (point A in Fig. 8.2).

*The CTE can be made more anisotropic through processing (e.g., chain alignment in polymers during extrusion), but the CTE is based on structural factors that are determined by the material more than the way it has been processed.

Thermal Conductivity of Common Materials			
Material	Class of Materials	Thermal Conductivity (BTU/h·ft·°F)	(W/m·K)
Diamond	Ceramic	1272	2200
Copper (pure)	Metal	226	391
Aluminum (6061)	Metal	104	180
Aluminum (2011)	Metal	82.5	143
Brass (30% Zn)	Metal	70	121
Nickel (pure)	Metal	43.3	75
Gray cast iron	Metal	28	48
Stainless steel (316)	Metal	9.4	16
Titanium (pure)	Metal	9	16
Inconel	Metal	6.5	11
Marble	Ceramic	1.2	2
Concrete	Ceramic	0.5	0.8
Glass	Ceramic	0.5	0.8
Polyethylene	Polymer	0.2	0.4
Calcium silicate	Ceramic	0.036	0.063
Fiberglass	Composite	0.026	0.045
Polystyrene (foam)	Polymer	0.02	0.03

TABLE 8.2 Thermal Conductivities of Some Common Materials

As temperature increases, the distance between atoms increases and the cumulative effect of the increased atomic spacing is for the part to expand. The shape of the curve is different for each material (actually, each atomic interaction). If the trough is very deep and narrow, the atoms will not change spacing much, even at high temperature. If the trough is shallow and wide, the material is more sensitive to temperature and will have a higher CTE. Increasing temperature causes the point to move vertically, and the shape of the curve determines how far apart they get for that increase in energy. For instance, the dotted line in Fig. 8.2 would have a lower CTE than the solid one.

Although there will be expansion in all three dimensions, the CTE is often cited as a single unit of length, not a volume. This is more appropriately called the *linear* CTE (denoted by α_L). Being a coefficient, we know it is a constant (based on the specific material) and that it modifies some variable in an equation. The equation for determining the amount a material will lengthen or contract for a given temperature is given by

$$\Delta L = \alpha_L \Delta T L \tag{8.1}$$

Equation (8.1) tells us that a part of a given length, L, will change in length, ΔL, when the temperature is raised or lowered by ΔT. The CTE will have units of *per temperature change* (i.e., 1/°F, 1/°C, 1/°R, or 1/K). It is good practice to use absolute temperature

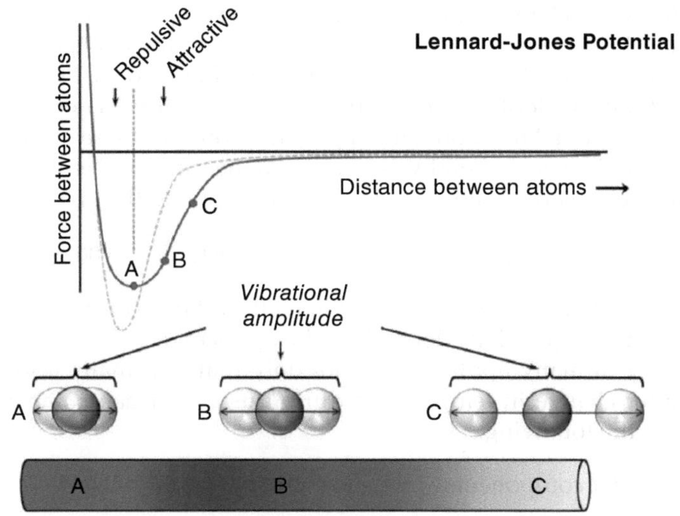

Figure 8.2 The temperature of a material is really a measure of energy of the individual atoms. The more energy the atoms have, the more they vibrate and the more space they need. This increased movement of the atoms results in an overall increase in dimensions at elevated temperatures. The shape of the energy curve determines the CTE. A material represented by the solid line would expand more than one given by the dotted line at the same temperature.

when calculating ΔT because negative temperatures can give erroneous values if signs are accidentally dropped. Done carefully, however, there is no difference. Recall that for a *change* in temperature, Δ°F is equivalent to Δ°R, and Δ°C is equivalent to ΔK. The general behavior with temperature is shown by Fig. 8.3.

For a single material, length increases with temperature, and the overall change in length increases with initial length (i.e., a 6-in bar will extend twice as much as a 3-in

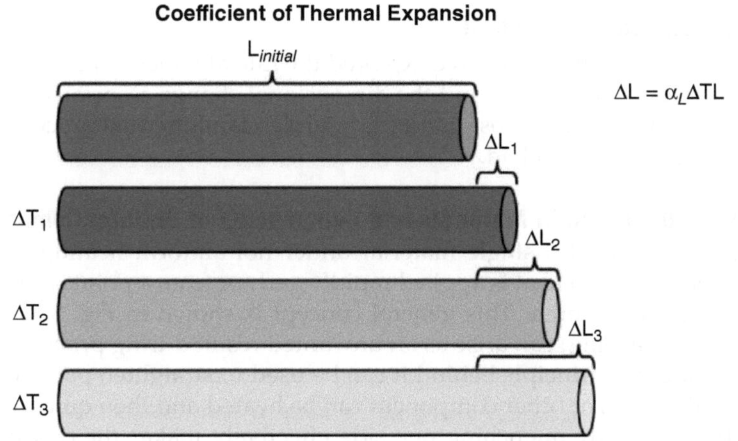

Figure 8.3 Materials tend to expand as the temperature is raised. The amount of expansion is determined by the original dimensions, the amount of temperature change and the CTE. For parts of equivalent size, a material with a larger CTE will expand more.

bar for the same change in temperature). For example, if we take a steel bar, 12 in. in length and heat it up by 100°F, we can determine the change in length. Referring to Eq. (8.1), we see we have ΔT and L, but we still need α_L. CTE values are compiled in tables and must be looked up. The use of "steel" is very generic as there are hundreds of steels (see Sec. 4.2 for more on that). When you find these values, however, you will discover that they are generally very close to the same. We will use 6.5×10^{-6} °F^{-1} for α_L (the units are sometimes given as in/in/°F). Applying the given values we get

$$\Delta L = (6.5 \times 10^{-6}\ °\text{F}^{-1})(100°\text{F})(12\ \text{in}) = 0.0078\ \text{in} \tag{8.2}$$

Is the value in Eq. (8.2) a small one? Will it affect the performance of a component? The answers to those questions depend on the application. Thousandths of an inch matter significantly in many applications. Not only that, but many applications will have a much larger change in temperature during use. Thermal expansion/contraction will *decrease* for the following:

- Smaller components
- Lower temperature variation
- Lower CTE

Because an application typically requires a fixed size and temperature, the easiest variable to change is the CTE. The CTE is most effectively changed by changing the material. Ceramics tend to have low CTEs, metals are moderate and polymers have large CTEs. As always, there are exceptions. To give some basic comparison, consider the example in Eq. (8.2), but now we will apply it to silicon carbide (SiC) and polyethylene (PE). SiC, with a CTE of only 1.5×10^{-6} °F^{-1} will extend by 0.0018 in, and PE, with a CTE of 111×10^{-6} °F^{-1}, will extend by 0.1332 in. These are significant differences, and although there are often properties that trump CTE as determining factors in material selection, it is still a very important one. The CTEs of a variety of materials are given in Table 8.3.

8.2.4 Thermal Distortion

In the last two sections we have explored the rate at which heat travels through a material (thermal conductivity) and the dimensional change associated with that change in heat (CTE). We can use those concepts to understand thermal processing and the effects of heat, both good and bad.

8.2.4.1 Distortion in Materials and Components of Uniform Composition

First, let us look at a single material under nonuniform heating/cooling. This commonly occurs when heat is applied rapidly and/or from a point source, such as a welding gun or blow torch. This general concept is shown in Fig. 8.4. Although thermal distortion (warping) can arise as an unwanted result during processes such as welding and casting, the principle behind it can be used to straighten parts as well. An already bent shaft, sheet, or other component can be heated and then quickly cooled to reverse the strain. Rapidly quenching one side effectively makes the *opposite* side at a much higher temperature, thereby causing the material to bend away from the side being cooled. This process is typically referred to as *flame straightening* since a torch is commonly used.

Coefficients of Thermal Expansion for Common Materials			
		Coeff. of Thermal Expansion	
Material	**Class of Materials**	**(10^{-6} °F^{-1})**	**(10^{-6} K^{-1})**
Polyethylene	Polymer	111	200
Nylon	Polymer	40	72
Polycarbonate	Polymer	39	70.2
Polystyrene	Polymer	38.9	70
Polyvinyl chloride	Polymer	28	50.4
Fiberglass[a]	Composite	11–22	20–40
Aluminum	Metal	12.3	22.2
Brass	Metal	10.4	18.7
Copper	Metal	9.3	16.6
Stainless steel (316)	Metal	8.9	16.0
Nickel	Metal	7.2	13
Inconel	Metal	7	12.6
Carbon steel	Metal	6.6	12
Gray cast iron	Metal	6	10.8
Glass	Ceramic	5	9
Titanium	Metal	4.8	8.6
Marble[b]	Ceramic	3.1–7.9	5.5–14.1
Sapphire	Ceramic	2.9	5.3
Aluminum nitride	Ceramic	2.9	5.3
Silicon carbide	Ceramic	1.5	2.8
Invar	Metal	0.8	1.5
Diamond	Ceramic	0.7	1.2
Quartz	Ceramic	0.2	0.3

[a]Value depends on polymer used for the matrix.
[b]Natural materials have higher variation in properties than engineered materials.

TABLE 8.3 The CTE of Common Materials

Thermal distortion or stress issues can also occur for a part where dimensions vary significantly over the component. This is very important in casting (see Chap. 11) and heat treatment (see Sec. 9.3). In both processes the material is quickly cooled from very high temperature. In casting, the rate of cooling is related to the mold's thermal conduction and starting temperature. It is not a fast cooling rate that is detrimental. It is *uneven* cooling that will cause distortion or cracking. Casting is a solidification process where volume change is a factor and mold restrictions on geometry also impacts quality. Uneven cooling causes uneven solidification which causes internal stresses as well.

In heat treatment fast cooling is often necessary. Part of the mechanism by which materials harden during quenching is time-dependent and cannot be slowed down

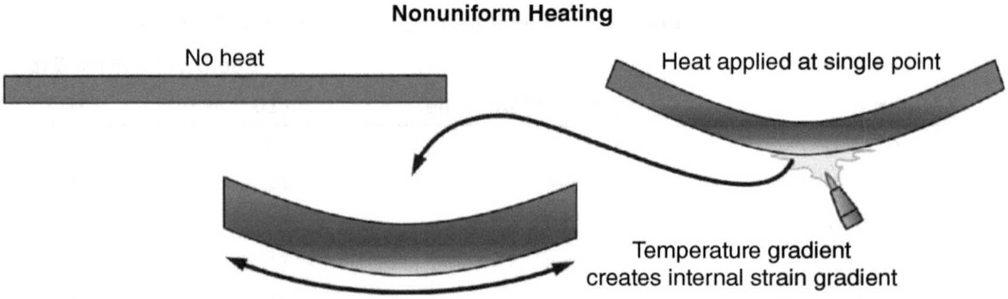

FIGURE 8.4 Thermal distortion arises when uneven heating exceeds the materials ability to resist the strain caused by thermal expansion. While this is often a problem, it can also be useful in straightening parts which are already bent.

without severely limiting or eliminating the effectiveness. If fast cooling is required, then sharp corners and uneven section thicknesses should be avoided, as noted in Fig. 8.5. These areas introduce stress concentration that magnifies the effects of temperature gradients. If a component requires small notches or protrusions, round-off corners, and if possible, separate into multiple parts that can be heat treated separately.

8.2.4.2 Distortion in Materials and Components of Nonuniform Composition

Thermal distortion is caused by uneven heating in a single material, but it can also be caused by uniform heating in two, rigidly connected materials. So what was it that caused distortion in a single material and how can we extend that to combinations of different materials? Recall, it is the difference in thermal expansion that is responsible for the development of strain. When combining two or more materials, the difference in their coefficients of thermal expansion will determine the difference in their size variation with temperature.

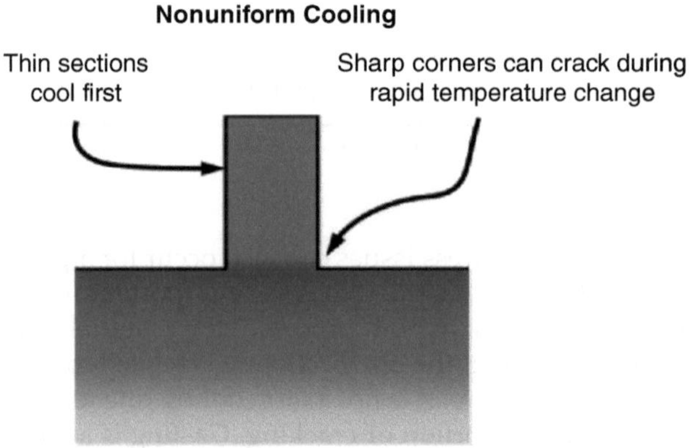

FIGURE 8.5 Causes for distortion and cracking of castings and heat-treated materials are often linked to uneven section mass and sharp corners. These should be eliminated if possible. Slower cooling will reduce the stress on parts, but many operations cannot be done efficiently or effectively under slow cooling.

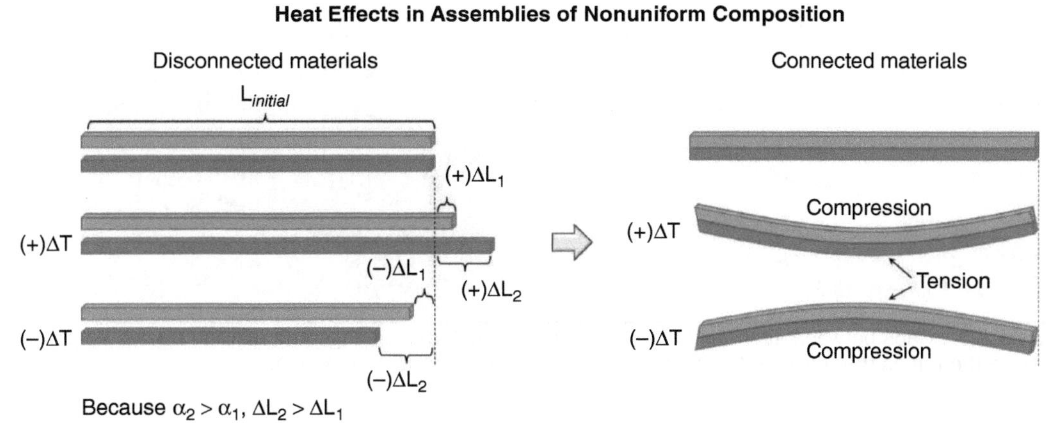

Figure 8.6 When materials with different coefficients of thermal expansion are combined, they will put stress on one another during changes in temperature. In this case the top material has a lower CTE and will undergo less dimensional change during elevation or depression of the temperature. This will either cause it to be under compression or tension, depending on the change in temperature from the equilibrium condition. While this property is useful in some applications, it is detrimental in many others. Materials should always be carefully paired when designing assemblies.

To illustrate, look at Fig. 8.6. Two materials with differing CTEs will differ in response to temperature changes. When the two are disconnected, or at least can move relative to one another, there is no strain gradient. When the two materials *are* rigidly connected, the unequal dimensional change results in a strain gradient similar to that developed under bending. One material is trying to stretch the other (the one with a higher CTE, or α) and in response, the opposite material is trying to compress the first.

If you refer back to Eq. (8.1), you will see that the magnitude of the forces is dependent on the difference in α, the length of the connected segments and the change in temperature. The distortion, which arises due to these differences, can be beneficial. One example is a thermostat that uses a bimetallic switch. The switch is opened (heat turns off) when the two metals are at a sufficient temperature to cause bending away from the contact (this is the room temperature you set). Upon cooling the strain is released and the material straightens, thereby making contact with the circuit and turning the heat back on.

It is infeasible to make every product out of one material. It is also unreasonable to expect a product will never change temperature. Because of these basic constraints, most design necessitates that we must account for differences in thermal expansion. Most of the time, the consideration will be minor. Recall, factors that limit the effect of thermal expansion are a small size, low CTE, and low temperature variation. Many products will not vary over a large temperature span during normal operation. In fact, most will not leave a reasonable range of ambient temperatures such as encountered outdoors (a range of less than 80°F (44°C) in most climates).

Also, most products are limited in size. Recall, the variation is considered between rigidly connected materials, and even if an assembly is large, most rigid connections will be relatively small. The last factor, the CTE of the materials, is something that can be addressed. Consider Table 8.3 again. You will see that polymers typically have the

FIGURE 8.7 Gaps between sections of large structures, such as buildings and railroad tracks, help reduce the forces that develop because of thermal expansion and contraction.

largest CTEs while ceramics have low CTEs. "Special" materials can have very low, or even negative* CTEs. Invar, for instance, is an iron-nickel alloy with an extremely low CTE (for a metal). Therefore, it can be paired with some ceramics, even in high-temperature applications. Typically thermal expansion is most critical to accurately account for in applications requiring:

- High-precision fitment
- Large changes in temperature during normal operation
- Very large structures

If we think of situations where these might arise, jet engines, cryogenic processing, and bridges might come to mind. All of these must account for normal operating conditions and changes in temperature during that period of operation. Engines, for instance, will often be "warmed up" before they are operated at full power to prevent damage or loss of efficiency. In general, connections between dissimilar materials should be avoided. Where this is unpractical, an analysis of the effects under normal operation should be conducted to determine suitability of the assembly. Adding allowance for expansion or contraction is common, and is easily observed in large assemblies like buildings and railroad tracks (Fig. 8.7).

8.2.5 Low-Temperature Effects

As temperature is reduced, materials tend to get more brittle. The magnitude of the change varies widely and in some cases is negligible. Each class of materials will have similar reactions to low temperature based on structure, but even within the classes there is variation. To illustrate the importance of low temperature, we will first look at metals, and steel in particular.

A popular example (among metallurgists, at least) of the consequences of low-temperature effects is the failure of Liberty Ships. These were cargo ships built during

*The mechanisms by which materials contract upon heating can vary and are beyond the scope of this text. The characteristic is not always displayed over a wide temperature range, but some materials such as cubic zirconium tungstate exhibit a negative thermal expansion over a wide range of temperatures. By paring negative CTE material with a positive CTE material, a zero CTE composite can be created.

World War II, and there were many instances of hull cracking, and even some cases of the ships sinking as a result of sudden fracturing of the steel. While there were claims that the rapid manufacturing of these ships by inexperienced workers were to blame, ultimately it was determined that the grade of steel (and improper design for that grade) was the cause. The steel, as it turns out, became brittle in the cold North Atlantic waters. There were design flaws as well that allowed cracks to propagate quickly and extensively, but the steel itself was not suitable for the application.

Why would the properties of steel suddenly change? One reason is that there is a critical temperature at which the ductility of some metals drops drastically. This is known as the **ductile-to-brittle transition temperature** (DBTT). The DBTT is a result of the material being unable to deform in the same way. The phenomenon is linked to microstructural features at the crack tip and involves concepts beyond our discussion, but it is vitally important to understand and account for its existence. There are many applications that require low-temperature operation, and material choice must be made carefully. An extreme case of this would be cryogenic equipment as shown in Fig. 8.8. The term **cryogenic** indicates a temperature below –238°F (–150°C), such as is encountered when dealing with some liquefied gases.

A DBTT is present in some metals (particularly pronounced in bcc metals, see Sec. 3.2.1) and polymers. When discussing polymers, it is referred to as the glass transition temperature (see Sec. 8.2.6.1). In both cases there is a precipitous drop in fracture toughness, or the ability of a material to withstand impact without breaking. The ability of atoms and molecules to move is impeded at low temperature, and instead of plastically deforming, the material will fracture. Recall that polymers are highly influenced by cross-linking and molecular structure (see Sec. 5.3.2.6) and thermosets may become brittle at elevated temperature also.

The temperature at which this occurs varies according to the material and the overall change varies likewise. An example of the way fracture toughness drops with temperature is shown in Fig. 8.9 (more on toughness and fracture in Sec. 7.4.3.2). You will notice face-centered cubic (fcc) metals are not effected significantly. These metals (e.g., copper, nickel, or austenitic stainless steel) are often suitable for cryogenic applications. Steel will vary drastically in its DBTT as a function of carbon content. A general DBTT temperature cited for steel is –60°C, but low carbon steel will remain ductile at much lower temperatures than high-carbon steel. This is reflected by the differences between their ductility at room temperature. The more brittle a material starts, the higher the temperature at which it is considered brittle. The effects of carbon in steel are shown in Fig. 8.9.

Consider a penny, for example. Copper was noted as a metal that does not become brittle at low temperature (it is fcc). If you take a penny and hit it with a hammer it will not shatter at room temperature. Some *will* shatter at very low (cryogenic) temperature. Why some and not all? Newer pennies (created after 1982) are copper-plated zinc, not mostly copper. That means they will behave more like zinc (hcp structure) and not like copper (fcc structure). Since zinc has a DBTT, it will crack apart when impacted at low temperature and copper will only deform. This is demonstrated in Fig. 8.10.

Ceramics are not subject to a DBTT or a glass transition temperature. Ceramics are characteristically brittle. Therefore, they cannot transition to being brittle. They are already there! For them, their DBTT may be thousands of degrees, and it is not exactly what would be considered a traditional definition of "low" temperature. There are means of toughening ceramics so that they are not as brittle, and one of these, transformation toughening, is discussed in Sec. 7.5.2.1.

FIGURE 8.8 Low temperatures require materials that maintain ductility and strength over a large range of temperature. The liquid nitrogen Dewar must be durable, avoid puncture, and consists of valves that must operate at –321°F (–196°C). 300-series stainless steels are popular for this type of application (see Sec. 4.2.6 for more).

8.2.6 High-Temperature Effects

Much of what was said in the previous section on low temperature can be said about high-temperature effects. It varies for each class and within each class. A major difference is that high temperatures eventually affect all materials. Although low temperature may have little effect on certain materials, all materials have a point at which they melt, decompose or otherwise cease to retain their "typical" properties. High temperature is a very nebulous phrase. It depends on which material we are discussing. Think of mercury, which is liquid at room temperature, and tungsten, which has the highest melting point of any pure metal. A bitterly cold day to us of –40°F (–40°C) is exceptionally hot to mercury. It is right about at its melting point [–37.9°F (–38.8°C)]. To tungsten, which melts at 6170°F (3410°C), it does not come close to qualifying as a "high" temperature.

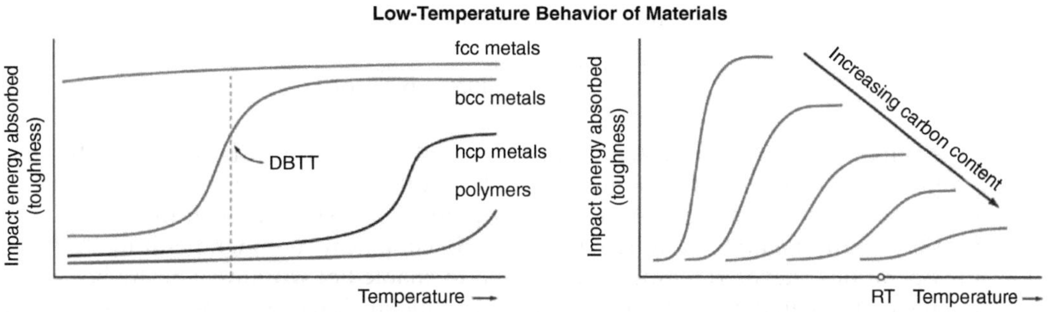

Figure 8.9 Materials tend to become more brittle at low temperature. Not all materials respond the same, however, and only some will exhibit a ductile-to-brittle transition temperature (DBTT). Body-centered cubic metals tend to be most susceptible. In polymers this phenomenon is known as the glass transition temperature. Properties vary with composition and adding carbon to iron will progressively reduce the toughness of the alloy and shift the DBTT to higher temperatures.

Hopefully it is apparent that some qualification is necessary to define what "high" means in regard to temperature. This can easily be applied by using a material's **homologous temperature**. The homologous temperature tells us the current temperature as a fraction (or percent) of the material's melting point. The equation used to determine the homologous temperature, T_H, is found by dividing the temperature of interest, T, by the material's melting temperature, T_M, using an *absolute* temperature scale:

$$T_H = \frac{T}{T_M} \tag{8.3}$$

So let us apply this to the example between mercury and tungsten from above. For mercury ($T_M = -39°\text{C}$) at $-40°\text{C}$, the homologous temperature would be

$$T_H = \frac{233\text{ K}}{234\text{ K}} = 0.996 \text{ or } 99.6\% \text{ of } T_M \tag{8.4}$$

Figure 8.10 The change in composition of pennies minted after 1982 means they will behave differently at low temperature. Older pennies will deform when impacted at cryogenic temperature, but newer pennies will crack apart.

For tungsten ($T_M = 3410°C$) at –40°C, the homologous temperature would be

$$T_H = \frac{233\ \text{K}}{3683\ \text{K}} = 0.063 \text{ or } 6.3\% \text{ of } T_M \quad (8.5)$$

This puts into sharp contrast the importance of considering the material's thermal properties when considering temperature. It is still difficult to define "high" temperature as a rock-solid fraction of the melting point. For instance, when machining, you typically want the material to remain relatively cool, near room temperature. In forming operations, elevated temperature is used to enable hot working. During hot working the material may be literally glowing hot, while during machining an unacceptable temperature will get nowhere near that hot. Unfortunately "hot" is a relative term for processing too, but we can now appreciate that each material has a different perspective on what temperature it considers hot. Typically, the higher the melting point the higher the acceptable service temperature, but it is not a perfect correlation.

Some of the basic changes that occur at elevated temperature include softening, an increased ductility, and increased reactivity. The basic effects of elevated temperature include the easing of atomic/molecular motion, reduction of internal strain, and a coarsening of the microstructure, particularly in metals. Some materials are better suited for high-temperature *service* than others. This means they still perform well at high temperatures. Ceramics are characteristically good at high temperatures. Although they maintain their strength well, they do not appreciably soften until very near their melting point. This makes them unsuitable for bending or tension forces (see Sec. 7.2.1) during service as brittle failure can still occur.

Some polymers are better at high homologous temperatures than others, but all have relatively low "melting" points. Recall that thermoset polymers (see Sec. 5.3.2.5) will harden at elevated temperatures which may or may not be desirable. Polymers will also degrade (decompose) at elevated temperatures, which limits their effectiveness. Metals fall between the other two classes. Most metals can serve under modest temperatures, but many will weaken at high temperatures such as in jet engines. As discussed in Chap. 4, there are certain categories of metals which perform well at high temperature, including refractories (Sec. 4.4.1) and superalloys (Sec. 4.4.4).

8.2.6.1 Glass Transition Temperature

The **glass transition temperature** (T_g) of a material is the point where it changes from being rigid and brittle to rubbery and ductile, and it only occurs in amorphous materials. It does not occur at a single temperature but progresses over a range of temperature. This phenomenon is most common in polymers, but any amorphous material, or the amorphous regions of a semi-crystalline material, will respond similarly.

When most materials melt (or solidify), there is a sudden change in properties. Liquids and solids are very different, and the transition between the two phases happens at a single temperature, so the characteristics are said to change *discontinuously*.* Ice and liquid water are two phases with distinct characteristics. There is no rubbery phase of water. On the other hand, amorphous materials do not change directly from solid to liquid with increasing heat. Rather than a distinct phase change, they decrease

*This is formally known as a first order phase transition. A change that is not abrupt, but takes place over a range of temperature is known as a second order phase transition.

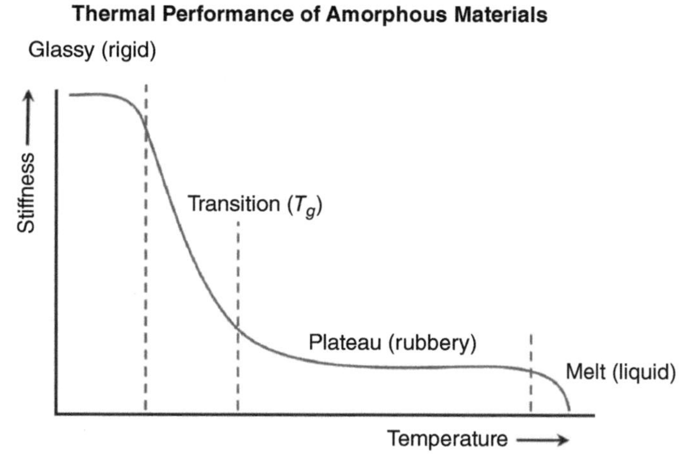

FIGURE 8.11 Amorphous materials, especially polymers, do not have a distinct melting point. Rather, they soften gradually over a range of temperature. The shape of the curve and temperatures at which the different regions begin and end is dependent on the material.

in strength and viscosity continuously as the temperature is raised. Even above T_g, the strength can be suitable for structural applications.

This is the case with some polymers, such as polyvinyl chloride (see Sec. 6.3.1.3), which are brittle below their T_g, but are ductile above it. Just because a material is ductile does not mean it is soft. By lowering T_g, a material can still be used in a rigid state, but it will be more resistant to cracking and impact. This also indicates that at lower temperature some materials may become brittle, as was discussed in Sec. 8.2.5. The general mechanical behavior with change in temperature is shown in Fig. 8.11. The shape of the curve and temperatures at which the different regions begin and end is dependent on the material. Even a single polymer composition will show variations based on molecular weight (see Sec. 5.3.2.3 for more on the molecular weight).

The temperature at which a material undergoes glass transition is widely varied. It can be experimentally measured using a differential scanning calorimeter (DSC), which measures heat absorption or release. For common materials, T_g is well known, but it can be measured as a quality control step, which is important since it will control processing and performance parameters. Some common materials and their glass transition temperatures are noted in Table 8.4. The T_g of a material is not directly related to the melting point of a material. Some polymers, for instance, will have a T_g very close to their melting point while others may not. Most processing, such as injection molding or extrusion of materials, occurs near to or above their melting point.

8.2.6.2 Creep

Materials will deform over time, even if the force on them is below that normally required to cause them to yield or deform permanently (see Sec. 7.4.1). **Creep** is the permanent deformation of a material over an extended period at a stress below its yield point. This behavior can be seen in items such as belts that stretch or cushions that compress without recovering their original dimensions. The process takes time, but the amount of time is dependent on the amount of stress (how close it is to breaking) and at

Material	Class of Materials	Glass Transition (T_g)[a]	
		°F	°C
High-density polyethylene (HDPE)	Polymer	–160	–107
Low-density polyethylene (LDPE)	Polymer	–130	–90
Polypropylene (PP)	Polymer	–10	–23
Nylon (6,6)	Polymer	120	49
Polylactic acid	Polymer	140	60
Polyethylene terephthalate (PETE)	Polymer	160	71
Polyvinyl chloride	Polymer	176	80
Polystyrene	Polymer	203	95
Acrylonitrile butadiene styrene (ABS)	Polymer	212	100
Polytetrafluoroethylene	Polymer	239	115
Polycarbonate	Polymer	300	149
Soda-lime glass	Ceramic	1022	550
Borosilicate glass	Ceramic	1040	560
Quartz	Ceramic	2192	1200

[a]Glass transition temperatures vary considerably depending on factors such as exact composition, molecular weight, additives, and processing. T_g occurs over a range of temperatures, and the listed values are representative only.

Note: The softening point of amorphous materials is known as the glass transition temperature. It is important for processing and applications, as it indicates the mechanical properties of a material at the specified temperature. Below the glass transition temperature materials are rigid and brittle. They become *rubbery* or *leathery* above that temperature and until they reach their melting point.

Table 8.4 Glass Transition Temperature for Common Materials

what temperature the stress is applied. Therefore, components under high stress at elevated temperature will creep the fastest.

There are two ways that time-dependent changes can cause problems. The first is through elongation, shortening, or warping, where the component changes dimensions enough to render it unsuitable for further use. The second is through **stress relaxation**, where an initial stress is diminished over time without a change of dimension. An example of this is a bolt being tightened to an initial specification, but over time the holding force reduces, even though the components are still held together. Creep is considered at a constant stress, so stress relaxation is technically not creep. The principles are similar, so they will be considered together.

The problem with creep is that it can change critical dimensions and ultimately leads to the failure of the component or larger assembly. For example, jet engines (Fig. 8.12) require tight tolerances and they operate over long intervals at high temperature and high stress. They have presented many of the challenges and have been the test bed for creep-resistant materials.

Ceramics are the most resistant to creep, metals are moderately resistant and polymers are quite susceptible (at least relatively speaking). Each category has a wide range of properties within it, but most materials will begin to creep appreciably at 30% to 40%

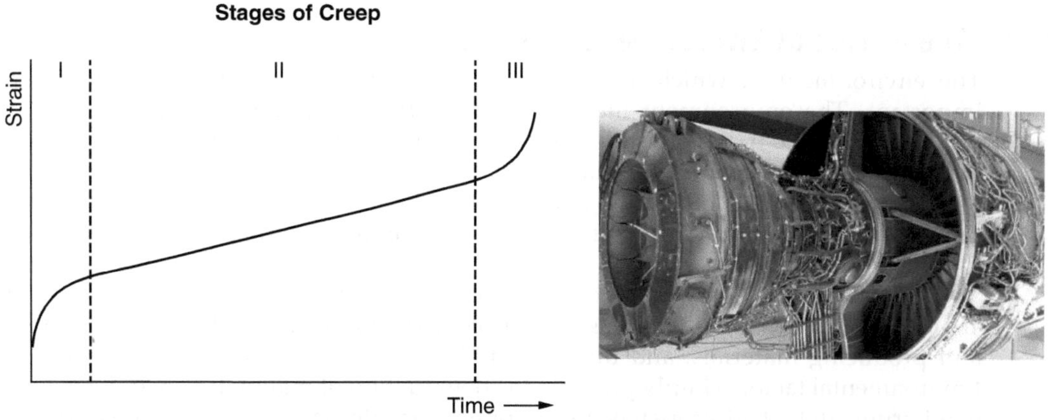

FIGURE 8.12 Creep involves dimensional change over time which is driven by stress and temperature. Components such as turbines blades (see engine on right) are operated over long periods of high stress and temperature yet must maintain tight tolerances. Understanding and controlling creep properties is essential to reliable operation.

of their melting temperature. Creep is attributed to microstructural changes from atomic diffusion (Sec. 8.4.1) and/or dislocation motion (Sec. 3.2.3.2). The reasons that creep occurs and the active mechanisms as a function of stress and temperature are important to design, but they are beyond our considerations. The results of creep are most important to us, and creep is considered to go through three stages:

1. **Primary creep** (Stage I)—relatively fast change in dimension occurring over a short duration and related to the initial loading of the object. Work-hardening often stabilizes the component.
2. **Secondary creep** (Stage II)—known as steady-state creep, the dimensional variation is relatively constant. Materials spend most of their useful lives in this stage.
3. **Tertiary creep** (Stage III)—final stage where deformation accelerates again due to stress concentration such as necking. This stage directly precedes fracture.

Stress relaxation is a problem when components that are supposed to connect assemblies fail to maintain the connection. For instance, a bolt may be tightened down, but over time the bolt permanently stretches enough that the pressure of the connection can reduce and possibly allow unacceptable movement. Likewise, a cable which tightly connects components may stretch and allow too much compliance. Guitar strings (or strings on any other instrument) will stretch over time and need retuning.

While there are other factors than simply the string material stretching, the relaxation of the metal contributes. Materials with a high elasticity and high strength will not be affected as strongly. The closer the material is to its yield point, the more likely it is to stretch, and again, temperature plays a decisive role. The main difference from creep is that stress relaxation involves a drop in stress over time and a minimal change in dimension.

8.3 The Basics of Environmental Effects

The environment in which a material, component, or product is placed is very important. The environment includes both thermal and chemical conditions surrounding the part. In the preceding sections we have considered the dimensional and mechanical impacts of temperature on materials. There are also important thermal implications for the composition and structure internally. Likewise, ambient conditions, such as light, humidity, salinity, and surface debris will also affect the useful lifetime of components.

This all goes back to the second law of thermodynamics (refresh yourself with Table 8.1 and the associated discussion if necessary). Basically, work goes into refining and producing materials, and they are rarely in their lowest energy configuration. Environmental factors simply promote the degradation of materials to a lower energy, equilibrium state. Unfortunately, most applications do not benefit from this process. Therefore, understanding and controlling environmental influence is an important undertaking.

8.4 Diffusion

The two main environmental impacts we will consider are diffusion and corrosion. These two are not necessarily linked, but both may be active simultaneously. Diffusion can be simply described as the movement of atoms *within* materials or between combinations of materials. Although most materials may seem quite inactive under normal circumstances, the atoms are still energetic enough to move around. It occurs in all materials, and it can be quite detrimental or it can be very useful. The rate at which they move is dependent on a number of variables which will be discussed in the next section. Corrosion will be covered in Sec. 8.5.

8.4.1 How Diffusion Works

The material's structure and the characteristics of the atoms or compounds diffusing through the material will determine the rate and effects of diffusion. This makes it difficult to generalize about diffusion in all materials. There are some general, governing concepts that we will establish. Before we get to that stage, let us start with examples to help establish what diffusion looks like. The easiest way to illustrate diffusion is in fluids. Materials diffuse quickly through liquids and gases. As shown in Fig. 8.13, if you put a drop of dye into water it will eventually disperse throughout the container, even without any stirring.* You can smell a meal cooking in rooms other than the kitchen. Atoms and molecules will go from an area of high concentration to low concentration. They want to achieve a uniform dispersion.

The tendency toward uniformity is present in solids as well. The biggest difference is that atoms and molecules in a solid state cannot move easily. They are relatively fixed in place, so transitioning from one location to another is greatly impaired. It is still possible for atoms to move in solids. The factors controlling diffusion include the atomic

*The random movements of atoms, known as Brownian motion, will aid in the distribution of matter in a fluid (liquid or gas).

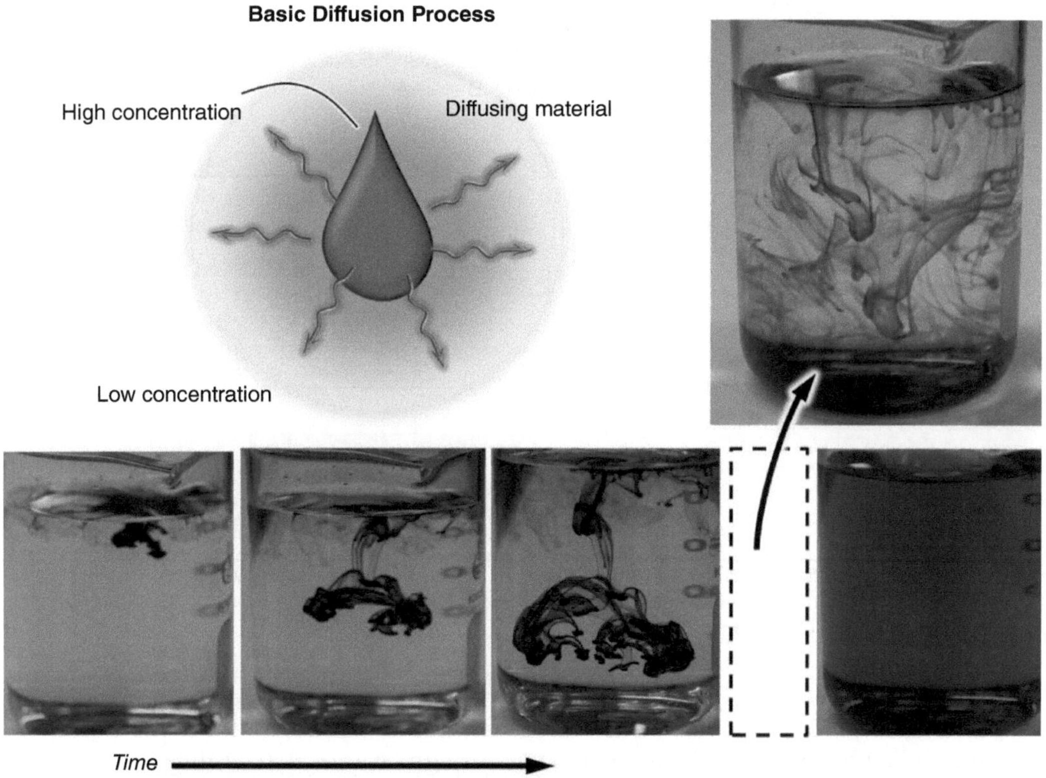

Figure 8.13 When the composition differs, there is often a tendency for matter to move to areas of lower concentration. Diffusion is the process by which concentration gradients are eliminated. This is the case in soluble mixtures, but some cases, such as oil and water, will not diffuse into each other.

bonding, the size of the atom moving relative to the other atoms, the temperature and others. We will begin examining the various factors, establishing a mathematical basis for our analysis, and then apply it at the end. This topic requires using some more advanced calculations than we have applied elsewhere, so take some extra time to understand where these equations are coming from, and what information they provide. There are also many introductory materials science texts with expanded discussion on diffusion and that examine it in greater depth. We will mostly cover the highlights.

Since diffusion is the movement of matter, we should first look at how we describe that movement. In the water and dye example, the dye moves from and area of high concentration to an area of low concentration. Wherever the dye and the water are in contact, diffusion will occur across the entire area of that interface. The dye has a very complex surface, but we can simplify our discussion by considering a flat interface between two metals. The process generally works in the manner depicted in Fig. 8.14, where atoms from the different materials exchange places.

The movement of atoms over time can be described by the diffusion flux, J. The **flux** is simply the flow (of anything) through a given area, and it is often described as

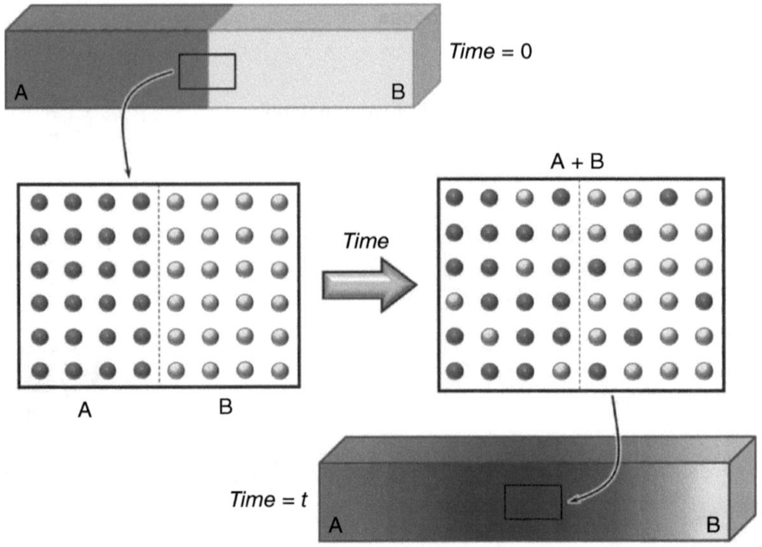

FIGURE 8.14 Diffusion involves the movement of mass through an area over time. This flux can be illustrated by two pure materials mixing over time through the movement atoms. Material further from the interface will stay pure longer.

a function of time. The flux can be calculated from the mass, M, (or number of atoms multiplied by their atomic mass), moving through an area, A, over time, t.

$$J = \frac{M}{At} \tag{8.6}$$

Looking at Eq. (8.6), you can see that time is in the denominator, so the flux is a rate. The higher the flux, the higher the diffusion rate. The units of flux are often given as kg/m^2s or $atoms/m^2s$. Other ways we can increase diffusion are by increasing the amount of mass flowing across the area, or by decreasing the area that a given mass is flowing through.

It is often important to know the concentration of a diffusing species at a given distance from the interface. This is exemplified by case hardening operations, discussed more in Sec. 9.3.8. If carbon is diffusing into steel from the surface of a component, the concentration will be very high at the surface and much lower in the center of the part. The concentration will vary from 0% to 100%. If we just think of iron and carbon as the two elements, the total of the two will add to 100%. If the carbon concentration is 2%, the iron concentration will be 98%.

Concentration is often expressed as mass per volume and has units such as kg/m^3. The variation in concentration, or the concentration gradient, is very important to the diffusion rate. The **concentration gradient** is the change in concentration of a certain substance as you change position. If there is no difference in concentration, there will be no driving force for diffusion and the rate will be negligible. If there is a sharp change in concentration, the driving force for diffusion will be high. Going back to the dye in water example, the dye will diffuse fastest when it is first dropped into the water. The rate at which it disperses will continually reduce until it is evenly distributed throughout

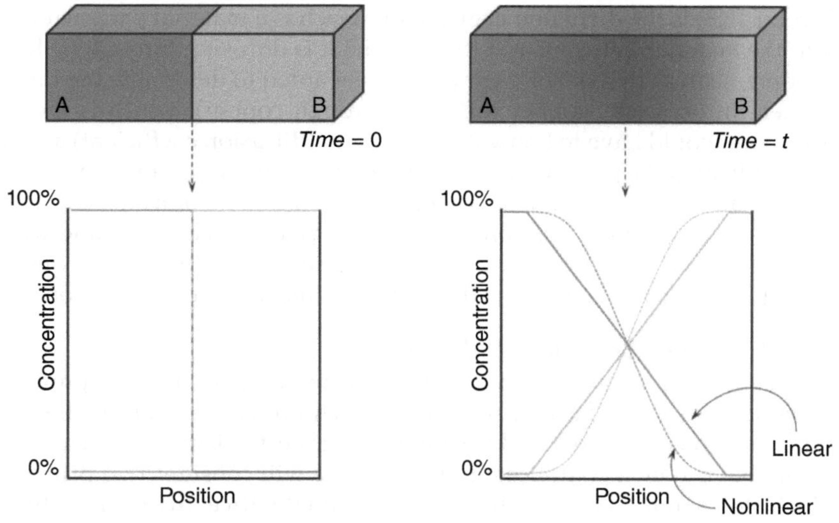

FIGURE 8.15 The concentration changes over time are shown. The extent of mixing reduces from the original interface outward. Far enough from the interface, the materials will be still be pure. The amount of time is critical to the distance over which mixing occurs. When assuming a linear concentration gradient, the mixing will exhibit a relationship shown by the solid lines in the graph on the right. More accurately, the concentration tends to show a nonlinear gradient such is represented by the dashed lines.

the container. The difference in concentration, C, from point 1 (x_1) to point 2 (x_2) is the concentration gradient. A linear* concentration gradient is given by

$$\text{Concentration gradient} = \frac{C_1 - C_2}{x_1 - x_2} = \frac{\Delta C}{\Delta x} \tag{8.7}$$

The way the concentration of atoms from materials A and B would vary is shown in Fig. 8.15. After time, t, the two elements will diffuse into one another and the concentration gradient will change from a flat, all-or-nothing interface to one that contains both elements and the original interface becomes indistinguishable. The lines, which depict the change in concentration as a function of position, will be straight for a linear concentration gradient. While this simplifies the calculation, a nonlinear relationship is more realistic. The general shape of the distribution tends to look more like the dashed lines in the graph on the right of Fig. 8.15.

If diffusion is occurring at a constant rate, also known as **steady state diffusion**, the rate of diffusion can be determined by **Fick's first law**:

$$J = -D\frac{dC}{dx} \cong -D\frac{\Delta C}{\Delta x} \tag{8.8}$$

---

*We will limit our discussion to linear circumstances where the *change* in concentration is constant over the distance in question. In most cases, there will be nonlinearity. In that case, use the differential form of concentration, dc/dx, where the variation in concentration can be described by any function, such as a parabola or square root.

D in Eq. (8.8) is the **diffusion coefficient**, which is a material parameter determined by both the material diffusing and the material it is diffusing through (D has units of area per time, such as m^2/s). For example, if we wanted to determine the rate at which copper self-diffuses (i.e., copper diffusing through copper), which is the flux of copper atoms, we would have to know the diffusivity (diffusion coefficient) and the concentration gradient at the point we are interested in. It is more common to approach the problem as a design parameter with a specific goal in mind. For instance, we want to have a specific concentration of carbon atoms in iron at a depth of 1 mm and we need to know how long it will take to achieve that concentration. The "how long" question is dependent on a number of conditions that we will look at in the next section.

8.4.1.1 Factors Controlling Diffusion

Diffusion is not a constant value. The concepts discussed in the previous section are dependent on many factors, not all of which are easily controlled. One of the most important considerations in diffusion is what mechanism or mechanisms are at work. While that can be a complicated question, we will consider two primary methods by which atoms move, both of which are considered **lattice diffusion**, or the movement of atoms through a crystal structure.

The first diffusion method is vacancy diffusion. In **vacancy diffusion** the atoms exchange regular atomic positions with vacancies, or unfilled atomic positions. Diffusion occurs by the exchange of a vacancy with the atom, so vacancies can be thought to diffuse in a direction opposite of atoms. If there are no vacancies adjacent to an atom position, it will not move from its position. The second method by which atoms diffuse is through interstitial diffusion. Interstitials are the spaces between lattice positions, and these are almost always open.

Interstitial diffusion occurs when the diffusing atom prefers to take an interstitial site, rather than a lattice position. This case is dominated by smaller atoms and will be negligible among atoms that form substitutional point defects (review these concepts in Secs. 3.2.3 and 3.2.4.1, especially Fig. 3.15). Because the number of available positions for diffusion are so much higher in interstitial diffusion, it tends to have higher diffusion rates than vacancy diffusion. Both are shown in Fig. 8.16.

Vacancies always exist in crystalline materials due to thermodynamic considerations, but their concentration increases with temperature. The vacancy concentration

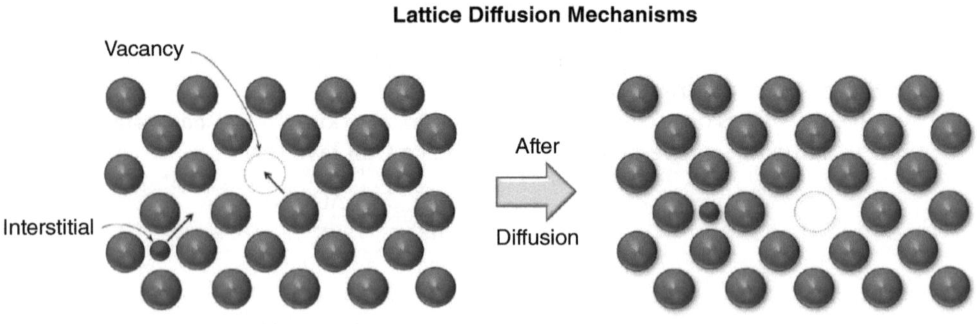

FIGURE 8.16 Diffusion can occur by atoms moving into vacancies or by the movement of interstitial elements from site to site. Interstitial diffusion is faster, but requires a small diffusing species. Vacancy diffusion is dominant in pure and substitutional alloys.

controls the rate at which diffusion can occur, therefore, vacancy diffusion is directly correlated to temperature. Interstitial diffusion also increases with temperature, but not because more interstitials form. Instead, the energy of the atoms is higher which makes them more likely to change positions. To formalize these concepts, we can relate diffusion flux through the diffusion coefficient (the diffusivity). **Diffusivity** is calculated by

$$D = D_o e^{\left(-Q_d/RT\right)} \tag{8.9}$$

where D_o is a pre-exponential constant, which is independent of temperature, Q_d is the activation energy for diffusion (the energy required to make the move), R is the ideal gas constant and T is the temperature using an absolute scale. The most important aspects of this equation are the trends we can determine from it. The ways the variables influence diffusion are summarized in Table 8.5.

Temperature is so important because it increases the opportunity and ability for atoms to move. The activation energy is not so direct. The activation energy describes the ease or difficulty with which atoms move through a material. It is the energy associated with a change in position. There are a variety of influences on atomic motion, but two basic ones are the elastic and chemical factors of diffusion. Essentially there are two questions to ask: "How much does the atom have to displace other atoms when it moves?" and, "Does the atom prefer its new position over its current one?"

Think of oil and water. Oil does not mix with water even if left for long periods of time. Elements in solids can have the same behavior. Most materials will have some limited solubility in one another, but for a material to diffuse from a high concentration to a low concentration it has to mix with another.

When atoms move, there will always be some stretching of other atoms' bonds. Refer back to Fig. 3.15. When a "foreign" atom is introduced, there will be an imperfect fit which always results in a change in lattice spacing. The magnitude of that mismatch determines the elastic energy required to accommodate the atom. Keep in mind also, the movement of atoms is taking place in a three-dimensional structure and motion must account for all the affected atoms. Atoms of much larger sizes will have a higher activation energy and will diffuse much slower. This concept is shown in Fig. 8.17.

Parametric Effects on Diffusivity			
Parameter	**Variable**	**Typical Units**	**Influence on Diffusivity**
Diffusion constant	D_o	m^2/s	Increases linearly with D_o
Activation energy	Q_d	J/mol, eV/atom	Decreases exponentially as Q_d increases
Ideal gas constant	R	J/mol·K, eV/atom·K	None, equals 8.31 J/mol·K or 8.62×10^{-5} eV/atom·K
Temperature	T	K	Increases exponentially as T increases

Note: The diffusivity of an element is determined by a few key parameters, those are related in Eq. (8.9), and their importance is described in this table.

TABLE 8.5 Parametric Effects on Diffusivity

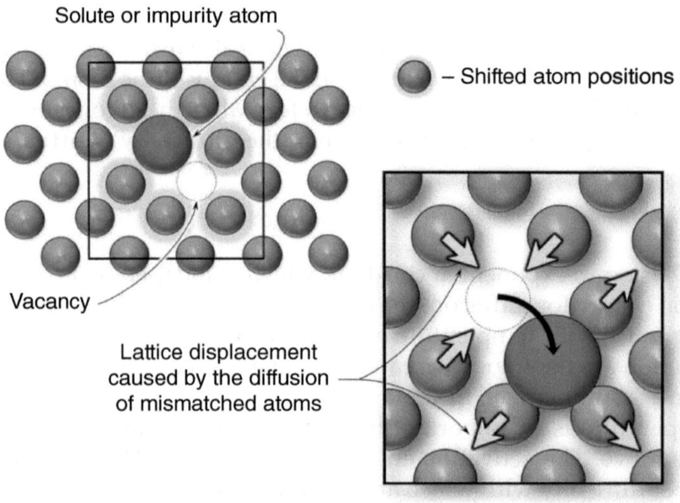

FIGURE 8.17 Impurity atoms or alloying elements will cause some degree of bond stretching in the lattice. Vacancy diffusion requires that the atoms which were previously around the vacancy to be pushed outward to accommodate the diffusing atom and the atoms previously around the impurity will contract inward.

Very small atoms, like hydrogen, can diffuse rapidly. Although some elements chemically repel one another, others just cannot fit together. Atoms come in a very wide range of sizes, so it is not always easy to get one element into the lattice of another. For instance, copper and zinc are highly soluble as they have a similar size and other parameters (see Sec. 3.2.4.2 for the rules of solubility, known as the Hume-Rothery rules). Copper and tungsten, on the other hand, have no appreciable solubility. Materials are sometimes coupled for their limited diffusion. For instance, in computer chips, tantalum can be used as a *diffusion barrier* to keep the copper wiring from diffusing into the silicon chip.

8.4.1.2 Calculating Diffusivity and Diffusion

The preceding discussion introduced the way diffusion happens and what conditions affect it. Now that we have a handle on what goes into the process of diffusion, let us apply it to a couple of problems. Looking at interdiffusion in the copper-zinc system, we can apply Eq. (8.9) to see how quickly zinc will diffuse through copper at 900°C (1173 K). First we need to determine the values of the pre-exponential constant, D_o, and the activation energy, Q_d. Here they are given to you, but you can look up these values in a variety of handbooks or research reports. Under the given circumstances, the values of D_o and Q_d are 2.4×10^{-5} m²/s and 189×10^3 J/mol, respectively. This gives us the following equation:

$$D = D_o e^{(-Q_d/RT)} = \left(2.4 \times 10^{-5}\,\frac{\text{m}^2}{\text{s}}\right)\exp\left(\frac{-189 \times 10^3\ \text{J mol}^{-1}}{(8.31\ \text{J mol}^{-1}\text{K}^{-1})(1173\ \text{K})}\right)$$

$$D = 9.11 \times 10^{-14}\,\frac{\text{m}^2}{\text{s}} \tag{8.10}$$

What does this mean? The 10^{-14} means the number is very, very small, but is that good or bad? To get a better idea of what our first result means, let us look at another example, tungsten diffusing through copper. In this system, D_o and Q_d are 1.69×10^{-4} m²/s and 226×10^3 J/mol, respectively. That yields a diffusivity of 1.44×10^{-14} m²/s, or about six times slower. Diffusivity is a useful number because of the information it gives us relative to other materials and for quantitative prediction. Although we can compare diffusivities of material combinations, this value is more useful for calculating the actual diffusion rate under specific conditions.

The diffusivity gives us insight into diffusion rates for a particular alloy system, and it can be plugged into Fick's first law (Eq. 8.8) to determine steady state diffusion. To do that, we need to know the concentrations at two points, and those concentrations cannot change with time. If we consider carburization (Sec. 9.3.8.1), where a high concentration of carbon is supplied at the surface of steel iron, there are some basic assumptions needed.

First, there has to be a sufficient supply of carbon at the higher concentration, and second, the ability for carbon to flow through must be maintained. That situation can be likened to a faucet and sink. The faucet must stay on, and the sink must not be clogged. That suggests there must always be a sufficient supply of carbon (faucet stays on), but there is never a saturation of carbon (no full sink).

The same applies to our copper-zinc example and the diffusivity we found in Eq. (8.10), and we can determine the diffusion flux with some other information. For example, if we know that point 1 (x_1) is 1 mm from the surface and point 2 (x_2) is 6 mm from the surface, and that the concentration at point 1 is 0.9 kg/m³ (C_1), and at point 2 it is 0.3 kg/m³ (C_2) we can calculate the flux. Using the diffusivity (D) found in Eq. (8.10), we get Eq. (8.11).

$$J = -D\frac{\Delta C}{\Delta x} = -D\frac{C_1 - C_2}{x_1 - x_2} = -\left(9.11 \times 10^{-14}\,\frac{\text{m}^2}{\text{s}}\right)\frac{(0.9-0.3)\frac{\text{kg}}{\text{m}^3}}{(1-6)\times 10^{-3}\,\text{m}} = 1.1 \times 10^{-11}\,\frac{\text{kg}}{\text{m}^2\text{s}} \tag{8.11}$$

This result is for zinc atoms diffusing through copper at a steady rate of 1.1×10^{-11} kg/m²s. The diffusivity is of primary importance. For interstitial diffusion, such as carbon in iron, the diffusivity is about 10,000 times higher under similar conditions.

8.4.1.3 Non-Steady-State Diffusion

Steady-state diffusion is not practical in most cases. It is much like turning on the faucet and simply measuring the flow of water. The room never fills with water and the water never runs out (hopefully). As diffusion occurs in real materials, there is typically a buildup of concentration with time. The concentration gradient changes with time, so we cannot simplify it like we did above. Like blowing up a balloon, the more air we put in, the harder it gets to keep going.

Unfortunately, by allowing the concentration gradient to vary, we also complicate the calculations, because now there are multiple variables that fluctuate simultaneously, namely the position and time. This requires partial differentials to solve, but we will limit our discussion to the results of the derivation. Specifically, if we are interested in a surface concentration, C_s, initial concentration, C_o, and concentration at a specific depth, C_x, we determine the time required to achieve that result. The equation takes the following form:

$$\frac{C_x - C_o}{C_s - C_o} = 1 - \text{erf}\left(\frac{x}{2\sqrt{Dt}}\right) \tag{8.12}$$

The "erf" indicates this is an error function, a special type of mathematical function that is often used to describe non-steady-state diffusion. The values for the error function can be tabulated in table form for reference. Because all solutions cannot be tabulated (there would be an infinite number of them), interpolation between values can be performed.

Interpolation is the process of estimating a value using known values associated with it. A table of values pertinent to the error function is given below (Table 8.6). You do this in a less rigorous way when you look down at the needle on your speedometer or the hands on an analogue clock. If the needle or hands are not exactly on a line, you guess what the actual value is based on where it falls between two marks that are known. The closer the needle or hand is to a known value, the closer it will be quantitatively. To perform interpolation mathematically, first rearrange Eq. (8.12) to solve for the error function as shown in Eq. (8.13). Then you can use that calculated value to determine what the fraction within the error function is

$$1-\frac{C_x - C_o}{C_s - C_o} = \text{erf}\left(\frac{x}{2\sqrt{Dt}}\right) \tag{8.13}$$

Let us say that we find the left-hand side of Eq. (8.13) to be 0.214 (i.e., erf $(b) = 0.214$). Looking at Table 8.6, we cannot find that number exactly, but we do see values for 0.1680 and 0.2227. If we assume the value to be the nearest of the two numbers, 0.2227, we know that b, and the expression $x/2\sqrt{Dt}$, is equal to 0.2. If we are looking for one variable, say the time, t, we can now solve for it using that value. This is sort of like looking at a clock and seeing the minute hand is between the 8 and the 9. Since it is closer to the 8, you assume it is 40 minutes after the hour instead of 45 minutes. Using the closest tabulated value obviously introduces error since we are knowingly using

b	erf(*b*)	*b*	erf(*b*)	*b*	erf(*b*)	*b*	erf(*b*)	*b*	erf(*b*)	*b*	erf(*b*)
0	0.0000	0.15	0.1680	0.7	0.6778	1.25	0.9229	1.8	0.9891	2.35	0.9991
0.01	0.0113	0.2	0.2227	0.75	0.7112	1.3	0.9340	1.85	0.9911	2.4	0.9993
0.02	0.0226	0.25	0.2763	0.8	0.7421	1.35	0.9438	1.9	0.9928	2.45	0.9995
0.03	0.0338	0.3	0.3286	0.85	0.7707	1.4	0.9523	1.95	0.9942	2.5	0.9996
0.04	0.0451	0.35	0.3794	0.9	0.7969	1.45	0.9597	2	0.9953	2.55	0.9997
0.05	0.0564	0.4	0.4284	0.95	0.8209	1.5	0.9661	2.05	0.9963	2.6	0.9998
0.06	0.0676	0.45	0.4755	1	0.8427	1.55	0.9716	2.1	0.9970	2.65	0.9998
0.07	0.0789	0.5	0.5205	1.1	0.8802	1.6	0.9763	2.15	0.9976	2.7	0.9999
0.08	0.0901	0.55	0.5633	1.2	0.9103	1.65	0.9804	2.2	0.9981	2.75	0.9999
0.09	0.1013	0.6	0.6039	1.3	0.9340	1.7	0.9838	2.25	0.9985	2.8	0.9999
0.1	0.1125	0.65	0.6420	1.4	0.9523	1.75	0.9867	2.3	0.9989	2.85	0.9999

Note: The columns of b values are the resulting values of the fraction $x/2\sqrt{Dt}$ in Eq. (8.12) and Eq. (8.13).

TABLE 8.6 Tabulated Values for the Error Function Allow for Simplified Calculation of Non-Steady-State Diffusion an incorrect number. If we want to be more exact (hopefully you do!), we can use the information provided in the table to get a more accurate value by interpolation as shown in Eq. (8.14).

$$\frac{b-b_1}{b_2-b_1}=\frac{\mathrm{erf}(b)-\mathrm{erf}(b_1)}{\mathrm{erf}(b_2)-\mathrm{erf}(b_1)} \tag{8.14}$$

For the example where $\mathrm{erf}(b)=0.214$, we need to find b. We can see in the table that the next lower value (b_1) is 0.1680, and the next higher value (b_2) is 0.2227. The b values for those two are given as $b_1=0.15$ and $b_2=0.2$. Solving with those known values, we get

$$\frac{b-0.15}{0.2-0.15}=\frac{0.214-0.1680}{0.2227-0.1680};b=(0.05)\left(\frac{0.0460}{0.0547}\right)+0.15=0.192 \tag{8.15}$$

We can see from Eq. (8.15) that the b value is much closer to the tabulated value of 0.2, but that there is an important difference. The error function is more sensitive to changes in the b value for smaller values. That means there is a big difference between 0.2 and 0.3, but there is a relatively small difference between 2 and 3. The general behavior of the error function curve is shown in Fig. 8.18. Also, the concentration is predicted by the error function, and the distribution of solute diffusing through or into the material is shown in as well. The shape of the nonlinear curve presented in Fig. 8.15 is time dependent, and the shape is follows the general trend plotted on the left of Fig. 8.18.

While the basis for this approach is very important, our purpose is to develop tools we can readily apply. Eq. (8.12) is particularly suited to solve for a variable in the error function when concentrations at the various locations are known (the left-hand side of the equation is a constant). If we know the position, diffusivity and time, we can solve for concentration by Eq. (8.12) or an approximation that does not require tabulated values. Milton Abramowitz (1915–1958) and Irene Stegun (1919–2008), known for their

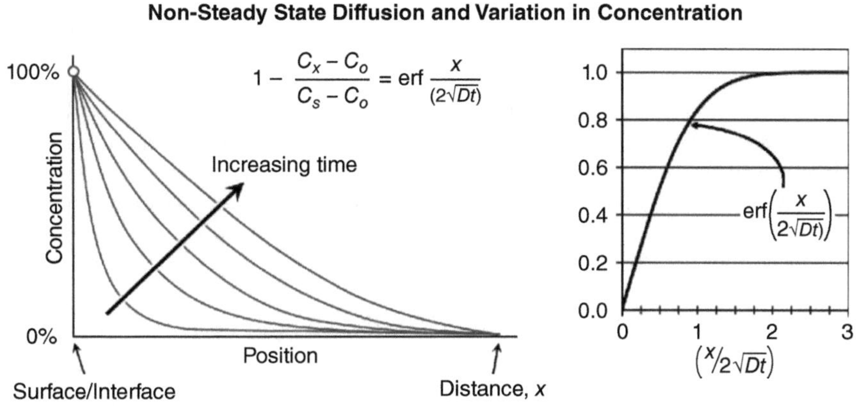

Figure 8.18 Non-steady-state diffusion is dependent on diffusivity, D, and time, t, in Eq. (8.12). Since diffusivity is relatively constant, the time will be the main parameter considered in processes involving diffusion.

handbook for applied mathematics, suggested a number of approximations, but the simplest (which does not introduce excessive error) is

$$\mathrm{erf}(b) \approx 1 - \frac{1}{(1 + a_1 b + a_2 b^2 + a_3 b^3 + a_4 b^4)^4}$$

$$a_1 = 0.278393; \quad a_2 = 0.230389; \quad a_3 = 0.000972; \quad a_4 = 0.078108$$

$$b = \frac{x}{2\sqrt{Dt}} \tag{8.16}$$

Although Eq. (8.16) can be solved directly, without the need for table values and interpolation, it does require diligence in entering many digits accurately. To compare methods, we will solve Eq. (8.16) using the same value found for $b = 0.192$ to (hopefully) get the same value for erf(b).

$$\mathrm{erf}(b) \approx 1 - \frac{1}{1.27231} = 0.21403$$

So the error function can be calculated within reasonable accuracy using Eq. (8.16). Diffusion, in general, is somewhat more involved of a concept than we have considered so far. It has specific applications in manufacturing, especially in surface conditioning (Sec. 9.3.8), but it is *always* an important consideration. Materials are very active, even if they do not appear so. Knowing the rules of the game, so to speak, will help you interpret changes and control the outcomes.

8.4.1.4 Example of the Diffusion Process

This discussion has probably been a lot to digest, but there are practical uses for it. Carburization is one of them (see Sec. 9.3.8.1). For instance, if we wanted to achieve a 1% carbon concentration at a depth of 1 mm in AISI 1018 steel by adding solid carbon at the surface at 700°C, we can determine how long it will take. First, you would need the diffusivity, which can be calculated using Eq. (8.9). We will use tabulated values for D_o and Q_d, which are 6.2×10^{-7} m²/s and 80×10^3 J/mol, respectively.

$$D = (6.2 \times 10^{-7}\ \mathrm{m^2/s}) \exp\left(\frac{-80 \times 10^3\ \mathrm{J/mol}}{(8.31\ \mathrm{J/mol \cdot K})(973\ \mathrm{K})}\right) = 3.13 \times 10^{-11}\ \mathrm{m^2/s} \tag{8.17}$$

Then you must determine your concentration gradient. You can use Eq. (8.13) and the specified concentrations to solve for the error function. The surface concentration, C_s, of carbon will be 100% since the part has pure carbon applied to the surface. The initial concentration, C_o, is 0.18% as discussed in Sec. 4.2.3. We have determined the concentration, C_x, must be 1% at distance, x, which is 1 mm. We then plug these values into the left side of Eq. (8.13) to get

$$1 - \frac{C_x - C_o}{C_s - C_o} = 1 - \frac{1 - 0.18}{100 - 0.18} = 0.9918 = \mathrm{erf}\left(\frac{x}{2\sqrt{Dt}}\right) \tag{8.18}$$

There are a couple of observations that can be made here. First, you can use the percentage (e.g., 25%) or decimal values (e.g., 0.25) for the concentrations as long as you are consistent. Whenever you have a pure material at the interface, such as the carbon, and a much smaller desired concentration within, you will end up with an error function value near 1. Now that we have the value of the error function, we can look back to Table 8.6 to find what the value within the parentheses, *b*, equals. You will note that the exact value does not appear in the table, so we will interpolate.

$$\frac{b-1.85}{1.90-1.85}=\frac{0.9918-0.9911}{0.9928-0.9911};\ b=(0.05)\left(\frac{0.0007}{0.0017}\right)+1.85=1.8706 \tag{8.19}$$

Since $b = x/2\sqrt{Dt}$, and we know D and x, we can solve for the time to achieve our specified conditions as shown in Eq. (8.20).

$$b=\frac{x}{2\sqrt{Dt}};\ t=\frac{\left(\frac{x}{2b}\right)^2}{D}=\frac{x^2}{4Db^2}=\frac{1\times10^{-6}\ \text{m}^2}{4(3.13\times10^{-11}\ \text{m}^2/s)(3.4991)}=2283\ \text{s or } 38\ \text{min} \tag{8.20}$$

What that means is that if you pack a piece of 1018 steel in carbon and heat it to 700°C for 38 min, you will achieve a 1% carbon concentration at a depth of 1 mm. By varying the concentrations or temperature, you can adjust the time required or determine the change in a part over time. Likewise, if you have steel with a high surface concentration, further heating without carbon at the surface will cause the carbon to diffuse further into the part and reduce the surface concentration but increase the concentration at greater depths.

8.4.1.5 Diffusion in Polymers and Ceramics

We have been considering diffusion in crystalline metals, and although many of the basic considerations remain valid, there are some differences in nonmetals. Ceramics are covalently bonded, and for atoms to move requires the breaking of very strong bonds. For this reason, interstitial diffusion is the primary mechanism. Some of the most beneficial and troublesome results occur by diffusion, and the "doping" of silicon is a good example.

Silicon is a semiconductor, but in its pure form it is not conductive enough to form circuitry. To create transistors using the silicon base material, the substrate is doped with other elements that change the electrical characteristics. Elements such as phosphorus introduce extra electrons to silicon, which enhances its electrical conductivity and makes it an n-type semiconductor. Other elements, like boron, change silicon to a p-type semiconductor by creating an electron deficiency (excess "holes"). Modern electronics depend on the exact concentration and depth of these impurities and diffusion is central. Other, damaging impurities, such as sodium, can diffuse through the silicon and short circuit the junctions. So diffusion of wanted and unwanted impurities must be carefully controlled.

In polymers, diffusion takes on a much different character. Due to their molecular structure, polymers are not as densely packed with atoms, and the diffusion of elements can often occur at a much higher rate. There is relatively little crystal structure, and bonding within chains is covalent. These factors lead to the majority of diffusion species to be distinct from the polymer itself. For instance, oxygen can diffuse through a polymer without really being part of the polymer chemistry.

The transfer of gas or other matter through a polymer is known as the **permeability**. This is vitally important to pressurized applications. If the fluid (gas or liquid) can permeate the polymer over time, the pressure will drop. If you blow up a balloon and set it in a corner, it will lose pressure over time. A soda bottle must prevent carbon dioxide from leaking through the polymer, so the choice of polymer and design of the bottle is important. The idea of permeability is important for a variety of applications.

In addition to permeability, polymers can absorb materials. For instance, some polymers, like nylon, should not be used in high-moisture conditions because they actually absorb water, which degrades their performance. The external substance diffuses in and can change the dimensions (polymers swell) and properties. The rate and extent of diffusion is dependent on a variety of factors which include the size of the element or molecule diffusing through the polymer (smaller is faster), the crystallinity of the polymer (higher crystallinity, lower diffusion) and the chemical compatibility (water will not diffuse readily through a hydrophobic polymer).

8.5 Corrosion

Corrosion, in the most basic terms, is rusting. At least that is the most obvious example. We do not often think of polymers "rusting," but they also corrode. Corrosion can take many forms and results in a variety of changes. Broadly speaking, corrosion is the conversion of a material's structure or chemistry to a more stable condition. It often results in the breakdown of engineered materials. We will consider corrosion and degradation to be equivalent terms for the purpose of this chapter. In natural materials we call it decay. Some corrosion can be stable and beneficial, but in other situations it can be quite detrimental. Either way, it is the law of the land (refer back to the second law of thermodynamics, Sec. 8.2.1). No material is immune, but we can make informed decisions to mitigate it or exploit it.

8.5.1 How Corrosion Works

Consider how materials start out. They begin in a "natural" form such as ore and precursor chemicals. Materials undergo significant refinement during production, all of which is aimed at imparting characteristics not found "naturally." These include structural and compositional characteristics such as crystallinity, molecular weight, purity, etc. Modern materials are tightly controlled for dimensional accuracy, and mechanical and physical characteristics. Corrosion is a universal enemy of these engineering efforts.

Corrosion occurs when the atomic or molecular structure of a material is modified by environmental factors. These factors include water, light, heat, chemicals, etc. This is intuitively understood when you bring items in from outside to prevent their damage, when you wash salt off your car in the winter, and when you seal a deck. There is a big difference between a vintage automobile that has been in a garage for 50 years and one that was parked in a field, even if neither were driven. You know there are actions you can take to reduce the rate at which materials and products degrade, but the exact concepts behind it are more complex. We will start by looking at the mechanisms of corrosion, and then we will investigate methods of controlling or preventing it.

8.5.2 Mechanisms of Corrosion

The way each material corrodes will vary. We will start our discussion with metals, perhaps the most recognizable material corrosion, and then extend the general concepts to polymers and ceramics. Corrosion is a chemical process. To understand it we need to understand chemistry, or at least some chemistry. Much of the quantitative analysis involved in corrosion will not be detailed. Our purposes are to understand the process and use that understanding to account for corrosion in production more than in design, but a product will experience the most corrosion during application, not manufacture. Corrosion can be critical to performance and safety in many applications, and detailed analysis and prediction is well worth the additional effort when needed.

8.5.2.1 Single Component Corrosion

The simplest version of corrosion involves a single material. The material's surface chemistry is changed by environmental exposure. The term **oxidation** technically describes a reaction resulting in the loss of an electron, but we will primarily consider it as the bonding of a material with oxygen.* This basic process is what causes an apple to turn brown after it is cut into slices. When a piece of steel is machined, there is a new surface created that is extremely "clean." If it is plain carbon steel, the only elements on the surface will be iron and carbon.† This does not last long, however, and the surface will quickly oxidize under normal, ambient conditions. The exposed iron has a natural tendency to bond with oxygen in the air, and the amount present in air is sufficient to coat the surface with an iron oxide layer. The higher the temperature and longer the time, the thicker the layer of oxide becomes.

Have you ever wondered why steel turns shades of brown, blue, and purple when heated to very high temperatures? Elevated temperature accelerates the oxidation process and the varying oxide thickness results in varied color since the oxide thickness determines how light is refracted from the surface. Most materials only form a very thin layer of "native" oxide (< 50 nm), which is transparent. Oxide thickness can be an important factor in production of certain products. For electronics, silicon is often coated with silicon dioxide as an insulating or capping layer, and like steel, the color indicates the thickness. Oxide formation is also used to anodize aluminum, which improves lubricity and corrosion properties and allows dye retention to color products (anodizing is discussed more in Sec. 9.3.8.2). In some applications, the exact thickness (within nanometers) of the oxide is critical (e.g., semiconductor electronics). Precision measurement of oxide thickness can be accomplished by reflecting a laser off the surface in processes such as interferometry and ellipsometry. Although oxidation is a naturally occurring process, it can be specifically controlled to produce enhanced properties. The general reaction where a metal, M, is oxidized is given by

$$M \rightarrow M^{n+} + ne^- \tag{8.21}$$

You will notice that there is no oxygen listed in the reaction of Eq. (8.21). Any reaction where an electron, e^-, is lost, as in the formation of the cation, M^{n+}, the reaction is termed an oxidation reaction. If an electron is gained, it is a reduction reaction. Any number of electrons, n, can be lost or gained and the number corresponds to the oxidation state. The variable, n, is commonly equal 1, 2, or 3 in metals with some metals having various possible oxidation states.

Typically, when an electron (or more than one) is liberated, another element will take those "free" electrons. When something is oxidized, something else is reduced. The

---

*Oxidation by bonding with oxidation is the most common example of corrosion of metals. The term reduction describes a reaction where an electron is gained.

†Not considering other minor alloying elements. See Secs. 4.2.2 and 4.2.3.

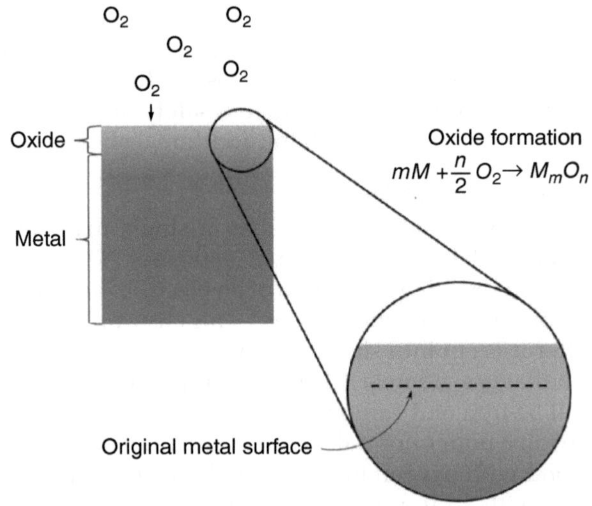

FIGURE 8.19 As an oxide layer forms, it takes up more volume than the metal alone. Although the oxide extends into the original surface, it also extends outward and causes the dimensions to increase. In some cases this change in dimension can harm performance of an assembly.

behavior of electrons in bonding was covered back in Sec. 2.3. To create a stable electronic configuration in oxidation, the metal bonds with oxygen in a ratio that satisfies the ideal eight valence electron rule. This varies for different metals and their oxidation states when bonded, and the general equation can be given by

$$mM + \frac{n}{2}O_2 \rightarrow M_mO_n \tag{8.22}$$

Oxides are a combination of metal atoms and oxygen, so they must occupy more volume than the metal alone. Although relatively minor, the formation of an oxide layer will increase the dimensions of a component. Changes due to native oxides are nearly imperceptible, but processes such as anodizing can create micron increases per surface,* and in precision fitments, this can inhibit proper function. The concept of oxidation is shown schematically in Fig. 8.19.

It is important to note that the amount and properties of an oxide are highly dependent on the host metal and temperature during exposure to oxygen. Gold, for instance, does not form any notable oxides,† but chromium quickly forms a stable, protective oxide layer, which enables the creation of stainless steels (see Sec. 4.2.6 for more). Even though the two metals respond differently to oxygen, both are resistant to corrosion. The differences in metals and multi-component corrosion will be discussed more in Sec. 8.5.2.2.

Different metals result in different ratios of oxygen. For example, CuO, TiO_2, and Li_2O all follow the same basic oxidation process but yield different results. Some metals

*A round part will increase in diameter by twice the oxide thickness. This occurs similarly for other shapes where the entire surface is coated.

†The most stable oxide of gold, Au_2O_3, decomposes around 160°C.

also have multiple oxides, such as iron, which may form FeO, Fe_2O_3, or Fe_3O_4. A simple example of an oxidation reaction for aluminum is given in Eq. (8.23).

$$2Al + \frac{3}{2}O_2 \rightarrow Al_2O_3 \tag{8.23}$$

Aluminum is assumed to be a 3+ valent element (it has three electrons in its outermost electron shell) and oxygen is 2– (it is two electrons short of having a filled valence shell). Al_2O_3 is a common ceramic that is discussed in Sec. 6.2.1.1, and you can see that metallic corrosion produces a ceramic. Because the product of corrosion is often a ceramic, ceramics are essentially impervious to corrosion, in the general sense, because they *are* corrosion, or at least the byproduct of natural processes involved in corrosion. They do not readily react with the environment because the reaction has already taken place (i.e., they are in a stable form). The reaction in Eq. (8.23) is very simple, but there are often other chemicals present and corrosion rarely produces a single, stoichiometric product. We can get a better picture of corrosion by considering reactions other than oxidation.

A single material can corrode in the presence of any chemicals that react with it. For instance, strong acids will dissolve many materials, but not all acids dissolve all materials. The chemistry of the two has to be correct. A metal that is stable in one environment may disintegrate quickly in a different one. The metal and the environment must be known before an accurate assessment of corrosion can be conducted. One acid-metal combination that results in a vigorous reaction is hydrochloric acid (HCl) and aluminum. The reaction for the two is given by

$$2Al + 6HCl \rightarrow 2AlCl_3 + 3H_2 \tag{8.24}$$

There are some important differences in Eq. (8.24) from Eq. (8.23), the most noticeable of which is the addition of an extra product. Our reactants are an element (Al) and a compound (HCl), so unless the product(s) consist of a compound containing all the reactants, there will be extra matter, in this case hydrogen. The hydrogen is given off in the form of gas, so you would see bubbles forming during the reaction. Corrosion often occurs from a combination of reactants and produces a variety of products. The chemistry in solution at the site of corrosion can change over time and actually become even more aggressive due to the byproducts of the process.

Figure 8.20 shows a schematic of a metal reacting with HCl. The HCl is dissolved in water and as the metal reacts with the chlorine the metal loses electrons to the rest of the metal and the chlorine splits from the hydrogen. The hydrogen proton then reacts with the excess electrons to form a hydrogen atom. This atom bonds with another to form diatomic H_2 gas, which is given off as bubbles. The reaction (corrosion) rate depends on the availability of the reactants to combine and then for the reaction products to leave the reaction site so fresh reactants can repeat the process.

8.5.2.2 Multi-Component Corrosion

Most processes that are considered corrosion are electrochemical. That means there is a chemical reaction involving the transfer of electrons. In the previous, single-component corrosion examples the material was oxidized, having lost electrons due to interaction with environmental chemicals. In many assemblies there are materials of differing composition which come in contact. The difference in materials can actually accelerate corrosion by allowing electrons to flow between materials, and it results in an electrical current.

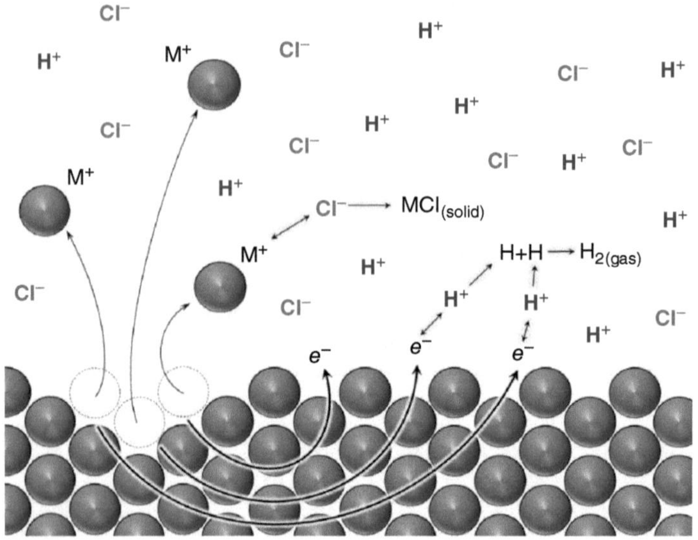

FIGURE 8.20 Single material corrosion by chemical attack can occur through the oxidation of metallic atoms and formation of compounds. The schematic illustrates a metal being corroded by a solution of hydrochloric acid (HCl). The metal atoms form a chloride compound with the chlorine anion, and the electrons released by the metal ions combine with hydrogen cations to form hydrogen gas. (*Note:* Single hydrogen atoms (H) quickly combine to form diatomic H_2 gas.)

The materials being paired determine the rate at which these electrons flow due to an electrical potential, or voltage, that develops. This is the basic way batteries work. There are two electrodes, one which supplies electrons (anode) and one which receives electrons (cathode). The flow of electrons (e^-) provides power for the device the batteries are used for. Batteries "die" because the anode is finite. It does not have an infinite supply of electrons, and when they are exhausted, it can no longer carry a current. See Fig. 8.21 for an example of this setup.

Batteries like these essentially operate on the basis of corrosion. Although the corrosion of metals is typically thought of as a bad thing, the chemical principles behind it are what make other beneficial processes possible. When different metals are connected by an electrolyte (e.g., saltwater), one of the two will corrode preferentially. To determine which one will corrode, we can look at a galvanic series. The galvanic series lists metals in the order that they will corrode in a given environment. Metals at the top of the list are **passive**, which means they are resistant to corrosion, and toward the bottom they are **active**, which indicates they readily corrode. The farther apart on the series, the greater the potential (electromotive force) and the more rapidly the anode will corrode.

Some metals and their relative positions on the galvanic series are listed in Fig. 8.22. When differing metals are combined, this can result in galvanic corrosion, where the anodic material is preferentially corroded. There is also a need for an electrolyte as discussed above and shown in Fig. 8.22. If the connection between differing metals is completely dry, corrosion will not progress. The corrosion rate is quantifiable through use of the standard electrode potential using the reduction of hydrogen as the standard. Based on this standard, some metals will have a positive electrode potential and some

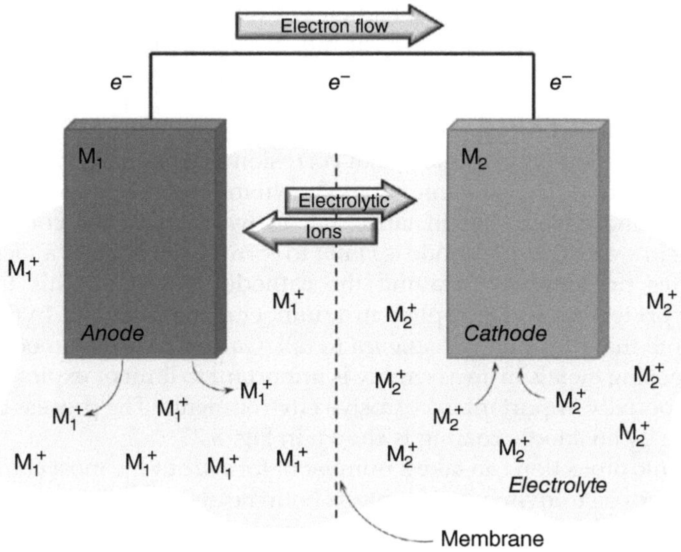

Figure 8.21 The basic components of an electrochemical cell consist of an anode, cathode, and electrolyte. The anode, metal 1 (M_1), will transfer electrons to metal 2 (M_2). During this process the anode corrodes.

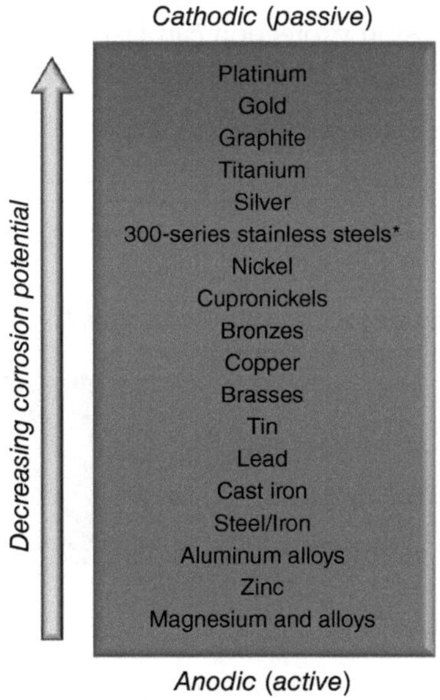

Figure 8.22 The galvanic series for listed metals in seawater is shown. Metals near the top of the list are passive and do not readily corrode. Metals near the bottom are active and can corrode rapidly. When two metals are paired, the one higher on the series will act as the cathode. The father apart they are, the faster the anode will corrode.

will have a negative electrode potential. The difference in the two determines the driving force for corrosion. We will limit our discussion to the qualitative importance of material couples in corrosion and corrosion prevention.

8.5.2.3 Galvanic Corrosion

Galvanic corrosion is the preferential corrosion of one material over another when in close combination. The galvanic series shown in Fig. 8.22 has practical implications for processing and design. For instance, if you want to control corrosion, you can pair materials in a way that the anode is *meant* to corrode. It becomes a sacrificial component. It corrodes preferentially, leaving the cathode protected. This technique is called **cathodic protection** and is applied in a number of applications. In this case, it is important to note that the anode is *designed* to fail. Galvanic corrosion occurs unwantedly as well. Choosing metals in an assembly is important to limit or exploit galvanic corrosion. This is especially important in corrosive environments. The process concept of galvanic protection by an anodic coating is shown in Fig. 8.23.

Galvanic protection can take a number of forms, but the most common is in coatings. Until the corrosive environment contacts both metals, the corrosion resistance of the coating alone determines passivity. For some coatings, a small scratch can actually "heal" itself. This is because as the metal oxidizes it expands and can refill the scratched area. This type of galvanic protection is very common in steel structures that are coated with zinc.

You may have heard of galvanized screws used in construction, but the concept is used in chain-link fence, guard rails, lights poles, and many other applications (see Fig. 8.24 for some examples). A steel component is coated with zinc because the zinc forms a fairly stable oxide that does not corrode quickly. If the zinc coating is damaged and bare steel is exposed, the large area of the zinc surface can supply electrons to the exposed steel, thereby protecting it from corrosion. Because the amount of zinc is typically much larger than the exposed steel, the corrosion protection can last for a significant length of time. This is a low cost way to achieve good corrosion performance.

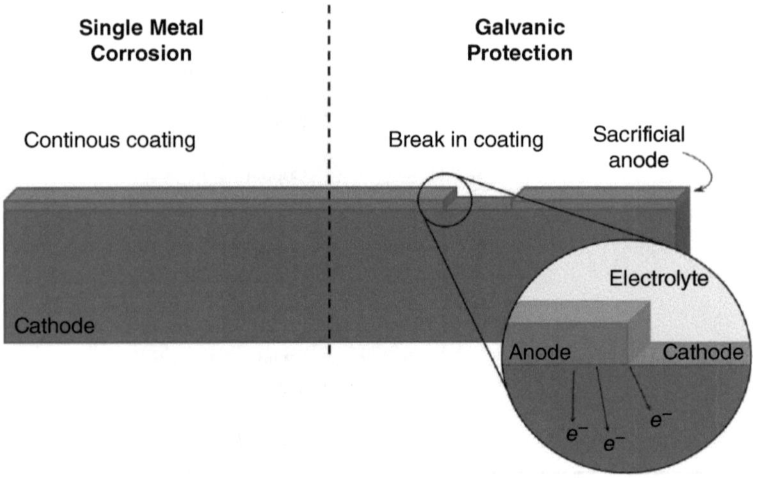

FIGURE 8.23 Galvanic protection can be used to stop corrosion in one metal by preferentially corroding another. In this schematic, a coating is shown which acts as the anode. The coating will corrode while the structural metal remains intact.

Figure 8.24 Galvanization is used to increase the corrosion resistance of steel in many applications. It can be identified by the geometric patterns on the surface, known as "spangle," which are actually zinc grains. Though not all galvanized structures appear the same, most will have this appearance.

The continuous supply of electrons is the key to corrosion protection in metals that are not inherently stable. This means corrosion protection can take other forms than coatings. For instance, on a large structure that must have a very long service life (decades), a sacrificial anode can be connected to the structure. For example, a large, magnesium block can be attached to a steel pier in the ocean. The magnesium corrodes quickly, and as it corrodes it protects the steel.

This can be achieved in a similar manner for underground pipelines which cannot be easily accessed. A magnesium anode in the form of a spike is driven into the ground and electrically connected by conductive wire or cable to the pipe. There is a minimum spacing between sacrificial anodes to ensure the base metal (cathode) is being protected, so they are spaced intermittently around or along the structure. Magnesium is a popular choice because it is so anodic, but other anodic materials, such as zinc, are available. For instance, a marine anode is shown in Fig. 8.25 that reduces corrosion, especially in saltwater environments. The fast corrosion of these sacrificial anodes requires them to be replaced periodically to maintain performance.

8.5.2.4 Concentration Cell Corrosion

Some coatings can be beneficial for corrosion protection, but others can be inadvertently damaging. Paint is one of the most common measures used to decrease corrosion. Can you imagine why it might end up causing more harm than good? Think of the discussion we just had about coatings and the coupling of materials in a galvanic cell. Paint, a polymer, is not actually cathodic in the traditional sense, but if a painted film is damaged, the underlying metal can undergo accelerated corrosion because the exposed metal will lose electrons and protect the metal still covered by paint. The scratched area starts to acts as an anode. This is depicted in Fig. 8.26. It is not only coatings, such as paint, that can cause this behavior. When rust is formed in solution it will deposit in the area and actually protect small areas itself. The area under the rust will then be cathodic. So rust itself can accelerate rusting.

Additionally, metals that are normally passive by forming a stable oxide layer can be susceptible to concentration cell corrosion. Those metals, such as stainless steel, require oxygen to resist corrosion. In situations where oxygen is depleted, such as under paint or decals or in small crevices, the metal will corrode rapidly and is often unseen. The exposed sections are passive leaving the hidden areas as the anode. For a concentration cell to occur, the electrolyte must be able to enter but not readily leave the area. A space of a few thousandths of an inch is usually adequate to create this condition. Examples of a detrimental coating and bolted assembly are shown in Fig. 8.27.

Figure 8.25 Sacrificial anodes are designed to corrode faster than the structure they are connected to. By preferentially corroding, they protect other metals higher on the galvanic series. (*Photograph courtesy of Jeff Heger*.)

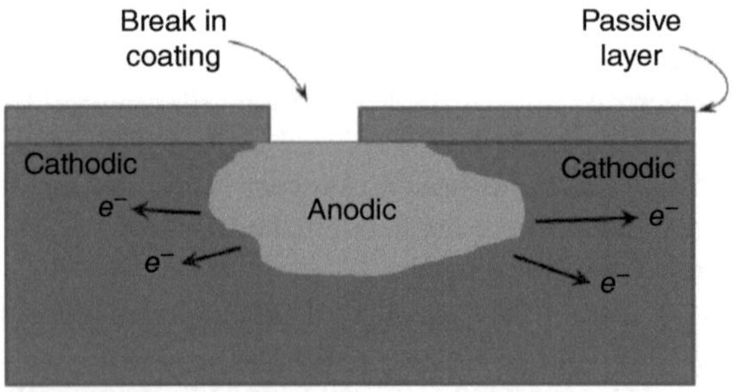

Figure 8.26 When a metal is coated with a passive material such as paint or even previous rust that has deposited on the surface, the area underneath the coating acts as the cathode and the exposed area acts as the anode.

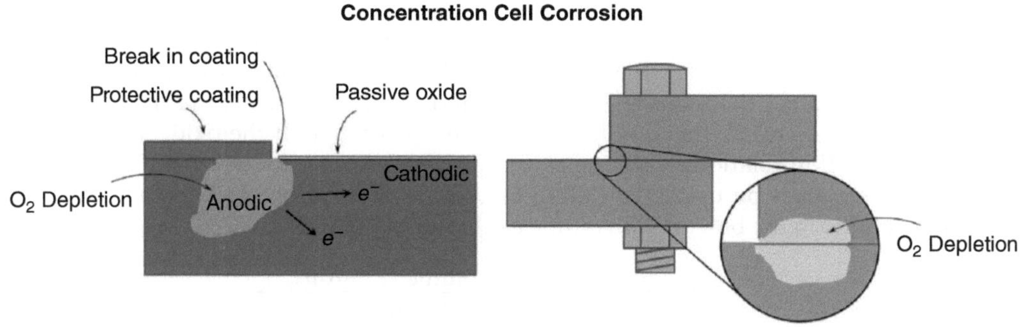

FIGURE 8.27 In small crevices, oxygen is not readily available and this can accelerate corrosion. Even in alloys which are normally passive, the lack of oxygen and high concentration of corrosive species will stop them from forming a protective oxide layer.

Any protective coating can become a problem, not just paint. For instance, in Fig. 8.28 a wheel from a car is shown. The wheel had a decorative and protective chrome coating applied, but the coating developed a scratch or break at some point. The wheel was continuously exposed to water, road salt and other dirt and debris that concentrated under the protective layer. Corrosion could progress unseen until finally the entire coating delaminated and peeled off. This is the sort of situation that must be avoided in critical structures. This wheel had been in service for 15 years, and is still structurally sound. In other circumstances concentration cell corrosion can be dangerous, not just inconvenient and unsightly.

8.5.2.5 Oxide Stability

The idea of passivation has been used to describe corrosion, but it is important to understand how passivation occurs. As discussed, all metals will form oxides, and most form stable oxides under normal operating conditions. For example, zinc and iron both form oxides, but zinc forms a passive oxide and iron forms a nonpassivating oxide.

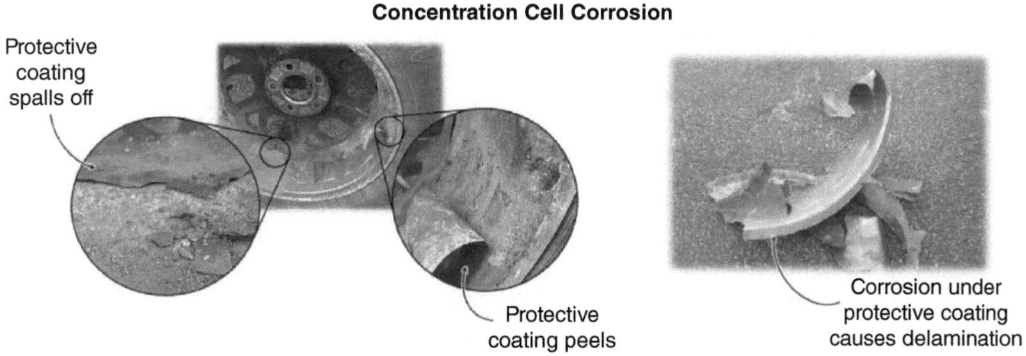

FIGURE 8.28 Concentration cell corrosion occurs where moisture, chemicals, and debris can get in or under, but they cannot escape. Protective films, such as the chrome plating on this vehicle wheel, can act to cover corrosion and accelerate the process. This form of corrosion is insidious as the coating still appears to be intact, as shown on the right.

This means as iron rusts, it will not stop. The oxide spalls off as it forms, thereby exposing fresh metal and the process repeats. The reason the oxide spalls off is because it has a much larger volume than the iron, and as the oxide becomes thicker, it develops a compressive stress that causes it to buckle and break up. Metals that have oxides of similar volumes to the metal will be stable and protective. If the oxide volume is too large or small in relation to the metal, it will fail. One way to describe the stability of oxidized metals is by using the Pilling-Bedworth (PB) Ratio, which is the ratio of the oxide volume to the base metal volume as shown in Eq. (8.25).

$$\text{PB Ratio} = \frac{\text{oxide volume}}{\text{base metal volume}} = \frac{M\rho_{\text{metal}}}{nA\rho_{\text{oxide}}} \tag{8.25}$$

In Eq. (8.25) M is the molecular mass of the oxide, A is the atomic mass of the metal, n is the number of metal atoms in the oxide and ρ is the density of the metal or oxide as labeled. As an example, we will consider ZnO. Zinc has a density of 7.14 g/cc, and an atomic mass of 65.38 g/mol. Zinc oxide has a density of 5.61 g/cc and a molecular mass of 81.41 g/mol. There is one Zn atom per oxide molecule. This gives us

$$\text{PB Ratio (ZnO : Zn)} = \frac{M\rho_{\text{metal}}}{nA\rho_{\text{oxide}}} = \frac{(81.38\ \text{g/mol})(7.14\ \text{g/cc})}{1(65.38\ \text{g/mol})(5.61\ \text{g/cc})} = 1.58 \tag{8.26}$$

All of the units cancel out in the equation, so we are left with only a number that we must interpret. If the ratio of oxide to metal volume is less than 1 or greater than 2, the oxide tends to be nonpassivating, because the film cracks due to excessive tensile (if <1) or compressive (if >2) stresses. Values slightly greater than 1 are most likely to be protective with a *slight* compressive stress, but it is not an absolute measure. There are many other considerations to make, but it does provide a helpful starting point. We can see that ZnO is expected to be passive. That means it will not readily corrode due to normal environmental conditions, and its lower position on the galvanic series helps protect iron which has higher PB ratios for its oxides, which are not protective. For iron in particular, the formation of varying oxides occurs in stages and by mechanisms that are beyond our scope. It is important to note that where there are multiple oxides possible, the type formed and sequence of formation are just as important to the corrosion rate. Of course, these factors are also controlled by the environment. The PB ratios for some common metal oxides are presented in Table 8.7.

8.5.2.6 Other Types of Corrosion

From our discussion so far, it should be apparent that corrosion can happen in many different ways. The way the metal is attacked is determined by the environment, both in regard to the electrolyte (e.g., saltwater, acid, etc.) and adjacent materials (e.g., galvanic factors, concentration cells, etc.). The way metals corrode can be divided into categories that include:

1. Uniform
2. Galvanic
3. Concentration cell
4. Pitting
5. Stress cracking
6. Intergranular
7. Dealloying
8. Erosion
9. Hydrogen damage

Metal	Oxide	Pilling-Bedworth Ratio
Aluminum	Al_2O_3	1.28
Cobalt	Co_2O_3	2.40
Copper	Cu_2O	1.67
Chromium	Cr_2O_3	2.02
Iron	FeO	1.78
Iron	Fe_2O_3	2.14
Iron	Fe_3O_4	2.10
Magnesium	MgO	0.81
Molybdenum	MoO_3	3.27
Nickel	NiO	1.70
Titanium	TiO_2	1.73
Vanadium	V_2O_5	3.25
Zinc	ZnO	1.58

Note: The Pilling-Bedworth ratio describes the difference in oxide volume and the volume of metal it is generated from. A ratio between 1 and 2 is indicative of a protective oxide, but there are many other important factors that should be considered as well.

TABLE 8.7 Metal/Oxide Relationships

We have discussed the first three at some length. We will now briefly explore some of the other forms. A schematic of these is shown in Fig. 8.29.

Pitting corrosion is a phenomenon that arises in normally passive metals and alloys. The mechanisms are similar to those discussed in concentration cell (crevice) corrosion. The passive film is destroyed locally, and corrosion begins to occur and creates its own crevice that concentrates corrosive elements and accelerates further breakdown. The process may be initiated by deposits of dirt, sand, previous corrosion products, etc. The concept is shown in Fig. 8.26, but the passive film is not continuous. Rather, it occurs as "islands" under which corrosion takes place.

As the name implies, **stress corrosion cracking** (SCC) occurs in components that are under high stress. This type of corrosion can be particularly troublesome because it occurs under environmental conditions that would be otherwise quite benign. The corrosion advances quickly, starting at locations of high stress. Since other areas are not affected by corrosion, the components may still appear to be in good condition. This type of corrosion is typically most pronounced in otherwise passive alloys.

Intergranular corrosion is the preferential corrosion along grain boundaries, usually due to a different composition at or near the boundaries. Impurities or alloying elements can "sensitize" the boundaries, and a metal that normally resists corrosion can disintegrate rather quickly. The most common example of this is in stainless steel. If enough carbon is present, the chromium (which makes it stainless) will form carbides at high temperature and can no longer prevent corrosion by forming a protective oxide. One of the most common ways this occurs is by welding, and welded areas can corrode

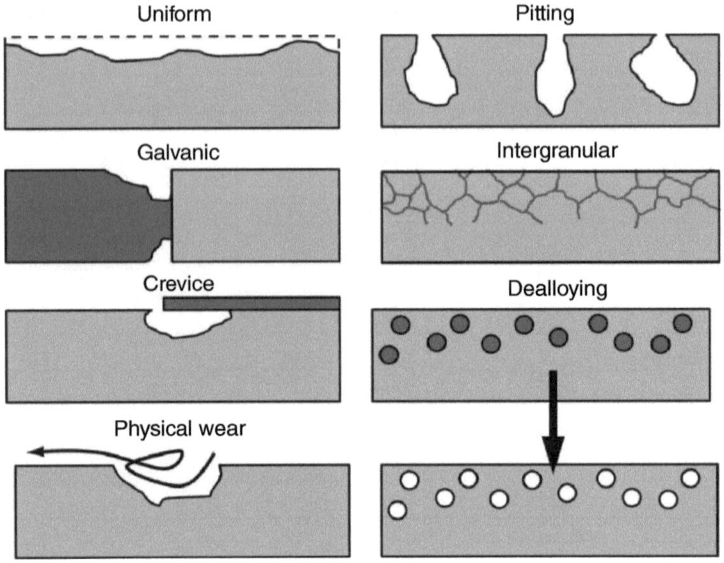

FIGURE 8.29 Corrosion can occur by a variety of mechanisms. In some cases more than one may be active at a time, or there may be multiple mechanisms over time. The type of corrosion depends on the material(s) and the environment.

and weaken very quickly. Specific alloys, like low carbon stainless steels, should be used when welded assemblies will be created.

Dealloying is a process by which a specific element in an alloy is leached out. This is most likely in alloys where an alloying element is much more anodic than the base metal. A common example is brass, where the zinc is more active than the copper (see Fig. 8.22). The zinc will be preferentially corroded out of the copper, leaving a porous and weak structure. This process is known as dezincification. The same basic concept applies to alloys of copper with other metals such as aluminum, and it can also occur in cast iron where the iron is leached out preferentially over the graphite.

Erosion corrosion and **fretting corrosion** occur by a combination of mechanical attrition and corrosion. For example, if a sand and seawater slurry is repeatedly washed over a metal component, the abrasive action of the sand will accelerate corrosion by removing any protective oxide that might form or by simply washing away corrosion products providing a continuous supply of fresh electrolyte. In some instances the flow might be very extreme, such as pressurized steam. The environment need not be liquid, as mechanical friction or vibration in assemblies can break up oxide passivation allowing further oxidation. The oxide spalls off and then acts as an abrasive and accelerates further breakdown of the surfaces.

Hydrogen damage is related to the tendency for dissolved hydrogen to embrittle metals. Hydrogen can be introduced during processing, but it can also be introduced from the environment the metal is used in. Especially at elevated temperatures, hydrogen will tend to form hydrides with metals and can bond with carbon or other alloying species, which weakens the metal. Steel, titanium, and magnesium are all susceptible to hydrogen embrittlement.

8.5.3 Corrosion in Ceramics and Polymers

As described in the previous sections, corrosion of metals results in a ceramic. So ceramics are essentially inactive by comparison. That does not mean that they are completely immune, but they can be considered completely passive for most applications. Situations that might compromise the structure of a ceramic include chemical changes. This might take the form of leaching of one of the elements, which results in a nonstoichiometric composition. For instance, when oxides are heated, they can lose some of their oxygen content but this is reversible, and once cooled they return to stoichiometry or near to it.

Another chemical change would be contamination, such as sodium impurities in silicon electronics, but the change does not compromise it structurally. In some advanced ceramics, such as partially stabilized zirconia (discussed in Sec. 6.2.1.6), the structure is not as stable. The application of moisture, heat, and stress can cause the zirconia to undergo a phase transition and impair the mechanical properties. In situations where ceramics are used in a stable form, there is little concern of environmental degradation under typical, ambient conditions. Ceramics are often used in demanding conditions of temperature, stress, heat, or chemical presence, so each circumstance should still be evaluated carefully.

Polymers are much different than ceramics, but they too are more resistant to corrosion than metals. In fact, polymers are commonly applied in place of metals as a corrosion-resistant alternative. Examples include sewer lines, storage tanks (see Fig. 8.30), decorative trim, vehicle body panels, and much more. Corrosion in polymers is often referred to as degradation, as it results in more structural than chemical change, and polymers are often applied to help prevent corrosion, such as in paint, films, and covers.

The reason polymers are relatively passive is that they are covalently bonded and less likely to form new bonds or exchange bonds like metals will. Polymers are not

FIGURE 8.30 Polymers degrade more slowly than metals in most environments, so applications where a long service life is desirable (such as the storage tanks pictured here), polymers (left) may be more desirable than metals (right).

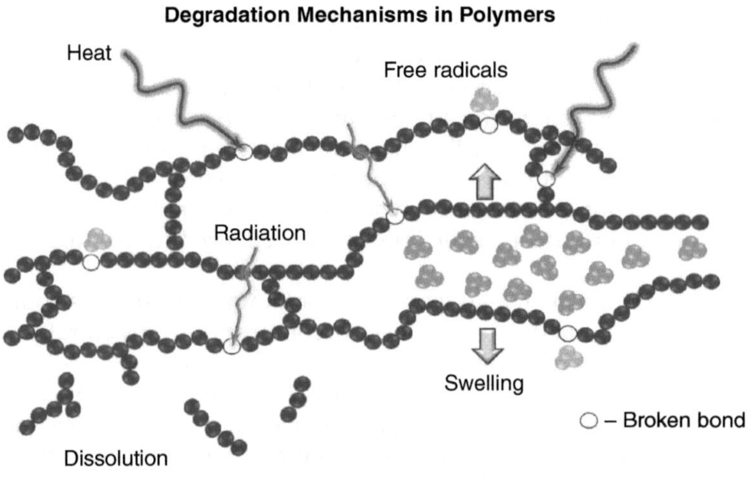

Figure 8.31 Polymers degrade in a variety of ways. Many mechanisms result in the scission of polymer bonds which negatively affects the mechanical properties. The mechanisms may come from normal environmental exposure (weathering) or they may be encountered as part of the material's application, such as radiation or high heat.

as stable as ceramics, and they will degrade more quickly. The long process of breakdown is something that has brought criticism to polymers. Their persistence in the environment is a concern when they are released as pollution. Although there are some biodegradable* polymers, such as polylactic acid (discussed in Sec. 6.3.1.5), most will not readily break down in the environment. They *will* degrade, however, and their appearance and performance are both affected.

The way polymers degrade is often linked to the breaking of bonds in the polymer backbone. The breaking of these bonds is known as **scission**. The results of scission are a lower molecular weight and a weaker structure. On the other hand, polymers can also cross-link as a result of environmental factors, which results in a more brittle mechanical behavior. The molecular breakdown of polymers is complex and can vary significantly for each circumstance, but we will focus on several key mechanisms. The various mechanisms are shown schematically in Fig. 8.31.

The thermal degradation of polymers is something that is important during processing and application. **Thermal degradation** is simply unwanted changes that occur at elevated temperature. When injection molding, for instance, if the melt is overheated or held at high temperature for too long, the polymer can decompose by evolving gases, such as HCl in the case of polyvinyl chloride. This indicates the long molecular chains are breaking down, and this is often accompanied by discoloration and weakening of the polymer.

Do not forget, however, that thermosets and thermoplastics have very different responses to heat (see Sec. 5.3.2.5), but the concept is the same in that degradation

*Bio-mechanisms are valuable for degrading polymers that are not normally considered biodegradable. For instance, certain bacteria and fungi can break down polymers that are normally persistent and that leach harmful additives into the environment.

occurs when they are *overheated*. Heat plays a critical role in most all degradation methods, and polymers which are constantly exposed to elevated temperature should be carefully chosen and monitored.

Radiation damage is another cause of polymer degradation. Although it can be related to high-energy radiation such as electron, x-ray, or nuclear decay products (α, β, and γ radiation), the most common cause is ultraviolet radiation that occurs in sunlight. This is one of the aspects in weathering. **Weathering** includes a collection of outdoor environmental factors that cause degradation of materials. A lawn chair will fade and weaken faster in the sun than in the shade, assuming all other factors are equal. Some of those other factors include chemical and moisture-based reactions.

Chemical degradation occurs when a polymer reacts with a chemical and that reaction causes a loss of integrity. For instance, many polymers are not suitable for use with petroleum products such as gasoline or oil. This is the situation for Buna-S rubber. Buna-N, which has a slightly different structure, is acceptable for those applications (see Sec. 6.3.3.2). The chemical and molecular structures of polymers determine what chemical compatibility it has. In some cases they are much more resistant than metals to chemicals, but in other cases, such as acetone, they are greatly inferior. Chemicals with a chemical structure similar to the polymer tend to be more likely to react.

It is not just liquids that are problematic. Gaseous molecules, such as free radicals, can also cause significant issues in polymers. **Free radicals** are chemicals with an unpaired valence electron that will react readily with molecules it encounters. For instance, ozone (O_3) is highly reactive, especially with polymer molecules containing double bonds. The ozone prefers to react with the bond and break it apart by individual oxygen atoms bonding to each side. A variety of pollutants will also accelerate the degradation of polymers.

In addition to chemical effects that result in bond scission, polymers may simply dissolve in a chemical or they may absorb the chemical, which causes **swelling**. Certain liquids will rapidly degrade the structure, a process known as **dissolution**. As mentioned above, acetone is a common example. Used in nail polisher remover and other applications, it attacks the polymer structure and dissolves the molecules. If you put polystyrene foam in acetone, the structure will melt away like ice in water. That is premise behind chemically welding polymers, so it can be useful if used judiciously. The compatibility between common polymers and solvents is provided in Table 18.2.

When a liquid is not reactive enough to dissolve the polymer but is still reactive enough to penetrate the structure, it causes swelling. A polymer component may absorb water, for instance, and the water forces the molecules farther apart, which increases the dimensions. Aside from obvious dimensional issues, the swelling also causes a reduction in strength. Some polymers, like nylon (Sec. 6.3.1.9), may absorb an appreciable amount of water and are therefore not recommended for submerged or high-moisture applications. The crystallinity and cross-linking of a polymer will both reduce tendency toward swelling by restricting space for the liquid to enter and by resisting the force required to absorb the chemical.

Although there are many ways in which polymers can degrade, there are also a variety of ways to avoid that degradation. This is commonly addressed by additives that resist free-radicals and absorb UV light without breaking chains (more on additives in Sec. 10.2.3). Again, polymers are often a good choice as a corrosion-resistant alternative to metals and a cheaper and more versatile alternative to ceramics.

8.6 Why It Matters

Heat is everywhere. All materials respond to heat, which is measured through temperature. Corrosion is everywhere. It is the natural tendency of all materials to degrade over time. Thermal and environmental concepts are at the center of processing and applications. While some applications will involve modest conditions of heat and environment, there are a host of them that must consider these factors very carefully. For example, when heat-treating steel, if no protective environment is used, a thick layer of scale will form, which may make a part unusable or at least unsightly. Aluminum, however, will look nearly unfazed even when heated to the same homologous temperature. There are fundamental differences in the way the two materials respond to temperature in an oxygen-containing environment.

If it was not for the second law of thermodynamics, we would not have to worry about it. Materials are constantly seeking the lowest possible energy state, which is not necessarily the most useful for us. Degradation has to be controlled, or at least considered, for reliability. We cannot afford to have a product change unexpectedly during use. We looked at the various ways heat can change the dimensions of parts, but it also changes the mechanical properties, sometimes in insidious ways.

Some materials are seemingly "immune" to temperature. They perform well at very low and very high temperatures. Others are unsuitable below a certain temperature. In polymers we call this the glass transition temperature, and it was the temperature below which the polymer behaves in a brittle fashion. In some metals there is a temperature at which they too become brittle, and it was called the DBTT. Ceramics are brittle nearly all the time. The exception is very near their melting point, which tends to be quite high, so they are used as structural materials at elevated temperature.

Ceramics were also quite resistant to corrosion. Although they are not entirely unaffected, they are quite stable in regard to metals and polymers. In many cases, ceramics are applied where corrosion resistance is of utmost importance, especially at elevated temperature. Many metals are susceptible to corrosion, so carefully choosing, pairing, and protecting them is part of design. The ones that are passive tend to share some common features about how they form oxides and whether that oxide coating is stable. It is not that metals like stainless steel and aluminum do not react to the environment, it is just that they stop reacting very quickly. In applications where painting or other passivation measures are undesirable, the use of more expensive, passive materials becomes a good option. Just remember, depending on how they are assembled and what other environmental influences are present, these passive materials can become active.

That is why heat and environment are so important. They are complex. There is not a single fix-all material. Applications typically require certain physical, mechanical, electrical, or other properties *primarily* to meet the needs of the application. Corrosion tends to be of secondary importance, but that does not mean it is not very important. Contrary to that, it can undermine whatever benefits of strength or weight a particular material may have. They must be considered together and assessed during operation.

Part of good design is based on an anticipated rate of degradation while maintaining a realistic expectation of service life. For instance, cell phones or other tech devices, which are constantly changing, do not have to meet any serious requirements on corrosion since they may only be used for a year or two. It is pertinent to balance durability with cost. The application will determine how critical it is. For instance, if a buried natural gas pipeline operating at high pressure fails due to weakening by corrosion, that can have devastating consequences. Alternatively, a shed for a residential property may be "in service" for many years, but it is much more apparent when the structure is losing integrity due to decay and has less of an impact if it fails unexpectedly.

8.7 The How and Why of Thermal and Environmental Concepts in Processing

8.7.1 Chapter Review

1. How does temperature affect a material? How does it affect corrosion?
2. What is the benefit of using an absolute temperature scale? What are the two scales?
3. What does the field of thermodynamics study? How it is useful to processing?
4. In your own words, list what the four thermodynamic laws state.
5. What is entropy? List three situations where its effects can be seen.
6. What does the thermal conductivity of a material dictate? How does it relate to what we perceive as hot and cold?
7. Which materials tend to have a high thermal conductivity? Which have a low thermal conductivity?
8. How is heat transferred through a material?
9. What does the coefficient of thermal expansion tell us about a material?
10. What leads to the expansion of a material as temperature increases?
11. Calculate the change in length for the following materials when heated by 450°F (use Table 8.3):
    a. A 6-in bar of fiberglass (use lower values)
    b. A 13-in rod of polycarbonate
    c. A 5-ft bar of gray cast iron
    d. A 0.125 piece of sapphire
12. Calculate the change in length for the following materials when heated by 600°C (use Table 8.3):
    a. A 50-mm bar of steel
    b. A 6-m rod of copper
    c. A 35-mm bar of silicon carbide
    d. A 50-mm bar of invar
13. Calculate the coefficient of thermal expansion for a material that
    a. Extends by 0.03 in after a 36-in section is heated by 350°F
    b. Extends by 0.013 in after a 21-in section is heated by 500°F
    c. Shrinks by 0.06 in after a 6-ft section is cooled by 200°F

14. Calculate the coefficient of thermal expansion for a material that
    a. Extends by 4 mm after a 2.6-m section is heated by 300°C
    b. Extends by 7 mm after a 242-mm section is heated by 650°C
    c. Shrinks by 0.21 mm after a 45-mm section is cooled by 100°C
15. In general, what design choices will result in thermal expansion being decreased?
16. What is the difference between thermal expansion and thermal distortion?
17. How does thermal distortion arise in a material of uniform composition?
18. How does distortion arise when dissimilar materials are combined?
19. Can thermal distortion be used to straighten parts? If so, how?
20. What manufacturing processes are most affected by thermal distortion? What unwanted effects arise in these processes?
21. In what applications is thermal expansion most important to account for?
22. How does a reduction in temperature affect the properties of most materials?
23. What is a DBTT, and what limits does it place on applications involving a material that exhibits it?
24. How does the composition of a material affect the DBTT?
25. What structural feature is most responsible for low temperature mechanical properties?
26. How can a material's room temperature properties inform you about its low-temperature properties? Think of ceramics and high carbon steel.
27. How does an increase in temperature affect the properties of most materials?
28. When referring to "high" temperature, why is the homologous temperature important?
29. What is the homologous temperature in the following circumstances (use Table 4.5 and Table 4.6)
    a. Aluminum at 600°C
    b. Zinc at 320°C
    c. Tantalum at 250°C
    d. Molybdenum at 2000°C
30. What is the difference between melting temperature and service temperature?
31. What is the glass transition temperature? What materials does it affect?
32. How is the glass transition different than melting?
33. When a material exceeds the glass transition temperature, is it unsuitable for use?
34. Compare two materials from Table 8.4, one with a glass transition temperature above room temperature and one below. How will their properties vary?
35. What is creep? How does it differ from stress relaxation?
36. Rank the classes of materials in order of their creep resistance.

37. What are the three stages of creep? In which stage does a component spend most of its time?
38. What interactions are included in environmental effects?
39. What is diffusion? Where can it be found?
40. What drives diffusion? When will it stop?
41. How does a concentration gradient change over time?
42. What does Fick's first law describe?
43. What is the important of the diffusion coefficient?
44. What are the two type of lattice diffusion? Which is the fastest? Why?
45. What is diffusivity, and how does it change with temperature?
46. How do differences in atomic size affect diffusion?
47. How does the diffusion rate in polymers and ceramics differ? When compared to metals?
48. What is permeability, and how does it affect everyday items you encounter?
49. What is the purpose of a diffusion barrier, and where would it be applied?
50. What is corrosion? What materials corrode?
51. How is corrosion a natural process?
52. What is oxidation? Does it have to involve oxygen? Why or why not?
53. What is a "native" oxide?
54. Is oxidation ever a good thing?
55. How does the volume of a component change during oxidation? Why? What effect can this have on its application?
56. Why are most ceramics nearly impervious to corrosion such as oxidation?
57. Why is the chemical environment so important to determining corrosion rate? Is it possible for a metal to stable in one and not another?
58. When two dissimilar metals are connected electrically what is created? How does this effect corrosion?
59. How does the corrosion rate of the anode and cathode differ?
60. What is meant by the terms passive and active in regard to corrosion?
61. Which metal will corrode faster in the following pairs:
    a. Copper and tin
    b. Zinc and nickel
    c. Platinum and gold
    d. Titanium and stainless steel
62. What is galvanic corrosion? How can this be a bad thing? A good thing?
63. How can a thin coating protect the underlying metal from corrosion? Name two ways.

64. What is the purpose of a sacrificial anode? What applications would it be useful in?
65. What is concentration cell corrosion, and how can it go undetected?
66. In what circumstances might concentration cell corrosion occur?
67. What is the Pilling-Bedworth ratio? What is desirable ratio?
68. What causes pitting corrosion?
69. Why can stress corrosion cracking be unexpected?
70. What problem can hydrogen cause in steel?
71. How can welding cause problems in stainless steel? What kind of corrosion is this known as?
72. What is the loss of a specific element from an alloy known as?
73. What is the combination of abrasion and corrosion known as?
74. What occurs in the corrosion of ceramics?
75. How do polymers compare to metals in their rate of corrosion?
76. What is the primary mechanism in polymer degradation?
77. What factors are included in the weathering of polymers?
78. How do chemical reactions effect the polymer structure?
79. What are free radicals, and what role do they play in the degradation of polymers?
80. Swelling and dissolution are both caused by liquids. What determines if one will happen over the other?

8.7.2 Concept Connections and Extensions

1. Recalculate Eq. (8.4) *and* Eq. (8.5) using the other absolute temperature scale. Do you get the same result?
2. Consider thermal expansion in multi-material assemblies. How could materials of varying CTE be used to eliminate the expansion without causing bending?
3. Look at Fig. 8.14, based on discussion of lattice diffusion mechanisms, what crucial feature is not present in the images of the lattice?
4. Consider Fig. 8.10 and the change in penny composition. Why do you think the composition was changed? Is the DBTT a significant disadvantage?
5. Use the thermodynamic laws listed in this chapter to describe the process of a metal casting cooling in a mold. More on casting can be found in Chap. 11.
6. Have you noticed water dripping out of a car's exhaust, especially after starting up? How could a longer commute actually help reduce corrosion in the exhaust system?
7. Describe moving through a crowd of people by vacancy and interstitial diffusion. Use that to explain why vacancy concentration is important.
8. If you connect dissimilar metals in a clean, dry environment, will galvanic corrosion be a concern? Explain.

CHAPTER 9

Conditioning Metals

If someone handed you a shovel with a steel spade, what do you know about that shovel? Of course you know the spade is made out of steel, but what more do you know? A quick glance at Table 4.2 will prove the point that there are many grades of steel. Suppose you are further told that the spade is AISI-SAE 1045 steel.

Going back to Fig. 4.4 and the associated discussion there, you can determine it is medium carbon steel containing 0.45% carbon. You can do more digging (no pun intended!) and find out from Table 7.1 that it has a yield stress between 50 and 162 ksi.* This is the stress at which it will permanently deform, which is obviously a bad thing for a shovel. You can compare it to other metals in the table and begin to get an idea of how it stacks up to other options.

This is getting closer to what we really need to know: how does it perform? You see, the performance is not simply linked to what material it is but also to what *condition* the material is in. Most shovels have been hardened and tempered. That gives us specific information on what conditioning steps have been taken, and what sort of applications it is suited for.

We could never determine that by simply looking at what elements are present. We need to know more about how they are arranged. We will look at the various ways both structure and composition affect the properties of metals and the way those properties can be tailored using mechanical and thermal processing.

9.1 How It Works

Metallic materials have two basic characteristics that determine their properties. The first is composition, or what the elemental and/or molecular make-up of the material is. For instance, steel is a mixture of iron and carbon, often with other elements in varying proportions. We can say that AISI-SAE 4130 steel has a composition of 0.28% to 0.33% carbon, 0.40% to 0.60% manganese, 0.15% to 0.35% silicon, 0.80% to 1.10% chromium, 0.15% to 0.25% molybdenum, and maximums of 0.035% and 0.040% for phosphorus and sulfur, respectively. While this is detailed and useful information, it only tells part of the story. The composition determines which characters are in the narrative (the elements), but it does not give us the plotline (what those characters do and how they relate). That information is determined by the second characteristic: structure.

The structure is the way in which elements or compounds are arranged, both at the atomic scale and over a much broader area. The principles of metallic structure are discussed in Chap. 3, so you will need to review and understand that information before

*1 ksi = 1000 psi.

conditioning makes much sense. That is because **conditioning** involves processes that modify the structure of a material in order to control the mechanical properties.

Those mechanical properties were discussed in Chap. 7, and again, it will be very valuable to understand them so that conditioning makes sense. Conditioning is achieved through mechanical means, involving force, and/or by thermal means through the application of heating and cooling processes. The temperature affects on materials are discussed in Chap. 8, and, as you may have guessed, you should be familiar with those concepts too!

Conditioning is, in many ways, the culmination and harmonization of outside influences on internal characteristics. For our purposes we will consider the starting composition of the metal to be fixed throughout the conditioning process, so we are left to determine the structural effects of conditioning. Understand that this is not strictly true. Sometimes elements are unintentionally lost or added during conditioning, and other times the change is intentional (consider carburization in Sec. 9.3.8.1, for instance). For simplicity, we will consider the base metal we start with to have a constant composition unless specifically altered as part of the process.

We will look at two ways to conditioning metals. The first is by mechanically altering the structure, often to increase its strength. The second way is through thermal processing. This involves a variety of heating and cooling techniques, which act to modify the structure and either makes it stronger or weaker. Thermal methods can change the way atoms interact by redistributing elements, through bonding (forming new compounds) or by reducing defects. Conditioning is a very deep topic, so we will limit our discussion to some of the most common methods and their underlying mechanisms.

9.2 Mechanical Conditioning

The first method of conditioning we will discuss is mechanical conditioning. Mechanical conditioning uses force and plastic (permanent) deformation to cause a change in the mechanical properties of a metal. The force can be applied through impact or as a continuous process, it can be repeated multiple times or achieved in a single step, and the amount and type of mechanical conditioning is material dependent.

Not all materials respond the same to mechanical conditioning. Some may respond by a significant increase in strength, while others may be relatively unchanged for the same process. Others may not be able to undergo as much deformation without breaking. There are fundamental limits on the degree of mechanical conditioning that can be applied. The amount of deformation and the resulting changes in the properties are dependent on a variety of factors, including temperature, but we will consider cold work as the primary means for this type of process.

9.2.1 Cold Work

Cold work (or work-hardening) is the general practice of deforming a metal well below its melting point, often at room temperature or just above. While this may not be "cold" in many senses, the reason it is considered cold work is because the metal accumulates damage,* which increases its strength but reduces its ductility. **Hot work** is the practice of deforming a metal above its recrystallization temperature (see Sec. 9.2.1.2), which is often around half of the melting point or less (see Table 9.1). Above this temperature, a

*"Damage" in this context is related to crystal imperfections such as dislocations, grain boundaries, etc. These concepts are described in Sec. 3.2.3.

Metal	Recrystallization Temperature °F (°C)	Homologous Temperature (approx.)[a]
Copper (99.999%)	250 (121)	0.29
Copper (OFHC)	400 (204)	0.35
Copper-5% zinc	600 (316)	0.43
Copper-5% aluminum	550 (288)	0.41
Copper-2% beryllium	700 (371)	0.47
Aluminum (99.999%)	175 (79)	0.38
Aluminum (99.0% +)	550 (288)	0.60
Aluminum alloys	600 (316)	0.63
Nickel (99.99%)	700 (371)	0.37
Nickel (99.4%)	1100 (593)	0.50
Monel (Ni 30% Cu)	1100 (593)	0.50
Iron (electrolytic)	750 (399)	0.37
Low-carbon steel	1000 (538)	0.45
Magnesium (99.99%)	150 (66)	0.37
Magnesium alloys	450 (232)	0.55
Zinc	50 (10)	0.41
Tin	25 (–4)	0.53
Lead	25 (–4)	0.45

[a]Homologous temperature calculated using melting point of major alloying element (e.g., Cu for all Cu alloys).

Note: The recrystallization temperature is an important processing as it defines the difference between hot and cold work, which have very different results on material properties. You may notice how sensitive the recrystallization temperature is to impurities, including alloying elements.

Source: Adapted from Dalton, William K. The technology of metallurgy. p. 71. New York: Merrill, 1994.

TABLE 9.1 Recrystallization Temperatures of Common Metals

material will stay soft as it is deformed. Cold working, then, is the opposite. The material will continue to get harder and stronger the more it is deformed.

This process has limits. We cannot achieve infinite strength by continuing to cold work a metal. The strength comes from a combination of decreasing grain size and increasing dislocation density (see more about these features in Sec. 3.2.3). As the strength increases, ductility decreases. Even though a material may get stronger, it becomes more brittle. Eventually the material will fracture as a result of becoming so brittle it cannot withstand more deformation. This is actually an intuitive and common process.

When you want to break a piece off of something, you may bend it multiple times until it finally snaps off. What you are doing is introducing strain until the structure breaks. You can take a paperclip and break it by bending it back and forth repeatedly. The reason this works is because you are mechanically working the point where it bends so much that it finally fails. You reduce the grain size and introduce so many dislocations and other defects that it becomes brittle.

So what about other processes, such as forming, that are used to create a product with specified dimensions more so than mechanical properties? Although the mechanical conditioning may be beneficial, it may also be detrimental. When purchasing metal stock such as rods and bars of steel, brass, aluminum, and other metals, the supplier may indicate certain conditioning states. These include cold rolled, hot rolled, full hard, ½ hard, ¼ hard, annealed, etc. The process for creating the stock may introduce a level of hardening that is undesirable, so you can specify stock that has a reduced hardness. More about these designations is given in Fig. 4.10, as associated with aluminum alloys, and the condition designations for brass, in percent cold work, are given in Table 4.4.

So what does it mean to have a certain "percent cold work?" To define this, we will reintroduce equations used previously to describe ductility (see Sec. 7.4.3.1). Specifically, Eqs. (9.1) and (9.2) describe the deformation of a material. This can be determined by a change from original length, L, to final length, L_f (for percent elongation), or by a change from original cross-sectional area, A, to final area, A_f (for percent reduction in area).

$$\%EL = \frac{L_f - L}{L} \times 100 \tag{9.1}$$

$$\%RA = \frac{A - A_f}{A} \times 100 \tag{9.2}$$

Referring back to Table 4.4, brass can have anywhere from 0% to 68.7% cold work as standard designations. Let us say we started with a piece of fully annealed (no cold work) brass with a square cross section, 1 in across. We can determine what it would take to achieve "spring" hardness condition (60.5% cold work) by solving Eq. (9.2) for the change in area.

$$A_f = A - \left(\frac{\%RA}{100}\right)(A) = 1\ \text{in}^2 - \left(\frac{60.5}{100}\right)(1\ \text{in}^2) = 0.395\ \text{in}^2 \tag{9.3}$$

We have solved for the final area, but we can now convert it back to length, l, per side, assuming it is square after the process, as shown in Eq. (9.4).

$$0.395\ \text{in}^2 = l^2 \rightarrow l = \sqrt{0.395\ \text{in}^2} = 0.628\ \text{in} \tag{9.4}$$

Once the dimensions are reduced to 0.628 in across, the cold work will have reached 60.5%. Some metals, like gold, can be reduced to a thin foil from a thick cross section, but many others cannot handle so much cold work without breaking. It is important to know the limit of cold work for a specific material, as it is common practice to significantly reduce the area of metallic components. The reduction of large stock into wire and small rod is a very useful process, but the way it is achieved will vary for each material. The simplest way to make large changes in area is to work at high temperature so that stress and strain do not build up, but no benefit to mechanical properties is achieved. Another way is to cold work and intermittently anneal the material (see Sec. 9.3.1 for more on annealing) and relieve the internal strain and allow further cold work. Alternatively, the reduction in size can be done by hot work and the final step or steps can be done by cold working. In one way or another, the stress and strain that accumulate in the material during cold work must be controlled.

9.2.1.1 Structural Effects of Cold Work

The way grain size and other defects affect the properties of a metal is discussed in Sec. 3.2.3, but a brief recap in the context of cold work will be helpful. Cold work modifies the structure by increasing the number of defects, which increases the strength. Dislocations, the mechanism by which many materials deform, cannot move as freely when the defect density is high. This makes deformation a more energy-intensive process, which results in a higher strength and lower ductility.

Consider a simple rolling operation (forming is discussed more in Chap. 12). The material enters at one thickness and leaves at a smaller thickness. The width of the material will increase if there is not a second set of rollers to constrain it laterally. Think about rolling out dough with a rolling pin and how the dough spreads in all directions. In the example of Eq. (9.3), the cross section started square and ended square, so there must be two sets of rollers, or the material will have to be rotated 90° and rolled again (less likely). During the process, a large amount of stress is applied to the material, which causes it to deform and internal strain to build (the basis of cold work). Although the material may start off with equiaxed grains that approximately the same dimension in all directions (like spheres), the orientation (or texture) of the grains may change to an elongated structure in the direction it has been rolled (the longitudinal direction).

Texture describes the collective crystalline orientations of grains within a structure. It is high if the grains exhibit highly aligned orientations, and it is low if the grain orientations are random (or equiaxed). Because of structural factors in deforming crystalline materials (i.e., slip planes as discussed in Sec. 3.2.2) there is a path of least resistance, so to speak, that naturally favors alignment during processing such as rolling. This concept is shown in Fig. 9.1.

9.2.1.2 Recrystallization

Cold work through various processes can reduce the grain size and introduce strain, which hardens and strengthens a metal. This built-up strain remains in the material and can change the performance. Have you ever seen a car's door window break? If you have, you will know that it shatters into many tiny pieces, rather than breaking into a

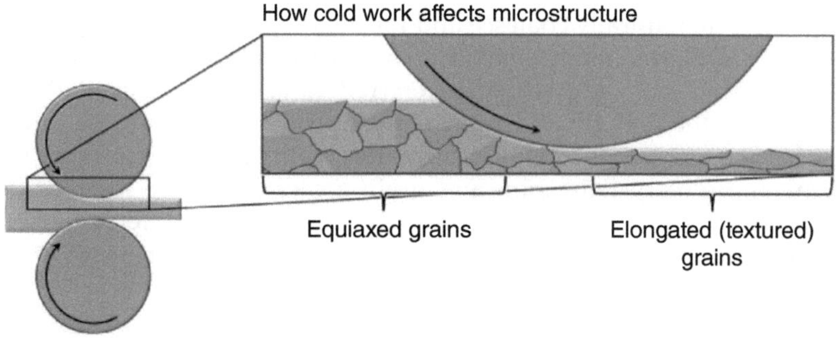

Figure 9.1 Large deformations, such as occur in rolling operations as pictured here, will cause the microstructure of the metal to align in a common direction. This alignment is known as texture, and it can significantly affect the material properties.

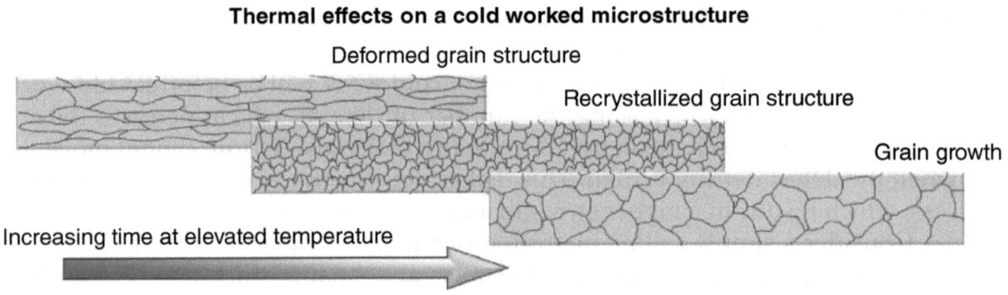

FIGURE 9.2 When heated above the recrystallization temperature, a cold-worked metal will undergo recrystallization and grain growth. The final structure and properties can be dictated by controlling the temperature and the time the material spends at that temperature.

few large ones. The reason for this is that the glass is tempered, which means that is has a built-in strain (see Sec. 10.3.2 for more on tempering glass). When that strain is released, it causes a sudden destabilization that results in a rapid fracturing.*

Internal strain can also be released through heating. While the principle behind it is different, the behavior in strain-hardened metals is similar. Long grains with significant strain stored in them will "fracture" into smaller, strain-free grains as the temperature is raised to a certain level. This phenomenon, known as **recrystallization**, is a means of reducing the internal energy of a material by eliminating strain in favor of smaller grains.

Each metal and alloy has a specific temperature at which this takes places, but it is also dependent on the amount of stored strain and the purity of the material. Table 9.1 shows the recrystallization temperatures for various pure materials and some alloys. You may notice that even a slight drop in purity results in a very significant increase in the recrystallization temperature. For example, 99.99% pure nickel has a recrystallization temperature of 700°F, but when the purity is reduced to 99.4%, the recrystallization temperature increases to 1100°F. Most pure metals and low concentrations of alloying elements have a recrystallization about 1/3 of the melting temperature (a homologous temperature of 0.33).

After recrystallization, many small grains exist that can still yield high strength. The presence of small grains is unfavorable from a thermodynamic perspective, and the grains will naturally coalesce and form larger grains if the temperature is high enough. This is similar to many small soap bubbles combining to form a few large bubbles over time. This grain growth, also referred to as **coarsening** of the microstructure, will decrease the strength and increase the ductility of the material, which may be favorable. The stages during heating above the recrystallization temperature of cold-worked metal are recrystallization and grain growth, and the final structure (and therefore properties) can be dictated through control of the temperature and time at that temperature. This concept is shown in Fig. 9.2. Heat provides the necessary energy to initiate these changes, and the higher the temperature, the faster the process.

*You may also be interested in Prince Rupert's drops that show this phenomenon quite drastically.

9.2.1.3 Considerations in Cold Working

We have discussed the effects of cold work has on the structure of metals, and the principal focus has been the way it affects strength. While high strength may be a good thing, it may also be undesirable, and there are a variety of other considerations as well. For instance, the elongated grain structure actually increases the strength more along the direction of the grains, but it becomes weaker across those grains. If the grains are equiaxed, the properties are isotropic, or the same in all directions. When they develop texture, they become anisotropic, where the properties now vary as a function of direction. Processing that introduces texture can be beneficial where the texture aligns with the forces during application (see Fig. 12.1 and associated discussion for an example of this).

Cold working is an alternative method to heat treating for increasing strength in a metal. Heat treating, or thermal conditioning, which is discussed in Sec. 9.3, can achieve similar or better results in mechanical properties, but the process is sometimes time consuming, requires multiple steps and elevated temperatures, and may have unwanted consequences such as oxidation during high-temperature processing. For those reasons, cold working can be a cost-effective method to achieve modest improvements, especially in regard to surface finish and dimensional stability. One of the greatest drawbacks for mechanical conditioning is that it requires a major capital investment and is usually only justified for large-volume production. More benefits and disadvantages are summarized in Table 9.2.

9.2.2 Burnishing and Peening

Cold work was described in the previous section in the context of forming operations where large deformations cause bulk changes in the microstructure (see Chap. 3 for more on microstructure). Mechanical conditioning need not be done to the entire volume of a material, and many times it is better to only apply it to a surface. That is the objective of methods such as burnishing, peening, and other related techniques. They all mechanically deform the surface of a component to improve surface strength and/or finish.

A smoother surface contains less flaws, which can be the starting point for fracture. A mechanically worked surface also develops a compressive stress, which acts to resist tensile forces. The treatment also helps improve fatigue resistance quite significantly, as the stress concentrations at the surface are removed and cracks cannot propagate as easily (fatigue is discussed in Sec. 7.6.5).

Benefit	Disadvantage
Increases mechanical strength	Decreases ductility
Suitable for large-volume production	May require large up-front investment
Excellent dimensional control and superior surface quality	Higher forces necessary than in hot working
Anisotropic properties (in strong direction)	Anisotropic properties (in weak direction)
Faster and operational costs lower than most heat treatments processes	Control and degree of mechanical enhancement lower

Table 9.2 Benefits and Disadvantages of Cold Working

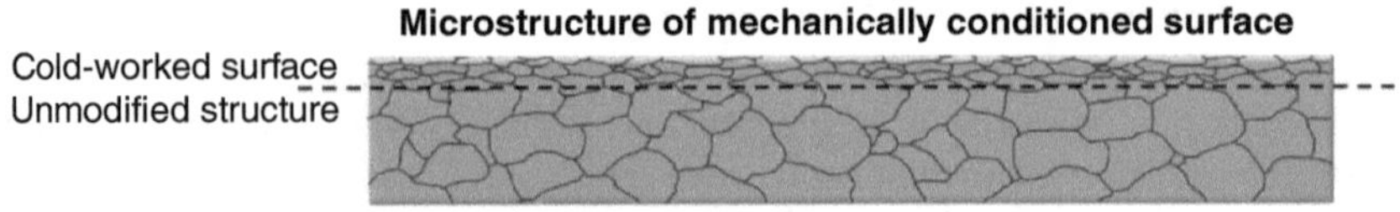

FIGURE 9.3 Mechanical conditioning can be applied to the surface of a component through burnishing, peening, and other methods. The common goal is to smooth and strengthen the surface by physically deforming it. The residual compressive stress at the surface makes it less susceptible to failure from flaws and fatigue.

Burnishing is accomplished by rolling or sliding a hard, smooth surface over the surface of the component. The fine irregularities in the workpiece surface are evened out and the considerable force required results in a mechanically worked surface layer. Burnishing can occur as a result of inefficient cutting or other processes, such as bending, where a material slides against another surface. Where burnishing happens as a side-effect, the parameters are not controlled and the quality and suitability are not guaranteed. Also, it is often localized and does not have equal effects over the entire part.

Peening is a process where impact is used to smooth a surface and cause the necessary cold working for increased strength. The process can be conducted using high-velocity shot, or it can be done using a reciprocating hammer tool,* which is similar in operation to a jack hammer. **Shot** is a generic term to describe the media used in peening. Shot can vary in size, shape, and properties such as density and hardness. The rate of impacts and the energy associated with them will determine the amount of surface deformation. The process is very similar to sand blasting (see Sec. 14.6.3.1), but instead of being abrasive and removing material, it is only intended to modify the surface structure. Peening can be done more easily to complex surfaces and in localized areas that are more vulnerable to surface defects.The general outcome of surface treatments, such as a burnishing and peening, is given in Fig. 9.3.

9.3 Thermal Conditioning

The previous sections discussed the way deformation of a material can result in a change in mechanical properties, and heat can act in the same way. **Thermal conditioning** is the controlled application of heating and cooling to achieve a change in the mechanical properties of a material. This is often referred to as **heat treatment**. Thermal conditioning can drastically change the properties of a metal. Some metals can be made very hard, very soft, or anywhere in between, just by applying and removing heat.

This ties into the opening discussion of the chapter regarding the spade of a shovel. Although the composition was known, it did not give us all of the information about the material. The spade was likely to have been produced using a forming operation, and even though the deformation may not have been great, it is likely to impart at least a small degree of added strength. In this case, however, thermal conditioning is likely to have played a much greater role in the strengthening required to serve the purposes of a typical digging implement.

*You may be familiar with what a ball-peen hammer looks like. Now you can better appreciate the name and the original use.

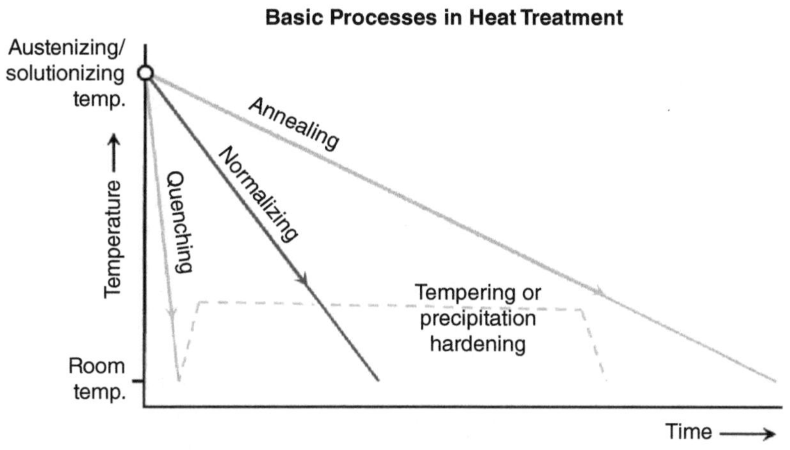

FIGURE 9.4 This graph compares the routes for achieving common thermal conditioning processes. Each of these processes will vary depending on the material and the desired outcome, and the trends shown here are for general comparison only. The labeled processes are described in more detail in the following sections.

We will now look at the most common thermal treatments applied to materials and their effects on mechanical properties. The heating and cooling protocols for common conditioning processes are shown in Fig. 9.4. Although this graph gives the general trends for such processes, there may be significant variation in the actual times and temperatures used. Heat treatment is a very involved engineering effort, and understanding and controlling the complex interactions of atoms and compounds are still active areas of research. The main processes we will be discussing are

- Annealing
- Normalizing
- Quenching
- Tempering (ferrous)
- Precipitation hardening (mostly nonferrous)

9.3.1 Concepts and Terminology in Thermal Conditioning

Before we examine specific processes in thermal conditioning, some basic terminology will be valuable to understanding the concept. As mentioned previously, thermal conditioning (heat treatment) involves controlled heating and cooling. The temperatures and rate of cooling are critical, so we should spend some time developing terms associated with them. Each will be discussed in more detail in the following sections.

Looking at Fig. 9.4, the graph starts at a point labeled "austenizing/solutionizing temp." This represents a critical temperature for ferrous and nonferrous materials, respectively. Many steels can be heat treated, but they must be heated above their **austenizing temperature** or else the process will not be successful. This temperature corresponds to the temperature at which the crystal structure transforms from bcc to fcc.

Iron prefers a bcc structure at room temperature, and unless it is raised to a temperature high enough to transform it to fcc, the subsequent cooling cycle will not achieve the desired structural change. The transformation of pure iron to an fcc structure (known as the austenite phase) occurs at 1675°F (913°C), but the temperature for alloys containing other elements will be different. Some alloys, like 300 series stainless steels (see Sec. 4.2.6), are fcc even at room temperature. The unique thermal conditioning requirements must be determined for each alloy.

Alternatively, some alloys require high temperature to evenly distribute their constituent elements. **Solutionizing** is the process of creating a single, well-mixed solution of elements in the solid state. Instead of a structural transformation, they have a compositional transformation where separate phases are dissolved into one another to create a homogeneous mixture. This is similar to heating up a pot of water with sugar in it. The solubility of the sugar increases at higher temperature where more can be dissolved (for phase diagram examples of this, see Sec. 3.2.4.3). This solutionizing process is most often used in nonferrous metals that can be precipitation hardened (see Sec. 9.3.7).

The basic protocols in heat treatment include annealing, normalizing, quenching, tempering, and precipitation hardening. These processes will have differences that depend on the alloy being treated. They do have some basic commonalities that are summarized in Table 9.3. When choosing a process, the cost (energy and time) and throughput (rate of processing) will be very important to determining the best way to get the necessary properties.

For instance, a low-alloy steel may achieve the required strength and ductility after quenching and tempering, but a more expensive alloy may achieve the same parameters without heat treatment. If the material cost is less than the heat treatment cost, the more expensive material is actually more economical. Each of the processes listed in this section will described in more detail in the following sections. The specific way they are achieved will be discussed in Secs. 9.3.6 and 9.3.7.

9.3.2 Annealing

Annealing is the process of heating a material to high temperature then slowly cooling it to reduce internal stresses. This results in a softer, more ductile behavior. Annealing can be done with varying times and temperatures, but the higher the temperature, the faster the material will soften. The temperature required for annealing is often a high

<table>
<tr><th>Thermal Process</th><th>Objective</th><th>Time Requirement</th></tr>
<tr><td>Quenching</td><td>Maximum hardness</td><td>Seconds</td></tr>
<tr><td>Normalizing</td><td>Moderate strengthening</td><td>Minutes</td></tr>
<tr><td>Tempering</td><td>Maximum toughness</td><td rowspan="3">Hours</td></tr>
<tr><td>Precipitation Hardening</td><td>Maximum strength</td></tr>
<tr><td>Annealing</td><td>Maximum ductility</td></tr>
</table>

Note: Thermal conditioning requires multiple processes, each of which has different requirements and outcomes. Heat treatment is a very versatile means for achieving specific mechanical properties.

TABLE 9.3 Summary of Thermal Conditioning

percentage of the melting point of the metal. This ensures that imperfections in the crystalline structure (Sec. 3.2.3) are given sufficient energy to dissociate and elements can diffuse (Sec. 8.4) as needed.

Elevated temperature can cause changes in the structure of the metal, a property known as allotropy or polymorphism. This concept was introduced in Sec. 3.2. This is important because it can put certain restrictions on the rate the material is cooled. Annealing requires very slow cooling, and it is often done by cooling the part in the furnace at a controlled rate or by simply shutting off the furnace and allowing the part to cool at that rate, which often requires hours to reach room temperature.

A metal may be annealed for many reasons, but its most common use is to simply reduce the hardness and increase ductility. This may be done to relieve stress built up by cold working (see Sec. 9.2.1), or it may be done to increase the machinability of a hardened component. After annealing, the part is easier to work with and may be thermally conditioned to higher hardness once the necessary processing is done. Therefore, annealing can be an intermediate or a final step in the manufacture of a component.

9.3.3 Normalizing

Sometimes a part can be modestly strengthened by faster cooling than used in annealing. Normalizing is the cooling of a part in air. The heating must be sufficient to reach the austenizing temperature of a ferrous metal, but the cooling process is relatively quick, depending on the size of the part that must cool. The process has a relatively low cost since no quenching medium is needed and the furnace is not occupied by parts for as long.

The increased strength is not nearly so large as that of quenching, but in many circumstances full hardness is not required or even desired. Normalizing, like all other thermal conditioning methods, is very sensitive to the metal's composition. For instance, certain tool steels (see Sec. 4.2.5) actually do reach full hardness even when cooling in air. Normalizing, then, can achieve significant strengthening, but the use of high alloy content (more expensive alloys) is required. A typical use of normalizing is to improve uniformity of hot-worked parts.

9.3.4 Quenching

Quenching involves heating a metal to high temperature then rapidly cooling it to a lower temperature, often at room temperature. The maximum temperature will be determined by the alloy being conditioned, but for ferrous alloys, it must be above the austenizing temperature, and for nonferrous alloys, it must be above the solutionizing temperature.

There is a very important difference between quenching for hardness and for precipitation hardening. In most ferrous alloys, the quenching process achieves maximum hardness. In nonferrous alloys, and any that are precipitation hardened, the quenching process does not achieve maximum hardness. Rather, it is an intermediate step before hardening. Precipitation hardening will be discussed in more detail in Sec. 9.3.7, and this section will primarily focus on the quenching process and outcomes in ferrous alloys.

Quenching must be done using a quenching medium that is at a lower temperature than the part. The **quench medium** or **quenchant** is the material used for quickly cooling a part from high temperature, often in excess of 1500°F (816°C). To be effective the quenchant must be able to transfer heat quickly, it must be relatively stable at the temperature the part being quenched is at (i.e., it will not explode, vaporize, or decompose excessively) and it should be cost effective. Fortunately, these conditions are met by common liquid baths such as water, brine (salt water), oil, molten salt, or molten lead. To reduce the stress from rapid cooling, it is preferable to use a quenching medium that cools the alloy at a rate fast enough to achieve full hardening (the **critical quench rate**), but also that is not so aggressive that it causes cracking. The avoidance of sharp corners and varying thicknesses will also reduce cracking.

The critical quench rate varies for each alloy, but in plain carbon steels, such as medium- and high-carbon steel, a water bath is preferable to achieve higher hardness. In alloy steels containing elements that prolong the hardening window, so to speak, the rate may be best achieved using oil. Other alloys may benefit by quenching to a relatively high temperature, such as in molten lead, which reduces the cooling rate. Some high-speed tool steels and other high-alloy steels use this type of process. The rate of cooling relative to plain water is given for some common materials in Table 9.4.

Looking at Table 9.4, you will notice that moving air cools a part at nearly twice the rate as still air. This can be true in liquids too. By circulating the oil, water, or other liquid quenchant, the cooling rate can increase significantly. This is partly a result of forced convection (e.g., moving air) being much more efficient than free convection (e.g., still air), but in liquids it also has to do with a phenomenon known as the Leidenfrost effect.

The **Leidenfrost effect** occurs when a liquid is rapidly vaporized, and that vapor prevents further contact of the liquid with the hot surface. For instance, you may see this effect if you put a drop of water onto a hot frying pan or liquid nitrogen on a table or floor. The droplets will appear to float around temporarily until they eventually boil off. The boiling process is greatly restricted, and when quenching, the formation of a vapor barrier reduces the heat transfer. This can be reduced by moving the part through the quenchant or by moving the quenchant across the part.

9.3.5 Tempering

Hardening is used to increase the stiffness and wear resistance of materials, but there are times when the final hardness is too high. The process for achieving full hardness is very simple. You just need to cool the part down from above its critical temperature (the austenizing temperature in most ferrous metals) at a sufficient rate, or the critical

Quench Medium	Quench Rate Compared to Pure Water
Brine 10% cone.	196%
Caustic Soda (NaOH) 10% cone.	138%
Fuel oil	36%
Machine oil	22%
Circulated air	3.2%
Still air	1.5%

Note: The quench rate is controlled by choosing the environment in which the part is cooled. Materials that conduct heat quickly result in a faster quench rate.

Table 9.4 Quenching Media Comparison

cooling rate. Though not impossible, it is much more difficult to achieve a combination of hardness and toughness directly during cooling. Rather, it is easier to **temper** a metal and achieve a balance of toughness and hardness by first quenching to full hardness and then systematically reducing the hardness in favor of higher toughness.

Tempering is achieved by reheating a quenched metal to a modest temperature and holding at that temperature for a prescribed period of time, usually a few hours. The time and temperature that each alloy should be tempered varies according to the composition and desired properties. Tempering should be performed soon after quenching because internal stress can cause cracking after quenching as the microstructure continues to be active, though at a much lower rate. Tempering is used extensively to achieve a higher hardness than the metal has when not thermally conditioned, but it also has a higher ductility than fully hardened parts.

9.3.6 Time-Temperature Transformation in Ferrous Metals

There are "recipes" for thermally conditioning metals to achieve certain properties. This concept was touched upon in Sec. 9.3.1 and illustrated in Fig. 9.4. The cooling rate is associated with varying mechanical properties in most ferrous alloys, but each alloy will have different requirements to achieve the same objective. This makes heat treatment a scientific endeavor, since there is no one-size-fits-all protocol. We will start digging into this topic by looking at a generic graph that displays the relationship between time, temperature, and the formation of new phases that result in new mechanical properties.

Even a simple iron-carbon alloy will have many possibilities for how those two elements interact. The **time-temperature-transformation (TTT) diagram** provides a detailed road map that describes the structure of an alloy in relation to the rate it is cooled. An example of a TTT diagram is given in Fig. 9.5. Even a cursory look reveals

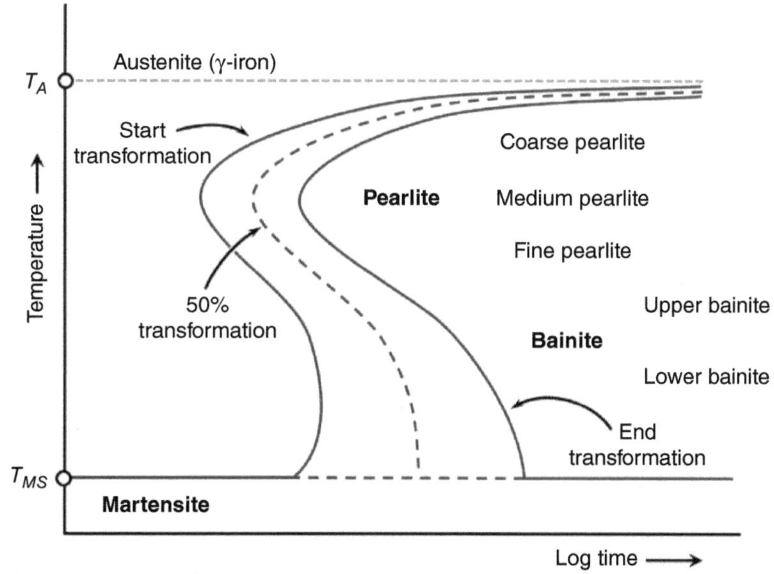

Figure 9.5 The time-temperature-transformation (TTT) diagram gives information on what is happening microstructurally in a metal upon cooling.

there are quite a few possible results after heat treatment. In order to understand why heat treatment affects ferrous metals the way it does, it will be worthwhile to first dissect the TTT diagram. After that we will examine what each phase in the diagram is and how that impacts the final properties of the metal.

9.3.6.1 Interpreting the TTT Diagram

In order to understand what Fig. 9.5 can tell us, we will start with the axes. The y-axis (the *ordinate* values) represents temperature. The two temperatures labeled in the diagram are T_{MS} and T_A, which are the temperatures at which martensite begins to form and the temperature at which the alloy transforms to austenite, also known as γ-iron. We will examine the martensite transformation a little later, but the austenite transformation is a very important temperature as it dictates the minimum (critical) temperature to which we must heat the metal to ensure the heat treatment is successful. T_A, then, is the minimum starting temperature for heat treatment of ferrous metals, but T_A is not the same for all alloys. Also, to ensure the conditioning is fully effective, the metal is typically heated somewhat above T_A as depicted by T_{start} in Fig. 9.6.

Next, we can see the *x*-axis (the *abscissa* values) is plotted as the log of time. That means the same physical distance at the left-hand side of the graph represents a much smaller number than on the right-hand side. For instance, 1/8 in on the left portion may represent 1 s, and on the right-hand side it may represent 100,000 s. This is a critical feature in heat treatment since some transformations require very fast cooling and other can occur over a long period of time.

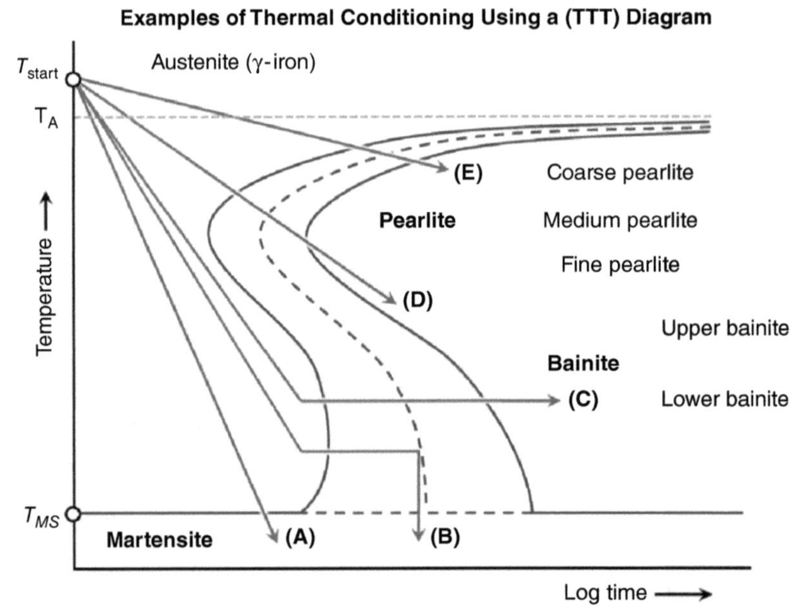

Figure 9.6 The cooling procedure will determine the phase(s) present once the thermal conditioning step is complete. Each phase has different properties and can be useful for different applications. Each thermal conditioning procedure labeled is discussed in Sec. 9.3.6.1.

The curved lines on the graph represent the "transformation" portion of the TTT curve. Time increases as we go from left to right, and once the "*start transformation*" line has been crossed, a new phase begins forming. The vertical position on the diagram (i.e., the temperature) determines which phase that is. The dashed line that mirrors the start transformation line is the point at which half of the metal has transformed to the indicated phase (e.g., pearlite).

After the cooling curve has crossed the "*end transformation*" line, there will be no more phase change. The reason for this is that at high temperature, before the cooling starts, the structure is entirely austenite, and austenite is the phase we change from. Austenite is unstable and will change to *something* as it is cooled below the T_A line. The rate of cooling determines which phase that is. Various cooling protocols are shown in Fig. 9.6.

Slow cooling will result in coarse pearlite (line "E" in Fig. 9.6). This is consistent with an annealing process that cools at a very slow rate. Normalizing, which cools at a modest rate, might give medium or fine pearlite (line "D" in Fig. 9.6). Based on line "D" in Fig. 9.6, we actually might have a mixture of all three types of pearlite. Do you know why? Remember, once the cooling curve crosses the start line, it transforms to whatever phase is most preferable at that temperature. Consider Fig. 9.7 as an illustration of how this occurs. You can think of it as a dinner buffet of sorts.

When you fill your plate, you are transforming open space into food. If you start loading your plate with dinner rolls, there will be less room for pot roast. After you add the pot roast, your plate may be full and you cannot transform any more of that open space. If there is a little room left, you might be able to add a desert, but it depends on the percent open space already transformed.

In line "D," in Fig. 9.6, the dinner rolls represent the coarse pearlite, as it is the first to appear. After continued cooling, the medium pearlite (pot roast) begins taking more available space, and it is quite possible that it could take up all of the remaining space. If the cooling is fast enough, there will still be room for fine pearlite (desert). Where this illustration breaks down is that there is always room for desert but not for fine pearlite!

In order to achieve bainite, a more complex heat treatment is needed. The *start transformation* curve is shaped so that it "overhangs" and bainite cannot be achieved by

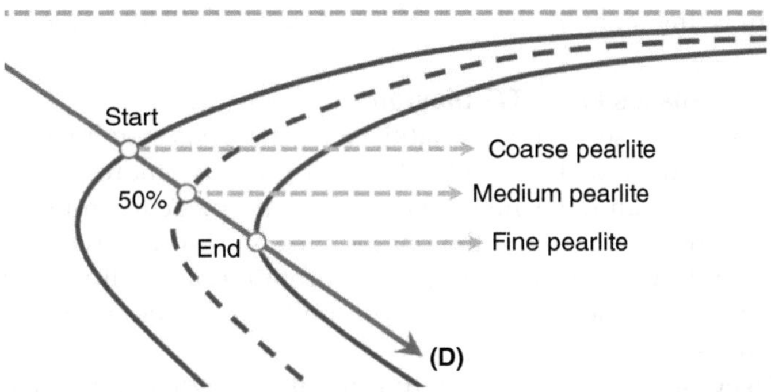

FIGURE 9.7 This enlarged area from Fig. 9.6 shows that line "D" has a complex transformation behavior. Transformations may include multiple phases forming during the cooling process. In the case pictured here, the transformation starts in the coarse pearlite region then transitions through the medium pearlite region until finally completing with fine pearlite.

a straight line without transforming some of the austenite to pearlite. Keep in mind that, on a logarithmic plot, a straight line is not a constant cooling rate. It is also important to note that the cooling rate of a material varies based on the difference in temperature between the part and the quenchant. A part will naturally cool faster in the beginning and more slowly at the end of a quenching procedure.

As shown by line "C" in Fig. 9.6, the cooling rate must be sufficient to miss the leading edge of the transformation start line, known as the pearlite "nose." One way to accomplish this is to cool a part very rapidly to an intermediate temperature. This is where quenchants such as molten salts or lead are useful. The part is still much hotter than the bath, so even what we consider as very hot is much colder than the metal. The part will then cool to the temperature of the heated bath rapidly and can be held there for sufficient time to pass until the transformation has completed. This is the process depicted by line "C" in Fig. 9.6 and results in a bainite structure. The intermediate temperature determines whether upper or lower bainite forms.

Because the transformation is time dependent, we can predictably achieve multiple phases. Line "B" in Fig. 9.6 is an example of this process. What it shows us is that by rapidly cooling a part and then holding it at an intermediate temperature we can begin forming bainite, as described above. Instead of holding it there until the transformation completes, we can rapidly cool it again while there is still austenite to transform. This can be done by a second quench in a different quenchant of lower temperature. Line "B" in Fig. 9.6 shows that the transformation starts in the lower bainite region, and after 50% of the austenite is converted, a second quench is performed to convert the remaining 50% to martensite.

The martensite transformation start is represented by the horizontal line at T_{MS}. It is notable that Martensite occupies the very bottom of the diagram, and requires very fast quenching to be formed. If we rapidly cool an alloy, we can form a structure that is entirely martensitic. This is shown by line "A" in Fig. 9.6. There are cases where the structure does not entirely transform even at room temperature and there may be a small percentage of "retained" austenite. This can cause distortion of heat treated part as the austenite will slowly transform over time under normal use. For critical assemblies and precision parts, the retained austenite can be forced to convert to martensite by **subzero cooling** [often < –100°F (–73°C)]. The lower the temperature, the faster the transformation will complete. Also, the greater the carbon or other alloying element content is, the more likely austenite will be retained after quenching.

9.3.6.2 Phases in the TTT Diagram

In the previous section, we established what phases occur for various heat treatment protocols, but it will be useful for us to discuss more about each of those phases. Although we can determine what phases are present, and even in what percentages, until we know more about the phases, we are unsure of what properties to expect. The phases and factors influencing their formation are still matters of ongoing research. Given their complex nature, we will limit our discussion to their general importance and effects.

Austenite is the fcc form of iron, and it is known as gamma iron (γ-Fe). When iron is heated above the austenizing temperature [1674°F (912°C) in pure iron], it changes from bcc to fcc. This structural change identifies iron as an allotropic or polymorphic material as it can exist in multiple, stable structures. The addition of alloying elements will reduce the temperature at which it changes to fcc, and in plain carbon steel, as the carbon

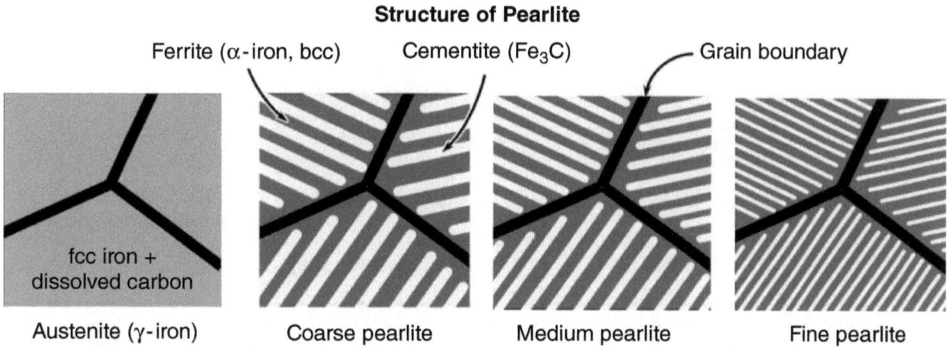

Figure 9.8 Pearlite is a mixture of ferrite and cementite layers. The cooling rate determines the size/spacing of each layer, with faster cooling (smaller spacing) resulting in a stronger component.

content is increased the temperature decreases to a minimum of 1341°F (727°C) at 0.76 wt% carbon, after which it begins increasing again. The reasons for this are related to the eutectoid with carbon that occurs at that concentration, but that depth of discussion is left for texts on metallurgy or materials science. You may see grades of steel that are referred to as "hypereutectoid," and this indicates the unique composition and structure of the steel.* For our purposes, the austenite phase is most important for the fact that it is a precursor to all other phases. When the austenite is formed, it essentially reverses any prior heat treatment or phase formation and gives us a uniform starting point.

Pearlite is actually a mixture of **ferrite** [bcc iron, or alpha iron (α-Fe)] and **cementite** (iron carbide, Fe_3C). Pearlite forms separate regions of ferrite and cementite, which are most often in a lamellar (layered) configuration. Ferrite is very soft and ductile and cementite is strong and brittle. The two phases appear as "stripes" when observed under a microscope (shown schematically in Fig. 9.8). The difference between coarse, medium, and fine pearlite is the size and spacing of the layers. When they are larger, the material is softer, and the thinner and closer they are, the stronger the material becomes. Also, the relative quantities of the two phases depend on the amount of carbon present, as ferrite can only hold a small quantity of carbon in solution (~0.02 wt%) and the rest will go to forming cementite or other phases, depending on the cooling rate.

Bainite is also a combination of ferrite and cementite, but it forms into needles or laths (upper bainite) or plates (lower bainite). The cementite forms into tiny elongated particles surrounded by a ferrite matrix. Once the formation of either pearlite or bainite occurs, that portion of the structure will not convert to the other. To maximize bainite formation, the pearlite "nose" of the TTT curve must be bypassed. If bainite formation begins at the high end of its formation temperature window, it forms needles that have a ferrite matrix that is low in carbon with what appear like broken strips of cementite within it. If it starts to form at the lower end of the temperature window, it forms plate-like structures where the ferrite matrix contains a large fraction of dislocations and the cementite is in the form of tiny particles.

*Hypereutectoid steels first transform to cementite (Fe_3C) instead of ferrite (α-iron) and results in a somewhat stronger structure.

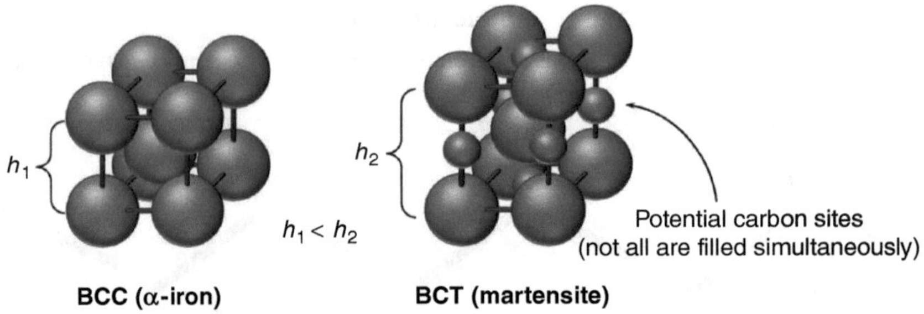

Figure 9.9 Martensite is the primary hardening phase in ferrous alloys. The fast quenching process prevents carbon diffusion into ferrite and cementite, so the carbon gets "trapped" during the transformation from fcc to bcc iron. Martensite forms as small needle-like body-centered tetragonal (bct) structures.

Martensite is the primary hardening phase in quenched ferrous alloys. The rate of cooling has to be fast enough to prevent carbon from diffusing and forming other structures like pearlite or bainite, which consist of separate ferrite and cementite phases. Because diffusion is limited, the transformation into martensite is considered a "diffusionless" transformation where atoms undergo coordinated movement to form a new crystal structure. Recall, iron and carbon are in solution above the austenizing temperature, but the fcc phase is not stable at lower temperature. This phase change results in the formation of a distorted bcc crystal structure where carbon can occupy spaces between normal iron lattice points (refer to Sec. 3.2 for more on these concepts). The distorted structure is no longer cubic, and is elongated in one direction, which forms a body-centered tetragonal (bct) structure (see Fig. 9.9).

The bct structure is not a stable structure, but it is "trapped" in place. This means that martensite will decompose when heated to a high enough temperature. When present, martensite assumes a needle-like structure, the extent of which is dependent on the initial carbon content. The "misplaced" atoms induce strain into the crystal structure as well as reducing the ease with which the crystal can deform (slip), which lends strength to the material but reduces ductility.

Due to the rapid cooling process, not all austenite is transformed to martensite. **Retained austenite** is untransformed γ-iron that may still transform over time. Retained austenite can be forced to transform by further cooling the structure below room temperature as described in Sec. 9.3.6.1.

9.3.6.3 Hardenability

The **hardenability** of a metal is a measure of how easily it can be fully hardened. This is directly related to the rate at which it must be quenched to achieve full hardness. Metals that can be fully hardened with slow cooling have a high hardenability.

The way ferrous alloys respond to heat treatment is dependent on a variety of factors. The TTT diagram gives us an indication of what time and temperature requirements we must satisfy to achieve a given microstructure and thereby a certain set of mechanical properties. Each alloy has its own TTT diagram, and this is related to its composition. Some alloys require extremely rapid cooling, while others do not need any quenchant other than still air. The hardenability of an alloy is a measure of how slowly it can be cooled and still achieve full hardness.

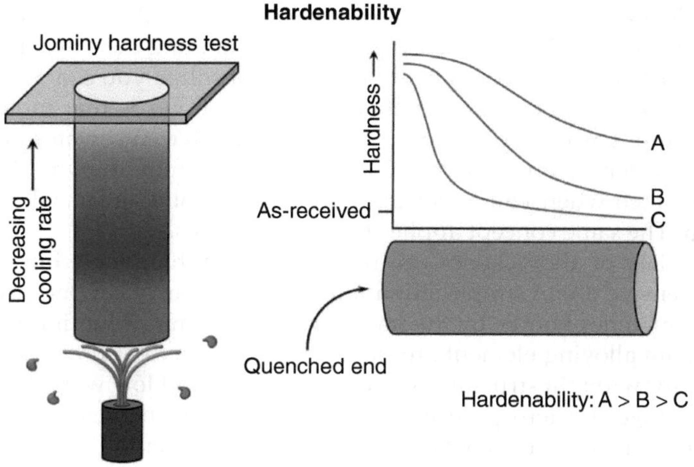

Figure 9.10 The hardenability of a metal is a measure of the depth of hardness achieved at a given quench rate. The hardenability can be tested experimentally using the Jominy test. Metals with a high hardenability will achieve a greater depth of hardness for a given quench rate. In the image above, alloy "A" has a higher hardenability than "B," which is higher than "C."

Medium carbon steels have to be rapidly quenched or they will not be completely hardened. Their surface may be hardened, but their interior, which cools more slowly, may be only modestly improved. While this may be beneficial in some circumstances, many times it is considered a quality control issue. There simplest way to increase the hardenability of a metal is to increase the alloying content. Looking back to Table 4.1, notice the purpose of many alloying elements includes increasing the hardenability (e.g., chromium and vanadium). Even for a specific alloy composition, the addition of carbon always improves hardenability as well.

The hardenability of a metal can be tested empirically by quenching on end of a hot bar until the whole test sample is cool. The end being quenched will cool the fastest, and the opposite end will cool the slowest. The hardenability is determined by the depth of hardness achieved, or the distance along the bar that obtains a desired level of hardness. The common test for this purpose is known as the Jominy test and is pictured in Fig. 9.10.

9.3.7 Precipitation Hardening of Nonferrous Metals

Our discussion so far has been primarily centered on the heat treatment of ferrous alloys, and they represent the majority of the materials that undergo heat treatment. There is a broad range of important alloys that do not undergo the transformations during cooling that characterize ferrous metals, particularly the iron-carbon interactions. Instead of trying to quench and "capture" an unstable phase-like martensite, the hardening is achieved by optimizing a stable* phase.

Precipitation hardening can only be done in an alloy consisting of elements that are insoluble (or immiscible) in one another. This concept was introduced in the context of

*Although not completely stable, it can be formed in more controlled manner.

phase diagrams for the sugar-water system in Sec. 3.2.4.2, and it may be helpful to review that material if you have trouble understanding the extensions in this section.

The primary emphasis is that the amount of solute you can dissolve in a solvent is almost always temperature dependent, and the trend is that more will fit at higher temperature. For example, more sugar will dissolve into water at high temperature than at low temperature. If you dissolve the maximum amount of sugar at high temperature, then when you lower the temperature, solid sugar will "precipitate" out of solution. The same concept applies to solid materials.

A variety of alloy classes can be precipitation hardened, but aluminum sees the most extensive use. A simple aluminum alloy consists of copper mixed into aluminum and is sometimes known by the trade name duralumin. Aluminum alloys with copper as the main alloying element are classified as 2000 series alloys (see Table 4.6) and are popular for wrought structures with high strength and low weight (e.g., sheet and plate in automotive and aerospace applications). To possess high strength they must be thermally conditioned using a process known as age hardening.

9.3.7.1 Age Hardening

Age hardening (alternatively known as precipitation hardening) is a process where a second phase is formed by the precipitation of an element or elements out of solution, and it results in enhanced strength in the alloy. The process is simple, and it mirrors the tempering process in ferrous alloys:

1. Heat to a critical temperature (the solutionizing temperature)
2. Quench rapidly to maintain the solution
3. Reheat to modest temperature and allow time for a second phase to form

The first step involves going above the **solutionizing temperature**, which is the temperature above which the elements are fully dissolved, like sugar in hot water. The second step requires fast cooling. Instead of forming martensite-like ferrous alloys, the quench does not impart significant strength to the alloy. Instead, it only acts to "trap" the solute in solution. For instance, copper will stay evenly distributed throughout aluminum, even though it prefers to cluster with other copper atoms in distinct regions. This is analogous to instantly freezing a solution of sugar in hot water. If there is not sufficient time for the sugar to "settle out," it will remain dissolved.

The last step is the one that is typically known as age hardening, but it cannot take place without the others. Age hardening, also simply known as **aging**, is the stage where the solute (e.g., copper) is allowed to diffuse and will form into clusters known as precipitates. This is very much like oil and water. The two do not like to be mixed and will separate out over time (think of Italian dressing). Copper and aluminum, and many other elemental combinations, act similarly. The biggest difference is that the rate at which they separate out is *much* slower. The dissolved (solutionized) and precipitated structures are depicted in Fig. 9.11.

Diffusion in solids was described in Sec. 8.4, and one of the main points to be understood here is that the diffusion rate increases exponentially with temperature [see Eq. (8.9)]. Elements will precipitate at room temperature (called **natural aging**), but the process takes months. What is interesting about natural aging is that a component will continue changing properties as time goes on with no external influence. The change can be significant too. To avoid unpredictability, the alloy is aged before the

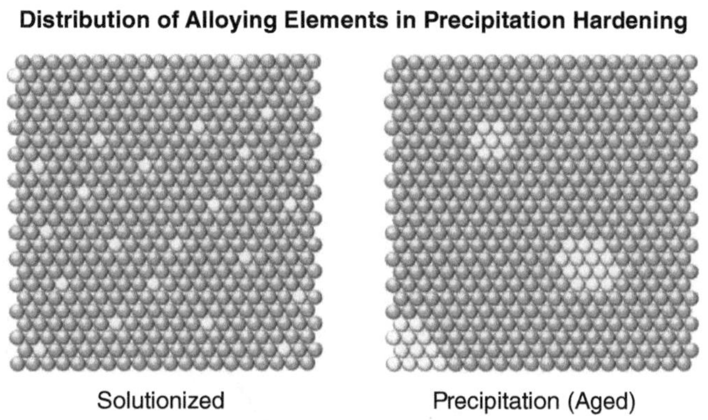

FIGURE 9.11 The process of age hardening requires that the alloy first be solutionized to create an even mixture. That mixture is then quenched to retain its structure and reheated to allow controlled precipitate growth.

material is used for manufacture. To reduce inventory required when waiting months for natural aging to stabilize, the much faster alternative of artificial aging is used.

Artificial aging is the process of applying elevated temperature to speed up the precipitation process. Instead of months, the precipitation hardening process can be completed in hours or even minutes, but there is a caveat. The higher the temperature used, the lower the maximum strengthening effect is. By increasing the temperature the diffusion rate increases, but the characteristics of the precipitates also change.

Precipitation is not as simple as copper particles forming in the aluminum. While that is the process, there is a series of steps that occur as the precipitates grow and how the precipitates interact with aluminum's crystal lattice. The moral of the story is that you can get good results quickly, but you can get the best results by taking more time. Of course time is money, so there is a trade-off to be made. If the application demands a greater strength, the extra aging time is worth it. The general behavior between time, temperature, and strength is shown in Fig. 9.12. Each alloy should be considered separately and specifically because the elemental additions will change the response to heat.

You can see in Fig. 9.12 that the age hardening curve reaches a maximum strength and then begins decreasing. This has to do with the effect the size of the precipitates have. In general, a distribution of many small precipitates will achieve a greater hardness than fewer large precipitates. For instance, if the alloy contains 4% copper, that quantity of copper can exist in different ways. This is similar to water vapor, which is like copper in solution. As the water vapor collects, it forms tiny droplets, such as fog. Once the droplets coalesce even further, they form rain.

The process of going from tiny precipitates to larger ones is where the strength begins to decrease. The larger the precipitates get, the farther apart they must be and the weaker the structure becomes. This is because the precipitates act as barriers to dislocation motion, which is the mechanism by which materials deform (see Sec. 3.2.3). This is like trying to hike through thick bushes or large trees. The trees are spaced further apart and easier to get through, even though they may contain more volume near ground level. Dislocations get tangled on small, closely spaced precipitates, but they can bow around larger ones. An alloy is considered **overaged** when it begins to decrease

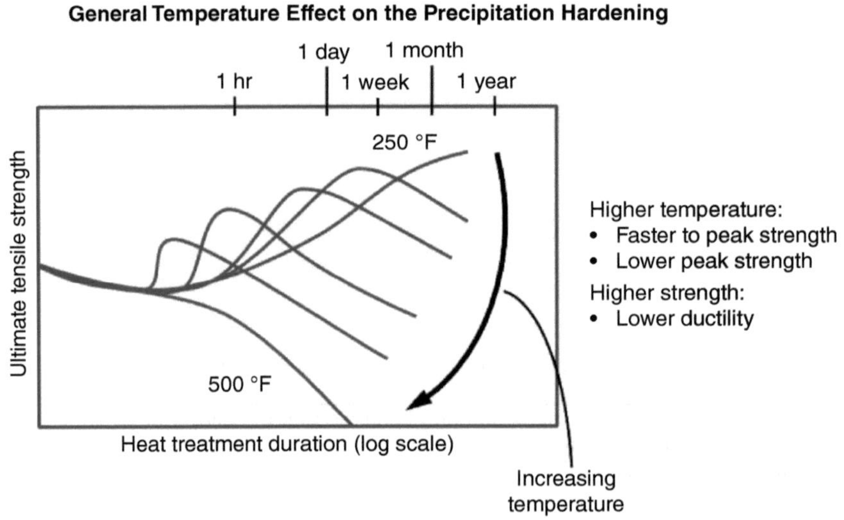

FIGURE 9.12 The temperature used during age hardening determines the maximum strength and the time required to reach that strength. The ductility will drop as the strength increases, but the effect is not as dependent on temperature. This allows for a "recipe" to be devised that achieves a desired balance of strength, ductility, and time.

in strength as a result of thermal conditioning that has been carried out for too long. The difference is depicted in Fig. 9.13.

9.3.8 Surface Engineering of Metals

The methods we have been looking at so far modify the mechanical properties of metals in bulk. That is, they are intended to harden, soften, or otherwise change the entire volume of a component. There are circumstances when a difference in properties may be

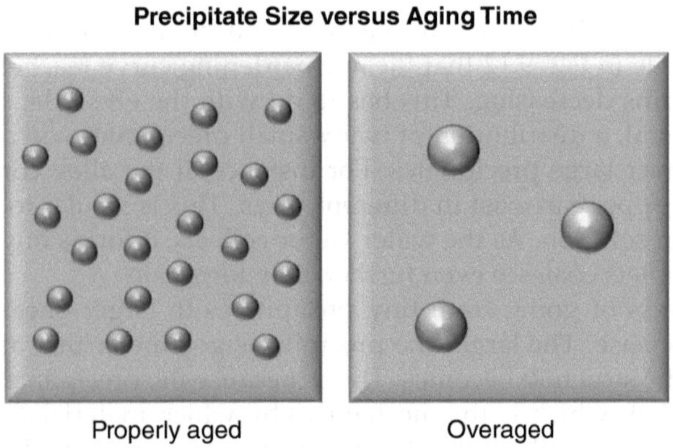

FIGURE 9.13 If the age hardening process is carried out for too long, the alloy begins to lose strength. This occurs because the precipitate size and spacing increases, which makes them less effective at blocking dislocation motion, the mechanism by which precipitation strengthens metals.

desired for the interior and exterior of a component. A property gradient is desired. While an assembly can certainly be created with two materials or two components of a single material thermally conditioned in two different ways, this adds complexity. When possible, reducing assemblies is beneficial.

The other option is to condition a single component in two different ways. This is accomplished by applying heat differently at different locations or by varying the composition at different locations. Both are effective and common practices to reduce cost and/or improve performance.

9.3.8.1 Case Hardening

The first method of selectively modifying a surface is by changing the composition. Recall that low-carbon steels do not harden well or at all with thermal conditioning. Their hardenability (see Sec. 9.3.6.3) is largely determined by their carbon content. Low-carbon steels are favorable for other factors such as low cost and ductility. To be able to get a hard surface in low-carbon steel is something quite useful, and it can be done by adding more carbon. The practice of selectively hardening the surface of a part is also known as **case hardening** (see Fig. 9.14).

Adding elements after a metal or alloy is in a solid form is not trivial, and depending on the element(s) being added, it may not even be possible. The reason is that it requires the elements to diffuse into the metal in the solid state (see Sec. 8.4 for more on diffusion). Diffusion is often a slow process in the solid state, even for relatively small elements like carbon, which enter as interstitial defects, but it is still practical. The reason is that the distance elements have to diffuse is fairly small, so even though they do not move quickly it can still be done within reasonable time.

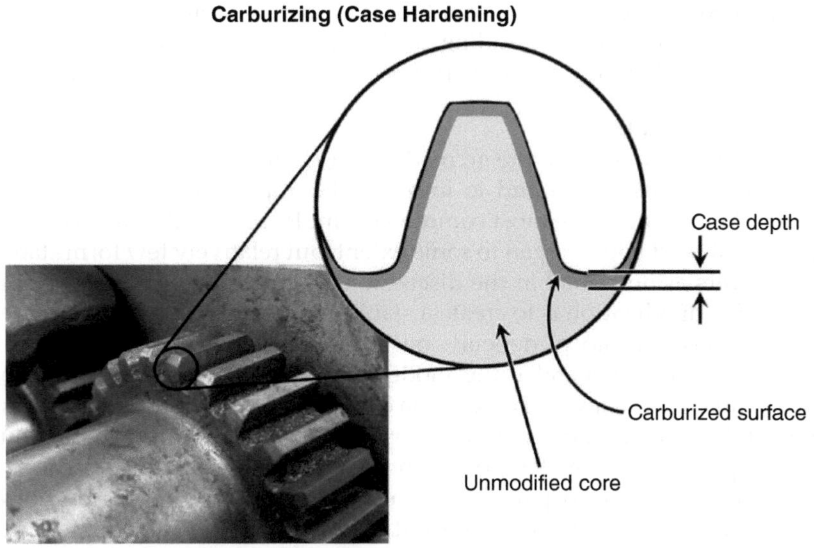

FIGURE 9.14 Carburizing, as well as other compositional surface treatments, modifies the chemical structure of the surface to directly harden or allow hardening through further processing. This combines a hard, wear-resistant surface with a tough, shock-resistant core. Such a combination of properties is ideal in applications such as gears.

Carburizing is the process of adding carbon to the surface of a component by introducing a carbon-rich environment to that surface at elevated temperature. The carbon-containing environment can be solid (powdered carbon, like crushed charcoal) or gas (e.g., carbon monoxide).* Carburizing is a diffusion process, and as discussed in Sec. 8.4, the time and temperature determines the depth of the "case," with an increase in either resulting in an increased case depth. The case hardening temperature is often between 1550 and 1750°F (843 and 954°C).

Nitriding is the process of adding nitrogen to a surface where it reacts to form a very hard nitride layer. It is similar to gas methods of carburizing to modify the surface chemistry, but it is not as diffusion dependent since a compound (nitride) is formed directly. This is generally accomplished by exposing the metal to ammonia gas at elevated temperature. Because the reaction involves the formation of nitrides, elements that readily form nitrides must be present. This requires the use of alloy steels with elements such as chromium, molybdenum, or aluminum. The process creates a very hard and thin surface layer and results in less geometric change than carburizing. Nitrides are popular for coating cutting tools (discussed in Sec. 13.4), especially titanium nitride (Sec. 6.2.3.4), because of their hardness.

Alternatives also exist, such as **carbonitriding,** where a combination of carbon and nitrogen is introduced to the surface. This can be achieved by using a combination of ammonia and carbon-containing gases. The part may or may not be quenched after treatment, but the minimum conditioning temperature is determined by that (i.e., the austenizing temperature must be exceeded if the part is to be quenched, see Sec. 9.3.6.1).

The areas treated for surface hardness can be removed by machining before the hardening step is conducted, or a diffusion barrier can be applied, which masks the areas and prevents diffusion. The second method is similar to taping off certain areas when painting; only the mask must be able to prevent atoms from getting through. The conditioned geometry must account for later material removal, and there may also be a small change in volume by adding additional elements (similar to oxidation discussed in Sec. 8.5.2.1) that may affect the performance of precision assemblies.

9.3.8.2 Anodizing

Anodizing is a process where an oxide layer is intentionally formed to prevent further environmental oxidation and to improve the appearance and/or function of certain metals. Aluminum is the most common example, as it readily forms an oxide layer. All metals will react with oxygen to some extent, but relatively few form stable, passivating oxide layers as described in the discussion on corrosion (see Secs. 8.5.2.1 and 8.5.2.5). The goal in anodization is to create a stable and useful oxide layer.

The process protocol depends on the material being anodized, but in general it consists of using the metal as the anode in an electrochemical cell where oxygen is produced and creates a layer of oxide. The chemistry of the electrolytic bath is controlled to prevent the formation of cracks and large pores. The anodized coating must be at least slightly porous (nanometer pores) so the underlying aluminum can continue to be oxidized. The anodized part is generally sealed so those pores do not allow further environmental damage (corrosion) and they also provide a surface for introducing dye. Sealing can be done by hydrating the oxide coating to cause swelling and closure of pores or the pores can be filled chemically.

---

*Liquids such as molten barium or sodium cyanide can also be used, but these will not be covered here.

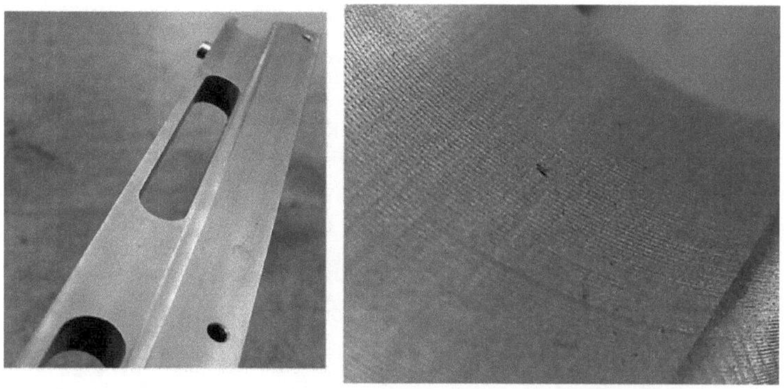

Figure 9.15 Anodizing is used for decorative and functional purposes. It involves the oxidation of a component, and only metals that form stable oxides are suitable. Dye may be added, as shown for the aluminum part here, but other metals, like the titanium, change colors based on the thickness of the oxide layer formed.

Brightly colored metals that still appear shiny and "metallic" have often been anodized to accomplish that appearance. Although decorative, it is also functional by increasing the wear resistance of a surface and as an electrically nonconductive layer. The anodization process increases the volume of the part as the oxidation occurs approximately equally outward and inward. A 4-μm thick coating will increase the surface height by 2 μm, and it will add 4 μm if opposing surfaces are both anodized. Some metals, like titanium, assume different colors as a function of oxide thickness, and they do not require dye. The oxide coating refracts light differently depending on the thickness, which changes the appearance. Anodizing does not cover surface roughness, and the part should be carefully prepared. The surface shown in Fig. 9.15 is anodized aluminum, but the machining marks are still clearly visible.

9.3.8.3 Surface-Specific Thermal Conditioning

In order to condition the surface of a component preferentially, the alternative to changing the surface composition is to localize the heat. Instead of allowing a component to heat entirely through, rapidly application of intense heat can be used to quickly bring the surface of a part to the critical temperature and then air or water cooling can be used to quench the material and achieve greater strength. This process is shown conceptually in Fig. 9.16.

There are a variety of ways to achieve this goal, but some of the most common are flame and induction heating. In flame heating, a part is subjected to intense heat from a blow torch, which rapidly raises the exterior temperature of the part. The flame or the part must be moved to ensure the entire surface area is heated, or multiple flames may be used to achieve a more even heating.

Induction heating uses a very high frequency current that flows through a coil which the part is inside of. The high frequency in the coil *induces* a current in the part and the electron flow raises the temperature through Joule (resistive) heating. The current, and therefore the heat, is concentrated at the surface of the part, but process factors, such as the frequency, and the material determine the depth of heating. The part is immediately quenched to achieve hardness, and medium carbon ferrous alloys are

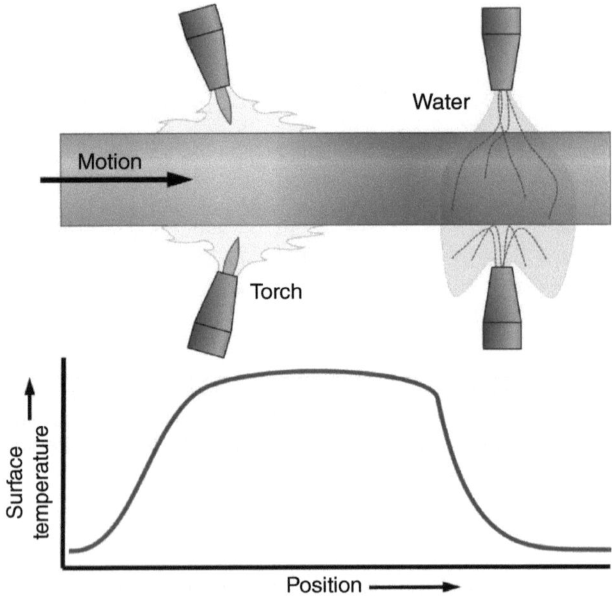

Figure 9.16 Surface hardening can be achieved by rapidly heating and quenching a part so that only the outside reaches a high enough temperature to be quench hardened. The heat may come from torch, induction heating or other suitable means.

commonly used. The ways (guides) on machine tools, such as lathes, are often induction hardened. Both flame and induction hardening can be done continuously by passing long parts through the center of a flame ring or induction coil followed by a water spray.

9.4 Why It Matters

Metals are versatile. Knowing the composition is important, but it does not tell the whole story. The properties of metals and alloys are also dependent on their *condition*. This can be compared to buying a used car. You may get an idea of the condition by simply viewing the car, but what has been done to it internally is of great value. Whether it has been driven hard is not always apparent. The same applies to metals. Whether a metal is mechanically or thermally conditioned is not obvious in most cases, but it has a major impact on how the material or component can be applied.

Mechanical conditioning is part of manufacturing and service. Whenever a metal is deformed, it has the potential to build up strain internally. High-temperature processing will immediately release the strain and no strengthening occurs. Cold work, where deformation is done near room temperature, will increase strength, which makes the cold-worked part capable of handling more force without failing (yielding). The downside is that it also reduces the ductility, so when a part fails it has a higher chance of fracturing quickly and without warning. Mechanical conditioning can be reversed through heating, also known as annealing or stress-relieving. Where this is undesirable, high-temperature should be avoided.

Temperature can have the opposite effect as well. It can harden ferrous and nonferrous alloys, but the processing is different for each. Ferrous metals rely heavily on iron-carbon interactions and nonferrous metals typically rely on the precipitation of a second phase. The procedures for achieving the desired balance between strength and ductility are specific to each alloy and must be stringently followed to achieve the desired outcome.

The "recipe" for achieving certain properties can be determined from an alloy's TTT diagram. The formation processes of various iron-carbon phases such as pearlite, bainite, and martensite are all functions of the cooling rate and whether intermediate temperatures are maintained before finally reaching room temperature. Thermal conditioning, also referred to as heat treating, is still an ongoing area of research, even though it has been in practice since the earliest metalworkers took up their trade.

Although it is often desirable to fully harden a component, there are also circumstances where selective hardening of surfaces is needed. There are two ways to achieve this: compositional variation and variation in thermal treatment. Both methods selectively modify the near-surface region of a component, and both result in a hard, wear-resistant exterior combined with a softer, tougher interior. Carburizing, nitriding, and anodizing were some of the compositional methods, and flame and induction hardening are popular thermal treatments. All methods are cost effective means for achieving a discontinuous structure.

Without thermal and mechanical conditioning many manufacturing processes would not be possible. The hardening of ferrous alloys brought about the earliest development of advanced tooling for improving manufacturing efficiency. These "tool" steels are still widely used in forming, stamping, shearing, and cutting operations. The ability to harden these materials is important because they can be machined or otherwise worked with *before* they are fully hardened.

Precipitation hardening has been instrumental to increasing the strength of strategic, lightweight alloys such as aluminum. Pure aluminum is very soft and mechanically unsuitable for its most popular applications, such as aerospace. By alloying the aluminum with elements that do not stay dissolved, the formation of discrete particles that strengthen the material is possible. The amount of strength and the time required to achieve it are dependent on the temperature used to age the alloy.

The application of heat can soften, harden, or toughen an alloy, and it will reverse mechanical conditioning. By carefully combining deformation and heating cycles, innumerable applications can be satisfied by only a handful of alloys.

9.5 The How and Why of Conditioning Metals

9.5.1 Chapter Review

1. What two characteristics of a metal determine its properties?
2. What is the purpose of conditioning metals?
3. Is it accurate to assume the composition of a metal is constant during conditioning? Why or why not?
4. What are the two primary means for conditioning a metallic material?
5. How does mechanical conditioning change metal's properties?

6. What process difference is there between cold work and hot work? What is the difference in the way it affects a metal?
7. What trade-offs are there to cold working a metal?
8. What is meant by "percent" cold work?
9. Calculate the percent elongation for the following scenarios:
   a. A 3-ft bar of steel extended to 3.5 ft
   b. A 2-in rod of brass shortened by 0.1 in
   c. A 6-m wire stretched to 8 m
   d. A 7-mm plate extended to 7.6 mm
10. Calculate the percent reduction in area for the following:
    a. A 1×2 in bar is reduced to 0.75×2 in
    b. A 2-in-diameter rod is reduced to 1.875 in. in diameter
    c. A 6-mm-diameter rod is reduced to 3 mm in diameter
    d. A 7×10-mm bar is reduced to 3×15 mm
11. Starting with a square 0.5 in per side, what is the final length per side after a 20% reduction in area if the final cross section is also square?
12. Starting with a rod of 20 mm in diameter, what is the final diameter after a 35% reduction in area?
13. Why must annealing be applied intermittently when extensive cold working must be done?
14. How do forming operations change the structure of a metal? How does this affect performance?
15. In your own words, describe recrystallization. How does heat allow this process to happen?
16. How do burnishing and peening reduce the likelihood that a component will fail?
17. How can burnishing occur unintentionally during processing?
18. How does shot affect the peening process?
19. How does thermal conditioning change the properties of a metal?
20. In general, describe the difference between annealing, normalizing, and quenching.
21. Why does steel have to be heated above the austenizing temperature during most heat treatment?
22. How do austenizing and solutionizing differ? When are they used?
23. Describe the balance between thermal conditioning, material choice, and cost. When is thermal conditioning the wrong choice?
24. What is the goal of annealing? In general, how long does it take?
25. What is the principal difference between annealing and normalizing processes?
26. How do the resulting properties achieved in normalizing differ from annealing and quenching?

27. What determines the rate of quenching?
28. What should the proper quenchant achieve?
29. How can the Leidenfrost effect be minimized? Why is this beneficial?
30. What step must precede tempering? Why is tempering used?
31. How does the maximum temperature differ in tempering from other heat treatments such as annealing or quenching?
32. What is the property balance that tempering seeks to achieve?
33. What is meant by a "recipe" for thermal conditioning?
34. How can a TTT curve be used to determine the outcome of thermal conditioning? In what type of metals is this used?
35. From left to right, what do the three curves on the TTT diagram indicate?
36. How can multiple phases from the TTT diagram occur in a single material? Give an example.
37. Why is a straight line on the TTT curve not a constant cooling rate?
38. How can a stepped curved like "B" in Fig. 9.6 be accomplished?
39. What is retained austenite? How can it be transformed?
40. Why is austenite important in thermal conditioning?
41. Describe the structure of pearlite.
42. What are the structural differences between upper and low bainite?
43. How does martensite form? What is its structure?
44. What compositional factors affect hardenability?
45. How can the hardenability of a metal be measured?
46. What characteristics must an alloy have in order for precipitates to form?
47. What variety of alloys are the most common in precipitation-hardened form?
48. Describe the process of age hardening.
49. Why is solutionizing an important step in precipitation hardening?
50. What is the difference between natural and artificial aging?
51. Why do the strength curves in Fig. 9.12 have a maximum somewhere in the middle?
52. When would you use higher temperature in precipitation hardening, and when would you use lower temperature?
53. What is the benefit of surface conditioning metals?
54. Name an application for case hardening and why it is appropriate.
55. How can carburizing be accomplished?
56. What differences are there between carburizing and nitriding? How can they be combined?
57. To what metals is anodizing often applied? Why is this done?
58. How is anodizing used as a decorative process? How can colors be achieved without the addition of any colorant?

59. A metal's surface can be selectively hardened without changing the composition. Describe how this is possible.

9.5.2 Concept Connections and Extensions

1. Describe the diffusion process for carburizing steel. Use the discussion in Sec. 8.4 to support your answer.
2. Consider the strength-structure relationships of coarse, medium, and fine pearlite. Why do the properties vary? Hint: Refer back to composite structure, especially Fig. 5.24.
3. How might a dull tool result in mechanical conditioning of a surface? When is this a bad thing for both the part and the tool?
4. What is the functional difference between sand blasting (Sec. 14.6.3.1) and shot peening? How is that difference accomplished?
5. What are the functional and cost benefits of surface hardening? Where are hard surfaces needed that surface hardening is not suitable for?
6. If carbon increases hardenability drastically, why are other alloying elements used so extensively?
7. Think about the cooling rate and the hardening effects in ferrous metals in regard to welding processes. How might different welding techniques result in different weld properties? See Sec. 18.2 for more information.
8. Might it be possible to cast and harden a component in a single setup? What advantages and disadvantages might there be?
9. Contrast the process for making metallic glass (Sec. 4.4.5) through rapid quenching and that used to form martensite.

CHAPTER 10
Conditioning Nonmetallic Materials

In the previous chapter, we explored the drastic difference that conditioning can make in metals. Polymers and ceramics are conditioned in different ways and with different outcomes. Unlike metals, simple heating and cooling cycles are often insufficient to increase their strength. Fortunately, there are different methods to strengthen or toughen nonmetallic materials, and we will consider some of the most common in this chapter.

10.1 How It Works

Nonmetallic materials are useful in many applications. The wide range of properties and applications of nonmetallic were described in Chap. 6, and in some cases, a single composition had different properties. This was the case for polyethylene (PE, Sec. 6.3.1.1), which is available as high density, low density, and other designations. Those polymers have the same elements in the same configuration, but their properties vary. This indicates that something can be done to change the properties through structural variations.

Also, as seen with polyvinyl chloride (PVC, Sec. 6.3.1.3), the rigidity of a polymer can be changed significantly, and a single polymer can be used as a rigid piping material as well as flexible sheeting used to create inflatable toys. Ceramics, on the other hand, are characteristically rigid and strong. Conditioning is often used to strengthen a material, but in their case, decreasing their brittleness is more useful.

10.2 Conditioning Polymers

Because polymers are made from individual molecules, their strength is related to the relationship between those molecules. The structural concepts of polymers are described in Sec. 5.3 and the interactions between polymer chains are covered in Sec. 5.3.2. Those sections are an important foundation for understanding conditioning processes.

Some of the primary conditioning methods we will examine are methods by which polymer molecules can be more intimately connected to increase strength, or they can be partially dissociated for increased ductility. This can be achieved by modifying the existing structure or adding other materials. Conditioning in polymers is not just for strength, either. They are often modified for increased environmental durability as well.

10.2.1 Mechanically Conditioning Polymers

Permanently deforming a material forces the atoms or molecules to realign. In polymers, this causes the molecules to orient in a similar direction. As they align, their rigid "backbones" carry the force and this increases their stiffness. This also reduces their ability to slide past one another, which decreases their ductility. Some processes, such as extruding, naturally induce chain alignment and can impart additional strength. Unlike metals, which are mechanically conditioned in the fully solid state, polymers can be given additional strength while they are molten. This is because the chains do not easily move by thermal action and will largely retain their orientation during cooling.

A word of caution, however, is that when a component is given higher strength due to favorable alignment in one direction, it will become weaker in directions perpendicular to that. This was emphasized in the discussion on composites (see Sec. 5.4.2.2, for example) and is known as anisotropy. This alignment is shown schematically in Fig. 10.1. In some situations, forces can be accurately predicted, and the properties can be optimized. For example, fishing line is always used under tension, and chain alignment adds strength where needed most.

You can tell when many plastics are deformed because they change color or appearance where the strain is the highest. This is most noticeable in clear polymers when they are bent or stretched and become opaque. This can be seen if you bend rigid plastic packaging, for instance. The reason it is no longer clear is because the molecules have been stretched into alignment which results in a higher density (smaller chain spacing) and consequently light cannot pass through as easily. This is related to crystallinity in polymers (Sec. 5.3.2.7), which increases strength. The harder it is for chains to align, the less effective mechanical methods will be at increasing crystallinity.

10.2.2 Cross-Linking

One of the simplest and most effective ways to achieve dramatic strengthening in polymers is through cross-linking. This topic is covered in more detail in Sec. 5.3.2.6, and here we will concentrate on the methods to achieve it. Cross-linking is a vital process in polymer engineering and can be controlled to achieve a desired balance of strength and ductility. Weakly cross-linked polymers will be more ductile, and as cross-linking is increased, the stiffness of a polymer also increases.

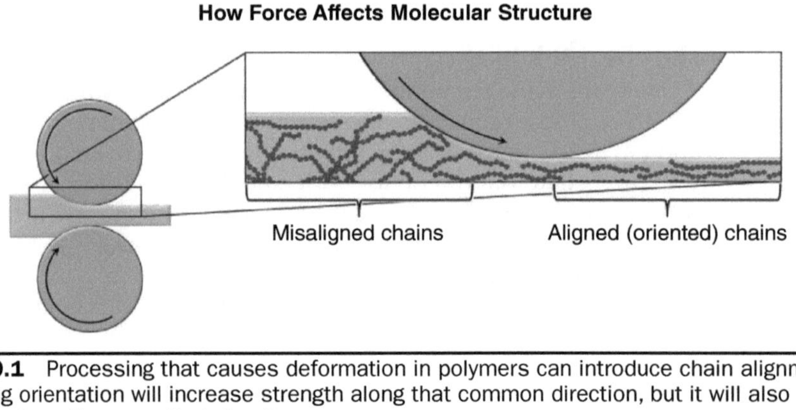

Figure 10.1 Processing that causes deformation in polymers can introduce chain alignment. Increasing orientation will increase strength along that common direction, but it will also decrease strength across that direction.

That is not to say that cross-linked polymers are always rigid. Polymers we consider "elastic" or "rubbery" can also be cross-linked, but their degree of cross-linking is minor. Although cross-linking can be controlled during processing, it can also occur in an unwanted way during service. This can be a byproduct of overheating, weathering, or other environmental exposures. The degradation of polymers is discussed in Sec. 8.5.3. We will now look at engineering controls to achieve it.

10.2.2.1 Thermal Methods

We discussed the two basic categories of polymers, thermosets and thermoplastics, in Sec. 5.3.2.5. The entire collection of polymers (thousands of them) fit into one of these categories. Their distinction is simply whether the polymer hardens or softens with heat. Heat, then, is very important. Based on the response of polymers to heat, it can also be very useful. Thermoset polymers harden by application of heat due to increased cross-linking.

Polymers that can be hardened, or *cured*, by heat have a chain chemistry that includes elements capable of forming additional bonds. When heated, the polymer chains are given sufficient energy to create new bonds between chains. Thermosets that cure with heat may be liquid or solid and can be strengthened during processing or as a later step, after the basic geometry has been created. Some of the polymers in this category include phenolics, epoxies, and polyurethanes (all of which are discussed in Sec. 6.3.2). Thermoset polymers are often better than thermoplastics at high temperature, and their applications often reflect that. Although they will not melt, they can still decompose if the temperature becomes too high.

10.2.2.2 Chemical Methods

Another method to increase the strength of a polymer is chemical curing . This is most often accomplished by mixing a resin with a hardener. In some cases, the hardener induces polymerization, but cross-linking is usually a part of the process. In this way, the short polymer chains are connected together by a chemical reaction and form longer chains and more cross-links. The chemical reaction can have several effects including the "opening" up of cyclic rings or exchanging a double bond for a single bond and a bond with another molecule in a cross-link. The chemical agent may provide additional elements used in the bond or it may simply act to break bonds and allow their reformation to happen with other molecules. Chemical agents can be equally harmful, and the uncontrolled breaking of bonds will shorten molecules, and it can embrittle polymers in some cases. This is a basic restriction on the environment in which polymers can be used. For example, it determines whether a polymer is suitable for use in contact with water or hydrocarbons such as gasoline.

10.2.2.3 Radiation Methods

Radiation is a useful means for curing polymers as it can be rapidly applied and localized. Heat and chemical methods will most often involve the whole polymer component. While this is usually most useful, there are applications where it is either impractical or undesirable to cure a polymer this way. Consider options for restorative dental work. Metallic amalgam, which contains mercury, silver, tin, copper, and other elements, has been used for filling cavities for years, but it has drawbacks related to its appearance and contents (particularly mercury). Other metals such as gold have been used as well, but this comes at a greatly increased price and still does not match the appearance of natural tooth.

Polymers can quite closely mimic the color and translucence of a tooth, but they must be durable and resist the corrosive environment of the mouth. Thermosets can be made hard, but you cannot partially fill a tooth and then apply intense heat or add a chemical hardener with ease. Chemical hardeners present a narrow window of time to sculpt the filling and heat must be very accurately applied or a patient would be burned.

Light-curing polymers were developed in the 1960s and have undergone a number of improvements over the years. The photocurable polymer, or **photopolymer**, is caused to cross-link by the application of light. Initially, the light had to be ultraviolet (UV), but advances have allowed visible light to be used, which has a deeper penetration depth into the polymer. The application of high-intensity light causes a polymer to cure within seconds to minutes and can be safely applied in a localized fashion. These same benefits allow the use of photopolymers for three-dimensional printing technology (see Sec. 21.2.2 for more details).

Other forms of radiation used to condition polymers include electron beam, gamma irradiation, and microwave radiation. Electron beam irradiation involves bombarding a polymer with high-energy electrons that interact with the polymer molecules and break existing bonds thereby allowing new bonds (cross-links) to form. The radiation can also cause chain scission to occur or the exchange of bonds to "rearrange" the molecular structure. This method can be used to induce cross-linking in thermoplastics, which then converts them to thermosets.

Radiation methods have advantages that include the ability to process a polymer by conventional means and then strengthen it afterward. This allows the simplicity and cost-effectiveness of a thermoplastic to be combined with the performance of a thermoset. The degree of cross-linking can be controlled by the energy of the radiation, and there is no concern of thermal degradation during cross-linking. The process also causes a controlled means of affecting the molecular structure, which reduces environmental influences during service resulting in a more stable product.

Radiation methods can be used for sterilization as well as cross-linking. Gamma irradiation is particularly popular for this purpose. Microwave curing, on the other hand, is typically a substitute for traditional thermal curing as it heats the polymer rapidly to initiate cross-linking. Polymer chemistries that work well with radiation include double-bonds, amine groups, and halogens (or halo-carbons). Some disadvantages of radiation include the high equipment cost, dangers of radiation exposure to people, and possible unwanted changes that may occur to the polymer.

10.2.3 Polymer Additives

The mechanical characteristics of a polymer are directly determined by the molecular chemistry and the relationships between chains. A single polymer chemistry, such as PVC, can be made to possess drastically different properties by mixing in additives. A **polymer additive** is a substance present in a polymer that does not occur in the backbone or side groups of the polymer chains, but which modifies the relationship of the polymer molecules with each other and/or with environmental factors. For example, a chemical may be added to soften a polymer (mechanical effect) by lowering its glass transition temperature (response to heat, an environmental effect).

Additives are used in nearly all polymer products for the purposes of adding color, protecting from environmental damage, and optimizing mechanical properties. The pervasive use of additives can make recycling more difficult since the polymer is often labeled but its "extra ingredients" are not. The used of additives can also affect the processing characteristics of the polymer, for better or worse, and must be carefully considered. We will look at some of the common uses for additives in the following sections.

10.2.3.1 Plasticizers

Plasticizers are additives to decrease the resistance to chain movement. They "lubricate" the chains by occupying space between those molecules and reducing chain interaction. They are available in the form of liquids or low molecular weight solids (like grease, oil, or wax) and they are used to improve the processing characteristics of polymers as well allowing new applications by making them more flexible.

Plasticizers do not bond to the polymer chains and can eventually "leak" out of the plastic. This process, known as **leaching**, is the cause of a characteristic odor of new products made from plastic. That "new car" smell is actually the scent of accelerated release of plasticizers. As time progresses the concentration is reduced and the rate of leaching is reduced. That also causes old plastics to crack and become brittle. Plastic restoration and protectant products can put plasticizers back into the product, but they cannot penetrate deeply and must be continually added.

PVC is one of the most common polymers that utilizes plasticizers extensively. It is brittle at room temperature when unmodified, but the addition of plasticizers can make it so flexible that it can be used in clothing and accessories such as purses. Some common polymer plasticizers include phthalate esters (especially for PVC), trimellitates (for elevated temperatures), adipates (low temperature and UV resistance), hydrocarbon oils, polyethylene glycol, etc.

A primary concern with the use of plasticizers is their long-term health effects as they leach out of plastics. For instance, you may see an electrical cord or package labeled as containing harmful substances (see Fig. 10.2). This is related to lead added to plasticize the PVC insulation. Lead and other heavy metals have been used to plasticize polymers for decades, but more recent attention and regulation has prompted the search for and has expanded the use of alternatives such as calcium-based materials.

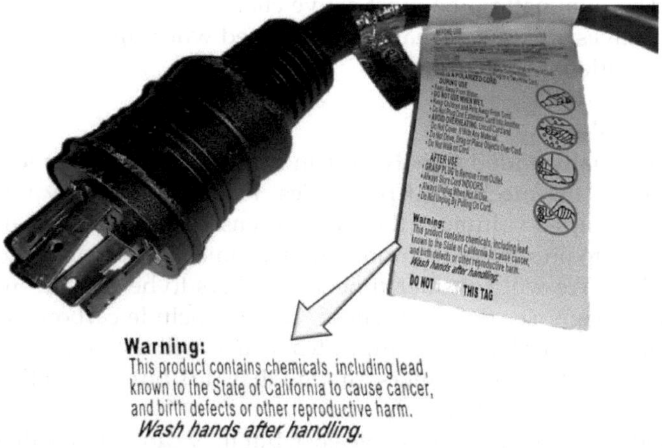

FIGURE 10.2 Additives are used to modify polymers for a variety of purposes, and some can be harmful chemicals, most notably lead used to plasticize PVC, and warning labels are mandated. The quantities used are quite small, but there are health risks in some cases of extended exposure.

10.2.3.2 Stabilizers

Stabilizers are additives designed to maintain a certain set of properties over time. A particular category of these, known as antioxidants, are used to prevent the damaging effects of free radicals. Those free radicals may be generated through direct chemical interaction, heat, or light interactions (photo-oxidation). The degradation mechanisms of polymers are described in more detail in Sec. 8.5.3. We will broadly consider that stabilizers such as antioxidants prevent chemical reactions from breaking down the polymer structure.

A more specific type of additive is an UV stabilizer. UV light can breakdown the polymer molecules or induce unwanted cross-linking. Carbon and titanium dioxide can be used to prevent this, but other common classes of UV stabilizers include benzophenones and hindered amine light stabilizers (HALSs). All of these materials act to safely absorb the UV light or neutralize free radicals that do form. The use of UV stabilizers prevents plastic from yellowing and becoming brittle, a characteristic quite important in applications such as signage, house trim, and other outdoor uses where exposure to UV and oxidizing agents is elevated.

10.2.3.3 Flame Retardants

Some polymers are naturally resistant to burning. For example, PVC gives off hydrochloric acid (HCl) as a byproduct. Although not a friendly chemical for other purposes, the HCl is heavier than air and displaces oxygen from the burn site. This allows PVC to self-extinguish a flame. Other polymers do not have this characteristic and can be dangerous if ignited as they melt and spread the flame with a long-burning fuel. To reduce the likelihood of a polymer igniting, a flame retardant additive is used.

Flame retardants are additives that act to reduce the potential for a polymer to sustain a flame by eliminating some elements of the burning process such as fuel, oxygen, or heat. Some reduce flammability by using a nonflammable filler to reduce the temperature by endothermic (cooling) reaction, by forming a charred, glassy layer that prevents heat from igniting further materials, by diluting the atmosphere surrounding the flame, or by eliminating the most active chemical radicals in the combustion process. These additives are used in many plastics used where ignition may occur, such as electrical applications.

10.2.3.4 Fillers

Fillers are usually added to polymers in significant quantities to reduce cost, but they also influence the mechanical properties. Fillers that simply take up space so as to reduce the required volume of the more expensive polymer are called **extenders**. Fillers that are used to impart some improved mechanical behavior are considered to be **functional fillers**. We will consider functional fillers to be reinforcement as described for composite design (see Sec. 5.4). These "fillers" include carbon, aramid, glass, or other fibers, particulate or laminar composites, which will not be considered as conditioning methods for modifying polymers (though they do), but instead these will be considered as distinct engineering materials.

Extenders are often used to save cost, but they can impart other valuable properties as well. Fillers can come from any category of materials (metal, polymer, or ceramic), can be natural materials or synthetic, and can possess any shape or size desired. That obviously leaves the definition of fillers very broad, but some common fillers, and their uses are provided in Table 10.1 to help give the concept more meaning. It is evident that

Filler Type	Benefit
Reinforcement (glass, carbon, etc.)	Added strength as composite
Glass, quartz	Reduced thermal expansion
Silver, copper, carbon	Increased electrical conductivity
Carbon, titanium dioxide	Improved weathering resistance
Wood flour, diatomaceous earth, carbon, silica, calcium carbonate, talc, etc.	Decreased cost
Molybdenum disulfide, fluorocarbons	Decreased friction
Microbubbles/hollow spheres	Decreased density, decreased conductivity

Note: The use of fillers can decrease cost and improve certain properties in polymers. Some common filler materials and their benefits are given here.
Source: Adapted from Walter E. Driver. *Plastics Chemistry and Technology*. p. 136. New York: Van Norstrand Reinhold Company, 1979.

Table 10.1 The Effect of Fillers on the Performance of Polymers

the range of materials and their impact on performance makes fillers a common asset in polymer design.

10.2.3.5 Foaming Agents

One of the most unique applications of polymers involves their lightweight and high elasticity. These properties are optimized in foamed polymers where they combine controllable stiffness and reduced density. Other properties include lower thermal conductivity and greater impact absorption. Devices for flotation, insulation, and cushioning are all made possible through the foaming of polymers.

There are two primary types of polymer foam: **rigid foam** and **flexible foam**. Both possess low density, but rigid foam maintains a much higher stiffness. It is used in applications such as insulation panels, whereas flexible foam is used in cushions for seating.

There are four ways to make foams. It can be achieved by mechanically frothing a liquid or molten polymer, by injecting gas into a molten or liquid polymer, by adding a blowing agent, or by adding hollow spheres (usually glass). The last technique is considered here more of a filler method described in Sec. 10.2.3.4. The nearest method to a conditioning process is the use of foaming additives known as foaming agents or blowing agents, which are physical or chemical additions that evolve gas and form bubbles within the polymer.

Chemical blowing agents give off gas during a chemical reaction, and **physical blowing agents** are liquids or solids that decompose at elevated temperature to form gas. Examples of chemical blowing agents include isocyanate and water, sodium bicarbonate (baking soda) and hydrazine. Physical blowing agents include fluorocarbons (some of which are being phased out for environmental reasons) and light hydrocarbons. These two methods can be used simultaneously as many chemical reactions are exothermic, and that provides the heat needed to decompose the physical blowing agents.

Certain blowing agents, such as azodicarbonamide, are banned or strictly regulated in some countries (European Union and Australia) while they are unregulated or even used in food products (as a dough conditioner) in other countries (like the United States). Blowing agents may decompose to form a variety of compounds that are trapped within pores of the foamed product and may be released over time (e.g., nitrogen, pentane, carbon monoxide, and ammonia).

10.3 Conditioning Ceramics

Ceramics are difficult to work with by mechanical means, so mechanical conditioning is a technique that does not lend itself to this class of materials. Most engineering ceramics are quite strong, so instead of conditioning adding strength, it often focuses on durability. The way this is accomplished is through decreasing their crack sensitivity (or by increasing their fracture toughness, see Sec. 7.4.3.2 for more detail). This allows ceramics to be less fragile and applied in more demanding conditions.

10.3.1 Toughening Methods

An increase in fracture resistance is an increase in toughness. Methods that act to reduce crack propagation in ceramics are known as toughening mechanisms. These increase the energy needed for a crack to increase in length and can be accomplished in a variety of ways. Among these are crack deflection, crack bridging, and transformation toughening. Transformation toughening is discussed in Sec. 7.5.2.1 and will not be repeated here. The mechanism can be applied in pure zirconia or in mixtures of ceramics, which include tetragonal zirconia (e.g., zirconia-toughened alumina).

Crack deflection and crack bridging both rely on a heterogeneous structure. That structure ensures the crack cannot progress without encountering a change in material properties that disrupts its growth. Crack deflection creates a nonlinear pathway on which crack propagation prefers to occur. This can be achieved by simply using polycrystalline ceramics where grain boundaries deflect the crack. This forces the crack to take a meandering path through the material. The longer the path, the more energy is absorbed during fracture. Areas of different composition or orientation will cause this type of behavior. The concept is shown in Fig. 10.3.

Crack bridging is a phenomenon that occurs in reinforced ceramics. Although a ceramic-matrix composite by definition, the material is entirely ceramic and contains small "whiskers," which impart a great deal of added toughness (whisker-reinforced ceramics are discussed more in Sec. 6.4.2.2). The reason they are able to resist fracture is because the whiskers can "bridge" the crack, as shown in Fig. 10.4. This is the same mechanism for any fiber-reinforced composite, but whiskers are particularly useful since they are very strong due to a low-defect density. Fiber-reinforced ceramics are used in applications where there is a high probability of impact or tensile loading such as in medical implants and cutting tools.

10.3.2 Tempered Glass

Another way to resist crack propagation is to reduce the tensile stress in a ceramic. This can be accomplished by engineering in a compressive stress. The compressive stress that already exists ensures that any tensile stress is reduced by that amount. When an applied tensile stress reaches the magnitude of the compressive stress, there is no stress

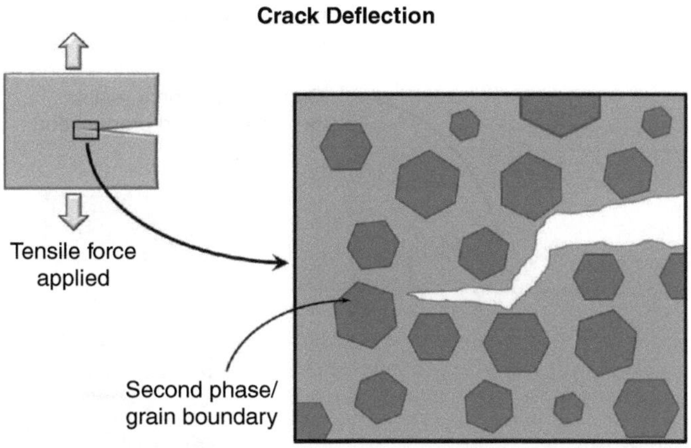

FIGURE 10.3 One method for toughening ceramics is to include inhomogeneities in their structure which cracks cannot easily pass through. These force the crack to divert and take a longer path which increases the energy needed to fracture the material.

on the material. The two simply cancel out and it is as if no tensile stress was ever applied.

This can be accomplished by tempering the ceramic, and is often applied to glass. **Tempered glass** is a glass component that has a residual compressive stress in the surface that resists crack formation and propagation. Tempered "safety" glass is used in applications where there is a likelihood of fracture, and the glass must not cause severe injury in the event that is does break. Car windows (Fig. 10.5) are tempered so that they shatter into many small pieces rather than into a few large pieces that would have a higher probability of causing severe injury. The exception is the windshield, which is annealed (cooled slowly) and has a sheet of plastic sandwiched between two glass sheets. The plastic holds it together on impact.

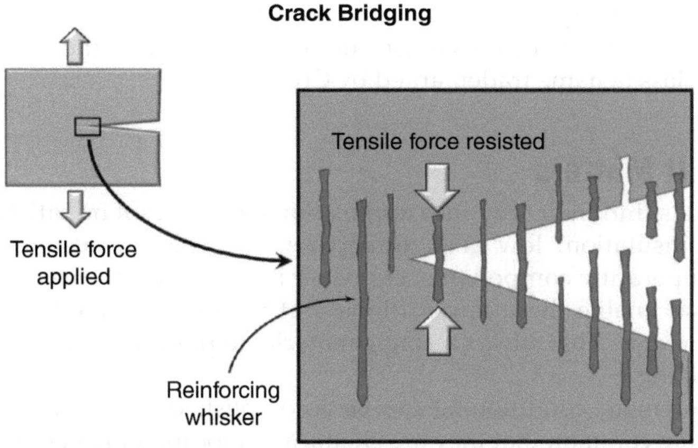

FIGURE 10.4 By incorporating a reinforcement phase, such as whiskers, a ceramic can be toughened through crack bridging. In this mechanism, the reinforcement carries a portion of the load even after the crack has formed in that area.

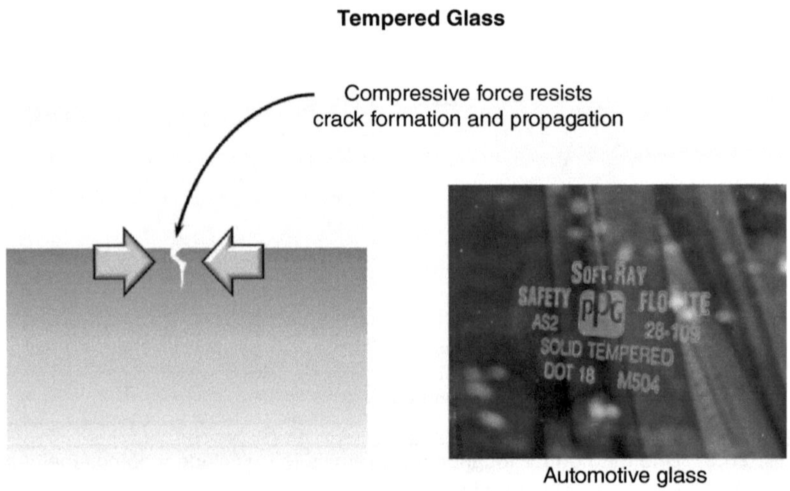

FIGURE 10.5 The use of tempering produces a surface with compressive stress that offsets an applied tensile stress. This allows for tougher, more impact resistance glasses such as tempered safety glass used in automobiles.

Glass is tempered by rapidly cooling the surfaces. The surface layer then shrinks by thermal contraction but the interior contracts later. When the interior cools (contracts) it cannot force the already solid outer layer to contract any further, so it produces a compressive stress in the surface. The reason tempered glass breaks so violently is that the built-in compressive stress is suddenly released across the surface resulting in rapid disintegration. That is, tempered glass can withstand more force before breaking, but when it breaks, it does so in a complete fashion.

Tempering can also be done by ion exchange, where the glass is treated in a high-temperature bath containing elements that have a larger atomic size. These elements exchange places with smaller elements already in the glass, and since they take up more space they add compressive stress to the surface. Glasses used in portable electronic devices such as smartphones use this type of process, the most well-known being Gorilla Glass (a name trademarked by Corning).

10.4 Why It Matters

Nonmetals find their way into applications that cannot be satisfied by metals. These include insulation, low cost, transparent packaging, cushioning, wear resistant, high-temperature components, etc. Their intrinsic structure is the source of the properties that enable these applications, but they can be modified to achieve further improvements. This allows a single material to possess a range of properties through conditioning.

In polymers, conditioning can be accomplished by mechanical methods, but it is much more common that they are chemically modified to alter their strength, ductility, weathering properties, or other aspects of their performance. Although there are a wide range of commodity and engineering polymers, as discussed in Chap. 6, the demand for low-cost performance has driven the market for conditioning.

The primary means for strengthening polymers comes from cross-linking. This is an integral part of polymer design and processing, and it is the basis of thermoset polymers. It can be achieved as part of the polymerization process or induced by various means during a later step. Thermoplastics can also be cross-linked, but this converts them into thermosets.

All polymers can be modified through additives. Additives are physical or chemical additions that reduce the cost or improve the performance of polymers. Fillers can be used to simply take up space, thereby reducing the volume of the more expensive polymer materials. Fillers can also reinforce the polymer forming a composite. In that sense, composites can be considered as "conditioned" polymers.

Additives are also used in small quantities to change the properties of polymers. This can be for the stabilization of polymers at elevated temperature or to withstand environmental conditions. Certain additives can also reduce the flammability of polymers or cause the formation of pores and result in foamed plastics. Although you might not expect that plastics need to be made more "plastic," that is exactly the goal of plasticizers. Additives can be used to modify polymers in many ways, and multiple types are often used together.

Ceramics are more difficult to modify. Their rigid structure and high melting point makes it difficult to introduce "additive" ingredients like those used in polymers. Even though elements can be added to modify the properties of ceramics, most elemental additions are considered an integral part of their structure, such as boron in borosilicate glass (see Sec. 6.2.1.3). That does not mean there are not ways of conditioning ceramics, and modifications to improve fracture resistance are most popular. This can be accomplished by introducing a compressive prestress, such as in tempered glass, or by increasing the energy required for a crack to progress.

Without conditioning we would not have the variety of materials we enjoy. Our everyday lives would be less convenient, more expensive, and less safe. Conditioning allows cheaper materials to possess superior properties and other materials to perform in ways they could not otherwise. For instance, smartphone screens benefit from the tempering of advanced glass compositions in order to have a much improved toughness and durability. Additives to polymers allow them to be used at greater temperatures and in outdoor environments without undue degradation. This has made common items from outdoor furniture to automobile trim last longer and to be more cost-effective. Without the research that has gone into developing new properties in nonmetallic materials, many indispensable products we use every day would not be possible.

10.5 The How and Why of Conditioning Nonmetallic Materials

10.5.1 Chapter Review

1. How is the strength of a polymer determined by its molecules?
2. How does the mechanical conditioning of polymers increase strength? What is the potential disadvantage of this?
3. What might prevent polymers from being strengthened through mechanical means?
4. How is crystallinity affected by mechanical conditioning?

5. What is result of cross-linking polymers?
6. How can heat cause cross-linking?
7. How does chemical conditioning work? When is this a bad thing?
8. What are the benefits of radiation conditioning in polymers?
9. What type of radiation is popular for sterilizing polymers?
10. What properties do additives modify?
11. What is the purpose of plasticizers? How do they work?
12. What health concerns are associated with plasticizers?
13. What properties do stabilizers stabilize?
14. Where are flame retardants often used?
15. What benefits are there with fillers?
16. What is the difference between an extender and a functional filler?
17. What applications do foamed polymers have?
18. What is the difference between chemical and physical blowing agents?
19. What is the primary goal with conditioning ceramics?
20. What is transformation toughening? How does it reduce crack sensitivity?
21. Describe how crack deflection and crack bridging increase toughness.
22. Why is glass tempered? Where is tempered glass used?
23. Describe the ways glass can be tempered.

10.5.2 Concept Connections and Extensions

1. How is conditioning central to so many polymers?
2. Why are there so many ingredients that are used to modify plastics? How is this more efficient than developing new polymers?
3. Describe how many adhesives (Sec. 19.3.2) are based on conditioning processes.
4. What the benefits of fillers? How is this a disadvantage at the end of the product life cycle?
5. Flame retardants are used in an extensive variety of common products. Do you think having these additives is important, despite the potential health risks?
6. Windshields are made from annealed glass held together by a plastic layer. What would happen if a windshield was made of tempered glass?
7. What properties make foamed polymers so attractive for insulation and packaging? What alternatives are there?

CHAPTER 11
Casting and Molding

Casting and molding operations are something most people have done in their daily lives. Have you ever made ice cubes? What about Jell-O? A waffle? Pour concrete? While most of these are not normally considered as casting and molding, they are instructive as to what the processes entails. In casting, a material in a "flowable" state is added to a form with the desired features and dimensions. That liquid—sometimes highly viscous—is then hardened through heating, cooling, dehydration, chemical reaction, or some combination of those. From the examples above, it should be clear that these operations can be very simple. Depending on the material and/or part, it can also be quite complex. In this chapter, we will look at the various considerations in metallic and nonmetallic systems and some of the casting and molding operations that are commonly used in industry.

11.1 How It Works

Complex shapes, thin sections, and fine detail are most easily produced using materials in a liquid state. This also allows the creation of internal features, something very few processes can create efficiently and accurately. There are quite a few methods for casting and molding, but the first distinction to make is the difference between those two terms. First off, the two terms are often used interchangeably, so defining the line between them is somewhat subjective. In general, **casting** is a process that utilizes materials in a liquid state and is most often applied to metals and ceramics, while **molding** is used with materials in a more viscous state, typically with polymers. Not to get too far ahead, but we will see that die casting and injection molding operate under the same basic principles, but one is used with metals (casting) and the other with polymers (molding), hence the difference in terminology. Both depend on a mold to give them the desired geometry.

11.1.1 Considerations in Casting and Molding

There are many similarities between casting and molding, so their commonalities will be discussed together. For instance, both are highly repeatable, can be fast, and are cost-effective methods of production. Although making the mold(s) can be complex and time consuming, the mold, or pattern, can be reused for each part. Casting and molding can also create highly complex parts that would be difficult, if not impossible, to make by other production methods. For instance, internal passageways and highly irregular surfaces with complex contours can be made, and the overall shape does not have to be easily held for material removal, such as occurs in machining. An engine block is a good example of the way complex shapes can be created. The block must have highly detailed coolant passages internally throughout the block and it has a complex external shape as well.

You might wonder how it possible to create internal geometries. Imagine trying to make ice cubes with holes running through them in all directions. There must be a material to prevent the water from filling those internal areas. If that material was a permanent part of the tray, it would be difficult or impossible to remove the ice cube from the tray or the material in those passages. Removing the cast material is a critical part of the process, and it emphasizes an important design consideration: mold geometry.

11.1.1.1 Generalized Process Description

The casting and molding processes involve the introduction of a material into a cavity, which the material fills and takes the size and shape of. The cavity, which represents the final part, is created by the **mold**. The material is introduced to the mold using a system of channels, known as the **sprues** (vertical pouring column), **runners** (horizontal channels), and **gates** (entryway to the mold cavity). A reservoir, known as a **riser,** may also be used in some designs. These features are intended to allow and control the flow of the material into the mold cavity. Not all molds will have all of these features, or they may take different forms depending on the need. Other features such as wells and recesses for capturing contamination may be included as well.

Likewise, additional materials may be added to the mold to create internal features or to include a component into the molded part, such as a metallic threaded insert being incorporated into a polymer. Although it is difficult to generalize such diverse processes, the basic features that may appear in a cast or molded product are shown in Fig. 11.1. For example, the design of a mold for processes which use pressure to force material into the mold will often differ significantly from those where gravity is used to move the material through the mold. More specifics are given in the sections describing individual processes.

11.1.1.2 Mold and Pattern Design

Designing molds and patterns can be a very complex undertaking. The intention here is not to provide the detail necessary to design complex molds, but to appreciate the considerations that go into it and present the basic features that are common in mold and pattern making.

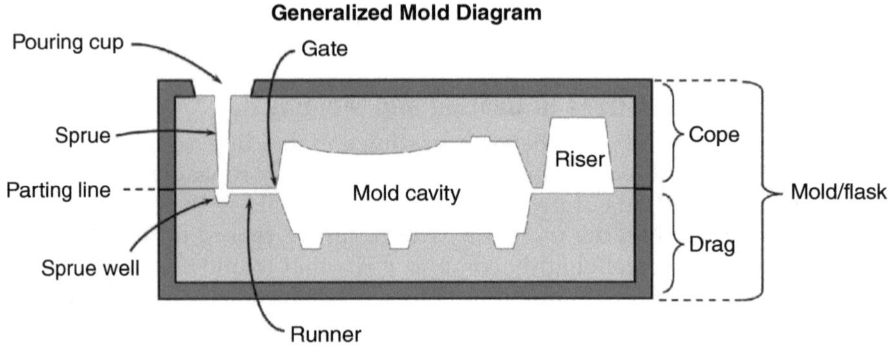

FIGURE 11.1 Although mold design can vary greatly for different materials and processes, in general there must be a way to efficiently get material in as a liquid and out after it solidifies.

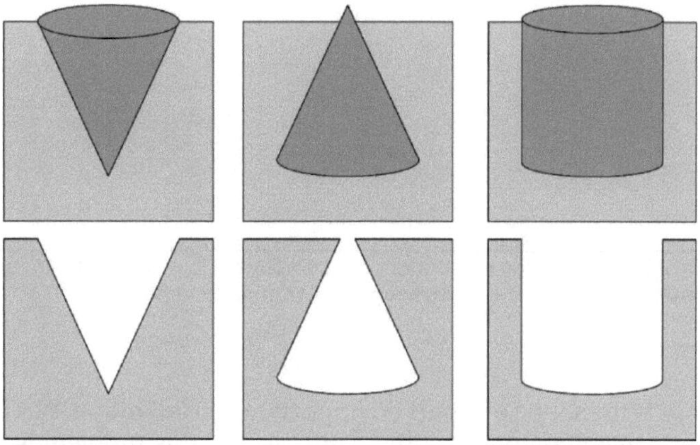

Figure 11.2 The sidewall angle is very important in mold design. Positive draft makes it easier to remove parts, negative draft can make it impossible, and zero draft is possible but often makes it much more difficult to remove parts.

The most important feature in designing a mold is to provide a way to remove the part. If you look an ice cube tray, a cake pan, a muffin tin, or any other mold, you will see they all have sloped sides. **Draft** is the intentional angle put on any vertical feature to allow it to be removed from the mold after solidifying. **Positive draft** angles the walls inward as you move to the bottom of the mold cavity, which gives it the shape of the letter "V." **Negative draft** slopes the wall outward toward the bottom of the mold and results in the shape of the letter "A." **Zero draft** leaves the walls vertical (if you want a letter to go with it, "H" would be one option). Imagine burying a cone halfway up in sand. Which way will it be the easiest to pull out of the sand? Of course, point down is the easiest when using sand, but when a mold is made out of a rigid material, it is the only way that will work (see Fig. 11.2).

That can actually be a drawback to casting and molding processes. They operate most efficiently when the mold has angled walls, which might not be desirable for the final product. Also, the part cannot have any features that would prevent removal from the mold, such as tabs or notches, unless they are very carefully placed. This indicates that orientation of the part in the mold is also important. Let us say you were going to cast the cone we thought of when illustrating draft. If we think of it like an ice cube tray, with one end completely open, there is only one way it will be removable. Most molding and casting processing use two halves to create a mold, so that would not work anymore. The geometry can be reoriented to work, but the cone would have to be rotated 90°. This is illustrated in Fig. 11.3.

Molds may be made up of two or more parts, depending on the complexity of the component being manufactured (Fig. 11.4). By using more mold sections, the number of features (nonsmooth surfaces) can be increased. The drawback is that the mold complexity must be increased as well as cost to manufacture. In real, three-dimensional parts, the mold cost and complexity can increase rapidly. Depending on the number of parts being manufactured this may not be a problem as the mold will pay for itself by reducing downstream work.

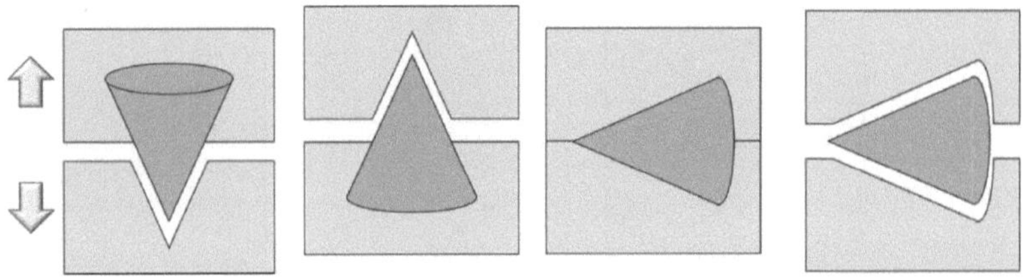

Figure 11.3 Multipart molds must orient the component in a way that allows complete removal. If this is not possible, the part must be redesigned or made another way.

Not only is the shape of a part an important consideration, but the size is as well. At first this may seem obvious, but the reality is that this problem is not always a simple one. Each material will have a different amount of shrinkage or expansion that occurs during solidification. Most materials contract as they solidify, but some expand, water being the most obvious example. **Shrinkage** is the volumetric change of a material during solidification, dehydration, and/or reduction in temperature. It is most pronounced during phase change (liquid to solid), but materials that are cast hot will also shrink while a liquid or a solid as they cool. The size effect of temperature on solids is discussed in Sec. 8.2.3.

Our main takeaway is that the mold should be a different size than the final part, and the size difference is dependent on the cross-sectional thickness. There will be more shrinkage in thicker sections, which can cause secondary problems due to stress build-up. This same concept was addressed in Sec. 8.2.4, where uneven heating/cooling results in disproportionate size changes. This can result in castings with cracking or warping. For this reason, cast and molded parts, especially those done at high temperature, should not have drastically different section thicknesses and avoid sharp corners where stress will concentrate during cooling and/or shrinking (Fig. 11.5).

Another consequence of shrinkage during solidification is that it creates cavities, both internally and externally. To reduce the effect of shrinkage, additional liquid should be allowed during the solidification process. This is most easily accomplished by providing a reservoir where liquid is retained while the part solidifies. A riser is a reservoir for additional material to be provided during the solidification of the casting. This is typically seen on metal castings that are gravity fed, so that as the component

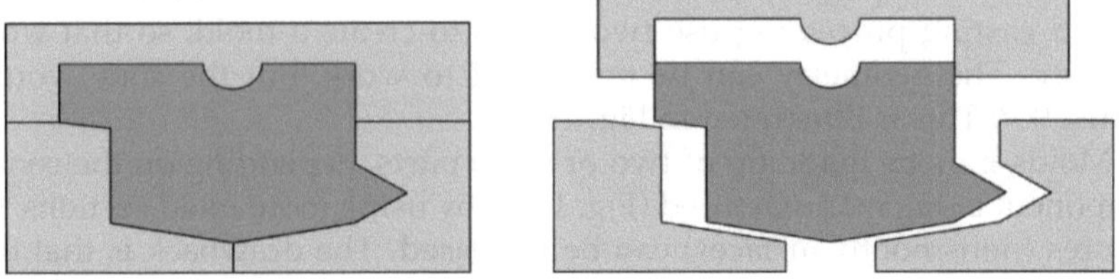

Figure 11.4 Molds made of more than two parts may be used for complex shapes with reentrant features on one or more sides. As the mold complexity increases, the cost also increases. Here, only two dimensions are shown. In real parts, sides that are featureless in this example may have even more geometric considerations.

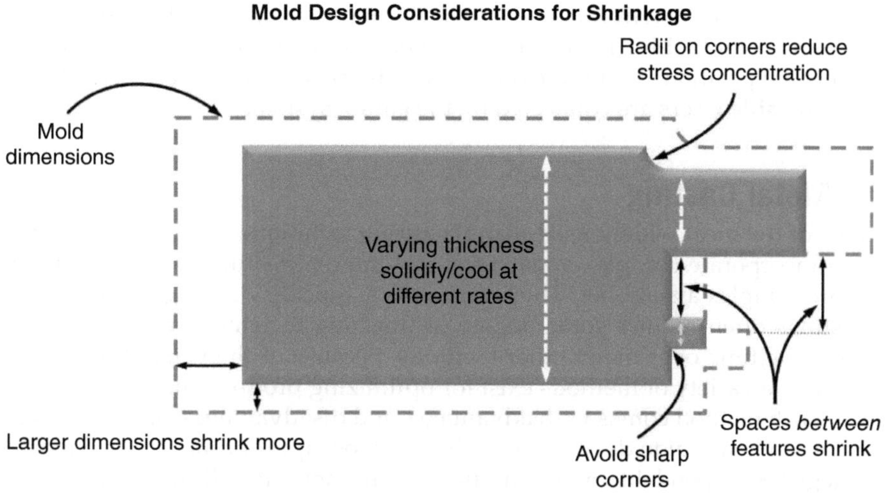

Figure 11.5 Mold design must take into account and volumetric changes in the material during solidification. While the changes may be minimal in some materials, they are pronounced in others, and each material should be evaluated individually.

cools, solidifies, and shrinks, additional molten metal can be provided. There are some basic requirements a riser must satisfy:

1. It must remain liquid until the component is fully solidified (last to harden).
2. It must be close to the component so that the liquid can be effectively supplied.
3. It should be as small as possible to reduce waste.
4. It should be shaped to minimize the solidification rate (minimal surface area).

In materials that solidify upon cooling, these requirements are straightforward. In other situations, such as the chemical hardening of polymers, there may be little that can be done by adding risers, but shrinkage is also much less of a concern. Different types of risers are defined by their positioning (Fig. 11.6). **Open risers** have a surface

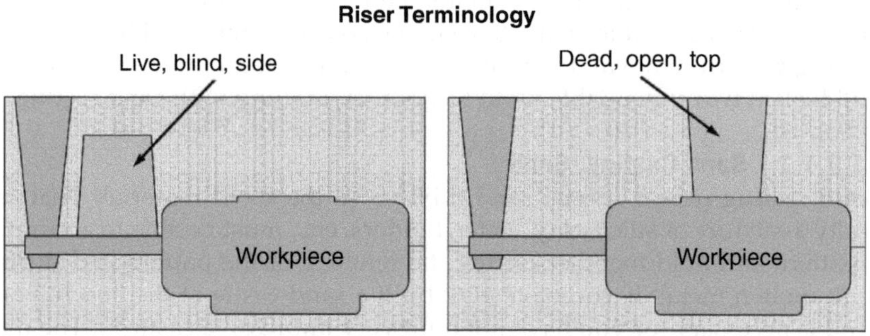

Figure 11.6 Risers can be configured with primary variables: whether they are positioned before the mold or after (live/dead), completely internal or exposed to the outside (blind/open), and whether they are next to or directly on the mold (side/top). The variations are illustrated here.

exposed to the environment outside the mold, whereas **blind risers** are contained within the mold. **Live risers** are positioned before the inlet of the mold cavity, and **dead risers** are positioned after the outlet. **Top risers** are directly connected to the mold cavity, and **side risers** are connected by the gating system.

11.2 Metal Casting

One of the most widely used manufacturing techniques for metals is casting. Castings can incorporate complex geometries and internal features that allow them to replace more complex assemblies. They can be made quickly with high repeatability. Although castings often require some degree of finishing to remove excess material, they still conserve time over many other methods. Because of the versatility and popularity of casting, a variety of methods exist for optimizing production.

Each method comes with advantages and disadvantages, and knowing more about each can help guide decisions on which is most appropriate. We will look at two main categories: expendable mold and permanent mold. Recall from Sec. 11.1.1.2, the mold must have the desired features of the final part and the necessary passage ways for introducing the molten metal to the mold. The mold material can be created from anything that satisfies some basic requirements:

- It must be stable and hold its shape before and during casting.
- It must have a melting temperature in excess of the pouring temperature of the metal being cast (preferably by a large margin).
- It should not react with the metal being cast or otherwise contaminate the molten metal.
- It must incorporate a means of extracting the casting from the mold after it solidifies.

11.2.1 Expendable Mold Techniques

As the name suggests, **expendable mold casting** involves a material that is destroyed at the end of the casting process. That is the way the part is removed from the mold. This can allow very complex parts, and even those with negative draft, to be cast, since the part does not need to be removed cleanly from the mold. Instead, the mold is broken to remove the casting. These expendable mold techniques must still be given shape, and the patterns used to create that shape have to be removed before or during casting. During pattern removal, the mold must retain enough integrity to hold its shape. We will look at two expendable mold techniques, starting with sand casting.

11.2.1.1 Sand Casting: Sand

Sand casting uses a special sand mixture as the mold material. That sand, which is really a mixture of silica, clay, water, binders, etc., must be able to accurately reproduce a pattern and hold together during the removal of the pattern and during infiltration with molten metal. If you have ever built a sand castle at the beach, you know moist sand is necessary (or you have not built very impressive castles!). The moisture acts to bond the sand particles together so they hold the shape they have been given. The same concept applies to sand used in casting. Casting sand can be bonded using water, known as **green sand**, or it may be oil-bonded. After casting, the sand is broken up to reveal the part contained within. The sand used in casting is developed to optimize the performance of the mold over repeated use. The sand can be remolded and cast again, which is a distinct benefit.

The sand does have a lifetime, however, as the repeated exposure to very high temperatures will degrade it chemically. It can be reconditioned to some degree by adding ingredients and mixing it, a process known as **mulling**. After the sand has been sufficiently mulled, it is then sifted through a screen in a process called **riddling**. This helps break up any clumps and gives a fine, free-flowing sand coating that is used, at minimum, to directly cover the pattern. For the sake of time and effort, after the initial layer is applied to the pattern, the sand is no longer riddled since it does not directly contact the pattern and will not impact the surface quality.

Sand for casting is much more intricately designed than you might expect. The sand should be able to meet requirements of temperature (not degrade or melt) and structure (hold the shape of the pattern and accommodate shrinkage during solidification). It should also allow vapors to escape as the binder material (e.g., water or oil) vaporizes and any dissolved gases are released when the molten metal is introduced. Additionally, the size, shape, and composition of the sand particles can affect the surface quality, permeability, and the likelihood of defects such as voids, cracks, sand entrapped in the metal surface, etc.

For example, fine, powdery sand will result in a smoother surface, but it has a lower permeability and can also result in the formation of voids. Silica-based sands are more likely to result in defects due to their high-temperature expansion characteristics, but alternatives such as zircon or olivine sands are more expensive. The material being cast and the required accuracy will determine if the cost is worth the benefits. It should be clear, however, that the quality of the part is dependent on the quality of the sand.

11.2.1.2 Sand Casting: Cores

In order to create internal passageways and features, a sand core is used. A **sand core** is a part of a mold for creating features not possible in a regular pattern and is often used to create holes or other internal features. The sand used for cores must be able to hold its shape without external support. Although regular casting sand can do this to some degree, it is too weak for many applications. Locating features for sand cores are added to the pattern for the mold, which allow the sand core to be precisely placed. The general process and function of sand cores are shown in Fig. 11.7.

Core sand incorporates binding agents that can be hardened by baking, which allows it to maintain its shape while handled. A good sand core should create a smooth surface, not contaminate the casting chemically or by evolving gas, it should be strong enough to handle before casting and not disintegrate during casting, but it should also allow the

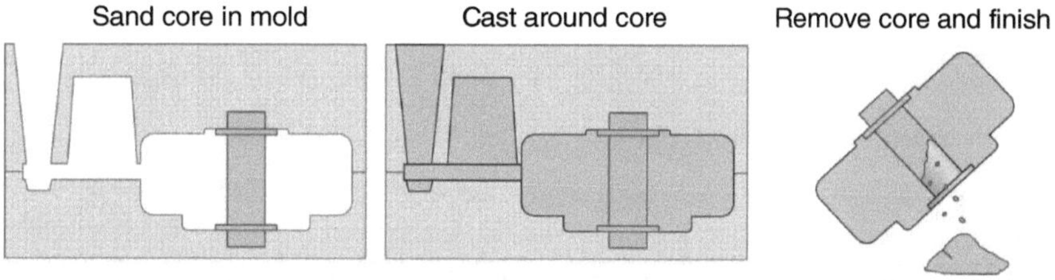

FIGURE 11.7 Sand cores are removable templates that create internal features in cast products. Common uses include passages for liquid and gas and holes for assembly at a later stage.

metal to contract without excessive restriction (it should be collapsible). The core should also be easily removed from the casting after it solidifies.

Although sand cores reduce the amount of machining required, they are still often followed by finishing operations to increase accuracy and improve finish. This is especially important in rotating objects, such as pulleys, where a slight misalignment of the sand core and mold would result in an unbalanced component. Sand cores require a separate mold to give them their shape, and the sand used in cores is more expensive. This results in cost increases, but they can save machining time later on, and they can achieve internal features that are otherwise impossible. These factors make sand cores a valuable asset in casting, both with expendable and permanent molds.

11.2.1.3 Sand Casting: Patterns

Sand can be given the desired shape using a variety of techniques. These share common design principles as outlined in Sec. 11.1.1. Referring back to Fig. 11.1, the mold is often composed of two halves. In sand casting, these are called the **cope** (top half) and **drag** (bottom half). Together they make up the **flask**. The pattern must be able to produce two separate halves that are precisely aligned. That way, when the cope and drag are combined, an accurate three-dimensional component is created. This may be done by using two separate patterns.

The pattern may be a solid piece (**solid pattern)**, which is patterned twice, two separate halves (**split pattern**), which can be molded separately, or a single piece (**match-plate pattern**), which can be used to create both halves at the same time. Simpler patterns have a lower cost, but for high-volume processing, the more efficiently a mold can be prepared, the more cost-effective it becomes to have a more sophisticated pattern. A solid pattern must use some backing fixture (known as a follow board) to align it in a mold and position the parting line. The split pattern eliminates the need for a follow board, but must still be carefully aligned. A matchplate pattern simply combines a split pattern onto a common plate so that they are necessarily aligned, and the top and bottom halves can be filled with sand without removing the pattern. These patterns are shown schematically in Fig. 11.8. Examples patterns are shown in Fig. 11.9.

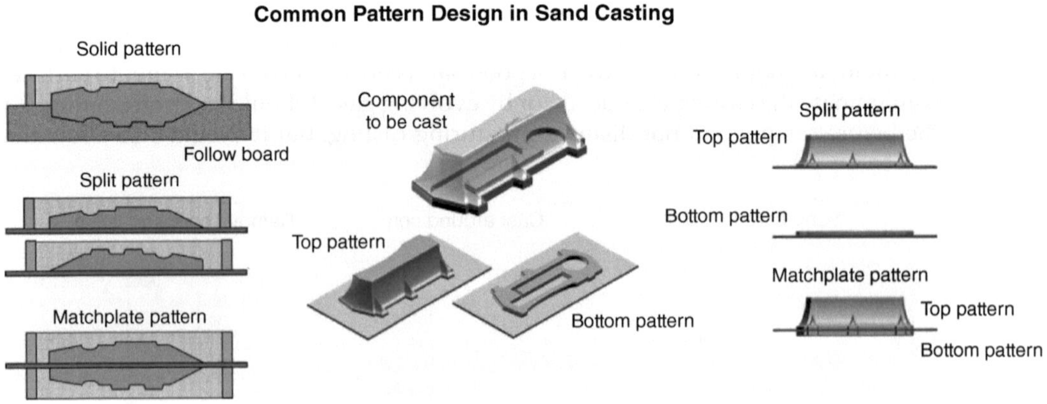

Figure 11.8 The same pattern can transferred to sand in a variety of configurations. The appropriate version is usually determined by the cost of fabrication, the number of parts to be made, and the required accuracy.

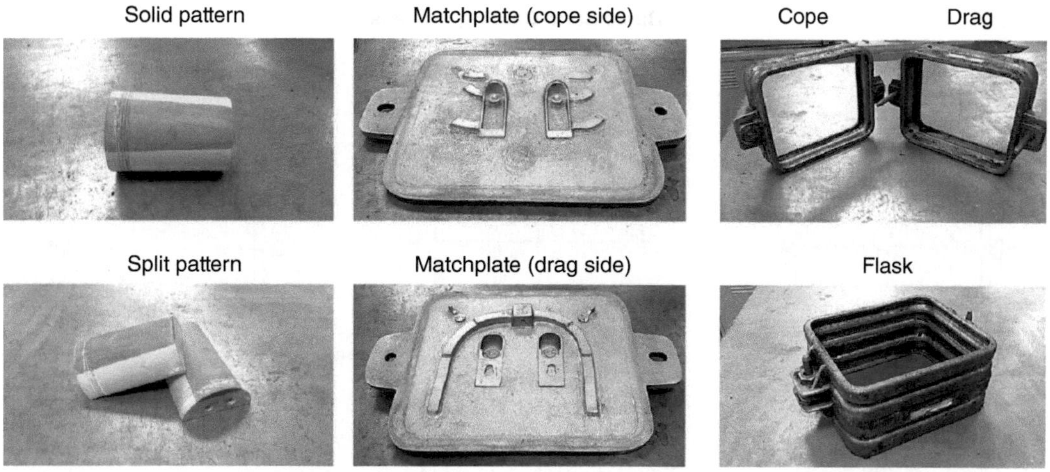

Figure 11.9 Casting equipment can be made from low-cost, readily available materials, such as wood, but for longevity and accuracy, it is more desirable to make them from durable and dimensionally stable materials such aluminum or steel.

The materials the flask and pattern are made out of will also determine their cost, accuracy, and durability. For instance, wood can be easily fabricated, but will wear down relatively quickly and can absorb moisture, which results in dimensional changes. Polymers can be used, but they are more difficult to shape and some are still sensitive to moisture and may be degraded by additives in the sand such as oil. Metals, typically aluminum, are used for making patterns with the highest durability and the greatest accuracy. Although the cost is higher, it is easily justified in high-volume processing. Wooden and metal patterns are shown in Fig. 11.9.

11.2.1.4 Sand Casting: Mold Techniques

So far we have discussed the sand and patterns used to create a mold, but we have not discussed how we put those two together. There are many variations, but the common theme is that the sand not only has to take the shape of the pattern, it must rigidly hold it during casting. The second requirement necessitates that the sand be tightly packed against pattern, and this packing process is known as **ramming** the mold. In its simplest forms, the sand casting process, shown schematically in Fig. 11.10, involves:

1. Combine pattern and flask (cope and/or drag).
2. Apply a powdered release agent to the pattern (i.e., parting dust).
3. Add mulled and riddled sand until the pattern is covered.
4. Incrementally compact the sand by hand and with implements as more sand is added.
5. Once the mold is filled above the top, use a flat, sturdy object to flatten the sand level to the top of the mold (known as **striking off** the mold).
6. The sprue and vent holes (if used) are then cut from the compacted sand.

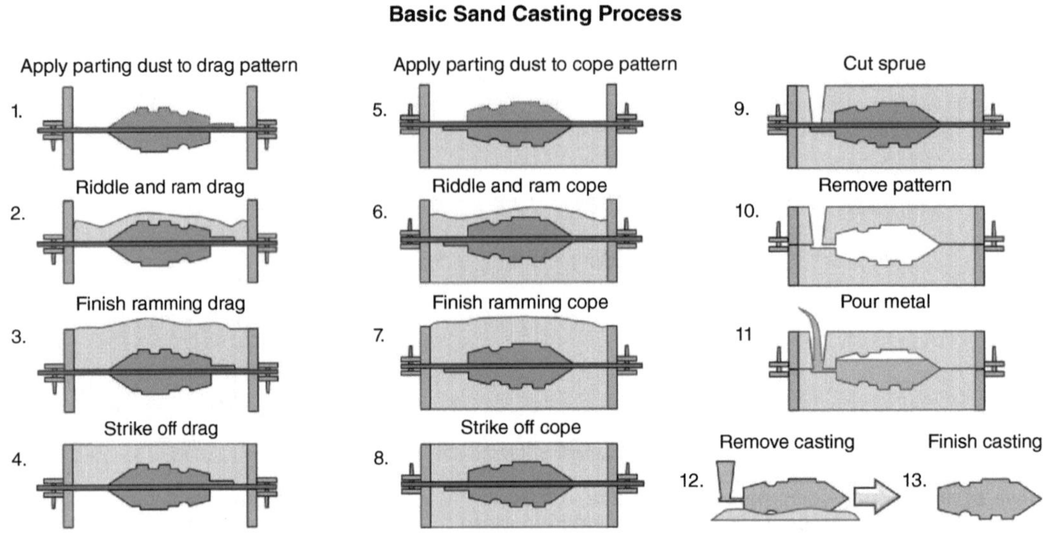

FIGURE 11.10 The sand casting process can be performed in a variety of ways, but the simplest method, done manually, takes the form shown here. In this case, a matchplate pattern is being used, and no risers or vent holes are included. See Fig. 11.1 for more on the features and terminology of molds.

7. The pattern is separated from the sand.
8. The mold is closed and the metal is poured.
9. The sand is broken up and removed from the casting.
10. The casting is finished by removing excess material.

The process just described is for manual preparation of sand molds. The amount of force used to compact the sand and the way and uniformity with which it is applied will determine the integrity and accuracy of the mold. The process is both time and energy intensive, and to increase productivity as well as repeatability, automation can be used. This can be as simple as compacting the sand by hydraulic pressure and/or vibration, or it may be completely automated where the pattern, flask, sand, and ramming process are all automated (known as automatic matchplate molding, shown in Fig. 11.11).

Some examples of improved methods for ramming a mold include jolt, squeeze, and when combined, the jolt-squeeze method. Permutations on the processes, such as gas pressure and or using a diaphragm to even out compaction pressure, are used to improve quality. Likewise, methods such as V-process casting, where a polymer sheet is shaped to the pattern through thermoforming (see Sec. 12.3.1; Fig. 11.12), and then backed by vacuum-packed sand.

Also, vertical flaskless molding can be used in continuous production lines for very high volume part runs. In this process, often referred to as the DISAMATIC process (after the company's name, DISA), the right half of one mold is also the left half of the next (see Fig. 11.13). Although automated processing improves quality and production rate, the equipment comes with a high capital cost.

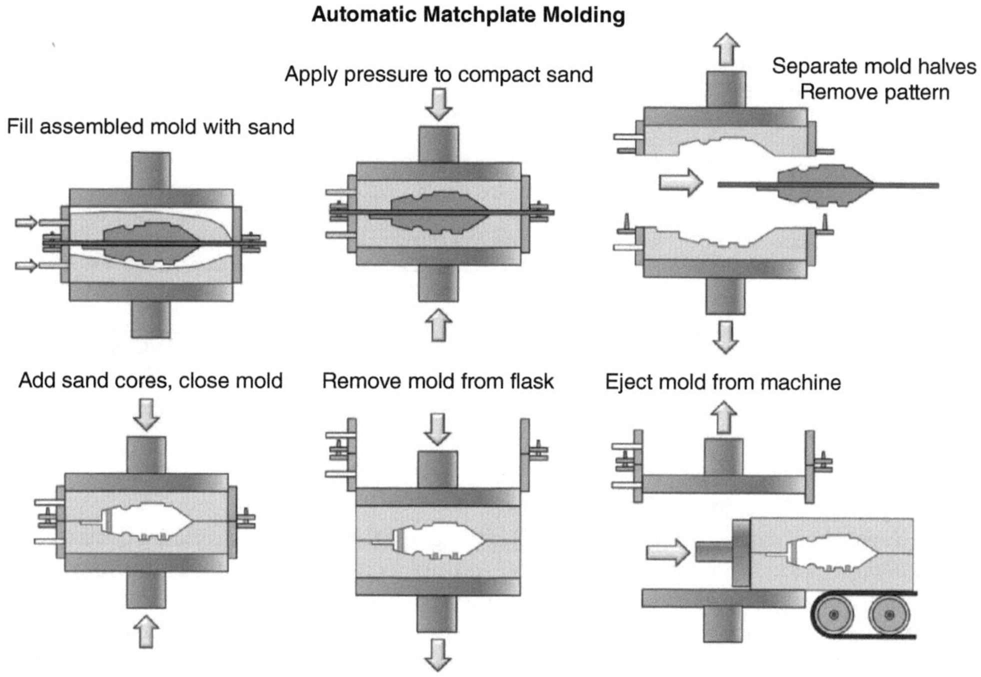

FIGURE 11.11 Automation can greatly improve product consistency and throughput. Here automated matchplate molding is used to introduce sand and compact and extract the mold.

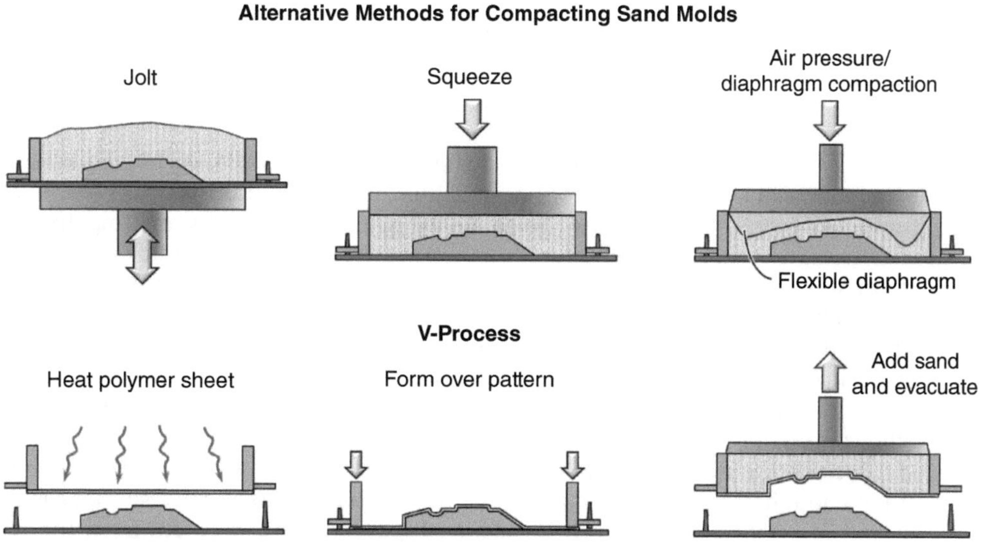

FIGURE 11.12 To improve uniformity, detail, and surface quality in sand casting, mechanized processes such as jolt, squeeze (and jolt + squeeze), air diaphragm compaction, and the V-process can be used to compact sand.

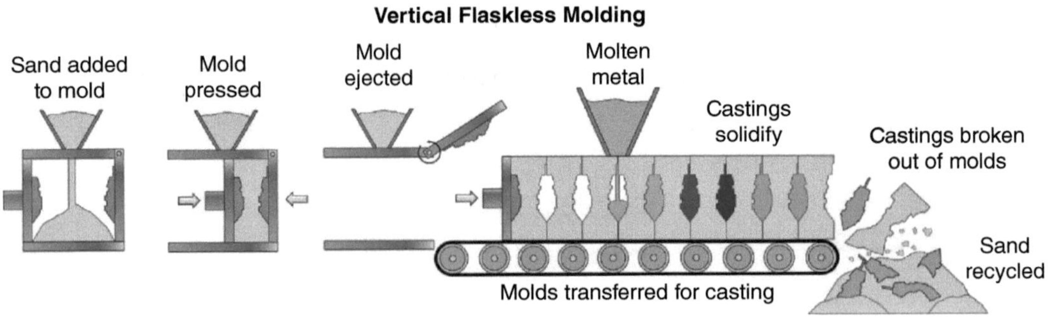

FIGURE 11.13 The vertical flaskless molding process, also known as the DISAMATIC process, is a highly automated casting technique suitable for very high volume products, such as automotive components.

11.2.1.5 Investment Casting

Investment casting, also known as **lost wax casting**, uses a sacrificial pattern and a sacrificial mold. As discussed in the preceding sections, sand molds are used to create a casting, are then destroyed to retrieve the component, and the mold material is recycled. Not all expendable mold techniques allow for reusing the mold or even the pattern. Most investment casting uses a sintered ceramic mold that cannot be reassembled for use after it removed. In general, the steps involved in investment casting include:

1. Create the pattern from a sacrificial material
2. Coat the pattern with a ceramic slurry to create a shell
3. Burn out the pattern
4. Sinter the ceramic shell for strength
5. Pour molten metal into the shell
6. After metal solidifies, break the shell off

Even the first step of the process can have several substeps, such as creating a metal mold in which to create the pattern (typically wax). The individual wax components are then connected to sprues and runners (the gating system) by melting the surfaces together. The ceramic slurry process will often use different ceramic particle sizes and chemistries for different layers and is done as a multistep process. The wax (or other pattern material) is then burned out at moderate temperature, but the ceramic must be fired at very high temperature to strengthen it. The process cannot be rushed or it may result in cracking or distortion. The molten metal is then poured into the ceramic mold and allowed to solidify. After it has cooled, the ceramic is broken off by vibration or other means.

Because the process is time consuming, multiple parts are cast at the same time (if small enough). The individual components are attached to a "tree" so that each is connected to a central sprue. In this case, each part must then be cut from the tree and excess material removed as part of the finishing process. Although more complex than sand casting, investment casting achieves superior quality dimensionally and in regard to surface finish. The process is shown schematically in Fig. 11.14.

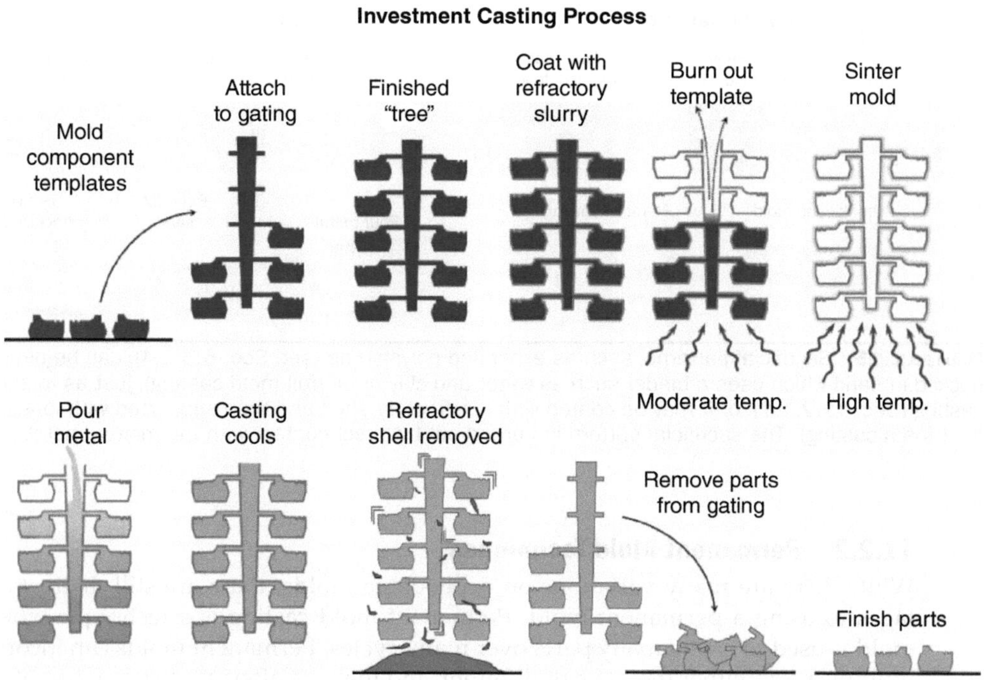

FIGURE 11.14 Investment casting is an expendable mold process with good tolerance and finish. The process starts with a sacrificial mold template and creates a durable mold around that template. The template is then removed and metal is put in its place. The general process is shown here.

A version of investment casting, known as **evaporative pattern casting,** uses a sand mold with a sacrificial pattern. The process may use a sacrificial pattern, most often polystyrene foam, directly packed in casting sand (the **full-mold process**) or it may use a ceramic coated shell that is backed by sand for support only (**lost-foam process**) (Fig. 11.15). The two are very similar, and require a pattern that will burn out as the molten metal directly contacts it. This is in contrast to investment casting as described above where the pattern is removed prior to pouring the molten metal. The pattern should burn out cleanly, not contaminate the metal or introduce porosity, and the surrounding sand should maintain the shape of the pattern as it is disintegrated.

11.2.1.6 Variations of Expendable Mold Techniques

Expendable mold casting is nearly limitless in the way it can be achieved. Any method that uses a mold that is nonpermanent falls under this category. We limited our discussion to some of the most common permutations, but many other techniques have been developed to simplify the process, improve quality, lower cost, or meet other specific demands. For example, the mold can be made from plaster, bonded ceramic, graphite, etc. The sacrificial pattern can be a polymer shell, sand shell, or unsupported ceramic shell. The mold material will determine the metal that can be cast (e.g., plaster only used with lower melting point metals), the detail that can be replicated, and the cost and complexity. A general overview of expendable mold techniques is given in Fig. 11.16.

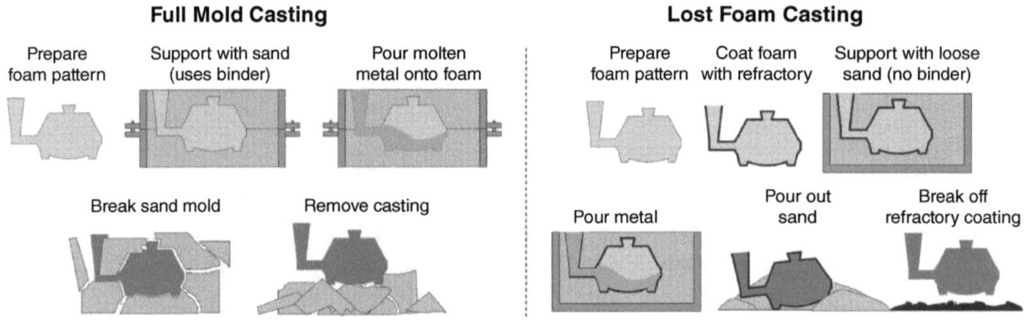

FIGURE 11.15 Sacrificial patterns, such as expanded polystyrene (see Sec. 6.3.1.4), can be directly packed in sand which uses a binder such as water and clay or oil (full mold casting), just as in sand casting (Sec. 11.2.1.1), or it may be coated with a refractory shell and then supported with loose sand (lost foam casting). The sacrificial pattern is burned out by direct contact with the molten metal.

11.2.2 Permanent Mold Techniques

While there are many variations on expendable molds, there are still distinct advantages to using a permanent mold. Permanent mold casting is a technique where the mold is used to create many parts over many cycles. Permanent molds can incorporate multiple parts, either the same or different, and they are often made from durable materials such as tool steel. Permanent molds have the same constraints on draft, cooling rates, and other design factors that expendable molds do, but since the mold remains intact, they must also allow for parts to be removed/ejected from the mold efficiently.

Permanent molds have distinct quality advantages over sand casting. The mold surface can be very smooth and finely detailed, the mold halves are precisely aligned and tightly clamped reducing material leakage into the parting line, which results in the formation of **flash**, or thin projections of material. Permanent molds also operate more efficiently as the mold never needs to be remade and no pattern for the mold is used as part of the process.

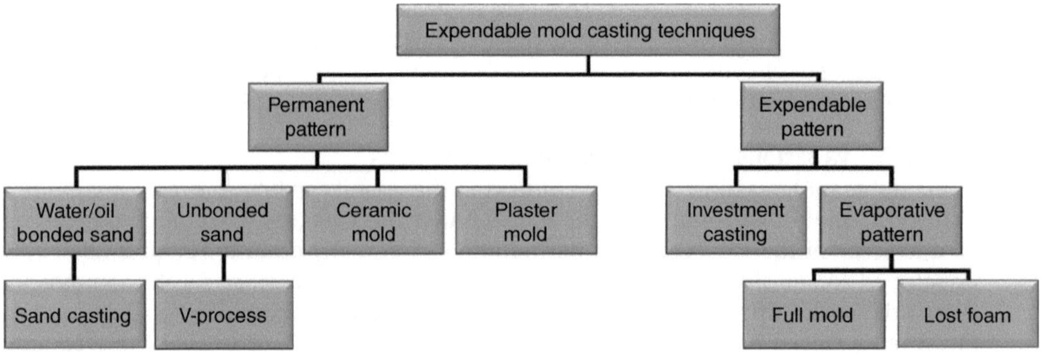

FIGURE 11.16 The variations on expendable mold casting techniques are many, and only a small sampling is provided here. Each method comes with advantages and disadvantages, so selection should be done carefully.

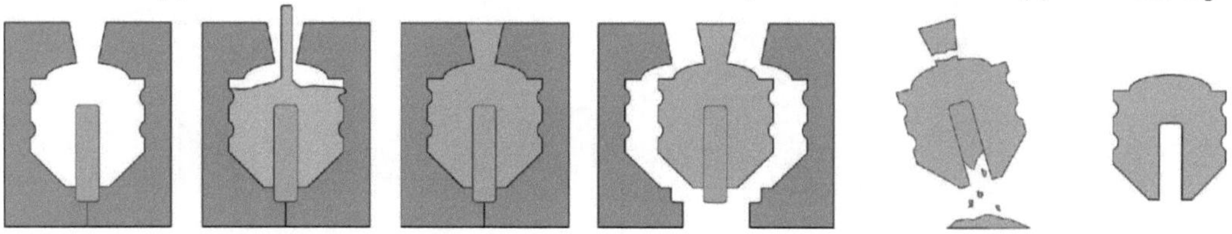

Figure 11.17 Permanent mold casting follows a process very close to expendable mold methods, but the ability to reuse the mold and to achieve higher quality make it desirable for precision casting.

The molds can be extremely expensive (tens of thousands of dollars), however, and changing the design of a component, even a little, often requires a new mold to be made. It is also limited in type of material being cast, since casting steel would destroy a steel mold based on the required temperature, and metals tend to be more reactive with each other than with ceramics used in sand or investment casting. Although all permanent mold techniques involve a reusable mold, they differ greatly in regard to how metal is introduced, how efficient the process is, and what materials can be used.

11.2.2.1 Permanent Mold Casting

Permanent mold casting is very similar to sand casting as it is often done with gravity acting to move the metal through the mold. Parts are typically of relatively simple geometries, and thin sections are avoided because of the difficulty in preventing premature cooling before the metal fills the mold, also known as **misruns** or **short shots**. The mold itself is often preheated to avoid cooling the metal too quickly and reduce thermal shock to the mold. The mold is heated significantly by the molten metal, so cooling is often used to regulate the mold temperature. After the casting is solidified, it is ejected from the mold while still hot. This reduces cycle time and any unwanted issues which might arise due to shrinkage. Sand cores can be used as the same way they are in sand casting,* which allows a variety of internal features to be created. The process is shown schematically in Fig. 11.17.

In a process known as slush casting, the metal is removed before it fully solidifies. In contrast to the permanent mold casting technique, the metal is simply poured out shortly after it is poured in. Since the mold walls are the first place solidification occurs, the metal quickly forms a solid "skin." If the molten metal is removed soon after this initial solidification occurs, it will leave a hollow casting with a well-defined exterior and poorly defined interior. This is acceptable in many decorative products or those which act as containers. The process is shown in Fig. 11.18.

11.2.2.2 Centrifugal, Semi-Centrifugal, and Centrifuge Casting

Centrifugal, semi-centrifugal, and centrifuge casting all use rotation as a means to force molten metal into the features of a mold. The differences are related to the location of the mold with respect to the axis of rotation. All of these methods use rotation as a

*When sand cores are used, the designation is often changed to *semipermanent mold* casting.

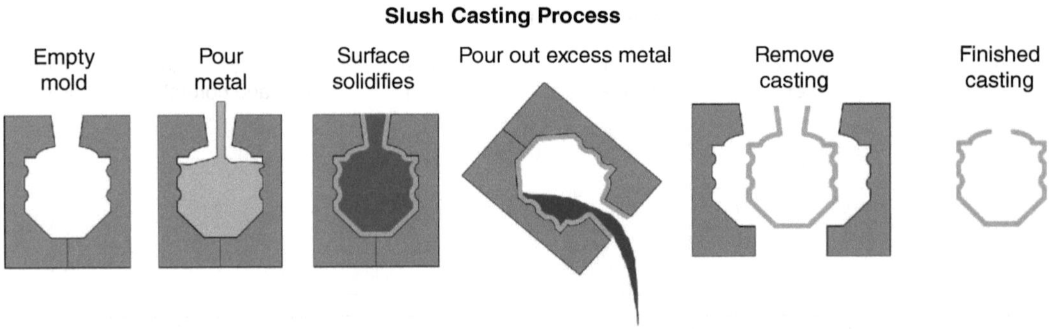

FIGURE 11.18 Slush casting mirrors the permanent mold technique except that it pours off excess molten metal after only the surface of casting solidifies. This leaves a hollow object with good external detail.

means to force the metal toward the exterior of the mold or molds. This is similar to a merry-go-round you might find on a playground. The faster it is spun, and the nearer to the outside edge you are, the more tightly you must hold on. In casting, the mold provides the containment for the metal, and the rotation ensures that the external features of the mold are filled. These three methods are depicted in Fig. 11.19.

In **centrifugal casting,** the mold is rotated around its own central axis. This causes the molten metal to be propelled to the outside walls by centrifugal force. This technique creates a hollow casting with the external features being dictated by the mold. Although the exterior does not need to be circular, the interior will be. This process can be done vertically or horizontally. In a horizontal configuration, the process can be used on small to very large structures (up to 10 ft in diameter and 50 ft long) such as pipe or pressurized tanks. The wall thickness depends on the amount of metal added to the mold, and therefore a single mold can be used to create castings of various internal dimensions.

Semi-centrifugal casting also rotates around the central axis of a mold, but the mold is filled with metal, whereas in centrifugal casting the center is hollow. The rotation then

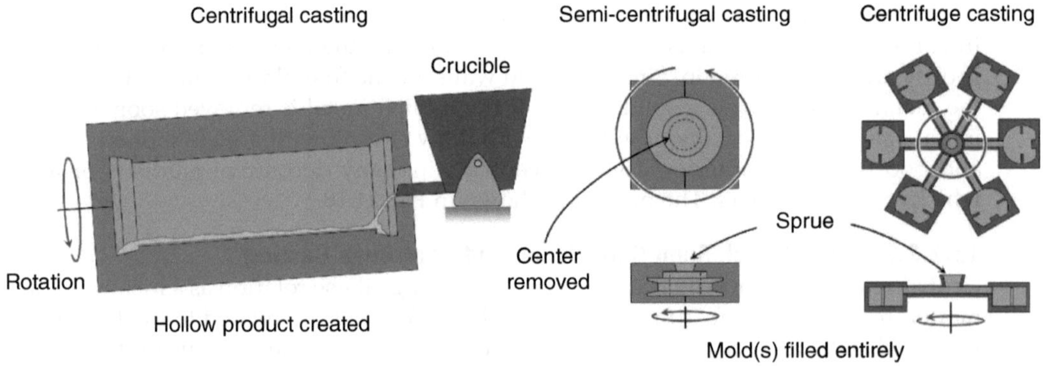

FIGURE 11.19 By rotating the mold, find detail can be replicated and impurities concentrate at the inner surface or center of the material. These can then be eliminated or later removed from the casting.

causes the metal to travel to the limits of the mold cavity. Impurities are most often lighter than the metal being cast, so when the casting solidifies they concentrate in the center (heavier objects "settle" on the outside rim during rotation). For this reason, semi-centrifugal casting is done around a vertical axis, as horizontal rotation will not separate out impurities as well. The concentration of impurities in the center presents a structural deficiency, so most semi-centrifugal cast parts are later machined to remove the center. Examples include pulleys, gears, or other objects that benefit from a hollow center and rotational symmetry.

Centrifuge casting uses molds which are placed away from the axis of rotation. Often multiple molds are placed around a central sprue, and a gating system is used to deliver the molten metal to the molds during rotation. This method provides high-purity castings, since the impurities concentrate in the center of the gating system and not in the molds. Because of the large rotational mass, mold size is more limited. This method may also be used to make evaporative patterns ("trees") for investment castings.

11.2.2.3 Die Casting

Die casting is a category of techniques that introduce a molten metal into a mold (die) using high pressure. One of the primary benefits of using pressure to introduce the molten metal is that very fine detail and thin sections can be accurately formed. The dies used in this method are often hot-work tool steel, and the metals being cast are typically nonferrous alloys with modest melting temperature. Like other permanent mold techniques, the mold itself can be very expensive, so maximizing lifetime is important. This can be achieved through maintaining an elevated temperature during processing to reduce thermal cycling and by minimizing turbulence in the metal "shot," which can erode the surfaces. The general process for die casting is

1. Mold is sprayed with a release agent (lubricant).
   a. Casting inserts added to mold (optional).
2. Mold is closed.
3. Metal is injected into mold.
4. Pressure maintained during solidification.
5. Mold opened.
6. Castings ejected.

The same considerations of draft, section thicknesses, and geometry (see Sec. 11.1.1.2) must be considered in die casting. Also, the die (mold) can be separated into multiple sections for complex parts. A limit to die casting is that sand cores cannot be used since they would be eroded by the forceful injection of the metal, but other retractable pins can be used for simple features. Also, due to the rapid rate of filling and impermeable nature of the dies, air cannot escape unless vent holes are provided. These are small grooves (0.010–0.020 in, 0.25–0.50 mm) placed along the parting line, and they must be located in any positions where air may become trapped.

Components such as threaded inserts and bearing surfaces can be cast into the part during the casting process, but the material must have a significantly higher melting point and adding inserts increases the cycle time. Ejector pins are also used to remove the castings after cooling. A sign that a component is die cast is a flat round impression left from where the pin is located.

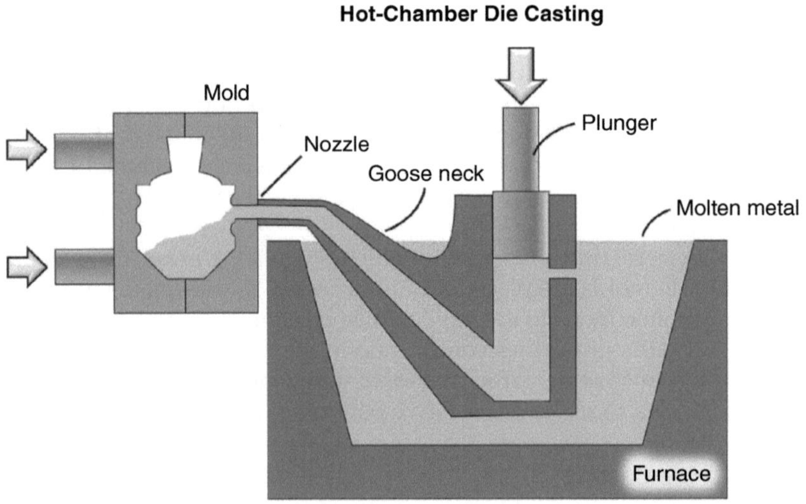

FIGURE 11.20 Hot-chamber die casting maintains a molten reservoir of metal so time between injections is minimized. This allows the process to be very fast, but it restricts the process to metals with very low melting points.

Die casting setups vary significantly in regard to the way the metal is prepared and introduced. The two main divisions are hot-chamber die casting and cold-chamber die casting. **Hot-chamber die casting** uses a plunger submerged in a reservoir of molten metal. As the plunger is advanced, it forces molten metal into the mold. When the plunger is retracted, it refills the chamber with molten metal and prepares it for the next shot. This equipment is shown schematically in Fig. 11.20.

This constant supply of metal allows for injection cycles to be very quick, reaching as many 1000 injections per hour for small castings. Larger castings cannot be cycled as quickly since solidification must occur and parts be ejected before the next cycle can begin. Remember, each shot can contain multiple parts, either of the same type or of complementary parts such as used in an assembly. This makes die casting capable of producing a hundred thousand small parts in a single day, all with superior quality and detail to most expendable mold casting methods.

Hot-chamber die casting is limited to metals with relatively low melting points and that do not react chemically with the mold equipment submerged in them. Although alloys of aluminum, copper, zinc, and other low-melting materials are commonly used, ferrous alloys are much rarer. To benefit from the quality and speed capabilities of die casting while avoiding thermal and reactivity issues, cold-chamber die casting is used.

In **cold-chamber die casting**, the metal to be cast is melted in a separate unit from the plunger and must be transferred to the molding equipment (see Fig. 11.21). In this case, the metal is poured into the injection chamber and the plunger is then advanced to deliver a measured shot volume to the mold. By melting the metal separately and introducing it shortly before injection, the risk of contamination and thermal degradation to the equipment is reduced. Ferrous alloys and more reactive metals can be cast by this method, but the compositions are typically similar to nonferrous alloys used in the hot-chamber process. Also, the time required to transfer the metal to the chamber increases cycle time, though the production rate is still very good.

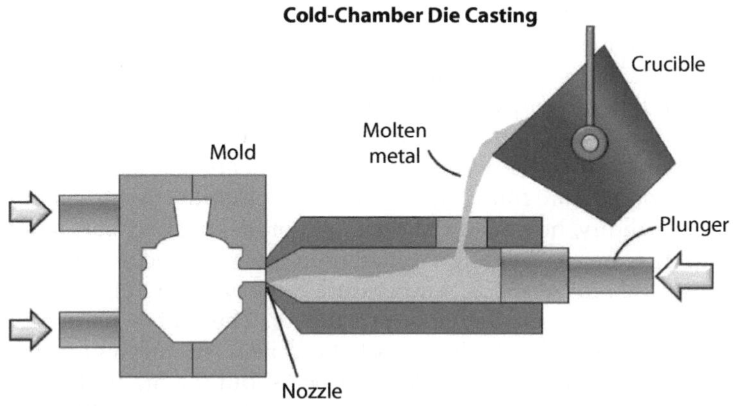

Figure 11.21 Cold-chamber die casting introduces the molten metal to a separate injection chamber right before the casting cycle. This allows for metals with higher melting temperatures to be used, but it increases cycle time.

11.3 Polymer Molding

Polymers present a distinctly different set of properties than metals when being processed in their "molten" state. Thermoplastic polymers do not undergo distinct melting, but instead they simply soften to a highly viscous form and thermoset polymers harden with temperature (review Secs. 5.3.2.5 and 8.2.6.1 about polymer characteristics at elevated temperature). This introduces unique benefits and disadvantages to molding operations for polymers. We will start by examining some aspects in molding polymers and then discuss common manufacturing operations.

11.3.1 Considerations in Polymer Molding

The first major distinction to be made is the difference in processing thermoset and thermoplastics. Thermosets cannot be poured or injected into a mold by heating the polymer, but that does not mean they cannot be shaped through molding. When heated, thermoplastics do not form a "traditional" liquid (a Newtonian fluid*), and they cannot be easily poured into molds, but they can be injected and otherwise molded. When discussing processing, then, we will have to consider these two categories separately, since they often have very different requirements.

Molding requires a transition from being soft and/or flowable to fill a mold to being rigid enough to maintain the mold shape after removal. Because temperature is a disadvantage to molding thermosets, they are often processed as a two-part mixture that hardens in the mold (e.g., epoxy, see Sec. 6.3.2.3), or they are dissolved in a solvent and hardened through evaporation of the solvent and/or the addition of heat. Therefore, polymerization can be an effective hardening mechanism in the molding of thermosets.

*Technically, whether a fluid is Newtonian or non-Newtonian is determined by whether the viscosity is dependent on the shear rate. For practical purposes, most low viscosity fluids (air, water, etc.) are Newtonian, while viscous materials (honey, paint, toothpaste, *molten polymers*, etc.) are non-Newtonian.

The relatively low temperatures used during molding also make thermosets more energy efficient to process, but not necessarily cheaper overall.

Thermoplastics can be significantly softened through heating, and they are most often molded significantly above their glass transition temperature (see Sec. 8.2.6.1). This makes them very formable and they can be injected through small passageways without prematurely solidifying. Heated molds are also used to prevent the polymer from solidifying before filling the mold. The mold must be sufficiently cool to allow the polymer to solidify, however. Polymers are often still very soft when removed from a mold, but they are rigid enough to maintain their molded shape.

As discussed in Sec. 8.5.3, temperature can be a mechanism of degradation to polymers. This presents a challenge in processing. The higher the temperature used with thermoplastics, the lower their viscosity becomes. This allows for the molding pressure to be reduced and aids in the molding process, but the higher the temperature and the longer the polymer is held at that temperature, the more it will degrade. This includes the breakdown of polymer chains, evolution of gas, loss of polymer additives, etc., which will negatively impact the properties of the polymer after it is molded.

11.3.2 Gravity Casting

The simplest method for molding polymers is gravity casting. This is functionally equivalent to permanent mold casting discussed for metals (Sec. 11.2.2.1) where a mold is prepared and the material is poured into that mold and allowed to solidify (Fig. 11.22). Although it can be done with molten thermoplastics, it is usually done with a thermoset that cures in the mold, sometimes by application of heat. This is really not much different from baking cupcakes or muffins. The ingredients are mixed, added to the mold, and then cured.

The process is very simple and low cost, but the time required for each component to cure can be relatively long, so large batches are often processed together or continuously, so more molds are needed for increasing the production rate. The process also allows for inserts, such as threaded components or any additional feature of a different composition, color, or property can be added. Likewise, reinforcement phases can be added to strengthen the final product.

Gravity casting can also be used to fill or encapsulate objects. For example, sensitive electronics can be covered with a polymer, and after it solidifies, the electronics will be protected from the outside environment (see Fig. 11.23). Some of the categories of casting include embedment, encapsulation, and potting. An embedment is when an object is completely covered with a transparent polymer. Encapsulation is the coating of electrical

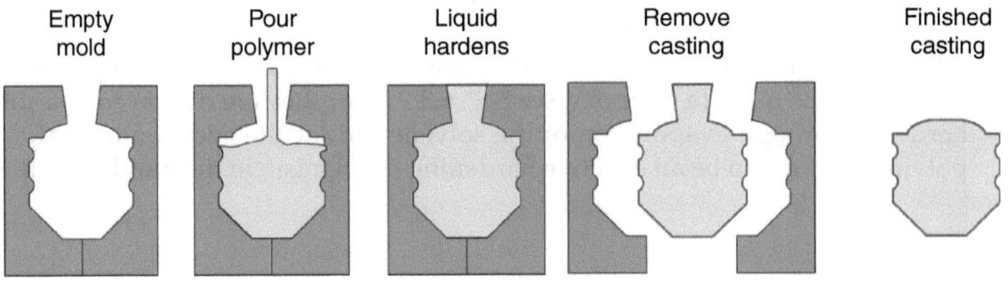

Figure 11.22 Gravity casting polymers is a very simple process. Objects can also be placed in the mold prior to casting to encapsulate them with the polymer.

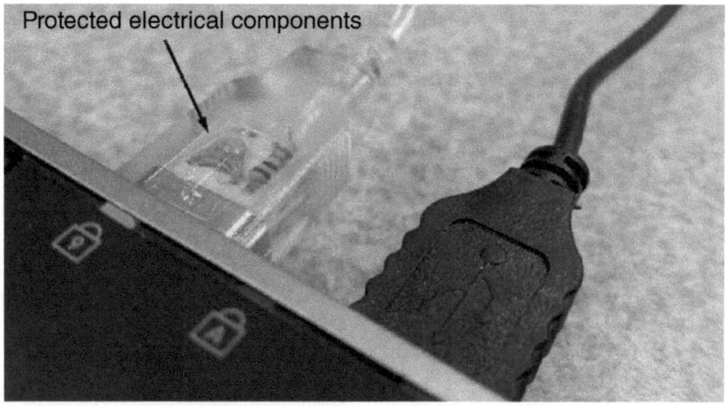

FIGURE 11.23 Because of the low temperature and stress, delicate components can be coated with a polymer to increase durability. Here, a USB cable made with transparent polymer is pictured alongside an opaque equivalent.

components, often by dipping. Potting is also used to coat electrical components, but the mold is part of the final product.

11.3.3 Injection Molding

Injection molding is the most widely used method for mass production of polymers and involves the forced introduction of a molten or liquid polymer to a mold chamber where it solidifies (Fig. 11.24). The process can be very fast and operate continuously for very high production rates and at low cost per part. The equipment and facilities for this type of processing are expensive and specialized, so only large runs of parts are justified. Once the major equipment is in place, the molds are the only investment, though they present a significant expenditure.

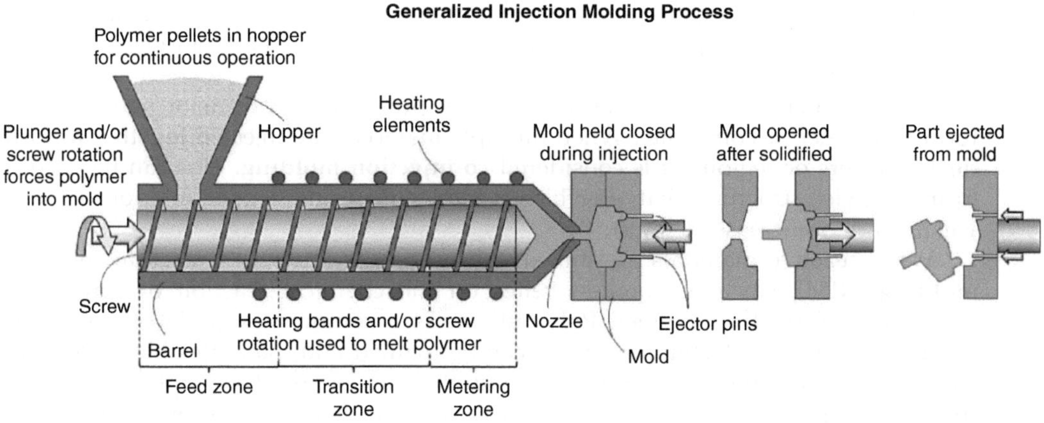

FIGURE 11.24 Injection molding may take many forms, but the general components and process are shown here. Some machines may use only a plunger and heating elements, others may only use a screw or it may be any combination. The commonality between them is that hard pellets are melted, injected, and solidified during the process.

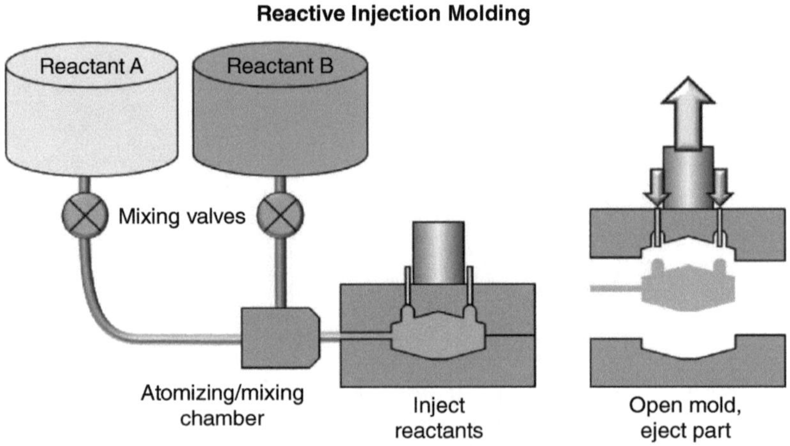

Figure 11.25 Injection molding can be done using multiple ingredients. Reactive injection molding combines reactants which polymerize when they mix.

The way the polymer is introduced to the mold varies, and some of the most common mechanisms for converting polymers, typically in the form of pellets, to molten plastic are by friction, external heat, or a combination of the two. The molten polymer is injected by screw (auger) action, by plunger, or by a combination of those. The polymer chemistry will determine the amount of heating, mixing, and force that are required to fill the mold efficiently. The machine size will determine the shot size (fractions of an ounce to 50 lb or more), clamping force (typically in tons), and the maximum mold size.

Injection molding is commonly done using thermoplastics, but it can also be done with thermosets. For thermosets, the polymer is heated once in the mold rather than in the barrel of the molding machine. Also, multiple materials (e.g., resin and hardener or polymer and foaming agent) can be introduced together in the mold or right before entering the mold (Fig. 11.25).

The multimaterial process can involve two polymers or precursor materials that form a polymer or composite. When multiple materials are injected together to form distinct layers or regions, it is considered **co-injection molding**. This can be used to form sandwich structures that often have a solid surface and foamed interior to reduce weight and/or increase strength.

When reactive materials are mixed, the process is known as **reaction injection molding** (RIM), since the process relies on the chemical reaction of constituent materials to form the polymer. Polyurethane is a common example of a polymer processed using this technique, and it can be foamed and made flexible or rigid by controlling cross-linking (more about polyurethane in Sec. 6.3.2.4). Applications range from beverage coolers to automotive parts, such as steering wheels and bumpers. To ensure proper mixing, the reactants are sprayed (atomized) together just before entering the mold. The multicomponent nature of this process also enables composites to be formed by mixing the reinforcement with the resin in the mold [sometimes referred to as **reinforced reaction injection molding** (RRIM)].

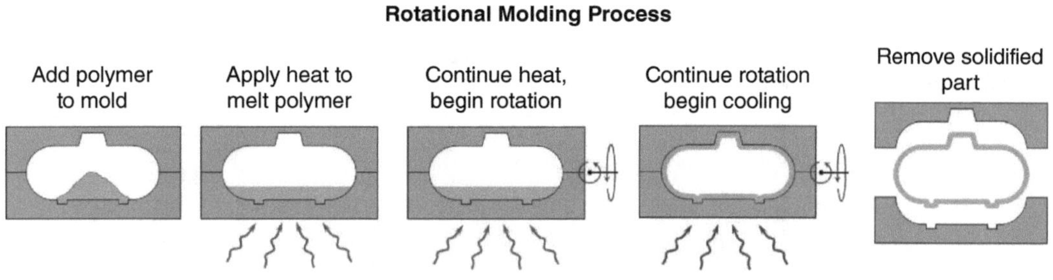

Figure 11.26 Rotational molding uses rotation around multiple axes to uniformly spread the polymer around the mold. This is well suited for large parts and those with complex surfaces.

11.3.4 Rotational Molding

Rotational molding uses a concept similar to centrifugal techniques (see Sec. 11.2.2.2) in that rotation is used to fill the features of a mold, but the rotation rate is much lower, and gravity is the dominant force that causes the material to spread (Fig. 11.26). Perhaps you are not much of a cook (that is what microwaves are for, right?), but you have probably at least seen someone drop butter into a hot pan. To spread the butter around the pan, it is tilted in various directions, and that is the general concept behind rotational molding.

A preset amount of polymer, as powder or liquid, is added to a mold and the mold is then closed. Depending on the type of polymer, heat will either melt or cure it as the mold is rotated about orthogonal axes. The mold is then slowly cooled while continuing the rotation to fully coat the interior of the mold. A benefit of this process is that it uses the entire amount of polymer and has no waste, but the cycle times tend to be longer (minutes rather than seconds). For these reasons, it is used in low-volume production and for large, complex parts such as boat hulls, canoes, large body panels, and storage tanks and bins.

11.3.5 Blow Molding

Blow molding uses pressurized air to expand material into the shape of the mold, very much like blowing up a balloon. The polymer must be very soft to achieve this, and the equipment must simultaneously secure the polymer at the air's entry point while allowing it to expand uniformly elsewhere.

In **injection blow molding**, a specifically designed preform, known as a **parison**, is molded first (Fig. 11.27). The parison is heated, clamped at top and/or bottom in the mold, and air (or inert gas) is injected to expand the polymer outward. The temperature must be correct or the polymer will be too weak or too rigid to correctly fill the mold. Also, since the entire polymer charge is contained in the parison, the wall thickness and distribution of plastic must be well controlled. There may also be mechanical displacement by the blowing tube where it physically stretches the parison before introducing the air (known as injection stretch or just stretch blow molding).

Plastic bottles are the most common product created this way, but other containers and hollow products may also be produced. Plastic bottles use a parison that looks very much like a test tube with a threaded top on it. During blow molding, the wall thickness decreases dramatically in the lower portion of the parison, but the threaded portion and neck, where it is held during blow molding, remain much thicker and stiffer.

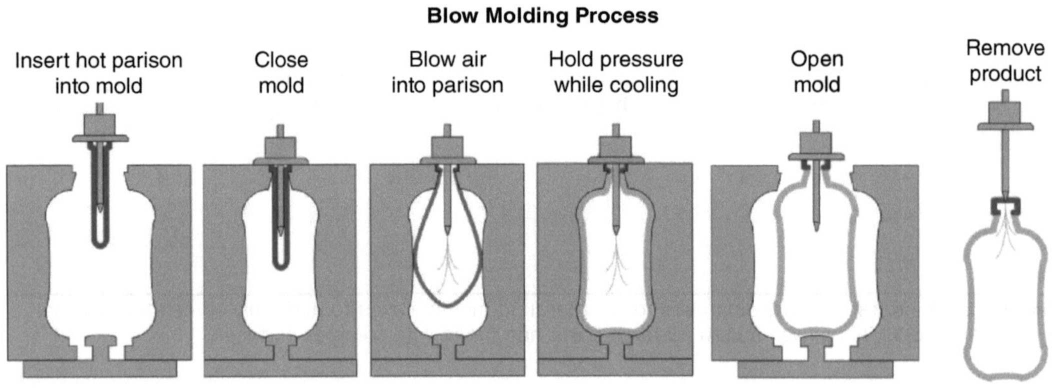

Figure 11.27 Injection blow molding is a common method for creating hollow products, such as containers. A preheated parison with appropriate dimensions is required.

In **extrusion blow molding**, instead of using an injection-molded parison, the polymer is directly extruded downward into the open mold as a thin-walled tube. Once the extruded plastic has reached the bottom of the mold, the mold closes, pinching off the top and bottom, and creating a seal. Air is introduced to expand the plastic into the shape of the mold. The general process is depicted in Fig. 11.28.

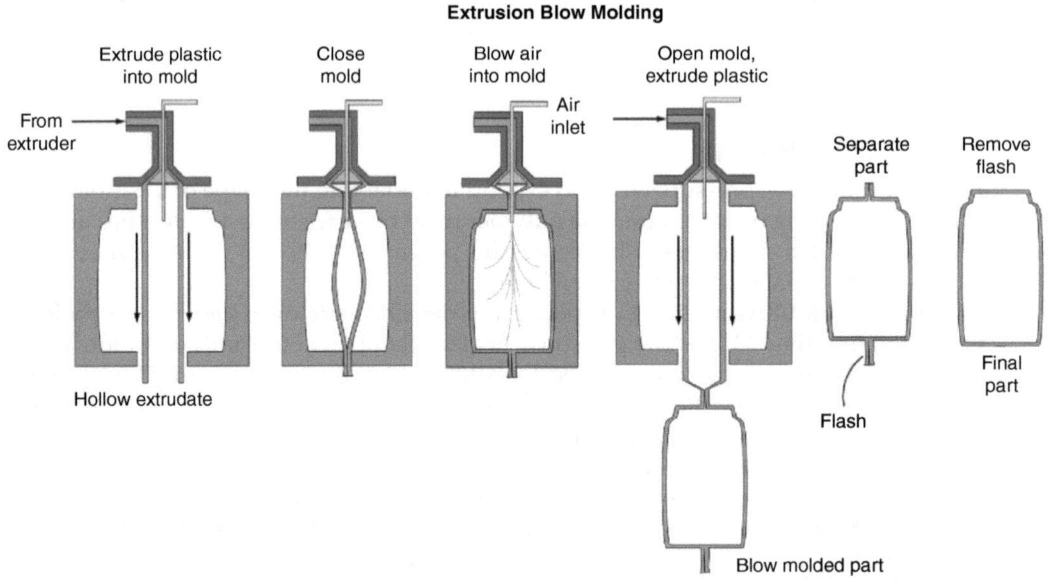

Figure 11.28 Extrusion blow molding is used to create hollow parts and can be continuously fed from an extruder. Unlike injection blow molding, it does not require a parison and can be used to produce very large parts.

The advantages of extrusion blow molding include the ability to make very large products and the elimination of injection molding as a prior step. Injection blow molding is a faster process, can create a more uniform wall thickness, and does not generate the scrap required in extrusion methods. The additional cost and complexity of injection blow molding is a drawback.

11.4 Ceramics

Many ceramics are characterized by a very high melting temperature and low ductility, which makes them difficult to melt and form. Some ceramics, especially glasses, can be molded in ways similar to polymers because they too go through a glass transition. Others have to be processed in a fundamentally different way. We will look at the most common methods for molding ceramics.

11.4.1 Glass Molding

Glass is similar to polymers in the way it behaves at high temperature. The temperature that is considered "high" is much different for the two (see Sec. 8.2.6 for more on that concept), but both become very viscous instead of melting to form a "traditional" liquid. When glass is heated (usually between 1500 and 2000°F or 816–1093°C) it gradually softens until it is quite formable, yet it still reasonably retains its shape.

Perhaps you have seen, or at least familiar with glass blowing. In that process, the glass is heated until it becomes molten, then a "gob" is collected on the end of a long tube. As air is blown into the gob, it expands, much like blowing a bubble with chewing gum. The glass can deform under relatively small amounts of force, yet it is stable enough that it does not collapse.

This makes glass particularly suitable for blow molding, just as used in polymer molding (see Sec. 11.3.5 and Fig. 11.27). This is the process by which incandescent light bulbs and other similar containers are made. Although the temperatures are higher, the process is identical.

Because of its brittle nature and high processing temperature, glass is sensitive to irregular cross-sectional thicknesses. As discussed in Sec. 8.2.4, stress will build up where uneven contraction occurs due to thermal gradients. In glass this results in cracking during cooling. To eliminate these stress glass products, especially those which combine varying section thicknesses (such as the handle on a pitcher), should be cooled very slowly, as process referred to as **annealing**. This can be achieved by cooling parts in a furnace at a controlled rate or by shutting off the furnace and keeping it closed until cool. Transparent parts with residual stresses can be identified by the way they refract light unevenly and possess a rainbow of colors upon inspection (in an otherwise colorless glass).

11.4.2 Slip Casting and Solid Casting

Where ceramics deviate most significantly is in their casting processes. Although the term molding is typically used for highly viscous, nonmetallic materials, casting is often applied to low viscosity ones. Since ceramics have such high melting points, they are processed at low temperature, and instead of being melted, they are dispersed in a liquid to form slip. **Slip** is a mixture of clays and other materials suspended in water to

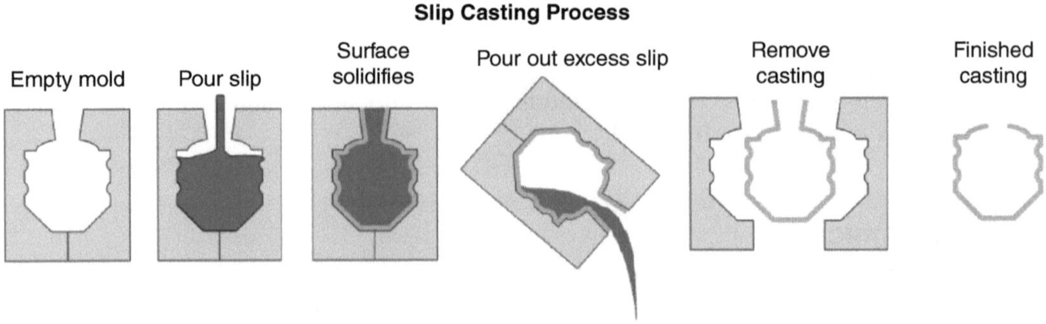

FIGURE 11.29 The slip casting process begins with pouring the slip into a plaster mold. The wall of the casting solidifies by moisture loss, and the excess slip is poured out.

allow those materials to be poured into molds. This mixture also often contains a **deflocculant,** which is an additive that helps keep the particles suspended.

The way slip castings solidify is through water loss (see Fig. 11.29). This is accomplished by a combination of evaporation and the use of a permeable mold, such as plaster. The plaster will absorb water from the slip and the surface in contact with the mold will eventually become denser and stronger. Until the water is sufficiently removed, the casting will be very soft, so the time in the mold is important. Too little time and the casting will collapse upon removal. Too long and the casting can crack (if constrained by the mold geometry) or the process will simply be inefficient.

The time required for the product to be ready depends on the initial water content, the moisture level in the mold, and the ambient conditions (temperature and humidity) during drying. Very thin (i.e., dilute) slip can be quickly and easily poured, but it will require more time to reach the same structural requirements as more water must be removed. On the opposite end, molding clay, like used in sculpture, can be thought of as very (very!) thick slip,* and will dry to structurally stable form more quickly but does not replicate mold details as well.

There are two basic ways that ceramics of this type can be produced. Either a mold is completely filled and dries or it is filled temporarily and the excess liquid is poured out leaving only a shell or skin of material along the mold walls. These methods are analogous to permanent mold casting (see Fig. 11.17) and slush casting (see Fig. 11.18), as discussed for metal casting (see Sec. 11.2.2.1). Products that are to be solid should account for the large amount of shrinkage that occurs during drying, and thick mixtures are often used. Products created by these methods are shown in Fig. 11.30.

11.4.3 Firing Ceramics

Ceramics produced using slip are not engineering ceramics like silicon nitride, aluminum oxide, or tungsten carbide (see more on these in Sec. 12.4). Rather, they are ceramics in the same sense that pottery, dishes, or decorative items are "ceramics."

*Clay has different ingredients than slip, so simply adding water is not equivalent.

Figure 11.30 This slip cast vase uses a plaster mold. Slip is poured into the mold, allowed to dry for a desired period, and then the extra slip is poured out leaving a hollow object. Solid casting uses the same basic process, but the mold is completely filled and allowed to dry.

After casting, these **greenware** products are removed from the mold, allowed to finish drying throughout and then fired for strength (known as **bisque firing**). During this firing step, the elements of the mixture fuse together (known as **vitrification**). Until fired, the ceramic has very little strength. After firing, however, the products are quite strong. The temperature of firing depends on the chemistry of the ceramic, and temperatures are often referred as cone numbers. These steps are shown in Fig. 11.31. Glazing, another firing step, is discussed in Sec. 20.4.1.1.

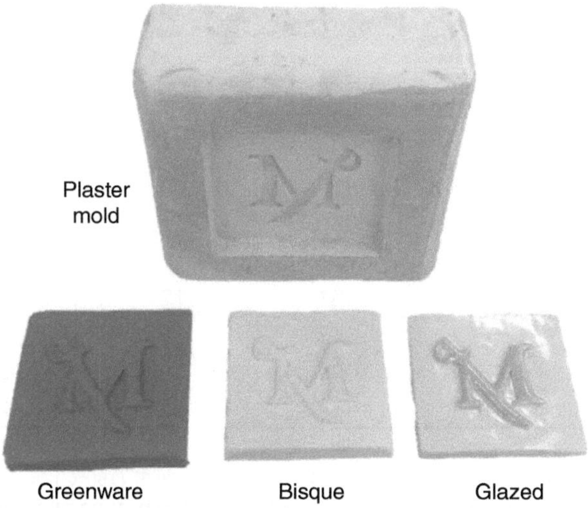

Figure 11.31 Ceramics are often created in a "green" state where they are still pliable. They are dried and fired to create bisque. Decorative and functional properties are included during glazing and firing again. These stages are illustrated in the making of tiles, shown here.

The **cone number** is the position on a temperature scale that corresponds to the point at which a **pyrometric cone** softens and slumps. These cones are simple devices that indicate the temperature visually, and multiple cones with varying softening temperatures can be used to determine a general temperature range. Though modern furnaces have precise temperature control, effectively making the cones redundant and unnecessary, the temperature controllers on **kilns**, or furnaces specifically for the purpose of firing ceramics, may still allow the user to choose a cone number rather than a specific temperature.

The relationship between cone numbers and actually temperatures is listed in Table 11.1. The manufacturer of the pyrometric cone will typically provide cone numbers and temperature equivalents for their products, and since the cone bends over slowly, the rate of temperature increase is also important. For a slow ramp rate (27°F (15°C)/h), the same cone number will indicate a lower temperature than at a high ramp rate (270°F (150°C)/h) since the temperature will not change as much during the period over which the cone deforms.

	Temperature Equivalent[a]			Temperature Equivalent[a]	
Cone Number	°F	°C	Cone Number	°F	°C
022	1094	590	04	1971	1077
021	1143	617	03	2019	1104
020	1180	638	02	2052	1122
019	1283	695	01	2080	1138
018	1353	734	1	2109	1154
017	1405	763	2	2127	1164
016	1465	796	3	2138	1170
015	1504	818	4	2161	1183
014	1540	838	5	2205	1207
013	1582	861	6	2237	1225
012	1620	882	7	2269	1243
011	1641	894	8	2295	1257
010	1679	915	9	2320	1271
09	1706	930	10	2336	1280
08	1753	956	11	2381	1305
07	1809	987	12	2399	1315
06	1855	1013	13	2419	1326
05	1877	1025	14	2458	1348
05-1/2	1911	1044	15	2523	1384

[a]The listed temperatures correspond to a heating rate of 270°F (150°C)/h using Orton brand, self-supporting pyrometric cones. Variations on the cone type and heating rate will affect the temperature equivalent.

Table 11.1 Pyrometric Cone Ratings and Equivalent Temperatures

11.5 Why It Matters

Few assemblies do not incorporate cast or molded parts. The use of casting and molding eliminates the complexity of assemblies by allowing for multiple parts to be combined in a single component with a more complex geometry. Although more complex, the use of a mold allows for rapid processing with relatively little waste. Whether a coffee pot, a computer, or a car, all benefit from the use of curved surfaces, thin sections, and rapid production allowed by casting and molding. Many applications would be difficult or even impossible to satisfy using other methods of production such as machining. Casting and molding operations are essential manufacturing techniques to provide quality and low cost together.

Casting and molding do have requirements regarding the shape of the parts being created. This is because the part or pattern must be removed from the mold with relative ease so as not to damage the part or the mold. For this reason, cast and molded parts will often have sloping sides to create positive draft and will have smooth transitions between surfaces to reduce stress concentrations. The degree to which these are required is determined by the material being processed and the method being used.

Expendable mold casting allows for very complex geometries to be created. Since the mold is destroyed, extracting the part is simplified. This is especially true in situations where a sacrificial pattern is also used. This includes sand cores, which make complex internal surfaces possible. Without these techniques, common items such as engine blocks would be much more difficult to produce. Sand casting also reduces up-front mold costs and is suitable for low volume and high volume production. Drawbacks include the surface quality and dimensional accuracy, but these are dependent on how the process is performed.

Permanent mold castings give a much better surface and dimensional accuracy, but the equipment and mold costs are often quite high in comparison to expendable techniques. For this reason, a large number of parts must be required to justify the up-front investment. Methods that use pressure to introduce the material to the mold, such as die casting and injection molding, are able to achieve very good quality and high production rates.

The way the material fills the mold can be adjusted in many ways. Hollow and solid products can be created through the use of rotation or air pressure. Some processes may be done under inert atmospheres (especially reactive metals), and temperature can be used to soften or harden a material (in the case of polymers). Glasses can be processed similarly to thermoplastic polymers, but crystalline ceramics cannot. The use of dehydration to set ceramics in the "green" state and subsequent firing to harden the product has been in place for thousands of years, but modern technology has transformed it from an artisan practice to high volume production.

You will be hard pressed to look far for a cast or molded product you rely on every day. Maybe it is a coffee mug, a water bottle, a toothbrush, a hair brush, your cell phone, or the steering wheel of a car. In part or in whole, we rely on casting and molding. Even something as humble as a toilet would be a matter of great complexity without casting and molding. The complex curves and hollow interior present a substantial challenge to create by other means, especially out of a hard and impermeable material like porcelain. Something this complex becomes quite "mundane" when good decisions and design are used. It is something to think about next time you are just sitting around.

11.6 The How and Why of Casting and Molding

11.6.1 Chapter Review

1. How are casting and molding operations some of the most common, even in daily life?
2. What is the difference between casting and molding?
3. What are the benefits of casting or molding parts?
4. Describe the function of the following mold features:
   a. Sprue
   b. Runners
   c. Gates
   d. Riser
5. What is draft in regard to mold design? Why is it so important?
6. How can draft be a disadvantage for part design?
7. How can complicated parts with multiple reentrant features be accommodated through mold design?
8. How does shrinkage affect mold design?
9. What type of defects may occur because of shrinkage?
10. Describe what properties a riser should have to function well.
11. Describe the position and relationship to the mold cavity of an open, live side riser.
12. Why is casting popular for complex metal parts?
13. Name two basic design considerations a mold for metal casting must satisfy.
14. What is meant by a mold being expendable?
15. In sand casting, what is special about the sand used for the mold?
16. What processes are used to prepare the sand in sand casting?
17. What is a sand core used for?
18. How does the sand for a core differ from mold sand? How is it given its shape?
19. Describe how the cope, drag, and flask are related.
20. What is the difference between a split pattern and a match-plate pattern?
21. Given some advantages and disadvantages of common mold materials.
22. In your own words, summarize the sand casting mold preparation process.
23. Describe how automatic matchplate mold differs from hand ramming molds.
24. How can sand molds be prepared faster and with greater quality than hand ramming?
25. Describe the DISAMATIC process and how it can save time in sand casting.
26. How is investment casting similar to sand casting? How is it different?

27. Why is investment casting also known as lost wax casting?
28. What are the primary advantages of lost wax casting over sand casting?
29. Describe the difference between full mold and lost foam casting.
30. Compare expendable and permanent casting processes in terms of cost and throughput.
31. What kind of parts does slush casting create?
32. Describe the differences between centrifugal, semi-centrifugal, and centrifuge casting.
33. Where do impurities concentrate in centrifugal, semi-centrifugal, and centrifuge casting?
34. Why is die casting popular for small, thin objects?
35. What are the material limitations associated with die casting (the molten metal)?
36. Are sand cores used in die casting? Why or why not?
37. What are the benefits and the disadvantages of hot-chamber die casting over cold-chamber die casting?
38. What differences do molten polymers exhibit when compared metals?
39. Polymers are cast (liquid) and molded (highly viscous). What is the difference in the polymers used for these methods?
40. What determines the soonest a polymer can be removed its mold after casting or molding?
41. Why are thermosets difficult to mold through heating?
42. Describe the gravity casting process.
43. What are the differences between encapsulation and potting?
44. What aspects make injection molding a popular option for polymer processing?
45. How is injection molding machine size determined?
46. How do thermoplastic and thermoset injection molding differ?
47. What is co-injection molding?
48. How does the RIM process work?
49. What is the principle behind rotational molding?
50. What types of objects are commonly created by rotational molding?
51. What types of products are made using blow molding?
52. What are the primary differences between injection and extrusion blow molding?
53. What is a parison? How does its design affect the final product?
54. Why are ceramics less common in casting and molding processes?
55. Which types of ceramics are most often molded?
56. What is the difference between slip casting and solid casting?

57. How does ceramic slip harden?
58. What is the purpose of firing ceramics?
59. Describe the greenware and bisque stages of ceramics.
60. What role does a pyrometric cone play in the firing of ceramics?
61. Why does the heating rate affect the cone number?

11.6.2 Concept Connections and Extensions

1. Notice the plaster mold in Fig. 11.31. The "M" is backward. Why? What does this indicate for mold design in general?
2. What the benefits of gravity casting over pressurized casting? Why do so many processes use pressure?
3. If you are tasked to design a pattern for sand casting, list some of the considerations you must make in its design. This should include geometry, material, and other factors.
4. Investment casting and sand casting are both expendable mold processes. Why is investment casting a better option for fine detail and surface finish?
5. What is the benefit of centrifugal and other rotational methods in regard to purity? What other strategies can be used?
6. Describe the similarities and differences between slush casting and slip casting.
7. Which method, slip casting or solid casting, do you think is better for fast production? Why?
8. Consider the structure of polymers (Sec. 5.3). How might this affect the properties of an injection-molded component with fine features?
9. Have you ever noticed all of the fine "hairs" on a new tire? What mold feature produces them, and how does this mold feature improve quality?
10. What do you suppose the origin of using "cones" to verify temperature in a furnace was? Are there better methods now?

CHAPTER 12
Forming

Forming was among the earliest methods of manufacturing developed. **Forming** is accomplished by simply bending, hammering, or otherwise permanently changing the shape of an item through deformation. It does not require high temperature or complex tools, although these can extend the utility and improve the production rate. It encompasses the work of a blacksmith or jeweler in centuries past as well as highly automated controls to produce aerospace components in the most modern engineering marvels. We will touch on the most popular methods and basic considerations in forming operations, but new techniques and deeper understanding of the principles are continually progressing.

12.1 How It Works

Forming is a vital process for modifying the shape of standard stock or previously processed materials. Forming processes may be performed as the first and only means of processing, or they may be used as a portion of multi-step processing regime. The basic requirement in this manufacturing method is that the material be formable. A **formable** material is one that can be reshaped by the application of force, and highly formable materials can undergo significant deformation at low to modest temperatures. As discussed throughout Chap. 7, there are various forces that arise during processing and materials will respond to them differently. Chap. 8 explored some of the thermal effects on materials, and both force and temperature are critical to achieving good results during forming. Although a variety of materials can be formed, some must be heated considerably, and others can be easily processed at room temperature.

12.1.1 Considerations in Forming

Most of the processing considerations associated with forming are dealt with in Chap. 7, so if you are unfamiliar with the concepts of stress, strain, and plasticity, some time spent reviewing those will be time well spent. Forming relies on plastic deformation. During plastic deformation material "flows" and atoms change bonds. Going all the way back to Chap. 2 (especially Sec. 2.3), it was established that bonds can be rigid or nonrigid, and rigid bonds do not allow atoms to move very easily.

Materials characterized by ionic and covalent bonds are typically not processed through forming processes due to their rigid bonds. The main exception is polymers, which have a covalently bonded "backbone," but they do not deform by the breaking of covalent bonds. Rather, they deform by the sliding of individual molecular chains (Sec. 7.4.3). That leaves most ceramics and extensively cross-linked polymers unsuitable for forming operations due to being brittle.

12.1.1.1 Temperature Effects (Cold Work and Hot Work)

For some materials, heat can be used to increase ductility. This is commonly used in metal forming operations, even in ductile metals. Cold work is discussed in Sec. 9.2.1 as a means to mechanically condition metals. The deformation process causes internal, atomic-scale defects to accumulate and strengthen the metal. This also limits the extent to which a metal can be deformed before breaking and increases the force required to conduct each subsequent forming operation (Fig. 12.1).

Hot work involves forming operations conducted above the **recrystallization temperature** of the metal. For steel this temperature is above 2000°F (1093°C), and, in general, it is done around a homologous temperature of 0.6 [see Eq. (8.3) and associated discussion]. Recrystallization is described in Sec. 9.2.1.2, and involves atomic restructuring of the metal to reduce internal strain. While cold work seeks to introduce strain, and the associated strengthening it produces, hot work allows for significant changes in shape to be accomplished without breaking. What constitutes "significant" is material dependent, with some metals, such as gold, being able to be hammered into foil without breaking and others, such as zinc, breaking after relatively little deformation. The concept of ductility and the measurement of it are discussed in Sec. 7.4.3.1.

Hot work allows for nearly unlimited dimensional changes and reduces the force required to deform the material. This reduces the rigidity and power requirements on machinery, but it also requires the material to remain hot during the process, which also requires energy input before and/or during the forming operations. Hot working also increases the reaction rate (proportional to the temperature) with surrounding air (or oxidation, as discussed in Sec. 8.5.2.1), so dimensional accuracy and surface quality is not as good. Also, since internal stresses do not accumulate during hot work, the final strength of the material is not as high, but the ductility is better. The application will determine if this desirable or not.

Also of concern is the temperature uniformity. Deformation processes result in heat generation. If you bend a piece of plastic rapidly back and forth, you can actually feel it get warmer at the bend. In forming operations the metal can get hotter during deformation, particularly for cold working, but in hot work it often cools down as it contacts the lower temperature tooling. This leads to a temperature gradient where the center of the

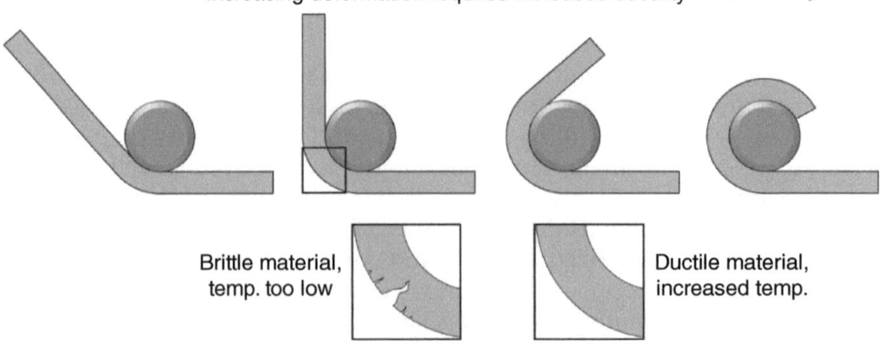

Figure 12.1 The amount of deformation a material can undergo is determined by its ductility. An increase in temperature can often increase ductility, but some materials, such as engineering ceramics and thermoset polymers are unsuitable under most conditions.

part has a higher temperature, and during final cooling this can result in distortion or even cracking. Process control to mitigate these results must be undertaken.

Finally, to achieve a balance of cold and hot working properties, an intermediate temperature, which is below the recrystallization temperature but well above room temperature, can be used. This is sometimes referred to as "warm working," but that name is a bit of a misnomer since cold working is technically anything below the recrystallization temperature and hot working is anything above, but it provides a way to distinguish the process. Using elevated temperatures allows for greater ductility than regular cold working and better dimensional and surface qualities than hot working. Similarly, forming operations may incorporate both hot working and cold working by starting with hot working for the bulk of the cross-sectional reduction and finishing the part with cold working to achieve a balance of efficiency and material properties.

12.1.1.2 Microstructural Benefits of Forming

Forming processes are volume conservative. They do not remove material, but rather, they reorient the material through force. The deformation of the material will cause a microstructural refinement (metals, see Sec. 9.2) or molecular reorientation (polymers, see Sec. 10.2.1). Ceramics are often only formed when distributed as a slurry or paste and the particles are disconnected and do not show as great of a dependence on forming.

In cases where the forming process does affect the microstructure, the resulting material displays and increased strength, either from an increased molecular alignment or decreased grain size and increased defect density. This gives forming operations an advantage over other processes such as machining and casting. For example, in closed-die forging (discussed in Sec. 12.2.1.2), the shape of a component is created by forcing the material, as a solid, into the desired shape. This forces the material to align microstructurally, and can be done in a way to orient the strongest direction with the force applied during use. The basic structural comparison among these methods is given in Fig. 12.2.

12.1.1.3 Deformation Rate

The deformation rate includes the strain rate, or how fast the material changes dimensions, as well as the overall deformation by each operation. Strain rate is entirely time dependent, while strain is dependent on the amount of deformation compared to the original size [see Eq. (7.4) and the associated discussion]. For instance, a high strain rate is achieved if a bar of steel is reduced from 1 to 0.95 in thick in 1 ms, but the overall strain is not large (5%). The reason this is important is because forming operations can be done at varying rates. If a large deformation is done quickly, it is more likely to cause the material to fail.

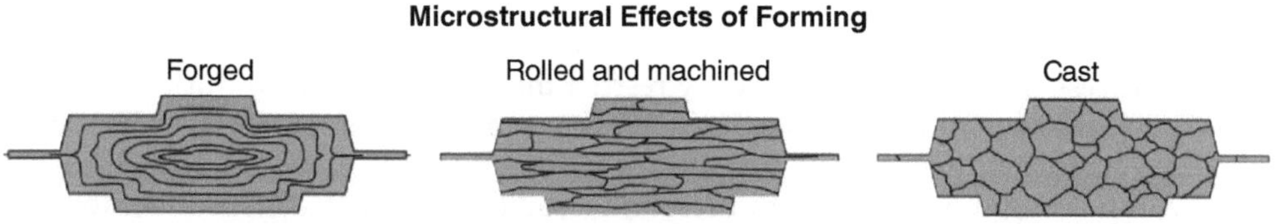

Figure 12.2 Forming processes are volume conservative, and they reshape the existing structure as a solid. This results in an alignment of structural features with direction of deformation and results in higher strength.

There are two common approaches to making large deformations. They are either done slowly or are done in multiple steps. Slow deformation gives time for defects to move through the material nondestructively, and small deformations have the benefits of requiring less force and give the material time to "relax." Temperature is still very important. Cold work will continue to accumulate internal stress and time is not enough to allow for the material to release internal energy. Higher production rates can reduce the cost to the consumer, but excessive speeds can be detrimental to the quality. In multistep forming processes, the steps typically occur within seconds of one another, so hot work is used to maintain ductility and reduce strain rate sensitivity.

If you want to reduce a bar from 6 to 1 in. in diameter, it can be done in a single step, but it will be difficult in practice and possibly detrimental, especially if the metal is not highly ductile. It will require much greater force to create that extent of deformation, and the accuracy achievable will likely be inferior to processes such as forging and rolling which are done more gradually. The equipment requirements, then, are a major driving force in using multiple operations. Although the equipment may require more space, cost is reduced and reliability increases.

12.2 Forming Metals

Much of the preceding considerations pertained to metals, though temperature and strain rates are universal processing considerations. Broadly speaking, metals are highly ductile and well suited to forming operations. Whether aluminum foil, automotive body panels or soda cans, products we use every day relay on the formability of metals. Despite the characteristic ductility of metals, there are many that are not highly formable.

This is especially true of alloys that have been thermally conditioned for increased hardness (see Sec. 9.3). For instance, 6061-T6 aluminum, a common structural aluminum alloy (see Sec. 4.3.1), has been precipitation hardened to increase strength, but in doing so it loses ductility. There are specific grades of aluminum that are very good for forming operations, or alternatively, a grade of aluminum that can be hardened should be formed *before* the hardening stage or the area of maximum deformation should be heated prior to or during the forming operation. Heat is critical to forming processes, and in metals, it may result in a variety of structural changes.

12.2.1 Metal Forging

Forging includes processes that use compressive force to deform a material in the shape dictated by the tooling. The tooling, or dies, can be very general in shape (flat, round, "V"-shaped, etc.) or they can have a specific geometry, very much like the molds used in permanent mold casting. Both methods start with an **ingot**, or workpiece with a general starting volume and geometry.

12.2.1.1 Open-Die Forging

A process that does not specifically restrict the flow of the metal during deformation is considered **open-die forging**. The metal can flow unconstrained laterally, but the dies may have some general shape. If the lower die cannot adequately support the workpiece, workholding must be used to hold and position the ingot. This can be done by hand (using tongs!) in small operations, or it may be done by industrial forklift in multi-ton ingot operations. Workholding also allows quick and accurate positioning of the ingot, which is especially important when drawing out a part (see Fig. 12.3). Open-die forging

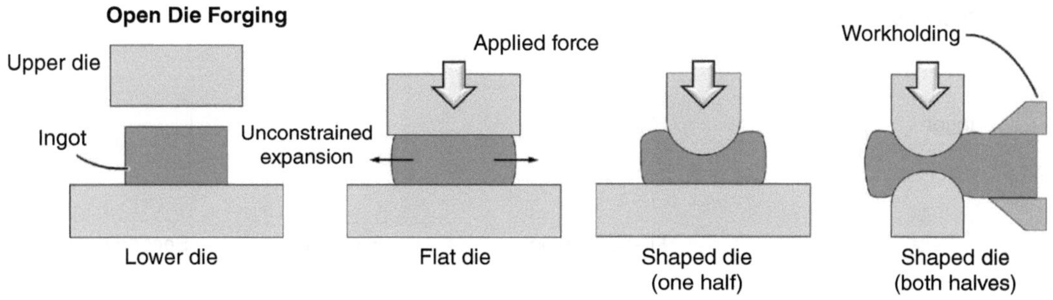

FIGURE 12.3 The two most basic classifications of forging are open- and closed-die forging. Open-die forging has simpler equipment and produces general shapes.

can accommodate significant variation in the initial ingot size and shape, and it can actually be used to *produce* ingots for further processing.

As you can see in Fig. 12.3, open-die forging can be used to create a variety of shapes, but none with a high degree of complexity. Some of the basic functions achieved in open-die forging are upsetting and drawing out. **Upsetting** is any forging operation that increases the cross-sectional area, and **drawing out** decreases it (see Fig. 12.4). Because forging is a volume conservative process, the length must decrease if the cross-sectional area increases and vice versa. The initial ingot size should take these into account.

Drawing out must be performed using multiple pressing operations. To create a part with a uniform cross section, these operations are conducted along the length, and the ingot must be rotated as well. For example, to create a square cross section, the ingot must be rotated in 90° increments. To create a round cross section, the ingot must be rotated in small increments, the exact angle of which is dictated by the required accuracy.*

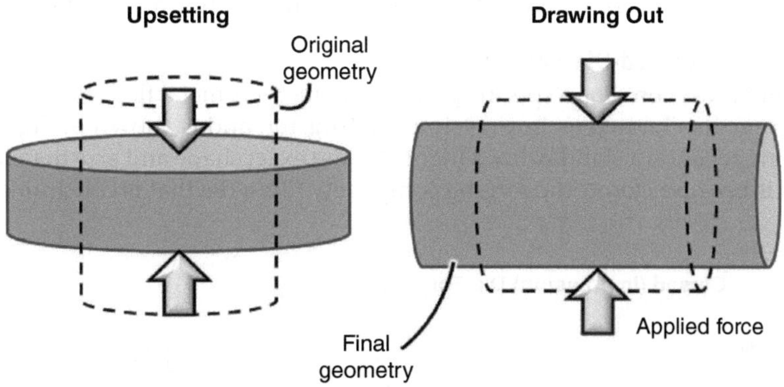

FIGURE 12.4 When material is deformed during forging, the cross-section changes. An increase in cross-sectional area during forging is known as upsetting the part, and a decrease is known a drawing out. Both methods result in a change of length inversely proportional to the change in cross-sectional area.

*It is not practical to create a smooth and highly round part by this method. Rather, a multisided polygon (e.g., octagon or decagon) can be created and later machined or otherwise processed if needed.

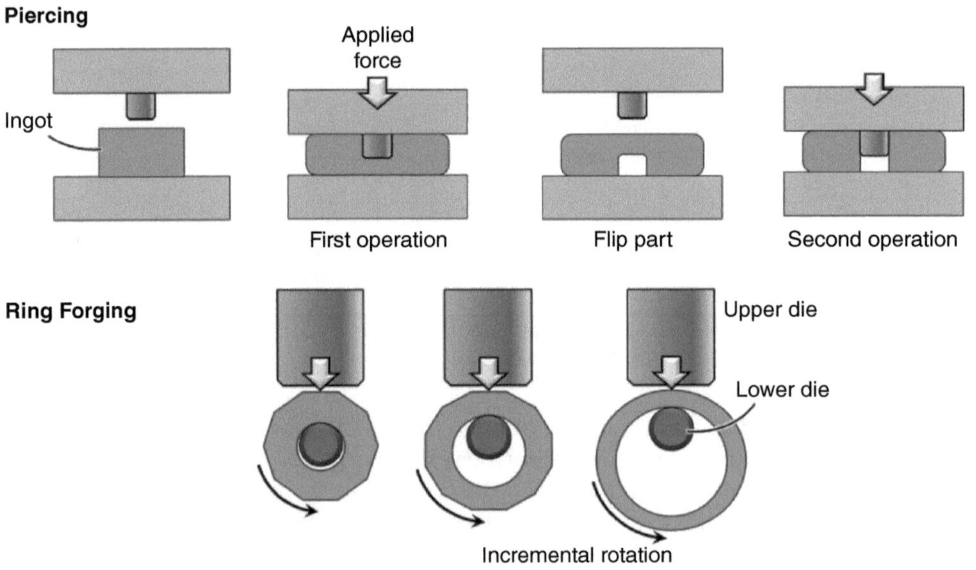

Figure 12.5 Some specialized processes in open-die forging include piercing and ring forging operations. Piercing must be performed before ring forging to create the hollow center.

Piercing produces a divot (two piercing operations on opposite sides create a hole) and after a hole is creating, a **ring forging** can be completed. Rings can also be created through **ring rolling,** which continuously applies force to the stock while rotating it, usually in a horizontal plane. These processes, shown in Fig. 12.5, can be done on very large or small parts. Open-die forging can accommodate parts that are many tons in weight, and most products created by this method require significant machining afterward.

12.2.1.2 Closed-Die Forging

When the die contains a specific pattern that restricts metal flow, the process is **closed-die forging**, also known as **impression die forging**, and is shown in Fig. 12.6. Closed-die forging relies on a standardized ingot of more exact shape and size than open-die forging. This is because closed-die forging completely fills a die that is constrained from all sides.

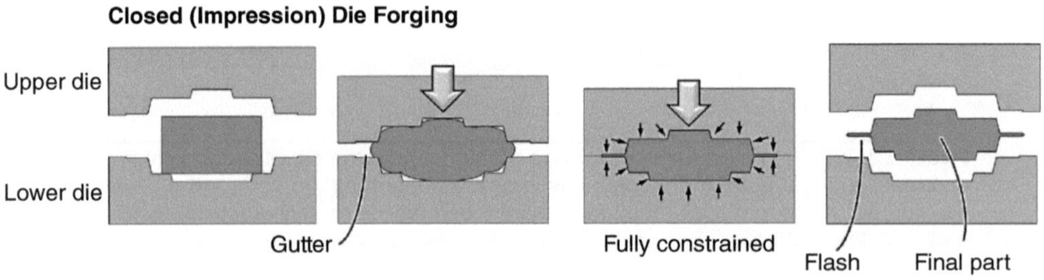

Figure 12.6 Closed-die (or impression die) forging forces material into a mold and must allow for excess material to squeeze out (in the gutter). This creates flash that must be removed during later processing.

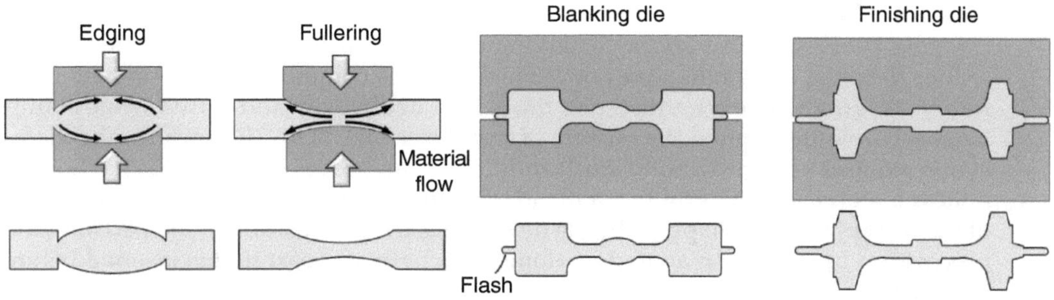

Figure 12.7 Closed-die forging typically begins with steps to preform the ingot before the final shape is imparted. Dies that gather (bulge) material are used in edging operations, and those which distribute it outward are used in fullering operations.

Closed-die forging can produce complex shapes, but the geometry has the same basic design considerations as casting (see Sec. 11.1.1) since the part has to be efficiently removed from the mold after it is forged. Forging is often a multistep process. Especially with complex impression dies, the force required to force the material into fine detail would be so great in a single process that it would prematurely wear the die. To accommodate greater detail, a die may contain several patterns which progressively increase the deformation and detail in a workpiece.

The first operation(s) in forging may include edging and/or fullering to distribute material. **Edging** operations pinch off an area and collect material within, and **fullering** distributes material outward to form a thinner section. Next, **blocking dies**, which have a rough, oversized shape to preform the metal, are used. The final forging step uses **finishing dies** to give the part its final geometry. These operations are shown in Fig. 12.7. After the forging process is complete, secondary operations such as machining and finishing are applied. These include **trimming**, or the removal of flash, which can be significant.

12.2.1.3 Forging Operations and Equipment

Forging can be accomplished by quick, repeated blows, known as **hammer forging**, or it may be achieved by the slow application of increasing pressure, or **press forging**. Open- or closed-die forging may be called **drop forging** if the operation involves an upper hammer (die) being "dropped" or otherwise propelled downward to impact the workpiece. The forging process involves massive force and the equipment must be large and capable of absorbing the transferred energy. The force may be developed by a rotating flywheel, compressed gas or steam, hydraulic fluid, gravity or a combination, and the way the force is transferred to the die may vary as well (i.e., piston, screw, etc.).

The equipment can be relatively simple with only small variations in processing conditions possible, or there may be significant computer control of motion, temperature, and force. While quick impacts, such as in hammer forging, will introduce heat from the rapid deformation, slower processes, such as press forging, will remove heat. Lubricants can help reduce heat loss and aid in metal flow and removal of parts from impression dies, but ingots may need to be heated during processing if they become too cool.

12.2.1.4 Swaging

Swaging processes can vary in method and result, but in general they are used to reduce the cross section of a part or taper the end. A common production method is **rotary swaging** where a multisection die is repeatedly open and closed while simultaneously rotating around the outside of a part (see Fig. 12.8). This ensures the diameter is reduced uniformly. Solid stock and tubing can be processed by this method. In hollow materials, such as tubing, a mandrel setup can also be used to expand the end of the workpiece. The shape of the die determines the final feature of the part, and the process can be started on an end or along the length of a part that is inserted before the machine is started.

12.2.1.5 Upset Forging

Upset forging is a process by which a section of workpiece can be upset (increase the diameter, as shown in Fig. 12.4). This process "mushrooms" out only a section of the piece, and is useful in the manufacture of components that have significantly varying diameters, such as valve stems in engines and bolts and other fasteners. Because fasteners are produced in such quantity, upset forging is responsible for more parts being produced than any other forging technique. It can also be done very quickly, often employing multistation dies where the parts are progressively formed until the final geometry is achieved.

Upset forging is often done in a horizontal direction to facilitate high-production rates, but it can be performed vertically as well. Some general process considerations include the geometry to be achieved and the amount of deformation required. For instance, if the unsupported material is too long, instead of cleanly deforming, it will buckle instead. This limits the change in diameter to about one and a half times the

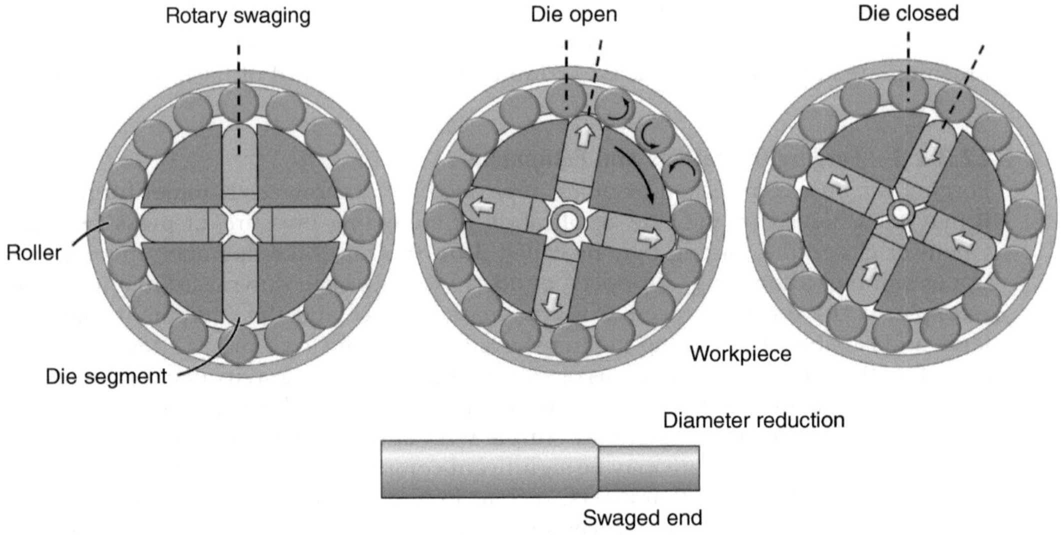

Figure 12.8 Rotary swaging is often used to reduce the diameter on the end of a component. It can be used to create tapers, straight reductions or radii as well as to increase the diameter of tubing using an alternative setup.

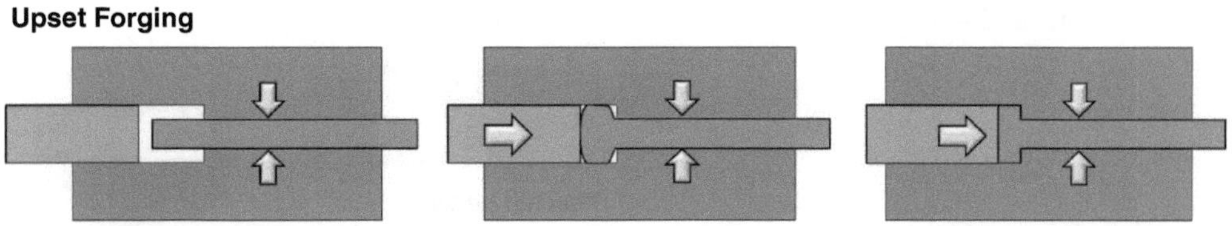

Figure 12.9 Upset forging is used to shape a portion of material to increase the cross-sectional area. The process may be done with single impact or it may be done incrementally using several dies of varied geometry.

length of the unsupported material, and likewise, the unsupported length should be no more than three times the diameter. This process and the design concepts are shown in Fig. 12.9.

12.2.1.6 Roll Forging

Roll forging is a process where a feature of limited length is added to a part using a rolling operation. Rolling is discussed separately in Sec. 12.2.3, but roll forging is distinct in that it does not produce a component of constant cross-sectional geometry. Instead, the rollers have an asymmetric shape, which advances the stock and simultaneously deforms it to create the desired feature(s), as shown in Fig. 12.10. The process can be performed multiple times (continuously) along a single piece of stock, or, more often, it is performed on discrete parts.

12.2.2 Metal Extrusion

The **extrusion** process for metals involves forcing a solid material through an opening smaller than the starting cross section. The shape of the opening determines the shape of the extruded material, also known as the **extrudate**. Many forms can be created so long as the desired part has a constant cross section. The process is very much like squeezing frosting through an icing tip or even aerosol whipped cream. The tip often has a star-shaped cross-sectional pattern, which is visible as ridges along the length of the "extrusion." In manufacturing, the process often starts with a billet of general shape and size that is then pressed by a ram until the force exceeds the amount needed to cause it to flow through the die opening.

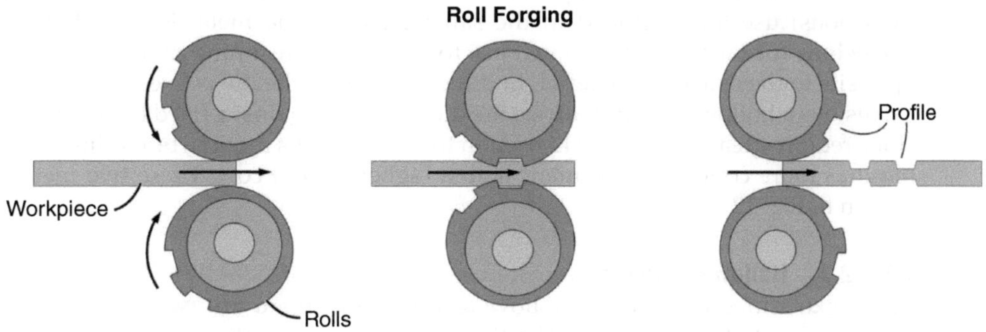

Figure 12.10 Roll forging can be used to create features over a short length of a single part, or repeatedly create the same feature in a long part.

FIGURE 12.11 Metallic extrusions can vary greatly in their size and complexity. In general, the softer the metal, the easier it is to extrude into complex geometries. Aluminum extrusions are found on many products as it is a ductile metal, lightweight, and relatively low cost.

The size of the billet and mechanical properties of the material being extruded determine the length of the extrusion and the amount of force required. Ductile materials, such as aluminum, can be extruded at room temperature (**cold extrusion**) and their cross sections can have a higher complexity. Materials such as stainless steel, which work hardens extensively, must be processed at higher temperature (**hot extrusion**), are more limited in what forms they can take and must be thoroughly lubricated to avoid damaging the ram, chamber, and die. Some examples of the complex cross sections possible with extruded aluminum, the most popular metal for extrusions, are given in Fig. 12.11.

All extrusion processes must consider the complex deformation of material during extrusion. Friction, temperature, and billet and die geometries determine the material flow characteristics. There are very large shear deformations during extrusion, and differences throughout the volume of the material can lead to cracking. Proper lubrication, temperature, and process design can improve quality and reliability.

There are two basic configurations of extrusions. **Direct extrusions** (or *forward* extrusions) use force applied in the same direction the metal is extruded. **Indirect extrusions** (or *backward* extrusions) use force (and the resulting motion) in a direction opposite to the extruded material. In direct extrusions the die is fixed, while in indirect extrusions a hollow ram that carries the die is moved against a fixed billet. Direct extrusions require greater force due to friction from the billet's motion, but indirect extrusion requires more complex equipment with a higher overall cost. These two methods are shown in Fig. 12.12.

12.2.2.1 Hollow Extrusions

Many extrusions incorporate hollow features, and there are two basic methods by which to create these features. The first utilizes a combination of a ram and mandrel where the mandrel creates an *internal* shape over which the metal flows. Because the mandrel movement is part of this extrusion method, it will be referred to as a **moving**

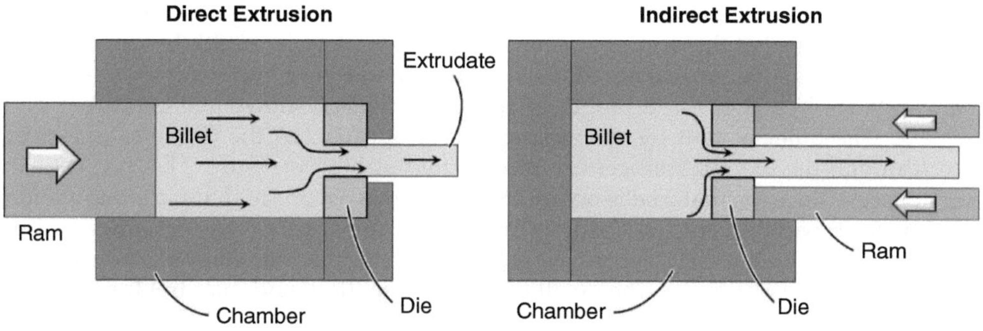

Figure 12.12 The two primary configurations of extrusion are direct and indirect extrusion. In direct extrusion the force on the billet and extruded material are in the same direction, and in indirect extrusion they are in opposite directions.

mandrel extrusion. The mandrel may move independently or in combination with the ram. The basic setup for this process is shown in Fig. 12.13.

The other method involves a more complex die assembly that has a **fixed mandrel** in the center. You may rightly wonder how a mandrel can be suspended in the center of a

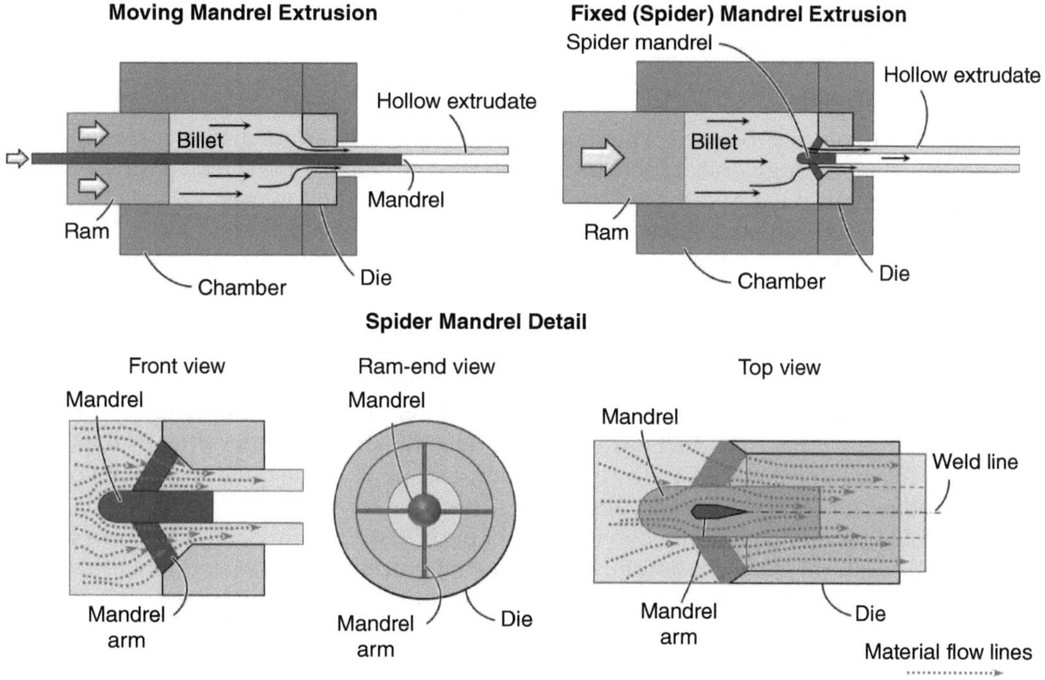

Figure 12.13 Hollow extrusions can be created with a variety of cross-sectional geometries. Simple sections can be created by using a moving mandrel which may be linked to the ram or move independently. The mandrel creates a void with the same shape as it has. For extrusions using a fixed or spider mandrel, the material flows around the mandrel and its support arms. Split surfaces are recombined within the extrusion die.

die to create a hollow section. The way it is accomplished is that the mandrel is positioned slightly ahead of the die opening and it is held in place by multiple supports. Due to its design, it is sometimes referred to as a **spider mandrel** or a spider (the supports or arms) and mandrel. The basic arrangement for this process is shown in Fig. 12.13.

The billet is split by the supports as it approaches the die, but as it progresses through the die the surfaces are recombined at high pressure. This is a cold-welding process since the material is not molten when this occurs, and any contamination of the surfaces will prevent full welding from occurring. For this reason lubricants cannot be used, and hollow sections are largely limited to materials that extrude well without them (e.g., aluminum). The detail of the material flow and mandrel function are given in Fig. 12.13.

Very complex cross sections can be created using extrusion. Extrusion allows shapes that would be inefficient, difficult, or just plain impossible by rolling, machining, or other operations (e.g., Fig. 12.11). Extrusion speeds can exceed 1.5 ft/s (0.46 m/s), dimensional tolerances are good and surface quality can be very good (all are dependent on the material type).

Advancements on the method, such as hydrostatic extrusion, can improve part quality and capability, but it also requires greater equipment complexity and upfront expense. Extrusion processing is still an active area of research, and more recent developments, such as equal channel angular extrusion (ECAE),* where a billet is extruded multiple times around a 90° bend, are also being investigated for possible applications in creating high-strength materials through microstructural refinement related to cold working, but this is still limited in output.

12.2.3 Rolling

Rolling involves processes where compressive force is applied to a material while advancing it through one or more sets of rollers. The cross section created can be almost any shape, but it is typically rectangular or circular. Rolling is used extensively in the production of standard stock such as bars and rods of aluminum, steel, brass, etc. Therefore, one of its primary uses is as an intermediate step in production, and most products have undergone a rolling operation at some stage of their production.

The main categories of rolling are hot rolling and cold rolling. **Hot rolling** is performed at elevated temperature (above the metal's recrystallization temperature) and allows for greater deformation to be sustained without breaking the material (see Sec. 9.2 for more on this concept). **Cold rolling** is done near room temperature† and results in a stronger product with better surface finish and tighter dimensional tolerance. To combine the workability at high temperature with the quality of cold working, hot rolling can be followed by cold rolling. These are shown in Fig. 12.14.

Rolling is relatively limited in the type of parts it can produce since the cross section must be uniform along the length. It is still a vital process, however, and much work has gone into improving the efficiency and quality of the materials processed by it. Rolling equipment, often called **rolling stands**, are stations that are used to perform one or

---

*Also known as equal channel angular pressing (ECAP). The path through the die and shape of the billet may vary.

†Rolling at elevated temperatures which are below the recrystallization point of the metal (too low to be hot rolling) is sometimes referred to as "warm" rolling.

FIGURE 12.14 Rolling operations are performed at elevated temperature and near room temperature. The differences include the amount of deformation possible and the quality of the end product. As shown here, a hot rolled bar will have a rougher and thicker oxide layer on the outside than one that has been cold rolled.

more passes where the material is gradually reduced in thickness and/or width with each pass.

The number of rollers and the ability to reverse direction determine the naming convention. Two rollers create a "two-high" roller stand, three rollers a "three-high," and so on. To add support to smaller rollers, they can be backed by larger ones (backing rolls) to increase the rigidity. These backing rolls decrease deflection of the roll in contact with the workpiece. Crowning of rolls can also be used to compensate for deflection, but the design becomes much more critical as deflection is determined by the size and strength of the workpiece.

Backing rolls are included in the naming convention, but where multiple backing rolls are used, the configuration is usually just called a **cluster mill**, though you may also see it called a 20-high cluster mill, for instance. Cluster mills allow for very small rolls to be used in contact with the workpiece while still being rigid and producing accurate material. Small rolls reduce the contact length with work, which results in lower forces and power consumption. Some common roller configurations are shown in Fig. 12.15.

The process is very much like rolling out a pizza crust. As the material is rolled, it decreases in thickness, but it also spreads out. A similar phenomenon happens when a piece of metal is rolled. As the metal passes through the roller the thickness decreases, but the volume must stay the same. This necessitates that the length must increase. A 4-ft bar of steel may enter the roller stand, but when it leaves it may be 4.5 ft long.

This also means the metal will be traveling faster when it exits than when it enters. Rolling processes often involve many separate stations that progressively reduce the cross section. The third set of rollers will be fed by the second set and so on. Each time the thickness is reduced, the speed increases. If the rollers are not operating at the correct speeds, the material will bunch up and buckle (in-feed too fast) or it will tear apart (in-feed too slow). In industrial rolling operations, computer monitoring and control is used to continuously adjust and optimize the process. A multistation process is shown in Fig. 12.16.

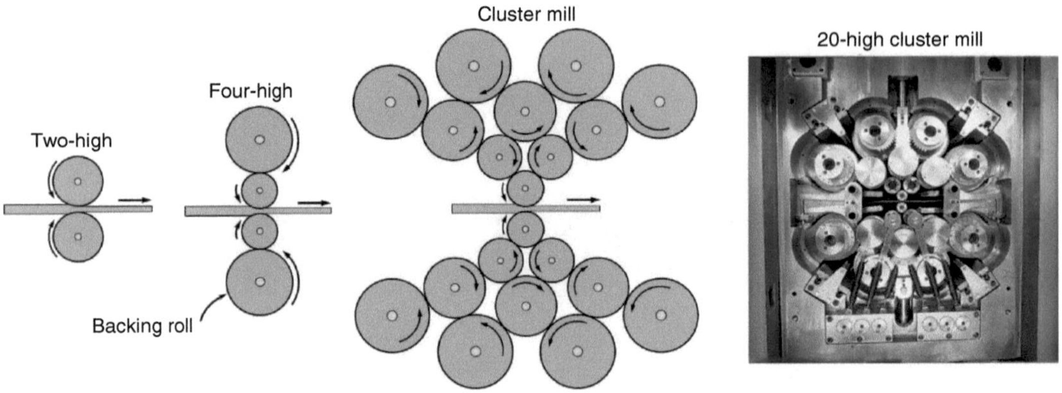

Figure 12.15 Rolling processes are accomplished using roller stands, which can have a varying number of rolls. Where small rolls are used in contact with the workpiece, backing rolls are added to decrease deflection and improve quality. (*Photo by The Theis Group, CC BY-SA 3.0.*)

The material must be ductile to undergo significant cross-sectional changes, and the reduction in area before fracturing is a measure of the material's ductility (see Sec. 7.4.3.1). As with other forming operations, the process is often done at elevated temperature to reduce the amount of force required and to extend the deformability of the material. Sometimes hot rolling is followed by cold rolling for the improvement in mechanical properties, dimensional accuracy, and surface quality.

To control lateral dimensions, a second set of rolls can be added to constrain the material as it is rolled. Also, the rolling may result in geometries that are more complex, and this is known as **shape rolling** or **profile rolling**. For instance, I-beams and C-channel can be rolled using a progression of rollers that incrementally form the material until it reaches its final shape. This is shown in Fig. 12.17.

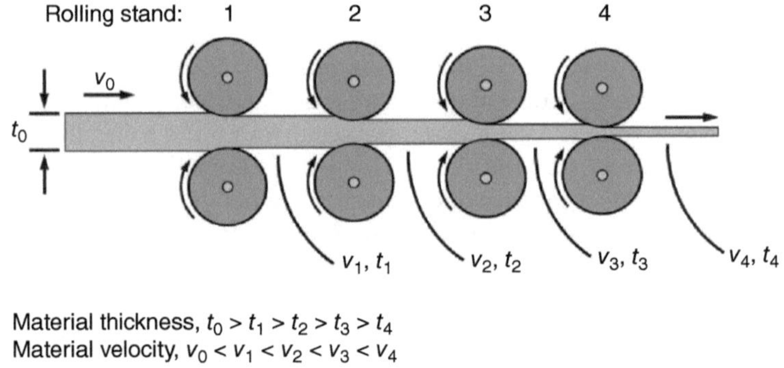

Figure 12.16 Rolling processes are often done continuously with a series of roller stands. The first set of rolls must be rotating with a lower surface velocity than the next because as the material is reduced in thickness, the velocity must increase (constant volume).

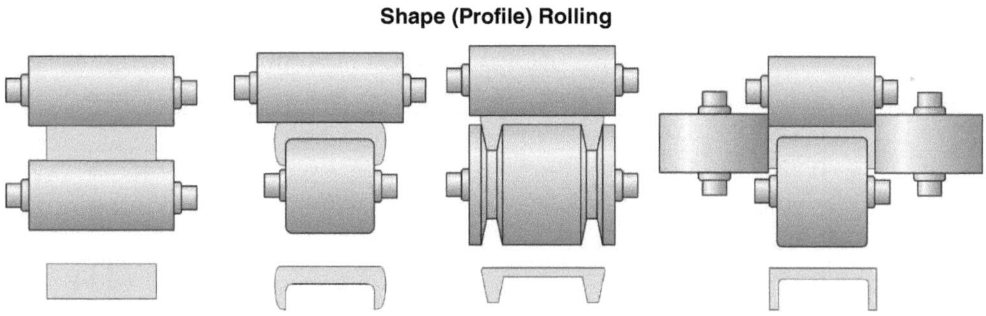

FIGURE 12.17 Example of a multistep rolling process to form a structural shape. Each step serves a specific purpose in shaping the material and reducing its thickness. The exact nature of the rolling progression will depend on the shape being made and the material being formed.

12.2.4 Sheet Metal Forming/Shearing

Sheet metal is processed in fundamentally different ways from "bulk" materials. Prior processing has already left it with a thin cross section, so forming operations such as forging cannot be done, but the sheet can still be formed in three dimensions. The major difference is that hollow objects are created. Compare this to the differences in the way wood and paper are processed. Though similar operations may be achieved, such as cutting, the tools and resulting objects differ. This makes sheet goods very useful in a variety of applications.

So what is a "sheet" of metal? This is a more difficult question than you might expect, or at least it will take more explaining. **Sheet metal** is material that has a thickness much smaller in one direction of its cross section than the other. That thickness can vary, but typically sheet goods come in 4 ft × 8 ft sheets and often have a thickness below 1/8 in and most often nearer to 1/16 in.

The thickness of sheet metal is given as its "gauge." For instance, 16 gauge (16 ga) mild steel sheet metal is about 0.060 in thick. Lower gauge numbers indicate greater thickness and higher numbers, thinner. For instance, 26 ga mild steel sheet has a thickness of about 0.018 in. This is only for steel, however, and aluminum is about 0.051 and 0.017 in for 16 and 26 ga, respectively. Likewise, the thicknesses for other metals will also vary for each gauge number. The relationship between gauge numbers and sheet thickness for some common sheet metals is given in Table 12.1.

Although very low gauge numbers approach ¼ in. in thickness and can be considered sheet, material that thick is relatively uncommon for wide areas, and stock of that thickness is often considered "strip" (if narrow) or "plate" (if wide), not sheet. The typical applications of sheet metal utilize thin stock for its formability or light weight for large areas (used for decorative and functional covers).

12.2.4.1 Bending Operations

First, we will consider simple bending operations. Just like folding a sheet of paper can result in complex shapes (think origami), sheet metal can be bent to create a variety of objects. The simplest is a straight-line bend. This can be achieved with a general purpose **press brake** or **hydraulic press** with appropriate tooling, which is a versatile forming tool which can create a variety of geometries and angles (see Fig. 12.18). The angle is determined by the amount of deformation, the tooling geometry, or both.

Sheet Metal Gauge Conversions								
Sheet Gauge (ga.)	Steel		Galvanized Steel		Stainless Steel		Aluminum	
	inch	mm	inch	mm	inch	mm	inch	mm
3	0.2391	6.07	—	—	—	—	—	—
4	0.2242	5.69	—	—	—	—	—	—
5	0.2092	5.31	—	—	—	—	—	—
6	0.1943	4.94	—	—	—	—	—	—
7	0.1793	4.55	—	—	—	—	0.14	3.56
8	0.1644	4.18	0.1681	4.27	0.172	4.37	0.125	3.18
9	0.1495	3.80	0.1532	3.89	0.156	3.96	0.112	2.84
10	0.1345	3.42	0.1382	3.51	0.141	3.58	0.1	2.54
11	0.1196	3.04	0.1233	3.13	0.125	3.18	0.09	2.29
12	0.1046	2.66	0.1084	2.75	0.109	2.77	0.08	2.03
13	0.0897	2.28	0.0934	2.37	0.094	2.39	0.072	1.83
14	0.0747	1.90	0.0785	1.99	0.078	1.98	0.064	1.63
15	0.0673	1.71	0.071	1.80	0.07	1.78	0.057	1.45
16	0.0598	1.52	0.0635	1.61	0.063	1.60	0.051	1.30
17	0.0538	1.37	0.0575	1.46	0.056	1.42	0.045	1.14
18	0.0478	1.21	0.0516	1.31	0.05	1.27	0.04	1.02
19	0.0418	1.06	0.0456	1.16	0.044	1.12	0.036	0.91
20	0.0359	0.91	0.0396	1.01	0.038	0.97	0.032	0.81
21	0.0329	0.84	0.0366	0.93	0.034	0.86	0.028	0.71
22	0.0299	0.76	0.0336	0.85	0.031	0.79	0.025	0.64
23	0.0269	0.68	0.0306	0.78	0.028	0.71	0.023	0.58
24	0.0239	0.61	0.0276	0.70	0.025	0.64	0.02	0.51
25	0.0209	0.53	0.0247	0.63	0.022	0.56	0.018	0.46
26	0.0179	0.45	0.0217	0.55	0.019	0.48	0.017	0.43
27	0.0164	0.42	0.0202	0.51	0.017	0.43	0.014	0.36
28	0.0149	0.38	0.0187	0.47	0.016	0.41	0.012	0.30
29	0.0135	0.34	0.0172	0.44	0.014	0.36	0.011	0.28
30	0.012	0.30	0.0157	0.40	0.013	0.33	0.01	0.25
31	0.0105	0.27	0.0142	0.36	0.011	0.28	0.009	0.23
32	0.0097	0.25	0.0134	0.34	—	—	0.008	0.20
33	0.009	0.23	—	—	—	—	—	—
34	0.0082	0.21	—	—	—	—	—	—
35	0.0075	0.19	—	—	—	—	—	—
36	0.0067	0.17	—	—	—	—	—	—
37	0.0064	0.16	—	—	—	—	—	—
38	0.006	0.15	—	—	—	—	—	—

Table 12.1 Sheet Metal Gauge to Decimal Conversion Values

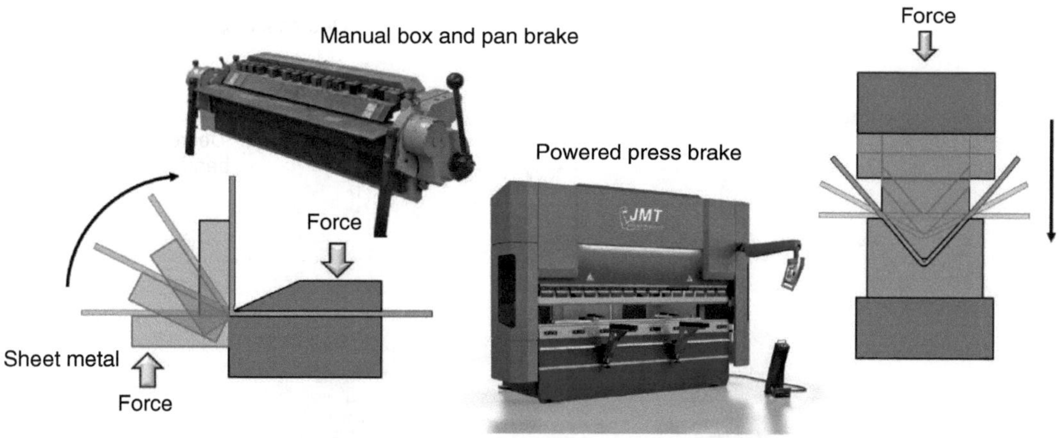

Figure 12.18 Bending operations can be done on thin, soft materials by hand using a manual box and pan brake, or for larger parts requiring greater force and repeatability, a powered press brake can be used. (*Powered press brake photo by JMTUSA, CC BY-SA 4.0.*)

Although simple versions, such as box and pan brakes can be used manually to create products, more complex, computer-controlled versions are available for increasing accuracy and productivity. The size and tonnage of the machine determines the maximum thickness and dimensions of the metal it can bend. The tooling for bending can determine the geometry completely or can be general purpose with more flexibility. Bending can be applied to many types and thicknesses of material, but it is commonly applied to sheet metal to increase its rigidity and create assemblies (more on assembling in Sec. 19.2.1).

In **bottoming** processes, an upper and lower die are brought together until they are only separated by the thickness of the sheet being bent. This limits their use to specific angles and material geometries. A more universal method is **air bending** where the upper and lower dies are used to create a variety of bend angles, the exact measurement of which is determined by the upper die (or punch) displacement. This makes air bending less accurate in high-production settings, but reduces the overall cost of tooling since one die set can be used for multiple materials and geometries. The concept of these processes is shown in Fig. 12.19.

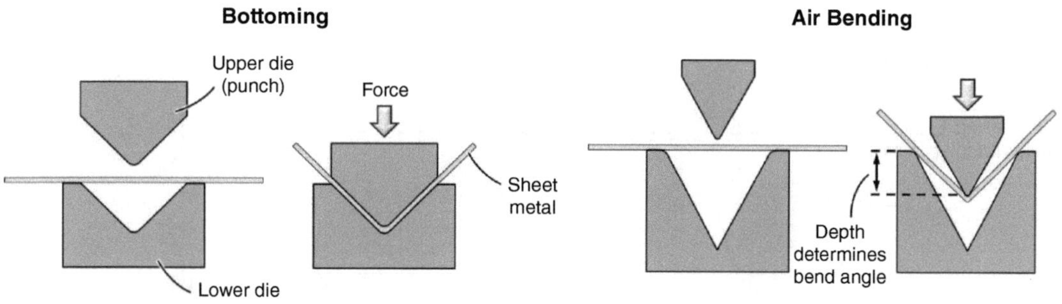

Figure 12.19 Making straight bends in sheet metal can be done using bottoming or air bending processes. Bottoming has higher repeatability, but it is limited to one angle. Air bending is preferable where bend angles may vary frequently and do not need to be as accurate.

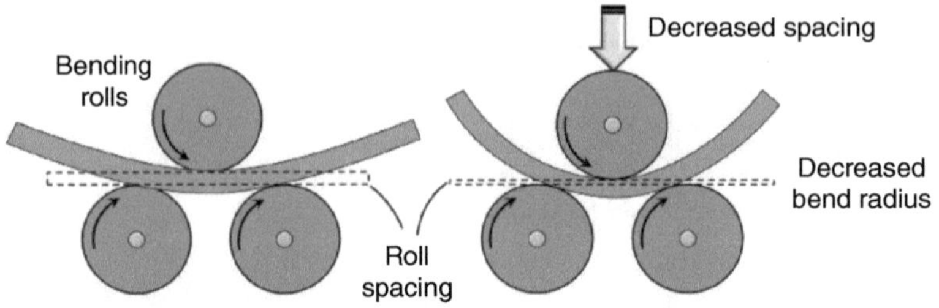

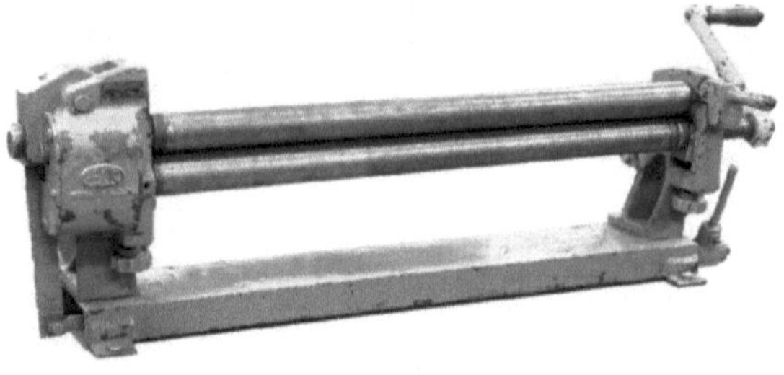

FIGURE 12.20 Roll bending uses a bending force combined with continuous advancement of the material to create smooth radii. The size of the radius is determined by the spacing of the rolls. A small hand-operated roll bender is shown, but large hydraulically powered versions are common for thicker materials.

To create larger, constant bends, or to introduce large radii, roll bending can be used (see Fig. 12.20). **Roll bending** uses multiple, offset rollers to apply bending stress to a piece of metal as it is advanced through rollers. The principle is identical to three-point bending (see Sec. 7.6.4), although in some cases more than three rolls are used. The curvature created is determined by the spacing of rolls, and smaller radii can be created by using multiple passes with decreasing roll spacing. Sheet, bar, and tubing can all be bent this way using various configurations. The material constraints (thickness, type, condition, etc.) are determined by the physical size of the machine, power (if motorized), and rigidity of the rolls and frame.

The same principle that creates bends by rolling can also be used to remove them. In **metal straighteners**, bending in opposing directions balances the internal stress and strain to remove unwanted bends (see Fig. 12.21). A common application is for stamping

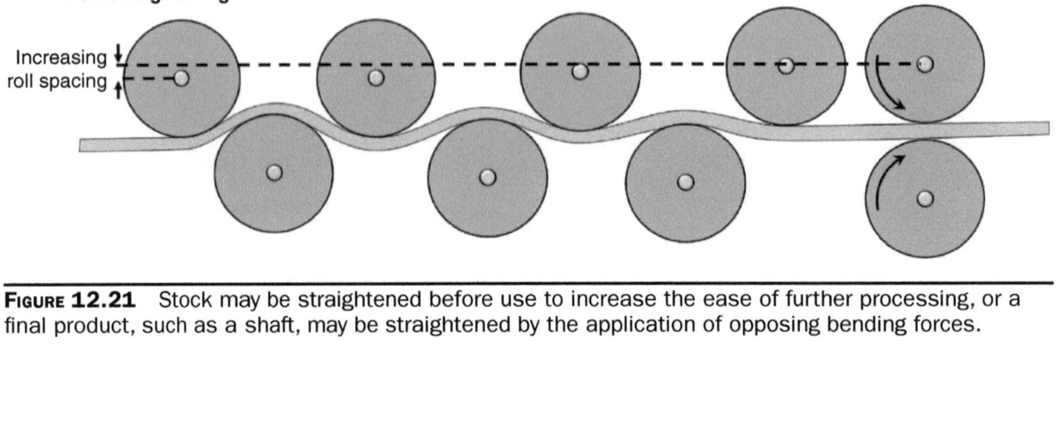

FIGURE 12.21 Stock may be straightened before use to increase the ease of further processing, or a final product, such as a shaft, may be straightened by the application of opposing bending forces.

operations (see Sec. 12.2.4.4), which use metal strip that is fed through a die. The metal strip is often supplied as a spool so the machine can be run continuously over long lengths of material. Being wrapped around a spool causes the metal strip to naturally curl, and this tendency can be removed by straightening. The stock can be straightened in more than one direction by the application of rolls in each desired direction (e.g., one set horizontal, one set vertical).

Uniform straightening of round cross sections can be achieved in a single step using cross roll straightening, which uses hyperbolically shaped rolls their with rotational axes aligned at an angle to rotate and bend the stock simultaneously. This process can be applied to solid rod or tubing, and is capable of straightening a wide variety of material sizes.

Not all bends are created along a straight line and there are sheet forming operations that create complex contours (i.e., a surface bent along more than one axis). There are many variations, but there are two principal ways in which to create complex shapes: using generalized equipment with skilled application or using specifically designed dies for high production. The latter will be described in more detail, but the former is at least worth mentioning.

Custom sheet metal work of great intricacy and accuracy can be completed by craftsmen to create one-of-a-kind components, such as motorcycle fenders, but the methods are unsuitable for high production. Manual tools, such as an English wheel, a shot bag and hammer, or even a shaped tree stump (used similarly to an anvil), require a great deal of skill to use accurately. Obviously, it does not take complicated tools to make complicated parts. An English wheel (shown in Fig. 12.22) is a common example of a basic tool that requires a skilled worker to produce accurate and aesthetic parts.

12.2.4.2 Drawing Operations

To efficiently create sheet metal components of intricate design, methods such as drawing use punch and die sets to form the metal to a desired shape. If the part to be created

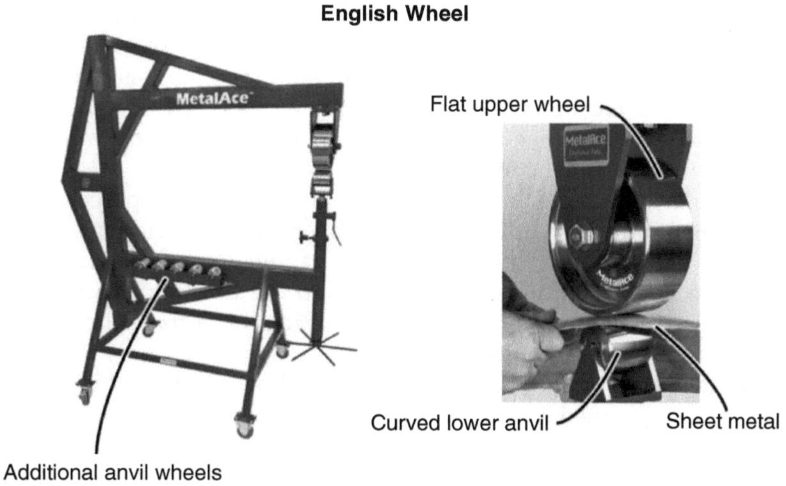

Figure 12.22 Sheet metal forming can be done using general-purpose tools and skilled labor, or using specialized production tools and automation. The emphasis on craftsmanship over productivity can be found in custom, contoured panels created by hand using tools such as the English wheel shown here. (*Courtesy of MetalAce English Wheels.*)

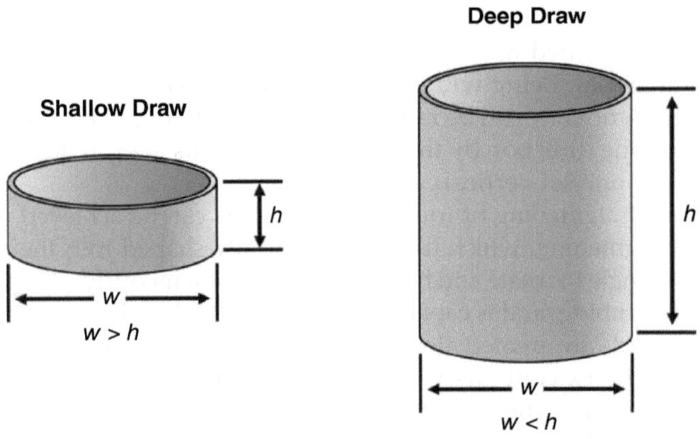

FIGURE 12.23 The depth of draw determines the designation of the process as shallow or deep drawing. Shallow drawing has a depth (or height), *h*, less than the smallest distance across the die, *w*. Deep drawing has a depth/height greater than the smallest distance across the die. Although cylindrical parts are pictured here, the same principle holds for any shape.

has a depth less than the diameter or smallest distance across its opening, it is a **shallow drawing** operation. The process is considered **deep drawing** when the component has a depth (or height) greater than its diameter or smallest dimension across the opening. Sometimes it is more informative to use the **draw ratio**, which is the depth of draw divided by the smallest opening width, to indicate the degree of forming. A draw ratio greater than 1 is considered deep drawing. Shallow and deep drawing are shown schematically in Fig. 12.23.

Try pushing a sheet of aluminum foil into a cup with the bottom of another cup. What happens? Aluminum foil is very ductile, which aids in drawing operations. Even so, you will most likely end up with wrinkling or tearing of the foil. To reduce that, a **blankholder** is used that applies pressure to the sheet blank and controls the material flow during drawing. The blankholder pressure must not be excessive or the material will tear. If it is too light, the material will wrinkle. This is due to the compressive stresses that develop when trying to convert a flat sheet into a more complex shape. A general process schematic for drawing is given in Fig. 12.24.

Where large changes in size or shape are needed, the process usually involves a progression of punch and die sets that each form the material further than the last. This allows for careful tuning of metal flow to avoid tearing or wrinkling of the metal. Lubrication is also applied to the punch and die to reduce friction and improve the surface finish on the part. Typically a draw ratio greater than 2 begins to reach the point of tearing, wrinkling, or both.

An example of a deep drawing process is shown in Fig. 12.25, where stainless steel blanks are being formed into the upper and lower halves of a keg barrel. Due to its tendency to work harden, (see Sec. 9.2.1), stainless steel requires special caution with drawing and can be very damaging to the dies. In food-contact applications, a corrosion-resistant alloy is essential, and a substitute material is not possible. As with cutting tools, coatings (see Sec. 13.4.4) are used to reduce friction and resist wear of the dies. In this application, after the part undergoes drawing, it is then rolled to create a bead for rigidity, and the bottom is trimmed for welding.

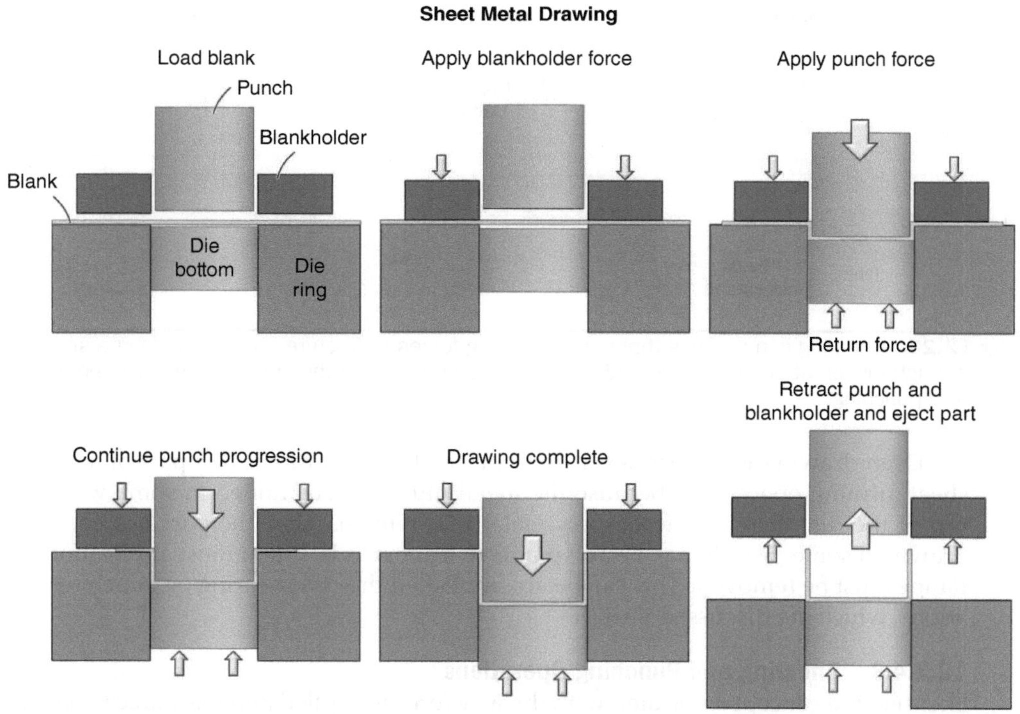

Figure 12.24 The drawing process follows the basic steps shown here. Depending on the depth of draw, material type and other considerations, the process may involve different equipment or multiple drawing setups.

Figure 12.25 The process of deep drawing is shown here for creating a component of a keg barrel. The sheet blanks are drawn, undergo secondary operations of rolling and trimming and are then assembled by welding. Note that this setup draws the material upward instead of down into the die. (*Courtesy of Brian Monteith and American Keg Company.*)

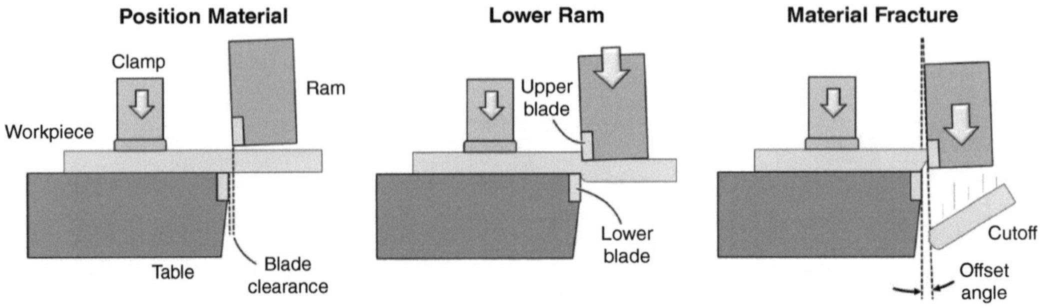

FIGURE 12.26 Shearing is a process that uses opposing forces to fracture a material along a specific path. Although this is often straight, curved cuts are also used. When the cutting path is closed, it is considered punching.

Deep drawing is almost never done in isolation, but rather it is part of a series of sheet forming operations. Because the metal must be constrained in some way at the top of the die, there is often excess metal that remains after the part has been deep drawn. In some cases it may be acceptable to leave it as is, but in most applications this flange must be removed. This can be accomplished though shearing or punching operations, which are discussed next.

12.2.4.3 Shearing and Punching Operations

Shearing is a concept associated with the way force is applied and mechanical deformation (see Sec. 7.3.2), but it is also a manufacturing process. Shearing and punching operations differ from "true" forming operations since they cut rather than deform material to create new shapes, but they do not generate chips as will occur in machining. Forming, shearing, and punching are often combined in sheet processing, so it is not a stretch to discuss them together.

Shearing processes use offset, opposing forces to cause the separation of a material into multiple sections as shown in Fig. 12.26. The simplest example of such an operation is the cutting action of scissors, which are also referred to as shears. The opposing blades cause the material being cut to separate along a well-defined path. Most scissors only cut straight, and to cut contours you must angle them. Shearing operations performed on metals often do not allow for manipulating the material or machine in such a manner, and the shape of the cutting edges will determine the shape of the final part.

A metal shear, as shown in Fig. 12.27, will most often consist of two straight edges equipped with a hardened steel or carbide edge. One edge is fixed and the material is secured against it. The second jaw is moveable and used to apply force to the material, often in a scissor-like pivoting action or similar to a guillotine. The material initially deforms sharply until it eventually fractures. The thickness and ductility determine how cleanly the cut edge is formed, with increases in both factors causing the edge to be rougher. To be successful, the opposing edges should be offset slightly to create clearance between the two. The amount of clearance needed is determined by the properties of the material being cut (e.g., thickness and ductility).

If shearing is similar to cutting with scissors, then punching operations are like a hole punch. The hole punch creates an internal feature where a section of material is removed. Just like loyalty punch cards where the store may use a uniquely shaped

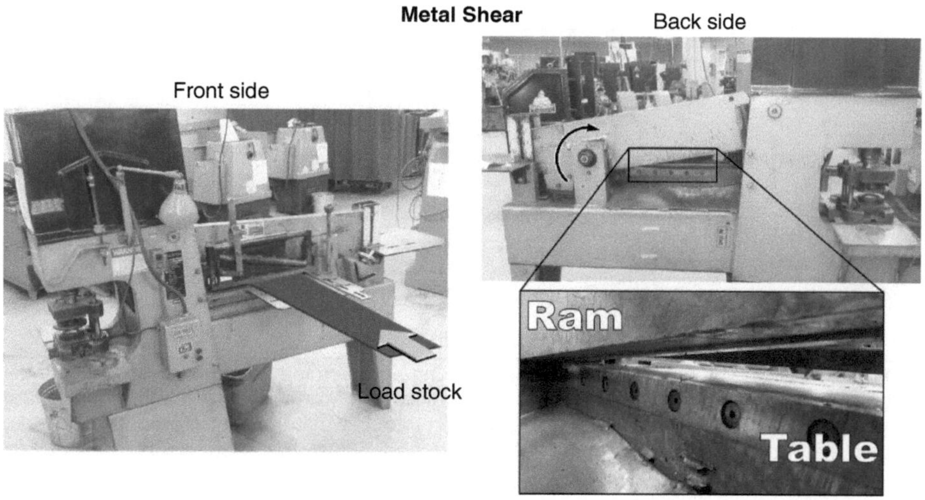

FIGURE 12.27 A metal shear is designed to create straight cuts, and is often used for sizing sheet metal and relatively thin bar stock. The workpiece is loaded on the table, clamped and then the ram is lowered until the part fractures.

hole punch, punching operations in metal can also be done with complex shapes. Punching operations are similar in many regards to shearing operations, but the cutting action takes place on curved surfaces that form a complete part or feature, as shown in Fig. 12.28.

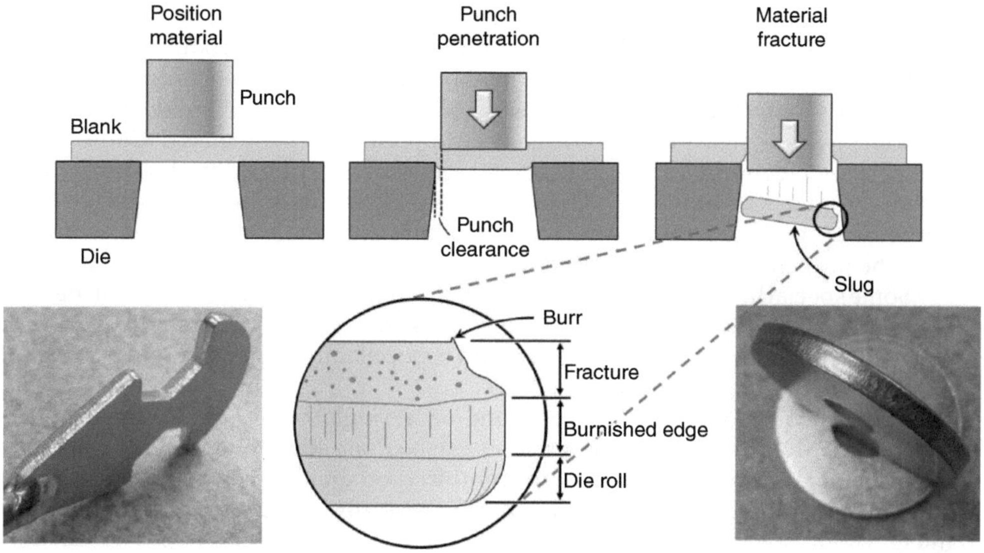

FIGURE 12.28 The general process of shearing and punching involves opposing forces applied to the material, which result in a combination of tool penetration and workpiece fracture. The material characteristics and tool geometry both affect the outcome of the process. Simple shapes, such as the circular part on the right, as well as more complex shapes, on left, can be created.

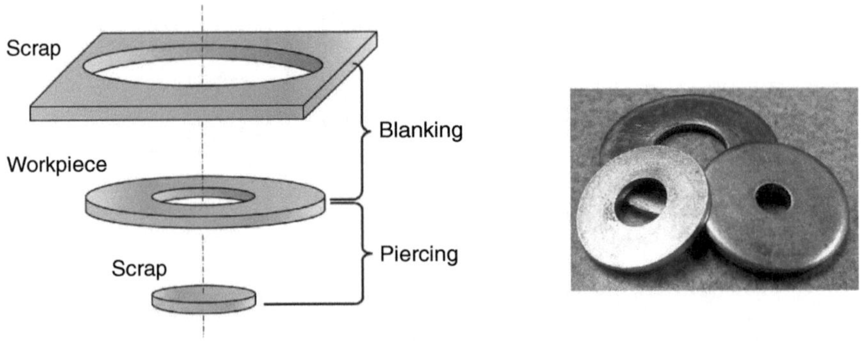

FIGURE 12.29 The difference between blanking and piercing operations is the relationship between the workpiece and the scrap. In blanking, the workpiece is punched from a larger blank. In piercing, the workpiece has a hole punched into it. Washers, as pictured here, are a common part that incorporates both processes.

A variety of features can be created on the edge or interior of the workpiece. If the area removed during punching is the desired part, the operation is called **blanking**. If the punching operation adds a hole to the desired part, it is called **piercing**. That is, a piecing operation *adds a feature* to a part, whereas a blanking operation *creates a part* (which may be further processed later). These concepts are shown in Fig. 12.29.

12.2.4.4 Stamping Processes

Stamping processes include any combination of punching bending and other forming techniques applied in sequential fashion. Stamping is a broader category under which those operations are included. For instance, a part may be created that requires punching and bending operations. The part is considered a stamped product.

Stamping may include a variety of methods. Bending and punching have been covered already, but a few more categories are worth mentioning. Coining and embossing are both processes by which patterns may be imparted to a flat material. **Coining** creates a pattern by indenting a shape into a solid section. It may be a simple pattern, such as letters and numbers, or it may be more complex, like designs used for coins (go figure!). **Embossing** stretches a sheet into a shallow depression, and is similar to drawing; however, the purpose is more often decorative.

Some special bending operations have unique names as well. For instance, **hemming** is a bending operation that folds the edge of a sheet over to add strength. **Curling** creates a circle shape with the edge of sheet and is popular in hinges. **Flanging** is a bending operation that is done along a curved line.

The punching operation can be done on a single blank, or the process can be automated. Multiple punching operations can be done simultaneously in one stroke of the machine using a **compound die** (or **combination die**), or they may be done sequentially using multiple stations. An example of processing using a compound die is given in Fig. 12.30.

When using a **progressive die**, which applies a variety of features to a part sequentially, the stock is automatically advanced through the machine, known as a press (or **press brake** such as in Fig. 12.18) with each stroke. During production, each cycle creates a finished part, even though it may take several strokes for the part to be made.

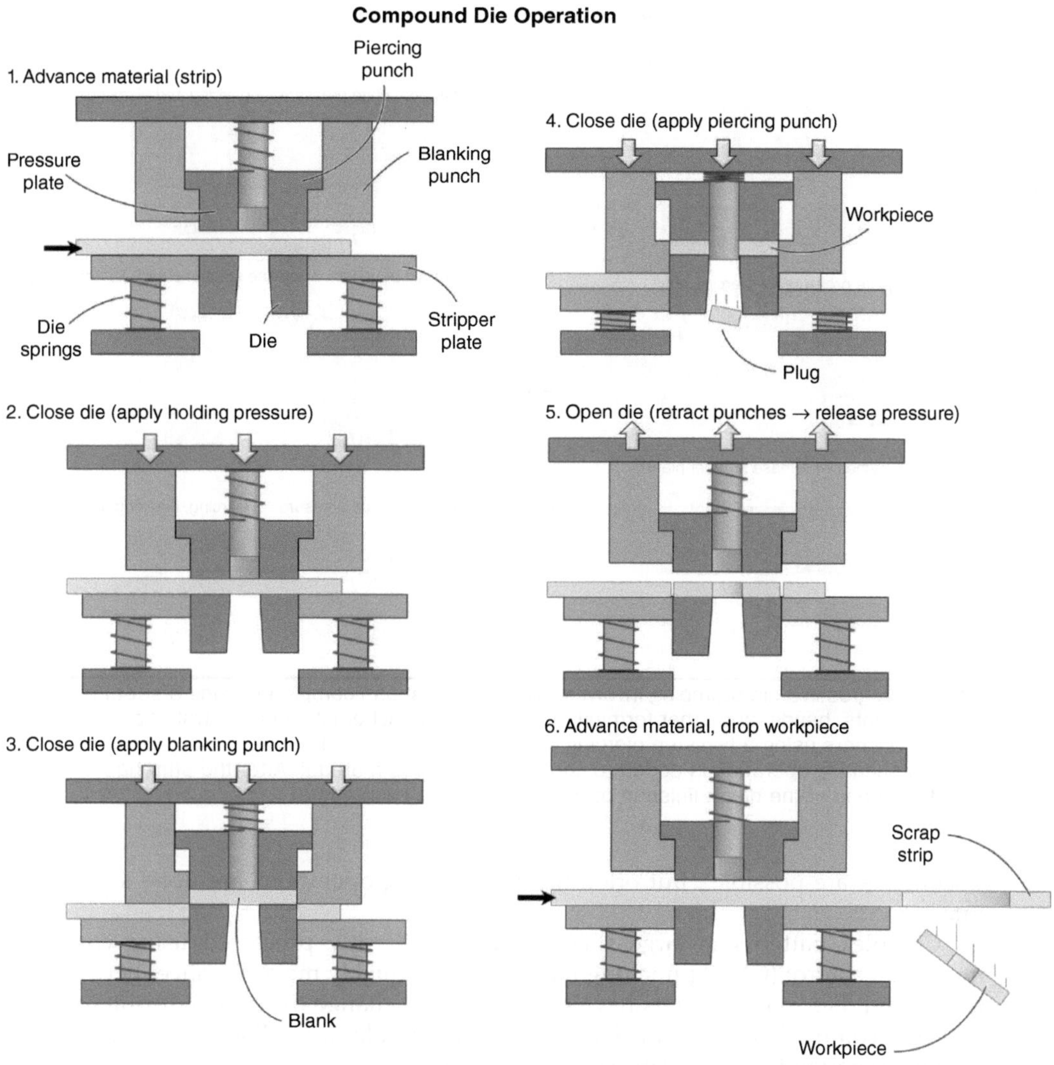

FIGURE 12.30 Compound (or combination) dies perform multiple forming operations with each stroke to create a part. This allows a part with multiple features to be created using a single station, but the complexity of die design is higher.

This makes each operation simpler to design than a combination die, and it allows for many operations to be completed with a single setup. An example of a progressive die operation is given in Fig. 12.31. Compare this to the combination die shown in Fig. 12.30, as both produce the same part.

The part may have any number of bends, punching, and shearing operations and the full range of processes can be done highly efficiently for certain parts, with some machines able to perform multiple strokes each second. High-tonnage machines are capable of processing thick materials (typically not much greater than ¼ in, but greater

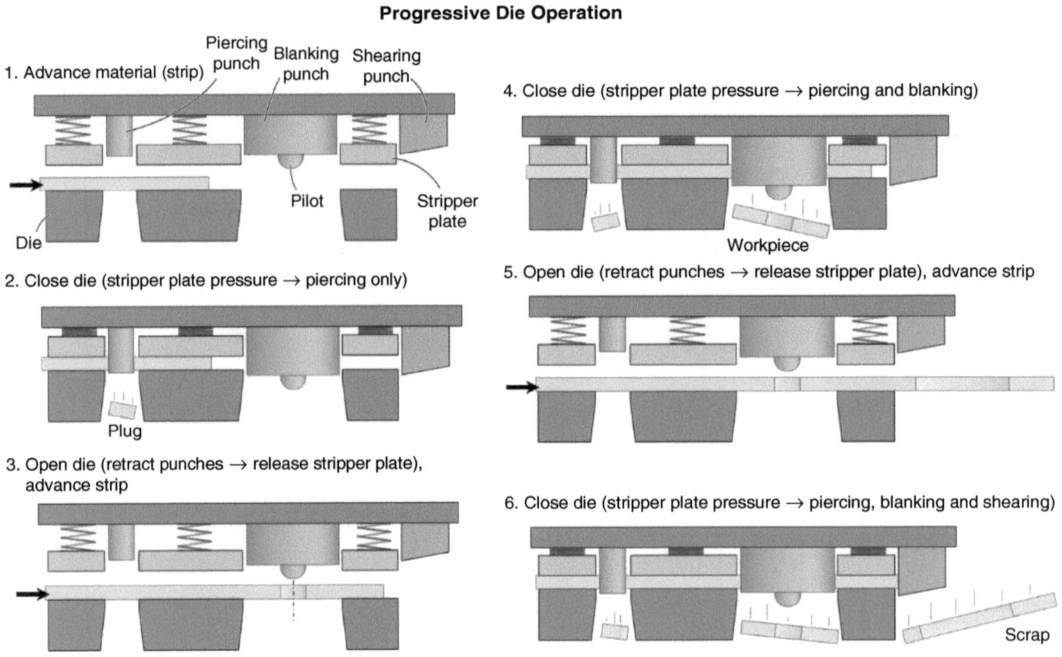

FIGURE 12.31 Progressive die stamping involves multiple operations completed using a set of punches to create holes, slots, bends, and other forming operations. This schematic demonstrates a method for manufacturing washers using a two-step piercing and blanking process. To make the scrap more manageable, a shearing operation is added to chop up the excess material. After the strip has progressed to the end of the die, a finished part is created with each stroke.

thicknesses are possible), but often heavy punching operations are done at a slower rate.

Complex patterns in large sheets can be efficiently processed using computer numerical control (CNC) punching machines that quickly move the sheet between the punch and die. Some versions have the ability to exchange the punch during this process to accommodate different shapes and sizes of holes as well as rotate the punches during operation to create features at any orientation.

12.2.5 Powder Metallurgy

Despite the impressive capabilities of methods such as forging, casting, machining, etc., there are some applications that pose problems related to efficiency or quality. Metals with very high melting points, that are difficult to machine, are brittle, or have a particular sensitivity to other manufacturing conditions can be processed using powder metallurgy (PM). **Powder metallurgy** is a family of processes that use a metallic powder (pure metal or alloy) to create parts through the application of pressure and heat. Products created this way are sometimes called **sintered metals**.

Have you ever built a sand castle? The processes share some characteristics. Specifically, by packing together many particles under suitable conditions, the particles can be made to stick together. With sand, moisture is a necessary ingredient (refer back to Sec. 11.2.1.1 in regard to casting). With metals, high pressure is the requirement

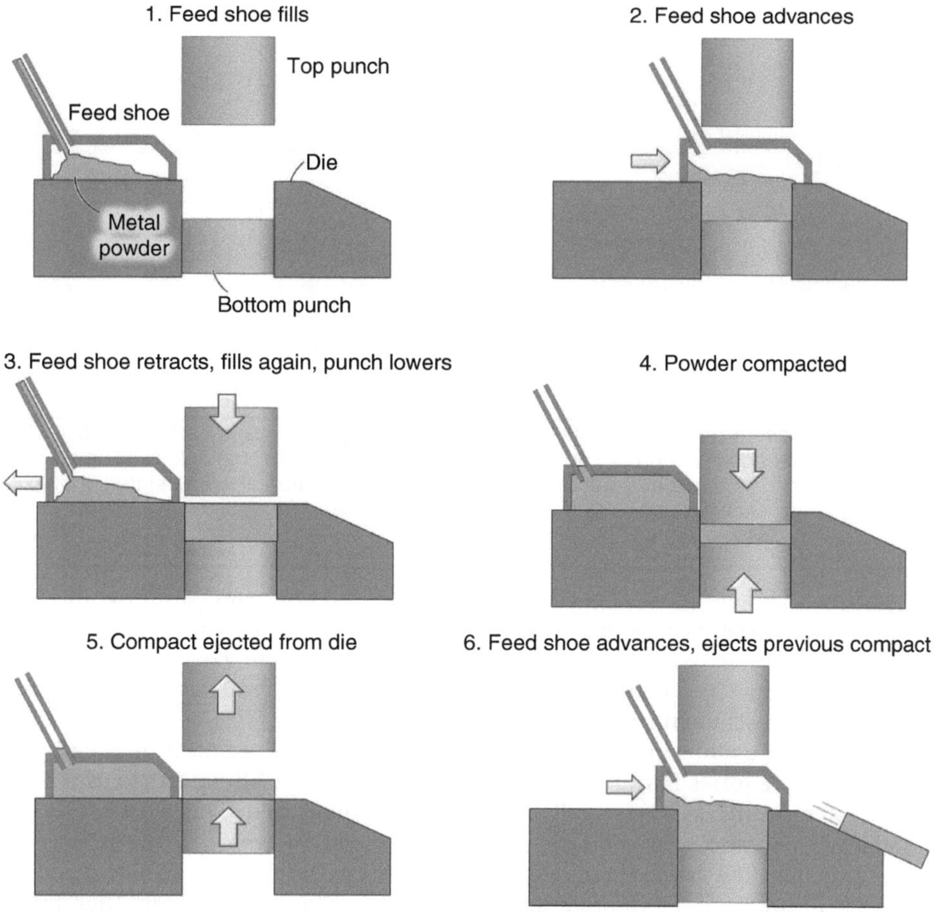

FIGURE 12.32 Powder metallurgy starts with compacting metal powder into the shape of the desired part. To do this efficiently, the process is automated so that every cycle produces a part. More complex punches can be used to create surface features and holes.

and binders are often added to maintain the strength in the **green** (pressed but not sintered) condition. The higher the pressure, the more densely the power will be compacted to a certain degree, but increasing pressure eventually plateaus in the increase in density it provides. This initial compaction is important to later steps that rely on a robust component. The general process steps for creating these compacts are given in Fig. 12.32.

Continuing with the sand castle concept, the wet sand may support its own weight, but the structure is not very strong. Likewise, when metal powders are pressed the particles may stick together, but the compacted powder is still weak. The condition where powder is compacted but not yet strong is known as the **green state** or a **green compact**. It can be broken apart with relative ease, but it is strong enough for light handling needed to move it through the rest of the processes.

After compaction, the parts are then sintered in a controlled atmosphere furnace containing an inert (e.g., argon) or reducing (e.g., hydrogen) gas. **Sintering** is the

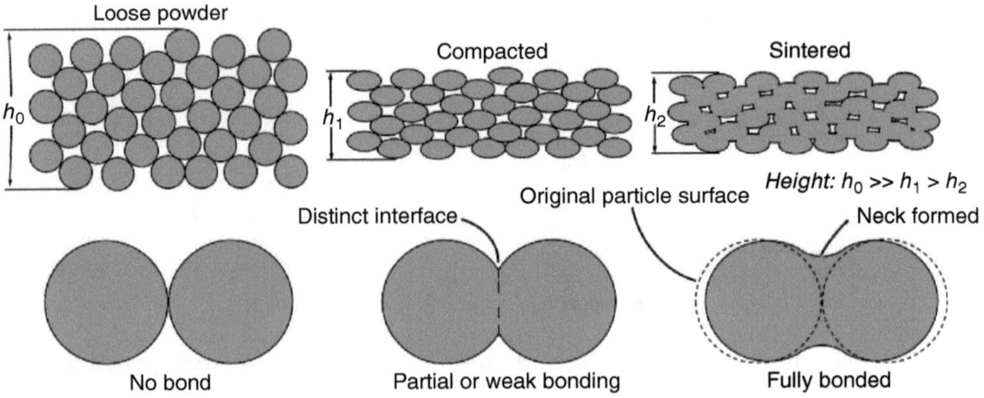

Figure 12.33 Sintering is a high-temperature process that causes particles to bond by material transfer. After compaction the power will hold together due to some cold welding caused by pressure and friction, but it will only be suitable for application after the sintering step. The dimensional changes during each step must be accounted for.

process of bonding metal particles together at high temperature, but without exceeding the melting point of the material. The atoms of the metal can move enough to diffuse (see Sec. 8.4) across the areas where the particles touch. These small connections are called "necks," and they form in order to reduce the overall surface area of the metal. A generalized schematic is shown in Fig. 12.33.

After the sintering process, density increases and strength is greatly improved. Sintering requires material to transfer from the bulk of the particle to voids, and this transfer results in an overall reduction in dimensions. This will cause the part to change size and must be accounted for. Because the mass of the material does not change and the volume decreases, the density will increase. Full, theoretical density is the ultimate goal, but often not all porosity can be removed, and therefore densities above 95% are common.* Full density can be achieved by using higher pressure and differing treatments, such as hot isostatic pressing, which applies fluid pressure around the entire part while at high temperature. The generalized process flow for PM is given in Fig. 12.34.

One of the main benefits of PM is that the material used is minimized. Unlike forging and casting, PM requires no removal of flash, gating, and often no finish machining. Some parts may require finish processing, but the degree needed is typically very low. The output of the process can be very high, accuracy is good, complexity is very good, and the resulting parts can be comparable in strength to alternative processes yet more economical. Almost any metal can be processed by PM (e.g., steel, copper, brass, bronze, stainless steel, titanium, etc.), and many products, from large to very small, from automotive components to electronics, benefit from this technique. Some of the possible parts are shown in Fig. 12.35. These include filters and self-lubricating bushings.

*Low density, porous parts can also be created intentionally. One application is oil-impregnated bushings, which possess better lubrication properties than solid bushings.

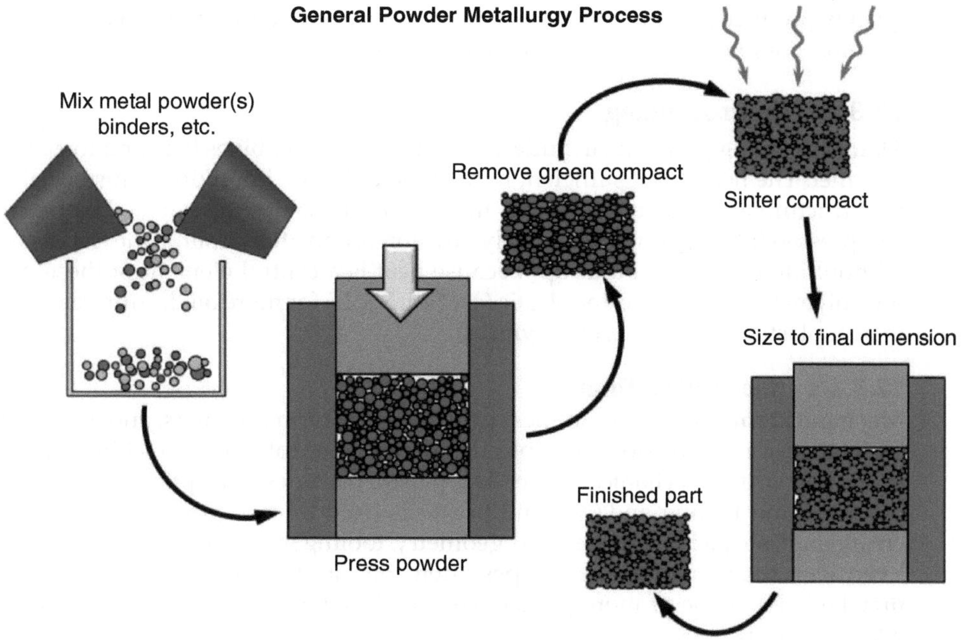

FIGURE 12.34 Powder metallurgy starts with loose powder and converts it into solid parts. The process varies based on application, but most parts will follow the stages shown here.

FIGURE 12.35 Powder metallurgy is used to make an assortment of parts from a variety of materials. The automotive industry is the largest consumer, but medical, electronic, aerospace, and other industries also rely on PM as a source of economical parts. Shown here are filters (left) and bushings (right). (*Courtesy of AMES.*)

12.3 Forming Polymers

Polymers can be formed through a wide variety of operations, most all of which involve elevated temperature to reduce the stress required for forming and to increase final quality. Polymers have a much lower softening point than metals, and the equipment used to form them is much simpler. For example, a strip of plastic can be softened with a hair dryer and bent by hand. Simplicity and accuracy are two different matters altogether, and in commercial production dedicated, specialized equipment is used to quickly produce polymer products with complex geometries. We will consider some of the most common processes in the following sections.

12.3.1 Thermoforming

Thermoforming processes include any method that combines heat and force to shape a polymer. The heat is not sufficient to melt the polymer, but it does cause weakening of the structure and aids in molecular motion. This enables the part to be stretched and compressed into shape with relatively little force, and after cooling, the shape is retained without internal stress build-up. Because heat is a central element of these processes, the polymer must be a thermoplastic (see Sec. 5.3.2.5 for more on the differences between thermoplastic and thermoset polymers).

12.3.1.1 Mechanical Forming

Mechanical forming is the simplest way to thermoform polymers, and is accomplished by heating a piece of stock and physically bending, rotating, stretching, or otherwise shaping it using mechanical force. The plastic becomes very pliable at elevated temperature, and tooling and/or hand power is used to create the desired geometry. To create multiple parts with the same geometry, tooling is used that produces the pattern. Even with the use of tooling, shapes often consist of basic curves or bends. The part may be fixed at one or more points to ensure it holds the shape. An example of this is given in Fig. 12.36.

12.3.1.2 Vacuum Forming and Drape Forming

Mechanical forming has limitations on the design complexity and is generally unacceptable for created large surface patterns. Vacuum forming and drape forming are similar processes that use a three-dimensional surface to form a polymer sheet by application of

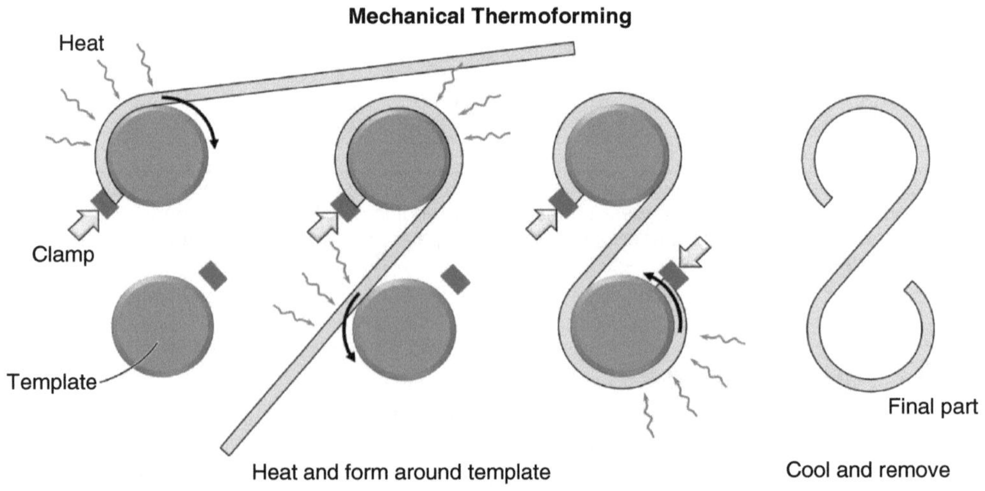

Figure 12.36 Mechanical thermoforming is the simple application of force and heat to create a desired shape. The process can be applied to very large and thick plastic components, and after the plastic cools, it retains the shape of the template.

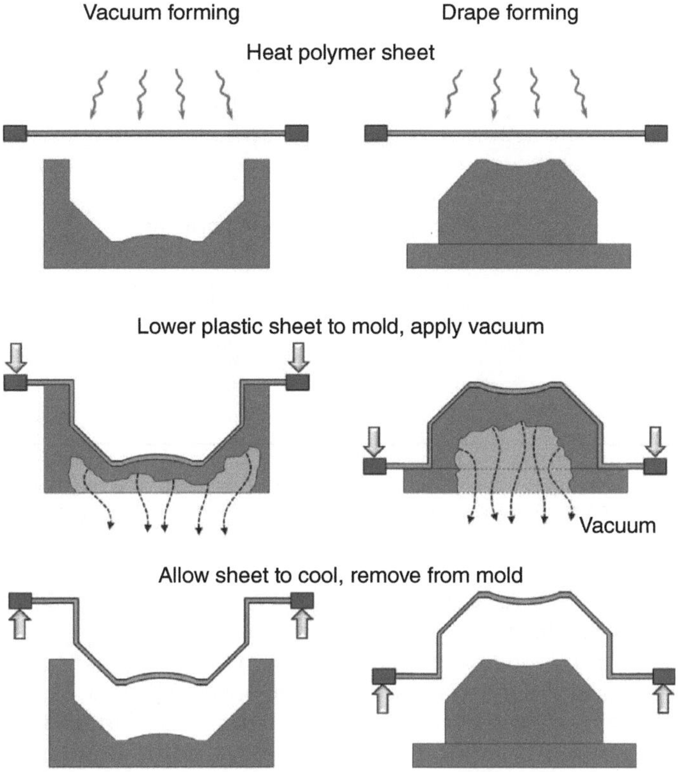

FIGURE 12.37 Vacuum forming and drape forming follow identical steps, but the mold differs between the two. Vacuum forming uses a cavity (negative) mold and drape forming uses a protruding (positive) mold.

pressure across the sheet surface (see Fig. 12.37). **Vacuum forming** uses an applied vacuum pressure to pull a sheet of softened plastic into the shape of a mold cavity (negative mold). **Drape forming** uses an applied vacuum pressure to pull a sheet of softened plastic into the shape of a protruding mold (positive mold).

Both methods use the same basic mechanisms, but positive molds are easier to make and negative molds typically yield better quality parts. Both processes can also be achieved by using positive pressure* from the top of the mold (known as **pressure forming**).

The aspect ratio of the mold is important, since it will determine the amount of stretching the plastic must do in order to follow the features of the mold. If the mold is too deep (or tall), the plastic may thin excessively, tear, or not completely follow the mold features. To assist in effectively molding in such cases, a plug assist can be used (see Fig. 12.38). A **plug** is a physical feature with roughly the shape of the mold that

*Vacuum "pressure" really uses atmospheric pressure to force the plastic against the mold. Pressure forming can apply higher pressure if desired.

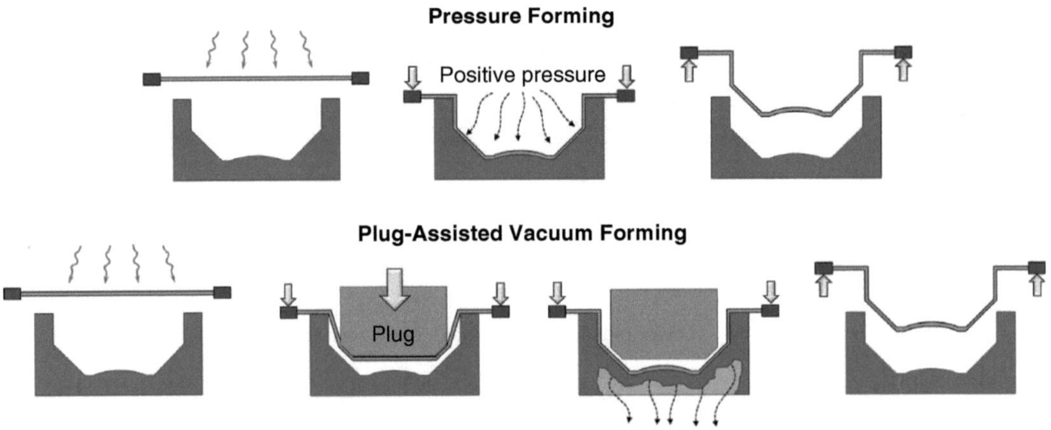

Figure 12.38 Pressure forming and plug-assisted vacuum forming are closely related to vacuum forming. In pressure forming, a positive pressure is applied instead of a negative pressure, and in plug-assisted vacuum forming, a plug mechanically stretches the plastic sheet into the mold before the vacuum is applied.

applies mechanical force to the plastic sheet before vacuum/pressure is applied to pre-stretch it. This results in a more uniform thickness and better detail in the formed sheet.

There are numerous variations on vacuum forming and drape forming. Some may include both positive pressure and vacuum, positive pressure may be applied through the plug, light pressure may be applied through the mold to "inflate" or bubble" prior to insertion of a plug and application of vacuum and so on. Despite the many changes that can be made, all of the methods use the same basic concepts. The differences exist to improve quality on a design-specific basis.

All molds should be designed with typical considerations, such as draft to ensure the molded part can be released. Unlike injection molding or casting operations, since the material is not molten or liquid, the stretching and deformation of the sheet must be considered. Common problems include distortion and "**webbing**," where the polymer sheet folds or creases due to sharp corners and/or features with excessive spacing or height differences. Consider draping a bed sheet over a chair. The sheet naturally forms these folds and creases, and a similar result occurs in softened plastic sheets.

To reduce distortion and webbing and increase detail accuracy, many, very small holes for vacuum should be positioned along corners of the mold and in any recessed areas. Although they can be placed across flat surfaces, and may need to be to aid evacuation of air, these holes should not be placed across areas that may be visually important, since small divots may appear in the surface. These divots and distortion are shown in Fig. 12.39.

Mold design is a combination of three-dimensional shape and positioning of vacuum points to optimize the evacuation of air and control sheet thickness and distortion. None of the mold design parameters will be effective without proper heating of the plastic sheet, and the temperature must be controlled for each type of polymer and sheet thickness. A common application of thermoforming is in packaging, and an example of a part used to store pretzels is shown in Fig. 12.40. The thermoformed sheet appears opaque until held up to light. It then becomes apparent where the thinnest sections are. In this case, it is at corners and within the deepest cavities, as is common.

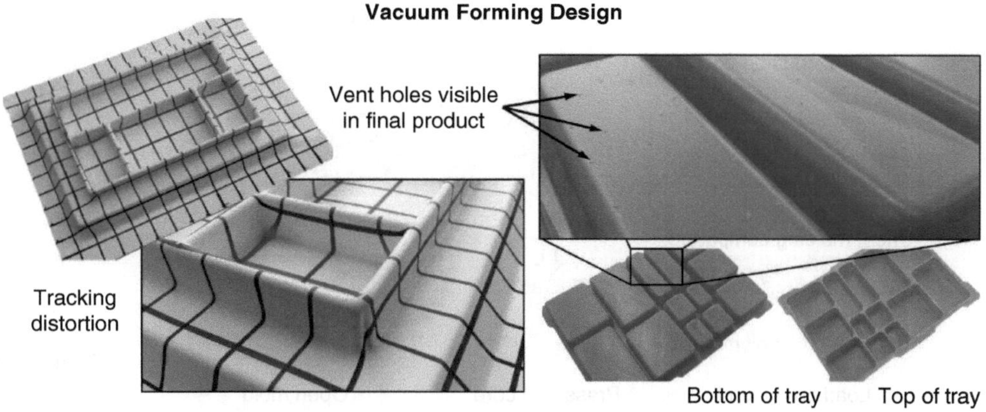

FIGURE 12.39 Vacuum forming molds should incorporate good design to prevent excessive distortion and to replicate detail well. Some common features of vacuum-formed products are shown.

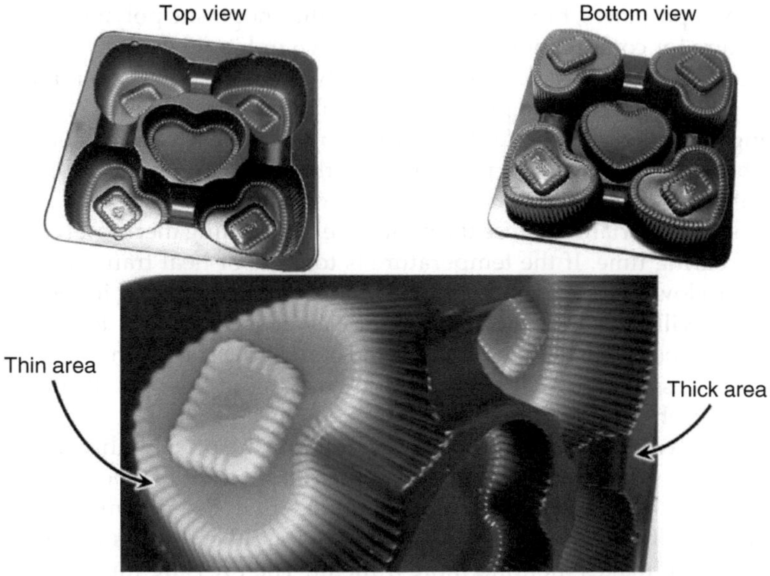

FIGURE 12.40 Vacuum forming of sheet is often applied in packaging. The product showed here houses chocolate-covered pretzels. The deepest recesses require the greatest stretching of the plastic sheet, which results in thinning. This is evidenced by the ease with which light can pass through the initially opaque plastic.

12.3.2 Compression Molding

Compression molding is a process that uses a combination of applied force and mold geometry to create a part. This process is similar to impression (or closed) die forging (see Sec. 12.2.1.2) as it starts with a predetermined volume of material, known as a **charge**, and presses it into a shaped mold. This process is particularly useful in molding

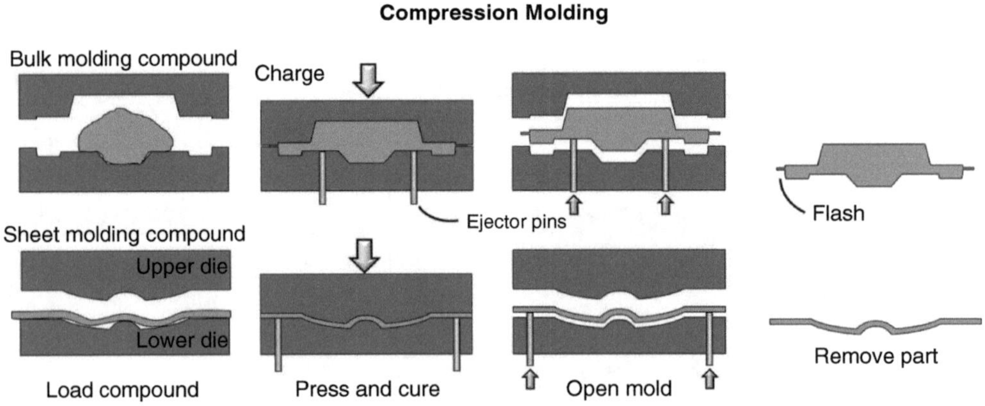

FIGURE 12.41 Compression molding is often done using thermoset polymers mixed with great quantities of reinforcement and fillers. Depending on the type of starting material (sheet or bulk molding compound), the final part can have varying degrees of thickness and geometric variation.

thermoset polymers, but can be used with thermoplastic polymers as well. The general procedure for compression molding is shown in Fig. 12.41.

Because thermosets are most commonly used, the polymer and/or mold are heated to set the material during pressing. The polymer may be heated by external sources or during processing to create the preforms used for each mold. For instance, extrusion may be used to process the polymer directly before molding, and the curing process is partially completed before compression molding is started.

This is important because the cycle time for molding thermoset compounds depends on the curing time. If the temperature is too low or heat transfer or chemical reaction rates are slow (see Secs. 5.3.2.5 and 5.3.2.6 for more on curing in thermosets), the molding time will be long. Since time is money, it is important to reduce the cycle time as much as possible. If the polymer is too hard at the start of molding, the force required will be excessive and will likely lead to cracking of the part. There must be a balance between preheating and curing within the mold.

The polymers used in compression molding are often highly filled (see Sec. 10.2.3.4 for more on fillers), and commonly contain a mixture of fiber glass and polyester or epoxy resin. There is so much filler added that the polymer has a thick, doughy consistency. The same properties that allow a baker to toss a pizza crust without it falling apart, makes molding more difficult. The pressure must be considerable to get such a viscous compound into the features of the mold. Because of this, fine detail is not used in compression molding.

The polymer materials used in this process are known as **bulk molding compound (BMC)** or **sheet molding compounds (SMC)**, which contain extenders, reinforcement, and other additives, and are technically composites. BMCs can be so filled that they have a fibrous, cotton, candy-like appearance. The fiber reinforcement in BMCs is shorter than SMCs, though both use chopped fibers. SMCs are produced in sheet form by pre-impregnating fibers with the polymer and other additives. This material is stored in roll form (like carpet) until ready for processing. SMCs have higher strength than BMCs, but are not as suitable for thicker products or those that have significant variations in thickness.

Figure 12.42 The process for creating a tractor suspension component is shown here. The compression molding process results in multiple metal plates being permanently combined with the rubber compound. (*Courtesy of Beau Kibbey, Superior Tire and Rubber Corporation.*)

An example of compression molding for making tractor suspension components is shown in Fig. 12.42. These parts are similar to motor mounts used to isolate vibration in cars. They use a rubber compound to bond and isolate metal plates that provide strength and mounting capability. The rubber preform is processed and weighed before being placed into a preheated mold. The mold must be assembled with the necessary plates before being loaded in the molding machine.

Once the preform and the mold are both ready, the preform is added to the mold and the mold is closed, which forces the compound into the features of the mold. To ensure a complete fill, more rubber is added than needed, and this results in "squeeze out," which is excess rubber being ejected from the cavity. After the molding is complete, the part will be trimmed to remove excess rubber.

12.3.3 Transfer Molding

Transfer molding is a process used extensively for thermoset materials where the polymer is injected by plunger either as a liquid or a solid (see Fig. 12.43). We will consider the molding of solid materials as a forming operation similar to compression molding (see Sec. 12.3.2) but with an enclosed mold cavity. Liquids, or what might be closer to nonrigid solids,* are our primary consideration here, particularly those which set in the mold either by simple chemical action or by modest heating.

Transfer molding has some similarities to cold-chamber die casting (Sec. 11.2.2.3) in that the material to be molded is introduced in a quantity for an individual shot. This is unlike injection molding which maintains a reservoir of material to inject from. Transfer molding is advantageous that complex shapes of thermoset materials can be molded, but the process is slow compared to injection molding and results in a greater quantity of waste material from the gating system.

12.3.4 Polymer Extrusion

Extrusion processes for polymers involve the melting of a polymer followed by passage through a shaped orifice (or die) and rapid solidification to maintain the shape created by that die. The equipment and process considerations are very similar to that of injection molding (discussed in Sec. 11.3.3), but extrusion differs in the final destination of the polymer. Instead of being injected into a mold, the polymer is ejected into free space, much like toothpaste leaving a tube. The polymer is often quickly cooled by air or in a

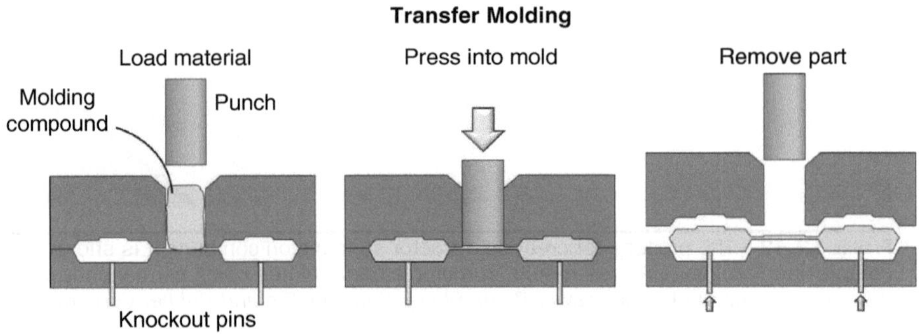

Figure 12.43 Transfer molding is a technique that can create detailed components from thermoset molding compounds in addition to thermoplastic materials.

---

*Molding compounds, which are not rigid, but do not flow without significant effort, are typical. These may be classified as highly viscous liquids, but in practice they act more like solids to the average observer.

water bath to help it retain its shape. We will consider two types of extrusion: profile extrusion and blown film extrusion.

12.3.4.1 Profile Extrusion

Profile extrusion uses a shaped die to create a component of constant cross section with indefinite length. Extrusion is a continuous process, so very long lengths of plastic rod, tube sheet, or other forms may be created. Alternatively, shorter lengths of material can be cut off intermittently, with the extrusion rate and frequency of cutting determining the component length. The basic configuration for profile extrusion is shown in Fig. 12.44.

The die used for extrusion determines the cross-sectional geometry of the extruded product, but the relationship between the die geometry and part geometry is not obvious. This is because polymers are subjected to greater dimensional change during cooling (see Sec. 8.2.3 about coefficient of thermal expansion and Sec. 8.2.4 about thermal distortion).

Because sharp corners and small sections will cool and harden first, the extruded section will be constrained in those areas first, and subsequent cooling cannot be accommodated there. This geometric phenomenon is known as **die swell**. The die shape should take this into consideration and remove such features if possible. Even for a shape as simple as a square, the die will look much different than expected. This concept is shown in Fig. 12.45.

Although the design of the die is the most direct way to address the geometry of the extrudate, postforming rollers or other forming may be used to enhance geometric accuracy. The application of forces to control dimension after extrusion is known as **sizing** or **postforming**. This process will be more critical if working under tight tolerances and with complex sections.

Profile extrusion is a common process and used to make a wide variety of products with simple or complex geometries. It is used for common food items, such as macaroni, and is also the basis for many low-cost 3D printers (specifically, fused deposition modeling, see Sec. 21.2.1). Extrusion can be used to create **film** (thickness 0.01 in or less) and **sheet** (greater than 0.01 in) as well as tube, bar, rod, and other hollow or solid profiles. Examples of the extrusion process are given in Fig. 12.46.

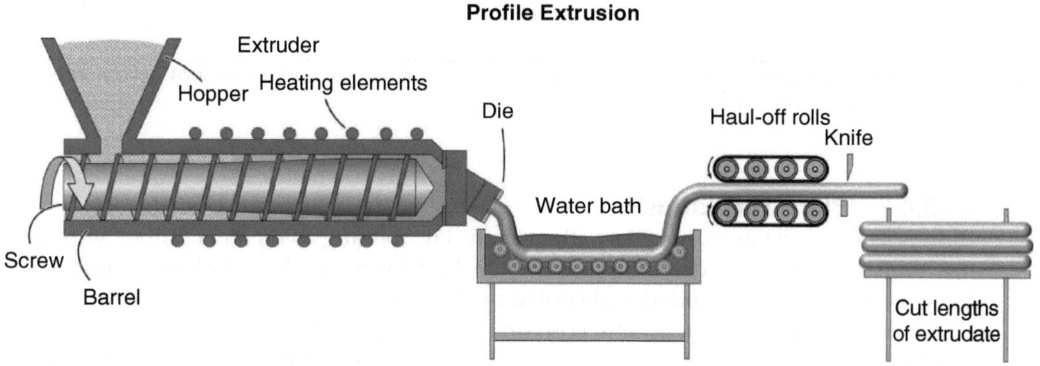

Figure 12.44 Profile extrusion can produce a wide variety of cross-sectional shapes. The geometry is quickly cooled using air blast or water cooling to help maintain the correct geometry. After the section is rigid, it can be cut into lengths.

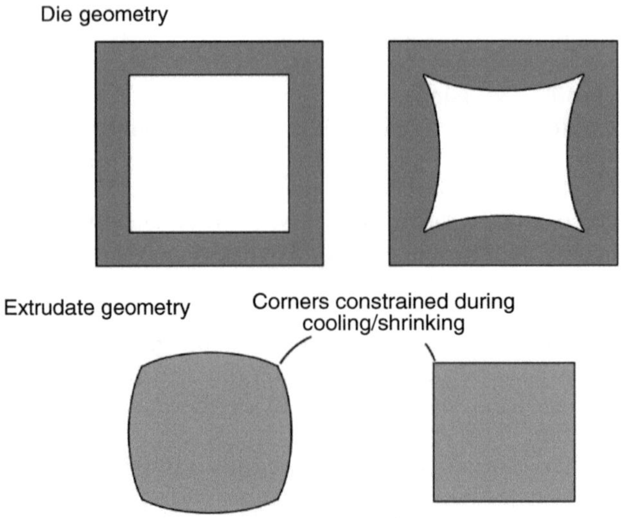

FIGURE 12.45 Due to the high rate of shrinkage during polymer extrusion the die geometry must take into account uneven dimensional change. Sharp corners cool quickly and will not accommodate the slower polymer contraction occurring at larger features.

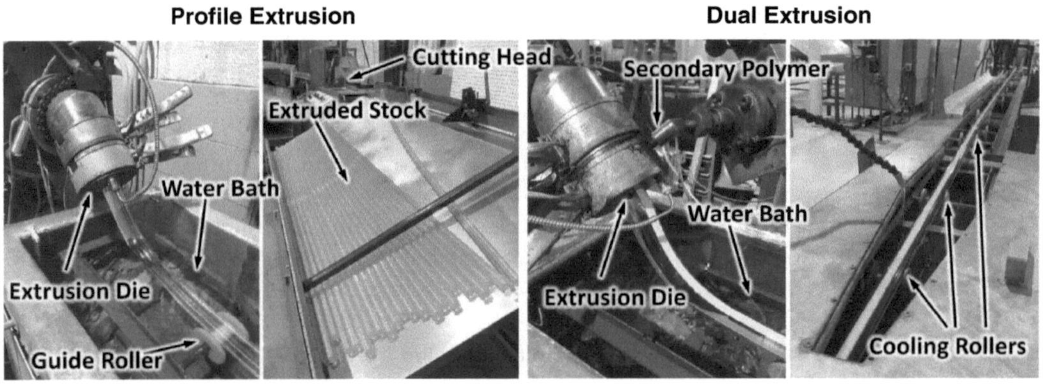

FIGURE 12.46 Extrusion processes are capable of efficiently creating an assortment of solid and hollow cross sections. (*Courtesy of Lance Nelson.*)

12.3.4.2 Blown-Film Extrusion

Blown-film extrusion uses a die with an annular opening to create a "tube" with very low wall thickness (see discussion in Sec. 12.2.2.1 for more about hollow profiles). The hot polymer can be expanded and cooled by flowing air into and around the extrusion. Guide rings and cages are used to contain the expanded plastic, which can be multiple feet in diameter. As the plastic leaves the die it cools as it travels upward. The point at which it cools is known as the **frost zone**. Above this point the plastic can be collapsed, processed through a series of rollers and finally wound up on a take-up roll where it is stored. This process is shown schematically in Fig. 12.47.

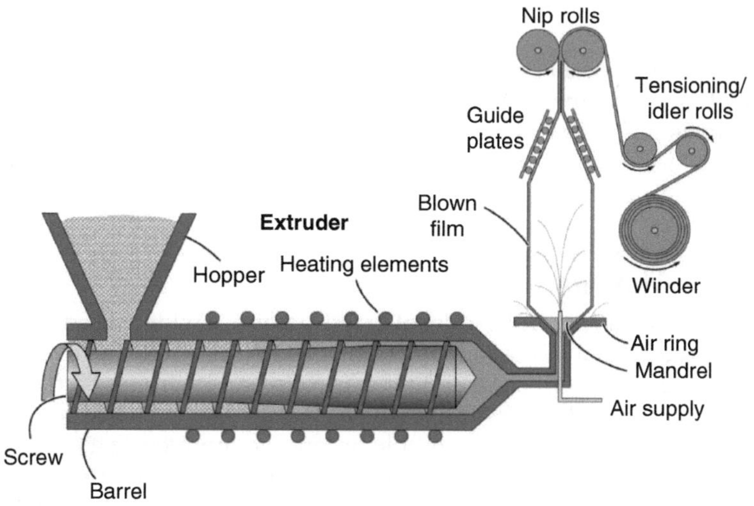

Figure 12.47 Blown-film extrusion is a quick process for creating thin plastic products, such as bags, thin-wall tubing, and other thin films.

Blown-film extrusions are used for a variety of film products, one of the most common being grocery bags. Controlling film thickness is a key concern as the alignment of the mandrel must be exact to ensure the plastic is extruded evenly. Air flow and roller conditions also play a definitive role in controlling quality.

12.4 Forming Ceramics

Ceramics are notoriously brittle, and cannot be plastically* deformed like metals and polymers. To undergo significant changes in size and shape required during forming, ceramics would have to be heated to very high temperatures. There are other methods of forming them, however, and these involve processing the ceramic as a slurry or "paste," rather than as a solid material.

A ceramic paste is a special formulation of ceramic particles and binders that results in a formable ceramic dispersion that must be strengthened through heating. This is the principle behind using clay to form a structure and then firing it to strengthen and solidify it. The clay state is easy to work, the fired state is strong enough to use.

As a paste, the ceramic is easy to form into complex shapes, but the geometry of the die must consider the changes that occur during sintering. As shown in Fig. 12.33, the sintering process closes porosity and results in an overall reduction in volume (increase in density). In nonsymmetric cross sections and those with varying wall thickness, the shrinkage during sintering will not be uniform.

The simplest method of creating ceramic parts is by **uniaxial pressing** where the ceramic is consolidated by applying pressure along a single axis, and constraining the exterior within a die. This is depicted in Fig. 12.32 for PM, and like PM, ceramic pressing is also followed by sintering to increase the strength of the compact. Any shape able

*"Plastic" in this context refers to mechanical properties as discussed in Sec. 7.4.3.

to be created by molding can be used for ceramics, but drying and sintering allowances must be included.

Extrusion is another method for creating ceramic parts, and the process mirrors that of metals and polymers. As the ceramic material is extruded, it must be supported to maintain its shape. The extrudate is then sintered to final density. Ceramic pastes do not need to be heated before extrusion, so thermal contraction is not an issue, but dimensional change during sintering is a concern.

12.5 Why It Matters

Forming covers a very diverse set of processes and is widely used with many materials. Forming is essential in the manufacture of nearly all products. Even parts that are cast or machined often start as stock that was rolled or otherwise formed. Forming is used extensively to create long-lasting, rigid, structural components (e.g., forged components), but it is also used to create disposable decorative products (e.g., vacuum forming).

One of the foundational considerations in forming is ductility. Knowing which materials can be subjected to forming will save a great deal of time, money, and frustration. Some of the considerations with plastic deformation (the basis for forming) are further described in Secs. 7.4.3 (mechanical concepts), 9.2 (metals and work hardening), and 10.2.1 (mechanically conditioning polymers). What we saw in this chapter is the way materials encounter these forces. Each type of process and the specifications for that process will result in varying amounts of strain on a material.

Some processes were universal, like extrusion, and are used with each class of material. Others are specific, like thermoforming, which is only applied to polymers. A material's properties and the practicality of the process are important. For instance, temperature is an important factor in forming, as most materials soften with heat. Most ceramics, however, have high softening temperatures and may never be truly ductile, so forming them is very limited. The right combination of process and material matters.

We first looked at some common metal forming processes, including forging, extrusion, rolling, sheet metal forming, and PM. Each is suited to different tasks, so knowing their strengths and limitations is important. A hollow component is best made by extrusion. A complex three-dimensional shape is best made through forging, but more specifically, impression-die forging is most suitable. For materials that are too brittle to bend and shape, PM gives us a way to create complex parts with minimal machining.

Most polymers are ideal candidates for forming. They are weak, ductile, and can be significantly softened at modest temperature (thermoplastics, at least). Thermosets harden with heat, but there are ways around that problem. For instance, molding compound which has a thick, dough-like consistency can be molded through compression or transfer molding and hardened chemically or by heating. A downside to polymers is that their molecular structure can complicate their response to forming operations. They may swell or distort when pressure is released, and this must be accounted for in the design of tooling, extrusion dies being a prime example.

Ceramics are much less common in forming due to their brittle nature, but they can be extruded and formed in a "green" state before they are rigidly bonded through heating. Although the ceramic paste used in extrusion does not swell and thermally distort as polymers may, sintering does result in dimensional change which must be carefully accounted for.

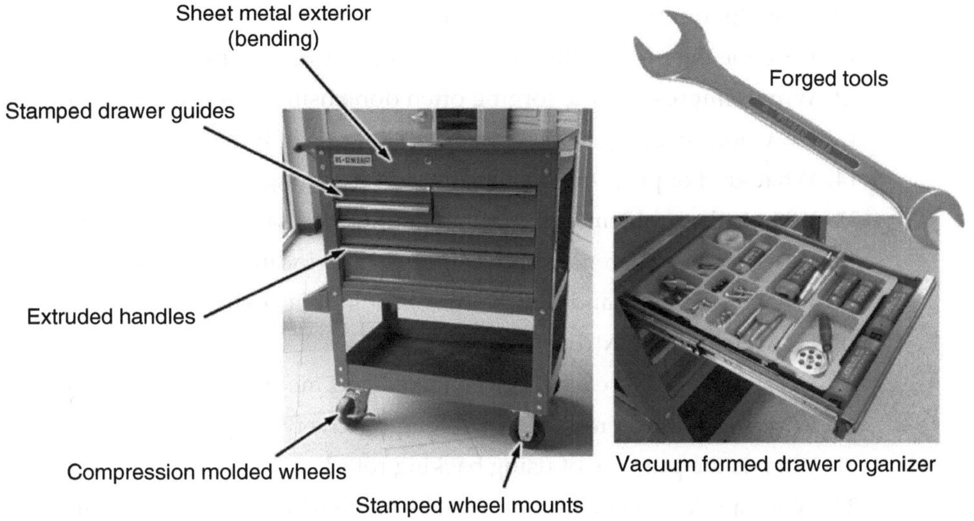

FIGURE 12.48 Forming is used to create many products, but there are also many products that use multiple forming processes. Some of the features of a toolbox are given as an example.

There are many products we use every day that rely heavily on forming operations. Without a basic understanding of forming, we could not begin to contribute to the manufacture of these goods. Consider a basic necessity in manufacturing, as well as in hobby and home repair, the humble toolbox (see Fig. 12.48). Though design varies, many are based on sheet metal as well as incorporating many other forming processes. By knowing the basics of forming, we cannot only appreciate what goes into making the toolbox, but we can also know why.

12.6 The How and Why of Forming

12.6.1 Chapter Review

1. Why was forming popular as an early form of manufacturing?
2. What properties make a material formable?
3. Why is plastic deformation so important in forming operations?
4. How does temperature play a role in forming?
5. How is hot working defined? How is this different from cold working?
6. Describe the microstructural differences between parts made through forging, machining, and casting.
7. How does the rate something is formed at affect the material's response?
8. Why are metals suitable for forming?
9. Describe the differences between open- and closed-die forging, and what type of parts they are suitable for?

10. How do upsetting and drawing out differ?
11. Describe the process for creating a ring by forging operations.
12. Why is impression-die forging often done using multiple forging operations?
13. How does swaging achieve such as high rate of impacts?
14. What kind of parts are made using upset forging?
15. What is the problem with a long unsupported section in upset forging?
16. Describe the differences between direct and indirect extrusions.
17. What are benefits and disadvantages of indirect extrusion?
18. How are hollow extrusions made?
19. Describe the differences between fixed and moving mandrel extrusions.
20. Why are profiles produced by rolling relatively simple?
21. What is the purpose of using backing rolls?
22. What special considerations must be made when rolling is done using multiple stations which all engage the workpiece simultaneously?
23. Why is sheet metal different than other forms of metal?
24. What is the relationship between gauge number and thickness of metal sheet?
25. How is bending used to create parts?
26. What is the difference between bottoming and air bending operations?
27. What kind of geometries does roll bending produce?
28. How can metal be straightened by bending it, such as with roll straighteners?
29. How can complex curves be created manually?
30. What kind of parts are created using drawing?
31. What factor determines if a drawing operation is deep or shallow?
32. What role does the blankholder play in drawing operations?
33. Why are shearing and punching not "true" forming operations? How do they fit well with forming operations?
34. How does shearing actually separate a workpiece into two pieces?
35. Why is clearance designed into shearing operations?
36. How is punching different than shearing? How is it similar?
37. Describe the difference between blanking and piercing.
38. Describe three processes which may be incorporated into stamping operations and how they are used.
39. Compare and contrast compound and progressive dies.
40. Considering design, what are the benefits of a progressive die?
41. Why is PM a popular process for brittle materials?
42. List the major steps in producing a part by PM.
43. What are some benefits of forming polymers over metals?

44. Describe what thermoforming involves.
45. How is the proper application of heating and cooling important in the mechanical forming of polymers?
46. Describe the design differences between vacuum forming and drape forming.
47. What is the role of a plug in thermoforming?
48. Why can pressure forming apply more pressure than vacuum forming?
49. Describe what major considerations must be made in vacuum forming design.
50. How is compression molding different than other polymer molding techniques discussed in Sec. 11.3?
51. What is a BMC?
52. When is transfer molding a good option for creating parts?
53. Why is a water bath often used directly after polymers are extruded?
54. When extruding polymer profiles and why is the die shaped differently than the desired profile?
55. What are sizing and postforming used for in the extruded products?
56. Why is the frost zone important in blown film extrusions?
57. Why are ceramics so difficult to form?
58. What are some ways ceramics are formed, and what makes it possible?

12.6.2 Concept Connections and Extensions

1. What specific advantage is there to hot working steel, which is allotropic? Why is this advantageous? (Hint refer back to Sec. 9.3).
2. What bonding characteristics of metals make them suitable for extrusion? How about polymers?
3. Given the microstructural development in forging (especially cold), what effect will heating at a later stage have? See Sec. 9.2.1.2.
4. Look at a plastic food storage container. What clues are there to how it has been made?
5. What challenges are there to molding thermoset materials? How are these overcome?
6. Using a stress-strain curve (Sec. 7.4.1), describe the minimum useful point on the curve to achieve forming processes.
7. How is forming more energy and material efficient than casting? How is it not?
8. How do the properties of ceramics make them less popular in forming operations? Where they can be produced this way and what post processing is required?

CHAPTER 13

Concepts in Material Removal

Machining covers numerous processes that involve the removal of material. Since it can be done using linear motion, rotation, chemical reaction, electricity, force, etc., it is difficult to define with much detail without describing a single process. Machining can be used to create products as the only type of processing used, but most often it is combined with other manufacturing techniques such as casting or forming. The reason machining is so ubiquitous is that it excels at holding tight tolerances and creating a very good surface quality. Therefore, it is used extensively as a method for creating and finishing products. It is an integral part of nearly all manufacturing operations, and understanding the basics of how it works is well worth the time.

13.1 How It Works

Machining is a **subtractive manufacturing** technique. That term has only recently come into use since the rise of additive manufacturing, which is discussed in Chap. 21. Machining starts with a part of general shape, either standard stock or a pre-shaped through casting or forming. The majority of work starts as a bar or rod of rectangular, hexagonal, or round cross section. That stock is then shaped through the removal of material to create flat or curved surfaces, holes, slots, notches, threads, etc.

Machining is exceptionally versatile in what it can create and how it can be accomplished. Methods that do not use the physical interaction of a tool with the material will not be considered here, but they are discussed further in Sec. 22.2. The remainder of this chapter considers the design and interaction of a tool with a material to remove the material by force in what are often considered "traditional" machining operations.

13.1.1 The Basics of Material Removal

Material is removed through the breaking of atomic bonds in the stock being machined, so the concepts of Chaps. 2 and 7 are going to be valuable, and they are worth refreshing your knowledge on or getting familiar with. In machining, a **cutting tool** directly contacts the workpiece and causes the separation of the material. A simple but important concept to keep in mind is that there must be some method of interaction between the tool and the workpiece. The two must be brought into contact continuously or repeatedly, and with sufficient force to cause the deformation and removal of material.

There are many ways to achieve this, and process design depends on the material and equipment available. Because machining involves one material (the tool) removing another (the workpiece), the tool must be stronger than the workpiece or else it will quickly breakdown and be ineffective. The tool must also be compatible with the machine being used, and, ultimately, the ability to create a part of suitable quality in a timely fashion is dependent on many aspects working together.

There are many cutting tool design considerations we will explore through the remainder of this chapter, but the basic tool requirements are as follows:

1. It should be stronger than the workpiece (not deform)
2. It should remain strong at elevated temperature
3. It should resist wear
4. It should not react chemically with the workpiece
5. It should have a geometry suitable for the process

We will assume that the cutting tool satisfies the general requirements listed above. In Sec. 13.4, we will see that there are significant differences in how cutting tools satisfy those requirements. The way material is removed is dependent on

1. The material characteristics
2. Tool type
3. The tool geometry
4. The tool material
5. The process conditions (temperature, rate of removal, etc.)
6. The relative motion between the work and the tool

We will consider each of these aspects, starting with the basic cutting process. The other considerations are discussed in the following sections.

13.2 The Cutting Process

When machining a part, the process is much different than you might expect. If you look at the tool being used, it probably will not seem very sharp. The material being removed may also look different from the stock it is being removed from. It is important to have a little background on how cutting actually happens so we can interpret feed-back such as noise, vibration, heat, and other clues about whether a process is efficient.

In general, machining requires the deformation and fracture of a material under applied stress. There are many possibilities in the exact progression of a machining operation, but we will look a simple case of a ductile material being cut by a single edge in continuous contact. You can think of this as peeling a carrot or whittling a stick with a pocket knife. A single, sustained stroke causes material to be removed until the material being removed breaks apart, the cut is stopped or the end of the material is reached.

In a ductile material, especially common structural materials such as metals like steel, the cut is not achieved by "slicing" or "splitting" a section of the material off. Instead the cutting tool acts to deform the metal and causes it to flow off from the

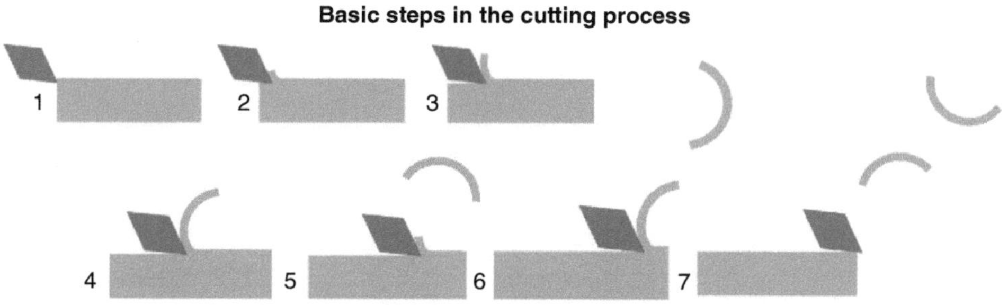

Figure 13.1 The process of cutting is very basic, but the phenomena that control the process are complex. That is, it is easy to remove material, but the physical factors that determine the most efficient parameters must be carefully considered for each tool and material combination. These stages of material removal are described in the text.

base material. The material that is removed, or the chip, is actually the product of deformation and shear. It does not take a sharp tool to efficiently cut many materials.

It is not always easy to visualize how this happens practically. The basic aspects of a cutting operation are shown in Fig. 13.1. The cutting tool applies force to the material, the material deforms, eventually fractures from the base material and flows over the cutting edge. More specifically, the process involves a sequence of steps where

1. The cutting edge contacts the workpiece
2. Stress increases until the material separates
3. Tool progresses and material "flows" over the cutting edge
4. A chip is formed, and for ductile materials, the chip can become very long
5. The chip eventually becomes unstable and fractures
6. The next chip is started and the process repeats
7. The tool exits workpiece and the cut is terminated

The characteristics of the cutting tool and material being machined will determine how the process progresses, and we will consider each in the following sections. Before discussing those specifics, however, keep in mind that the tool and the material affect the way the cut occurs. A geometric parameter, known as the **rake angle** (Greek lowercase letter, theta, θ, see Sec. 13.2.1.2), will determine the force and flow characteristics of the material. The material, especially in terms of ductility and work hardening, determines the **shear angle** (Greek lowercase letter, phi, ϕ). Materials that work-harden (see Sec. 9.2.1) will have a lower shear angle since the deformation is spread out ahead of the cut. This results in a larger **chip thickness**, which is always greater than the **depth of cut (DOC)**. These features are shown in Fig. 13.2.

The chip thickens when cut because the shearing action acts to "pile up" the material before it exits over the cutting face. Think of it as a deck of cards spread out in a line. As you push the cards together to stack them up, the stack gets thicker. The cards do not take up any more volume; they are simply rearranged due to the force (your hands) and their ability to slide. This is analogous to the process in ductile metals. If this seems confusing, see Sec. 3.2.2 for more on the process of shear (or *slip*) in metals.

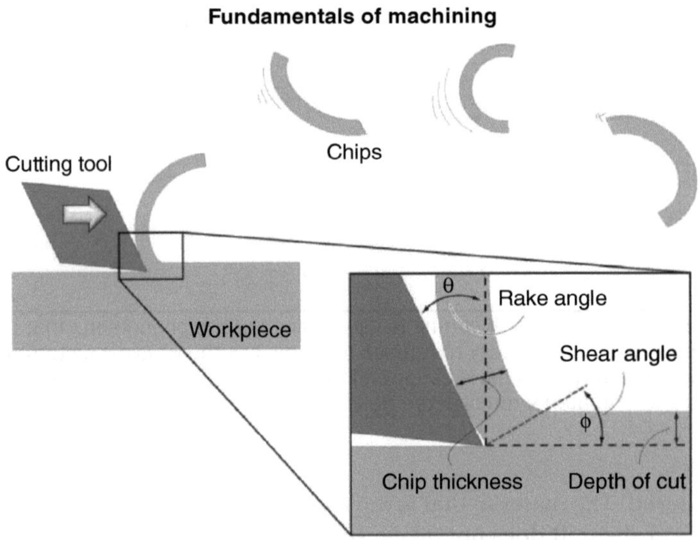

FIGURE 13.2 The general features of a machining process are shown. The tool and material will determine the exact nature of the interaction.

One reason metals are processed extensively through machining is that their atoms have the ability to be "shuffled" around in this way. Brittle materials, which cannot be deformed easily, will not form long, continuous chips. Instead they create very small, discontinuous chips. They also resist building up on the cutting edge. When a material deposits on a cutting edge, it is known as a built-up edge (BUE), and this case is discussed more in Sec. 13.2.2. Because ductile materials tend to form a BUE, they also come with continuous chips. Although there are some variations, these cutting results are often categorized as three types (Fig. 13.3):

1. Continuous chip
2. Discontinuous chip
3. Continuous chip with built-up edge

13.2.1 Tool Geometry

Every cutting tool has a geometry set by the manufacturer. The geometry of a tool is simply its shape with key aspects considered in the way it interacts with the material it cuts. Thinking in the simplest terms, imagine you are scraping paint off a surface with a putty knife. It has a flat edge that is not particularly sharp. How do you hold it? How do you apply pressure? You would not hold it perpendicular to the surface, and you probably will not hold it parallel either. You find somewhere in between that balances the force and ease of getting below the paint layer. Cutting tools work similarly. They interact with the workpiece and force some of that workpiece to separate. If the angles and forces are correct, the process is efficient. If the angles are not right, it will quickly breakdown the tool and result in poor quality.

While cutting tool design is a complex topic, we will concern ourselves with only the basic features that control machining operations (Fig. 13.4). Each material has a

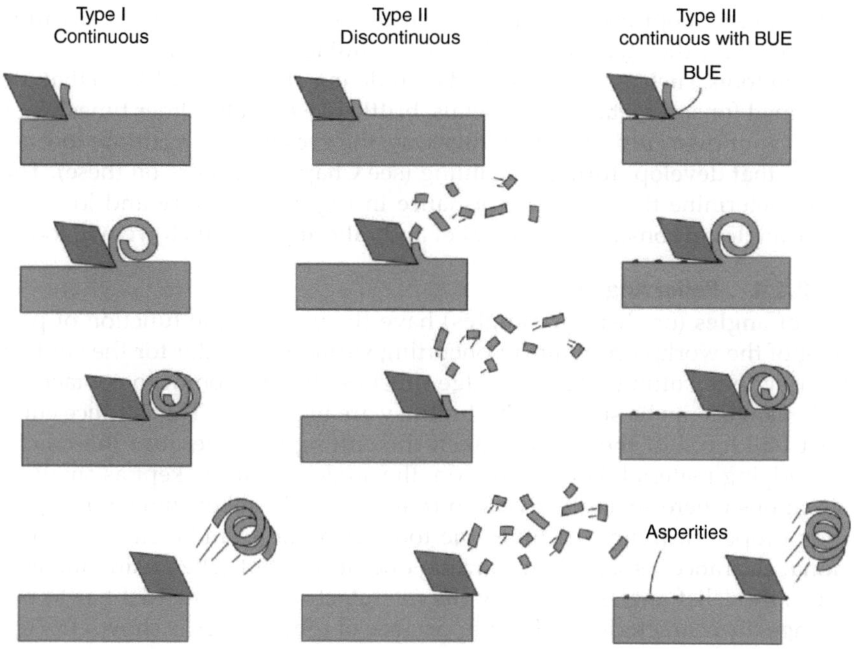

FIGURE 13.3 The way material is removed depends on several factors including the material type and conditions of machining. Three basic chips are formed during machining, and these are shown in this figure.

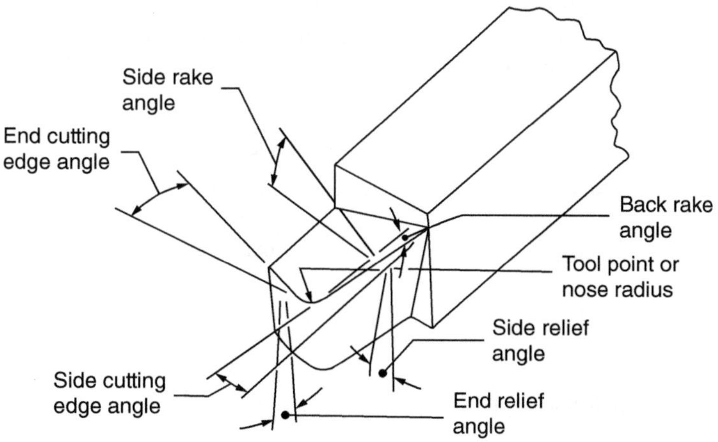

FIGURE 13.4 Common cutting tool geometry includes angles and features that dictate the cutting characteristics, strength, and lifetime of the tool. Some basic features are given here. (Courtesy of Industrial Press, Inc., from *Machinery's Handbook*, 29th edition, Oberg, Jones, Horton, Ryffel. Fig. 1, pg. 757. ISBN 978-0-8311-2901-9.)

different set of optimal conditions regarding tool geometry, and many times the design is something dictated entirely by the manufacturer. For instance, you can purchase cutting tools, such as end mills and carbide inserts (see Sec. 13.4.2) that are specifically designed for soft metals, hard metals, brittle plastics, etc. Other times you may actually grind your own cutting tools. Either way, the geometry of a cutting tool determines the forces that develop during machining (see Chap. 7 for more on these). Those forces, in turn, determine the tool's performance in regard to quality and longevity. The most basic angles to consider are the relief and rake angles, which we will look at next.

13.2.1.1 Relief Angles

Relief angles (or **clearance angles**) have the very simple function of preventing rubbing of the workpiece against noncutting surfaces. In order for the cut to be controlled by a definite cutting edge, the edge must be the only point to contact the workpiece. Relief angles can be small (5–7°), but they are important. They reduce cutting force and heat buildup, but they also weaken the cutting edge because the relief removes the underlying material. For that reason, the angle should be kept as small as possible. In situations where an aggressive feed rate is used, the relief angle must be greater to provide proper clearance. At times the tool orientation can be changed to provide additional clearance, especially in turning operations, but when using tooling such as end mills, the relief angles are set by the manufacturer and the tool has to be reground to change the characteristics. The importance of relief angles is shown in Fig. 13.5.

13.2.1.2 Rake Angles

The rake angles on a tool determine the way the cutting edge enters the material. A tool with what most people would consider a "sharp" edge has a positive rake angle. A neutral (zero) rake angle meets the material at a 90° angle. A negative rake angle has a cutting surface tilted toward the material, making the included angle of the cutting edge larger. This is shown in Fig. 13.6.

Rake angles can control the size, shape, and way chips flow from a material. Increasing the rake angle decreases the cutting force, but it also weakens the cutting edge as it gets narrower. The smaller volume at the edge also means heat buildup will occur more

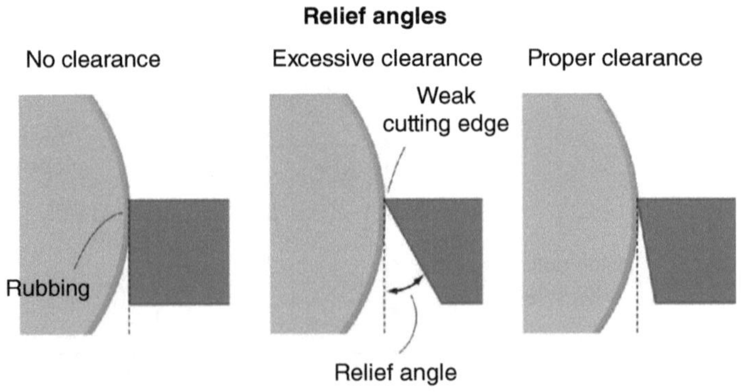

FIGURE 13.5 Relief angles allow the material interaction to happen only at the cutting edge. Without relief (clearance), the cutting tool will rub against the workpiece and cause unnecessary heating and loss of surface quality. In some operations, like turning, the tool clearance is at least partially determined by how the tool positioned and not only the tool geometry.

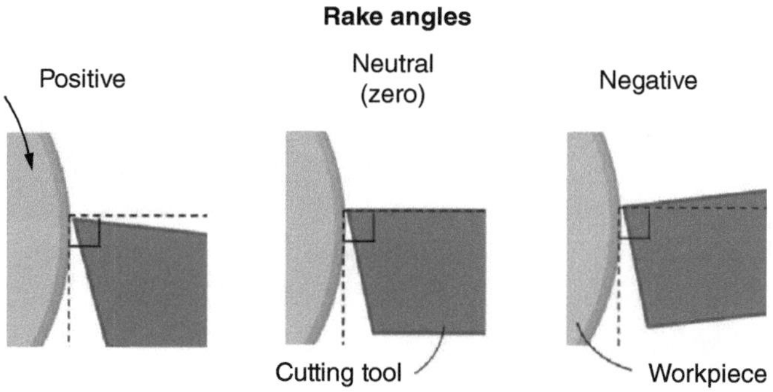

Figure 13.6 The rake angle determines the cutting force and strength of the tool, both deceasing with a larger positive rake angle. As the rake angle increases, the tool's cutting edge becomes "sharper," but the cutting characteristics may not improve, depending on the material.

quickly, too. Rake angles are usually optimized within a certain range for each type of material.

13.2.1.3 Nose Radius

The nose radius of a cutting tool refers to the leading point of the cutting edge and its geometry. For instance, a square has corners with no radius. By rounding the corners, a radius is created until that radius is so large a circle is formed. A sharp corner represents the smallest nose radius and a circle the largest.* Carbide turning tools will specify the exact nose radius, and a larger radius is both stronger and creates a smoother finish on the workpiece. The disadvantage is that it has higher contact area and more force during cutting. Smaller nose radii also act to concentrate heat, which can prematurely degrade the tool. Varying nose radii and their effects on surface finish are shown in Fig. 13.7.

13.2.2 Material Characteristics

The material being machined determines all other facets of process design. Soft, ductile materials like copper will be processed under very different conditions than a brittle material like cast iron. The primary difference in their processing is the way the material separates from the workpiece. Think about breaking off a piece of taffy versus a piece of a candy cane. One will snap while the other stretches and tears. Machining is not so different. When sawing a soft, pliable material such as a polymer, the material that comes off, known as **chips**, will be much different than when polishing glass. The processes (the tool, the motion, etc.) are different and the way the material responds is also different.

Because polymers are so easy to process by forming and molding, and ceramics are difficult to process by machining, we will focus our attention on the machining of metals, which dominate the machining industry. That does not mean there is not significant time and attention paid to polymers and ceramics! Many components are machined from polymers, including state-of-the-art medical devices that carry stringent quality requirements. There are also machinable ceramics and processes, such as abrasive machining (see Sec. 14.6),

*Assuming the circle is not exceptionally small.

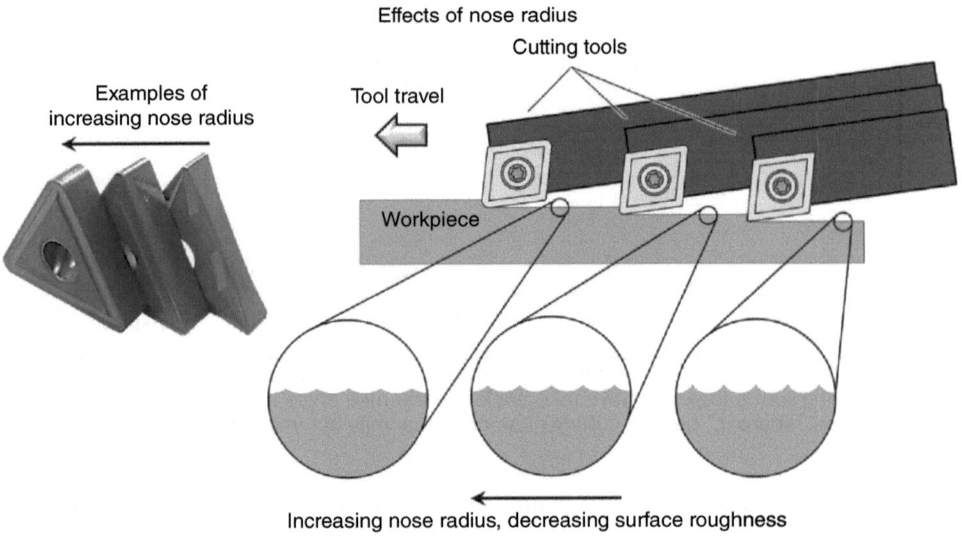

Figure 13.7 The nose radius is an important geometric feature as it controls the surface finish. A larger nose radius produces a smoother finish and is stronger, but it creates larger cutting forces.

which are well suited to dealing with these hard, brittle materials. Our focus on metals will provide the basic insights necessary to understand other materials as well.

When dealing with metals, most display some degree of ductility, but the level of ductility will vary. The way this affects the machining characteristics is most clearly displayed in turning operations (see Chap. 16). Turning operations, also known as lathe work, utilize a rotating workpiece from which the outside is stripped off by a cutting tool. If you have used a small, handheld pencil sharpener, you have done this type of process. As you rotate the pencil, a layer is sliced off and it creates a sharp point.

The way the layers are removed is determined by the material's mechanical properties (Sec. 7.4). Brittle materials will result in small, discontinuous chips, but ductile materials will create long, continuous chips. As noted above, these are known as Type I and Type II chips, respectively. Sometimes the material being cut "smears" onto the tool tip creating a **built-up edge (BUE)**. When a BUE is formed in conjunction with continuous chips, it is a Type III chip. These conditions are shown in Fig. 13.3.

The BUE can be beneficial to tool life as it protects the cutting edge. This is most often to the detriment of surface finish, since the BUE intermittently breaks off and reforms. The material that breaks off is welded to the surface and is noticeable as "speckles" along the surface. In extreme cases, the surfaces will be quite rough. This is shown conceptually in Fig. 13.8.

13.2.3 Thermal Characteristics

Machining creates heat. The rate of machining will determine the amount of heat created. The material being machined may be a good conductor of heat, such as copper or aluminum, or it may be very poor, such as stainless steel or titanium. This is important because heat is a significant obstacle to production. Faster production means lower cost per part but results in more rapid generation of heat and higher machining temperatures.

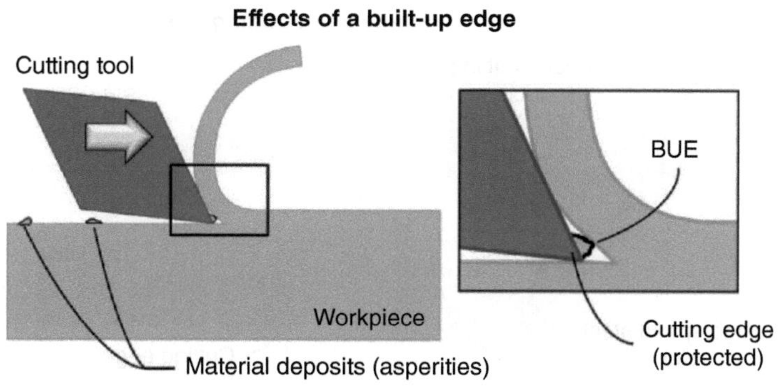

Figure 13.8 A built-up edge (BUE) can help protect the cutting edge, but it also negatively impacts surface finish.

To a certain degree heat can soften the material ahead of the cut and reduce the cutting forces, but materials that do not conduct heat well concentrate the heat at the cutting edge and this can quickly wear down the cutting tool. Cutting tool materials that maintain their strength at elevated temperature allow higher machining rates, a concept discussed more in Sec. 13.4. To further enhance the cutting properties, fluids known as **coolants**, which reduce friction and remove heat, can be added.

Coolant is often a water-based liquid that contains lubricants, though pure oil can also be used. The chemistry of the coolant will determine its efficiency at removing heat and reducing the friction at the cutting edge. The use of coolant should be done judiciously. If coolant is not supplied in a great enough quantity, intermittently, or if it does not reach the cutting edge, it will cease to be useful and can actually be detrimental. For instance, if coolant is added in short spurts during a machining operation, the temperature will increase greatly between applications. The coolant will then rapidly cool the part and tool and this causes thermal cycling which can be destructive to the tool. Proper application of coolant is important to controlling temperature, and the process parameters, as discussed in Sec. 13.5, must also be chosen to balance productivity with tool life.

13.3 Tool Types

With so many types of machining operations, there is an equally vast array of cutting implements. Saw blades and sand paper are both cutting tools since they remove material, but they have distinct uses and characteristics. It is useful to categorize tool types by their general design, and it gives us a basis for discussing more specific interactions later on. Cutting tools can be categorized based on the number and placement of the cutting edges.

13.3.1 Single Point Cutting Tools

Single point cutting tools, as shown in Fig. 13.9, only have one cutting edge that interacts with the workpiece. These are most common in turning operations, as discussed in Chap. 16. The reason they are common in turning is that the workpiece rotates, so it continually brings new material into contact with the cutting edge. Even with only one cutting edge, the process is still efficient. Other tools may be considered to have one

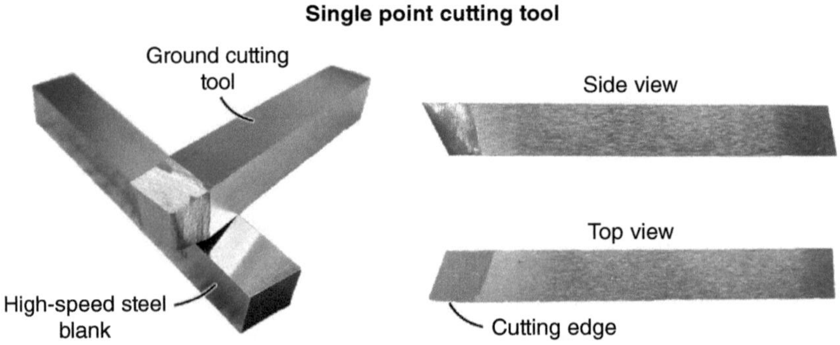

FIGURE 13.9 Single point cutting tools can be created from high-speed steel blanks by grinding, or carbide tooling can be used (see Fig. 13.12). Note the features appearing on this tool in relation to those listed in Fig. 13.4.

cutting surface, such as a chisel, knife, or an axe, but they are not tools for machining in a traditional sense. Single point tools are not limited to turning either. Others include fly cutters and single-flute end mills that rotate against a part as described in Chap. 17, which covers milling operations. The only limitation to the configuration of single point cutting tools is that all material removal is caused by a single cutting edge.

13.3.2 Multi-Point Cutting Tools

Multi-point cutting tools have two or more cutting edges positioned in a definite pattern, such as the end mill shown in Fig. 13.10. The term "definite" will make more sense in the next section when we discuss random point cutting tools, but for now we will assume that the cutting edges are in a fixed pattern. That pattern is usually well defined from the manufacturer. For instance, a saw blade will have a specific pattern to the location, size, and number of teeth (sawing is discussed more in Chap. 14).

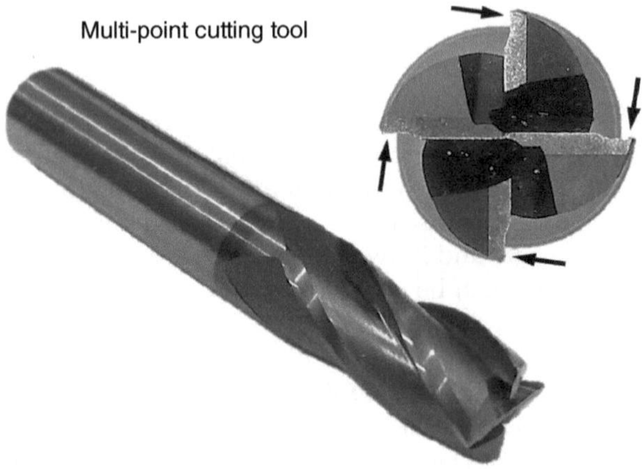

FIGURE 13.10 Multi-point cutting tools distribute the work of cutting over two or more surfaces. A four flute end mill is shown here.

Random point cutting tools

Figure 13.11 Random point cutting tools are nearly always abrasive tools such as the grinding attachments shown here.

A common example of multi-point tooling is the end mill. End mills will most often have two or four flutes unless they have a large diameter (1 in or greater), then they may have six or more flutes. The flutes are regularly spaced around the diameter of the end mill. Note that flutes and cutting edges are not the same thing, but they come in equal numbers. Flutes provide clearance for material to be removed by the cutting edges. This principle is also seen in circular saw blades where many teeth (usually around 30–80) are positioned around the perimeter of a large diameter cutting tool. More about the individual cutting tools will be presented in their related processing chapters.

13.3.3 Random Point Cutting Tools

A random point cutting tool is one that has multiple cutting surfaces that are irregularly positioned. They often have little control over the cutting surfaces except some basic factors such as size, spacing, and material composition. Random point tools are exemplified by abrasives, such as grinding stones (see Fig. 13.11) and sand paper (see Sec. 14.6 for more on abrasives). The grit attached to the backing does not have a specific pattern, but it does have a grit size, spacing, and composition determined by the manufacturer.

There are some other tools with random patterns as well, but they are less common. These include hand-sharpened saws and files, which cannot be as perfectly spaced as mechanically produced versions. Some newer end mills also intentionally design cutting edges to engage differently (spacing, helix angle, etc.). The small inconsistencies are beneficial since it reduces harmonic vibrations that can degrade tool life and surface finish. These types of tools are not truly random, however, because the cutting edges are intentionally placed.

13.4 Tool Materials

There are many choices in tool material, but there is no clear, one-size-fits-all solution. The tool must be stronger than the workpiece, otherwise the workpiece will not be machined; the tool will simply fail. Tool type depends on the needs of the customer and the capabilities of the manufacturer. Perhaps the most qualified source for information on cutting tools is the cutting tool manufacturer, and they are often quite happy to assess your individual needs and make recommendations. It is important to have a general knowledge, however, so you can be a smarter "shopper." The same objective may be achieved in a variety of ways, but not all methods are equally efficient and/or cost-effective.

A simple illustration of this is chopping a carrot. To reduce the carrot into slices, a cutting tool (a knife in this case) is used. The size, shape, and material the knife is made of will determine the efficiency of the process, the quality of the cut, and the longevity of the tool. If the knife is made of sharp, hardened steel, the cut can be made easily and repeated quickly without wearing the knife out and while maintaining quality. If a plastic, disposable knife is used, the task will require more force and likely result in the breakage of the knife while resulting in a lower quality cut. A ceramic knife could also be used. It will maintain a sharp edge longer, but it has a higher initial cost, is more likely to chip if mishandled, and cannot be easily re-sharpened.

This brings into view some of the considerations in choosing a cutting tool material. There are some materials that offer excellent high-temperature properties and can last a long time when used properly. The trouble is that they tend to be more sensitive to misuse and have a significantly higher initial cost. The necessary production rate, the machine rigidity, the operator, the workpiece material, the required quality, and many other factors contribute to the best choice.

We will not explore these concepts exhaustively, but rather it is more efficient for us to focus on the key factors and do our best to account for them in our decisions. Even under ideal conditions, cutting tools wear down over time. They have a demanding job and their performance can be greatly extended through proper use, but it is inevitable that they will fail. The goal is to have it occur less frequently and more predictably. We will discuss what causes failure and how we can prevent it in Sec. 13.4.5.

13.4.1 High-Speed Steel

High-speed steel (HSS) is a type of tool steel (see Sec. 4.2.5) that is well suited to material removal due to its high hardness and wear resistance. It is also relatively tough and can withstand impact better than some of the "advanced" tool materials. It is an all-around good performer for general machining work. HSS is a cost-effective choice as well, and it is used extensively in end mills, drills, and turning tools.

One of its most attractive qualities is that it can be reground (re-sharpened) or custom ground (see Fig. 13.9) using basic abrasive processes (see Sec. 14.6). Because HSS hardens deeply during conditioning (see Sec. 9.3.6.3), this allows the machinist to re-sharpen tools multiple times and to create tools with custom geometries at will without losing the cutting performance. Drill bits can be given new life easily as well, but end mills require specialty fixtures to regrind. Once an end mill is reground, its use should be carefully noted and the reduced diameter must be compensated for.

The disadvantages for HSS are that it is not as hard as alternative, ceramic materials, and it has a lower allowable cutting temperature. If a cutting tool is sharpened, the geometry is permanently changed, which is problematic in automated processes. These problems are largely addressed by indexable carbide inserts, which will be discussed in the next section.

It is worth noting that not all HSS is equal. There are a variety of grades with varying compositions. As noted in Table 4.2, two grades of HSS are designated by T and M, for principal alloying elements of tungsten and molybdenum, respectively. That does not tell the whole story, however, since there are further designations, such as M42, a high-cobalt (8%) HSS that has greater wear resistance and high-temperature performance than other grades. These are often sold simply as "cobalt" cutting tools.

13.4.2 Carbide Cutting Tools

Carbide is a rather generic term since many types of carbide exist. Carbides are simply carbon-containing compound, often with just carbon in conjunction with a metallic or semiconductor element.* For instance a tool material might be silicon carbide, tungsten carbide, or titanium carbide. The most common is tungsten carbide (WC), however, and the term "carbide" is often used as short form for that particular compound. Carbides contain a mixture of covalent and ionic character in their bonds, depending the elements present, but they are rightly categorized as ceramics, though there are other "ceramic" cutting tools (see the next section).

Carbide cutting tools are also known as cemented carbide or sintered carbide, and they have superior properties of hardness, wear resistance, and high-temperature performance when compared to HSS. One of the principle benefits of carbides is their ability to retain their strength and wear resistance, even at elevated temperature, a property known as **hot hardness** or sometimes as *red hardness*. This allows carbide cutting tools to be run at faster speeds (discussed in Sec. 13.5) and without the need to cool the cutting area with coolant. Although coolant is often used, it must be applied carefully to avoid thermal shock where heat can build up intermittently, which might occur if the coolant is blocked by a feature of the workpiece or flow is interrupted for some other reason.

Carbide tooling consists of small particles of carbide (powder) that are held together by a binder. This is similar in concept to a Rice Crispy Treat, which consists of crisped rice adhered together by marshmallow. Even though the crisped rice (carbide particles) are crunchy (brittle), the marshmallow (binder) is strong and ductile enough to create a relatively tough composite material (particle reinforced composites like WC are described in Sec. 5.4.3). The binder in carbide tooling is often cobalt since it "wets" the surface of WC and strongly adheres the particles together. It also has a high melting temperature and strength, which make it suitable for high-temperature operation and allow a minimal amount to be used.

Carbide tooling allows for greater productivity, but the cost of tooling increases as well. The material being cut, the number of parts to be made, and the general machine set up will determine whether the benefits of carbide tooling outweigh the additional cost. It cannot be sharpened as easily (if at all) and is more prone to chipping and fracture, especially in interrupted cuts. As with HSS, there are a variety of grades of carbide and the tooling design also varies. We will look at the two most common tool types: solid carbide and inserted carbide.

13.4.2.1 Solid and Brazed Carbide Cutting Tools

Solid carbide tooling is often limited to drills and end mills, which have a fixed cutting geometry. This is different than HSS, which sees extensive use in custom-ground cutting tools, especially for turning operations. Although the color of a carbide tool is often a

*The element will be less electronegative than carbon (see Sec. 2.3.1.3, particularly Fig. 2.5).

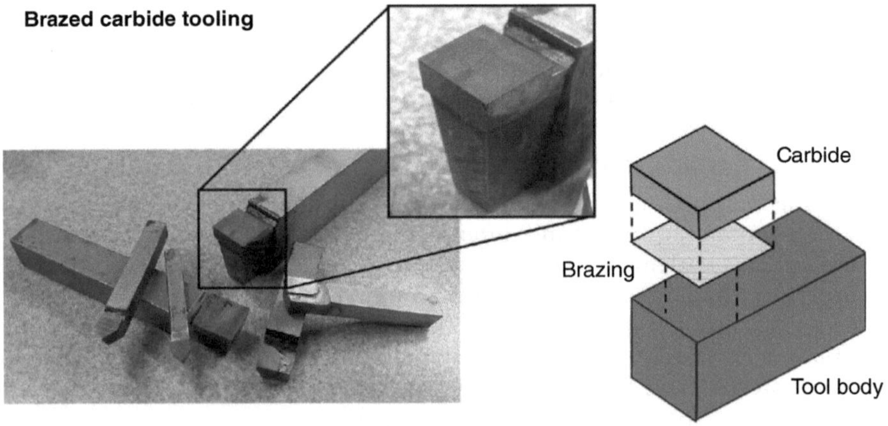

FIGURE 13.12 Brazed carbide tooling uses a carbide cutting surface brazed to a steel tool body.

darker gray and the weight is greater than for HSS, there is little observable difference otherwise in their basic characteristics. It is during cutting that the most significant differences are realized.

In turning operations a cutting surface made from solid carbide may be adhered to a low-cost, steel tool body. This allows less of the more expensive carbide material to be used. It is permanently attached to the body through brazing (see Sec. 19.3.1 for more on brazing). This creates a robust tool that can be re-sharpened to a limited degree, as long as a diamond sharpening wheel is available. Typically this type of tool will use a tougher grade of carbide since even a small chip may ruin the tool. Brazed carbide tooling is shown in Fig. 13.12.

13.4.2.2 Inserted Carbide Cutting Tools

Inserted carbides (or just **carbide inserts**) are small cutting tools that must be held (inserted) in a larger tool body for support. The primary advantage with inserts is that they can be quickly replaced without significant loss of position, so a machine's coordinates do not need to be reset except on critical tolerance work. The insert can have a variety of shapes (round, square, diamonds, or parallelograms of different angles, triangular, etc.), sizes (thickness, width, center hole diameters, etc.), and geometries (rake angles, relief angles, chip breakers, etc.). This makes it possible to optimize tool choice for specific needs, but it also creates a complex decision-making process, especially when deciding on new tooling. The tool holder can only accommodate small changes to the insert specifications, such as a different nose radius or including a chip breaker. If the cutting tool geometry needs to change more significantly, a new tool holder will be required.

Due to the diverse offerings in carbide inserts, they are designated by a series of numbers and letters which indicate their features. Designations may be present in ANSI or ISO format,* and manufacturers may also have specific features to their inserts that no other has. Figure 13.13 shows the general alphanumeric identification system used for carbide inserts under ANSI B212.4. The corresponding tool holders have a similar

*ANSI is the American National Standards Institute, and ISO is the International Organization for Standardization.

1 2 3 4 5 6 7 8 9 10

Ex. D C M T 2 1.5 1 * * *

Carbide Insert Identification ANSI B212.4																
Feature Described	**Common Designations**															
1. Shape	**A**—85° parallelogram; **B**—82° parallelogram; **C**—80° diamond; **D**—55° diamond; **E**—75° diamond; **H**—hexagon; **K**—55° parallelogram **L**—rectangle; **M**—86° diamond, **O**—octagon; **R**—round; **P**—pentagon; **T**—60° triangle; **V**—35° diamond; **W**—80° trigon															
2. Relief Angle[a] (deg)	**A**	**B**	**C**	**O**	**E**	**F**	**G**	**H**	**J**	**K**	**L**	**M**	**N**	**P**	**R**	**S**
	3	5	7	15	20	25	30	0/11	0/14	0/17	0/20	11/14	0	11	11/17	11/20
3. Tolerance (× 0.001 in.)		**A**	**B**	**C**	**D**	**E**	**F**	**G**	**H**	**J**	**K**	**L**	**M**	**N**	**U**	
	Inscribed	1	1	1	1	1	0.5	1	0.5	2–5	2–5	2–5	2–4	2–4	5–10	
	Thickness	1	5	1	5	1	1	5	1	1	1	1	5	1	5	
4. Type	**A**—with hole; **B**—hole and one countersink; **C**—hole and two countersinks; **F**—no hole, chip groove both surfaces; **G**—same as F with hole; **H**—hole, countersink, chip groove on rake surface; **J**—same as H with all features on two sides; **M**—hole, chip groove on one rake surface, **N**—no hole; **Q**—same as C; **R**—no hole, chip groove on rake surface; **T**—hole, one countersink, chip groove on one rake face; **U**—same as T with features on both sides; **W**—same as B.															
5. Size	Regular polygons—the size of the inscribed circle in eighths of an inch; rectangular and parallelograms—the first digit is width in eighths and second is length in quarters of an inch															
6. Thickness	In sixteenths of an inch (1 digit for whole numbers, 2-digits if not)															
7. Cutting Point	Numbers **0–8** indicating nose radius in sixty-fourths of an inch, or letters indicating a *square insert* with *chamfer* of **A**—15°; **D**—30°; **E**—15°; **F**—3°; or *double chamfers* of **K**—30°; **L**—15°; **M**—3°, or **N**—truncated triangle, or **P**—flattened triangle															
8. Special Cutting Point	Size of facet (in sixty-fourths of an inch) measured parallel to its edge.															
9. Hand	**R**—right hand tool; **L**—left hand tool															
10. Other	Edge treatment or surface finish. Honed (in thousandths) to **A**—0.5–3; **B**—3–5; **C**—5–7, or **J**—polished on rake face; **T**—chamfered, manufacturer's standard negative land (rake face)															

[a]Where applicable, second number is facet angle.

Figure 13.13 Carbide inserts follow a standardized identification scheme as shown here.

system for designating the size, shape, and special features. A single insert type may be mounted differently and used for different applications, depending on the design of the tool holder. Similar guidance for identifying a tool holder can be obtained from tooling companies and ANSI B212.5 designations.

Carbide tooling inserts are used for turning, drilling, threading, boring, grooving, and much more. Again, the insert allows for the cutting edge to be quickly and accurately changed. Depending on the insert geometry, it may be able to cut on top and bottom surfaces. If the insert has any rotational symmetry, then it can also be spun around to a new tip. For instance, a square, one-sided (top only) tool has four cutting surfaces and a triangular tool with cutting edges both top and bottom has six cutting surfaces (if it is an equilateral triangle). A carbide insert that can be repositioned to utilize another cutting edge is known as an **indexable insert**. Other tools, such as some threading and parting tools, may only have one cutting point.

13.4.3 Advanced Tool Materials

Despite the advancements in cutting speeds allowed through the use of carbides, the demand for better and faster still exists. Tooling that is more robust, wear resistant, and temperature resistant is an ongoing area of development. Tooling represents a major cost in manufacturing, and it is one of the greatest limitations to production rate. By engineering new materials and tool designs, there are still opportunities to reduce cycle times and increase profitability. The following sections describe some of the other materials used in tooling, all of which are based on ceramic materials.

13.4.3.1 Ceramics and Cermet Cutting Tools

Ceramics are characteristically hard materials (see Sec. 6.2 for some examples). Wear resistance is generally proportional to hardness, so it makes sense that reducing tool wear would be best achieved by using harder materials. It is not quite so simple, however, but hardness is the main advantage of ceramic cutting tools. Their principal disadvantage is their lack of toughness (susceptibility to fracture), a concept discussed further in Sec. 7.4.3.2.

Hopefully the term "ceramic" has caught your attention. You should be wondering how such a general term is used and why carbides, which are ceramics, are not included in this category. **Ceramic cutting tools** are completely nonmetallic (no metal binder) and are often based on oxides, such as alumina (Al_2O_3).

Cermets are cutting tools that consist of nonmetallic particles held together by a metallic binder (ceramic/metallic), though some definitions focus on having a combination of ceramic particles such as titanium carbide and titanium nitride. Because traditional carbides use a metallic binder such as cobalt or nickel, they are technically cermets. The term cermet, however, is often reserved for carbides other than tungsten carbide. They are also sometimes called "hardmetals," which may refer specifically to those based on titanium carbide. The crossover of materials and design can make cutting tool designations confusing at times, so focusing on performance metrics is more important than terminology.

For our purposes, ceramic cutting tools will be used primarily to refer to alumina-based cutting tools. They may use a nonmetallic binder or be sintered as a pure material under high temperature and pressure. Alumina is a very hard material, but it tends to be brittle. Additions of other ceramics such as zirconia or titanium carbide are used to increase the toughness of the tool (called **mixed ceramics**). Silicon carbide "whiskers"

Figure 13.14 Carbide inserts are made with a wide variety of size, shapes, and features. The thickness, hole size, chip breaker, etc. will all determine the performance of the insert in a given application. The tool holder must be matched to the insert so that it fits tightly and supports the cutting edge.

can also be used to create **whisker reinforce cutting tools** that possess greatly enhanced toughness (these are discussed in Sec. 6.4.2.2).

Another common ceramic material in cutting tools is silicon nitride (Si_3N_4), and it is most often paired with other elements to create **SiAlON** (silicon-aluminum-oxynitride, trade name Syalon) and uses ceramics (usually yttria, Y_2O_3) as sintering aids. This and other variants based on Si_3N_4 have hardness values comparable to alumina-based ceramics but with higher toughness and thermal shock resistance similar to carbides.

In general, ceramic cutting tools perform well at very high temperatures produced during high-speed machining, but they are often sensitive to thermal shock. They are used without coolant to prevent cracking due to fluctuations in cutting edge temperature. The extremely high speed (rpm) that ceramics can be run at actually melts the chips it removes and gives the appearance of sparks showering off of the cut area during machining. A key aspect of successfully using ceramics is that machining conditions result in heat being absorbed by the chip and not the tool or part. These conditions are discussed more in Sec. 13.5.

13.4.3.2 Cubic Boron Nitride Cutting Tools

Cubic boron nitride (CBN) is a special form (allotrope) of boron nitride that has exceptionally high hardness* and makes for a very effective cutting material. Because of the cost to produce CBN, it is often an insert brazed onto a carbide insert. This often means that only one cutting edge on a CBN insert is actually made from CBN (see Fig. 13.15,

*Of common materials in commercial use, only diamond is harder. There are some "superhard" materials being developed that may approach or even surpass diamond, but these are not currently available as cutting tools.

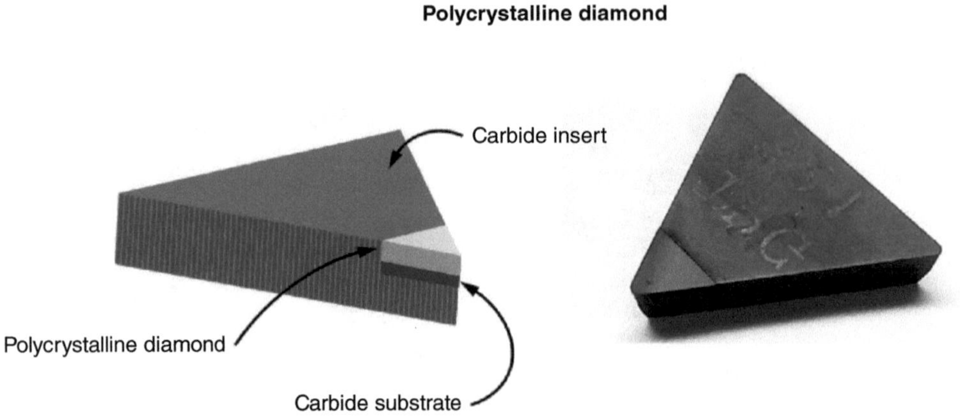

Figure 13.15 Advanced cutting tool materials are often expensive and difficult to produce. To reduce the cost, a small portion of the material is used only at the cutting edge. The material, such as polycrystalline diamond shown here, is brazed onto a carbide insert.

for instance). Despite its high cost, CBN is cost effective because of its longevity and the machining rates it allows.

CBN maintains it hardness up to 1800°F (982°C) and can be run at much faster speeds than carbide but often at lower depth of cut (DOC). Because of its heat stability, it can be run without coolant, but flooding with coolant is often applied for improved results. Unlike diamond cutting tools, it is fully compatible with carbon-containing ferrous alloys (which are ubiquitous). This makes CBN a very attractive tool material. It is used for machining all varieties of steel, but specifically suited for high hardness steel (> HR_C 45, see Sec. 7.6.1.3) and super alloys.

13.4.3.3 Diamond Cutting Tools

Diamond is the hardest naturally occurring material and most effective at resisting wear. Some of its most common machining applications include nonmetallic materials such as wood, plastics, and ceramics. When machining abrasive materials such as fiberglass or carbon fiber, diamond will yield the longest tool life.

Diamond cutting tools consist of a cutting edge made from diamond and can be single crystal or polycrystalline. **Single crystal diamond** cutting tools are often small and used for specialized, ultrahigh precision work [down to tens or hundreds of nanometers (1 nm is ~0.00000004 in)]. This type of tool is very much like diamonds used in jewelry, but instead of using facets for maximizing decorative effects, they are sharpened for optimizing material removal.

Polycrystalline diamond (PCD) cutting tools are made from many, small synthetic single crystal diamonds that are sintered together at extremely high temperature and pressure. This creates a solid tool with excellent abrasion resistance. Because the individual diamonds that make it up are small, the cost is much lower than single crystal diamond. It does not have the same anisotropy a single crystal has, which also reduces variability in wear and eliminates fracture along certain crystal orientations.

The small PCD insert is bonded to a carbide substrate to create a PCD compact, or PDC (note the reversal of the D and C in the abbreviation when it refers to a compact). PCD has a very low coefficient of friction and can produce a very fine surface finish. The single most important limitation with diamond cutting tools is the inability to machine carbon-reactive metals, primarily ferrous alloys. At high cutting temperature, the diamond will actually decompose and the cutting tool wears rapidly. Since steels and cast irons are used in such a large fraction of manufacturing operations, this is a significant disadvantage. Diamond tooling is still an important option for highly abrasive and most nonferrous materials.

13.4.4 Tool Coatings

Throughout our discussion of tool materials, the advantages, disadvantages, and applications have varied. As a general rule, more advanced tooling costs more. It is used because it increases output and reduces long-term cost by lasting longer and being more efficient. In an effort to combine high performance with low cost, tool coatings can be used. A **tool coating** consists of one or more thin layers (usually < 0.0002 in or ~0.005 mm) of a different material applied to the surfaces of a cutting tool. This is often similar to the process of painting an automobile, where a base coat (primer) acts to improve adhesion to the metal, a color coat is applied to add a decorative effect, and a clear coat is used to improve scratch resistance.

In cutting tool manufacturing, the coatings serve to

- Reduce wear
- Reduce friction
- Increase the allowable machining rate
- Change chip flow characteristics
- Reduce chemical reactivity

Coatings are primarily used to reduce the wear rate. Other factors listed above contribute to that reduction. Coatings are often used on HSS and carbide tools to enhance their cutting properties. A coated tool can often be identified by its color. The whole tool may be coated, but more often it is only applied to the cutting areas.

Cutting tool coatings are a very advanced science now, with multiple layers and chemistries used to optimize the durability and performance of the coating. Because the properties of each coating vary widely, there is no "one-size-fits-all" solution. We will look at some common tool coatings more specifically.

13.4.4.1 Titanium Nitride

Titanium nitride (TiN) is the most commonly used coating in tooling, and it is used in many other applications as well (see Sec. 6.2.3.4). It has become increasingly cost effective and is applied to HSS as a means to combine wear resistance with high toughness. It is easily identified by its bright, golden color, and is used in all varieties of tooling such as drill bits, end mills, carbide inserts, punches, etc. TiN allows for significant increases in cutting speed and the coating can prematurely wear at speeds that are too low. TiN oxidizes at modest temperature, so cutting speed should not be excessive, however, and the manufacturer's recommendation should be followed. The addition of this coating can increase cutting speeds up to 50% while maintaining the same tool life.

13.4.4.2 Titanium Aluminum Nitride and Aluminum Titanium Nitride

Titanium aluminum nitride (TiAlN) and **aluminum titanium nitride (AlTiN)** are common alternatives to titanium nitride. These coatings are harder than TiN, and the addition of aluminum increases the oxidation resistance of the coating (or at least oxide stability), so very high cutting speeds can be used. The difference between TiAlN and AlTiN is the ratio of aluminum to titanium, with the primary element being listed first in the name. Because of higher aluminum content available for the creation of stable aluminum oxide, AlTiN has a higher oxidation resistance than TiAlN. This allows for greater speeds to be used and higher production rates to be achieved. Visually, the color of these coatings depends on the ratio of titanium to aluminum, but TiAlN often has a dark violet hue and AlTiN has a dark gray to black color. Coating color can also depend on the thickness of the layers, so appearance alone is not a guaranteed method of identification.

13.4.4.3 Titanium Carbo-Nitride

Titanium carbo-nitride (TiCN) is harder and more wear resistant than TiN and can increase tool life by two or more times. It is also harder than TiAlN and AlTiN, but without the same oxidation resistance. TiCN is not as widely used as a primary tool coating, but it may also be used in multilayer coatings. TiCN is also tough, and a good performer for cutting ductile and "gummy" materials. Another benefit of TiCN is that it can be removed and reapplied relatively easily. This is particularly useful in applications such as molds, which are often very expensive to replace, since the coating can be renewed with little to no compromise to the underlying mold material.

13.4.4.4 CVD Diamond and Diamond-Like Carbon

The properties of diamond have been discussed in Secs. 6.2.4.2 (in general) and 13.4.3.3 (cutting tools). These properties can be applied to tools at lower cost by using coating methods. The two primary coatings are chemical vapor deposited (CVD) diamond and diamond-like carbon (DLC), also referred to as tetrahedral-amorphous carbon (ta-C). Applying diamond as a coating allows for larger tool sizes and complex geometries to be employed economically. The increased tool life over bare carbide can be as much as 50 times or more when machining abrasive materials and ceramics.

CVD diamond coatings are created by placing the tool in a vacuum chamber at high temperature and introducing the coating vapor that coats the surfaces. This process typically creates a thicker layer with a coarser structure. DLC coatings are applied at lower temperature, which creates a fine structure that is categorized as amorphous.

This may seem counterintuitive since amorphous materials have no crystal structure, which is the very thing that makes diamond, diamond. The reason the coating is "diamond-like" is the way the carbon is bonded and the properties it results in, including very high hardness, low coefficient of friction, higher toughness, and overall resistance wear. The exact structure of the carbon and the coating surface structure is determined by the deposition parameters, or "recipe," used.

13.4.4.5 Other Tool Coatings

A variety of ceramic materials can be applied as coatings. These are most often single element nitrides such as zirconium or hafnium nitrides, mixtures of elements such titanium chromium aluminum nitride (TiCrAlN) and titanium silicon nitride (TiSiN), or others such as boron carbide (B_4C). Table 13.1 gives some common coating materials and their properties.

13.4.5 Cutting Tool Failure

Cutting tools fail. It is inevitable. The goal is to make them last as long as possible while also being as productive as possible. While tools can fail in a countless number of ways, there are some basic mechanisms that cause failure in each of those situations. The failure mechanisms, summarized in Table 13.2, are

1. Deformation or fracture of the cutting edge
2. Thermal decomposition or softening
3. Abrasion causing excessive wear
4. Chemical reactivity
5. Material buildup on cutting edge

The most obvious standard of failure is fracture. When the drill breaks into two pieces, it has definitely failed. Plastic deformation, where the cutting edge becomes more blunt, is less obvious, but it is certainly noticeable. It will result in increased force requirements and excessive heat buildup. That is part of the next mechanism as well.

As the tool wears it cuts less efficiently and more heat is generated. Very high temperature can cause the binder used in carbides (often cobalt, see Sec. 5.4.3 for more on this structure) to weaken and the particles to dislodge. In HSS, excessive heat causes the metal to anneal* (see Sec. 9.3.2) and become softer. This leads to plastic deformation that increases the friction and heat that leads to further breakdown and so on. Once tool failure starts, it progresses quickly. Keeping tools in good condition should be a priority.

Abrasion can arise from a variety of sources, but abrasive particles within a material are often associated with sand casting (see Sec. 11.2.1.1). Sand can be embedded during the casting process and causes wear to the tool. At times the tool can actually react with the material being cut. For instance, diamond tooling will react with steel at elevated temperature because of the iron content and its affinity for carbon, but ceramic cutting tools are essentially inert and will be less reactive than carbides.

Even carbides can vary in reactivity based on the amount of cobalt binder they use, with micrograin versions having more. Some coatings can be problematic when cutting materials like aluminum, because it will build up on cutting edges due to increased surface roughness and/or chemical composition of the coating. In general, coatings are beneficial, and titanium diboride (TiB_2) and diamond coatings work well because of their low affinity for aluminum. Cutting tool and coating materials, as well as cutting conditions should be carefully chosen and monitored to account for and reduce the effect of these mechanisms.

*The term "anneal" as used here simply refers to softening by heat, rather than a more traditional austenizing and slow cooling procedure.

Material	Thickness Range (μm)	Micro hardness (HV 0.05)	Coefficient of Friction vs. Steel	Max Service Temp. (°C)	Deposition Temp. (°C)	Color	Coating Benefits	Applications	Trade Name
AlCrN	2–4	3000	0.55	1050	450–500	Gray	Heat resistance, wear resistance	Hard machining at high speeds, milling, hobbing of materials with medium to high hardness, application on tools with high impact load due to geometry	Crosscut™
AlTiN/ TiSiN	2–4	3500	0.4	1100	450–550	Copper	Wear resistance and oxidation resistance at cutting edge temperature of over 1000°C	Hard machining at high speeds, general high-speed machining applications and aerospace alloys, machining of gears	Hardcut™
AlTiN	2–4	3500	0.6	850	450–550	Blue-black	Heat resistance, wear resistance	Gear cutting, cast and nodular iron, hard machining at medium machining speeds, titanium machining, saw blades, and at medium lubrication conditions	Maximizer
AlTiCrN	2–4	3200	0.55	850	450–550	Silver	Heat resistance, wear resistance	Hard material machining at moderate speeds	Dominizer™
ta-C	0.5–2	> 5000	0.15	750	300	Dark gray	Wear resistance, galling resistance	Machining graphite, composites, plastics, and aluminum alloys	Tetrabond
TiN	2–4	2800	0.5	500	450–550	Gold	Wear resistance	Low-end machining applications, hobs	Ionbond™ 01
TiCN	1–3	3000	0.2	300	450–550	Blue-gray	Wear resistance, galling resistance	Machining alloy steels, stainless steels and cast irons at slower speeds, hobs, machining under good lubrication conditions	Ionbond™ 10
TiCrN	2–4	2200	0.55	600	450–550	Gray	Toughness	Moderate cutting speeds	Ionbond™ 14
TiAlCrN	2–4	1800–3000	0.6	800	450–550	Gray	Hardness, toughness	Moderate cutting speeds, interrupted cutting	Ionbond™ 15
TiAlN	2–4	3200	0.6	800	450–550	Gray	Heat resistance, wear resistance	Is similar to Maximizer, but better at lower cutting speeds and good lubrication conditions	Ionbond™ 16

TiAlCN	2–4	2800	0.4	700	450–550	Gray	Reduction of adhesive wear in comparison with TiAlN		Ionbond™ 18
CrN	3–5	2300	0.55	700	450–550	Silver	Toughness	Moderate cutting, Cu alloys	Ionbond™ 30
a-c:H:W + a- C:H	1.5–3	2800	0.15	300	200	Dark gray	Resistance against adhesive wear in soft alloys	Machining of Al, Ti alloys, wood	Ionbond™ 41
B4C	1–2	3400	0.55	800	200	Black	Wear resistance, galling resistance, corrosion resistance	Nonferrous machining	Ionbond™ 50
ZrN	2–4	2600	0.55	500	300	Gold	Hardness, resistance to pick up		Ionbond™ 60
Ti(C, N) + MoS_2	2–4	3000+ 2000	0.03	500	450 for Ti (C, N), 150 for MoS_2	Black	Galling resistance	Machining titanium, stainless steel and some aluminum applications	Ionbond™ 62
Ti(B,C,N)	12–16	3500	0.2	1000	1000	Gold	Low friction, high wear resistance, material compatibility	Rough turning of stainless steel and carbon alloy	Bernex™ 08
TiN-TiCN-TiN	4–6	2800	0.3	800	1000	Gold	Abrasion resistance, impact resistance, material compatibility	Turning of carbon alloy and stainless steels, cast steels, cast iron	Bernex™ 10
Ti(C) N-TiC-TiN	8–10	2600–2800	0.3	800	1000	Gold	Hardness/ toughness	Turning, grooving, moderately to highly interrupted cutting, milling super alloys and stainless steels	Bernex™ 13
Al_2O_3/ Al_2O_2	8–15	3000	0.3–0.4	1200	1000	Black/ maroon	Chemical, metallurgical and thermal stability, resists metal pick up, durability under low impact conditions	Turning, milling, roughing of various Fe materials and alloys	Bernex™ 29

Note: Tool coatings can dramatically improve production rate and tool lifetime, but the proper coating must be chosen.
Source: Table data and trade names for coatings used with permission, IHI Ionbond AG.

Table 13.1 Tool Coatings, Their Properties, and Applications

Cutting Tool Wear Mechanisms	
Wear Mechanism	**As Speed (Cutting Velocity) and Temperature Are Increased**
Deformation/fracture	Always present, but no significant change with speed. Primarily dependent on cutting force (feed).
Thermal decomposition/softening	Starts at high speed (high temperature). Increases modestly as speed is further increased.
Abrasion of cutting edge	Active at all speeds when abrasive particles are present. Increases with increasing speed.
Chemical reactivity	Starts at very high speed (high temperature). Increases rapidly as speed is further increased.
Workpiece material build-up	Most active at modest speed. Decreases at elevated speed and temperature.
Overall wear	Increases rapidly at low speed. Plateaus at moderate temperature. Increases rapidly at elevated temperature.

Note: Cutting tools can fail by a variety of wear mechanisms, which can be operating simultaneously. These mechanisms are directly related to temperature.

Table 13.2 Cutting Tool Wear Mechanisms

13.5 Cutting Parameters (Speeds and Feeds)

Section 13.4 describes the wide variety of cutting tool materials and tool coatings available to manufacturers. This smorgasbord of tooling can be a great asset in achieving more efficient production of high-quality parts. With great power comes great responsibility, and that responsibility manifests itself in the form of choosing the best settings for the machine/tool/material combination. The most advanced tooling available is still only as good as the operator. Here, we will look at what goes into determining the right settings for common machining operations.

13.5.1 Speeds and Feeds

When you hear of the right (and wrong) way to machine a material, you will often hear it termed as adjusting your "speeds and feeds." If you hear a tool chattering, squealing, squeaking, squawking chirping, or most other high-pitched noises, it usually means something is wrong with the setup (tool position or workholding) or the speed is not correct. If you encounter cutting tool chipping, breaking, or workpiece deflection, the feed is probably not right.

All that racket and destruction is pretty easy to identify, but the solution can take a variety of forms. For instance, squealing during cutting can mean that the cutting speed is too high, the feed is too low, the cutter is dull, the spindle is rotating backward, etc. There is a set of parameters to consider when dialing in the most efficient machining operation. The simplest factors are speed and feed.

13.5.1.1 The Importance of Cutting Speed

Cutting speed is a measure of the rate at which the surface of the tool and the surface cutter pass by one another. If the tool rotates, it is related to the rotational rate (rpm) of the tool. If the workpiece rotates, it is dependent on the workpiece rpm. In a band saw, it is the linear rate of the blade.

The reason speed is so important is that it generates the most heat. There are several tool/material interactions that control the generation of heat during machining, but speed is the number one contributor. As the speed increases, heat increases and the demand on the tool is greater.

That is why carbides can run at higher *speed* than HSS tools. They can withstand higher temperature and still perform well. Higher speed means a higher production rate, which means more profit. Although carbide is more expensive, it pays off in the long run.

So how do we determine the right speed? Fortunately, our job is easy. We look it up! Cutting speeds for many workpiece materials, tool materials, and processes are tabulated in handbooks (a few are given in Appendix D). For instance, if you are going to drill a hole in half-hard cartridge brass (see Sec. 4.3.2) using a 3/8-in-diameter HSS drill bit, you can determine the correct rpm. You will find that the tool/workpiece material combination has a constant value known as the **surface speed**, or simply the **cutting speed**. The cutting speed in this case is the velocity of the cutting edge in relation to the workpiece.

When a process utilizes a rotating cutting tool, the diameter of that tool is very important. Imagine the wheel on a bicycle rotating. If you look close to the hub, the spokes appear to be moving more slowly than toward the rim. The velocity of a point near the hub (small diameter) is lower than near the rim (large diameter). Large diameter workpieces or cutting tools must be run at lower rpm to maintain the optimal cutting speed. For every revolution, their outside surface covers more distance (higher velocity).

For this reason, small parts and tools are run at higher rpm, and the equation to determine the best rpm is given by

$$N_s = \frac{12V}{\pi D}; \quad \text{rpm} = \frac{12(\text{cutting speed})}{\pi(\text{diameter})} \tag{13.1}$$

For example, if you want to drill a ¼-in hole in AISI 1018 steel (see Sec. 4.2.3 for more on alloy designations) with an HSS bit, you can precisely determine the appropriate rpm. The only missing information is the cutting speed for the workpiece-tool combination, and that can be found in a metalworking handbook or electronic resource. Depending on the source of the information, you may be able to further refine the choice based on material properties such as hardness of the workpiece or use of coatings in tool design. For soft mild steels, such as AISI 1018, the cutting speed is around 100 sfpm (surface feet per minute). The correct rpm, then, is calculated as

$$N_s = \frac{12V}{\pi D} = \frac{12(100\ \text{sfpm})}{\pi(0.25\ \text{in})} = 1528\ \text{rpm} \tag{13.2}$$

The units in the equation for rotation rate include feet, inches, and min^{-1} (per minute). The reason these can be mixed is due to the presence of the number 12 in numerator, which is really a conversion from units of feet to inches. The diameter must be entered in inches for this equation to work as presented. In its most basic form, the equation for rpm is simply a velocity (V) divided by circumference (πD). The appropriate units and conversion of those units can vary. Using SI units, the equation for rotation rate is different since the surface speed is expressed in meters/minute and the diameter is expressed in millimeters. The equation is then given by

$$N_s = \frac{1000V}{\pi D} \tag{13.3}$$

Although you should be perfectly comfortable solving these equations, it is sometimes beneficial to use mental math on the spot and the approximated equation is generated by assuming $12/\pi \approx 4$ so as to result in the simpler version:

$$N_s \cong \frac{4V}{D} \tag{13.4}$$

This, simpler version can be used to solve our previous drilling problem where 4 multiplied by 100 is 400, and when divided by ¼, is equivalent to multiplying by 4. Finally, 4 multiplied by 400 is 1600 rpm, a result not too far from the 1528 rpm that we calculated using the full version of the equation. The difference can be important in some applications, but for short run, job-shop type operations, it is usually an acceptable error, especially if means the difference between making the calculation and not.

13.5.1.2 The Importance of Cutting Feed Rate

Ideally, material removal will be done as fast as possible. Each cut that does not result in a finished surface will never be seen by the customer and does not *directly* contribute to the final geometry and accuracy of the dimensions. For that reason, the bulk removal of material is accomplished using **roughing cuts**, which are cuts designed to maximize the rate material is removed with surface quality and accuracy as secondary considerations. **Finish cuts**, which are used to achieve high accuracy and acceptable finish, must also be efficient. Choosing the right feed rate is integral to both.

Feed rate is the velocity of a cutter relative to the workpiece as it progresses through the material. It is often expressed as a distance per unit time (e.g., inch per minute or meter per minute), but it can also be expressed as a distance per revolution (usually for turning, as discussed in Chap. 16) or a distance per cutting edge or per "tooth" on multi-point cutters as in sawing (Chap. 14) or milling (Chap. 17).

A high feed rate will increase the rate excess material will be removed, but it will also result in a poor surface finish, higher stress on the cutting tool, and higher temperature during cutting. When possible, it is best to use cutting tools that are designed for rough cuts, or "roughing." Although heat is a concern, it is not as closely linked to feed as it is to speed. The primary concerns with high feed are the strength of the tool, the rigidity of the machine, and the rigidity of the workpiece and workholding.

Just as with speed (rpm), the feed rate is important to get right. Too much feed and the tool will break. Too little feed and the cut will be inefficient, result in excessive rubbing, and the tool will wear quickly. The best value is determined by factors such as the tool material, tool design, and workpiece material. Tabulated data is available in handbooks and other resources to aid your decision, but the tool manufacturer will also provide this information as well. A few examples are given in Appendix D.

The way feed is tabulated is often in inches per tooth (for milling) and inches per revolution for lathe. As discussed more in the chapters on those processes, the tooling can vary quite a bit, but the calculation is changed only slightly.

For example, most milling machines will be programmed (if CNC, see Chap. 22.3) in inches per minute, but the suggested feed rate is in inches per tooth. The reason is that the rpm of the tool will affect the chip load, or the amount of material removed by each cutting edge. If the tool rotates very fast and moves very slowly through the material, each tooth will do relatively little work *per revolution*.

This is like a slow-moving tornado, which causes significant destruction over a smaller area than a fast-moving one. The difference with tooling is that when run too slow, it is not doing much work, and at too great a velocity, it will simply break the tool. Based on an optimal chip load, the linear feed rate (f_m) has to consider rpm [N_s, see Eq. (13.1)] with the chip per tooth, f_t, as well as the number of teeth, n, on a cutting tool. The equation for this is

$$f_m = f_t n N_s; \quad \frac{\text{feed}}{\text{min}} = \left(\frac{\text{feed}}{\text{tooth}}\right)(\text{number of teeth})(\text{rpm}) \tag{13.5}$$

What we see in Eq. (13.5) is that the more teeth we have, the greater the feed rate can be while maintaining the same chip load. A cutter with four teeth can be fed at twice the rate as a two flute cutter. This is not a good reason to make all cutters with as many teeth as possible. More teeth for the same size cutting tool will reduce the space for chips to flow out of the cut area and will clog with material more easily. The increase in teeth also results in an increase in teeth being engaged during the cut and that results in an increased requirement for machine power and rigidity.

For example, if we borrow the information used in Eq. (13.2), where 1528 rpm was calculated, and assume a 3/8-in-diameter end mill (same diameter in the previous calculation), 0.005 in/tooth chip load and four flutes, the rate we would feed the cutter is

$$f_m = (0.005 \text{ in/tooth})(4 \text{ teeth})(1528 \text{ rpm}) = 30.6 \text{ in/min} \tag{13.6}$$

So what about lathe? Lathe tooling is often single point. The equation is the same with the exception that n is equal to 1. Also, what was feed per tooth is now feed per revolution.* Because lathe tooling is in continuous contact during cutting, the tool can be rigidly held and there are no flutes to get clogged with material. Therefore, higher feed rates are often used successfully.

Imagine our lathe is set at 0.015 inch per revolution, but we want to know what the equivalent inch per minute would be at 2500 rpm. We need to give some leeway with the units of "teeth" and "revolution." One or the other does not cancel, but it is only for convenience sake that we add those units in the first place. Strictly speaking, n has no units and N_s has units of min^{-1}. That said, all we need to do is plug the known values into Eq. (13.5):

$$f_m = f_t n N_s = (0.015 \text{ in/revolution})(1 \text{ tooth})(2500 \text{ rpm}) = 37.5 \text{ in/min} \tag{13.7}$$

*With a little consideration you will find these are equivalent for any tool with one cutting edge. For instance, a one flute end mill will also have equivalence between the units of inch per revolution and inches per tooth, just like a single point turning tool does.

In that case, if you had a 37.5-in-long cut to make, it would take 1 min to complete the pass. Just enough time to grab a cup of coffee while the lathe runs. But, of course you would not. Would you?

The calculation of feed is much more useful than planning coffee breaks. It is used to calculate the time required to produce parts, to maximize tool life, to determine the power requirements of the machine, and to ensure the tool, part, and workholding are adequate to handle the force. Even with all the factors we have considered with speeds and feeds, there is yet another fundamental consideration in material removal which we will consider next.

13.5.2 Climb versus Conventional Milling

The direction a tool or workpiece turns during machining is very important. Most tools only cut well in one direction, but would you expect that the tool's path is also important? If you are milling, the cutter will almost always need to be rotating clockwise (as viewed from above), but cutting from left to right will give distinctly different results than from cutting right to left.

First, consider the illustration in Fig. 13.16. It shows that although the rotation direction does not change, the feed direction can, and feed direction affects the way the tool removes material. In **conventional milling** (also called *up* milling), the cutting edge starts at the thinnest point of the chip being formed and progressively increases in thickness. This results in an initial stage where stress must build up before the tool "bites" into the material. This is comparable to peeling off a sticker by starting in the center. It would not make much sense and requires more effort. It is easier to start at the edge and peel it up, which is analogous to climb milling.

In **climb milling** (also called *down* milling), the cutting edge engages the material at maximum depth and thins as it exits. This means the cutting edge is immediately engaged in the material, thereby removes the inefficient friction and stress buildup. Additionally, climb milling ejects chips away from the cutter's path, and conventional milling ejects chips into the cutter's path. Because climb milling reduces the pressure

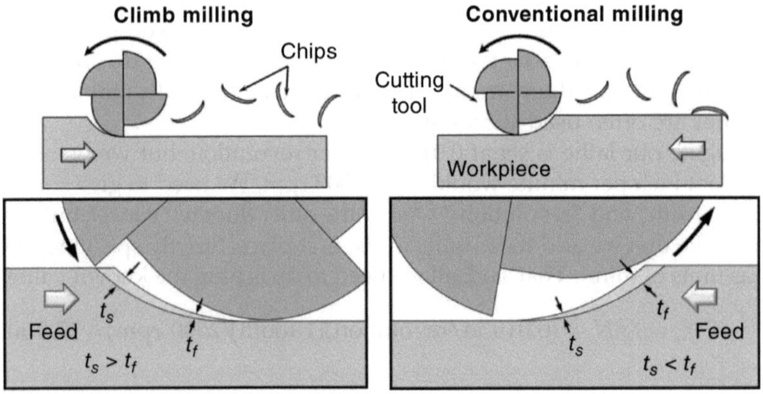

FIGURE 13.16 Climb milling has several advantages over conventional milling. This is mostly because the cut starts (t_s) thick and finishes (t_f) thin. Also, climb milling ejects chips from the cut area, but conventional milling ejects them into the cutter's path.

and friction during cutting, it also works well for work-hardening materials. The limited benefits of conventional milling include a less rigid setup being required, and in cases where the workpiece surface may be harder or abrasive (such as sand castings), since the cutting edge does not directly impact surface materials. The benefits of climb milling include

- Improved tool life
- Improved surface finish
- Works well on work-hardening materials
- Reduces power requirement

13.5.3 Machining Productivity

Having considered tooling types, materials, wear mechanisms, speeds, and feeds, we can consider the limitations on productivity. Manufacturers want to produce parts as fast as possible for the least amount of money possible. The obvious choice for this would be to increase the machining rate, but there are significant drawbacks to that. As the speed, feed, and DOC are increased, the amount of heat generated and tool wear increase rapidly. The general response to the factors is shown in Fig. 13.17.

Productivity has a limit. Finding the balance between output and processing cost is where optimization is focused. The time and cost required to replace tooling and scrap parts of inadequate quality will eventually outpace increased productivity (parts/hour, for example). The limiting factors include machine and tool properties. Tool materials, which are discussed above, are certainly important, but there are some basic machine considerations that deserve at least brief discussion.

High-speed machining (HSM) is a technique for removing material quickly by running the spindle at higher rpm (typically greater than 15,000 rpm) and taking lighter cuts at higher feed rates. Because the spindle operates at such high speed, the chip per

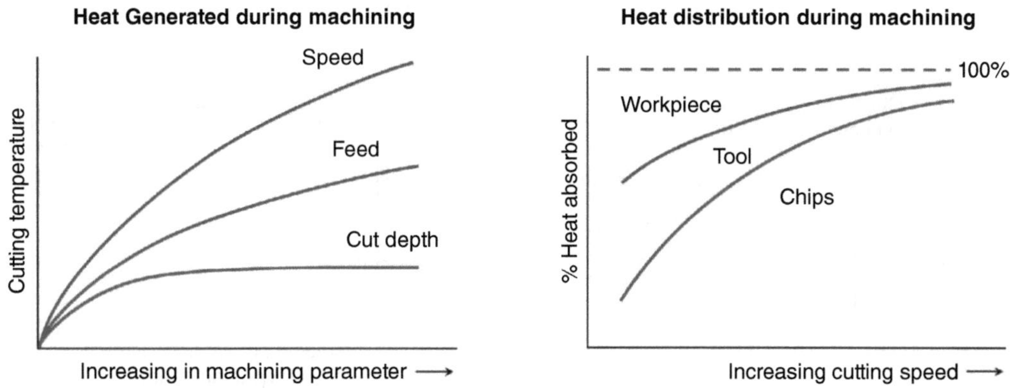

FIGURE 13.17 The cutting parameters used during machining will directly determine the conditions experienced by the tool and the workpiece. The generation of heat is a primary concern to part quality and tool life, and some basic relationships between material removal and heat are shown.

tooth can be kept relatively low while still resulting in an acceptable **material removal rate (MRR)** and surface quality. The MRR is simply the volume of material removed per unit time (e.g., in^3/min or mm^3/min).

The volume of material intersected during machining is fully determined by multiplying the DOC (in or mm), width of cut (in or mm), and feed rate (in/min or mm/min). The speed acts to determine the chip load. The benefits of HSM on MRR, quality, and tool life depend on the machine and tool, but for HSM to be effective, the machine should be more rigid, have faster feed and rapid rates and more accurate control over movement to compensate for the increased velocities and forces encountered.

It does not take expensive, purpose-built equipment to maximize production. Techniques such as plunge roughing and other tool and feed choices take advantage of the excellent axial rigidity of most all machines. Lighter cuts (smaller DOC) at high feed can also improve MRR without introducing vibration. This can actually be problematic if the speed is not increased sufficiently.

Chip thinning is a problem that arises on light peripheral cuts, since the chip never reaches the full thickness of what would be predicted by simple calculation, such as that in Eq. (13.5). If 50% of the cutter is engaged, Eq. (13.5) is accurate. Unfortunately, the way the cutting edge enters and exits the cut is less efficient at 50% than using 25 or 70% engagement. The effective chip thickness will decrease as the cut decreases, and the shallower the cut, the more drastic it becomes. This is shown in Fig. 13.18. To maintain the correct chip per tooth feed rate, the linear feed rate will have to be higher. This has the benefit of reducing finishing times, since finishing passes tend to have lower cut depths.

Regardless of the machine you are using, optimizing feeds and speeds will extend tool life, improve part quality and increase output. Machining can be approached in different ways, but generally lighter cuts performed at higher spindle speeds and feed rates result in the most efficient part production. Some common machining parameters and their effects are given in Table 13.3.

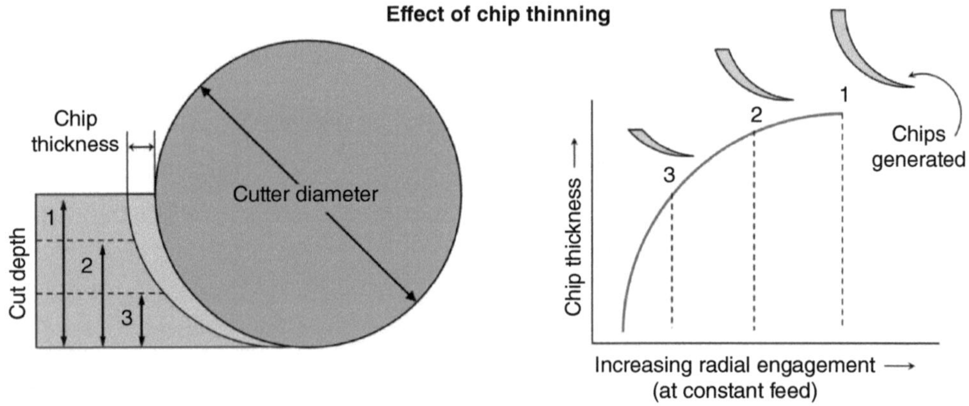

FIGURE 13.18 The radial engagement (cut depth along the perimeter) will determine how thick the chip is. For a 50% cut depth (half the radius) the chip thickness will be equal to the linear feed rate, but shallower cuts will result in a thinner cut.

Parametric Effects in Machining	
Parameter	**Effects When Increased...**
Speed	Temperature rises rapidly, tool wear accelerates, material removal rate increases
Feed	Temperature rises, cutting forces rise, chip flow changes, surface finish decreases
Depth of cut	Temperature rises modestly, tool deflection increases, material removal rate increases
Radial engagement (step over)	Tool wear decreases (between 25% and 70%), chip thickness approaches nominal, material removal rate increases, cutting power increases

TABLE 13.3 Summary of Cutting Parameters and Their Effects on Cutting

13.6 Why It Matters

Machining is a fundamental capability in manufacturing. Any **subtractive manufacturing** operation (one that removes material to produce a desired geometry) that uses a physical tool/material interaction can be considered under machining, and there is no limit to the materials and configurations possible. Even with many other ways of fabricating goods, machining is present in almost every one. For instance, even techniques that are alternatives to machining, like casting, molding, forging, and other operations that use shaped dies or molds, rely on machining to create the necessary tooling.

We have discussed the basic concepts in the cutting process, and without a solid understanding of the conditions during cutting, it is very difficult to diagnose a problem with quality or tool life. It is also difficult to optimize processing without a working knowledge of tool materials, speed, feeds, and other factors in machining. Machining is simple to accomplish, but it requires skill to do well. Manufacturing is a highly competitive industry, and reducing cost for the company and the customer must be a central goal.

Some ways this can be done is through the proper choice of tooling. The material of the tool, surface coating (where applicable), geometry, and cut conditions will determine the time required to produce a part and the longevity of the cutting tool. Advanced cutting tool materials, like carbides and ceramics, have allowed higher cutting speeds and that has improved tool life while decreasing cycle time. Fortunately, tool manufacturers study machining processes very closely to design new tools that can improve productivity, but with so many new choices, you too must be knowledgeable. It is not as simple as just buying the most expensive tooling to get the best results. Each tool/material combination should be considered together. The example given of diamond tooling with ferrous metals proves the point. The best tool might come at a lower cost.

No tool can survive abuse. Running the tool at too high (or low) of an rpm will increase the tool wear rate and lead to premature failure. Taking cuts that are too deep and too fast will result in the same. There is a right balance between aggressive

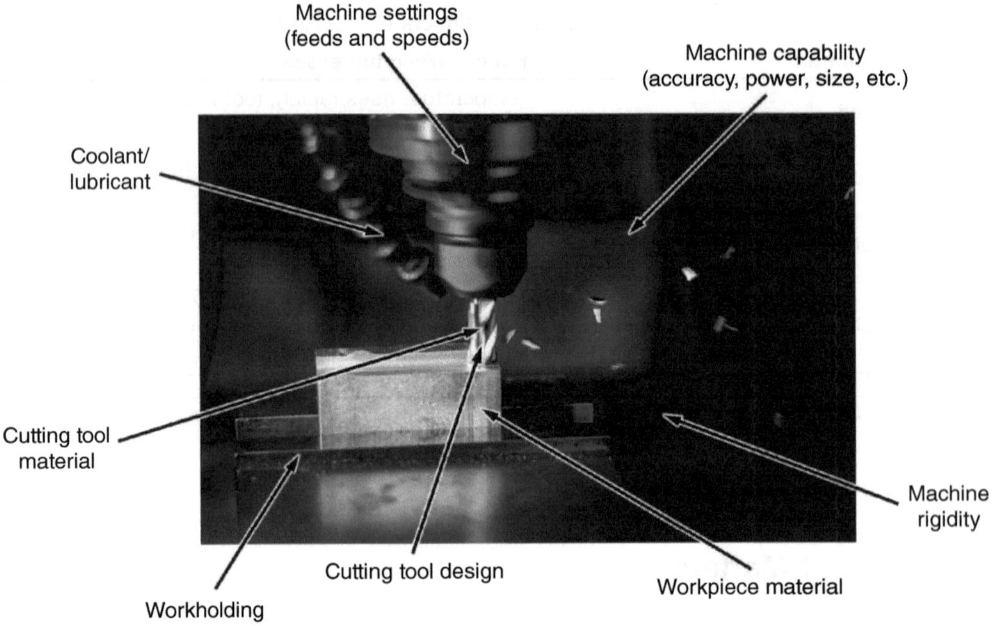

FIGURE 13.19 Machining productivity and efficiency relies on many factors simultaneously. Some of those factors that should be considered are listed here.

material removal and maximizing tool life. Good machinists and machine programmers find that balance. This is especially important in high-production scenarios where hundreds or thousands of parts must be machined, and every second of down time required to replace tooling costs money, not to mention the cost of tooling itself.

If you want to be effective at machining, you need to know the process from every angle. The machine capabilities, the tool holder, the tool geometry, the tool material, the workpiece material, and the machine settings all work together to determine the best result (see Fig. 13.19). Any weak link will limit productivity. In some cases a new machine with more power and speed is the only answer, but many times there is more to be achieved with what is available. Knowing how to achieve that is what makes you valuable.

13.7 The How and Why of Concepts in Material Removal

13.7.1 Chapter Review

1. What is meant by subtractive manufacturing? How does this relate to machining?
2. Describe the basic interactions that must occur during machining.
3. Why do you think cutting tool design remains a challenging topic after so many years of development?

4. What factors affect cutting characteristics?
5. How does a tool actually remove material from a workpiece, and why are some cutting tools not even sharp?
6. What determines the shear angle during cutting? What difference does it make to chip thickness?
7. Why are metals particularly suited to machining?
8. What is the primary difference between a Type 1 and Type 3 chip?
9. What does tool geometry affect during machining? Who determines it?
10. What is a relief angle? What does it control during cutting?
11. How are the nose radius and surface related? Why?
12. What is a built-up edge? Can it ever be a good thing?
13. Why is heat an important consideration in machining processes?
14. What purposes do coolants serve? How can improper use of coolants be damaging?
15. What application is a single point cutting tool most common in?
16. What is the relationship between the number of flutes and the number of teeth on a multi-point cutting tool?
17. Describe the design of a random point cutting tool. Where are these tools used?
18. What general considerations are necessary in choosing a cutting tool material?
19. Why is HSS still a popular choice for machining?
20. What benefits do carbide tools have over HSS? What are their limitations?
21. Why do brazed carbide tools use a tougher grade of carbide?
22. What benefit do brazed carbide tools have over carbide inserts?
23. Describe what the following carbide insert designations indicate:
    a. CNMG432
    b. TPMR321
    c. RNMG43 (why is there no seventh digit?)
24. What are the benefits of an indexable insert?
25. How have advanced cutting tool materials allowed for greater machining productivity?
26. What are the desirable features of ceramic and cermet cutting tools?
27. What makes CBN a good cutting tool material?
28. What disadvantage does diamond have given the high volume of ferrous processing?
29. What benefits do tool coatings provide?
30. How do tool coatings reduce cost for improved performance?
31. What are the performance differences between TiAlN and AlTiN?
32. In what applications is TiCN a better choice than TiAlN and AlTiN?

33. How do DLC coatings allow greater flexibility than inserts such as PCD?
34. Why is it inevitable that a cutting tool will fail?
35. What happens to the wear rate of a tool as the tool begins to wear?
36. Can multiple wear mechanisms be active at the same time? Explain.
37. How does temperature affect tool wear mechanisms?
38. Describe what "speed" is in reference to cutting.
39. How is heat related to cutting speed?
40. What is required to determine the appropriate cutting speed for a cutting tool and workpiece materials combination?
41. Calculate the appropriate cutting speed for the following situations:
    a. 2-in-diameter cutting tool and cutting speed of 150 sfpm
    b. 0.75-in-diameter cutting tool and cutting speed of 80 sfpm
    c. 0.25-in-diameter workpiece and cutting speed of 350 sfpm
    d. 20-mm-diameter workpiece and cutting speed of 100 mpm
    e. 10-mm-diameter workpiece and cutting speed of 30 mpm
42. What are the differences between roughing and finishing cuts?
43. What does feed rate measure?
44. How can you determine the appropriate feed per tooth for a cutting tool?
45. Calculate the feed rate for the following situations:
    a. 0.005 in/tooth, 1200 rpm, four flute end mill
    b. 0.005 in/tooth, 1800 rpm, single point cutting tool
    c. 0.1 mm/tooth, 3200 rpm, two flute end mill
    d. 0.15 mm/tooth, 600 rpm, six flute end mill
46. Calculate the feed rate for a 0.5 in, four flute end mill for each speed calculated in Question 45.
47. How does the number of flutes affect the feed rate for a tool?
48. How are conventional and climb milling different?
49. What are the benefits of climb milling? Is there any situation where it may not be the better choice?
50. Describe the balance between productivity and tool life. Specifically, why can we not maximize both?
51. How can tool life and productivity be optimized?
52. What is HSM? How does it differ from traditional machining?
53. Explain chip thinning and how it can affect cutting efficiency. Why might finish passes be done at a higher feed rate?

13.7.2 Concept Connections and Extensions

1. Carbide inserts that have cutting edges on both the top and bottom are used in negative rake tool holders. Why?
2. Show the appropriate conversions to justify the final units of *per minute* in Eq. (13.1). How is this information lost in Eq. (13.4)?

3. Describe how machining metal is less like true "cutting," and more like plowing? How does this explain that cutting tools with nearly square edges work so well?
4. Describe the benefits of an end mill with many cutting surfaces. What are the disadvantages?
5. If a cutting tool costs the same with or without a coating, which one do you choose? Explain your answer.
6. Describe the relation between the tool geometry and the workpiece material. How does each determine the cutting characteristics?
7. Recent developments have shown that cryogenic machining can be more effective at increasing tool life on difficult materials than traditional coolant. Describe the reasons for this and how it may also be environmentally beneficial.
8. What are the benefits of HSS over advanced tool materials? Do you think it will ever be phased out? Why or why not?

CHAPTER 14

Sawing, Broaching, Filing, and Abrasive Machining

Familiarity breeds contempt, or so the saying goes. Some processes are so common, you might not think twice about them. You just do what "works." Methods such as sawing and filing are used in just about every home, and most people will not stop to consider whether they have the best saw or file for the job or are using it most efficiently. The same can be said of sanding, an abrasive machining technique. Most know the general differences between sanding grit, but there is more to be considered.

This is not a great problem since the desired outcome might only be a few cuts or parts, but when producing hundreds or thousands of parts, it becomes very important. Quality, cost, and output are critical, and small errors can accumulate into big problems. In this chapter, we will examine sawing, filing and abrasive material removal, along with a related process you might not be as familiar with, broaching. Some of the facets we will consider are the way these processes work, how the tools are designed, and when and how to apply them.

14.1 How They Work

Sawing, broaching, filing, and abrasive processes are similar in their designs. Specifically, they use numerous cutting edges to distribute material removal, and the resulting features are dependent on the shape, size, and characteristics of the layout of those cutting edges. We will consider all of these processes to be **polyoperative cutting processes**, or those which engage *many* cutting surfaces in a single tool. This is a subcategory of multipoint tooling and distinct from operations such as milling, which engage only a few cutting surfaces simultaneously. In this chapter, the cutting tools have tens or even hundreds of cutting surfaces, and in milling operations, this is comparatively rare.*

The cutting tool may have a **definite cutting edge layout**, where each cutting surface is specifically placed with consideration to the body of the tool and to the other

*Some operations, especially those in horizontal milling, will have more teeth than the majority of processes, and certain operations, such as slit milling, are really sawing operations performed using a mill.

cutting edges, or it may have a **random cutting edge layout** where cutting surfaces have no specific cutting geometry, as described in Sec. 13.2.1. A random layout is only typical to abrasive processes, but that does not mean the tooling is not designed to perform. The size, spacing, material, and other factors are all engineered for specific uses. In definite layouts, the design is just easier to see and has a more recognizable geometry.

Despite the similarities between sawing, broaching, filing, and abrasive machining, there are significant differences, and we will consider each technique separately. Together these techniques can be used to create a variety of external and internal features, but each serves specific purposes in manufacturing. They may be used anywhere in the production sequence, from the first operation to the last, and they may be encountered multiple times along the way.

14.2 Sawing Basics

Sawing is one of the most common manufacturing techniques applied. It is often used to separate smaller workpieces from a larger piece of stock as the first step in a series of processes, but it may also be employed multiple times throughout production. In some disciplines, such as woodworking, sawing operations make up the majority of the processing protocol. Manual and powered versions of saws can be found in almost every manufacturing environment and even in most homes. Despite how common saws are, there are important aspects of their design and use that are commonly unknown or simply ignored. We will take time to examine these aspects so that you are more informed on decisions regarding the right saw for the job and the best way to use it.

14.2.1 Saw Blade Design

Just as with any cutting tool, saws are designed for specific uses. Some saw blades may work in a variety of conditions, but they are always designed with particular applications in mind. For instance, a saw designed to cut metal will cut wood, but the cut will not be as efficient. The spacing, size, and material of the saw blade will vary for each purpose. In cases where a blade is designed to be "multipurpose," it will not be the most efficient at any particular task, but provides additional convenience where a tool is not dedicated to a specific material or operation.

To better understand the differences, we will consider the basic features each saw blade will have. The most basic features and terminology of saw blades are shown in Fig. 14.1. Most blades have a neutral rake (hook) angle, which means the tooth face is

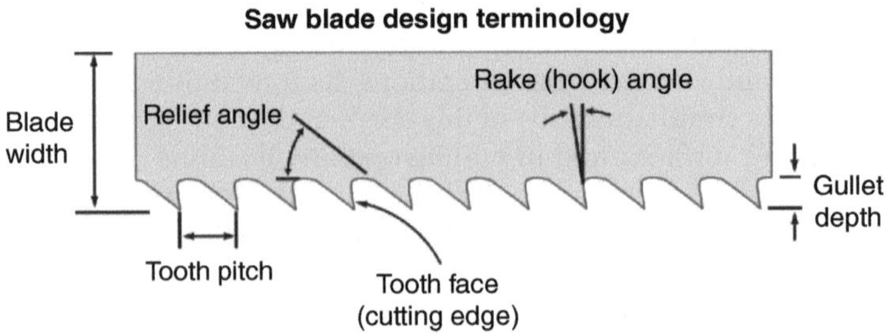

Figure 14.1 The general design of saw blades is largely dictated by geometry. The terminology for geometric design is given here. Further design designations are given in Fig. 14.3.

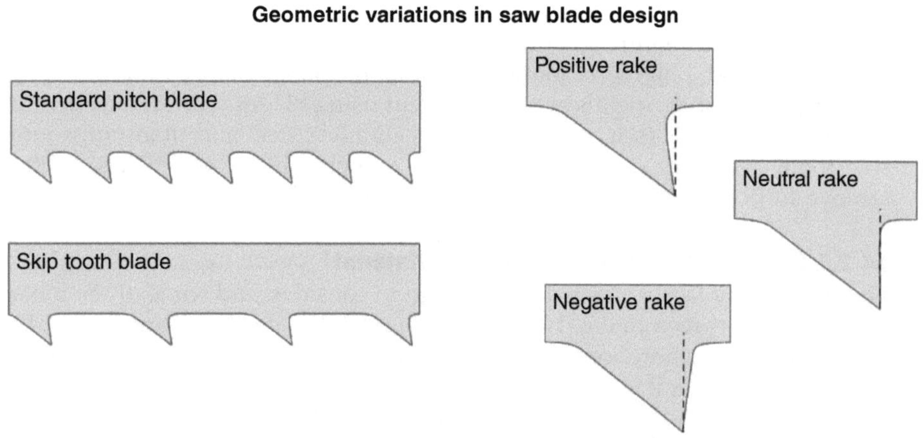

FIGURE 14.2 Blade geometry such as pitch and rake angle determine the cutting characteristics. In general, larger pitch and positive rake blades will remove material more aggressively.

90° to the workpiece surface, but for more rapid material removal a "hook" blade may be used, which have a positive rake angle.

14.2.1.1 Cutting Edge Geometry

Every saw blade must have multiple cutting edges (teeth). The size and spacing (pitch) of the teeth will determine the rate of material removal possible. The deeper the gullets and farther apart the teeth are, the more material can fit between the teeth without "clogging" the blade. During a cut, there are a specific number of teeth engaged with the workpiece at any given time. That number will be determined by the spacing of the teeth and the thickness of the material being cut. For instance, if a blade with a tooth spacing of 10 teeth per inch (tpi) is cutting through a 1-in-thick section, 10 teeth will be in contact with the workpiece simultaneously.*

Material is being progressively removed and fills the gullets, so the chip load in the gullet will increase as the part thickness increases (since it is a longer cut). Because of this, blades with fewer tpi should be used for large workpieces. In some applications a "skip" tooth blade is used where the tooth geometry is identical, but they are spaced twice as far apart (e.g., 4 tpi instead of 8 tpi).

More aggressive cutting is also achieved by using a **hook tooth** blade, which has a positive rake. Most blades have a **neutral** or **zero rake** geometry. With larger spacing and greater positive rake, the teeth are more likely to "snag" the workpiece and result in poorer surface finish. In applications where the cut is not well supported or the blade is used in a climb cutting configuration (see Sec. 13.5.2), a higher tooth count and negative rake is beneficial. A common example of this is a radial arm saw (see Sec. 14.2.3.4). The various saw tooth configurations are shown in Fig. 14.2.

---

*Technically a "tooth" includes the space behind the cutting edge also, so there may momentarily be 11 cutting *edges* engaged based on a points per inch designation. This is a minor consideration, but defines the difference between teeth per inch and *points per inch*, another measure of blade pitch.

More teeth improve surface quality, but also reduce chip space, increase cutting temperature and require greater machine power (more cutting per inch). Most soft materials (wood, polymers, some nonferrous metals) are cut with lower tpi and larger blade teeth. As such, metals are typically cut using blades with fine teeth, but in work-hardening materials (see Sec. 9.2), such as stainless steel and titanium, a more aggressive tooth geometry can be beneficial to reducing blade wear since each tooth can remove sufficient material to avoid the cut-hardened surface layer.

14.2.1.2 Saw Blade Tooth Design and Material

There are many facets of saw blade design to consider, and some of the most common parameters are given in Fig. 14.3. The tooth pitch has just been considered, but it is not necessarily a constant. Some blades will have a variable pitch (or **variable geometry**, since gullet depth, rake angle, and other factors can also vary), which helps reduce harmonic vibration and results in a smoother surface finish and faster cutting action.

The material the blade is made of also determines productivity. As discussed more thoroughly in Sec. 13.4, advanced cutting materials can increase cutting speed and therefore reduce the cutting time per part. With most saw blades there are three common materials: high-carbon steel (Sec. 4.2.3.5), high-speed steel (Secs. 4.2.5 and 13.4.1), and carbide (Secs. 6.2.2 and 13.4.2).

Carbon steel blades are a low-cost option, but they are not suitable for cutting most metals or in high-production environments with wood, plastics or composites. For continuous blades used in band saws (see Sec. 14.2.3.3), a hybrid carbon steel/high-speed steel construction, known as a **bimetal blade**, is often used. The low-cost carbon steel is used as the majority of the blade, but the tooth tips are made of more durable high-speed steel. Although this increases the cost, it also increases blade life and productivity.

Most carbon steel blades are made from a solid piece of steel, and the teeth are "set" in order to give the blade body space so as not to rub against the workpiece. **Tooth set** is a blade design feature where each successive tooth is given a varying direction and degree of offset from the blade centerline. In this way, the **kerf**, or the width of cut, is larger than the blade body and gives the blade clearance. The kerf width depends on the set of the blade and whether any anomalies are present, such as a wobble or bend in the blade.

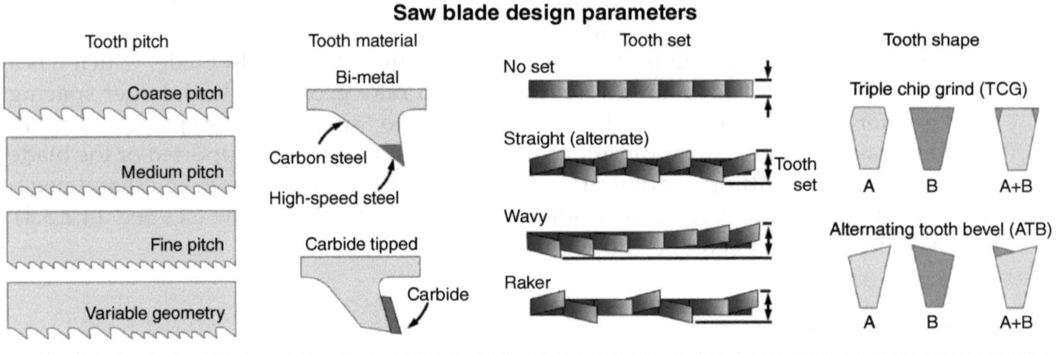

Figure 14.3 Tooth pitch (spacing), material, set, and shape all affect the performance of a saw. Each of the design parameters illustrated is discussed in the text.

Also available are blades that add carbide tips. Carbide, being much harder, will last a long time under demanding conditions. In some cases, especially in circular blades (Sec. 14.2.2.3), these brazed on cutting edges can be re-sharpened multiple times to increase the lifetime of the blade. Circular saw blades often have no set to the teeth since the carbide inserts are wider than the blade body and give the necessary clearance.

The shape of the teeth determines the strength of the insert against chipping and the ease with which it cuts through the material. Hard and abrasive materials benefit from a more robust cutting geometry, such as the **triple-chip grind (TCG)**, whereas an **alternating tooth bevel (ATB)** will require less force and "slice" through fibrous materials such as wood more cleanly, especially where the bevel angle is large. All carbide must be well supported to prevent chipping.

14.2.2 Types of Saw Blades

Because sawing is used in such diverse applications, there are equally diverse designs, not just in tooth geometry, but also in blade dimensions and configurations. For instance, in some applications a wire with embedded grit will be used to saw through hard materials such as granite. In the lumber industry, chain saws are plentiful, and in metalworking friction saws, which are essentially toothless, use frictional heat to melt through material. These variations will not be considered here, and our focus will remain on traditional manufacturing environments and blades with a general tooth design as described in the previous sections.

14.2.2.1 Straight Saw Blades

A **straight saw blade** is simply a linear blade design of finite length that has a limited **stroke length.** The stroke length, or the distance traveled by the blade during a single cutting motion, may be determined by the blade length or machine design. The blade length can be anywhere from an inch or two to multiple feet in length, and the design of these blades is described in Sec. 14.2.1. Examples of straight saw blades are given in Fig. 14.4.

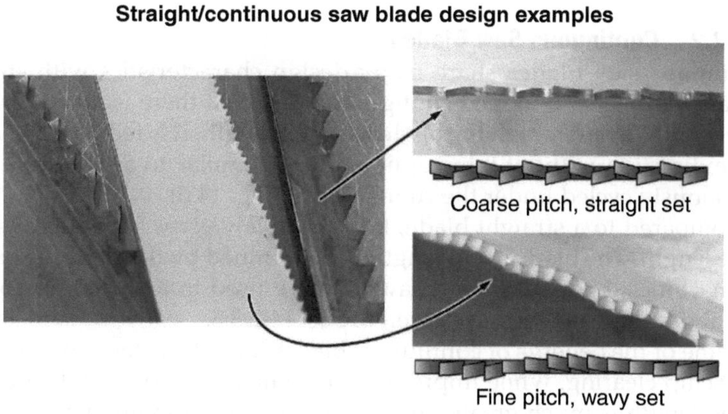

Figure 14.4 Straight saw blades are used in variety of applications. Shown here are blades with various designs used for cutting metal, wood, and plastic.

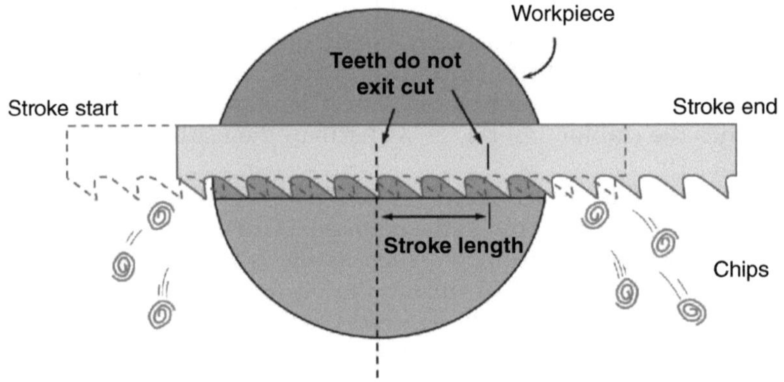

Figure 14.5 Reciprocal saws have a limited stroke length, and in cuts longer than the stroke, material may collect and reduce cutting efficiency. Orbital (elliptical) blade motion can be applied to assist in clearing material more efficiently.

The stroke length is restricted by blade length, but a stroke length less than the full blade length is often applied. For instance, a blade that is 6 in long may become clogged while cutting a 2-in part if the stroke length is only 1 in. The blade is much longer than the cut, but the cutting motion is less. This is common in reciprocal sawing operations.

The teeth that engage the center of the workpiece will not travel far enough to allow the chips to be ejected, and they will continue to build and increase resistance to cutting. If the stroke length is less than the cut length, there will be a chance of material buildup. This concept is illustrated in Fig. 14.5.

Straight blades are the type used in handsaws and reciprocal saws, as described in Sec. 14.2.3.1. Their use is largely limited to low production activities, especially where manually operated. Continuous cutting (no defined stroke length) is achieved using more efficient blade design, such as continuous and circular blades, which are discussed next.

14.2.2.2 Continuous Saw Blades

Continuous saw blades share many design characteristics with straight blades, but they are defined by a looped configuration, so that there is no end or beginning to the blade length. Even though the blade has a set length, it is fed past the workpiece surface in one direction without interruption. This is similar to a treadmill, where the belt is continuously cycled under the runner's feet (Fig. 14.6).

Compared to a straight blade, this effectively gives a continuous blade an infinite stroke length. The blade's real length is determined by the machine on which it is used. The machines, known as band saws, can be used in a variety of configurations, and more about these machines is given in Sec. 14.2.3.3.

Some of the benefits of continuous blades include faster material removal rates and better chip clearing, while improving blade life. Cost for each blade is higher, but the cutting length is much greater, meaning the work is shared between more teeth. This allows for a greater time for the teeth to cool down after each pass and the number of passes per volume of material removed will be lower.

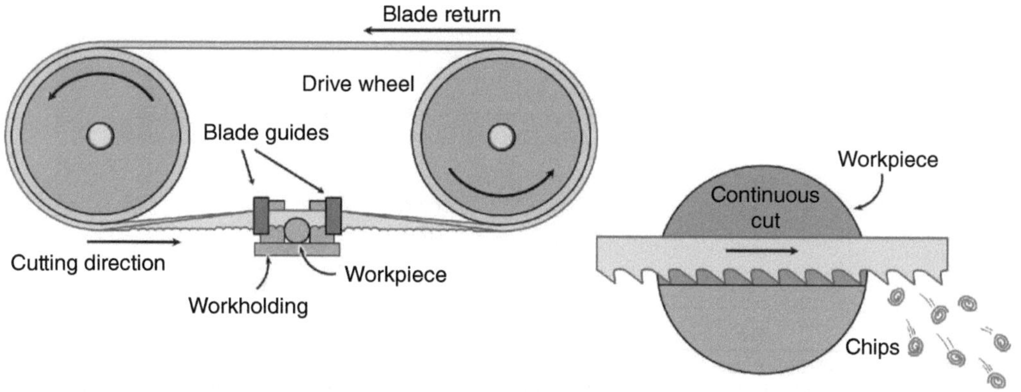

Figure 14.6 Continuous saw blades are capable of efficiently cutting parts both large and small since the blade motion is constantly cutting. The blade is shown in a horizontal configuration, but vertical processes are also common. Band saw machines are discussed more in Sec. 14.2.3.3.

Although the blades may be very long (10 ft or more), the maximum workpiece size is not nearly so large. The exact dimensions are determined by the machine design, but in general, the blade will have to pass over large wheels and will return on a path similar to the cut. The greater the cut length, the weaker the blade becomes, since it is unsupported over that length, and the less accurate the cut is.

Because of this, blades 12 or 13 ft in length may be restricted to only 6 to 12 in. in cut length. The maximum cut length is determined by the blade width (rigidity), the intended cut type (straight or curved), and the machine design (power, workholding, blade guides, etc.). Band saw equipment is discussed more in Sec. 14.2.3.3.

Band saw blades are often sold as a continuous loop with a specific length to fit a given machine (e.g., 159 in), but band saw blade stock is also sold in coil form that can be cut and welded by the user as needed. The process of welding often uses a dedicated blade welder (electrical resistance welding, see Sec. 18.2.6), usually attached to the band saw itself, to connect the ends and grind the joint smooth.

14.2.2.3 Circular Saw Blades

Circular saw blades have teeth that are positioned on the perimeter of a disk. These teeth have the same basic design geometry as found in straight and continuous saw blades. Like continuous blades, circular saw blades are run in a single direction and can be used to cut parts of any size that the machine can accommodate. Circular blades are rotated around a fixed axis, and work is fed through the rotating blade. The workpiece size must be less than the blade radius, and usually significantly less to accommodate the arbor on which the blade is mounted and the mounting structure that holds the arbor. These concepts are shown in Fig. 14.7, where a table saw configuration is represented. Other machines are discussed in Sec. 14.2.3.4.

As with other cutting tools, the material and tooth geometry of the blade will affect performance. These characteristics are described in 14.2.1, but the most defining characteristic of circular blades is the tooth shape. In particular, the triple-chip grind (TCG) and the alternating tooth bevel (ATB). These are especially suited for durability and cut quality, respectively. That is, the TCG shape has blunt corners that are strong and resist

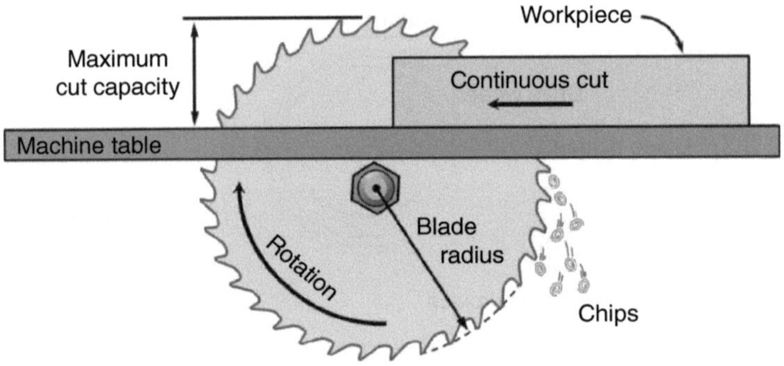

FIGURE 14.7 Circular saw blades are commonly used in wood working, but they also find use in other materials and industries. Here the basic setup of a table saw is presented, but the machine configuration can vary significantly.

chipping, especially when cutting materials such as laminates (see Sec. 6.4.3 for some examples), nonferrous metals (usually aluminum or other, soft metals), and abrasive materials such as polymers and engineered wood products (see Sec. 6.4.1.6).

ATB saw blades are especially good at leaving a smooth finish in wood since the sharp edges can cut cleanly through the wood fibers. The sharper the tip angle, the better the finish, but the weaker the cutting edge. The same basic tooth design is also used in steel blades, but as cost has come down for carbide blades, the performance benefits, such as longer tooth life, outweigh the upfront cost savings for steel. Examples of circular saw blades are given in Fig. 14.8.

Unlike most metals and plastics, there is a key difference when cutting wood. The wood has a fibrous structure that is aligned with the growth of the tree vertically. Lumber will have this grain running more or less parallel to its *length*, and will cut more easily with the grain than across it. Cuts made with the grain direction are called **rip cuts** and those made across the grain are **cross cuts**. Certain saw blades are dedicated to one or the other (known as rip blades and cross-cut blades, respectively), but many blades are known as **combination blades** and are designed to cut acceptably in either direction. Teeth on a rip blade are spaced farther apart and flatter, while cross-cut blades have closely spaced teeth that have a sharp edge to "slice" through the wood fibers.

14.2.3 Sawing Equipment

Saw blades are often general-purpose cutting tools. One blade may fit on a variety of different machines, and one machine may accommodate a variety of blades. As always, there are limitations, but what is important to note is that there are options for completing sawing operations. By changing the blade to one with a different number of teeth, of a different type of material, or any other factor, the cutting process can be optimized. Also, there are a great number of machines dedicated to sawing, so proper equipment selection is equally useful. We will look at common configurations used with each type of blade discussed in Sec. 14.2.2, namely straight, continuous, and circular saw blades.

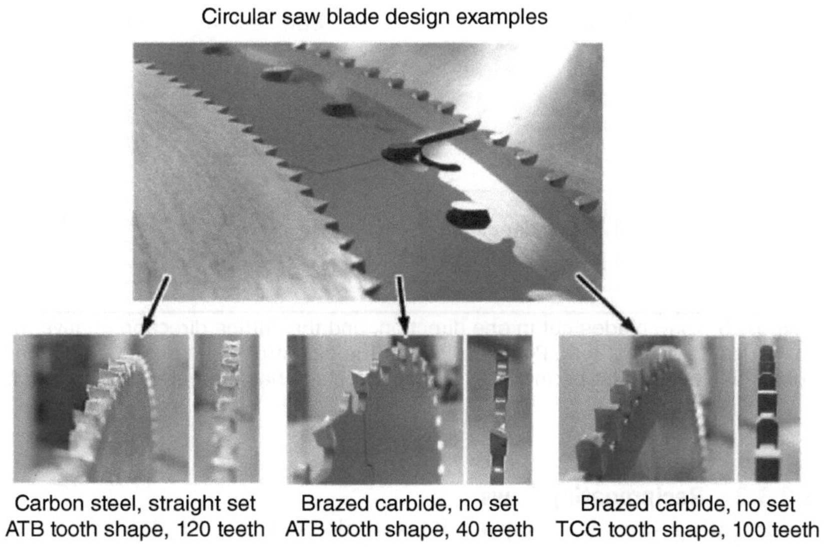

FIGURE 14.8 The circular saw blades pictured here are all 10 in. in diameter, but they have differing pitch, materials, and shapes for their teeth. These are illustrated in Fig. 14.3. The design of the saw blades determines its optimum function, and although there are multipurpose blades, the most efficient processing will use dedicated geometry optimized for the material and method of cutting.

14.2.3.1 Handsaws

Straight blades see extensive use in hand-operated saws and powered reciprocal saws. **Handsaws** are advanced through physical force to create a cut of limited distance. One consideration that should be carefully considered with this type of operation is the cutting direction. Saw blades are designed to cut in one direction, but some saws allow a blade to be mounted so that the cutting direction can be toward the user or away.

Most saws are designed to cut in the forward direction and are known as **push saws**. Saws designed to cut with a stroke toward the user (or machine support) are known as **pull saws**. Push saws must be more rigid, and may have a stiffening rib support along the noncutting edge. This type of saw is also known as a "backsaw," and cannot be used to cut beyond the full depth of the blade because of the back support. Backsaws are typically used for precision detail work, especially in woodworking. Push and pull-type handsaws are shown in Fig. 14.9.

Since pull saws create tension in the blade during cutting, they can be thinner and weaker, and are also often used for detail work. Many machines cut in a "pulling" direction since it naturally stabilizes the blade under tension. Neither saw should have force applied in both directions, as any noncutting return stroke will dull the teeth. There are many types of handsaws, such as keyhole saws, coping saws, hacksaws, etc., that we will not discuss, but all follow the basic cutting characteristics described here.

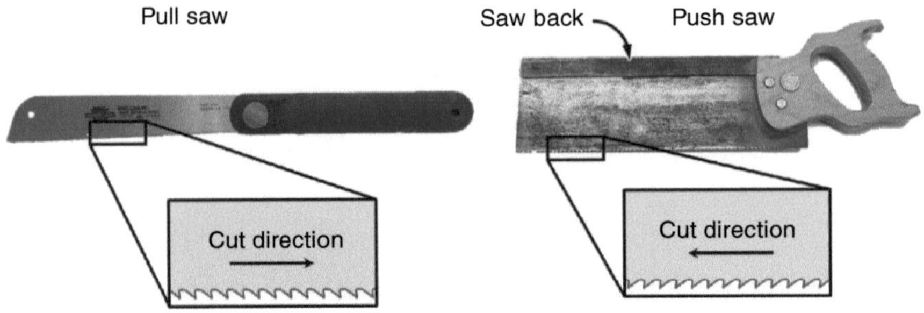

FIGURE 14.9 Saw blades cut in one direction, and the cutting direction relative to the handle determines the type of saw. Pull saws cut on the pull stroke and push saw cut on the forward stroke. Push saws must be more rigid since the blade has a greater tendency to flex.

14.2.3.2 Reciprocating Saws

Reciprocating saws (or reciprocal saws) also cut in a single direction, but they use a powered stroke at a much higher rate than what can be achieved by hand. Popular reciprocating saws are **jig saws, scroll saws,** and what are simply known as **reciprocating saws** (Fig. 14.10). The stroke length is defined by the machine design. A disadvantage of reciprocal saws is that the stroke length determines the ability to remove chips, regardless of the blade length.

Reciprocation may not be strictly linear, however, and some motion perpendicular to the cutting action may also be incorporated. Saws with this additional, elliptical motion are often referred to as **orbital saws,** and the orbital motion may be used optionally since it can negatively affect cut accuracy. A benefit of orbital cutting is that it reduces pressure on the noncutting return stroke. This will assist in clearing debris, reduce heat and blade wear, and increase cut efficiency.

14.2.3.3 Band Saws

Band saws use a continuous blade that loops around wheels that propel the blade. As a part comes in contact with the blade, cutting can be performed without interruption, even on very long cuts. This allows for a blade to quickly cut through a workpiece while reducing wear on the teeth in comparison to reciprocating methods used with straight

FIGURE 14.10 Common examples of saws that use reciprocating motion to cut through materials.

Band saw configurations

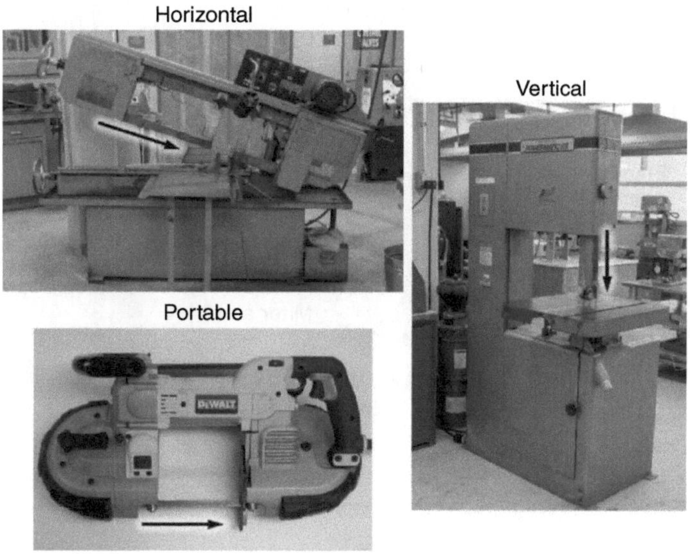

FIGURE 14.11 Continuous blades are most often used on band saws. These machines are available in horizontal, vertical, and handheld configurations.

blades that must reverse the blade direction after each cut stroke, and chip clearance is much more effective also. The configuration of band saws may take several shapes, and some common examples are provided in Fig. 14.11.

Band saw size is designated by the wheel diameter. The maximum workpiece size, the power, and the blade size all scale with that measurement, but not necessarily exactly. That means a 20-in band saw from one manufacturer may have different capabilities than from another, but in general they will be comparable. The major components of a vertical band saw are given in Fig. 14.12. A horizontal band saw is shown in Fig. 14.13. In this case, the vertical band saw is used for cutting wood, and the horizontal band saw is used for cutting metal.

The workpiece may be held by hand (common for vertical saws) or by a vise or other fixtures (used in horizontal band saws). Horizontal band saws are generally used for production work with metallic materials since the saw can apply more force and hold parts tightly. Cuts on bands saws may be straight (horizontal and portable band saws) or it may be contoured (vertical band saws) by turning the workpiece while advancing it through the blade.

14.2.3.4 Circular Saws

Circular saw blades are used in a variety of configurations. The blade may be fixed, may move about a hinged point, may slide or may be operated free hand. The blade rotation allows efficient material removal to be contained in a compact space. The most familiar applications are related to woodworking and construction, and the home hobbyist is likely to have a circular saw in some configuration for home repair or other work. Some common examples of equipment using circular saw blades are given in Fig. 14.14.

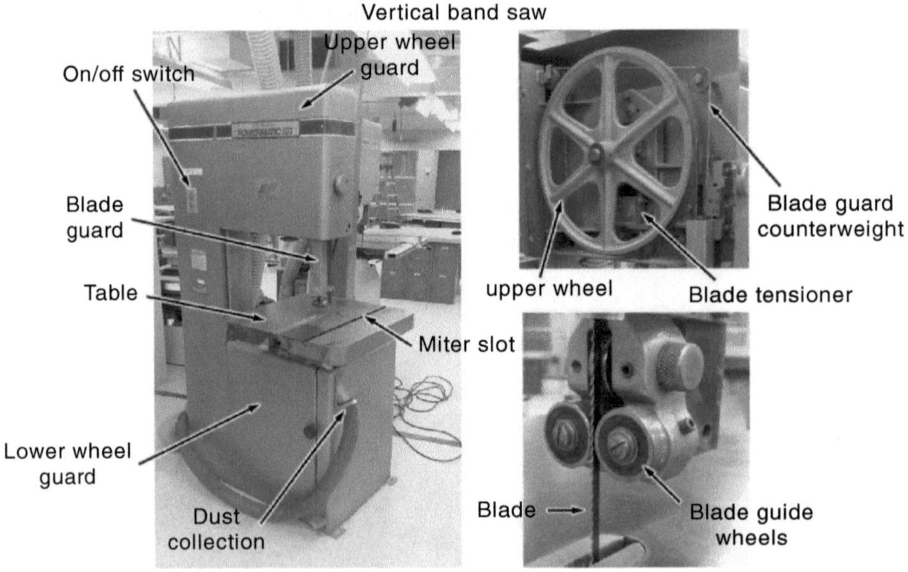

FIGURE 14.12 A vertical band saw is pictured with major components labeled. Each manufacturer's design will vary, but most band saws will incorporate these elements. This band saw is used to cut wood.

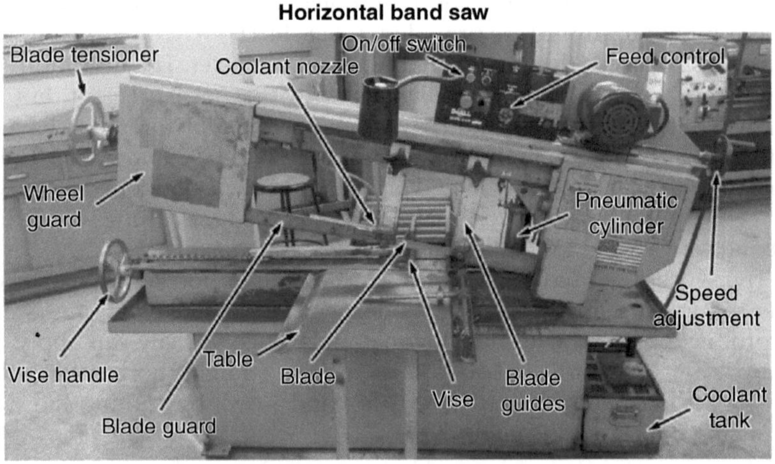

FIGURE 14.13 A horizontal band saw is pictured with major components labeled. Each manufacturer's design will vary, but most band saws will incorporate these elements. This band saw is used to cut metal.

FIGURE 14.14 Circular saw blades are used on a variety of machines. Some of the most common are the table saw, miter saw, and handheld circular saw shown here.

A **table saw** has a blade that is rigidly positioned during cutting, and material is passed through to cut the desired parts. The blade may be raised, lowered or tilted, but it does not move during the cut (other than rotating). Large sheet materials, dimensional lumber and most any other part can be cut with bevels, tapers, grooves, shoulders (known as rabbets), and much more by using the various features on the saw. A general part diagram is shown in Fig. 14.15.

Complex parts can be made efficiently by adding jigs and fixtures to hold the workpiece in specific orientations. The most common among these is the **miter gauge**. This is a small attachment that rides in a slot (a miter slot) and helps support the workpiece when cutting across the width rather than along the length. The miter gauge can be held perpendicular to the blade or can be angled (Fig. 14.16). It is also often used with a large fence attached to provide additional support. Angled cuts can also be made using miter saws or radial arm saws.

Miter saws and radial arm saws are related in their functions, but the way they achieve those functions varies. A **radial arm saw** uses a circular blade that is guided

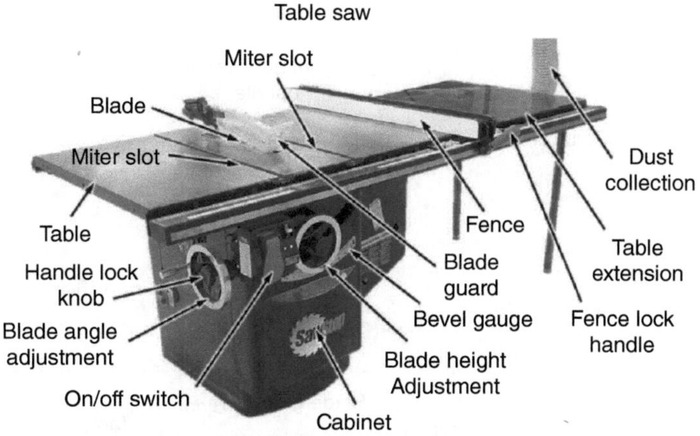

FIGURE 14.15 A table saw is pictured with major components labeled. Each manufacturer's design will vary, but most table saws will incorporate these elements.

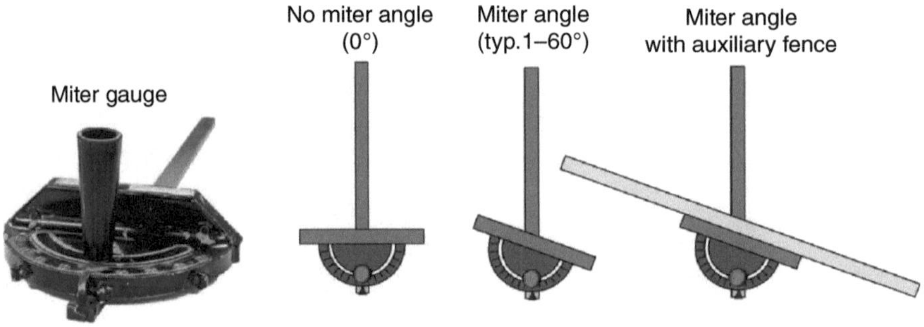

FIGURE 14.16 A miter gauge is commonly used on band saws, table saws, router tables, etc. It provides support while cutting angles, but in many cases an auxiliary fence is beneficial as it can be cut to support the workpiece up to, and even into the blade. The auxiliary fence can be made from low cost material safe to be cut.

along an overhanging support arm (Fig. 14.17). The blade and support assembly (including motor and adjustments) are fed across a part that is held stationary. The arm can be rotated about its main connection point to the machine base, but the motor and blade assembly can also be rotated to cut at various angles. In theory, the radial arm saw can achieve the same operations as a table saw, but in practice, this is not as efficient or as accurate.

Miter saws provide greater ease when creating angled cuts (miters) since it is purpose-built for the job (Fig. 14.18). Compound miter saws can be tilted in multiple axes to create compound angles. Like a radial arm saw, the motor and blade are fed through a stationary workpiece, but miter saws incorporate a blade pivot to lower the blade into the work in addition to the ability to slide through the work.

The base has angle graduations and quick release handles to allow it to be rotated to a desired angle (left or right). This allows for additional control over the cut and

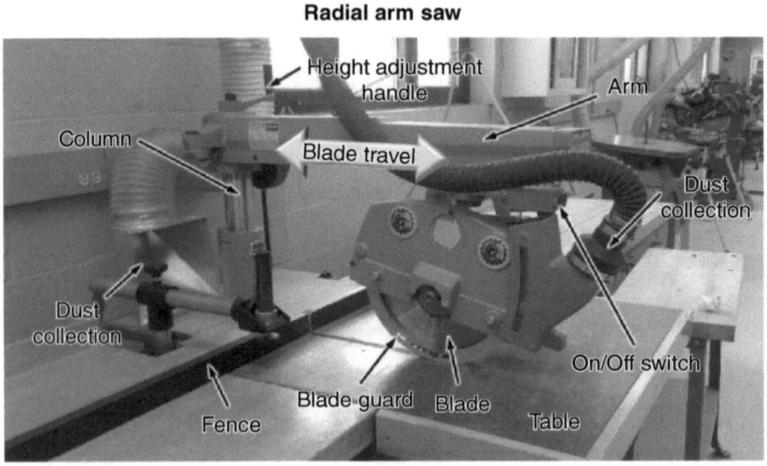

FIGURE 14.17 Prominent components of a radial saw are listed here.

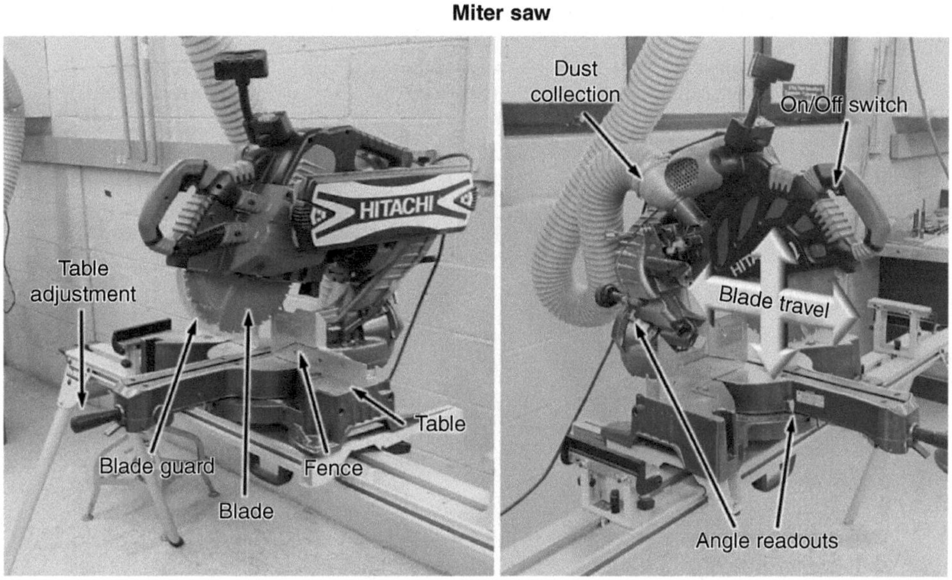

Figure 14.18 Prominent components of a compound miter saw are listed here.

makes the process faster in high-production settings, such as construction, where many cuts must be made and accuracy is not especially tight. That is not to say that miter saws are not capable of high precision, but care must be taken to ensure the angles represented are accurate, and many saws have positive locks, which "click" the saw into place, at common angles, such as 45° and 90°.

14.3 Sawing Techniques

As indicated throughout the discussion of various saw blades and equipment, this is a versatile technique. Each type of equipment will determine the techniques it is capable of, but, in general, there are three main types of cuts: straight, angled, and contoured.

A **straight cut** is used to cut material to length at the beginning of processing, but it can also be used multiple times throughout the manufacturing protocol. To differentiate "straight" and "angle" cuts, we will restrict straight cuts to those which are mutually perpendicular to the workpieces surfaces in a rectangular part, or to the longitudinal surface and the face of a circular part. Horizontal band saws, miter saws, radial arm saws, and table saws are preferable for making these cuts. Any handheld sawing technique must be carefully conducted and/or use a guide (fence) to make a straight cut.

An angled cut is similar to a straight cut but the blade may be tilted in one or more axes. For example, when using the table saw, the blade may be tilted and the part fed through as normal. This creates a **bevel**. Alternatively, the blade may remain vertical and the part may be angled to create a **miter**. Examples of methods to create miters and bevels are given in Fig. 14.19.

A **contour cut** is achieved by rotating the part (e.g., vertical band saw) or blade (e.g., jig saw) during the cut, thereby creating a curved feature. The contour capability

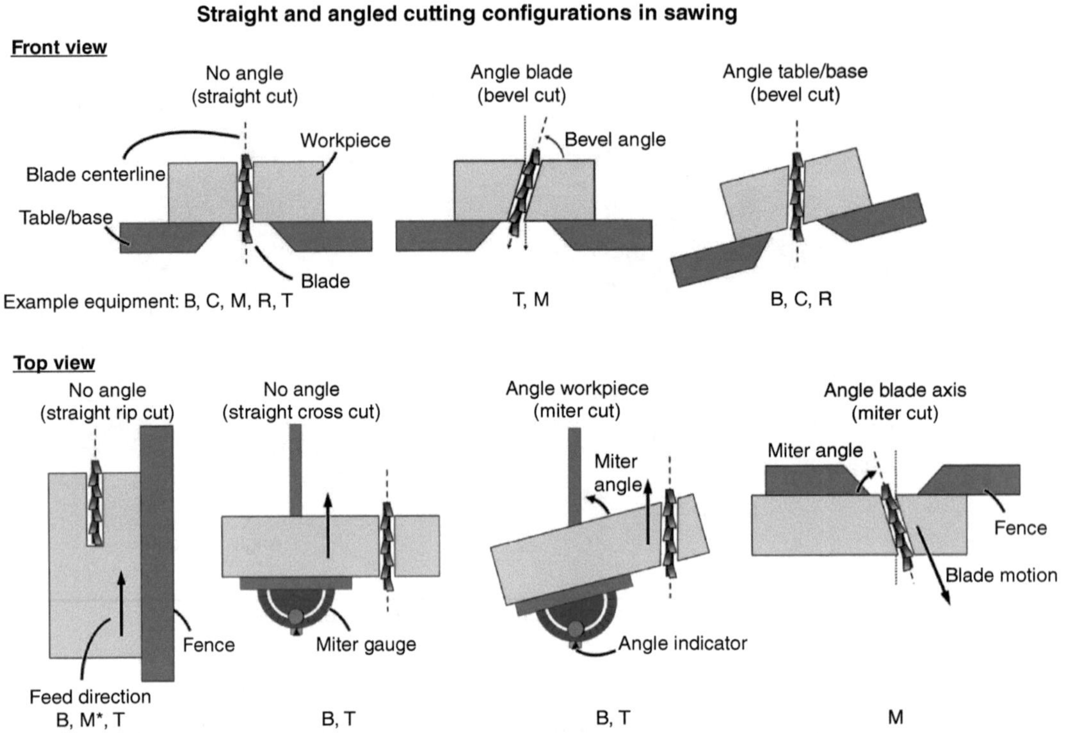

Figure 14.19 Straight and angled cuts can be made in many ways using saws. The equipment and blade characteristics determine the ease and accuracy with which the cuts can be made. Equipment examples are generalized and may appear differently on specific machines. Equipment lettering is as follows: B = band saw, C = circular saw (handheld), M = miter saw/radial arm saw, R = reciprocating saw, T = table saw. *Only as radial arm saw with blade rotated 90°.

is determined by the blade size and rigidity. The more rigid the blade is, the smaller the curvature must be (larger radius). To create tight curves, a thin blade with lower tension is needed, and excessive torsion on the blade will result in blade breakage. To reduce binding of the blade during cutting, a series of slots, called **relief cuts**, can be created up to the cut line so that material behind the cutting edge of the blade fall away intermittently in small pieces. Contouring considerations are shown in Fig. 14.20.

Internal circles and contours can be made, but there must be an entry point for the blade. This can be achieved by drilling a hole of sufficient size to accommodate the blade. This works well for reciprocal sawing equipment, such as jig saws or scroll saws, since the blade can be easily located in the hole, but for equipment such as the vertical band saw, the blade must actually be cut, inserted and rewelded. This is only justified for low production parts that cannot be efficiently cut by other means. To assist in cutting circles, circle cutting jigs can be used. This can be simple as locating a pin a recess at the center of the part to be cut and rotating the part against the blade at the proper distance (i.e., the pin is located a distance equal to the part radius). This is shown in Fig. 14.21.

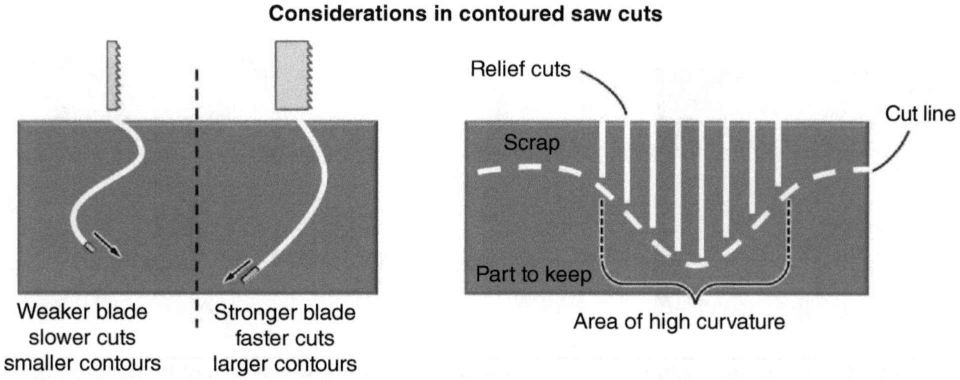

FIGURE 14.20 Contoured cuts are a useful feature possible with straight and continuous blades. The equipment and blade characteristics will limit the contouring capability, but in general, a thinner blade can provide greater maneuvarability. Relief cuts can be used to further enhance contour capability.

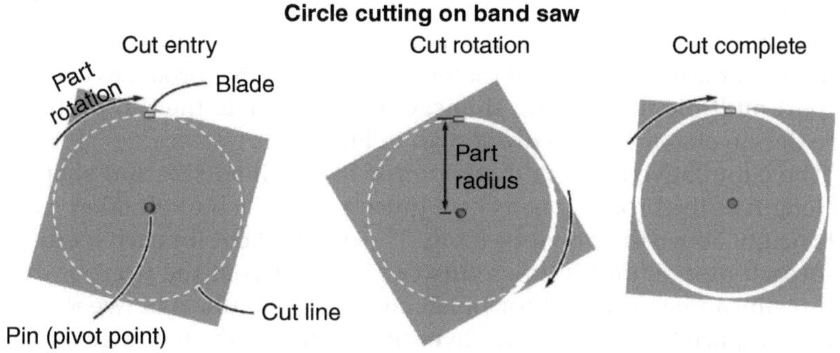

FIGURE 14.21 Cutting accurate circles can be accomplished by using a jig which controls the cut. An external circular cut is shown here, but internal circles can be made by drilling an entry hole for the blade to start in. Internal circles are much more cumbersome on a band saw. Other methods, such as reciprocating saws (e.g., jig saws and scroll saws) are preferable.

14.4 Broaching Basics

Broaching shares some key aspects with sawing, but it is used much differently. The primary application of broaching is to create noncircular holes, but it relies on holemaking operations to precede it. This is because broaching is most often used to modify a hole profile. One of the most common applications is to create a keyway to prevent a shaft from spinning relative to a pulley or gear. Likewise, a spline is used for similar reasons, but does not require a separate key to lock the shaft and gear (or wheel, pulley, etc.). Both of these features are shown in Fig. 14.22.

14.4.1 Broach Design

To create features internally requires a tool that can cut with linear motion, very similar to a saw blade. Reciprocal, continuous and other saw blade designs will not be able to achieve the same result, so broaches are uniquely designed to complete the entire profile

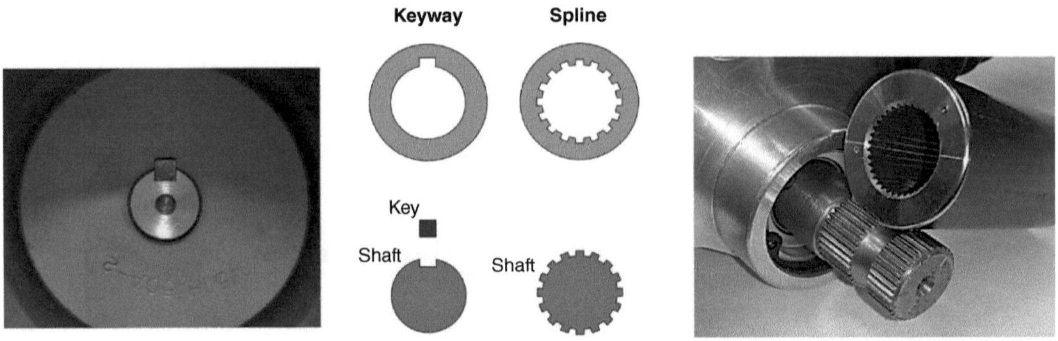

Figure 14.22 Broaching is often used to create internal profiles which cannot be made by other holemaking operations. Shown here are examples of a keyway and splined connection. Both are used to eliminate slipping of components during rotation.

in a single pass. Each tooth on a broach has a slightly different height and/or geometry, depending on the profile. The first teeth on the broach to contact the workpiece will take a small amount of material, and every tooth after will also take a small amount of material. For large amounts of material removal, some broaches can be passed through the part multiple times using shims. Others complete the full profile in one pass. The basic design characteristics are shown in Fig. 14.23.

The cumulative effect is that a profile of specified size and shape is created as the full length of the broach passes by a given point. Each tooth takes a cut equal in depth to its height above the previous tooth. The depth of cut for each tooth is engineered into the broach design, and heavier cuts can be created at the front of the broach and the final teeth can be designed for finish cutting. The gullets between teeth must provide sufficient space for chips to form over the entire length, and in general, broach teeth are larger with greater spacing than saw blades since chips do not clear until the broach traverses the entire workpiece. The general broaching process and common broach profiles are shown in Fig. 14.24.

Broaches have a constantly changing geometry. The teeth size, depth of cut, and profile geometry may change along the length of the tool. At the entry, the cut is light and has a profile with minimal detail. The further back along the broach you observe, the more defined the profile becomes. This change in cutting form is shown in Fig. 14.25.

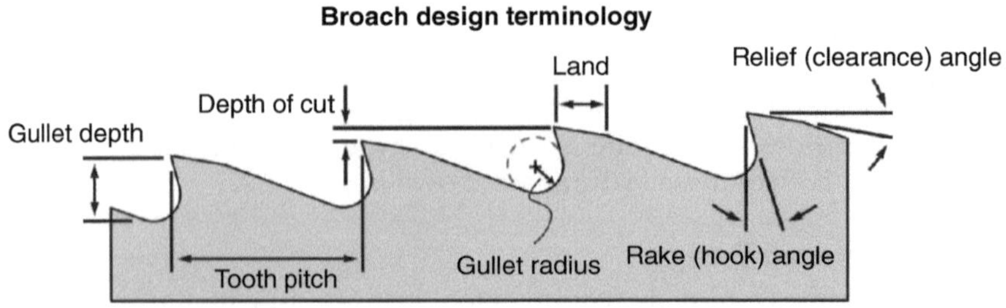

Figure 14.23 Broach design is similar to saw blade design, but the profile and cutting characteristics are entirely determined by the geometry, not machine settings such as feed rate.

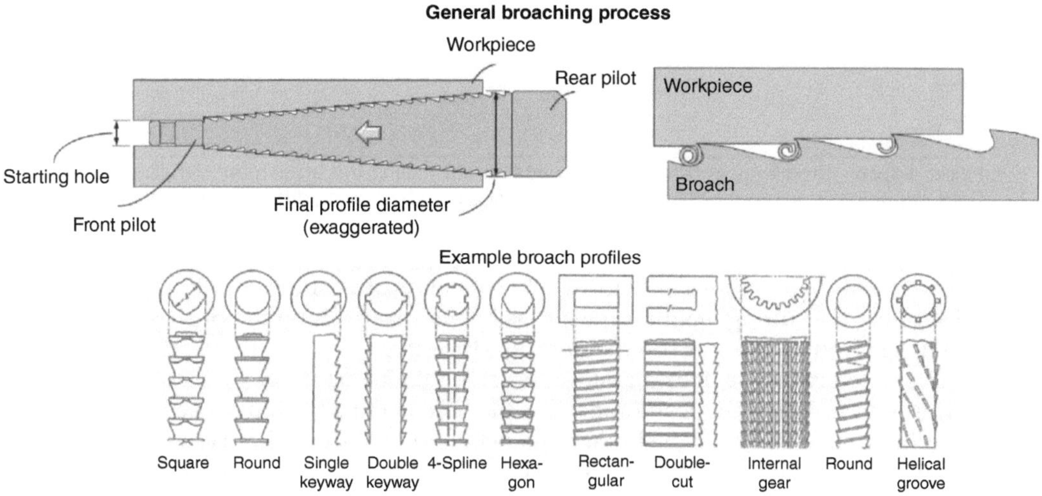

Figure 14.24 Broaches complete the entire cutting operation in a single pass. This is accomplished by a constantly increasing cross section, which can be found in a variety of profiles. Note the size change over the length of the broach is greatly exaggerated for clarity. (Bottom image courtesy of Industrial Press, Inc., from *Machinery's Handbook*, 29th edition, Oberg, Jones, Horton, Ryffel. Fig. 1, pg. 980. ISBN 978-0-8311-2901-9.)

14.4.2 Types of Broaches

To create the profile, a broach can be pushed or pulled through the starting hole. A **push-type** broach progresses through the workpiece by force applied from the back side, while a **pull-type** broach is drawn through the workpiece from the front (Fig. 14.26). Push-type broaches can be identified by a lack of attachment point needed to pull the broach through a hole. For instance, a thread, hole, slot, or other means of securing one end of the broach must be present to hold the broach. Also, the broach will have a

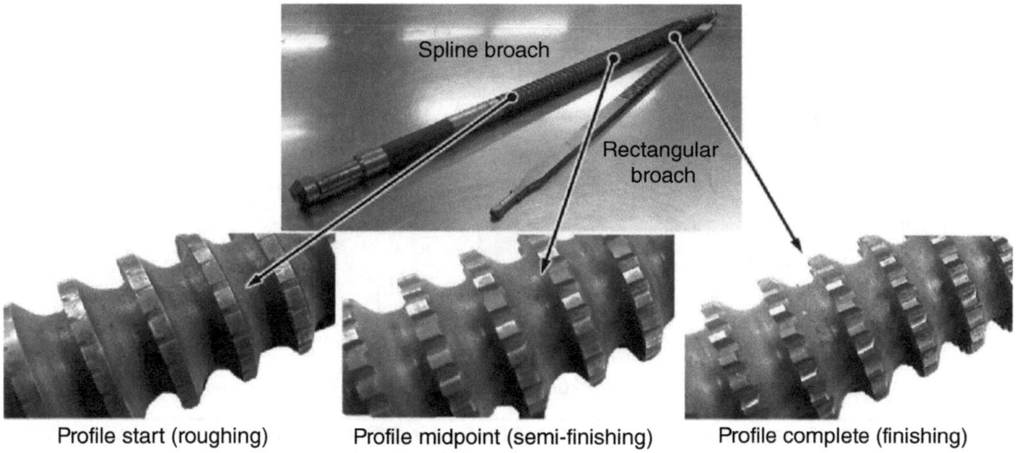

Figure 14.25 The profile change along the length of a broach along is a design choice built-in to the tool. Here the profile along the length of a spline broach is detailed. The profile becomes more pronounced toward the finishing end of the broach.

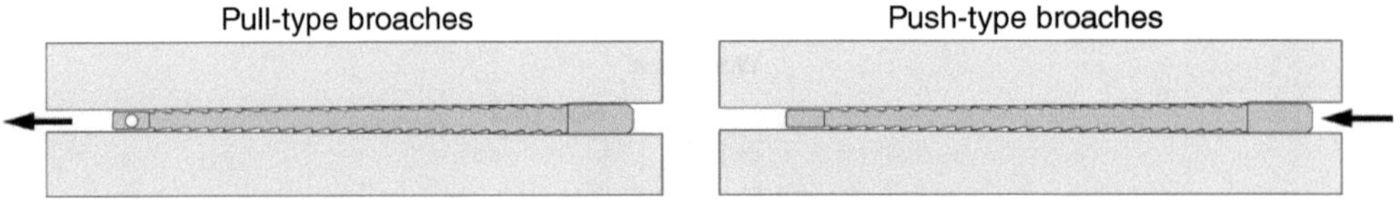

FIGURE 14.26 Broaching is done by either pushing or pulling the broach through the workpiece. Both result in the same features, but they are accomplished using different equipment.

majority of its noncutting length toward the direction of pushing or pulling. For instance, most of the noncutting length (no teeth) is behind the teeth on the push-type broaches, and it is in front of the teeth on the pull-type broaches.

A single broach, such as a keyway broach, may be used on a variety of hole sizes. This is accomplished by positioning it in the hole using a bushing. The bushings come in multiple sizes and hold the broach against the side of the hole. In cases where the cut is to be made in multiple passes, a shim is added before each pass to remove additional material (Fig. 14.27).

Broaching is often done using linear motion, but it can be also be combined with rotation to make helical profiles as well, a process known as **rotary broaching**. For instance, the rifling on gun barrels can be created using a rotary broach. In automated processing, broaching can be done as a reciprocal-type operation. Because the position and depth of cut are accurately controlled by a computer in CNC broaching, a single cutting edge can be used and is often made from carbide.

When applied to external surfaces, broaching is known as **surface broaching**. This process can create profiles along the length of a part, and can be done over many pieces efficiently. Although similar features may be created through other methods, such as milling, the use of broaching can be more efficient since the cutting is distributed over many more teeth. Also, surface broaches may be comprised of multiple, side-by-side broaches, so that a profile may be created *across* the length of a part. Surface broaching is shown in Fig. 14.28.

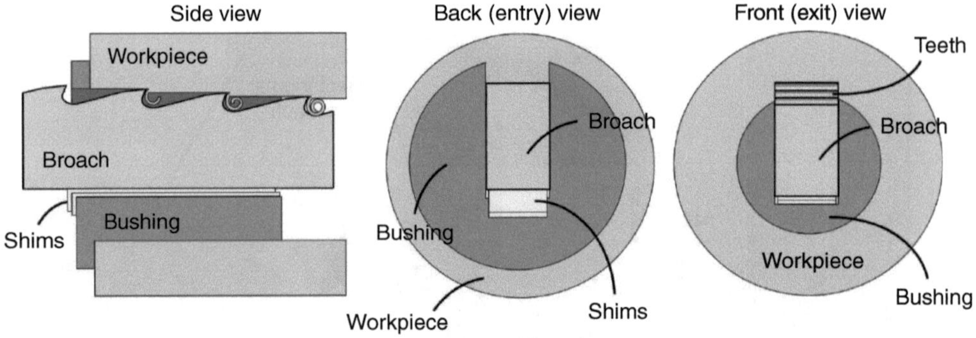

FIGURE 14.27 Broaching can be in a variety of ways for different purposes, but the simplest way is vertical, push-type broaching where the broach is forced through the workpiece using a hydraulic press or arbor press. One of the most common profiles is the keyway and kits can be purchased to allow creating keyways of varying depth in a number of hole sizes. Bushings and shims are used to achieve the desired keyway.

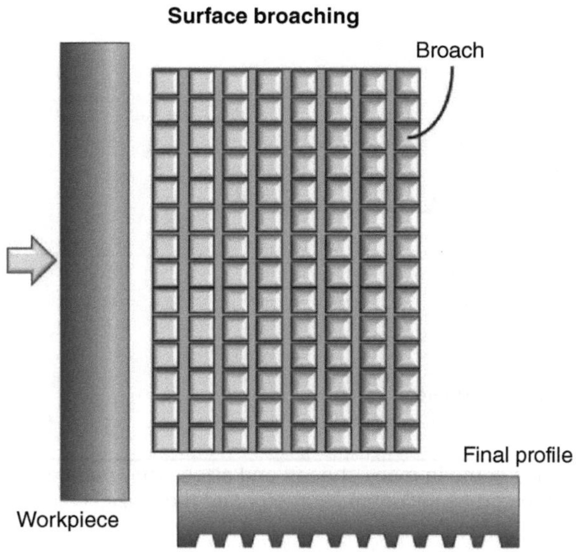

Figure 14.28 Surface broaching can be used to remove material along the length of a component, or, as shown here, it can be used to create a profile across the length.

14.5 Filing Basics

Filing removes a small amount of material using numerous small teeth. It often used for smoothing surfaces, rounding or shaping corners and for minor material removal to create features such as chamfers.

Filing is often used to remove burrs from a workpiece. **Burrs** are sharp extensions on the edge of a workpiece caused by deformation during cutting. Ideally, they will be minimal when cutting edge condition is good and machining parameters are correct, but some finishing work is often required, especially in small lot production or prototyping.

14.5.1 File Design

To allow a variety of applications, files come in many sizes, shapes, and degrees of coarseness. The size of a file is designated by its length (e.g., 8 or 10 in). The shape of a file refers its cross-sectional geometry. These shapes include square, flat, round (or rattail), triangular (or three square), half-round, etc. The coarseness refers to the size, spacing and aggressiveness of the teeth. For instance, a rasp is a very coarse file, while a smooth file is used for a finer finish but with lower material removal rate (Fig. 14.29).

14.5.2 Filing Techniques

Filing is most often done by hand. Files are designed to cut on the forward stroke, and reciprocating a file across a surface will wear it out prematurely. Filing machines are less common, but they can be used to increase productivity. For instance, a band, similar to that for sawing (see Sec. 14.2.3.3), can be used to continuously file. There are special techniques for filing features such as saw teeth or cross-hatching, but these are not common to production work. Another example is **draw filing**, where smoother

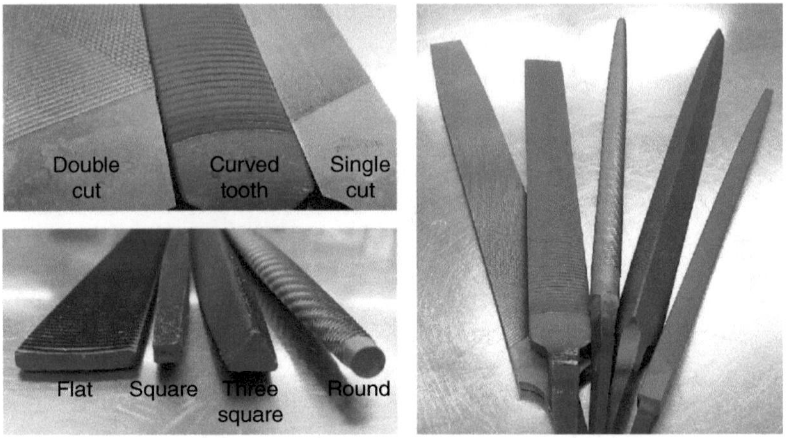

FIGURE 14.29 Files come in many shapes and sizes. They can be described by their tooth characteristics, their cross-sectional geometry and their size.

surface is achieved by advancing the file across a surface perpendicular to the stroke to increase the contact area. The shape of the file and the workpiece will determine the appropriate technique, but in production environments, filing will be avoided to reduce time and cost, and removal of burrs is often done using abrasive techniques, which are described next.

14.6 Abrasive Machining Basics

Abrasive machining relies on an abundance of randomly oriented cutting surfaces, which are not geometrically controlled, to collectively remove material from a workpiece. Abrasive processes include more familiar terms such as sanding, grinding and polishing. This type of machining is unique since there is no defined cutting edge geometry. Instead, it relies on a random distribution to ensure that enough of the cutting edges are functional to be effective. These edges are created by small particles, the type and size of which determines the outcome of the abrasive process.

Abrasives are most often adhered together and/or to a substrate like a cloth or paper backing. The abrasives may be bonded together rigidly or by a flexible material. The abrasive "tool" may be in the form of a sheet, strip, wheel, disc, belt or other conformations, some of which are shown in Fig. 14.30. Each type has specific purposes and benefits, and we will look at some examples of the processes in Sec. 14.6.2. To help understand why those processes work, we will first look at how abrasives are designed.

14.6.1 Abrasive Design

Even though the orientation of the abrasive particles used in abrasive processes is not precisely controlled, there is still significant design work. Close examination of a bonded grinding wheel, for instance, reveals similarities to Rice Krispies Treats. The particles (Rice Krispies) are bonded together by a coating of material (not marshmallow!) that connects the particles and creates a rigid structure. The properties of each

Figure 14.30 The size, shape, and configuration of abrasive tooling vary widely. The most common production methods use abrasive wheels, but surface preparation and finishing processes often use abrasive sheet or loose abrasive as in san blasting.

will determine the cutting characteristics. This bonded-particle concept was also applied to carbide cutting tools (Sec. 13.4.2), but in that context there was no open space between particles and a smooth surface was preferable.

The parameters that are central to the performance of the abrasive are material type (**abrasive type**), the size of the abrasive particles (**grit size**), the spacing of the particles (**structure**), and the rigidity (**grade**) of their connections (**bond**). These factors are included in the standardized designation established by the American National Standards Institute (ANSI). The designation (ANSI B74.13-1990) with explanations is given in Fig. 14.31.

The abrasive material, grain (or grit) size, the grade, and structure are all key factors in determining the performance of the abrasive and suitability for a given application. These are shown schematically in Fig. 14.32, and can be summarized as

- Type—ceramic the grains are made from
- Grit size—size of the abrasive particles
- Grade—hardness the particle bonds
- Structure—spacing of the particles

14.6.1.1 Abrasive Performance

The abrasive material is quite important because it determines the cutting characteristics. For instance, a very hard and strong abrasive will last longer during machining, but the abrasive grains wear down during the cutting process. That means that the abrasive may start to round over and become less effective at removing material from the workpiece. This condition is known as **glazing**, and it can be remedied by removing a thin layer of abrasive. This process is known as **dressing** when using a

1 2 3 4 5 6 7
Ex. 39-C-60-J-8-V-K

Abrasive Identification (ANSI B74.13-1990)		
Identification Position	**Feature Described**	**Common Designations and Explanations**
1	Prefix	Manufacturer-specific information about the abrasive. Not required.
2	Abrasive Material	**A**—Aluminum Oxide, **C**—Silicon Carbide, **Z**—Alumina Zirconia, **Other**—Specified by Prefix (Position 1)
3	Grain (Grit) Size	Increasing number (e.g.,80,120,220) indicates decreasing size. Grit number indicates the mesh size through which the largest particles will pass.
4	Grade	Indicates the hardness of the bonds connecting particles. Increasing hardness indicated by letters **A–Z** (soft-hard).
5	Structure	Increasing spacing indicated by numbers **1–16** (close-open spacing)
6	Bond	Indicates the material connecting grains. **B**—resinoid (synthetic resin). **BF**—reinforced resinoid, **E**—shellac or resin, **O**—oxychloride, **R**—rubber, **RF**—reinforced rubber, **S**—silicate, **V**—vitrified
7	Manufacturer Record	Used for internal reference or manufacturer's use. Not required.

Figure 14.31 Abrasive designation according to ANSI B74.13-1990. Applies to grinding wheels and other bonded abrasives.

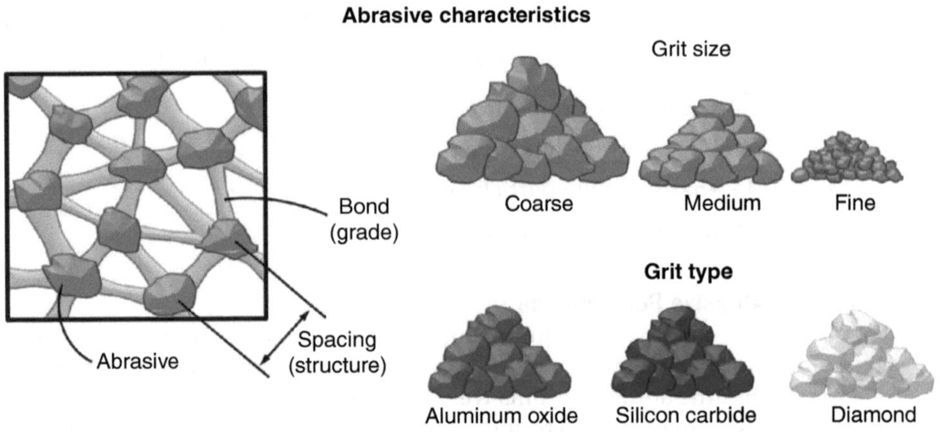

Figure 14.32 The performance of an abrasive is determined by several factors. These include the strength of the bonds and the size, type and spacing of the particles.

grinding wheel (or **truing** to restore the correct geometry). Alternatively, the abrasive can be exchanged when using a consumable type such as sand paper. This effect can be reduced by choosing a more **friable** grit, which means it is more likely to fracture under pressure and expose new, sharper cutting surfaces.

Abrasive cutting assumes normal wear of the cutting tool as part of the process. The design of the abrasive tool, like a grinding wheel, can be modified to adjust cutting characteristics. For instance, the abrasive type can be varied, as just mentioned, but the way the abrasive is held together and spaced will also determine the appropriate applications for the tool.

The grade can be varied to resist the loss of abrasive particles, or to enable a higher wear rate of the wheel. It may seem counterintuitive to want a wheel to wear down faster, but the faster the wheel wears, the fresher the abrasive particles will be and the cooler the cutting process remains. The material removal rate (MRR) of the workpiece with respect to the material removal rate of the wheel is known as the **grinding ratio** or **G-ratio** and is defined in Eq. (14.1). Values for this number are often 20 (20:1 ratio) or higher.

$$\text{Grinding ratio} = \frac{\text{MRR}_{\text{workpiece}}}{\text{MRR}_{\text{grinding wheel}}} \tag{14.1}$$

The way the abrasive is spaced will allow for more rapid material removal with less clogging (large spacing) or a better surface finish (closer spacing), which is also reliant on the grit size. The flexibility of the bond will determine how brittle the grinding wheel is and whether is best used on a machine or can be used by hand.* There are many factors that must considered in the determination of the appropriate grinding tools, and in large-scale production environments, small differences can have a great impact on quality and productivity.

14.6.1.2 Abrasive Cutting Mechanisms

The abrasive particles are randomly oriented in a cutting tool, and there are some particles that are not oriented for efficient cutting. The particles remove material by a combination of rubbing, ploughing, and cutting (depending on orientation), and more abrasive particles contribute to the first two interactions than to cutting. Because only a small portion of the cutting tool cuts, there is inefficiency that results in heat. This makes low melting point materials unsuitable for many high-speed abrasive operations.

It is quite typical for abrasive machines to rotate, oscillate or otherwise move the abrasive at high speed. This is often done by rotating an abrasive wheel at high rpm (thousands of rpm). As illustrated in Fig. 13.17, the speed (velocity at the cutting edge) of the cutting process most directly contributes to heat generated during machining. Coolant is almost always used in commercial grinding operations, and even in brief operations involving hand-grinding, the part is often intermittently cooled to avoid excessive heat buildup, which results in dimensional change, heat effects in the material (such as annealing), and surface quality issues such as burn marks.

---

*Abrasive tools operated by hand must be able to withstand a variety of loading configurations without shattering and require greater toughness.

14.6.2 Abrasive Machining Processes

The processes that use abrasive cutting tools are diverse. Flat, round, soft, hard... all materials and features can be machined using abrasives, but the details vary greatly. There are common applications for industrial grinding processes, and these are largely limited to metallic materials. It is not exclusive, however, and keep in mind that the applications are widespread. The most common limitations with grinding processes are associated with soft materials that "load" (clog) the abrasive.

Abrasive machining can be highly precise or terribly imprecise. It can be very fast, or it can be one of the slowest standard machining practices. As described in Sec. 14.6.1, the tool design determines the outcome of the process, but the method in which the abrasive is applied is equally important. We will consider some of the most common methods, but there are other variations on what we will discuss and some that we will not consider. The principles of these methods are quite similar, however, and learning about some of them will aid in understanding all of them.

14.6.2.1 Surface Grinding

Surface grinding operations encompass a category of processes in which the surface of a part is ground down to create flat or profiled features (Fig. 14.33). The shape of the wheel will determine the geometry of the feature created, and since the wheel breaks down during the grinding process, the profile must be restored (called truing) before final grinding, or a separate wheel may be used for the finishing operation.

The part is most often a ferrous metal and held by a magnetic chuck. The chuck is precisely ground for flatness so the surface created by the grinding operation is highly parallel and flat. The part is positioned on its surface and may be accompanied by fixturing or blocking to help position and secure it. The magnetic pull of the chuck can distort thin parts, and after grinding the part can "snap" back to its original shape, thereby removing the flatness created during grinding.

Surface grinding is best performed on hardened steels as they cut more cleanly and will not load the wheel as quickly. The grinding process often follows a hardening heat

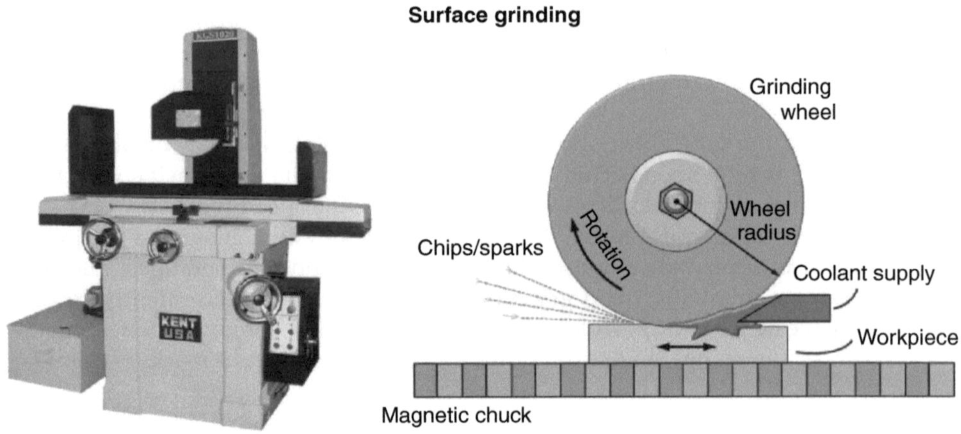

FIGURE 14.33 Surface grinding is used to create flat surfaces and straight profiles. (*Photograph courtesy of Kent Industrial USA, Inc.*)

treatment. The grinding process removes oxidation from the surface and is suitable for creating highly precise dimensions (routinely 0.0001 in tolerances). The grinding process is primarily a finishing process, so major features are first created through traditional machining such as turning (Chap. 16) or milling (Chap. 17) and then ground down by a few thousandths of an inch to final dimensions.

Surface grinding leaves a very smooth surface finish and can be used to create crisp edges on holes used in punching operations (see Sec. 12.2.4.3), or to sharpen other tooling such as shears. Therefore, grinding can be used to create new parts and recondition used ones.

14.6.2.2 Cylindrical and Centerless Grinding

In order the create round parts, the workpiece must be rotated against the grinding surface. Since the most efficient grinding tool is the grinding wheel, this entails rotating two circular sections against one another. This can be accomplished by holding the workpiece between centers (see more on this setup in Sec. 16.4.5), known as **cylindrical grinding** (Fig. 14.34). Power is supplied directly to the workpiece to rotate it against the grinding wheel at slow speed. The workpiece or the grinding wheel (whichever is smaller) is moved longitudinally to remove material along the length of the part.

In **centerless grinding**, the workpiece can be positioned between two wheels and supported from below (Fig. 14.35). One wheel does the grinding, and the other is the

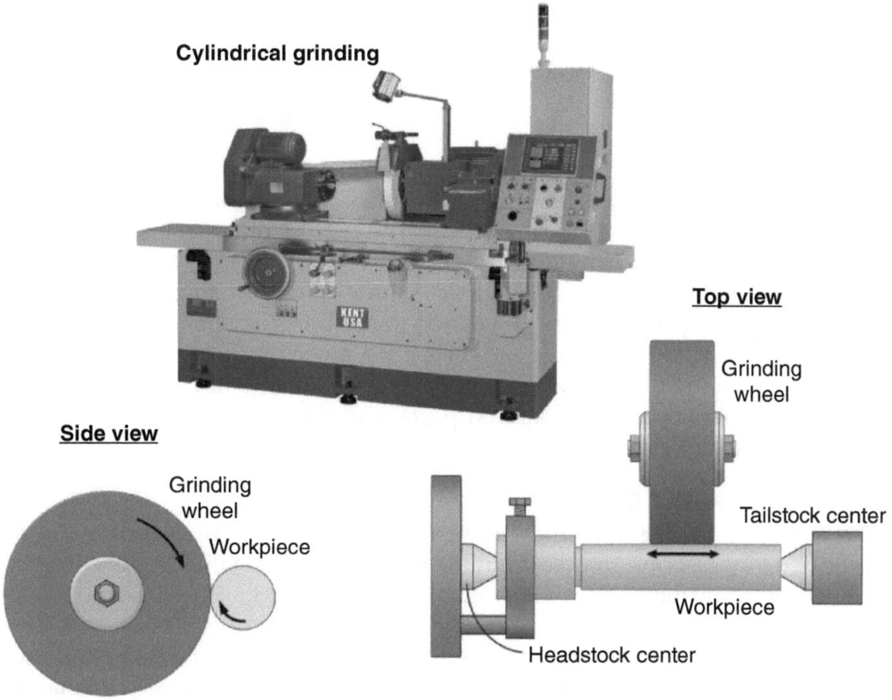

FIGURE 14.34 Cylindrical grinding rotates a workpiece against a counter-rotating grinding wheel. Either the grinding wheel or the workpiece is translated along the length of the feature. (*Photograph courtesy of Kent Industrial USA, Inc.*)

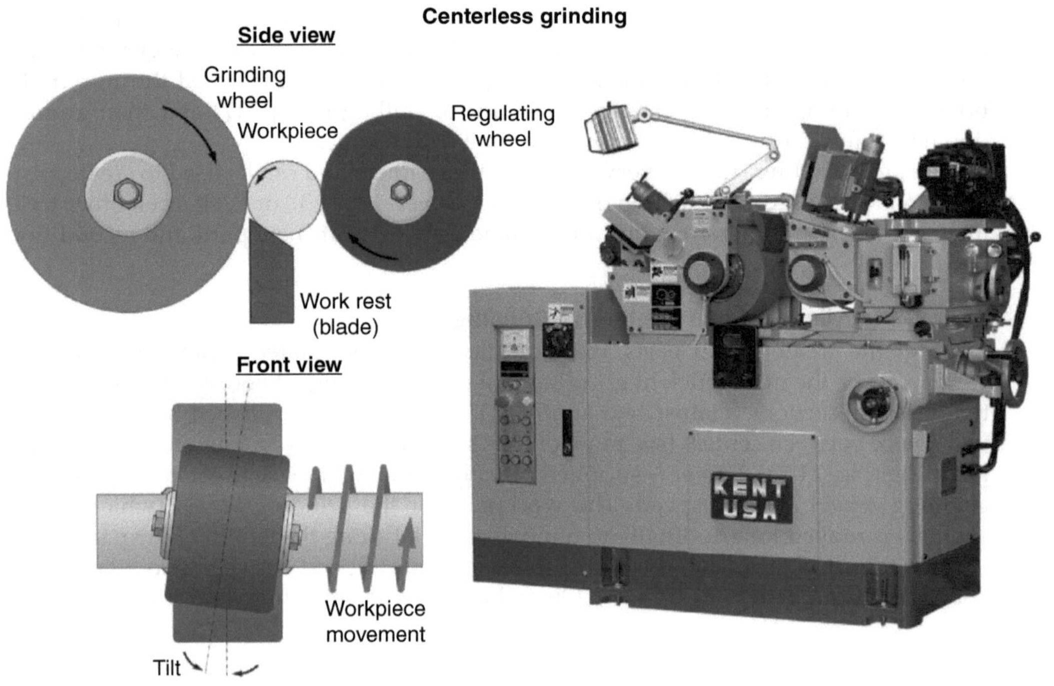

Figure 14.35 Centerless grinding uses a grinding wheel and regulating wheel to control the movement of a cylindrical workpiece. The workpiece rotates at a rate slower than grinding wheel, and progresses forward at a rate controlled by the tilt of the regulating wheel. (*Photograph courtesy of Kent Industrial USA, Inc.*)

regulating wheel. The **regulating wheel** controls the rotation of the workpiece as well as advancing it along its length. This is possible as the regulating wheel is tilted at a slight angle in relation to the grinding wheel. The regulating wheel is often rubberized and does not act to remove any material directly.

The benefits of centerless grinding include being able to grind the diameter of very long workpieces, less skill in processing parts (parts can be fed into one side and they automatically progress through to the other), more rapid processing, and the ability to take heavier cuts due to better support. The machines are not very versatile, cannot grind shoulders or process tapered parts, and they must grind fully cylindrical parts only (no flats or key ways).

14.6.2.3 Abrasive Cutting

All abrasives cut the workpiece in some way, but **abrasive cutting** will be considered unique to operations that section a workpiece into small components. That is, even though abrasive methods can be used to create features or otherwise modify a workpiece, abrasive cutting creates a new workpiece that is often further processed. These types of operations have much in common with sawing operations, including the equipment.

One of the most common abrasive cutting operations is very similar to the radial arm saw or miter saw discussed in Sec. 14.2.3.4. The reason an abrasive blade is used is to cut very hard materials, such as ceramic tile, or to accomplishing the cutting faster than with a saw blade.

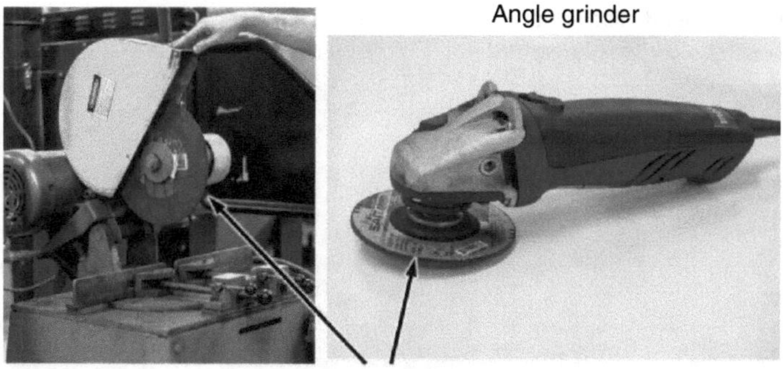

Figure 14.36 Abrasive cutting can be accomplished using a chop saw, which is similar to a miter saw (see Fig. 14.18) or by handheld tool such as an angle grinder. Both require sturdy abrasive wheels which can handle large loads under "imperfect" cutting conditions.

For instance, a chop saw (Fig. 14.36) is functionally similar to a miter saw (see Fig. 14.18), just with a thin abrasive blade. This allows angled cuts to be made in parts (often tubing), which are difficult or time-consuming to make using a band saw. Likewise, a cutoff wheel can be used on a handheld grinder, such as an angle grinder, to make cuts through thin sections, such as sheet metal. These tools are designed similarly to other abrasive wheels, but are generally reinforced to resist fracturing under variable loading conditions.

14.6.3 Abrasive Finishing Processes

Abrasive techniques are very diverse, and there are a variety of methods for improving the surface of a part without significant material removal. These are rightly considered machining operations, but the goal of most is not to create a feature or reduce the workpiece size. Rather, **abrasive finishing** operations enhance the surface characteristics of a part for aesthetic or functional purposes. Methods of preparing surfaces include sanding, sand blasting honing, and lapping.

14.6.3.1 Sanding and Sand Blasting

Sanding is a generic term that encompasses a range of processes used to smooth or remove the surface of a part. This includes smoothing wood, preparing metal for paint and removing old paint. Sanding is often done using a belt, disc, or sheet with abrasive adhered to one side. Equipment for this varies widely and includes large, commercial versions and smaller, handheld models. Belt sanders, disc sanders, spindle sanders, and hand sanding are all included under this category and serve specific purposes. Some examples of equipment in this category are given in Fig. 14.37.

Sanding can range widely in surface roughness, depending on grit size (e.g., 40–2400 grit) and often lack precise control. Although sanding can be used to remove significant amounts of material, it is not an efficient way to do so in manufacturing. In custom or limited production, it can be used to shape parts. Sanding (sometimes referred to as *grinding* in heavy material removal), must be done with great care and skill in these instances.

Figure 14.37 Sanding can be done mechanically or by hand using a variety of equipment to create flat or contoured surfaces.

Sanding is often done as a multistep process in which progressively finer abrasive is used. The abrasive should remove the scratches from the previous step and leave finer scratches. If the difference between the prior grit size and the current grit size is too great, it will take much more work to remove the previous scratches and prematurely wear out the finer abrasive. Moving to successively finer abrasive grades in small steps to create a smooth finish is known as following the **sequence of abrasives**.

Although it requires additional steps of changing abrasives, it is more efficient than using a fine abrasive to remove heavy surface imperfections. The first abrasive grit size used will be determined by the starting condition of the surface, and the final abrasive used will depend on the intended surface finish. An example of various abrasives is show in Fig. 14.38.

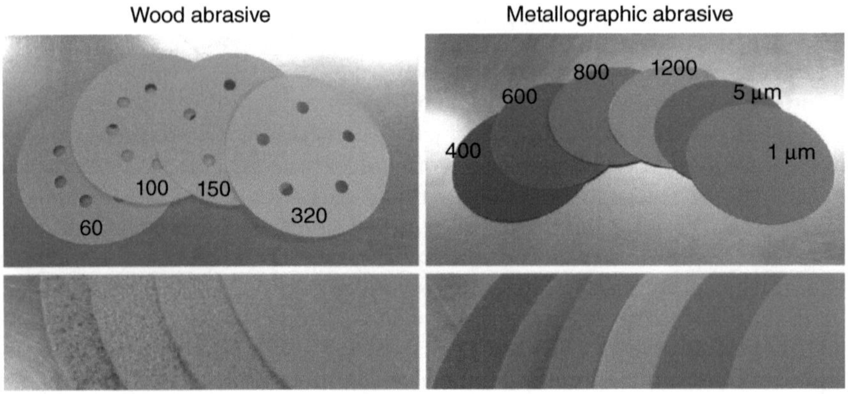

Figure 14.38 Abrasives are to be used in order of increasing grit *number*, which represents a decreasing grit *size*. Very small grit sizes are often represented by the particle size in micrometers (μm) instead of a mesh size number.

Grit Number to Particle Size Conversion*					
	Actual Particle Size (Average)			Actual Particle Size (Average)	
Grit Number	Micrometers (μm)	Inches	Grit Number	Micrometers (μm)	Inches
8	2210	0.0870	100	122	0.0048
10	1854	0.0730	120	102	0.0040
12	1600	0.0630	150	89	0.0035
14	1346	0.0530	180	76	0.0030
16	1092	0.0430	220	63	0.0025
20	940	0.0370	240	50	0.0020
24	686	0.0270	280	39	0.0015
30	559	0.0220	320	31	0.0012
36	483	0.0190	400	22	0.0009
46	356	0.0140	500	19	0.0007
54	305	0.0120	600	16	0.0006
60	254	0.0100	700	14	0.0006
70	203	0.0080	800	12	0.0005
80	165	0.0065	900	9	0.0004
90	145	0.0057	1000	7	0.0003

Note: As the grit number goes up and the grit size goes down, the surface will be smoother.
*Mesh grit size standards and conversions may differ depending on the manufacturer and standard referenced.

TABLE 14.1 Grit Number to Particle Size Conversions

Abrasives with a grit number less than 400 (and often only up to 220 or so) are used to finish wood. Due to their reflectivity, grit numbers in excess of 400 are needed to give metal a satisfactory appearance. Grit numbers can be very large (e.g., 2400), but many manufacturers change to a particle size measurement in micrometers (μm) for these. See Table 14.1 for conversion between micrometer and grit number abrasive sizing. It is difficult to see the individual abrasive particles of fine grit, but this also makes the scratches they leave hard or impossible to see with the unaided eye.

Sand blasting uses loose abrasive particles by propelling them at the surface at high velocity. This is accomplished by mixing loose abrasive powder into a high velocity air stream (Fig. 14.39). As the abrasive leaves the nozzle of a gun, typically operated by hand, it hits the surface with enough energy to abrade away material. Sand blasting can be used to remove rust, paint or to create a clean surface with sufficient roughness to allow for good adhesion by paint.

The abrasive particles follow the same compositional and structural considerations discussed in Sec. 14.6.1, and the roughness of the surface and material removal rate are controlled by abrasive characteristics. To yield a less aggressive process, materials other than ceramics, like silicon carbide and aluminum oxide, can be used. For instance, walnut shells can be used as a less damaging way to remove a layer of paint without damaging the underlying metal.

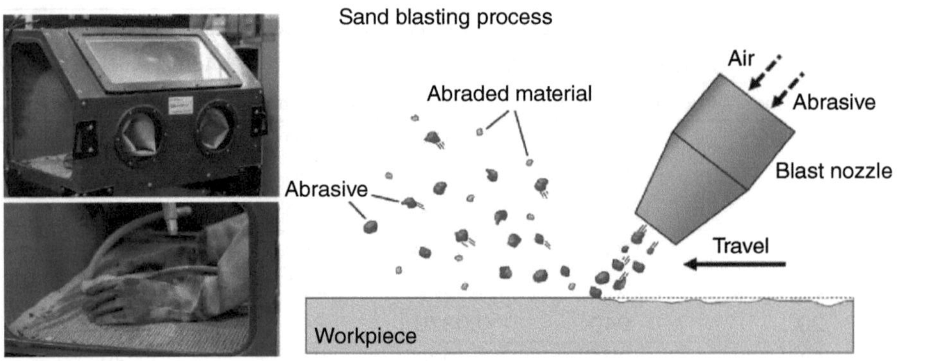

Figure 14.39 Sand blasting uses air to propel loose abrasive grit out of a nozzle and toward the workpiece surface.

Because the particles are not necessarily "sand," though they may be glass beads, the term **blast media** is more appropriate and encompasses the gamut of possibilities. Although sand blasting can be done in an open environment, it is often contained in a blast cabinet (for small components) or a blast room (for large items) to recover the media and reuse it. The particles are continuously collected and recycled into the stream. Additionally, the blast media must be kept clean and dry as foreign materials and oil, water or other liquids can cause equipment issues.

14.6.3.2 Honing and Lapping

For fine surface finishes with greater predictability, honing can be used. **Honing** is an abrasive process that used "stones" or blocks of abrasive similar to bonded abrasive used in grinding. The honing operation is performed at much lower speed and with less heat generated. Honing operations are typically accomplished through a combination of rotation and linear translation. A common application is in the preparation of engine cylinder walls. The walls must be smooth, but very fine scratches in the surface also help retain a film of oil that prevents wear.

For an especially fine finish, lapping is used. Lapping uses a fine abrasive embedded in a soft material, known as a lap, to remove fine scratches and create a highly polished surface (Fig. 14.40). The principle behind lapping relies on the lap being softer than the material being lapped. This ensures the abrasive embeds in the lap and does

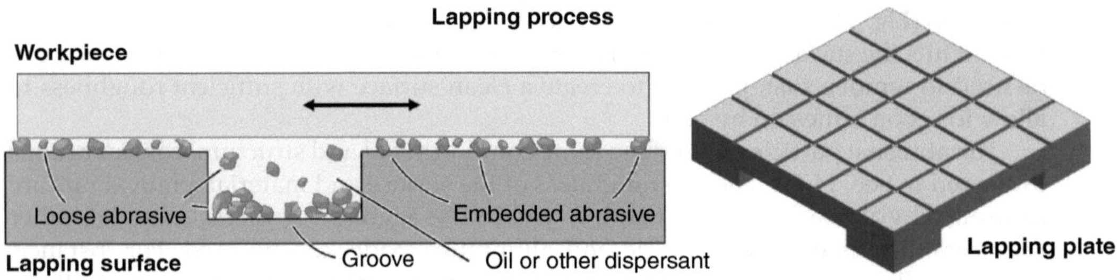

Figure 14.40 Lapping produces a very fine finish using a slurry of abrasive particles.

not wear it away. This "charged" lap then acts like a bonded abrasive surface to abrade the workpiece. The abrasive particles can also be added between the workpiece and lapping surface, usually as a paste or slurry. The process can be applied by hand or with automated equipment. The lap is often a cast iron plate, but it can be any suitable material, and hand laps may be leather.

14.7 Why It Matters

This chapter covered many processes, so it should be no surprise that they are vital to many products. Although these multi-toothed cutting methods can be used to create parts, it is much more common that they either begin or end the manufacturing sequence. For instance, sawing is often used to create roughly sized parts before other processing, and abrasive processes can be used to improve the surface finish of parts after the majority of processes are complete. Whether at the beginning, end, or in between, these cutting processes are essential to efficient production. Efficiency and capability matter.

We started this chapter by considering the basics of saw design. Saws are a common part of daily life, not just in manufacturing, so it is valuable to know what makes them different. We saw that blade design effects the material removal rate, strength and finish of a cut, so hopefully you can appreciate the difference between a hacksaw for metal and a handsaw for wood. We also discussed that saws only cut in one direction, so that might just save you time, effort, and money if you will take a little time to check the tooth orientation. Blade choice and proper use matter.

Sawing can be very versatile depending on the equipment and the way that equipment is used. Whether straight cuts or contours, the rigidity of the blade makes a difference. The use of a fence or other accessory, such as a miter gauge, allows for accurate angles, bevels, or profiles to be created. Whether by hand or by power equipment, the pressure, blade speed and rigidity of the setup is important.

Broaching is a topic that was perhaps less familiar, but there is no better way to cut complex internal profiles. Whether a simple keyway, a spline or other design, broaching provides an efficient way to work on the inside of a part. The cut can be made in a single pass, thanks to the tooth offset, or through the use of shims, a deeper cut can be made with multiple passes.

Filing is another process that is common to most people, but it is something that is often done incorrectly, or at least inefficiently. Files are designed to cut in one direction and are created with a specified coarseness. Filing should be kept to a minimum. Depending on the design of the file cross section, many features can be created or worked on. It can be used to smooth edges and surfaces and to create radii, chamfers, etc., but it is not an efficient production method. Filing is an important technique, but to ensure uniformity and efficiency, it should be used sparingly where high-volume production is needed.

Abrasive processes are another method that should get particular attention. These may be very fast or very slow, create rough surfaces or smooth surfaces, be done by hand or by precision equipment. Abrasive machining is versatile! We saw flat, round, and other profiles can be created by manipulating the part or by using grinding wheels that have the desired profile. Keeping the wheel, or any other abrasive tool, in good condition is essential to achieving good results.

In finishing, a sequence of abrasives should be used to efficiently and effectively smooth a surface. It can save time and money by choosing the coarsest abrasive

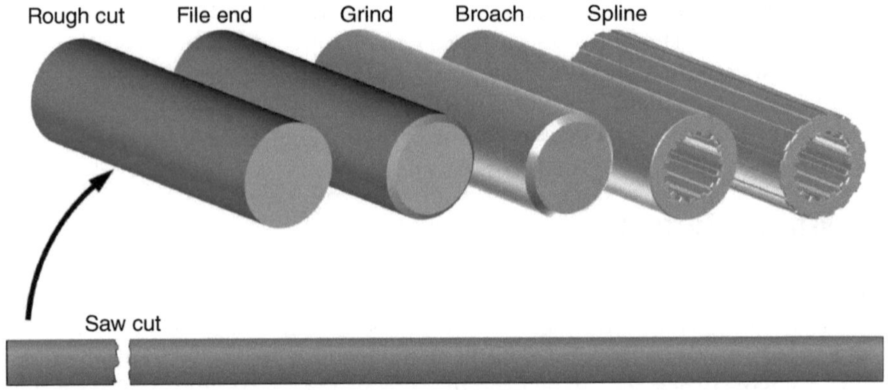

FIGURE 14.41 Sawing, broaching, filing, and abrasive operations are often used to supplement other operations or they can be used together to create parts, such as the splined shafts shown here.

carefully. Specifically, you do not want to start with an overly aggressive abrasive grit as it will create extra steps. Too fine of an abrasive will take much longer to get the surface smooth. Highly polished surfaces should only be used where it is a customer specification. Methods such as honing and lapping can achieve exceptionally smooth surfaces, but the time investment is significant. In cases where a finish such as paint will be applied, coarser abrasive is preferable and sand blasting is an especially useful method for quickly preparing an uneven surface or for areas difficult to reach by other methods.

Any one of the processes covered in this chapter may be applied in conjunction with other techniques, but they may all be applied to the same part (Fig. 14.41). These methods are very general and common practice in all manufacturing environments. Broaching is a technique that may not be as familiar, but without many assemblies we take for granted (like wheels!), would be impractical or at least more difficult. They matter because they are essential practices even if they are in a "support" role for other methods.

14.8 The How and Why of Sawing, Broaching, Filing, and Abrasive Machining

14.8.1 Chapter Review

1. What do sawing, broaching, filing and abrasive machining have in common?
2. How is abrasive machining different from other multi-point cutting methods?
3. Why is tooth pitch important in saw blade design?
4. Describe the role of tooth rake on performance.
5. What is the importance of tooth set?
6. Describe the difference tooth material and shape make.
7. How are straight saw blades used to cut? How do they limit production?

8. How can continuous blades cut without interruption?
9. What are the benefits of continuous saw blades over straight blades?
10. Why are continuous blades that are many feet in length limited to only small cut lengths?
11. What are the key design differences between rip and cross-cut circular blades?
12. How are push saws and pull saws designed differently?
13. What are benefits of orbital motion in reciprocal saws?
14. How is the size of a band saw measured?
15. Why are horizontal band saws only used in metalworking?
16. How are cut dimensions set on the table saw?
17. What is the function of a miter gauge?
18. Describe the differences between a radial arm saw and a miter saw.
19. List three ways to make angled cuts with saws.
20. What is the difference between a bevel and a miter?
21. How does blade size affect minimum contour size?
22. What kind of features are created using broaching?
23. Broaching requires what feature before it can be performed?
24. How is the cut depth of each tooth determined in broaching?
25. How can a pull-type broach be identified?
26. How can broached depth be increased for a second pass?
27. What applications are files used for?
28. How does tooth design affect cutting performance? Give two examples.
29. How does file shape effect the applications it is used for?
30. Why is the cutting geometry not controlled in abrasive cutting tools?
31. Describe how abrasive wheels are constructed, including what is meant by structure and grade.
32. What is the purpose of dressing a wheel?
33. What is the G ratio? Why might lower ratio be beneficial?
34. In what case is an open structure needed?
35. How is material removed by abrasive machining? Why is heat a key consideration?
36. Describe how abrasive processing can be slow, fast, precise, or imprecise.
37. What kind of features are created using surface grinding?
38. What are the differences between cylindrical and centerless grinding?
39. How is a chop saw similar to a miter saw?
40. What differences in abrasive tool design are needed to be used in manually operated equipment?

41. Why is the sequence of abrasives important?
42. What determines the minimum grit size?
43. How are grit number and grit size related?
44. What are the benefits of sandblasting?
45. When are honing and lapping used?

14.8.2 Concept Connections and Extensions

1. How is the cutting geometry of saw blades related to other cutting tools (Sec. 13.2.1)? What is the difference in how they are used?
2. When using a reciprocal saw to cut a wide object, what should be done to clear chips? When is this not practical?
3. A saw back is used to make the blade more rigid. What disadvantage does this have and why might a flexible saw be useful?
4. Consider the similarities between handheld circular saws and table saws. What situations are each better for?
5. Table saws, miter saws, band saws, etc. can make the same type of angled cuts. Which single saw would be the best for a small shop and why?
6. How is a broach with cutting surfaces on all sides more limited than a broach that only cuts on one side? Explain.
7. Why is the starting hole dimension so important when broaching? Would it be better if it was slightly undersize or oversize?
8. Why do you think magnetic chucks are popular for surface grinding, but not as much for other machining processes?
9. When removing significant material, how should the wheel condition be considered to ensure the proper dimensions and quality? Is this true for other cutting tools?
10. Pretend you are an abrasives salesperson. How are you going to convince a customer to buy a wheel that wears away faster, and therefore has to be replaced more often? Imagine it costs the same as other wheels.

CHAPTER 15
Holemaking

Perhaps the most common process in manufacturing is the creation of holes. These features are used to locate two parts, allow for fasteners to be attached, provide passage for other components, and so on. Although holes can have any shape, the profile is typically round. That is no mistake, since a rotating tool of any shape will leave a round hole, and most holemaking operations involve a rotating cutting tool.

There are some exceptions, such as punching operations (Sec. 12.2.4.3), which use shearing to create a hole, and broaching operations (Sec. 14.4), which requires a round starting hole and only modifies that hole. The majority of hole production is accomplished through drilling, and there are a variety of tools and methods associated with it. We will also look at ways of creating high-precision holes, such as by reaming, and methods such as offset boring can be used to create accurate holes of any size, not just standard dimensions.

15.1 How It Works

Holemaking operations are most often completed by rotating a tool with cutting edges on the end, although the cutting edges may also be on the perimeter. Because the end of the tool does the majority of the cutting, even very deep holes are completely dependent on a relatively small cutting area to do all of the work. The condition of the tool tip is vitally important then. Drills, reamers, and related cutting tools can be made of any material (Sec. 13.4), but the most common is high-speed steel. Like all cutting tools, the material will determine the appropriate cutting conditions, productivity, and expected lifetime (see Sec. 13.5.3).

Holemaking tools may use one or more cutting edges to do the work, and their diameter may be fixed or adjustable. By far, the most common setup is to have two cutting edges and a fixed geometry. Holemaking tools may work by removing material at the end of the tool (e.g., drills) or along the perimeter (e.g., taps). Because of the diversity within this category, the specifics will be addressed as each process is discussed. Some of the material may seem very familiar, but as with most processes, there is still more to be learned to get the most out of your efforts.

15.2 Drilling

Drilling is not limited to manufacturing production. We drill for oil, we drill teeth and we drill cores from trees and ice for scientific examination. Drilling is one of the simplest and most ubiquitous methods for modifying structures. It is a simple concept,

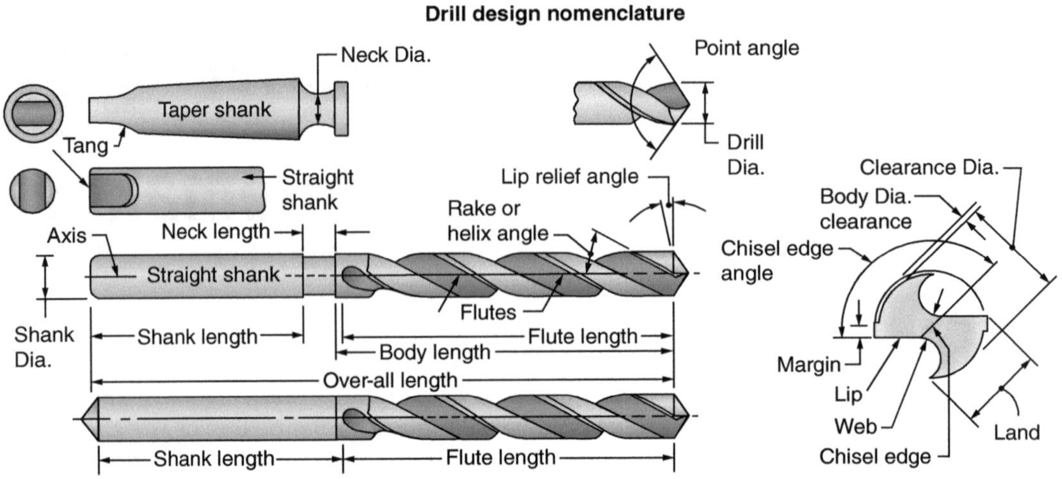

FIGURE 15.1 Twist drills have a number of design features that affect their performance. Here the various characteristics are labeled. (Courtesy of Industrial Press, Inc., from *Machinery's Handbook*, 29th edition, Oberg, Jones, Horton, Ryffel. pg. 867. ISBN 978-0-8311-2901-9.)

but rest assured, there has been significant thought and design put into the tools and manufacturing methods.

Drills cut at the forward end of the tool, and they must then transfer the chips out of the hole. This can be accomplished by designing the drill with an auger-type geometry that advances chips outward as it rotates. Alternatively, chips can be removed by air or coolant (liquid) being forced through or around the drill.

It is important that chips are removed, especially in deep holes, because they can collect and create pressure and heat by rubbing against the wall of the hole. This eventually results in breakage of the drill, and possibly loss of the part. Proper design and use are both important. To discuss these, we will focus our attention on the most common type of drill, the twist drill.

15.2.1 Twist Drills

Twist drills are used for all materials, though variations in their design are necessary to make them effective. They can be purchased in sets or singularly at home centers for relatively low cost. In manufacturing settings the type and size of the drills used will be more diverse, and we will start by establishing some vocabulary. To help identify the various terminology regarding twist drills, refer to Fig. 15.1, where the design features are labeled.

15.2.1.1 Drill Design: Diameter and Length

The most basic feature of a drill is the diameter. Twist drills can be purchased in an assortment of sizes, both standard and metric. Standard sizes up to 1 in include nearly 200 different drills (not including metric). That is a lot of choices. Drills are available in the following designations:*

*The details of drill nomenclature and geometry are specified in the standard ANSI/ASMEI B94.11M-1993.

- Numbers (97–1, 0.0059–0.2280 in)
- Letters (A–Z, 0.2340–0.4130 in)
- Fractions (in increments of 1/64th of an inch)
- Metric (from 0.15 mm up in increments 0.01–0.50 mm, depending on size)

Some drill sizes vary by less than 0.001 in, and there are even two designations that are exactly the same size (i.e., ¼ in and letter size E are both 0.250 in). You can see the way these sizes are distributed and their decimal equivalents in the drill chart inside the back cover of this book. In general, size increases as the drill number decreases (80–1), letter increases (A–Z), and fraction increases (1/64th and up). Millimeter designations do not require a decimal equivalent as they are given in decimal form.

There are so many drill sizes because there are so many needs that drills satisfy. Drills are used to provide clearance for fasteners, to create passages for fluid and air to flow, to allow locating pins to seat tightly and to provide a starting point for threading, reaming, counterboring, and other operations. Drills must be able to accommodate different material types and slight variations to maximize the efficiency of secondary processes, so having many choices is critical to maintaining productivity.

Though often not a common consideration for most, the length of a drill is also important. The "standard" length drills are referred to as **jobber length** drills. Shorter drills are referred to as **screw machine length** drills, a name derived from their original application (see Sec. 16.2.3 for more on screw machines). **Aircraft extension drills** are also available and have an extended length necessary to reach difficult locations. These varying lengths are shown in Fig. 15.2. A taper length drill is longer than a jobber drill. These drills have a flute length similar to taper shank drills but with a straight shank.

To compare these three types, consider a ½-in drill. As a jobber length drill, the flute length is 4.5 in and the overall length is 6 in. A ½-in screw machine length drill has flute and overall lengths of 2.25 and 3.75 in, respectively. A ½-in taper length drill has dimensions of 4.75 and 7.75 in, respectively. Depending on the depth of the hole and the space available, longer or shorter drills can be substituted for the standard jobber variety. See Sec. 15.2.3 for more on techniques to ensure proper holemaking in cases of deep or large diameter holes. Many other combinations of length, flute geometry, and other factors are also available, but they are less common.

15.2.1.2 Drill Design: Tip Geometry

Because drilling relies on the tip to do all of the cutting, the tip condition is very important. A side note: the tip *should* do all of the cutting! Drills are not meant to cut along their sides (laterally). You may try to use them this way, but that is not their intended use.

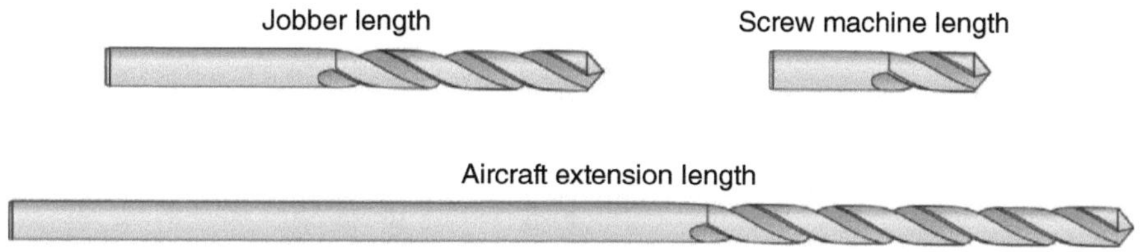

FIGURE 15.2 Twist drills can be purchased in varying lengths. Common naming conventions are shown above.

The tip angle is very important because it dictates the way the drill interacts with the material. The most common angle is 118°, but other angles are available (e.g., 90°, 135°, etc.). The reason the tip angle controls the cutting characteristics is due to the cutting process. The first part of the drill to contact the workpiece surface is the chisel edge (see Fig. 15.1). The drill must be advanced into the workpiece, overcoming the friction and plastic deformation at the center until the lips of the drill can engage the material.

A broader angle (e.g., 135°) will result in greater cutting engagement sooner in the depth and result in a thinner chip being formed, due to a shorter cutting edge. Larger tip angles are often cited as being preferable in harder materials (such as alloy steels) and smaller angles for soft materials (such as polymers), but it is not always so simple. For instance, a broad range of angles (90–140°) may be cited for a class of materials, such as aluminum alloys. In low-production environments, a general-purpose 118° tip will serve most applications well, but if you experience issues related to dulling, heating, and quality, consult a tool manufacturer on material-specific design.

In general, increasing the tip angle reduces the cutting force but increases the likelihood of the drill "**walking**" or deviating from center. More complex drill geometries such as **split point drills**, **dubbing** (changing the cutting edge rake angle) and other faceting techniques can be used to enhance cutting characteristics and lifetime. As with all cutting processes, the rake and clearance angles will have the greatest effect on performance (see Sec. 13.2.1 for more on these aspects). A traditional drill tip geometry and a split point drill are compared in Fig. 15.3.

15.2.1.3 Drill Sharpening

Drills are capable of being re-sharpened multiple times and still function well. It is imperative that sharpening retains (or restores) the original, factory geometry. The drill was designed to perform with the given tip, rake and clearance angles, and deviating from the original design may result in premature failure. This is not an absolute restriction, however, and some liberty may be taken in modifying the geometry for a specific purpose. Just do not get carried away!

Some of the most troublesome results of drill sharpening are the improper change in tip angle, the modification of the clearance angle, and an uneven final geometry,

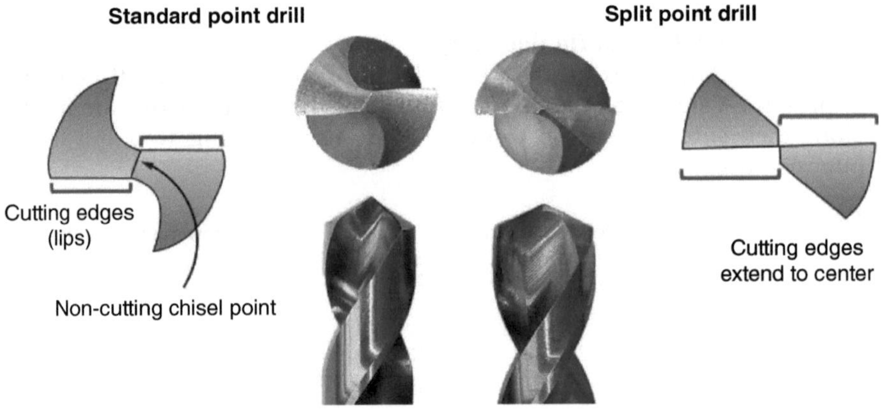

FIGURE 15.3 Standard drill tip design includes a chisel edge which does not cut. A split point drill cuts all the way to center allowing it to engage immediately and reduce deflection.

FIGURE 15.4 Drills can be sharpened freehand and verified by drill point gage or the geometry can be precisely controlled using a drill sharpener which gives a faster, more repeatable result but with less flexibility.

where the angle, cutting edge lip, or other feature is not centered and symmetric. This results in the cutting being unevenly distributed and causing loss of quality, accuracy, and accelerated tool failure. Also, the web of the drill thickens toward the shank, and web thinning is necessary to reduce the chisel edge size after repeated sharpening.

Depending on what issue has been caused, it may be easy to detect. For instance, if the tip angle is not symmetric, chips will only be observed exiting one flute of the drill, or at least the majority of chips will come from that flute. This is caused by the fact one cutting edge is doing most of the work. The drill may wobble, walk, chatter or otherwise result in additional noise and heat during cutting.

Drills may be sharpened by hand and their accuracy confirmed by eye (if you are experienced), or the geometry may be checked by a drill point gauge, which has measurements and indication for the angle and lip length (Fig. 15.4). Alternatively, a drill sharpener can be used. This may be a dedicated attachment to a tool grinder or a small, single-purpose unit that is used solely to recondition drill points.

You can "get away" with imperfections to drill geometry when working by hand and in small part numbers. Any inefficiency in the drilling process is magnified during large production runs and automated processes. Each time that drill has to be removed from the machine to be sharpened is time the machine is not producing parts. In these circumstances, the cost of the drill may be less than the downtime due to sharpening errors, especially smaller drills. A new, factory-conditioned drill is preferable for the consistency it provides.

15.2.2 Other Drill Types

Twist drills are the most common type of drill, but there are certain cases in which they are not the best suited. Instances of this include very deep holes, very large diameters, when drilling wood, drilling ceramics, etc. Even with the many variations that twist drills come with, there are dozens of other types of drills for every application imaginable. We will look at some of the more common variants you will find in manufacturing and what their advantages are.

15.2.2.1 Center Drills (Combination Drill and Countersink) and Spot Drills

One of the most common operations in drilling is to prepare a location for the drill to start. Strangely enough, this also requires drilling! The difference is that a shorter, more rigid drill is used to create a small hole in the material that guides and stabilizes the drill used

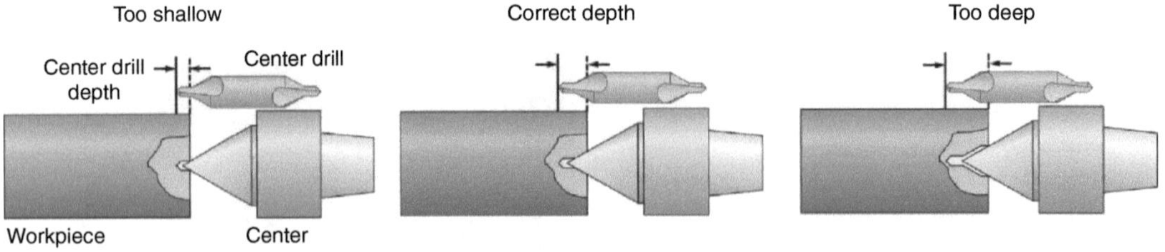

Figure 15.5 Center drills are design to produce a small hole in the end of a part to accommodate a center. If the appropriate depth is not used, the center will not seat properly.

in the following step. The most common drills for this are center drills and spot drills. A **center drill** (more accurately known as a combination drill and countersink) is a short drill designed to create a tapered hole (60° angle) in which a center is located to support a workpiece, often in the lathe (see Sec. 16.4.5 for more on turning between centers). The depth of the drilling operation is critical to ensure the center is properly seated. This concept is shown in Fig. 15.5.

Center drills are also used to create a small starting hole that provide clearance for the chisel edge of a twist drill so that the lips can begin cutting immediately (see Fig. 15.1 for more on the terminology in drill design). This prevents the drill from walking. Center drills are used widely for this purpose, but their size specifications are not always precisely known by the user. They are often designated by trade numbers, not sized by diameter, so many people will choose one that is proportional to the hole size they need.

For instance, you may find a #3 center drill, but unless you measure it, you cannot be sure what the outside diameter is, what the tip diameter or length are, etc. Figure 15.6 provides a reference with these dimensions, and this is particularly useful when deciding on processing to ensure the center drill provides adequate clearance for a drill when used to a specified depth. As the workpiece or drill size increases, a larger center drill is required.

Center drill (combination drill and countersink) geometry specifications			
Trade number	A: Pilot diameter (in.)	B: Body diameter (in.)	C: Pilot length (in.)
000	0.0200	0.1250	0.0300
00	0.0250	0.1250	0.0300
0	0.0313	0.1250	0.0380
1	0.0469	0.1250	0.0469
2	0.0780	0.1875	0.0780
3	0.1090	0.2500	0.1090
4	0.1250	0.3125	0.1250
5	0.1875	0.4375	0.1875
6	0.2190	0.5000	0.2190
7	0.2500	0.6250	0.2500
8	0.3125	0.7500	0.3125

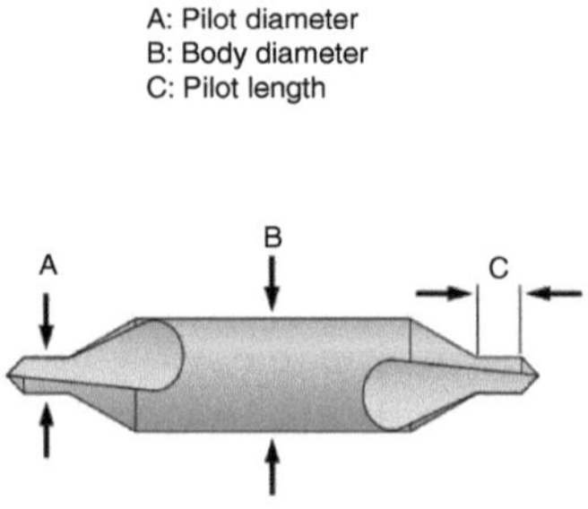

Figure 15.6 Center drill (combination drill and countersink) specifications regarding size in relation the trade numbers they are given.

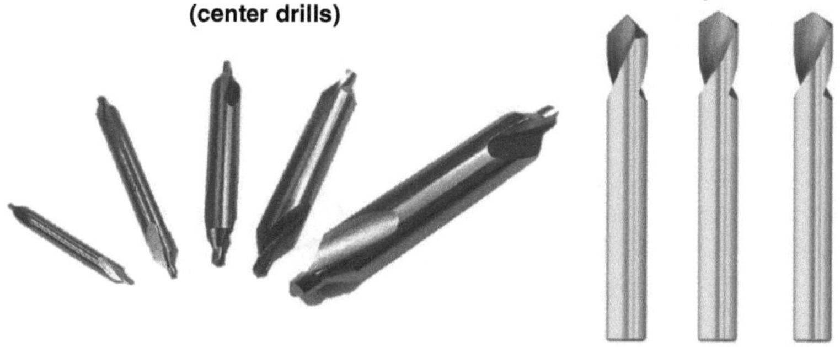

Figure 15.7 Center drills and spot drills are both used to create pilot holes for twist drills. Center drills have the additional purpose of creating a locating surface for a center. (*Spot drill images courtesy of Atlas Cutting Tools, LLC.*)

Knowing the depth/diameter relationship is especially important for automated processing where you must program a depth instead of producing it manually.

A **spot drill** (or spotting drill) serves a similar purpose of creating a hole that a twist drill later follows. Spot drills are usually sturdier as there is no increase in diameter for a countersink. Also spot drills do not need to have a 60° angle to match the taper on a center (although they can be 60°), and they are typically 90° or 118°. In general, a spot drill is preferred over a center drill for creating a starting hole. They are more expensive and less common in the average shop, but if you are planning a production drilling process, plan on using a spot drill. Center and spot drills are shown in Fig. 15.7.

15.2.2.2 Gun Drills

Gun drills are straight, often single-flute drills with integrated coolant passages used for drilling very deep holes. Although "deep" is an imprecise term, **deep hole drilling** is typically defined as a hole that is three times deeper than its diameter. So a 0.5-in-diameter hole is considered deep if it is 1.5 in or more into the part. They can be used to drill holes with very high **aspect ratios**, or the ratio of the depth to the width (diameter), even more than 300:1 with specialized setups. Their construction makes them more rigid and efficient than twist drills for these applications. The design allows for efficient cooling and lubrication at the drill tip, and it also enables the continuous removal of chips from the hole to further reduce friction, stress, and heat. As the name implies, gun barrels are a good example of the kind of applications these drills can serve in. These drills are shown in Fig. 15.8.

15.2.2.3 Masonry (Hammer) Drill Bits

A **masonry drill bit**, sometimes referred to as a hammer drill bit, is designed to drill through hard materials such as concrete, brick, stone, and other abrasive structures that would quickly dull other types of drill bits. The primary functional aspect that allows for these bits to be used is their carbide tip, which can resist wearing down even when applied against other ceramics (see Sec. 13.4.2 for more on carbide tooling). Drilling ceramic materials is demanding due to their high hardness and brittle nature. If you have a basement, the walls (foundation) are most likely concrete block, and a regular

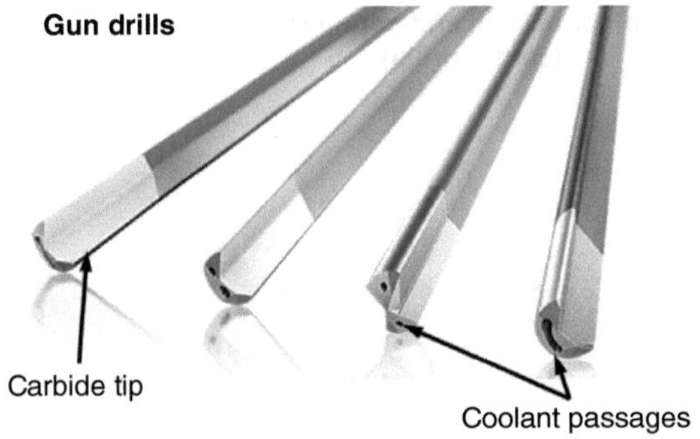

Figure 15.8 Gun drills are especially suited to drill very deep holes. This is due to the ability to flow coolant through them and continually flush chips out of the cutting area. (*Courtesy of Sandvik Coromant.*)

twist drill will not perform well at drilling this material. A masonry bit can drill reliably, and is often used with a rotary hammer drill that combines rotation with an intermittent forward impact to help fracture the material being cut. This type of drill is shown in Fig. 15.9.

15.2.2.4 Brad Point, Forstner, and Spade Drill Bits

Brad point, Forstner, and spade drills are used to cut flat-bottom holes in relatively soft, fibrous materials like wood. These include a pilot, which keeps the drill centered to help compensate for the large flat, cutting faces. Otherwise, the drill would "wander" across the surface causing the hole to be misplaced and to drill oversize. Drilling through fibrous materials is also supported by slicing (or "scoring") the fibers on the outside of the hole before the bulk of the material is removed.

Brad point bits are quite similar to standard twist drills except for the cutting face geometry. Forstner bits are used for larger diameter holes and have cutting edges around their circumference. Spade bits (less commonly known as "paddle" bits) are simpler in design and cheaper, but do not give as good quality of a cut. Their main

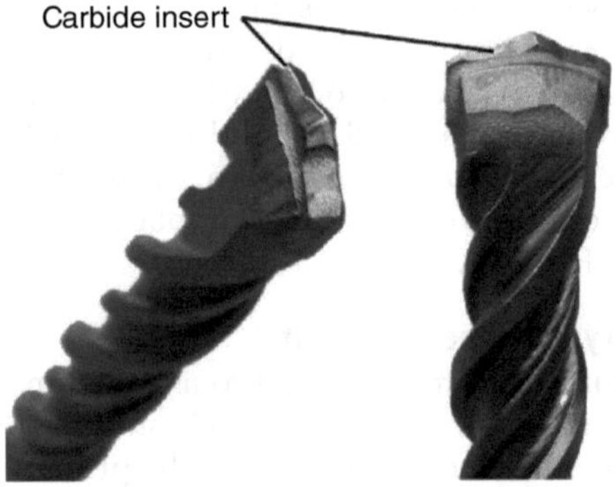

Figure 15.9 Masonry (or hammer) drill bits are used to drill through hard, brittle materials without wearing out. They are often used in conjunction with a hammer drill that rotates the bit as well as reciprocating to create impact force that breaks up the material. The cutting demands are met by using a carbide tip.

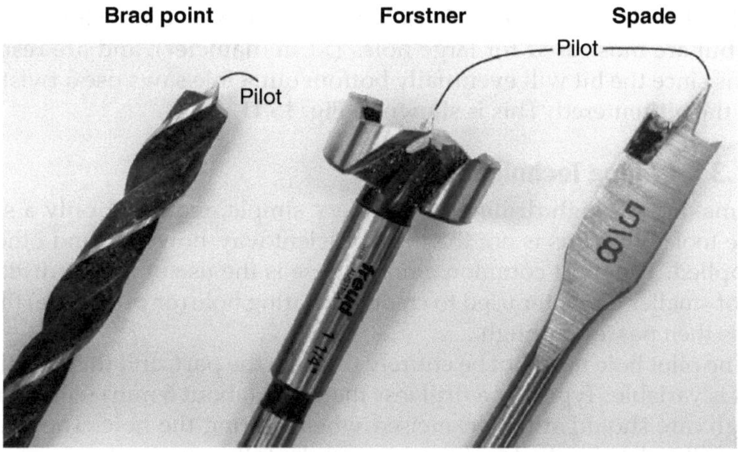

Figure 15.10 Brad point, Forstner, and spade bits are used to drill flat-bottom holes in relatively soft, fibrous materials, most often wood. Their design allows for a clean cut to be made by guiding the drill bit with a pilot and cutting through fibers at the edge of the hole before the material is removed.

advantage is that they can be sharpened into custom profiles with relative ease. These drills are shown in Fig. 15.10.

15.2.2.5 Step Drills and Hole Saws

Step drills, sometimes called a *unibit,* are drills with an incrementally increasing diameter. They have very short distances along which that diameter is set, so they are most useful in thin materials such as sheet metal. The deeper the drill is progressed, the larger the diameter, and the diameter size is marked on the bit. Step drills are also convenient for drilling large diameter holes since they incorporate multiple pilot drills (see Sec. 15.2.3) into a single tool. Step drills are shown in Fig. 15.11.

Hole saws are related to saws in that a perimeter of saw teeth is used to cut through material. They have a fixed diameter for each bit, but the size can be changed for a single tool. That is, the cutting diameter may be varied on a single shank as needed,

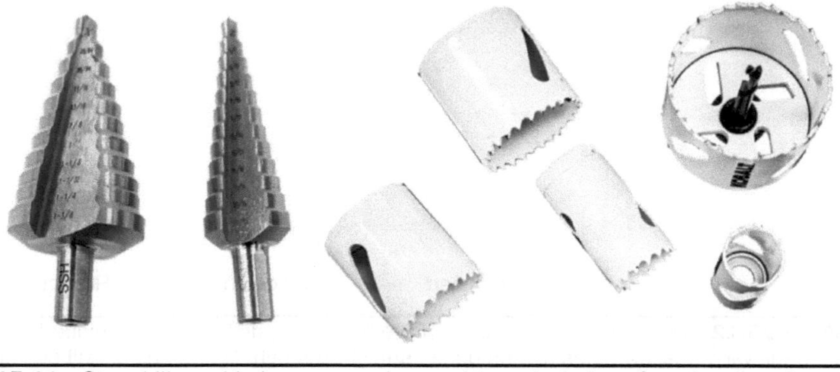

Figure 15.11 Step drills and hole saws can be used to make holes of varying diameters. Neither can drill very deep holes, but both are relatively quick at making large diameter holes.

usually through the use of a threaded collar and set screw. Hole saws vary widely in size, but are most used for large holes (>1 in diameter), and are restricted to shallow depths since the bit will eventually bottom out. Hole saws use a twist drill as a pilot to keep the bit centered. This is shown in Fig. 15.11.

15.2.3 Drilling Techniques

Holemaking through drilling can be very simple, requiring only a single step with a single tool. Often this is not the most efficient way, however, and other techniques can be applied. The most common among these is the use of a pilot drill. A **pilot drill** is a drill of smaller diameter used to create a locating hole (or **pilot hole**) that another, larger drill is then passed through.

The pilot hole need not be entirely through the part, and the size difference between drills is variable. Typically a drill less than ¼ in (about 6 mm) will not need a pilot hole, though care should still be exercised when starting the hole. The minimum size for a pilot drill is the chisel edge dimension on the following drill.

To prevent heat and chip build-up, especially in deep holes, it is good practice to peck drill. **Peck drilling** is the process where multiple smaller depths are cut with short retract motions between each cut to "break" the chip and help remove waste material. This also gives the drill a short time to cool and/or for coolant to better reach the cut area.

A peck motion can include a retraction of only a few thousandths of an inch, or, in the case of deep hole drilling, the retraction can entirely remove the drill. This is most important where the aspect ratio of the hole is greater than 3 or the chips cannot evacuate well otherwise. If chips are not observed leaving the hole during drilling, it is especially important to retract the drill entirely so as to clear the drill flutes and reduce friction. Retracting the drill entirely between each cut is known as **deep hole drilling**. Simple, peck and deep hole drilling are shown in Fig. 15.12.

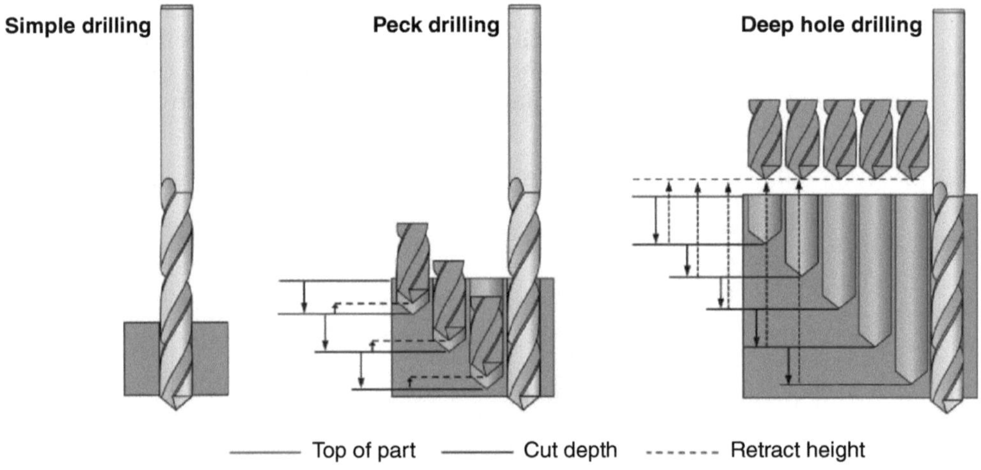

Figure 15.12 Depending on the depth of the hole being drilled, the technique for drilling the hole will vary. Shallow holes can be drilled straight through. Deeper holes will benefit from peck drilling where the drill is retracted slightly to break the chip or from deep hole drilling where the drill is entirely retracted between each cut.

15.3 Reaming

Reamers are used to precisely enlarge a hole and produce excellent surface finish. The reaming operation only removes a small amount of material from a hole created by another holemaking operation, usually drilling. Although the amount of material removed will vary, reamers are designed to remove less than 1/32 in, and often less than 0.010 in (0.3 mm), since they cut with a small chamfer on the front edge. If the preexisting hole is not created with at least minimal care, and the hole is very rough, out of round, etc., the reamer may not be able to fully clean up the hole. The basic design features of a reamer are shown in Fig. 15.13.

Reamers are used in circumstances where a tight tolerance and smooth finish are needed. Reamers can be used in soft or hard materials and can be operated by hand or machine (more on the types of reamers in the next section). The tolerance advantages are most often justified in precision fitment and mechanisms requiring smooth motion without being loose.

15.3.1 Types of Reamers

Reamers come in a variety of designs for various applications. Some are straight, some are tapered. Some are operated by machine (i.e., chucking reamers), others are operated by hand. They can be a fixed diameter or adjustable. They may have flutes that are straight or helical. Obviously, there must be a reason for this diversity!

Although reamers serve the principal purpose of enlarging and improving hole quality, they can accomplish that in many ways. **Hand reamers** are spun and progressed through the hole by hand, and to ensure proper alignment and easier starting, the beginning portion of the reamer is tapered. **Expansion reamers** allow for the hole size to be adjusted by a few thousandths of an inch using a screw located in the end of the reamer to accommodate variations in mating parts or for repair work. **Adjustable blade reamers** have a greater variation in diameter modified by means of adjusting nuts on the outside diameter of the reamer. Although, the actual cutting size of each shell is fixed, **shell reamers** allow for variation in size to be achieved with a single shank as the cutting end can be exchanged. A variety of reamers are shown in Fig. 15.14.

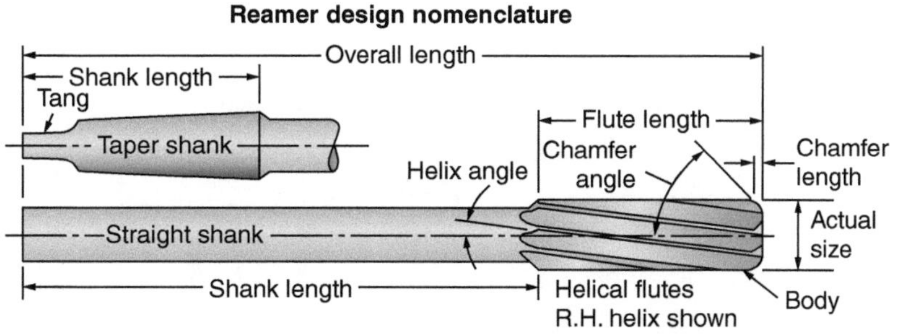

FIGURE 15.13 Reamers have a number of design features that affect their performance. Here the various characteristics are labeled. (Courtesy of Industrial Press, Inc., from *Machinery's Handbook*, 29th edition, Oberg, Jones, Horton, Ryffel. pg. 849. ISBN 978-0-8311-2901-9.)

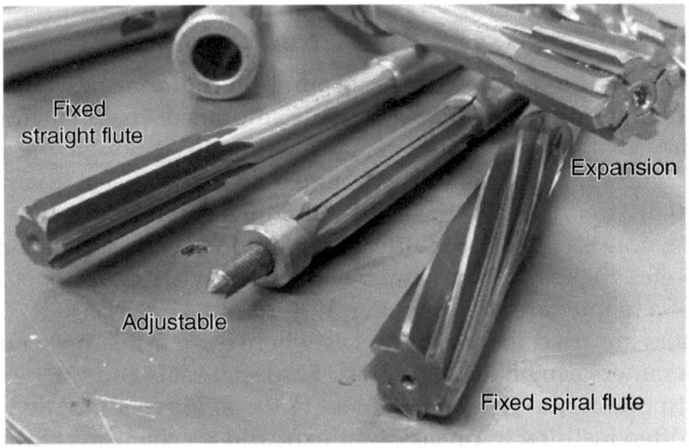

Figure 15.14 Reamers come in a variety of configurations to allow for diverse hole types to be accurately sized.

15.4 Counterboring, Spot Facing, and Countersinking

Counterboring, countersinking, and spot facing are all methods of modifying an existing hole. These operations are often used to prepare a hole for a fastener used to create assemblies. The resulting features of these methods are shown in Fig. 15.15. The methods to create each are described next.

A **counterboring** operation creates a concentric flat-bottom hole of larger diameter and lesser depth in relation to a preexisting hole. Counterboring is often used to create space for a bolt head to be recessed within a part. If used for such as purpose, the depth and diameter of the counterbore will be dictated by the geometry of bolt head.

Counterboring tools may have a locating pilot to help ensure the counterbore is concentric with the preexisting hole. If the pilot is too small, it will allow the counterbore to be off center, but if the pilot is too large, it can bind in the hole. Some counterboring tools have adjustable pilots (see Fig. 15.16), but others have fixed pilots that are part of the tool itself. If the workpiece is rigidly held and position can be maintained between tools (such as in a milling machine), no pilot is necessary, and an end mill can be used in place of a dedicated counterboring tool.

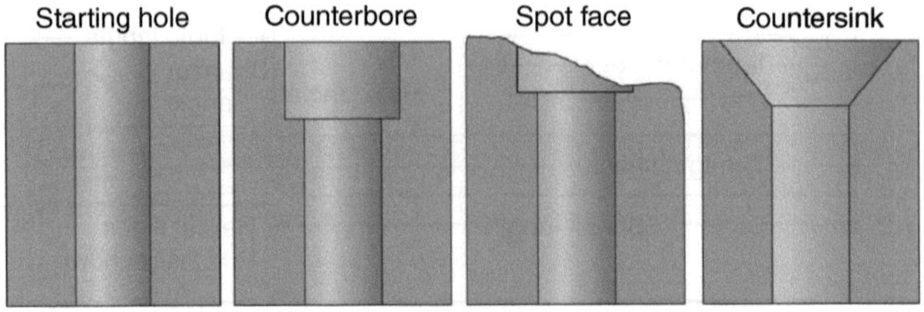

Figure 15.15 Holes may be modified in a number of ways to accommodate fasteners or otherwise change the hole characteristics. Some of the most common modifications are shown here.

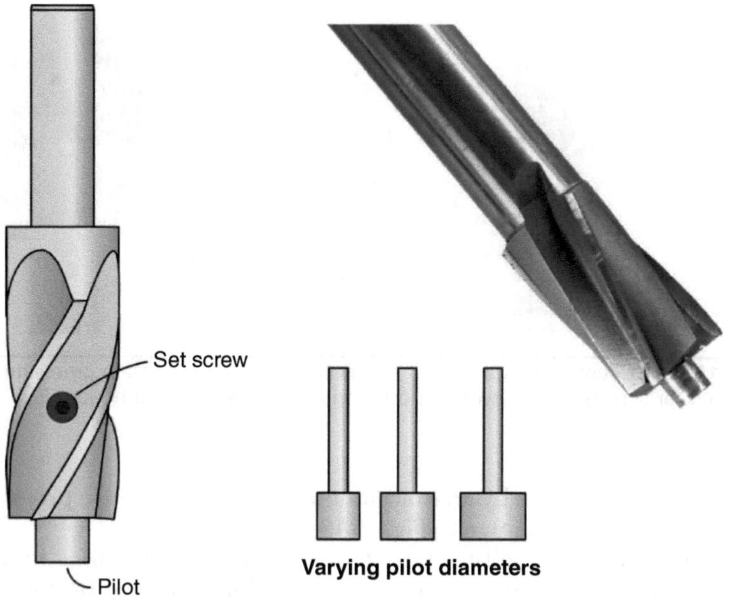

Figure 15.16 Counterboring is often used to enlarge holes to create clearance for a bolt or screw head to be recessed into the surface of a part. To help center the counterboring tool, a pilot with matching diameter to the hole being modified can be used.

Spot facing uses the same tooling as counterboring, but it is only used to flatten a surface concentric with a prior hole. This is most often used when a bolt head must sit flat on a surface that is angled or uneven, which is common with cast or forged parts.

Countersinking is a process where a tapered section is added to a hole. A countersink, the tool used to create the feature, is tapered with a specific angle designed to match the screw or bolt and allow it to be fastened flush with the surface. The diameter of the hole will increase with increasing countersink depth, but the angle is constant. Countersink angles are usually 82° or 90°, but other angles are also available.* Countersinks may be a standalone tool or combined with a drill to reduce the number of setups (e.g., center drills). The more flutes a countersink has the faster it can cut, but it also leads to chatter (vibration). Countersinks are often run at relatively low rpm with a heavier feed. Varying types of countersinks are shown in Fig. 15.17.

15.5 Tapping

Tapping operations create threads within an existing hole for later use with fasteners or other threaded components. Tapping is not the only way to create threads, but it is the most common. Like drills and other hole making operations, the tap is rotated while advancing it through the part. The biggest difference is that taps must be fed at the exact rate that matches the thread they produce. Just as one turn of a screw will move the

*These are most commonly 60°, 100°, 110°, and 120°, depending on the application.

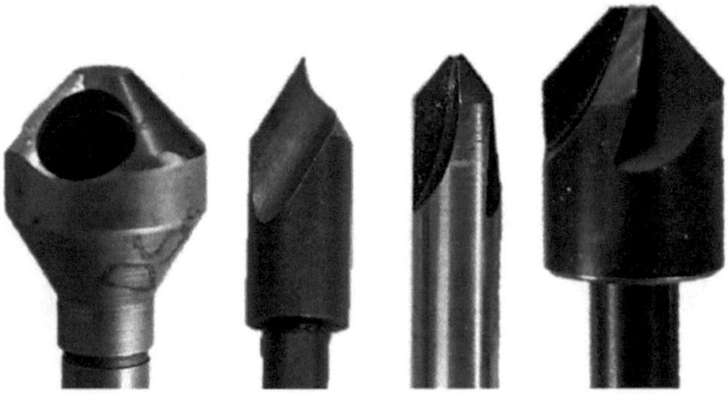

FIGURE 15.17 Countersinks are used to create tapered recesses in preexisting holes as clearance for screws, bolts, rivets, or other fasteners with angled heads.

screw in or out of a hole by an exact amount, the tap must also progress at a given rate. The design of a tap is shown in Fig. 15.18.

Because the tap is self-feeding, tapping can easily be done by hand or machine, as long as there is nothing to restrict the motion in or out of the hole. As the tap is rotated it rides in its own track and progresses at a rate determined by its thread pitch (see Sec. 15.5.1.2 for more on the anatomy of threads).

The most important stage of tapping is during the start of the cut. If the tap is started crooked or not enough pressure is applied, the hole can be damaged and the tap can be broken. To help align the tap, a type with a tapered end is preferable (see Sec. 15.5.2 for more on the types of taps). Taps can be fed through a hole using a drill press, mill, or lathe with a practice known as "power tapping." This is done at low rpm and must be carefully setup so as not to break the tap by overloading it.

When tapping by hand, it is common practice to feed a tap a partial turn into the cut and then reverse the direction to break the chip and reduce cutting force. Ample lubrication is also suggested. Caution should be exercised when tapping blind holes as the

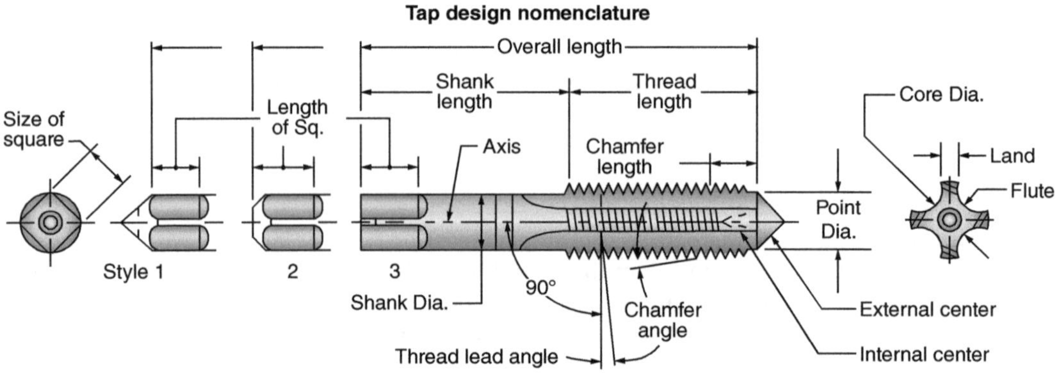

FIGURE 15.18 Taps have a number of design features that affect their performance. Here the various characteristics are labeled. (Courtesy of Industrial Press, Inc., from *Machinery's Handbook*, 26th edition, Oberg, Jones, Horton, Ryffel. pg. 873. ISBN 0-8311-2625-5.)

tap will bind if it reaches the bottom and result in the tap breaking. Broken taps are notoriously difficult to remove from holes and can result in the loss of a part. To better appreciate these considerations, we will first look at thread specifications and then more specifically at taps and related tooling.

15.5.1 Screw Threads

Thread specifications are standardized to ensure universal fitment of parts. This is why you can buy a nut and bolt from different manufacturers and have them work together. Although there are some limited uses for customized thread geometries, the vast majority of operations create threads that conform to some basic standards of size and fit. We will look at the general considerations next.

15.5.1.1 Thread Designations

There are many types of threads, but we will focus on Unified Screw Threads (some others are given in Sec. 16.5.6). Most nuts and bolts following this system and are classified in the format of XX-YY UNC/UNF, where XX is the diameter, YY indicates threads per inch (tpi), and the UNC/UNF designates the thread as being Unified National Coarse (UNC) or Fine (UNF).*

Coarse threads are fewer in number and deeper, which can prevent stripping the internal threads of a softer material. Fine threads have a higher count per inch and increase the unthreaded area of the bolt, thereby increasing the tensile strength. This is also beneficial for applications where the wall thickness around the threaded portion is thin or where fine adjustment of the threaded component is needed. Note that as bolt diameter decreases, it requires fewer threads per inch (tpi) to be considered coarse. For instance, a ¼-in screw is considered a fine thread with 28 tpi (i.e., ¼-28 UNF), but a #5 bolt is considered coarse with 40 tpi (i.e., 5–40 UNC). It is relative to the diameter.

15.5.1.2 Anatomy of Threads

Now that we know a little about how threads are classified, we should look more closely at their design. There are some basic geometric considerations that threads must follow to be universal. The diameters, angles, and spacing of the threads are all critical. The major features of threads are shown in Fig. 15.19. Although this provides a valuable basis for understanding, the exact determination of these features differs depending on the thread geometry.

The **thread pitch** is the spacing between two peaks (or troughs) of the thread. This also means that the pitch is equivalent to one divided by threads per inch (1/tpi). For instance, if a screw has 40 tpi, the pitch would be 1/40 = 0.025 in. This is true for single start threads (the most common), but will vary for multiple lead threads (see Sec. 16.5.6.1).

Most standard threads have a **thread angle**, θ, of 60°. For that to stay true, the thread depth, or the distance between the peak and trough of the treads, must get smaller as the pitch gets smaller. This change in depth also affects the way that the diameter is measured. For example, ¼-20 UNC and ¼-28 UNF bolts both have a ¼-in outside diameter, or **major diameter**, but their minimum, or **root diameters** will be different, as will their **basic**

*Additional designations include UNEF (extra-fine) and UNS (special), which typically (but not always) has more threads per inch than fine or extra-fine designations and therefore higher strength.

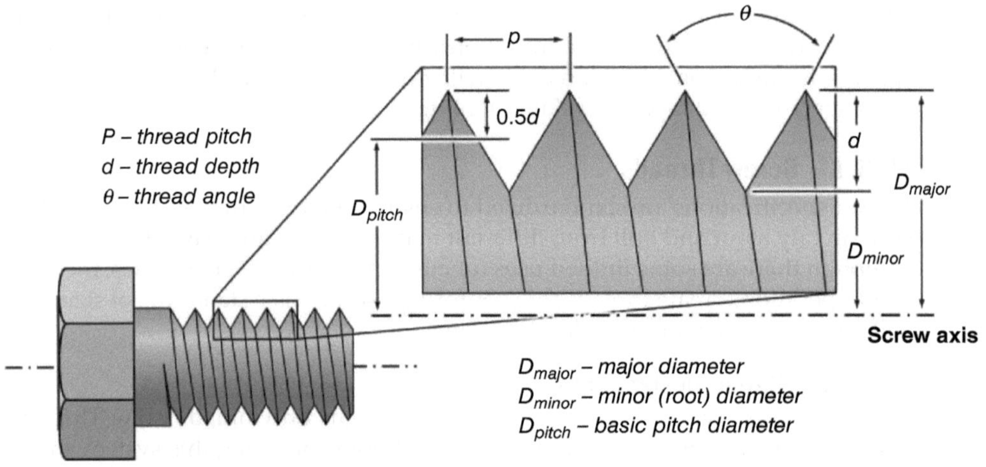

FIGURE 15.19 The characteristics of threads determine their applications and performance. Some basic features of threads are shown here.

diameters, which are determined by the midpoint of the thread depth. In practice, perfectly sharp threads are rare, so the way these are defined can change as well.

15.5.1.3 Class of Fit

There are tolerances in thread size specifications. External and internal mating threads cannot be the exact same size or they would bind together. For that reason, they need to be of slightly different size, but if they are too far off, the fit will be sloppy, and the strength of the assembly will be reduced. It is wasteful, however, to use a precision fitting thread in an application not requiring it. For these reasons, classes of fit are specified in addition to the basic thread designations.

The class of fit of a thread is based on the maximum deviation allowed from nominal values related to the pitch diameter and nominal diameter. The classes of fit include class 1A, 1B, 2A, 2B, 3A, and 3B, with "A" indicating external threads and "B" indicating internal threads. This designation is often added to the end of a standard thread designation (i.e., ¼-28 UNF 2B).

Class 3 threads have the highest tolerance and class 1 threads have the lowest (loosest fit). The benefit of looser fitting threads is the ability for quicker assembly/ disassembly or slightly dirty or damaged threads to be assembled. Class 2 threads are the most common in general-purpose applications.

15.5.2 Types of Taps

Taps are available in a number of designs. The most common are taper, plug, and bottoming. All three types can be used with hand tapping or machine tapping. The biggest differences are the ease of use by hand and suitability for threading blind holes. The design differences are shown in Fig. 15.20.

A **taper tap** is used to start threading a hole and improve alignment. This is especially important when starting a tap by hand to help align it in the hole and reduce initial cutting force. The main disadvantage with taper taps is that the full thread depth

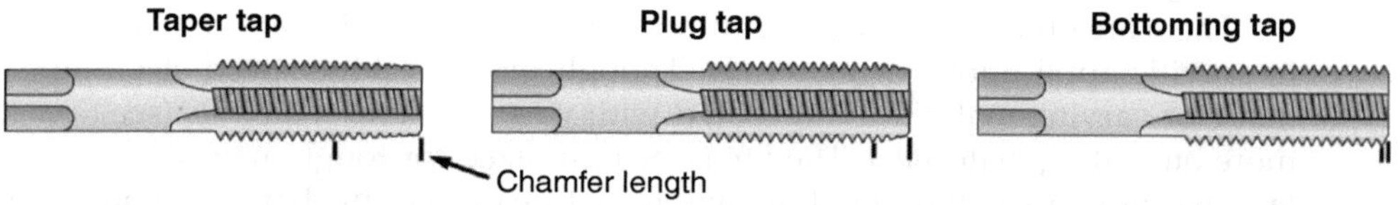

FIGURE 15.20 There are three, very common tap designs for a single-thread specification. The main difference is the number of partial depth threads they cut to get a hole started by hand and the effect that has on the depth they can create fully formed threads.

is not achieved in shallow blind holes since the first 7 to 10 threads are chamfered and do not create a full depth cut.

A **plug tap** is a tap with a short chamfered section (about four threads) that can be used to extend the threads further into a blind hole. Plug taps require more care when starting a hole since there are fewer "starter" threads, but they can be used in blind holes where there is sufficient depth that the screw or bolt does not need to reach the bottom.

A **bottoming tap** only has one or two threads chamfered on its end and is difficult to start by hand, but it is the most suitable for fully threading a blind hole. A bottoming tap can follow a taper or plug tap to increase the thread depth of a previously tapped hole, and this makes starting the tap in the hole a simple task as the threads are already formed.

In addition to the taper on the end of a tap, other features will also determine their performance. One of the most common types of taps is the spiral point tap, typically used on plug taps. In a **spiral point tap** the cutting edges of the first few threads are angled forward to push chips ahead of the cut and prevent clogging of the flutes, so these taps are best for through holes.

For blind holes, spiral flute taps are a better choice. **Spiral flute taps**, not to be confused with spiral *point* taps, have flutes similar to twist drills. This improves the chip flow out of the hole and is beneficial for high-speed tapping, such as in CNC machining. **Straight flute taps**, as pictured in Fig. 15.21, do not clear chips effectively, but depending on the depth of the hole, this may not be problematic as long as the chips can be removed from the tap between holes.

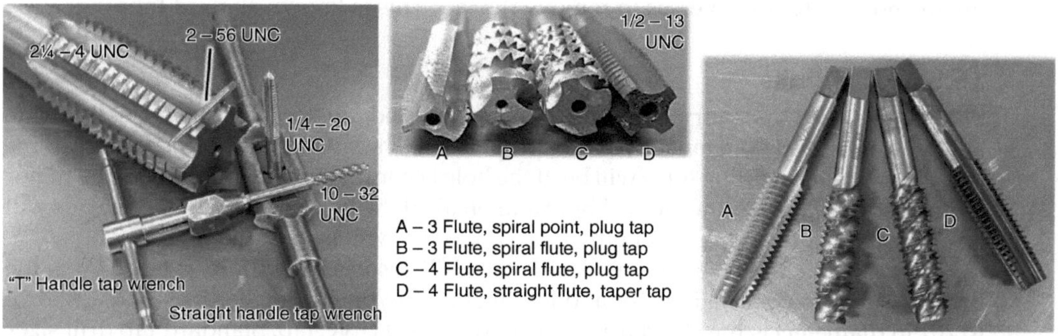

FIGURE 15.21 Taps and tap handles come in a variety of styles and sizes. Some common examples are shown here.

There are other variations in cutting geometry of taps that effect their performance in specific applications or materials. Though most are high-speed steel, taps can be made of varying materials and coated with wear resistant materials (see Sec. 13.4 for more on cutting materials). They may be optimized for tough materials such as stainless steel or they can have no flutes and form threads (i.e., push material out of the way) rather than cutting them. The greatest variations in tap design are based on thread specifications. Tapered threads such as pipe threads will require fundamental changes in the tap geometry, but the cutting process will be similar.

15.5.3 Tap Handles and Tool Holders

Taps can be held for machine or hand operations. Hand tapping requires a **tap handle,** which comes in T-handle and straight handle varieties. The tap is secured against slipping in the handle by utilizing the square end on the tap (see Fig. 15.18). Tap handles are typically only suited to a certain range of tap sizes. Small tap handles are preferred with small taps to aid in feeling for problematic resistance. Common tap handles are shown in Fig. 15.21.

When tapping by hand, it is common practice to advance the tap by a partial rotation and then reverse direction by a rotation to break the chip. The tap should not have excessive force applied or it may break, often more easily than expected. Care should always be taken to avoid applying off-axis force during tapping, and forward pressure should be applied judiciously. The tap will advance into the hole by its own cutting action. Be careful when tapping blind holes not to tap too quickly, and cautiously proceed to the bottom of the hole without building up unnecessary pressure.

When machine tapping, the feed may be determined by the tap's own cutting action or it may be rigidly held. When the tap self-advances, any rpm (within reason) can be used, but the rpm is typically slow (a few hundred rpm or less). When the tap is advanced by the machine, the rpm of the spindle must be exactly matched to the feed rate to achieve the correct rate of advancement (i.e., to match the thread pitch).

To accommodate any error, a **floating tap holder** can be used. This holder allows the tap to vertically slide in and out of the holder slightly but resists rotation. Any variation in the feed and rpm from the ideal case can be "absorbed," but the amount of travel is typically small and the feed rate must be close. Alternatively, in computer-controlled machines with precision feedback, the tap can be rigidly held. The spindle and feed rate are precisely coordinated to allow the feed to be adjusted as the rpm slows down and eventually reverses to remove the tap. This is known as **rigid tapping**.

15.5.4 Tap Drills

Because the exact geometry of a thread is standardized, the tap must be precisely made and the hole it is used in must also be of proper size. The smaller the hole is relative to the tap, the greater the cutting force will be. If the hole becomes too small (the size of the minor diameter of the thread), the tap will bind and break. If the hole is too large, the threads will be of limited depth and the strength of the connection will be reduced (Fig. 15.22).

Each standard tap size has a specific tap drill associated with it. A **tap drill** is the drill size used to create a hole for a following tapping operation. For instance, a #7 drill (0.201 in diameter) is the standard tap drill for a ¼-20 UNC tap. Standard tap drill sizes are listed in many handbooks and in some drill charts, which list drill designations and decimal sizes (see inside, back cover for a list). Alternatively, the appropriate size for a tap drill can be calculated using Eq. (15.1), where D_{drill} is the drill diameter, D_{tap} is the

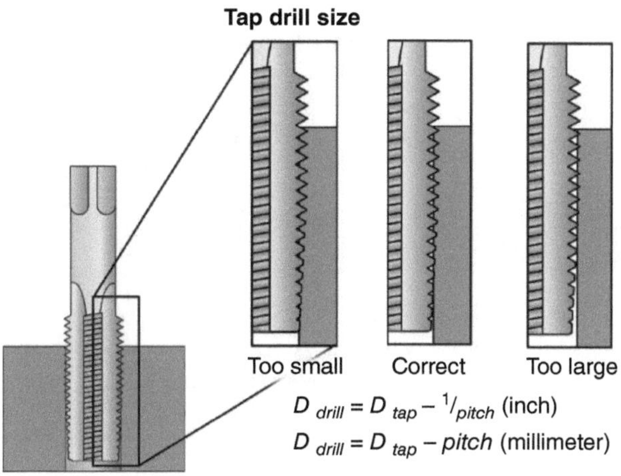

FIGURE 15.22 Tap drill size is very important to the threading operation. If the tap drill is too small, the tap will bind and break. If the tap drill is too large, the threads will be cut very lightly weakening the hold of a bolt fastened in the hole. The proper tap drill size can be calculated using the equations given.

major diameter of the thread (outside diameter of tap) and *pitch* is the thread pitch (see Fig. 15.18).

$$\text{a)}\qquad D_{\text{drill}} = D_{\text{tap}} - \frac{1}{\text{pitch}} \quad \text{(inches)}$$

$$\text{b)}\qquad D_{\text{drill}} = D_{\text{tap}} - \text{pitch} \quad \text{(millimeters)} \qquad (15.1)$$

A standard tap drill results in approximately 75% thread depth. Although a single size is specified for a tap drill, some variation is acceptable. For instance, a larger tap drill may be desirable in an application where a difficult material is being cut and less force during cutting is desired. As long as the threaded hole is not being used in a strength-critical application, the loss of thread depth is not necessarily a disadvantage, and it can save time and tool cost if used judiciously.

15.6 Tool Holders for Holemaking

The way holemaking tools are held during processing varies widely. The tool may be held directly in the spindle by means of a friction fit or is may be clamped in place using other workholding. Each method has benefits and disadvantages, and some of the most common techniques will be highlighted here. As they are the most common, drills will be used to highlight the differences.

15.6.1 Straight Shank Tools and Drill Chucks

Many tools are held on a **straight shank**, which is a cylindrical surface with constant diameter. For instance, many drills, end mills, reamers, measuring tools, etc. are of this type. Because the shank is straight, the tool holders are often adjustable and can accommodate a range of tool sizes, and the tools can also be held by a variety of tool holder types. This ability to interchange efficiently is a key advantage over taper shank tool holders.

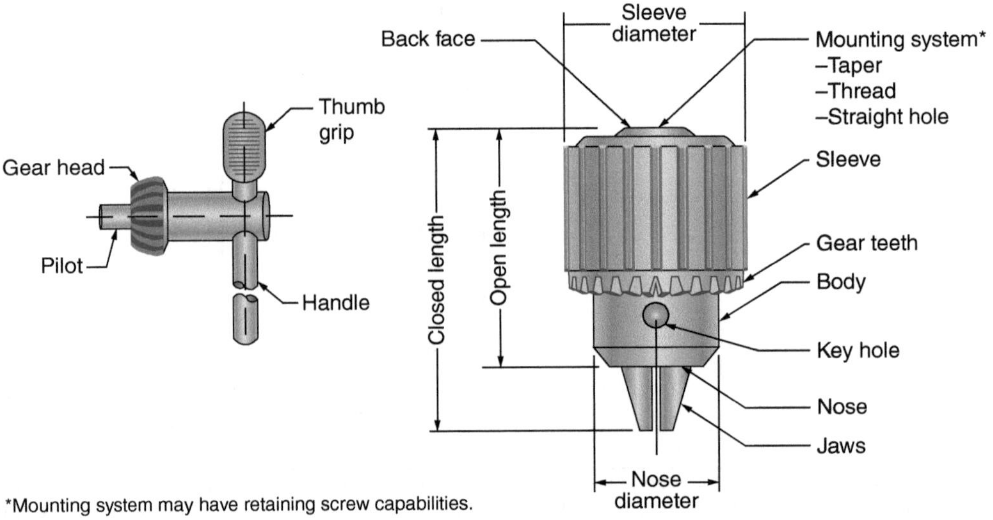

FIGURE 15.23 The features of keyed drill chuck and corresponding chuck key are listed here. (Courtesy of Apex Tool Group, LLC, manufacturer of Jacobs® drill chucks.)

All straight shank tool holder types will not be detailed here since they are discussed in other sections related to specific equipment (e.g., mill, lathe, etc.), but the most common holemaking tool holder, the drill chuck, will be given attention.

A **drill chuck** is an adjustable tool holder, typically with three jaws, which clamps the exterior of a drill, reamer, tap, or other tool in order to center it with the spindle and secure it during cutting (Fig. 15.23). As the sleeve (or collar) of the chuck is rotated, the jaws are either drawn in and move farther apart, or they extend out and closer together. The jaws are designed to move at the same rate and the chuck should be self-centering, but the accuracy is dependent on the initial quality and the amount of use the chuck has had.

Drill chucks are used on a great assortment of equipment. They may be attached to a spindle using a taper or threads, and their size varies according to machine and tool size. Chucks are used on equipment ranging from small handheld drills to large metalworking machinery (Fig. 15.24). The main difference is the size of the chuck and the durability of its design. Many drill chucks will hold a maximum of a ½-in shank since tooling in excess of ½ in is more likely to have a tapered shank, but chucks can hold larger sizes and smaller drills can be purchased with tapered shanks.

The smallest drill a chuck will hold is also important. A large chuck will often have a gap at the center that is larger than small drills, usually below 1/8 in. Very small drills require small chucks designed for the purpose, and some small drill chucks can even be mounted in a larger chuck to reduce the setup time (known as micro drill adapters).

Many drill chucks are tightened or loosened using a **chuck key**, which is a pinion-type tool that matches teeth along the perimeter of the chuck. The size of the chuck determines the size of the chuck key, which is designated by a number or by the drill chuck's maximum capacity (e.g., 3/8 in, ½ in, etc.).

Some drill chucks do not use a key, and rather they are tightened by twisting the chuck by hand. These keyless chucks are beneficial in that there is no chuck key to keep track of and tool changes can be made very quickly. Although typically very good,

FIGURE 15.24 Drill chucks come in a variety of sizes and styles.

these chucks are sometimes noted to be insufficient for high loading where the tool may slip and counterclockwise operations, which tend to loosen the jaws. The quality and design of the keyless chuck can largely negate these problems. Keyed and keyless chucks are shown in Fig. 15.24.

15.6.2 Taper Shank

The simplest method to hold a drill or other holemaking tool is to have a direct connection to the spindle, or in the lathe, the tailstock quill (see Fig. 16.7 for tailstock terminology). A shank with a slight taper (a few degrees) will hold very tightly but can still be removed relatively easily. Tapers are very common in machine tooling, and it is beneficial to discuss them further.

15.6.2.1 Types of Tapers

A **standard taper** is an industry standard designation that ensures interchangeability between manufacturers. There are a variety of standard geometries, however, and knowing the basic differences is beneficial. The type of taper influences the uses it is applied in and the equipment it is used with. In some cases, multiple tapers will be used in a single workholding setup. An example of this is shown in Fig. 15.25.

Many tapers were developed by individuals and companies and still bear their names (e.g., Brown and Sharpe, Morse, etc.). Most of these tapers come in various size designations indicated by a number. For instance, a machine spindle may be designed for a #4 Morse Taper (i.e., MT4). Standard tapers are now covered by ANSI/ASME* specifications. The taper geometries and typical uses for common taper types are given in Table 15.1. It is important to note that all aspects of the taper geometry are controlled (diameter, length, thread specifications if applicable, etc.), not just the taper angle.

*ANSI = American National Standards Institute; ASME = American Society of Mechanical Engineers.

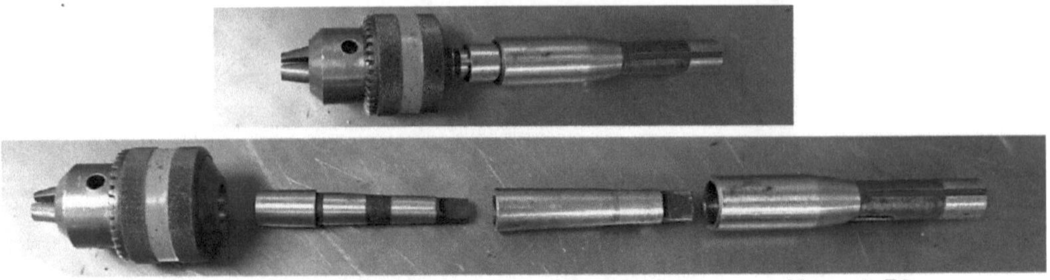

Figure 15.25 A variety of tapers may be used even for a single tool. Here a drill chuck is mounted to a shank that is inserted into a sleeve to increase its size. This is then inserted in another sleeve to change the taper type. Whenever possible, the fewest adapters should be used.

Tapers can serve two primary purposes. They locate a tool by guiding it to a given depth and by maintaining it on center with a spindle or other attachment, and they can also act to hold the tool rigidly in place through friction. Tapers that have a shallow angle, like Morse and Jacobs, are **self-holding**, which means they will generate a tight enough fit with sufficient friction to keep the connection together and resist movement

Taper Specifications (ANSI/ASME B5.10-1994 (R2008))				
Taper Type	**Taper Slope (in/ft)**[a]	**Taper Angle (deg)**	**Self-Holding?**	**Common Applications**
Morse	~0.6–0.63	~3	Y	Drill presses, lathe spindle, lathe tail stock quill
Jacobs	~0.62, 0.76 or ~0.95	~3–4	Y	Mounting for drill chuck on shank
Brown & Sharpe	~0.5	~2.4	Y	Milling, grinding and lathe spindles
Jarno	0.6	2.9	Y	Profiling and die sinking machines, some lathes
NMTBA (aka CAT[b])	3.5	16.3	N	Milling machine spindles, especially for CNC
British Standard (BT)	3.5	16.3	N	Milling machine spindles, especially for CNC
R8	3.63	16.85	N	Milling machine spindles, especially bridgeport-type

[a]In some cases, taper depends on size designation. Refer to ANSI specifications or a manufacturing handbook for more detail.
[b]More common reference. Named after the manufacturer, Caterpillar.

Table 15.1 Taper Geometry Plays a Definitive Role in the Function of Workholding during use. Steeper angles, such as the R8 taper, are **self-releasing**, which means they are easier to disassemble making tool changes more efficient.

It is common practice to adapt from one taper size to another, or from one taper type to another, using a **taper sleeve**, which has the internal taper of the tool desired and an external taper of the machine spindle or other connection. Because self-holding tapers are so effective, they must be disconnected from one another by mechanical force. For instance, a drill drift must be used to separate a drill sleeve and cutting tool. A **drill drift** is simply a wedge that is hammered into the gap between the drill sleeve and the tang of the tool. The concepts of adapting between sizes and taper types through the use of drill sleeves are shown in Fig. 15.26.

Self-releasing tapers are useful for accurately locating tool holders and making tool changes much easier (Fig. 15.27). For instance, in CNC milling, where tool changes can be automated, the tool holder may be drawn and held in the spindle by a power drawbar or retention knobs. When the force holding the tool in the spindle is removed, it must not remain seated. CAT and BT tapers are very effective for this purpose. R8 tooling, as is normally used on manual knee mills (i.e., Bridgeport-type mills), may remain tightly held in the spindle after the drawbar is loosened. To remove it, a modest tap on the top of the drawbar with a brass or lead hammer will overcome the friction holding it in place, but make sure the drawbar is still threaded into the tool holder by a few turns (still very loose) or else the threads may be damaged.

Although many of the tapers discussed here have been in use for decades, there are new challenges that have spurred innovations in this area. For instance, to meet the ever-increasing demands of accuracy and speed in machining, the **HSK taper** was developed to address dimensional changes that occur at very high machining speeds. The HSK taper is very shallow (1.2°) and has additional locating surfaces to ensure the tool position is highly accurate. The design also improves holding efficiency at high rpm thanks to a *weaker* design that relies on the spindle to carry the force and results in lower distortion and high holding force.

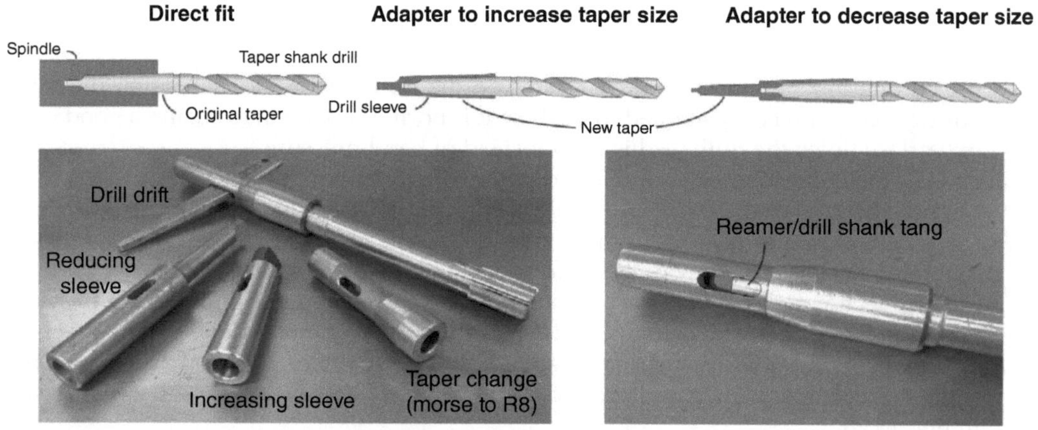

Figure 15.26 Tapers come in many types and sizes, and to accommodate the differences between a tool and a machine, taper sleeves can be used to adapt between sizes and types. The connection between the sleeve and tool is separated by forcing a drill drift in the opening of the sleeve.

Figure 15.27 Common self-releasing taper tool holders. Note that CAT and BT tooling is not identical although they have the same taper angle and similar general appearances.

15.7 Offset Boring

The boring of a hole is a process to enlarge a prior hole to a size determined by the position of a cutting edge, not the diameter of a cutting tool, such as a drill (this is discussed more in Sec. 16.5.3 in regard to lathe work). Offset boring uses a single-point cutting tool mounted to a special holder that can be precisely adjusted to dictate the hole diameter.

Although drills and reamers come in many sizes, there are circumstances where a size that is not readily available is needed. Even though boring is used to remove material, it is inefficient as a rough cutting tool, and it should be preceded by a drill only slightly smaller than the desired hole size. Enough material to allow for the hole to be completely cut by the boring bar is essential.

15.7.1 Offset Boring Tool

An **offset boring head**, the tool used for rotary boring on a mill, is similar in many regards to that used for lathe. Both use a boring bar, but the greatest difference is that the cutting tool and entire holder rotates on the mill. This necessitates that the offset boring head is precisely aligned to the hole it is boring out. Also, to ensure a good quality surface finish the quill feed is used instead of hand pressure.*

The offset boring head has a micrometer adjustment on it that sets the diameter of the hole (in conjunction with the tool geometry). In addition to this fine adjustment, there are often multiple mounting holes on the head that offer greater changes in diameter. Because of the way the head is designed, it cannot be as rigid as other hole-making techniques, and cuts are typically light, so finishing operations are the most common use. One of the primary benefits is that large diameter holes can be precisely sized with a smooth finish. To ream such holes would be impractical and expensive, if the correct size reamer was available. The setup of an offset boring head is shown in Fig. 15.28.

*Although technically possible, boring on a drill press is impractical in most all circumstances.

Figure 15.28 Hole boring can be performed on the mill using an offset boring head. The position of the cutter is moved linearly to increase the cut diameter. Feed is typically controlled using the quill feed.

15.8 Why It Matters

Holes are an integral part of our lives. When two pieces need to be connected, there is usually a hole created for the fastener to pass through. Where one part must rotate within another, there is usually a hole. This leaves many applications dependent on holemaking, and manufacturing must be able to provide these features efficiently.

We started by considering drilling operations, which are the most common holemaking techniques. Drilling is done extensively in manufacturing, but it is also very common in home and hobby work. Hopefully the wide array of sizes and styles of drills now make more sense. There must be such variety since there are so many needs in this process category. Picking the right drill is good practice at home, but it is essential in a production environment.

Reaming is one way to improve hole quality, but, along with tapping, it is dependent on the right hole to start. Reamers come in a variety of styles, but in general, their work should be limited to removing very little material. Only enough should be left to allow the surface to be cleaned up and properly sized.

Counterboring, spot facing, and countersinking also rely on a preexisting hole, and the size and depth of the features depends on the purpose. For instance, counterbores are usually used to provide space for the head of a bolt to recess into the surface of a part. Depending on the bolt geometry, the depth and diameter of the counterbore will vary. Getting the right fit is important.

Screw threads are standardized, and so taps, which create them, must also follow these design standards. That does not mean every tap is the same. There are important differences including the flute and tip geometry. If you are threading a shallow blind hole, it would be a mistake to pick a spiral point taper tap. If you do not know why, maybe you should review that section!

The right tap and procedure, especially the right tap drill, is essential to getting good quality and good tool life. When deciding on a thread specification, consider the use. Finer threads allow for more precise adjustment and higher strength of the connection, but they are also more susceptible to dirt and damage and are not as well suited for low precision fitments.

The way the tool is held is equally important. The first consideration is the tool shank. Straight shanks can be held by a drill chuck, and this provides a convenient and fast way to change out tools. Especially with larger holemaking tools, a tapered shank may be a better choice since it self-aligns and holds, and it can provide a more rigid setup.

Holes go into just about every product, whether it is an oven or an automobile. Not all holes are created by drilling or other holemaking operations, but knowing the processes is important to determining the best way to make these when needed. It just might even make you think twice the next time you need to hang something on the wall.

15.9 The How and Why of Holemaking

15.9.1 Chapter Review

1. In general, what are holemaking operations? How do these differ from punching and broaching?
2. Name three applications for drilling that are not related to manufacturing. What does this suggest about drilling operations?
3. What are the four twist drill size categories? Why do twist drills come in so many sizes?
4. Name three lengths of twist drills from longest to shortest.
5. What is the importance of the tip angle of a drill?
6. What is meant by "walking" in regard to a drill? How does a split point help?
7. What are the benefits of sharpening a drill by hand? The disadvantages?
8. What is the full name for a center drill?
9. What are center drills used for?
10. What is a spot drill? How does is differ from a center drill?
11. When drilling, what is the definition of a "deep" hole?
12. What makes gun drills well-suited for deep hole drilling?
13. How is the cutting edge unique on a masonry bit? How is the drilling process different?
14. What common features do drill bits designed to cut wood have? Name three of these drills.
15. What is the role of a pilot drill? How does a step drill provide this?
16. What are the differences between peck and deep hole drilling in regard to the drill movements?
17. What are the benefits of a reamer? Why must another holemaking operation precede reaming?
18. How does the size variability differ between an expansion reamer and an adjustable reamer?
19. What is the primary difference between a counterboring and spot facing operation?

20. Why is countersinking used?
21. What type of features do taps create?
22. Why are standard design designations important for threads?
23. What is meant by the designation ½-13 UNC 1A?
24. What benefits do fine threads have over coarse threads? Disadvantages?
25. Why is the class of fit important for threaded assemblies?
26. What is the difference between taper, plug, and bottoming taps?
27. Why might a spiral point tap be a bad choice for a blind hole?
28. What applications are spiral flute taps good for?
29. Describe the process of tapping by hand, and why it is done that way.
30. What is a tap drill? What percent thread depth does this result in?
31. What kind of holemaking tools can a drill chuck be used with?
32. How is the size of a drill chuck determined?
33. What are the benefits of a keyless chuck?
34. Where are tapers used in holemaking?
35. What determines if a taper is self-holding or self-releasing?
36. What is the function of a drill drift?
37. Why are self-releasing tapers often used in automated machine tools?

15.9.2 Concept Connections and Extensions

1. Why does drilling seem so simple? How is it more complicated?
2. What reasons are there for grinding drills rather than replacing them?
3. Consider the drilling techniques in Sec. 15.2.3 and the drill pictured in Fig. 6.9. What purpose do the holes in the end serve? How might they increase productivity?
4. Taps have a tendency to break off in the workpiece. Why is this difficult to fix, and what method is most successful (see Sec. 22.2.1.4). What other strategies can you find to remove a tap?
5. What are the benefits of using smaller thread depth when tapping a difficult material? How does this affect the strength of the connection? What other factors affect strength (see Sec. 19.2.2.3)?
6. Describe how an offset boring head is used. How is the task of making large diameter holes achieved when using a CNC mill (see Sec. 22.3.1)?
7. What is the difference between a spot drill and a countersink? How could they be used for similar purposes?
8. Why do steep tapers release more easily than shallow tapers?
9. What are the benefits of tapping holes by hand? Why is it normally done with a machine?

CHAPTER 16

Turning and Boring

Turning is one of the most popular and versatile techniques in production. **Turning** can be broadly defined as a method by which parts can be created with an external, circular cross section. For instance, a highlighter, which I am sure you are using now (hint, hint), has a circular cross section. If you were to cut the highlighter in half (across the width), you would see that it is circular. It is easier to just look at the back end of it, and this indicates that it is circular at any point. Turning operations, by nature, create circular objects.

Take a look around you. Can you identify anything that was or can be manufactured using turning? The cylindrical shape may be modified by contours or conical (cone shaped) features when viewing it from the side. The legs of a coffee table, a candlestick holder, a vase, rolling pin, beverage container, etc. are examples of what *can* be turned, but this does not mean that they *are* turned. Figure 16.1 shows a simple cylinder on the left with the transverse (lateral) cross section shown. The **transverse cross section** is the shorter dimension (width of a part). If we were to look at a part across the longer dimension (length), it would be considered the **longitudinal cross section**. On the right of Fig. 16.1 are contoured wood pieces. Although it is not as obvious with more complex shapes, a transverse cross section at any point where turning has been applied will be circular. Note that the wooden workpieces are square on the ends, where turning has not been applied. Those sections are unmodified, and this demonstrates that any starting geometry will be reduced to a circular final product with sufficient material removal.

The process of boring is closely related to turning. The difference between boring and turning is that **boring** is done *internally* and turning is done *externally*. The same features and geometric considerations apply to boring. Boring creates circular cross-sections, but the internal surface is circular while the external surface can have any geometry. Given that internal circular or cylindrical features can be created by holemaking operations (Chap. 15), you may ask, "What is the purpose?" That is a good question. It will become apparent as this chapter progresses; boring is a more complicated technique, but it is also much more versatile. Drills can create cylindrical holes, but boring can create any type of contour. Just as with turning, the longitudinal cross section can be complex, but the transverse cross section is always circular (Fig. 16.2).

16.1 How It Works

One of the most unique aspects of turning and boring, commonly referred to as "lathe work," is that the workpiece is the primary source of motion. Sawing, milling, abrasives, drilling, etc. all share the common feature that material removal is principally controlled by a moving cutting surface. With turning, the part is the principal source of

Figure 16.1 Turning creates an external, circular cross-section. Although the part may be a simple cylinder as shown on the left, the exterior may feature a more complex shape that does not change the cross-sectional geometry. The right image shows external contours applied by turning wood from square stock. Although these shapes are not simple cylinders, they still have circular cross sections where they have been turned.

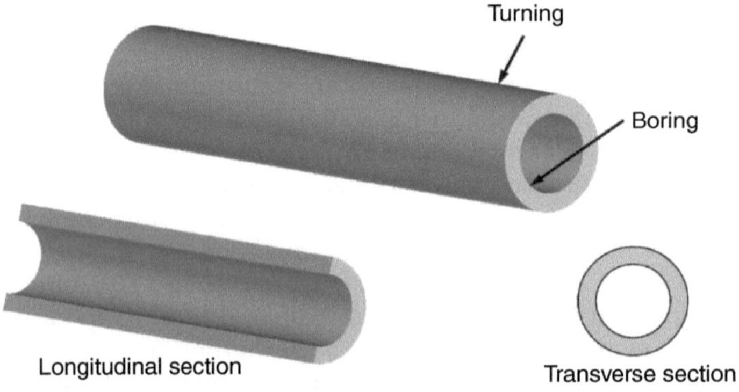

Figure 16.2 When look at the longitudinal cross section of a turned or bored part, it can have any shape. The transverse cross section will always reveal a circular outside diameter (turning) or inside diameter (boring). Tubing (as pictured) has circular outside and inside diameters. This configuration can be achieved by a combination of turning (external) and boring (internal).

movement, and the cutting tool is stationary. Of course, the tool is not entirely stationary, otherwise no cutting would occur. The material removal is a result of rotating the workpiece into the cutting tool.

For processes such as milling and drilling, a key parameter is the speed with which the cutting tool moves. In turning and boring, the workpiece needs to be considered in the same way. Because workpiece diameter will vary, it is important to choose the correct rpm. Larger workpieces require a lower rpm than smaller ones, so each time material is removed from the diameter, the best choice in rpm changes (see Sec. 13.5). The optimal rpm value is constant in other methods, such as milling (Chap. 17).

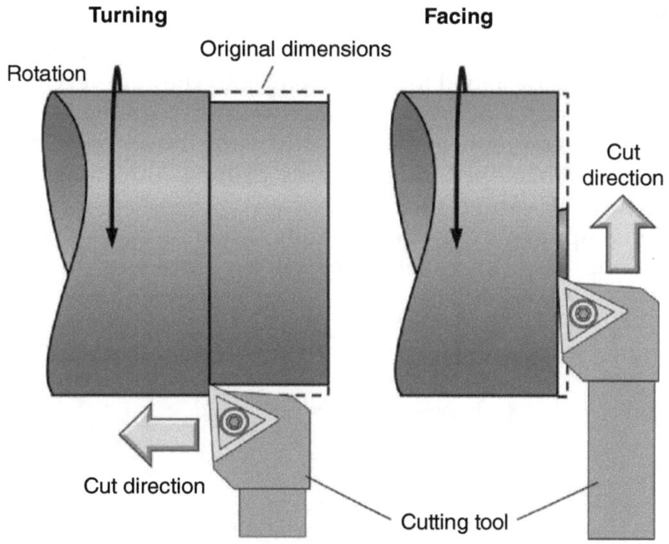

Figure 16.3 The two main operations in lathe work are turning (left) and facing (right). Turning is performed parallel to the axis of rotation. Facing is performed perpendicular to the axis of rotation. Notice the depth of cut and cutting surface of the tool are determined differently in the two.

There are two principal operations in turning. The first bears the name, **turning**, and it is the process of removing material along the length of the part, parallel to the axis of rotation. This operation will act to make the workpiece skinnier. The second method in turning is called **facing**, and it removes material using motion perpendicular to the axis of rotation. This process will make a workpiece shorter.

The workpiece is always rotating, and the type of operation is determined solely by the motion of the cutting tool. You may notice, however, that *both* parallel and perpendicular motions are needed to remove material. One direction will determine the depth of cut (DOC) and the other will be used to remove the material. Figure 16.3 shows the two turning operations and their relation to the cutting tool. Note that different portions of the cutting tool are used in the two operations. Typically, cutting tools are designed or oriented such that they are more efficient at cutting in one of the two directions, but most can be used in either. For heavy cuts, the design and orientation becomes very important.

16.2 Types of Lathes

Different materials and tasks may be turned using lathes of differing designs. For instance, wood turning is done using a simple *speed lathe*, which has control for rpm, but the tooling is held and manipulated by hand. We will not consider this type of operation. Though it is important, it represents a rather small segment of manufacturing. We will look at the most common type of lathe, the engine lathe, and the original production lathes, the turret lathe and the screw machine. Though these are still in use, computer technology has largely replaced mechanical control (see Sec. 22.3 for more on computer-controlled machining).

16.2.1 Engine Lathes

Engine lathes are a general-purpose class of turning equipment. They are generic in the sense that they are not designed for any specific job. They are very versatile. Nearly any type of machining can be performed using the basic ingredients found on an engine lathe, but the most common operations deal with round parts that are turned to a desired diameter. Other operations such as drilling, threading, and boring can be accomplished through various tooling.

The major components that are generally found on an engine lathe are shown in Fig. 16.4. It is important to realize there will be differences in each manufacturer's

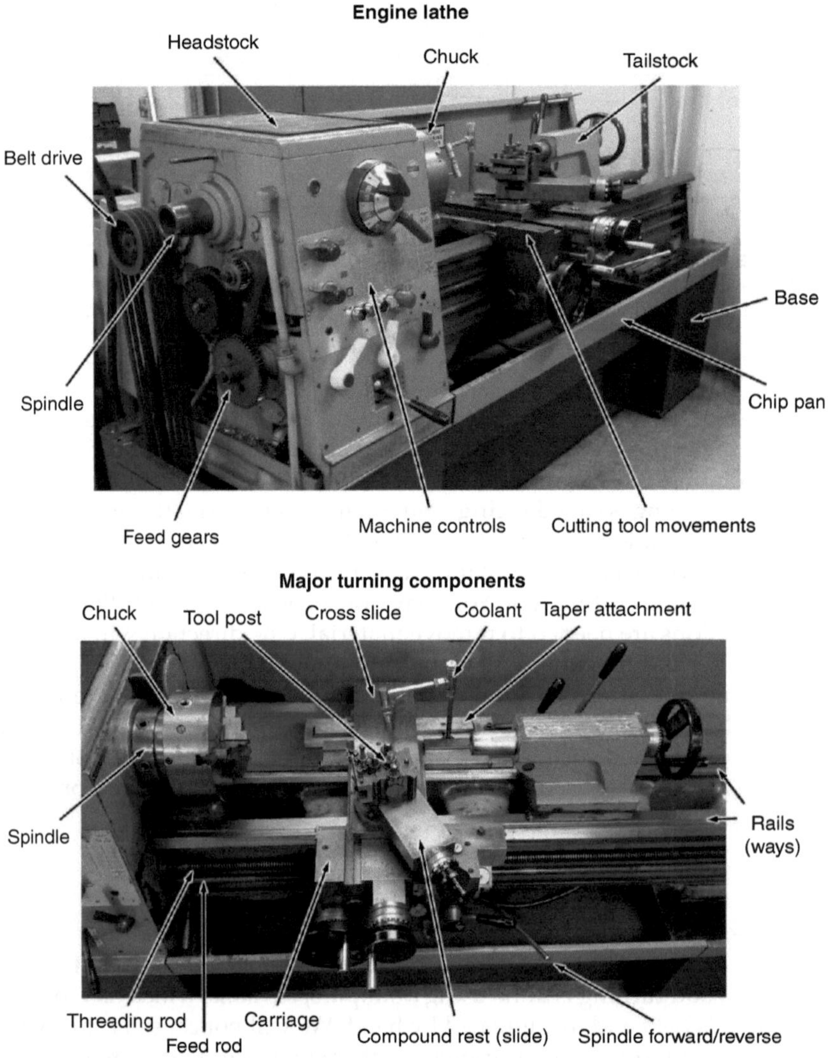

FIGURE 16.4 A general engine lathe is depicted. This is the most common type of lathe and found in most any shop. Although there are variations among lathes, the features labeled in the schematic are often found on lathes in some form.

FIGURE 16.5 The spindle is directly connected to the drivetrain of the lathe. Power is converted to rotation and transferred from the spindle to the workholding.

setup and even between models from the same manufacturer. Some of the most important and universal items are the spindle, carriage, cross slide, tool post, and tailstock. All features are important, but to understand the basic operation, these will be most pertinent.

Take note of the location of the spindle in Fig. 16.4 and the magnified view in Fig. 16.5. The spindle is the rotating component of the lathe that makes all other work possible. The spindle is commonly used to mount some means to hold the workpiece. That is, the spindle provides motion, but another component actually holds the workpiece. Some popular workholding methods are discussed later in this chapter (see Sec. 16.4). The spindle is usually hollow, which allows workpieces with a smaller diameter to be fit through the center of it. Workpieces can be fed from the back and do not have to be cut individually before machining. This is usually preferable as it reduces metal scrap and the associated cost.

The spindle creates rotational motion of the workpiece, but the next two components we will look at create linear movement of the cutting tool. These components are the carriage and cross slide (see Fig. 16.6), and the two are needed to provide longitudinal and transverse movement, respectively. You will see that the arpon and saddle are both part of the carriage. The largest handle on the lathe is usually the carriage handwheel, and you will find that it moves the cross slide, tool post and tool, and compound rest since they are all mounted on the carriage.

The carriage moves longitudinally and is used to reduce the diameter of a long workpiece. The cross slide moves transversely and is used for facing or for setting the DOC when turning. By combining motion in these two axes, angles, radii, or any type of contour can be achieved. Because hand control of angles is unreliable at best, the compound rest can be used. It has precision graduations that can be used to set a specified angle. By feeding the tool using the dial on the compound rest, the angle, or taper, can be accurately cut. Most work will utilize only the carriage and cross slide motion to make straight cuts.

The tool post also rides on the carriage, and it is where the tool holder and cutting tool are mounted. Because there is a great difference in the tooling used, the tool post and tool holder will allow for changes in the height and angle of the cutting tool. The angle of the

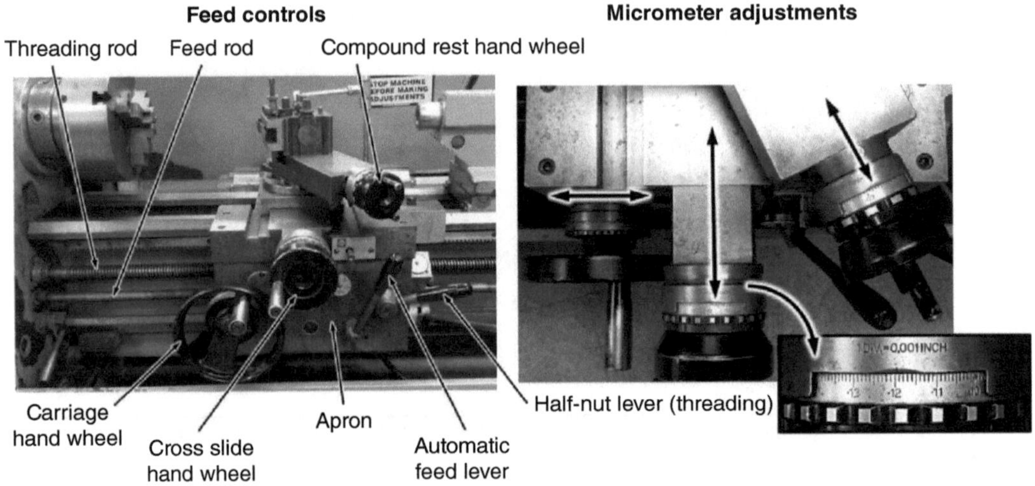

Figure 16.6 The motion of the cutting tool is primarily controlled by the carriage (longitudinal) and cross slide (transverse). The compound rest can be positioned to move linearly in some combination of these two (a specific angle).

cutting tool will not necessarily change the geometry that is cut. This is because the tool is moved by the other mechanisms. The angle of the cutting tool *will* affect the cutting action, as there must be proper clearance. If not secured sufficiently, the tool post, holder, or cutting tool itself may shift. If cut quality is poor, always check the tool setup and that the cutting edge is in good condition. See Chap. 13 for more on tool geometry and cutting characteristics.

The final piece we will look at is the tailstock. The major components of the tailstock are the anvil (the body), a ram or quill, and levers and handles to move or lock the motion of these parts (Fig. 16.7). First, the anvil houses the other parts and is designed to slide along the rails over the length of the lathe so it can be precisely positioned for various machining operations. The body of the tailstock can be locked to the lathe rails (also referred to as the "ways") by applying the tailstock (anvil) locking lever. This still allows for the ram to be extended or retracted. The ram (quill) is extended/retracted by turning the hand wheel and can also be locked by applying the ram locking lever.

In many cases, the tailstock is used to support long workpieces. Long is a relative term, as thin workpieces will tend to deflect more over shorter distances than thick workpieces (see Chap. 7 for more on this concept). If extra support is needed, a center will be used. A center is designed to fit into a mating recess in the part. This usually requires an a hole with the correct 60° angle. A tool called a combination drill/countersink or simply a "center drill" is used (see Fig. 15.5 for the proper technique), but any existing hole may be used, though not as effectively.

The tailstock is used to drill holes as well. A drill chuck is mounted in the ram, and the handwheel is used to advance the drill into the end of the part. Graduations on the ram can be used to determine the depth of the hole. Tapping, reaming, and other hole-making operations can also be achieved using the tailstock. This basic process for center drilling and supporting a workpiece is shown in Fig. 16.8. The use of centers will be described in detail in Sec. 16.4.5.

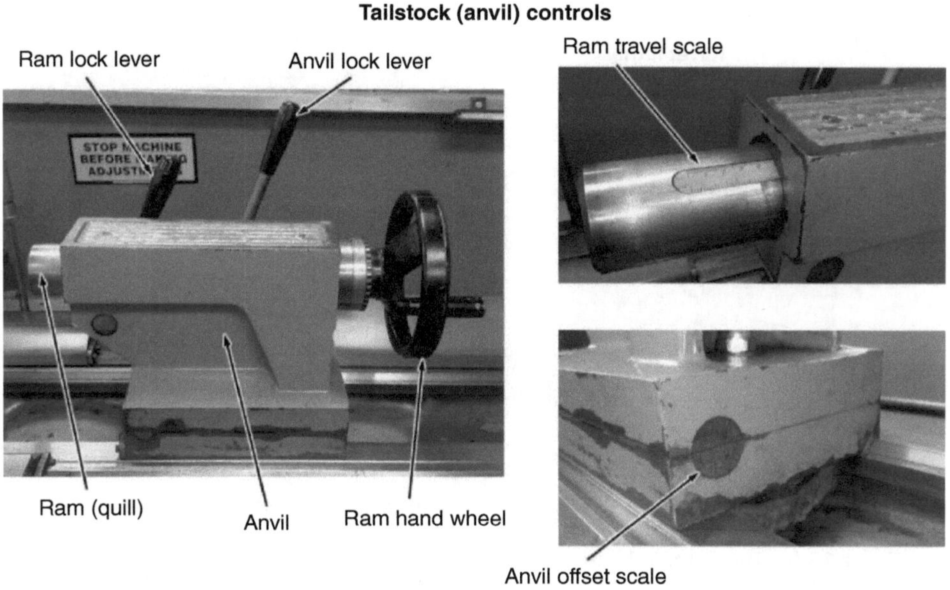

Figure 16.7 A tailstock is vital to many operations on the lathe. It can support workpieces or facilitate mounting and feeding tools such as drills.

Figure 16.8 The tailstock performs a number of important functions. It can be used to support the workpiece or to complete cutting operations. After facing the workpiece (left image), a center drill is often used to create a recess (center image) that is used with a lathe center (right image) to add support.

16.2.2 Turret Lathes

Turret lathes were developed early in the history of mass production. The general shortcoming of engines lathes is that tool changes and tool positioning are often time consuming or inaccurate for complex parts and between parts. Though there are many methods of enhancing repeatability on an engine lathe, the tool changes cannot be easily overcome. That is the principal advantage of **turret lathes**. They combine the ability to simultaneously mount many tools and set the travel during machining. This dramatically cuts down on tool changes and allows for precise, repeatable operations to be done very quickly. The basic structure of a turret lathe is shown in Fig. 16.9.

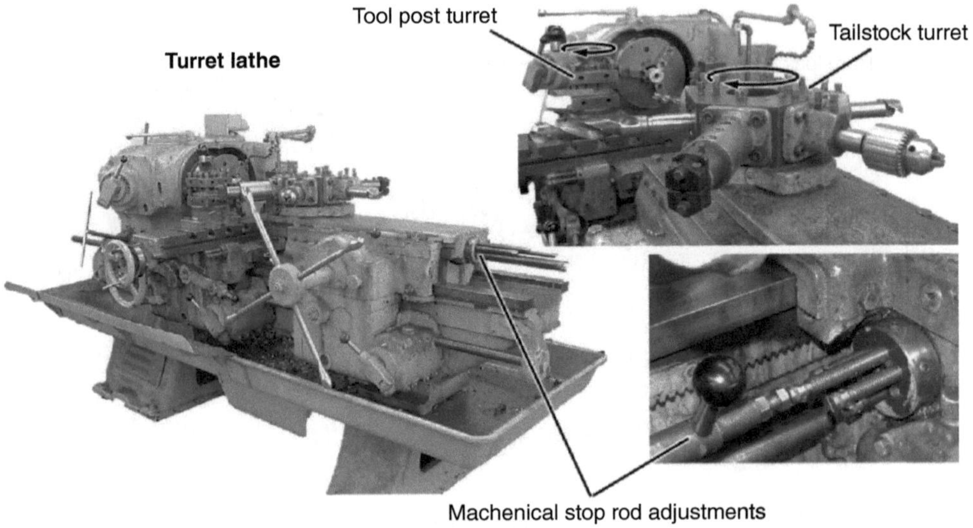

Figure 16.9 The turret lathe is a mechanically controlled production lathe. It houses multiple tools with preset travel stops to make complicated parts quickly and reliably. (*Courtesy of Cindy L. W. English, MFA.*)

There are two main differences between a turret lathe and an engine lathe. The tool post can hold multiple tools, and the tools can be moved into position by a ratcheting motion where the tool post will rotate into the proper position. The tailstock can also be cycled in this manner. Instead of holding one tool, four, six, or eight tools can be mounted. The tailstock is moved by a separate handle, and the tool turret is selected by retracting until the rotary gear is engaged. The travel allowed for each tool is determined by a mechanical stop. A set of rods rotates with the tool turret and the length can be adjusted to set the full cut depth.

16.2.3 Screw Machines

In a similar capacity to turret lathes, **screw machines** are production lathes, and they use mechanical control to set tool travel. Unlike turret lathes, their motion is automatically controlled. The spindle speed and direction, the turret rotation, tool motion, and other functions are all controlled mechanically. Instead of stops, they use a rotary drum and cams that change, advance, and retract the tools (see Fig. 16.10). The geometry of these cams determines the distance and duration of travel for a particular tool, as well as the feed rate. The cams for each tool must be offset so tools that may interfere with one another advance and retract without causing a collision. A single cam may be responsible for multiple tools if used for a turret. Because of the complexity and importance of the cam profile, they should be produced by a skilled machinist who understands the implications of their design. Cam blanks come with degree markings to assist in creating their profile.

The tools are oriented around the spindle to approach the workpiece from multiple angles. This allows them to operate simultaneously, or more commonly, to cut down on wait time as a tool retracts. An example of a screw machine is shown in Fig. 16.11. The part must be automatically advanced, and cut by tools from multiple stations, then parted off and the cycle repeated. Like the turret lathe, screw machines are limited to

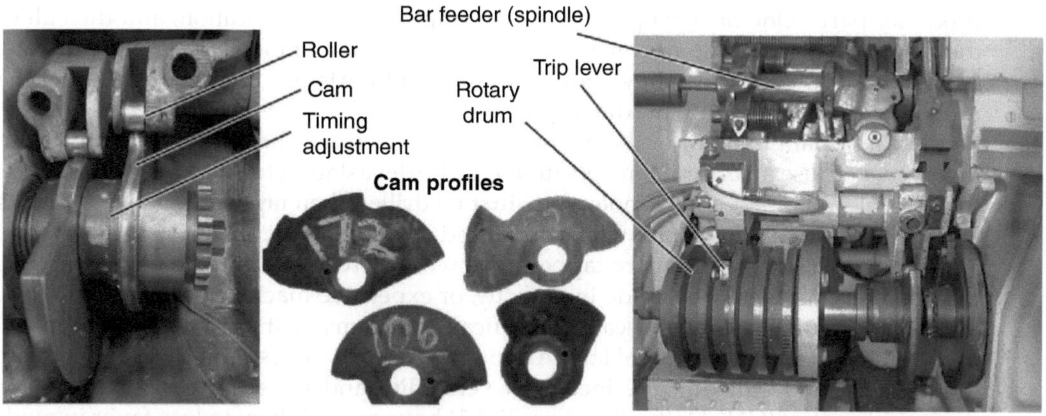

FIGURE 16.10 Screw machines are controlled by rotary drums with trip levers to control machine functions and cams which advance and retract tools. The timing and profile of the cams determines the rate and magnitude of tool motion. (*Courtesy of Cindy L. W. English, MFA.*)

FIGURE 16.11 Screw machines are able to perform multiple operations on a part by housing and controlling multiple tools simultaneously. The motion of the tools and machine functions are all controlled mechanically. (*Courtesy of Cindy L. W. English, MFA.*)

linear motion. Any contours must be created by the tooling, so custom tooling is commonly required unless only basic features are needed.

Screw machine are used primarily for small, fairly simple parts (think screws!). The operations required to produce parts can be done at one spindle with time given to each tool (see Fig. 16.10), or there can be operations at multiple spindles simultaneously. The multiple spindle screw machine is similar to a Gatling gun, which uses multiple barrels to quickly fire ammunition. Instead, the multiple spindle screw machine quickly processes parts. One or two operations are completed at each location, and then all of the spindles rotate to the next station. Each revolution completes one part, but each spindle holds a part. This allows for the number of parts per revolution to equal the number of spindles during production.

It is important to monitor the operation of the machine when multiple spindles are involved, because an error in one position may be devastating to later operations. Take, for example, a tapped hole. A hole must first be drilled then tapped. If the drill breaks and no hole is created, the tap will be advanced into solid material and will also break. After these tools break, no more tapped hole will be created, but the part may still be cut off. Many parts may be made incorrectly, or expensive machine damage can occur. This compounding failure may cause significant problems, so it must be caught early to prevent down time and material loss. Although screw machines served very effectively for decades (and still do), the increased versatility and lowering cost of computer numerical control (CNC) lathes (see Sec. 22.3.1) have caused them to lose favor in modern manufacturing environments. These machines were built solidly and widely used in their prime, and World War II era machines are still in service in some shops.

16.3 Lathe Tooling

As discussed in Chap. 13, tooling decisions are determined by the type of material, geometry, and support of the cutting surface. There is no one-size-fits-all solution to cutting tools. Some operations require fine detail, but some require heavy material removal. These operations are best served by tools with different characteristics. In general, there are a number of geometries that are suited to specific tasks such as cutting threads, radii, chamfers, grooves, etc.

Many general tool geometries, such as the 60° angle used in threading, are available commercially. It is difficult and time consuming to form complex contours by hand using a generic tool shape, so the desired geometrical features are commonly cut into the tooling. This takes significant time and skill up front, but the increased productivity and accuracy later in the manufacturing process make up for the initial investment. A variety of cutting tools and features created are shown in Fig. 16.12.

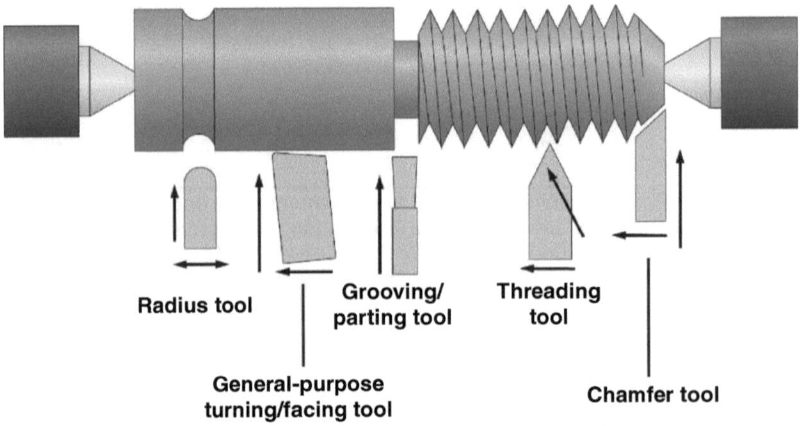

FIGURE 16.12 The cutting tools used in turning operations may be general purpose or have specific applications.

FIGURE 16.13 A tool post and related adjustments allow the tool height and angle to be varied over a wide range. This allows for many different tools to be used and also enables a single tool to be adjusted for a variety of tasks.

Not only can tool geometry be varied significantly either by the tool manufacturer or by custom grinding, the tool post and tool holder allow for adjustment as needed. For instance, the tool can be tilted to mimic a chamfer tool or to add clearance for facing or turning. While operations such as grooving and threading have limited room for error, other, general-purpose tools offer a great deal of variability. A tool post is shown in Fig. 16.13.

16.3.1 Lathe Work and Cutting Tool Choice

Given that there are a number of options regarding lathe tooling, which one is right for what application? While you can typically get away with using a single, general-purpose tool for simple projects, to increase productivity and quality, the best approach usually requires more than one tool. For instance, consider the part in Fig. 16.14. There are many different features that require different tools. We will now look at each.

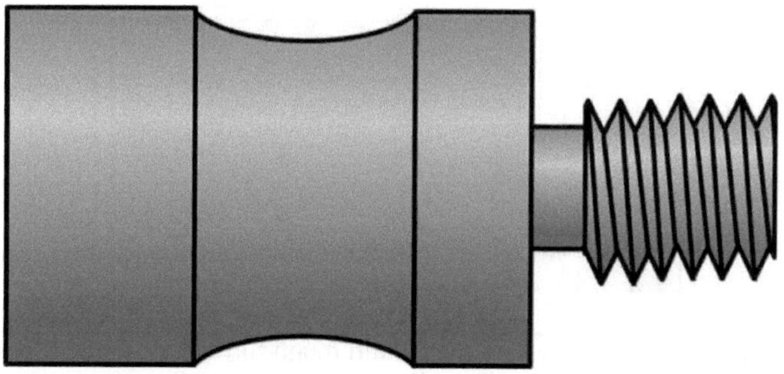

FIGURE 16.14 Even a relatively simple part like this one requires multiple tools.

The most common starting point for turning is bar stock. This has a specified external dimension that must be reduced to create the part. The stock size should always be greater than the part size so that at least *some* material is removed. More specifically, do not use ½ in stock for a ½ in part. Although very good, the as-received tolerances and surface quality are not sufficient for most customer specifications. This necessitates the first steps be related to reducing the overall diameter and length. This is known as rough turning. Of course, the closer the starting material is to the final dimensions, less scrap will be generated. This means less money will be wasted on material and less time will be invested in removing it.

Rough turning, or roughing, usually starts with turning along the length of the part, or facing longitudinally. Either process can be done first, as they are not dependent on one another, and both are required. If a center is needed, facing will be the first operation, as demonstrated in Fig. 16.8. Turning will create a truly circular part and facing will true the end so the two are perpendicular. In order to get an accurate measurement, reference surface(s) should be machined.

The material to be removed from the part should be done as quickly as possible (time is money!). Cut depth and feed rate should be maximized as much as possible. This results in a large amount of stress on the cutting tool. A rigid cutting tool and tool holder are essential for this (see Fig. 16.15). The larger the angle of the cutting point, the stronger the tool will be (review cutting tool geometries in Sec. 13.2.1).

The trouble is that clearance is needed so that the tool does not rub. For this reason, a 90° tool should not be used to machine a shoulder that meets at a 90° angle. There is no room for clearance. The tool choice should be justified by the amount of time saved in using a more rigid tool over the time required to change it later on. The fewest possible cutting tools should be used.

After the stock is roughed out to the outer dimensions, the shoulder can be machined. This too requires a significant amount of material removal. Unlike the outer diameter, there must also be consideration given to the corner at which the shoulder and minor diameter meet (the 90° angle). Because clearance is required, a tool of dimensions less than 90° is needed (see Fig. 16.16). The angle can be almost anything, but a balance between rigidity and maneuverability can achieved.

This part requires threads, and in order for threaded objects, such as a nut, to fully seat on the shoulder, a relief groove is required. The groove should be slightly lower than the root diameter so that the threads do not bind. It is preferable to use a tool with the correct width so that the only motion required is plunging (transverse feed using

FIGURE 16.15 Machining this part starts with rough turning. A sturdy tool with large tip angle is capable of doing the work most efficiently. A larger nose radius also increases the strength of the cutting edge.

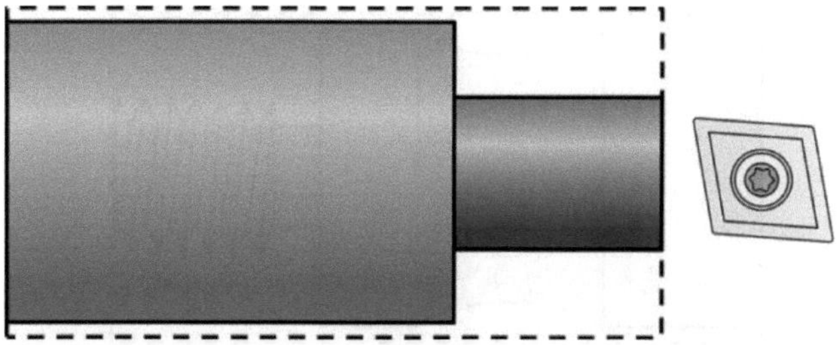

Figure 16.16 After rough turning the outside diameter, the smaller diameter can be turned. A sturdy tool with large tip angle can support larger cut depths, but a smaller angle helps ease the process of cutting the shoulder.

the cross slide). Again, the shoulder to the left of the groove requires that the tool has side clearance. An example of such a tool is shown in Fig. 16.17.

After the groove is cut, the threading can be done. Threads have to be precisely cut to match the pitch, major and minor diameters, and thread angle required for the mating nut (or internally threaded part). The thread angle is the only feature controlled by the tool itself, the other features are dependent on tool positioning and feed rate. Specialized, threading tools are used for cutting threads, and the general shape is shown in Fig. 16.18.

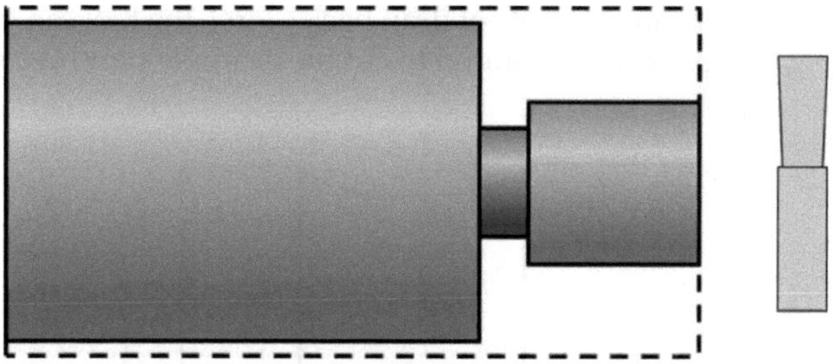

Figure 16.17 In order to create clearance (relief) for threading, a groove slightly deeper than the threads should be cut. This is done most efficiently with a grooving tool that is already the desired width.

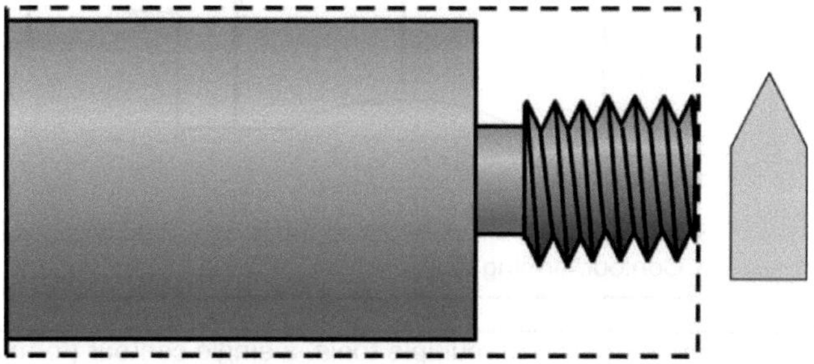

Figure 16.18 To cut common threads, a 60° tip is used.

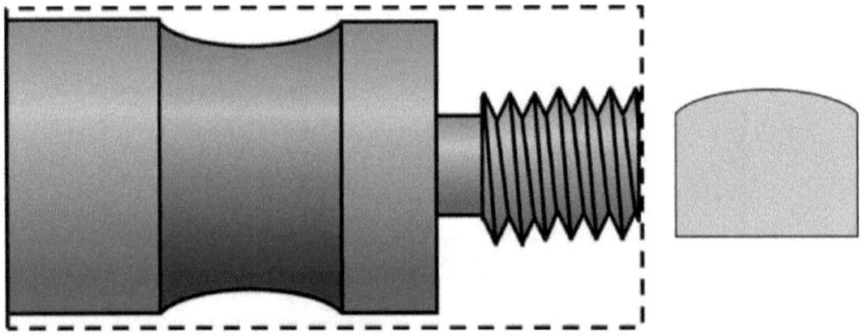

Figure 16.19 To create the radius in this part, a cutting tool with same shape is used. A custom cutting tool like this can be created from a high-speed steel blank.

The last operation specified for this part is cutting the radius in the diameter. A uniform, accurate radius is very difficult to cut free-hand, and even using computer-controlled machines, it requires a very fine tip tool, or left-hand and right-hand tools. This is because the sides are steep, and to maintain side clearance, a narrow tip is needed. Alternatively, the tool can be large and have the specific radius already formed in its geometry (see Fig. 16.19). Radius tools are not uncommon, and they can be created by hand from tool blanks. If used, the cutting motion is entirely a plunging action, just as used in grooving. Likewise, this can be considered a grooving operation.

In certain cases, particularly high-volume production, it may be desirable to create one tool that will take the place of many other tools. These specialty tools are commonly plunged into the part and transfer their profile to the workpiece. This greatly simplifies the processing, but there are a couple of drawbacks. First, the tool can be very complicated. As shown in Fig. 16.20, the tool required to cut the example part here would be very difficult

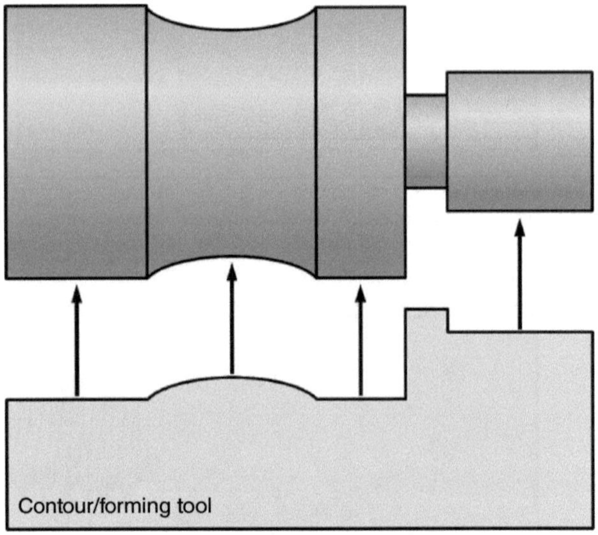

Figure 16.20 As an alternative to multiple tools, a single contour (forming) tool can be used to create the part (except the threads) in one plunging motion. It is important to consider that the force will be very high once the tool is fully engaged, and the part must be rigid to resist bending.

to create and maintain due to the complexity. As a result, the tool itself has a very high cost. This can be easily offset by the time savings if many parts will be made. Another downside is that there is a tremendous amount of force applied to the part. The cutting surface is very large and that will act to deflect the workpiece and may actually break it off if the cutting is done aggressively. Small diameter parts and weak materials can be problematic.

The steps to make the simple part detailed here are basic to many other parts made by turning. Small radii and chamfers commonly added on a part can be made by simply "breaking" the corners with a hand file. Otherwise, they could also be made by using tools with the appropriate geometry or by tilting straight-sided tools in the case of the chamfers. The cutting tools shown in Figs. 16.15 and 16.16 are made from carbide, but high-speed steel (HSS) is also very common. Chapter 13 discusses the cutting characteristics and considerations of using different cutting tool materials. Custom tooling, such as a form tool, is much easier to create from HSS blanks than from carbide. Although the process to create a feature may be very basic, there are many other factors influencing the process.

If the part needs to be cut off from the piece of stock being held in the machine, a parting tool may be used. This is nearly identical to a square grooving tool, but general has greater side clearance for reaching deep recesses. Parting off is a good way to conserve material as another part can be cut from the same stock. This saves scrapping the rough stock being held in the machine.

This brings us to a point where it is pertinent to discuss how the workpiece is held. The following sections will describe some common workholding methods and considerations for using each.

16.4 Workholding

Turning requires that the workpiece be held in some manner. There are numerous ways to achieve the basic function, but we will focus on some of the most common. When choosing a method, the basic considerations include simplicity, precision, speed of part change, durability, cost, etc. Workholding can be done using universal, off-the-shelf methods, or it can be customized for a specific type of part. Typically, the decision is based on cost. The number of parts, the rate of production, the part features, and the required accuracy will all dictate which method is most effective. We will look at general workholding duties first. These include supporting and locating the workpiece.

16.4.1 Workpiece Support

Workholding must prevent the part moving longitudinally or rotationally. When you cut a steak (assuming you eat steak), you must hold the steak (workpiece) in place while you cut it. Depending on how tough that steak is, you will have to hold it more or less aggressively. Usually you will use a fork (workholding) to hold the steak in place while you cut it with a knife (cutting tool). The same basic features are present when cutting material on a lathe.

The workholding will support the work while it rotates, but it must also keep the workpiece from slipping under cutting forces, as shown in Fig. 16.21. The pressure used to firmly hold the work during cutting should not cause damage to the part. By evenly distributing force, rather than concentrating it, the proper support can be provided without detriment. In a production environment, there must be a balance between firmly gripping a part and preventing surface during heavy machining. If the gripping surface will later be machined or scrapped, these considerations are greatly simplified.

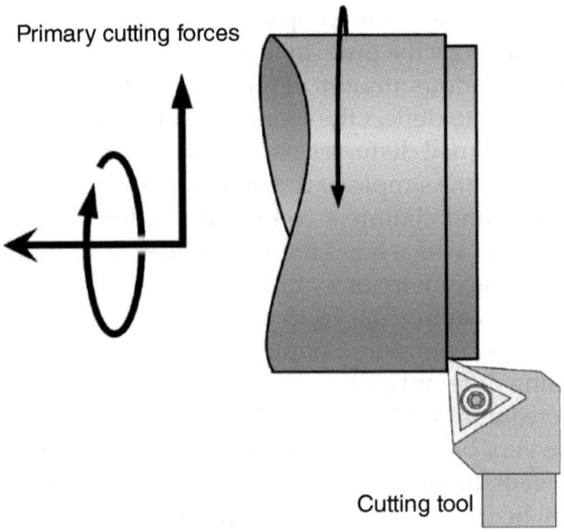

FIGURE 16.21 Workholding used in turning operations must resist translational and rotational motion. Most workholding relies on frictional force to secure the part.

The aggressiveness of machining, the material being machined, and the condition of the tooling will determine the workholding requirements. Materials that are difficult to machine, large cut depths, improper feeds and speeds, incorrect tool setup, and dull tooling will all act to increase the required force for material removal. The tooling and machine setup should be optimized to maximize efficiency, regardless of workholding. If the workpiece cannot be thoroughly supported due to unique geometric considerations, the cut depth should be reduced as a last resort.

16.4.2 Workpiece Positioning

Accuracy and repeatability are typically the most important aspects in workholding for turning operations. Second to that is speed. Turning creates circular cross sections, and if your workholding cannot hold the part on center throughout the machining process, it will not be accurate. The greatest problem with repeatability is maintaining **concentricity**. This entails positioning a part perfectly on center every time the workpiece is secured. Imagine if you were to create a pin with a shoulder on it that was to be received in a counterbored hole (see Sec. 15.4 for more on counterbores). If one diameter of the part is not aligned with the second diameter, the part will not fit into the receiving hole. If two diameters are not concentric, they are eccentric. Although eccentric features do have a multitude of uses, concentricity is more commonly required. The differences between these two conditions are shown in Fig. 16.22.

Most workholding in turning should position a part of any size diameter in the exact same position on center. It is the relative location of the tooling that determines the final diameter. Some methods are better than others, and this will be discussed in context with the various methods. An easy way to check the runout (wobble) of a setup is to take a light cut until the surface is completely machined. After this, take the part out, put it back in and cut again,* just barely touching the surface.

*Note that if you reposition the workpiece in the same orientation, you will not notice a difference. It becomes noticeable when you rotate the workpiece between clamping.

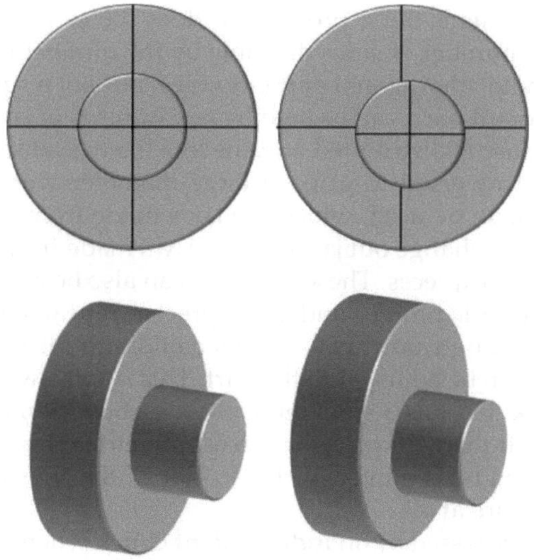

Figure 16.22 The ability of a workholding method to center a workpiece is essential when turning multiple diameters. Two circles exactly aligned are considered *concentric*. Misalignment is known as *eccentricity* and can lead to a part being scrapped. As many operations as possible should be completed without removing the part.

If you notice the cutting tool is only touching intermittently, there is runout. Layout die (bluing) can be used to aid the visibility. Runout is an important feature that is determined by the machine spindle and the workholding. If a precision ground bar is used, the amount of runout can be determined by measuring the deflection of a needle on a dial indicator during one revolution. The magnitude of change is the runout.

Some workholding methods can also position the workpiece at a specified longitudinal (lengthwise) position. If the workholding is a custom setup for specific part, then allowances can be made to add this extra control, but it is often at the expense of generality. Specialized tooling is less adaptable. An alternative method for positioning work along its length uses a previously turned diameter and associated shoulder.

The caveat with using a shoulder to position work is that the both ends must be machined. The shoulder provides a physical interlock that allows for heavier cuts to be made without sliding the workpiece farther into the spindle, and this can be a benefit during heavy machining. If there is no need to cut a shoulder, the added support must be weighed against the added time and possibly material, as well as concentricity concerns.

16.4.3 Lathe Chucks

We will start specific discussion of workholding with chucks. These are the most common method due to their ease of use, speed of part change, adjustability in workpiece size, and overall suitability. Although they are very user-friendly, they do suffer slightly in accuracy. Some chucks can be exceptionally precise, but in general they are not as accurate as collets (next section), but they are more versatile. The two main classifications of chucks are universal and independent as described next.

Universal chucks have jaws that move together in synchronized motion. This is dictated by a scroll within the chuck, so they cannot be operated individually.

With most parts this is not a problem. It does mean, however, that the part must be round or have a number of sides divisible by the number of jaws. For instance, a three jaw chuck can hold a hexagonal part (six sides) but not a square part (6 is divisible by 3 but not by 4). Round parts can be held by any number of jaws. The more jaws, the more evenly the pressure is distributed and the less likely surface damage is. Some jaws can be removed and reversed to grip on large diameter workpieces as well (Fig. 16.23). Also, **soft jaws** may be used, which are jaws made from a softer material, like aluminum. The ability to change out jaws allows jaws made from a softer material to be used to hold delicate workpieces. These soft jaws can also be machined while in the chuck to accommodate specific profiles and allow precise, repeatable positioning.

It seems as though universal chucks could suit all workholding needs since the number of jaws can be adjusted to the part. This is not always the case since some parts are not bar stock that has a symmetric cross section. Remember, in turning the workpiece is turning! If the workpiece has an asymmetric cross section, like a rectangle, that is a major problem for a universal chuck. Not all the jaws would apply the same pressure or any pressure at all.

This can be addressed by an **independent chuck** where each jaw can be exactly positioned to apply equal pressure (Fig. 16.24). Typically independent chucks have four jaws. The disadvantage to independent chucks is that they must be manually (and carefully) centered in a manner similar to checking for runout. A precision measuring device (usually a dial indicator) is used to check if a part known to be round is wobbling.

The ability to manually set the center can reduce runout, but, more importantly, it allows the workpiece to be intentionally positioned off-center. The downside is in time,

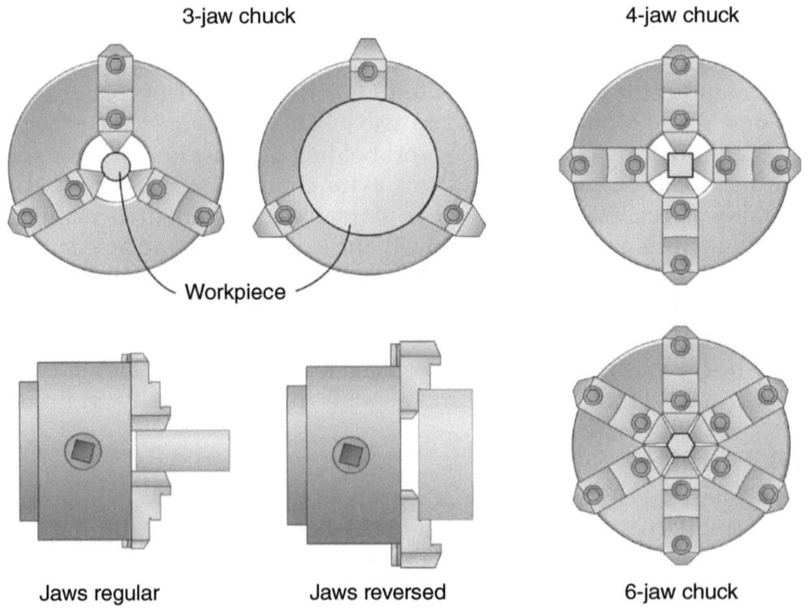

FIGURE 16.23 Lathe chucks come in a variety of configurations. The main difference is whether a chuck is universal or independent. In a universal chuck, the jaws all move in unison. In an independent chuck, the jaws are moved individually.

Figure 16.24 Independent chucks are useful for nonstandard stock or for doing operations on oddly shaped workpieces. This could not be achieved using a universal chuck.

as setup can be extensive, especially if the operator is not experienced. Safety should also be duly considered, as unbalanced (off-center) work is more likely to eject from the chuck at high speed. Lower speeds and extra caution should be applied if the workpiece is large, far off-center, or oddly shaped. The part's center of mass (where it balances) must be on the spindle centerline for maximum stability. Additionally, some workpieces are not suitable for chucks at all. These may be plates or some other, completely different configuration. They can still be done using the lathe, but that is territory for the faceplate method (see Sec. 16.4.6).

16.4.4 Collets

A popular alternative to a chuck is a collet. Collets are typically used to clamp on the outside diameter (OD) of a workpiece. Like universal chucks, they are self-centering. A common collet type, known as a **spring** or **split collet,** is shown in Fig. 16.25.

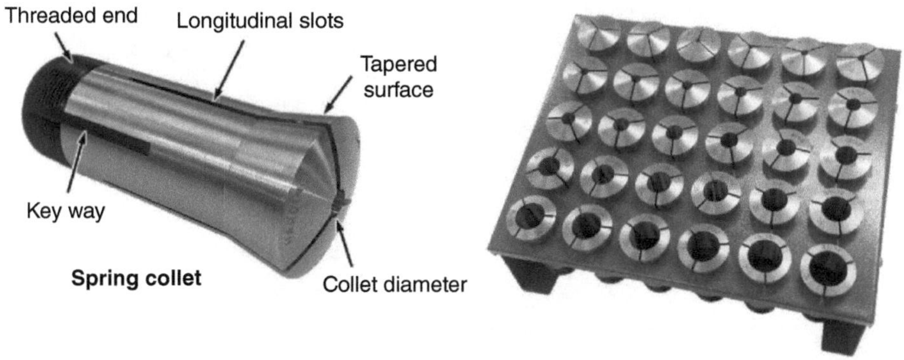

Figure 16.25 Collets operate by flexing open and closed in small amounts. Because spring-type collets do not have a large amount of adjustability, a set is needed to accommodate various workpiece sizes.

To open and close a collet, the body of the collet is shifted into or out of the spindle. The spindle and the collet are tapered so that when the collet is forced in, it closes. If you reach into a jar of pickles and grab one, you may not be able to get your hand out. When you have something in your hand it takes up more space. When you try to remove your hand, the opening of the jar squeezes it down on the pickle. This is a general picture of how collets work.

With collets, this is made possible by the spring-like behavior of the collet. The collet acts like a spring because it is solid at one end and slotted along the length. This cantilever behavior allows for some adjustability, but it is *very* small compared to chucks. For this reason, collets are available as sets with a range of sizes. Another limitation is due to the shape of the collet opening. A round collet cannot securely hold a hexagonal part, even if it is the same size. A three-jaw chuck can. Collets with special shapes (hexagonal, square, etc.) have to be purchased to hold parts that are not round. A benefit of collets is that they can accurately center parts very quickly.

Not all collets are created equal. In fact, there is quite a large variation between different types of collets. As shown in Fig. 16.26, collets can vary in the way they connect to the spindle, the slot configuration, overall size, etc. All the collets in Fig. 16.26 are designed to hold round parts or tools, but the similarities are few beyond that.

A notable difference is the way the collet taper is actuated. An R8 collet uses an internal thread (R8 configuration shown in Fig.15.27), a 5C collet uses an external thread, an ER-type collets use a ring that applies force from the front of the collet (pushing it in rather than pulling). There are also a number of similarities. These include slotting, tapers, and keyways. Not all collets have these features, but they all must be secured to machine spindle and must transmit force to a part or tool. It is not uncommon for other workholding to have a standard taper such as the R8 (standard tapers are discussed more in Sec. 15.6.2.1). This allows the direct use of the spindle for many tool holding configurations. Many times, the holder for an ER collet will use an R8 or other mill spindle adapter, such as the BT20 taper shown in Fig. 16.26.

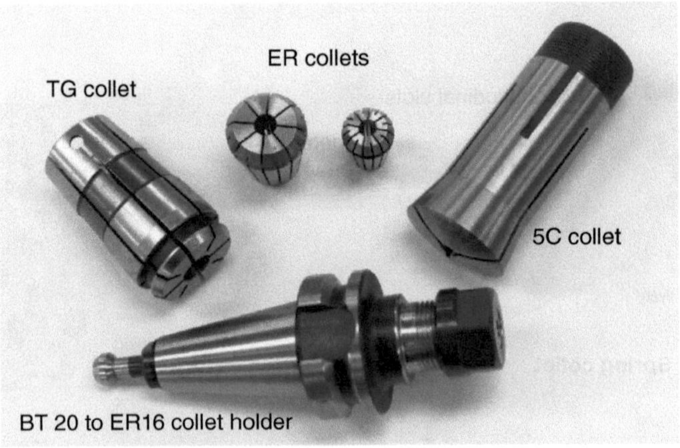

FIGURE 16.26 A variety of collets are shown, and they are not interchangeable. The image on the right shows an ER-type collet holder, which has an R8 taper shank. Combining tapers of different types is common in machine tooling.

Figure 16.27 Split collets are very limited as to the part size each one will securely hold. Adjustable collets have a much wider range of size, but they require dedicated workholding.

An example of the wide variety of collet adaptions can be seen in specialty varieties too. For example, if more adjustability is needed, collets with metal jaws connected by rubber can be used. These collets allow for greater variation in workpiece geometry while maintaining a secure hold on the part. Adjustable collets made entirely of metal are also used and have more variability in size. These have spring-loaded jaws and use a dedicated chuck designed to work with them (Fig. 16.27). Additionally, like soft jaws for chucks, some collets can be machined to a specific shape. These are commonly called emergency collets and are made of brass, plastic, or steel.

16.4.5 Between Centers

Another, common method of workholding in the lathe is between centers. A center is simply a cone-shaped component that has a well-defined angle (typically 60°). This is important because the workpiece must have hole of matching angle to seat securely. The center is aligned so that its tip is on the axis of the spindle. This ensures the workpiece being machined is concentric.

There are two major distinctions in regard to centers: live or dead (that is a major distinction in many other categories too!). It is somewhat intuitive to remember the difference. A **live center** moves with the workpiece rotation. A **dead center** is fixed in place (motionless). The live center rotates because there are bearings housed behind the tapered tip that allow free motion, even under large stresses. The dead center is simply one, solid piece which must be lubricated to prevent excessive friction between it and the workpiece. The two types of centers are shown in Fig. 16.28.

Because parts come in all shapes and sizes, centers are available in a wide variety of geometries. All achieve the same basic result: support and center the workpiece. Hollow parts of large diameter, such as a pipe, require a larger "pipe" center, which also has a steeper angle. A female center is used to support a workpiece that is tapered to point and cannot have a mating hole in the end. The half-male center is available exclusively as a dead center. The cut out on the end is to allow facing of a part held between centers,

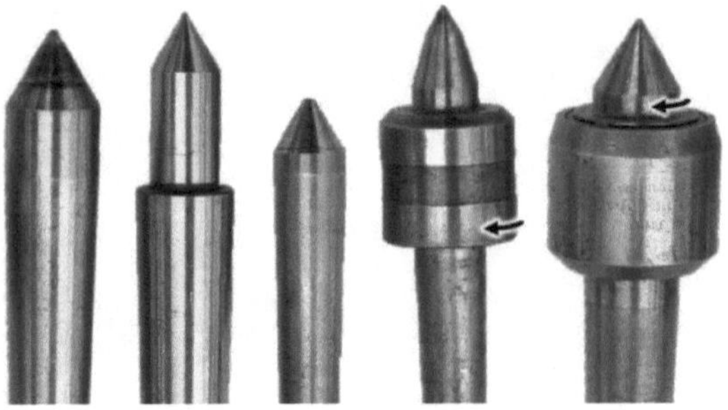

Figure 16.28 Lathe centers come in dead or live versions with different designs. A live center rotates with the work, a dead center does not. Note that the live center may rotate at the conical end only or the entire end may rotate.

and the cut out is positioned toward the tool. Turning between centers is a versatile and valuable workholding technique for turning.

Turning between centers necessitates two mounting points. This is accomplished by securing one center in the spindle and the other in the tailstock. There is a vital difference between these locations. The spindle is the source of rotation, but the tailstock does not rotate. Choosing the proper center for the each end is important. If a live center is used in the tailstock, there is no problem. If a dead center is used, the part will be rotating while the center is not. This can cause excessive heat and deformation between the tailstock and center. Dead centers can still be used, but it is crucial to supply lubrication between the workpiece and center. This commonly consists of white lead and oil. Despite the extra setup considerations, dead centers are more accurate than live centers, so they are used for precision work.

Even though a center will provide a significant amount of friction against workpiece rotation, it is usually not sufficient to hold the workpiece during cutting. That is, even with a dead center in the spindle and a live center in the tailstock, the workpiece can still slide against the dead center. This is where a **lathe dog** is added. Lathe dogs are integral to securing a workpiece held between centers. The workpiece is inserted through the dog and clamped. The dog is then keyed into a faceplate that keeps it from rotating (faceplates are discussed in more detail in Sec. 16.4.6). This concept is shown in Fig. 16.30. If a **self-driving center** is used, as shown in Fig. 16.29, it has a serrated cone that can resist cutting forces enough to allow uninterrupted cutting of the entire workpiece length. In this case, no lathe dog is required, but cut depth and aggressiveness of cutting are more limited.

Because the dog is bolted directly to the part, care should be taken to avoid marring the workpiece surface. This can be achieved by clamping on the raw, unmachined stock material or by using a soft shim, such as brass. The heavier the cutting, the more securely the workpiece must be held. By compromising and taking lighter cuts, the clamping pressure can be reduced. The dog also prevents cutting from the point where it is clamped onward. The part must be taken out and reversed to complete a full-length cut. As mentioned above, a self-driving center can be used to alleviate this problem, but a lathe dog provides better support and allows for more aggressive machining. In cases

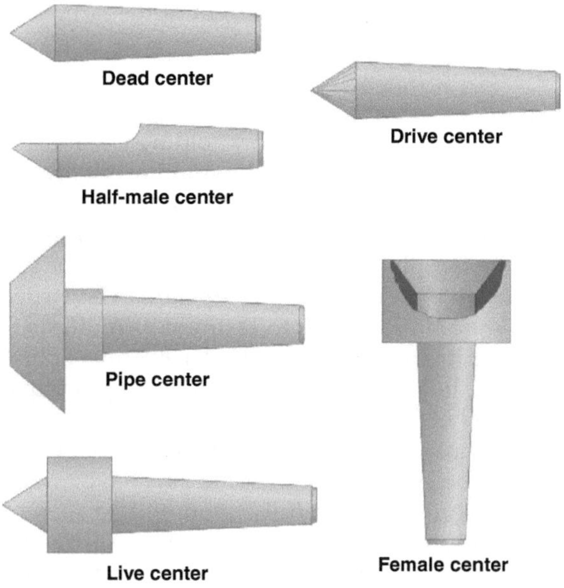

Figure 16.29 Lathe centers come in a variety of geometries. This allows for many types of workpieces to be cut between centers, or supported over long lengths.

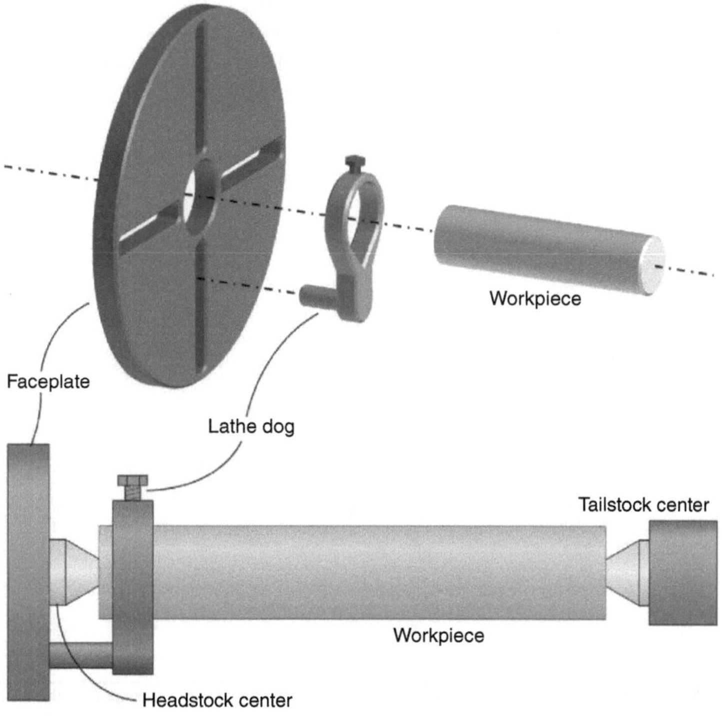

Figure 16.30 To securely hold the workpiece during machining, a lathe dog is used in conjunction with a faceplate to lock the part with the spindle.

where an uninterrupted cut along the length of the lathe is not needed, the lathe dog will be preferable.

16.4.6 Faceplate

The final method we will address in workholding is the use of a faceplate. A faceplate is simply a flat plate that is mounted to the spindle. A workpiece is held to the faceplate by some means of clamping. This is commonly done by using nuts, bolts, and clamps. These are often used in milling (Chap. 17) where work is fastened to the table of the machine. Ideally, the faceplate will be perpendicular to the spindle axis so that any cuts are also perpendicular. If necessary, the faceplate can be "trued" by making a facing cut on the plate. This should be done with restraint as multiple cuts will reduce the thickness of the plate until it is no longer usable. An example of how a faceplate may be used to create an off-center hole is shown in Fig. 16.31. Step clamps, which are integral to the process, are described in Sec. 17.4.4.

Because the workpiece is clamped to the faceplate, there is no absolute reference to "center." That is, the spindle axis and the part have no direct relation. For this reason, the workpiece must be carefully mounted to ensure that the feature being machined is in the appropriate location (on center). This unconstrained mounting is also a benefit. The workpiece can be machined in any area, and not only about its central axis.

This allows for eccentric features to be produced. In Fig. 16.31, you can see the hole is not centered on the part. It is centered on the spindle axis, as it must be, but the part is shifted. This allows holes to be drilled in a grid pattern or any other pattern. It is important to remember that just because it can be done, does not mean it should be done. In a shop with other equipment (hopefully all of them!), it would be easier to do something of that nature on the drill press or mill. The lathe can be very versatile, but it is not always the most efficient method.

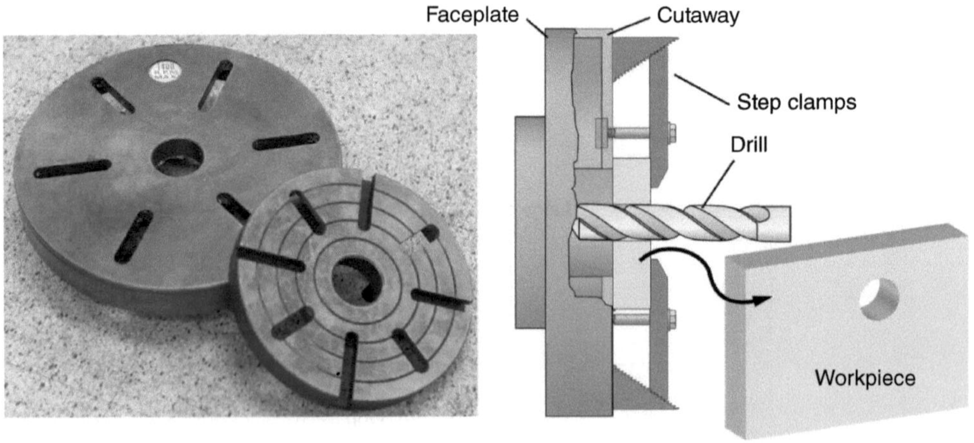

FIGURE 16.31 A variety of nontraditional workpieces can be held in the lathe using a faceplate. The method of clamping the workpiece to the plate differs greatly from other lathe workholding methods, but it also allows unique features to be created.

Another benefit of holemaking on the lathe by using a faceplate is that very large holes can be made. The tool does not rotate like on a drill press or mill. Rather, the workpiece rotates, which, at times, makes large holes easier to machine. A serious caveat to faceplate machining is mass distribution. If a workpiece is held off-center, there will be a corresponding force that acts to shift the part further off-center. Try taping a roll of quarters to the side of a drill and mount it in a hand-held drill (OK, do not *actually* try it). It will not take long to realize how much force uneven mass distribution can create. At the high rpms available on most lathes, there is a higher likelihood that the part can be thrown from the faceplate. That is extremely dangerous. It also puts significant stress on the lathe spindle. A counterweight can be used to alleviate the balance issue, but the workpiece must always be held securely and the speed reduced accordingly for massive, unbalanced parts.

16.5 Turning Processes

Turning operations can accomplish many tasks, and the lathe is a very versatile machine. There are many specialized processes to create features such as grooves, threads, and knurls. We will look at a few of the most common processes in this section.

16.5.1 Turning

Turning has been discussed in general terms so far, but some other considerations remain. Turning can be as simple as cutting in a straight line, parallel to the spindle axis. This creates a cylinder with a constant diameter. This is the most common process used on a lathe as raw material rarely has a useable exterior. Typically, to improve surface quality, accuracy, and meet required dimensions, the outer diameter must be turned.

To set the DOC, the cross slide must be advanced toward the part. The distance traveled beyond the OD is the DOC. There is an important caveat in turning when reaching a dimension. In reality, the amount of material removed from the diameter is *twice* the DOC. This occurs because the workpiece rotates. As it rotates, both "sides" pass the cutting tool and are removed. If you move the cutting tool in by 1/8 in, the change in diameter will be ¼ in.

On some lathes this is compensated for. That means when you turn the handle it reads out that you have moved 1/8 in (125 thousandths or 0.125 in), the tool actually moves 1/16 in so that the true cut is 1/8 in. Confusing? Look at Fig. 16.32 for a visual representation. If you are unsure whether a lathe is compensated, you can take a small cut and compare the diameter change to the handle readout, or you can mount a dial indicator to see if there is a difference between the two. If the dial indicator moves half the distance indicated on the cross slide handle, the lathe is compensated. If there is only a small difference, either the lathe or the dial indicator is inaccurate. It should be different by a factor of two or none at all.

16.5.2 Facing

Facing is simply a cut using cross slide motion only. This creates a flat end (face) on the part. The general procedure is shown in Fig. 16.3. Of the most important use for facing is to prepare the surface for future operations. If a piece of stock is cut in a band saw

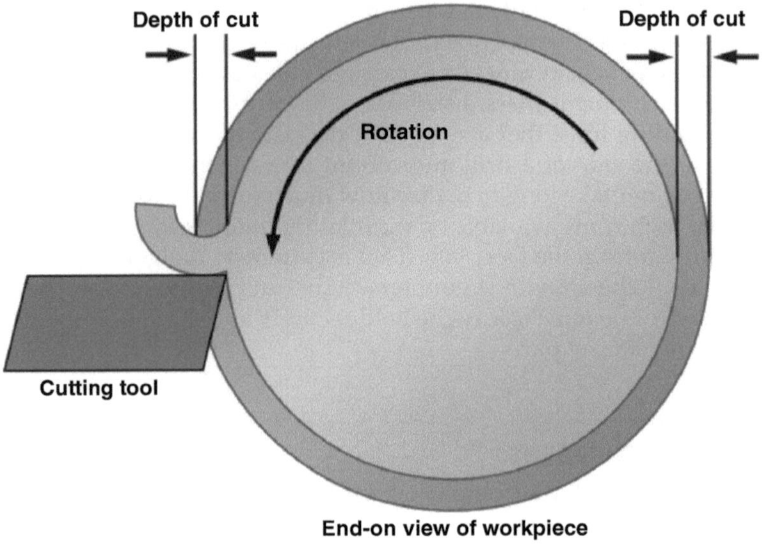

FIGURE 16.32 The depth of cut on a lathe is taken from both "sides" of the part. The diameter is reduced by twice the distance the tool is advanced. Some lathes compensate for this, but it is always best to check before making a cut.

before being transferred to the lathe, the ends will be uneven. The level of unevenness will be determined by the quality of the saw, the blade, the workholding, etc., but it is never "good enough." Facing is a simple, fast process that should always be applied. It creates a smooth, flat plane on the end of the workpiece (see Fig. 16.33). This is needed for a drill to start on (a common next step). Uneven surfaces will cause a drill to wobble, cut oversize, or even break (see Chap. 15).

To make the face flat, the tool has to be properly positioned. If the tool is too high or low, the center of the part will not be flat. This is where a drill will start, so improper tool height can be just as bad as not facing at all. If the tool is above the lathe (spindle) centerline, there will be a rounded nub. If it is too low, a sharper nib will remain. These features are shown in Fig. 16.33. The tool height can be accurately set by comparing the point (cutting edge) of the tool with a center mounted in the tailstock. The tip of the tool should match the tip of the center. If a face cut has been made, the tool can be fairly accurately centered on the remaining nub or nib.

Another consideration is how fast to face the part. The workpiece turns at a fixed rate, which is the rpm, but this is not the whole story. If you look at a spinning bicycle wheel, the spokes are more difficult to see at the outer edge than in the center. The speed (or velocity) at the OD is higher than the inside diameter. For this reason, the feed for facing should be reduced toward the center. Ideally, as discussed in Sec. 13.5, the rpm should be increased instead of reducing the feed. This is unrealistic in manual machining,* so be less aggressive when cutting at the center. This will save the tool from breaking or excessive wear.

*Some lathes can automatically increase the rpm based on manual movement of the cross slide.

Figure 16.33 Facing is a common starting point for other operations. The tool must be correctly positioned; otherwise the surface will not be entirely flat.

16.5.3 Boring

Boring is essentially the same operation as turning, only it is applied to an internal diameter. This means boring requires a preexisting hole. This can usually be achieved by simply drilling a starting hole. This begs the question, "Why not just drill the final hole?" The answer is multifaceted. Boring is more accurate than drilling and can produce a smoother surface finish. Also, boring can be done to create complex geometries that are impossible to replicate with a drill. Very large holes are impractical to drill as they require many drilling steps to progressively increase the hole size, but boring, with one tool, can create a hole of any size larger than the size of the bar itself.

The starting hole must be large enough to fit the boring bar. This can be tricky at times because the boring bar should as thick as possible for rigidity. Again, drilling a large hole can require multiple steps with several drills, so a very wide bar is impractical. Rigidity is also an issue for very deep holes. The larger the unsupported length, the greater the deflection [see Eq. (7.9)] and increases undesirable cutting chatter. As a general rule, the bar should only extend as far as needed for the cut being made. This means it is usually just slightly longer than the hole is deep. As the boring bar diameter decreases, so does the rigidity. For this reason, DOC should be reduced for thin bars. The process of boring and some common boring bars are shown in Fig. 16.34.

Boring can produce any contour, so long as the tooling allows it. Manual movement of the boring bar must be informed by *external* indicators since the cutting is *internal*, where visual information is not available. Micrometer readouts on the machine handles or dial indicators mounted to display the change in depth and diameter can be used. This is especially important when doing operations such as backboring. **Backboring** is a method where a larger diameter is cut deeper in the hole (Fig. 16.35). Normal boring bars will rub against the smaller diameter, so a special backboring tool must be used. Getting accurate dimensions during this type of operation is not trivial. It can save a second operation with a separate setup, but if backboring can be designed out of a part, that is advisable. Backboring is typically applied where boring from the opposite side is not practical.

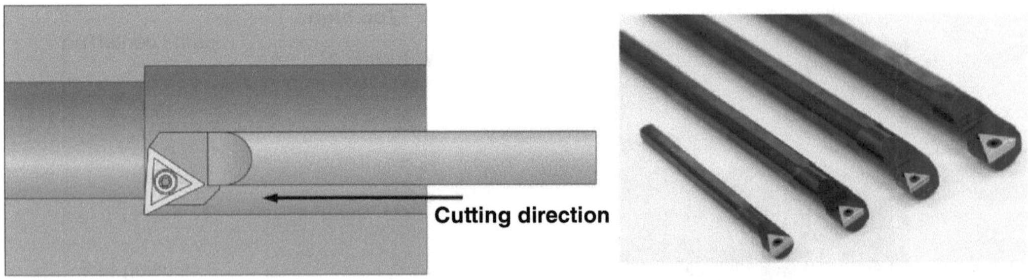

FIGURE 16.34 Boring on the lathe requires an internal cutting device. This is similar to external turning, but new considerations must be made based on rigidity.

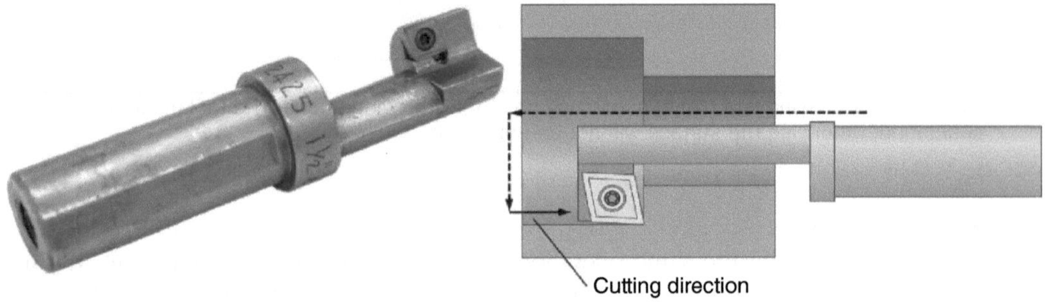

FIGURE 16.35 Holes can be backbored to produce a larger diameter deeper in the part. Specialized processes in boring typically require specialized tooling.

16.5.4 Tapering

Tapering creates angles on a part. The simplest example of this is a cone. A cone is simply a round part that has a constant taper. Not all tapers must come to a point. Rather, it is more common for a section of the part to be tapered while others are straight (cylindrical). Tapers are ubiquitous in manufacturing. As discussed in the collet section (see Sec. 16.4.4), machines tend to utilize tapers for holding parts or tools. The lathe tailstock usually uses a Morse taper to hold the center, drill chucks, etc. The holding is largely due to friction between the two surfaces because of the wedging action (see Sec. 15.6.2.1 for more on tapers). Being able to accurately create tapers is important.

There are a variety of methods to create tapers. The method is primarily determined by the part. For instance, the workholding choice will directly impact the ability to taper by certain methods. If working between centers, a conical tip cannot be made because the center is supported. That does not mean a taper cannot be created. When turning between centers, the tailstock can be offset to create an angle. This means that the part is actually tilted so that a straight cut will create an angle. To stay securely in the centers, the taper angle must be shallow. The benefit of this method is that very long tapers can be created. Tapering between centers is shown in Fig. 16.36.

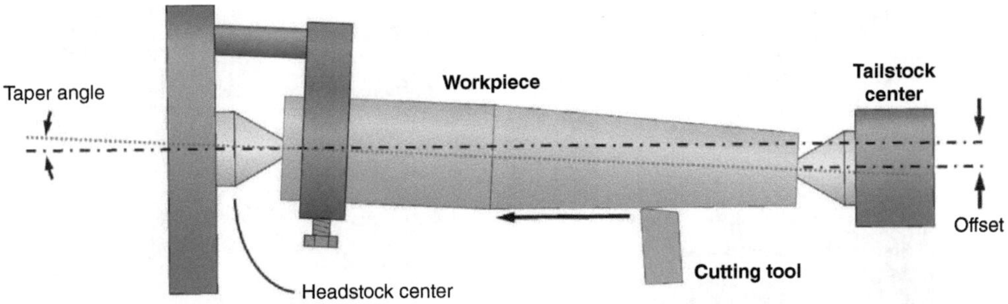

Figure 16.36 Tapering between centers can be accomplished by offsetting the tailstock so that the center is not in line with the spindle axis.

In other methods, the taper can be created by feeding the tool at an angle. There are two main methods for doing this. If you consult Fig. 16.6, the compound rest (or top slide) is capable of being tilted to any angle. The small handle associated with compound rest can then be advanced or retracted to create a taper. Although this method is very simple, it is not capable of creating long tapers. The travel is limited by the compound rest, not the size of the lathe. Another method of angling the feed is using a taper attachment. What the taper attachment does is forces the cross slide in or out as the carriage moves up or down the ways. This allows long tapers to be created, but steep angles are not possible. These feed-based methods can be used when appropriate without changing the workholding. This allows for a faceplate, collet, or chuck to be used. Both methods are shown in Fig. 16.37

The final and possibly the simplest method of creating tapers is to use a tapered tool or a tool at an angle. This is limited to very small tapers, but any angle can be created. Typically the accuracy and quality are not as good using this method. The tapers created in this manner are more commonly referred to as **chamfers** because of their small size. Usually these small tapers are used to help locate mating part (such as a pin) or to "break" edges so that they are not sharp. Methods using a file and cutting tool to create chamfers are shown in Fig. 16.38.

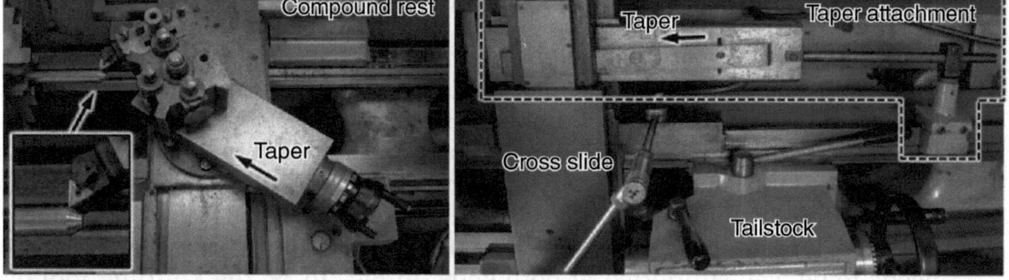

Figure 16.37 Tapering can be done by feeding the cutting tool at angle. For steep, short tapers, the compound rest can be used. For longer, shallow tapers, a taper attachment can be used.

FIGURE 16.38 Simple, short tapers called chamfers can be created by angling the tool. The cutting tool can be a file or regular lathe tool. Tapers made by these methods tend to be of a lower accuracy and quality than by other methods.

16.5.5 Knurling

To improve grip on certain items, the handle, knob, etc. will be knurled. **Knurling** simply roughens the surface with a defined pattern. There are two main categories of knurling: pressure knurling and cut knurling**. Pressure knurling** is not a material removal process so much as it is a forming process. You will not remove much if any material when creating the pattern by pressure knurling. Rather, the knurling tool applies sufficient pressure to indent and deform the material.

Cut knurling does remove material. This tends to provide a better finish and allows for faster processing. Also, because it is not deforming the metal, it will not increase the diameter like pressure knurling. Forming is a volume conservative process. In order to displace the metal, pressure knurling must squeeze the metal out. The diameter of a knurled part will increase after processing. Where this is undesirable, cut knurling should be used instead.

Knurling tools have rollers that form the desired pattern. Some knurling tools will only have one roller, but many tools have two rollers that form the pattern. Because the pattern is determined by the combination of two separate rollers, the rollers must engage the part equally. This is problematic if the tool is too high or too low (uneven pressure) or if the tool is tilted too much (partial pattern).

If the tool is not properly centered height-wise and square to the part, the pattern will not properly form. The resulting knurl will look like a chewed-up mess (like something your dog got ahold of). Although it may serve its function of improving grip, it will not do so as well and will not be aesthetically pleasing. Many times the longest process in knurling is getting the tool setup properly, but it is worth the initial effort.

Because form knurling does not cut the part, there are a couple of unique considerations. First, the force required is quite large. To displace the metal, it must deform the surface. This can lead to excessive, transverse bending of the workpiece. The workpiece needs to be well supported at both ends. In Fig. 16.39, the knurling tool only has two rollers that are positioned adjacent to one another. This **"bump-type" knurling tool** forms knurls by feeding the cross slide into the part. The force is concentrated in the transverse direction. This not only makes it very stressful on the part, it also puts a lot of stress on the machine.

Figure 16.39 Knurling creates a pattern in a part to allow for better grip. There are a variety of patterns and sizes, and the knurling tool will determine which pattern is formed into the exterior. The most common form is a diamond-shaped pattern. The revolving head bump-type knurling tool shown on the right can make fine, medium, and coarse knurls (right).

To reduce this one-directional stress, a **"scissor-type"** or **"straddle-type" knurling tool** can be used (see Fig. 16.40). In this case, the rolls are located across the part from one another and this can reduce or eliminate the bending force as only direct compression of the part is used. Cut-type knurling tools are capable of producing knurled surfaces on thin parts and with minimal pressure on the lathe and part. Also, because the pressure and deformation is reduced, many materials can be successfully knurled. Form knurling

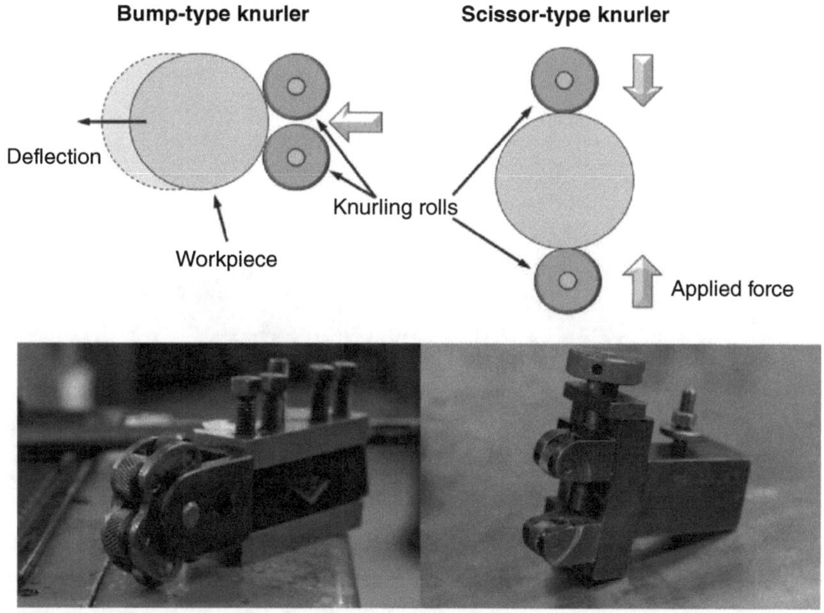

Figure 16.40 Knurling tools come in a variety of configurations. Shown here are bump-type (left) and scissor-type (right) knurling tools. Scissor-type tools balance the force and decrease part deflection and stress on the machine.

will cause surface hardening of materials, that work harden (e.g., stainless steel) and does not work well with very hard, brittle materials such as cast iron.

Although knurling seems like a fairly basic operation, there are many technical specifications that will determine the outcome. To make the various patterns in a variety of part types there are different wheel pitches, wheel materials, numbers of wheels, etc. The major considerations are the required pressure and production rate. Cut knurling will reduce pressure and increase productivity, but typically requires more setup time and is only justified for large-scale production. Most general-purpose knurling can be done well with a pressure-type tool. A variety of available knurling tool configurations are shown in Fig. 16.41.

16.5.6 Threading

Threads are one of the most common features machined into cylindrical parts made by turning. Think of screws, bolts, fittings, etc. Any part that screws into another must be made in some way. Although threads are often formed (i.e., a *forming* operation, as discussed in Chap. 12), we will look at how they can also be cut. Most lathes allow for adjustment of the feed rate to create threads with any common pitch.

As discussed in Sec. 15.5.1, the thread pitch is simply the distance between "peaks" in the thread. The cutting tool geometry and feed rate are critical to getting the appropriate thread geometry. Also, the OD is determined by the preceding turning pass. If the OD is too large, the threaded end will not screw into the mating part. If the OD is too small, the fit will be sloppy and weak. The preparation steps before threading are critical to the final quality. Universally, the diameter must be reached before threading commences. To create threads, there are two common approaches: single-point threading and die threading. We will look at each.

FIGURE 16.41 Knurling tools come in a variety of configurations.

16.5.6.1 Single-Point Threading

Single-point threading is done with a special cutting tool that must be fed along the part to create the threads. Because a single point of contact is used, this gives us a lot of freedom, but that needs to be used responsibly. Threads are standardized (see Sec. 15.5.1). The standardization ensures that the nut from one manufacturer will fit the bolt from a different manufacturer. It is good convention to choose standard screw threads when cutting them on a lathe. This does not limit the lathe, however, and nonconventional threads are easily created (maybe too easily!). Nonconventional threads will deviate in diameter or thread pitch.

For example, a very common screw type has a 0.250 in OD and 20 threads per inch (tpi). The true diameter can vary somewhat and still work. The closer the fit to the nut or internally threaded part, the tighter the fit and stronger the connection. When performing manual operations in turning, the accuracy of the OD will vary. It should always be slightly less than 0.250 in (the inside diameter of the nut) to avoid interference. The caveat is that the lathe will put 20 tpi on *any* diameter. The threading setup on a lathe only affects the feed rate, so the prior dimensions are entirely dependent on the operator. The freedom to match the pitch with any diameter can be quite useful, but the most common goal is to match existing thread specifications.

The most common geometry for a single-point cutting tool is a 60° tip that comes to a sharp point (see Fig. 16.42). This forms a unified screw thread, but there are many other types of threads. The profile may be very different, depending on the application. Acme threads, for instance, are much flatter at the major and minor diameters. This requires different tool geometry, but the other major considerations such as feed rate and diameter are the same. A variety of profiles are shown in Fig. 16.43, but still others exist. Some threads are better under high load or more accurate and can reduce backlash.

Backlash is a universal consideration for machine screws. There must be some space between the mating threads or the friction would be excessive. That space means when you change a direction, there is a certain amount of rotation before the screw engages on the opposite face. As the threads wear, the backlash increases. By choosing a different thread type or using preloading, the backlash can be minimized. One of the most

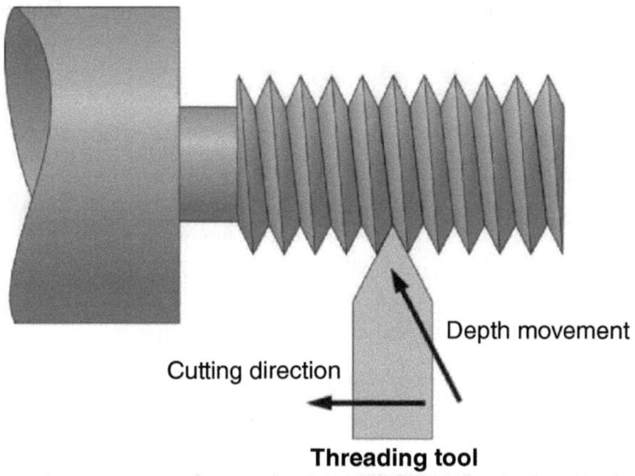

FIGURE 16.42 Turning by using a single-point tool requires precise feed and tool geometry. The tool should be fed in at the thread angle, not in a purely transverse direction.

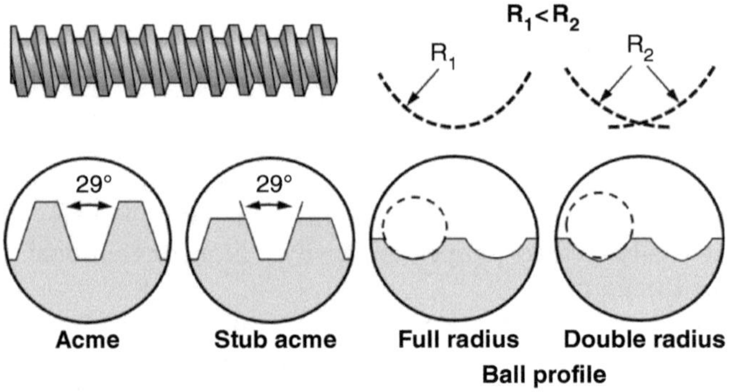

Figure 16.43 Thread profiles may vary greatly depending on the intended application.

common screw configurations which reduces backlash is the **recirculating ball screw** (or simply, *ball screw*). It uses ball bearings as the mating "thread." This allows for very small clearance to be provided since the ball bearings will roll and not bind even when tightly held. As shown in Fig. 16.43, a ball screw can have a fully round profile, but a double radius or "gothic arch" profile will reduce the contact to four points and reduce the friction further. These concepts are shown in Fig. 16.44.

As a foolproof approach, regardless of the amount of backlash, always advance the tool in the direction you will cut before reading the micrometer measurements on a machine. This ensures repeatability and rigidity. If you go too far or only need to make a small adjustment, make sure you back out significantly (0.100 in, for instance) and then return to the point you are trying to reach, again going the direction you will be cutting. This process accounts for the backlash by taking the "slop" out of the gears before using the positioning measurement. **Digital readout** (DRO), which measures and displays actual movement, not the screw rotation, is available on many machines, and it is immune to backlash since it does not use the thread pitch to determine position. DRO is shown in Fig. 17.7 for milling, but it is available for lathes and other machine tools as well.

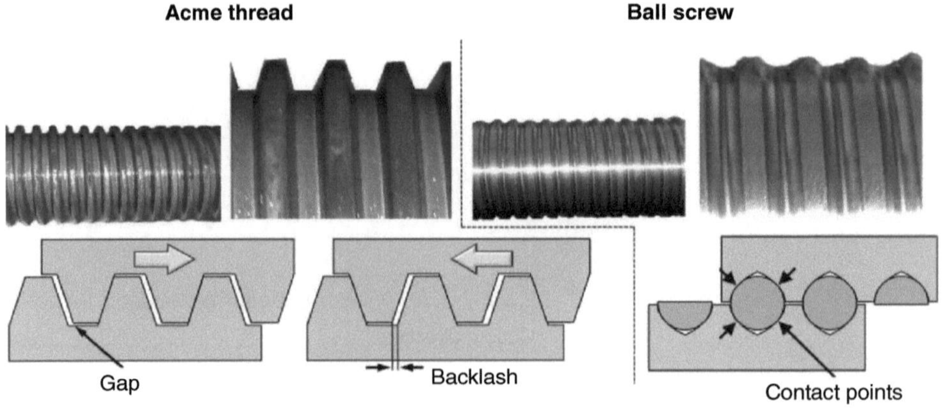

Figure 16.44 Common machine screws, such as those using an acme thread (left), will inherently have some backlash. Updated designs such as the recirculating ball screw (right) can considerably reduce backlash.

Among all these types of threads, there are other defining features (yes, there is more to it!). These universal features include the **handedness** of the thread and the lead number of the thread. The most common type of thread will be right-handed and single lead. Right-handed means the screw will tighten with clockwise motion ("righty tighty, lefty loosey").

Left-handed screws are cut in the reverse direction of right-handed screws, but follow the same considerations. This mean turning the screw clockwise will *loosen* it (righty loosely, lefty tighty, so to speak). Left-hand threads are common on rotating shafts where the nuts must be "self-tightening." This design is used on bench and pedestal grinders. The nut on the left side has a left-hand thread and the right side has a right-hand thread so that the nuts will both tighten when the grinding wheels are rotating in the same direction.

The **lead number** (e.g., single lead) indicates the number of passes a thread was cut in. If you were to wrap a string around the screw by following the pitch, one string would outline the entire thread pattern. This is not always the case. A multiple lead thread will have to be cut using multiple passes. Each pass will use a feed that is too fast for the final thread.

For example, an 18 tpi screw can be cut by spacing three 6 tpi cuts equally. The final thread is 18 tpi, but when you turn a nut, instead of advancing at the rate of 18 tpi (slow), it advances at the rate of a 6 tpi thread (fast). The nut must also have three leads to fit. Multiple lead threads require more careful setup to cut properly, but they can drastically increase the efficiency of threading components on and off or moving something by screw motion.

Some of the most common applications include soda bottles or other commercial, screw-on caps. Instead of having to spin the lid five or six times to get it closed, you may only need to spin it twice (or less depending on the number of the leads). Multiple lead threads are typically stronger since they achieve the same rate of movement without having to reduce the screw's minor diameter as happens when decreasing the pitch. Handedness and multiple lead threads are depicted in Fig. 16.45.

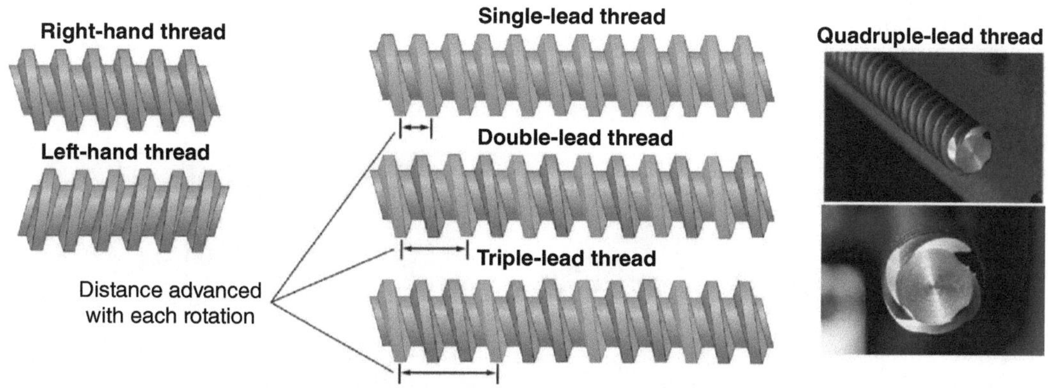

Figure 16.45 Threads can be made with different handedness so that it can tighten or loosen for a given rotation. The distance the thread engages for one turn can be changed for a single pitch by increasing the number of leads. This effectively allows motion at the rate of a much coarser thread while maintaining the strength of a finer thread.

There are many types of threading tools and techniques for cutting threads on the lathe. Like the other turning methods we have discussed, threading can be done on the inside of a part as well by using a boring bar (not to mention through tapping, as discussed in Sec. 15.5). Thread cutting may be done with a single point, multiple points, and with tools in various orientations.

16.5.6.2 Die-Cut Threading

We talked about taps in Sec. 15.5.2. They are used to cut internal threads in a pre-drilled hole. A **thread cutting die** is essentially the inverse of a tap. It will cut the standard threads, but it cuts them on an external surface. Just like with single-point threading, the diameter must be properly sized to achieve a standard thread. Because the die engages the full diameter of the part, it is extremely important that the part is not oversize. If it is, the die will bind and either the part or the die (or both!) can be ruined. Unlike single-point threading, die threading cuts threads based on the die itself. The freedom to cut any thread on any diameter is restricted. Although custom dies can be made, when purchasing a die set, it will come with common, standardized thread specifications (Fig. 16.46).

Some dies are fed on by hand. This requires careful approach so that the die is perpendicular to the length of the part. If the die starts crooked it will get jammed, cross-thread, or otherwise ruin the part. Plenty of lubricant should be used to prevent excessive wear or breakage. Threading is a heavy material removal process, and a die must cut the full depth of the thread over a very short distance, so the forces are high.

Rapid application of a threading die can be used in a production environment by using a **self-opening** (or **geometric**) **die**. These use separate inserts mounted to jaws, similar to a lathe chuck. The cutting can be spread out farther, and at the end of the cut, instead of having to reverse the feed to remove the die, the jaws snap open. This is triggered by the die being pulled (not fed) forward. That is achieved by stopping feed of the assembly and continuing spindle rotation. The die self-threads further, pulling the jaws forward. After it extends a pre-set distance, the jaws are triggered open. The desired thread length must account for the stopped feed and the length it will self-thread until opening. The jaws must be closed before the next part can be threaded. The self-opening die is shown in Fig. 16.47.

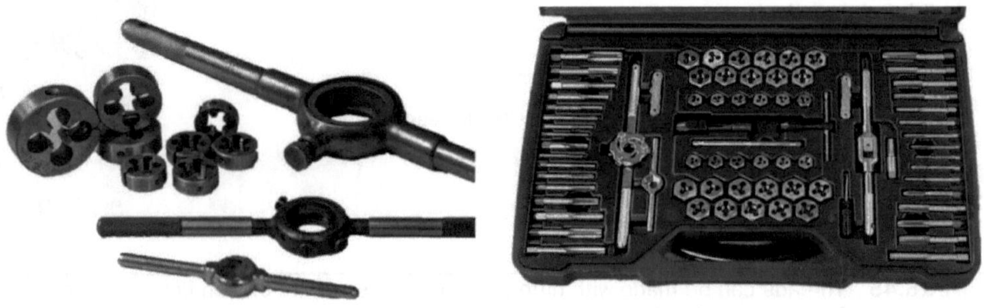

Figure 16.46 Dies used to cut external threads in common, standardized sizes. Usually a set (right image) is purchased to accommodate a range of options.

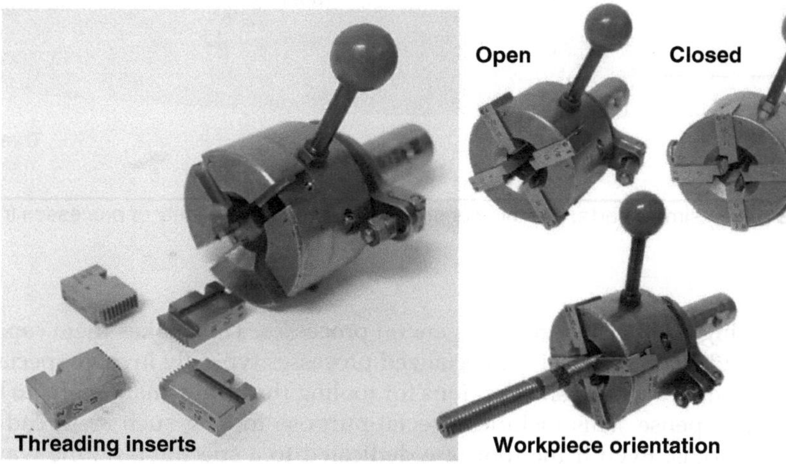

FIGURE 16.47 Self-opening (geometric) dies use separate inserts for cutting threads. They are closed and threaded onto a part. The feed is stopped and the die continues threading on as a nut would. After it has extended a small amount, the jaws snap open and it can be retracted. (*Courtesy of Cindy L. W. English, MFA.*)

16.6 Why It Matters

We have examined a number of facets of turning. The versatility of a lathe is one of its defining features and the reason you will find one in any general-purpose shop (job shop). There is simply no better way to create cylindrical parts. That does not mean there are no other ways, but the method of workpiece rotation is unrivaled for speed and accuracy. The fact the tool is not rotating allows for the use of a single tool to achieve a wide variety of geometries. The two most common processes are turning and facing, but many other processes are important as well.

We saw that there are a number of lathe types. These lathe types are essentially general-purpose and production varieties. Remember, whenever you focus a machine on a single task to increase productivity, you tend to lose versatility. The same can be said of tooling. Specialized tooling for operations such as creating radii has little use outside of that. General tools still come in a variety of geometries. The choice will depend on the amount of material you need to remove (roughing versus finishing) and the features to be replicated.

Workholding has some key considerations. These are support and positioning. The options we looked at were chucks, collets, between centers, and faceplates. The method of workholding must be chosen based on the operations to be done and vice versa. Your options are only limited by your equipment. Many methods will achieve the same basic functions, but there are some features that are characteristic of each. We looked at some of the techniques used on the lathe, and these can inform the workholding choice. For instance, if you need to bore a hole, working between centers is not going to be an option. If you want to cut a taper, there are different methods that excel at steep or shallow angles and long or short lengths. Other options such as threading and knurling are mostly dependent on the tooling, but there are many considerations within that realm as well.

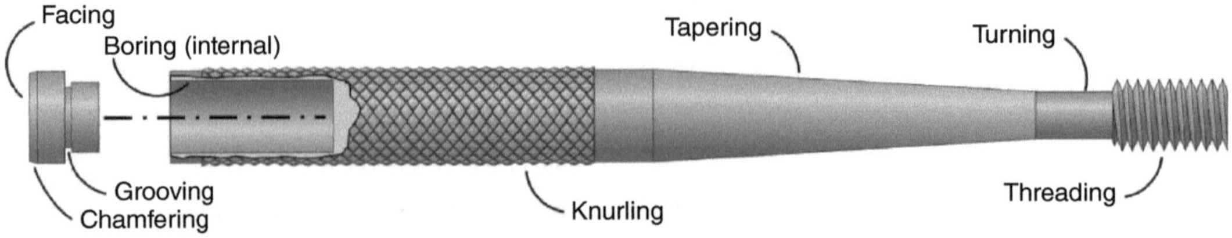

Figure 16.48 Even simple parts, like this hammer handle, require a variety of processes involving turning.

Turning and boring are very general processes. This makes them capable of generating a wide variety of parts. Specialized processes typically involve specialized tooling. There are many off-the-shelf options for tooling that allow many parts to be made with minimal expense. This includes special-purpose tooling such as threading tools and knurling tools. Even if the tools are dedicated to a specific task, the way they can be applied still allows a great deal flexibility.

One example of this is threading. Although screw threads generally follow a standard convention, the ability to create any diameter part with any desired thread pitch is beneficial. That design freedom is valuable when creating prototypes or custom manufacturing, while still allowing for conventional parts to be made as well. Most parts created on a lather incorporate a variety of processes (Fig. 16.48), so knowing what is available and how they can complement each other is valuable. The possibilities using the lathe are only as limited as your imagination. Hopefully the material we have covered in this chapter gives you some good ideas.

16.7 The How and Why of Turning and Boring

16.7.1 Chapter Review

1. What is turning? What is boring?
2. What is the difference between longitudinal and transverse cross sections? Draw a rough sketch of what this would look like for a passenger car.
3. What are the two principal operations in turning?
4. Describe why motion in two axes is needed to remove material.
5. What is the spindle of the lathe and what is it used for?
6. Why is a spindle most commonly hollow?
7. What are the cross slide and the carriage? How do they differ?
8. What is the purpose of the compound rest?
9. Why is the tool post adjustable?
10. When facing, why does the angle of the tool not affect the shape of the cut?
11. Name two operations that the tailstock can be used for.
12. Why is a thin part more likely to need support from the tailstock?

13. How is a turret lathe different from an engine lathe?
14. Name two advantages of using a turret lathe over an engine lathe.
15. What advantage does an engine lathe have over a turret lathe or screw machine?
16. Why would you use a multiple spindle screw machine instead of a single spindle version?
17. Why should stock for turning be oversize?
18. Describe why there is a compromise between cutting tool strength and ability to reproduce detail.
19. What are the first machining operations when starting from standard stock?
20. Why is there a groove at the end of a threaded section?
21. Why is the angle of the cutting edge important for threading?
22. What tooling issues are associated with turning radii?
23. What is a form tool? Name an advantage and disadvantage of using a tool like this.
24. Is workholding always required? Why or why not?
25. Name five considerations that determine the type of workholding used?
26. Turning operations put a variety of forces on a part. Draw the primary forces generated during facing, turning, and grooving.
27. What balance must be reached for securely holding a part?
28. Name three factors that may increase the forces during cutting. How can these be avoided?
29. Describe concentricity and eccentricity. Why are these important?
30. What is runout? How can this be measured?
31. What are some attractive features of chucks?
32. What is the difference between universal and independent chucks?
33. Why is the number of chuck jaws important?
34. How are soft jaws used with a chuck?
35. What advantage and disadvantage do independent chucks have?
36. What safety concerns arise when turning unbalanced parts?
37. How does a collet apply pressure to hold a part?
38. What limitations do spring collets have?
39. What are some similarities and differences between R8-, C5-, and ER-type collets?
40. What is the benefit of using a collet with jaws connected by rubber?
41. What is an emergency collet used for?
42. What are the two main types of collets? How are they different?
43. What must be done to the end of the part before a center can be used?
44. What are the two main functions of a center?

45. Name three types of centers and why they would be used.
46. What is a lathe dog? Why is it used?
47. When would you not want to use a lathe dog? What is the alternative?
48. What is a faceplate?
49. Why would you use a faceplate over other methods?
50. What are some benefits and disadvantages of faceplates?
51. When are holemaking operations efficient using a faceplate? When are they not?
52. How should unbalanced parts be addressed?
53. What is important to note about the DOC when turning in regard to the actual amount of material removed? What does it mean if a lathe is "compensated?"
54. Why is facing used?
55. What happens when a tool used for facing is not on center?
56. How should feed be adjusted when approaching the center of the part?
57. What operation must precede boring?
58. How is boring more versatile than drilling?
59. Why should a boring bar be as thick as possible? What is the trouble with very deep holes?
60. Why is it difficult to manually create bored holes?
61. What is a taper?
62. How can a taper be created between centers?
63. What limitation does tapering with compound rest present?
64. What does a taper attachment do? What advantage does it have over the compound rest? What disadvantage?
65. What are the two main categories of knurling? How do they differ?
66. Why is workpiece support so important to bump-type form knurling? What alternative configurations can help?
67. Where is threading needed?
68. Why is tool geometry so important for threading?
69. How is the thread pitch adjusted when threading?
70. What are the two main categories of thread cutting?
71. Why is the starting diameter so important when threading?
72. What is backlash? Why is it important?
73. How can backlash be reduced? What process is used to remove it from measurement?
74. Discuss the handedness of a thread. Why would you use different types?
75. What is meant by the lead of a thread?
76. What advantages do multiple lead threads have over single lead threads?

77. How can threads be cut internally on a lathe?
78. How is a thread die different from a tap? How are they the same?
79. How is single-point threading more versatile than die threading? When is it an advantage? When is it a disadvantage?
80. Why must lubrication be significant when die threading?
81. How does a self-opening die (also known as a geometric die) work? How can this aid production rate?
82. What special considerations are required when using a self-opening die?

16.7.2 Concept Connections and Extensions

1. The engine lathe is commonly cited as the most versatile piece of equipment in a shop. How is that justified (use concepts throughout the chapter)? What operations would you not choose the lathe for?
2. Turret lathes and screw machines are "programmable" in the sense that they are set to run a specific task. How does this differ from modern programming (Sec. 22.3)? Which is more difficult?
3. One tool can create various geometries and sizes on the lathe. How is this different than the mill (Sec. 17.3), and why are multiple tools typically used then?
4. What challenges are encountered with long workpieces, and how are these overcome? How can that solution interfere with other processes?
5. Why is a low rpm preferred when using a faceplate? Why not use a mill instead?
6. Why should a part be left in the chuck for every possible operation? How can this potential problem be used to create unique features when controlled?
7. When is a collet preferable over a chuck and vice versa? What role might an emergency collet play in addressing this?
8. How is boring similar to turning? What unavoidable downside can compromise efficiency in removing material and/or cut quality?
9. Compare pressure knurling with cut knurling using the concepts in Fig. 12.1. How might the properties differ, and what are the advantages of each?
10. When threading a round object, does it make more sense to use a die or to do it using single-point threading? What circumstances would warrant each? How might it vary between a screw machine and a CNC lathe (Sec. 22.3.1).

CHAPTER 17

Milling

The terms "mill" and "milling" can be unclear. There are saw mills, rolling mills, paper mills, pepper mills, ball mills, etc. None of these are related to what we will consider in this chapter. In woodworking, reducing boards to finished dimensions is also sometimes referred to as milling. Though getting much closer to the intended processes, this too is outside the scope of this chapter. Though the concepts of milling show up in many places, we will focus our attention on common milling equipment often used in metalworking, though it is not limited to metallic materials.

17.1 How It Works

For our purposes, **milling processes** involve multipoint cutters that can remove material in variable quantities along multiple surfaces simultaneously. They are different than saws, which have multiple teeth, but are only used to cut through a material in one direction with a single cutting surface, and this leaves a fixed kerf width (see Sec. 14.2). Milling cutters are different from drills in that they can cut around periphery of the tool, not just at the tip. Milling tools can create straight surfaces or contours, but the profile is dictated by the tool geometry and tool path.

You may be feeling at this point that "unclear" is all the more you will get from this chapter. Do not worry; a picture is worth a thousand words, so we will save some space and consider Fig. 17.1. The figure shows the important distinction with milling. A single tool can be used in multiple ways, cutting on multiple surfaces. This makes a single setup versatile. One size tool can make any size feature, so long as it is not smaller than the tool and not larger than the machine can accomodate.

Possible and practical are two very different things. For instance, it is impossible to create a ½-in hole with a ¾-in end mill. On the other hand, it is *possible* to cut a 1-in groove with a 1/16-in end mill, but it is impractical. Likewise, it is impractical to use a milling cutter to drill holes, since drills are more efficient. Milling is very versatile, but that is not a good excuse to be lazy and try to accomplish every task with a single tool. The ability to cut multiple ways with a single tool must be practiced responsibly. This will reduce time, increase tool life, and improve part quality. Before we can detail these considerations, some general background on the equipment and tooling will be important.

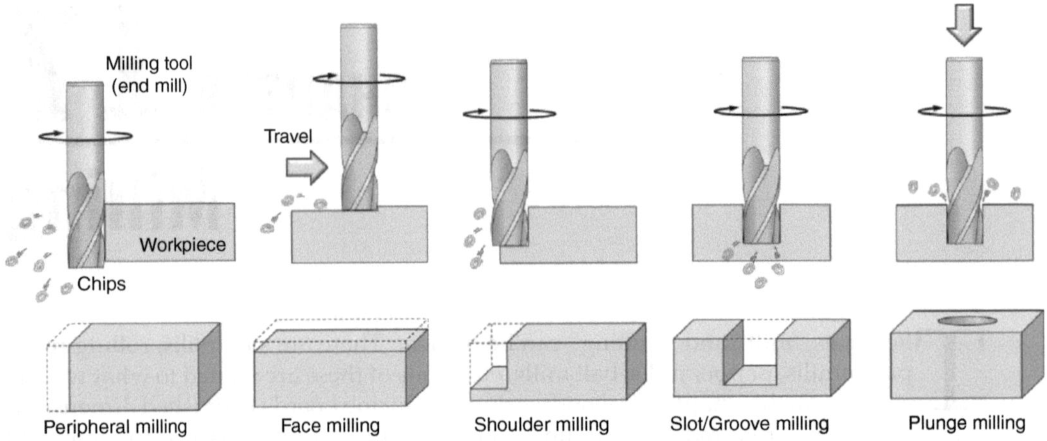

Figure 17.1 Milling can be done using multiple cutting surfaces in different ways. Some common techniques are shown here.

17.2 Types of Mills

Milling machines are general-purpose devices that can hold a variety of tools and workpieces to accomplish many tasks. The general theme is that they rotate the cutting tool and the workpiece piece is secured to a table that moves the part into the tool. Most manual machines will have three axes in which the part can be moved. The part height is adjusted (z-axis), then the workpiece is moved left or right (x-axis) or in and out (y-axis) as needed. All axes have micrometer adjustment on them, which means that precision movement is possible to accurately machine features in all three axes (Fig. 17.2).

Mills come in various sizes (from benchtop to multistory versions) and with differing capabilities (horsepower, maximum workpiece size, maximum spindle rpm, etc.). A hobby mill for the home machinist may cost a thousand dollars (or less, especially if purchased used), or multimillion dollar versions can be had for large part production. Despite the diversity, the main differentiation between mills is based on the orientation of the cutting tool. The two main layouts are vertical and horizontal mills, with the name indicating the spindle axis orientation. We will consider each next.

17.2.1 Vertical Mill

Vertical mills have a spindle with its axis of rotation orientation vertically (perpendicular) to the table where the workpiece is mounted. The primary cutting motion comes from a rotating tool, and the primary workpiece movement comes from linear motion of the table it is mounted to. The workpiece can be mounted in various ways, and these will be discussed in Sec. 17.4. Vertical mills are the most common type of mill and come in further subdivisions based on design.

The most common vertical mills are the knee-type (also known as ram-type or by the manufacturer trade name of Bridgeport) and bed-type. Knee mills have vertical adjustment that raises and lowers the table. Bed mills have vertical adjustment on the head of the mill. The added rigidity of a fixed table makes bed mills suitable for heavy-duty

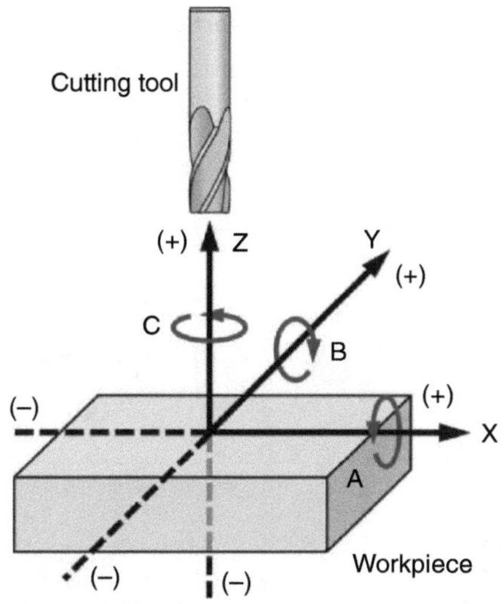

Figure 17.2 Milling machines offer precision movement in at least three axes. These axes correspond to Cartesian coordinates of X, Y, and Z, and have positive and negative values as shown here. Rotation about *x*-, *y*-, and *z*-axes are known as A, B, and C axes, and have positive values as shown here. Negative values (not shown) would rotate in the opposite direction.

milling operations, and this is the primary configuration for production mills like computer numerical control (CNC) versions. Knee mills are suitable for light- and medium-duty work and are lower cost than bed mills, so they are the primary type of manually operated machine. Differences between the two styles are shown in Fig. 17.3.

Some terms, such as table, head, etc., have been used so far, and it is useful to further define the components of a vertical mill. A typical knee mill, as shown in Fig. 17.4, has

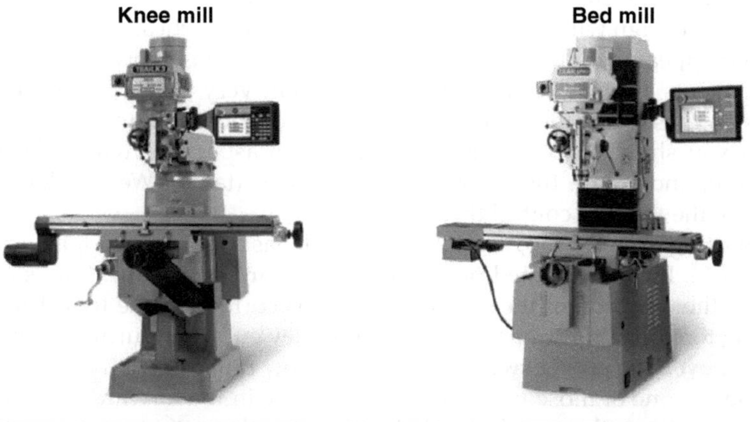

Figure 17.3 Vertical mills are often further divided into knee- and bed-type versions. Bed mills have a head that moves in the *z*-axis, and this provides a more rigid machine but greater cost than a knee mill. Both of these models have computer control included, but bed-type mills are much more common in these CNC applications. (*Courtesy of TRAK Machine Tools.*)

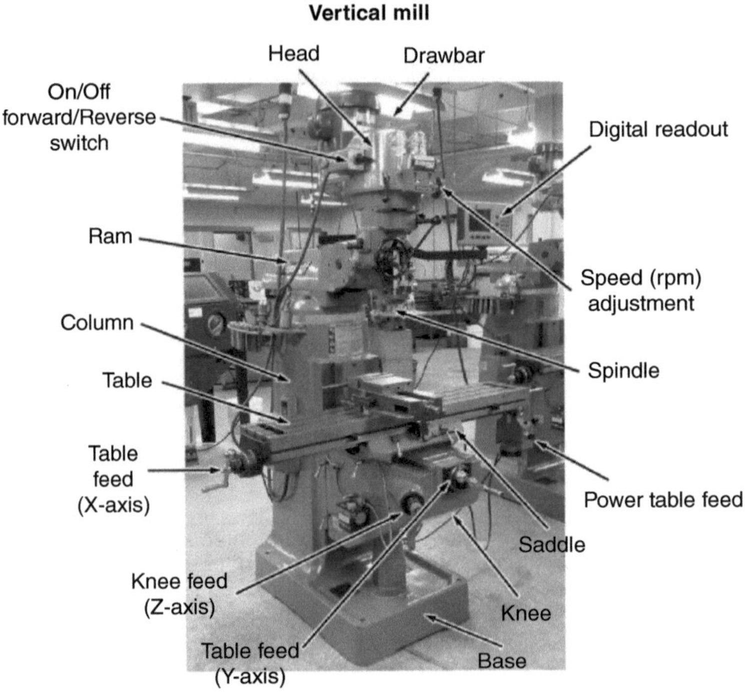

FIGURE 17.4 Some of the major components of a typical vertical knee mill are given here.

many features and adjustments, but for general-purpose use, most are not needed. We will highlight some of the most common of these to help understand the way mills work.

17.2.1.1 Features and Functions of Knee Mills

Although design will vary between manufacturers, vertical mills are relatively uniform in their appearance and function. A typical knee mill is shown in Fig. 17.4, and various components are labeled. It should be apparent very quickly that there are many features on these machines, and only a portion of those features are labeled in Fig. 17.4. What you should focus on now are the main aspects of where the tool is mounted (spindle) and where the workpiece is mounted (table). We will look more closely at some of the ways to control these next.

Starting from the top, the head of the mill has controls for adjusting the tool rotation (Fig. 17.5). It can be turned on and off in forward or reverse. Care should be taken to ensure the direction of the spindle rotation is correct for the tool. The vast majority of cutting tools are "right-handed," meaning they cut when turning clockwise, as viewed from above. The power switch is labeled as low range or high range, and it is off when it is not on one of those two. To run the spindle in the forward direction as needed for most operations, the switch should be turned to the setting corresponding to the gear it is in (i.e., when in high gear, the switch should be turned to high range).

The choice between high and low gears depends on the rpm desired. If you are not familiar with the way to determine the optimum rpm, it would be a good time now to

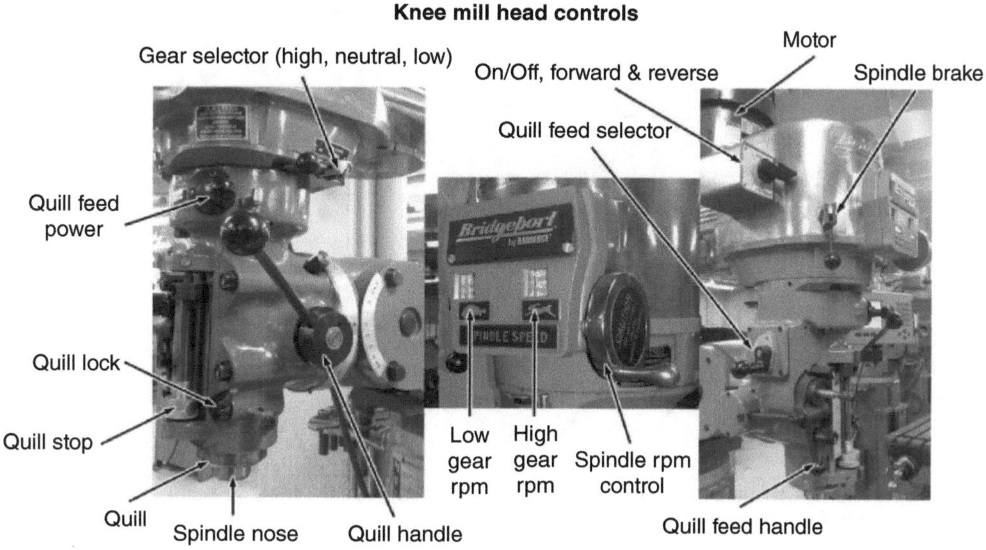

FIGURE 17.5 A variety of machine controls are positioned on the mill head, where the motor is located. These control the main components of the quill and the spindle rotation.

learn (see Sec. 13.5.1). Each gear has a range of speeds with low gear running from 500 rpm and lower and high gear running from about 500 to 3000 rpm. At lower speeds the motor has higher torque for large cutting tools. Smaller tools, which can be run faster, should be operated in high gear for greater cutting efficiency. Both ranges should be adjusted *while the machine is running* until the desired rpm is set on the dial (turtle for low gear, rabbit for high gear). Some older mills will require changing a belt from one set of pulleys to another, but these models are becoming less common, though the need to change a belt between pulleys may be found on a variety of production equipment, especially drill presses.

Aside from speed control, the **quill** is the other major feature of the head, and it can be controlled using multiple methods. The first, and most basic, is manual operation, very much like a drill press. The quill lock should be disengaged and the quill handle rotated counterclockwise to lower the spindle. Only holemaking operations (see Chap. 15) should be done with the quill down. Although milling operations such as those in Fig. 17.1 can be done with the quill lowered, it is not best practice since the longer the distance is between the tool and head, the greater the deflection will be [see the term, *L*, in Eq. (7.9)].*

The distance the quill will lower is dictated by the quill stop. To repeatedly drill holes to same depth, the quill stop can be precisely adjusted and locked in place. As shown in Fig. 17.6, the full length of travel of the quill is about 5 in, which can be used

*This is probably as good as any place to note that there are a variety of opinions about what is "best" and what is "good enough." Do not be surprised when you see others taking shortcuts, because you will be tempted to as well.

Figure 17.6 By unlocking the spindle, it can be operated similarly to a drill press. This most common for holemaking operations.

to create most hole depths, but the quill travel can be combined with raising the table to achieve even greater depths. Of course, longer tooling can be substituted as well.

The quill can be engaged with a transmission to lower or raise it at a constant rate (i.e., auto feed). For holemaking operations where surface finish is especially important, like boring operations (Sec. 15.7), this is quite valuable. The feed rate and direction (down or up) can be set as needed. When not in use, the quill feed should be disengaged to avoid accidental motion.

The major positional adjustments of a mill include movement of the **head**, **ram**, and **table**. The head and ram are not moved during cutting, but rather they can be positioned prior to cutting. Adjustments on the ram (rotation on the column and linear movement) are typically used to adjust the **work envelope**, which is a hypothetical volume in which the tool can reach all points. In most cases, these will be left at orthogonal angles so that the spindle (and cutting tool) is orthogonal to the table in all directions.

In fact, it is important to check for any tilt in the head as it will result in unwanted ridges in a workpiece and can affect the accuracy of parts. The process of setting the spindle perpendicular to the table in the x- and y-axes is known as **tramming** the head, and it should be checked periodically even if the machine has not been adjusted. A dial indicator is used to measure any deviation in height as the spindle is rotated. When the head is properly trammed, none will be present.

Precision is important in milling operations, and the many adjustments on the mill are intended to aid the operator in setting it up quickly and achieving high accuracy. The table movements are measurable in increments of one thousandth of an inch (0.001 in) as labeled on the hand wheels that control the motion. Rotational movements of the head and ram are indicated in 1° increments, but precise angles should be verified by precision measuring equipment.

Likewise, digital readout can be used to verify linear movements of the table. This is important because as the machine is used, the lead screws, which the table moves along, wear out and the movement becomes less accurate. Digital readout measures

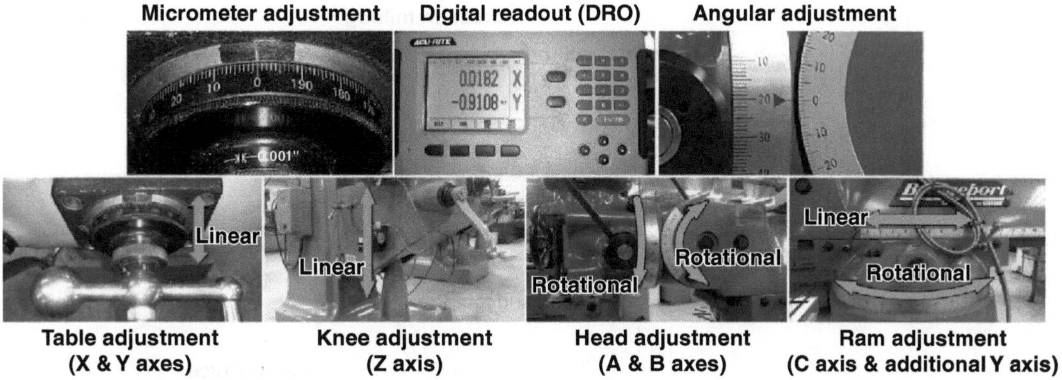

FIGURE 17.7 Knee mills can be adjusted in a variety of ways for specific operations. Most operations do not require the head and ram to be tilted or otherwise adjusted, but most will use table and knee adjustments. These can be precisely positioned using the micrometer readouts or digital readout. For the relationships of axes, see Fig. 17.2.

actual displacement, not rotation of the hand wheel, and will account for the inaccuracies. The various adjustments of a knee mill are shown in Fig. 17.7.

17.2.2 Horizontal Mill

Horizontal mills have a spindle with its axis of rotation orientation horizontal (parallel) to the table where the workpiece is mounted. Although table movement is similar to that of a vertical mill, the applications tend to be different. For example, horizontal milling is often used for creating gears (straight and helical teeth), slots, and profiles. The cutters used in horizontal milling are often of a larger diameter than used in vertical milling, and can have an appearance similar to circular saw blades (Fig. 17.8).

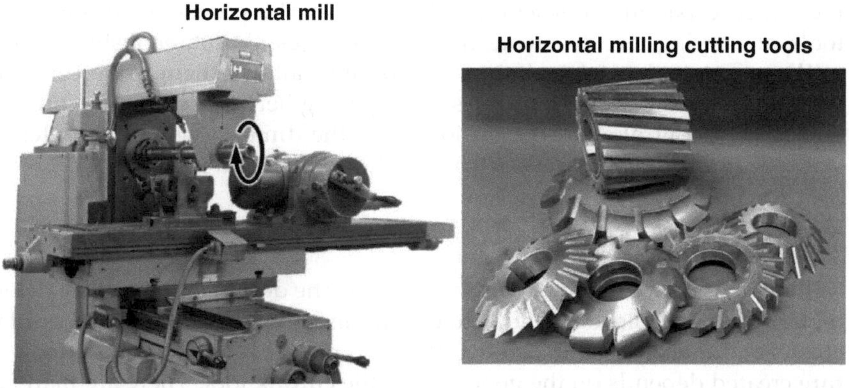

FIGURE 17.8 Horizontal mills are often used to create profiles not practical on a vertical mill. The cutting tools may have a variety of sizes and cross-sectional profiles. (*Photograph on left courtesy of Cindy L. W. English, MFA.*)

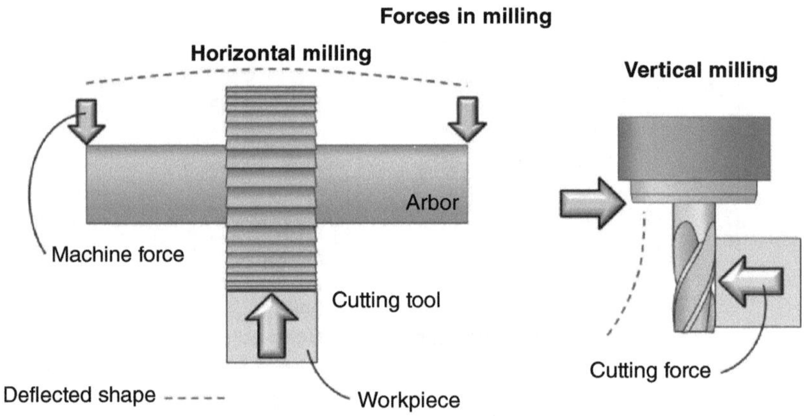

Figure 17.9 Horizontal mills support the cutting tool at two points, and deflection is resisted by a large arbor. In vertical milling the tool is not so well supported, and forces in the cutting tool are higher and resulting deflection limits the accuracy at high load. In general, cutting forces are higher in horizontal milling as the cutting tools tend have greater contact area with the workpiece.

Horizontal mills are more rigid and able to cut more aggressively than vertical mills, since they have a cutter mounted on an arbor supported at both ends to distribute the force. In a vertical mill, the cutting tool is only supported from one end, and deflection will be greater as the force is primarily carried by the cutting tool, and not a large arbor (Fig. 17.9). Because the cutter contact area tends to be greater in horizontal milling, the forces developed during cutting can be quite high, and methods such as climb and conventional milling (see Sec. 13.5.2) become very noticeable.

Many of the same features can be created on a horizontal or vertical mill. Cutting tools used in horizontal milling can be straight, angled, or contoured, and may or may not have cutting edges on the side for shoulder milling. Tooling for horizontal milling and the resulting profiles are shown schematically in Fig. 17.10. Examples of these tools are shown in Fig. 17.8.

Horizontal milling is more commonly applied for heavy material removal and has the added capability of stacking cutting tools. The process of using multiple cutting tools of varying size to create multiple features simultaneously is known as **gang milling**. The variety of tool sizes and profiles allows them to be combined in many ways to create a variety of profiles. The spacing between cutters becomes very important since it controls part dimensions, and the dimensions can be determined by the stacked cutters or by spacers between cutters, as shown in Fig. 17.11.

17.3 Mill Tooling

Milling requires tooling that cuts as it rotates. The design of the tooling determines what material it is suitable to cut, how quickly it can remove material, the type of profile it can create, and much more. In milling, the workpiece is moved into the tool, so every feature created depends on the geometry of the cutting tool. There are many (thousands!) of different cutting tools available. Fortunately, they have basic similarities, and most of the differences are manufacturer's variations on common design. We will only highlight the major aspects common to mill tooling, but be aware there are many options, and tool manufacturers probably have a tool designed for your specific application.

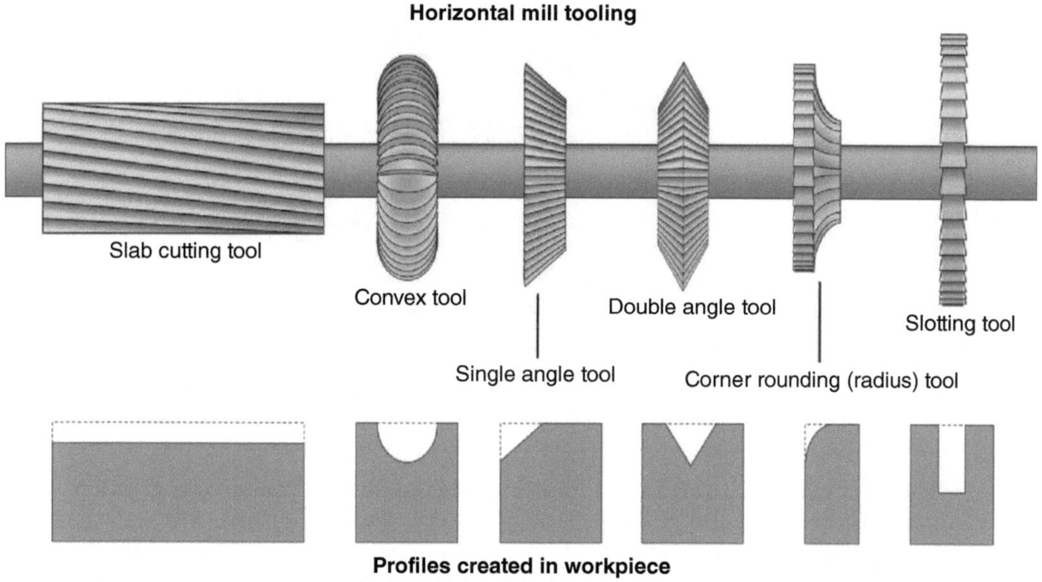

Figure 17.10 Horizontal milling cutters come in various profiles to create the corresponding geometry in the workpiece.

17.3.1 End Mills

End mills are cutting tools that are designed to cut radially and axially as needed in the process of milling. They are the most common mill cutting tool since they are general-purpose tools (Fig. 17.12). Because they can cut on the side (periphery) and bottom, they may be used to create a variety of features (refer back to Fig. 17.1). They come in many standard sizes. They are most commonly purchased in 1/16-in increments, but smaller increments and custom sizing are available. As discussed in Sec. 13.4, they come in many material types and coated varieties are also available.

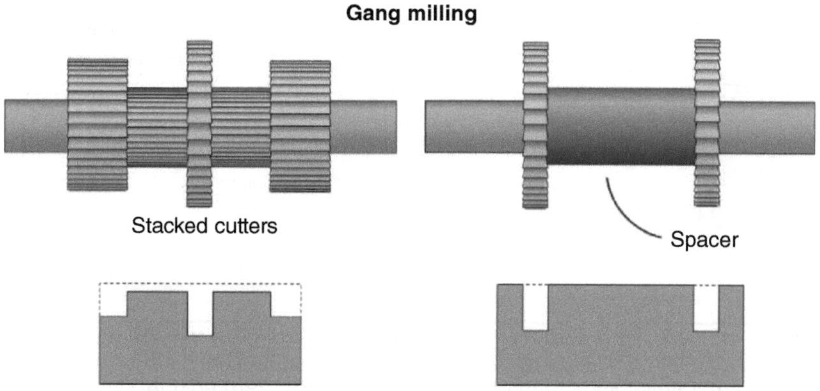

Figure 17.11 By combining multiple cutting tools, a complex profile can be accurately created in one setup. Tools may be stacked together or spacers can be used to precisely separate them.

FIGURE 17.12 End mills come in many sizes and geometries. Most are general-purpose cutting tools able to create a variety of features, but some are especially suited to particular applications.

The key design features of end mills include the flutes (number and geometry), the diameter and length, the cutting edge geometry, and the profile. We will continue discussing these in the following sections, and some features of end mills are shown in Fig. 17.13. Note that these features are for solid end mills and design will be different for inserted carbide end mills.

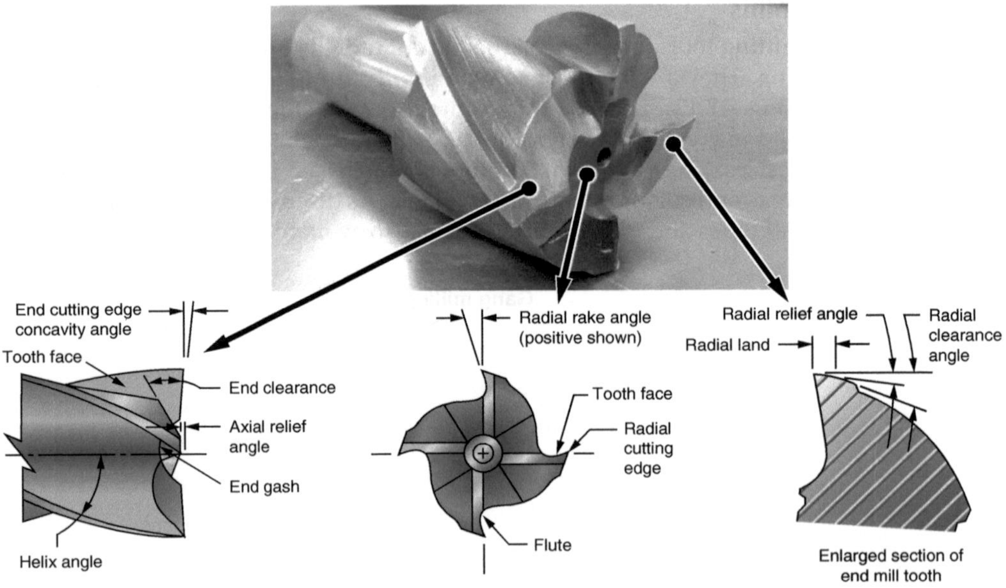

FIGURE 17.13 End mills share many common design elements even though they may vary significantly in appearance and function. The common design elements of end mills are shown here. (Bottom image courtesy of Industrial Press, Inc., from *Machinery's Handbook*, 29th edition, Oberg, Jones, Horton, Ryffel. pg. 815, 816. ISBN 978-0-8311-2901-9.)

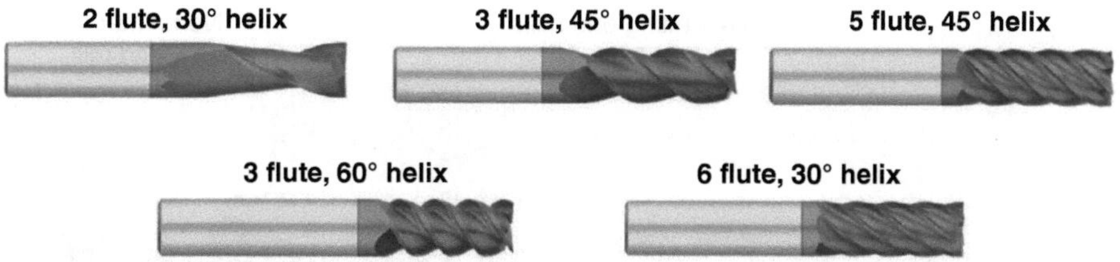

Figure 17.14 The number and helix angle of end mills determines the cutting characteristics. More flutes and a higher helix angle distribute the cutting force more efficiently, but the ability to clear chips from the flutes is reduced. (*Courtesy of Atlas Cutting Tools, LLC.*)

17.3.1.1 Number of Flutes

A major distinction in end mill design is the number of flutes, which are equal to the number of cutting edges (or *teeth*). For instance, an end mill that is ¼ in. in diameter may be referred to as a two, three, or four flute end mill. The number of flutes is important because it is the primary factor that determines the cutting characteristics (Fig. 17.14). More flutes will spread out the cutting force and tend to create a better surface finish. Also, based on Eq. (13.5), the more flutes an end mill has, the higher the feed rate can be with all other factors remaining equal.

One limit to this is that the more flutes there are on a given size of end mill, the closer they must be and the less space there will be for chips to be cleared. Imagine going through a revolving door on a hotel. If there are three spaces to enter, you may be able to fit with a suitcase quite comfortably. Now imagine there are six spaces in the same door. Each space must get smaller. With end mills, high feed rates will more easily clog the flutes of end mills with many teeth. In soft materials where cuts can be aggressive, a two flute end mill may be preferable.

The **helix angle**, which is the angle of the flute with respect to the centerline of the end mill (see Fig. 17.13), will determine how the chip is created and can reduce chip flow at high angles. An increasing number of flutes and a high helix angle are often used for light finishing passes.

Another important distinction is whether or not an end mill is center cutting (Fig. 17.15). A **center-cutting** end mill has cutting edges that traverse the entire face of the end mill and allow for the end mill to be safely plunged vertically into a workpiece. Some end mills are not center cutting, and the noncutting area of the face will create excessive heat and force if it is plunged into a workpiece. Two and three flute end mills are always center cutting, but four flute and higher end mills may not be. If in doubt, check!

17.3.1.2 End Mill Profiles

End mills come in various geometries to create specific profiles in a part or to change the cutting characteristics of a tool. **Square end mills** have a 90° edge that leaves a square edge in the part. These are the most common type of end mills. A **corner radius** (or **bull nose**) end mill has side and end cutting edges that are 90° from one another, but the corner has a small radius. Whatever that radius is, it will be transferred to the workpiece. A corner radius end mill may also be used as a first operation tool that is followed by a finishing end mill. The benefit of using a corner radius end mill is that it can cut more aggressively since it has a stronger cutting edge than a square end mill (see Sec. 13.2.1.3), and is more rigid than a ball end mill, which will be twice the radius in diameter.

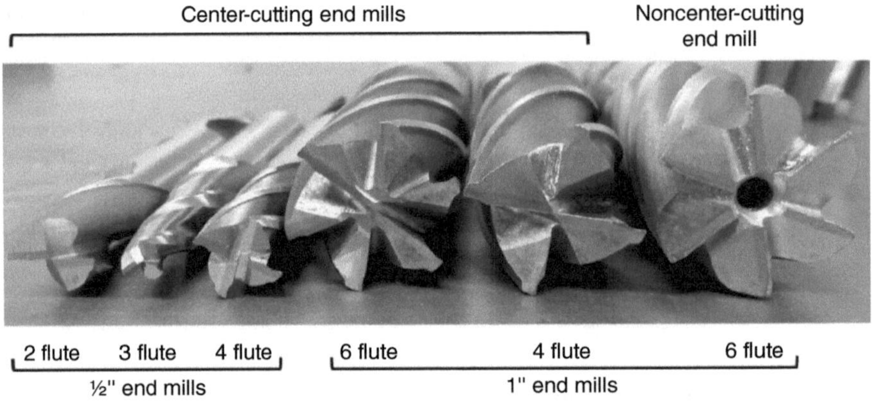

Figure 17.15 An end mill can only be lowered directly into a workpiece (plunged) if it is center cutting. Many end mills are, but some are not. Always check that the cutting edges go all the way to center of the tool before plunging.

A **ball end mill** has an end with a semicircular profile, the diameter of which is determined by the cutter width. This round end is particularly useful for creating smooth contours, which cannot be achieved with the sharp corner of a square end mill. Ball end mills are primarily used for finishing contoured features, and any heavy material removal is more efficiently achieved using a square or corner radius end mill. These end mills are shown in Fig. 17.16.

17.3.1.3 Special-Purpose End Mills

Even though general-purpose end mills can achieve a variety of tasks, there are some specialty end mills that are designed for particular work (Fig. 17.17). For instance, **roughing end** mills are very similar to regular, square end mills, but they have a scalloped cutting edge that reduces the cutting force by creating smaller chips. This makes heavy cutting easier on the machine and tool, but the surface created by these end mills is rather rough and should be followed by a finishing pass with a suitable end mill.

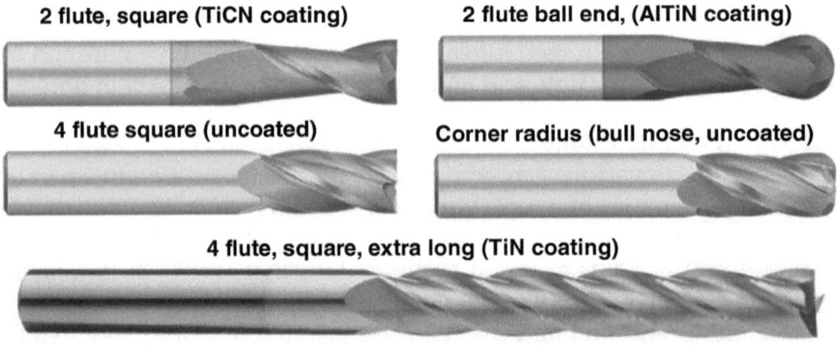

Figure 17.16 End mills come in a variety of materials, designs, and profiles for particular applications. Some common, general-purpose designs are shown here. (*Courtesy of Atlas Cutting Tools, LLC.*)

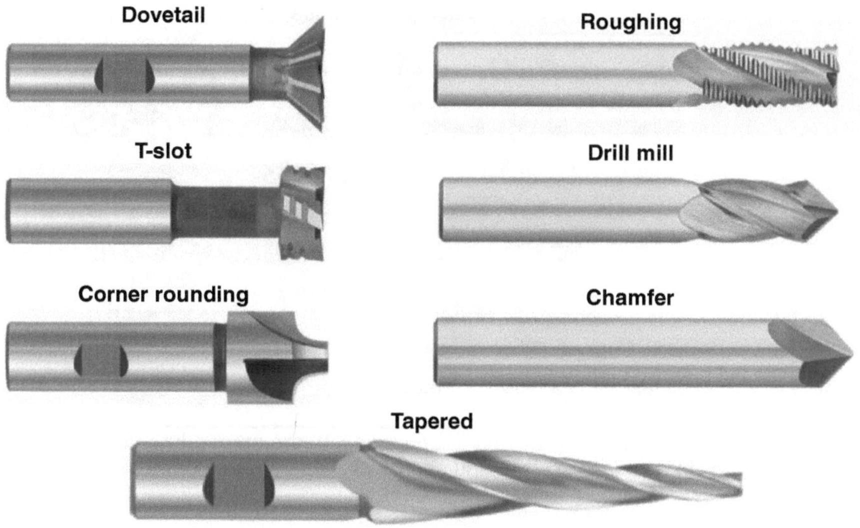

Figure 17.17 Specialty end mills serve a limited number of applications that are not possible with general-purpose end mills or not efficient. (*Courtesy of Atlas Cutting Tools, LLC.*)

At first glance, roughing end mills may appear like taps (Sec. 15.5.2), but they are designed much differently.

Other special-purpose end mills include **dovetail end mills** and **T-slot end mills**. These create an internal profile that requires a preexisting slot, which can be made by traditional milling. The size and angle (for dovetail end mills) varies for specific applications.

Another profiling end mill is a **corner rounding end mill**, which is a cutting tool designed to create a radius on the outside corner of a workpiece. This is different from the corner radius end mill, because it is a dedicated tool, unsuitable for general-purpose work. A **chamfer end mill** is similar in that it is used to create a tapered (chamfered) edge along the profile of a workpiece and sometimes to create shallow angled slots in the face of a workpiece. It, too, comes in varying sizes and angles but is only used in finished operations. A **drill mill end mill** can be used to cut on the periphery or to plunge through a part with greater efficiency than a square end mill. A drill mill takes the middle ground between an end mill and drill. Angles similar to a chamfer mill are available. For larger tapers at a shallow angle, a **tapered end mill** is used. This is useful in creating molds where draft angles are necessary.

17.3.2 Face Mills

Face mills remove material from the top of a workpiece and are designed to cover a large area at a modest cut depth. Facing operations are often used to flatten a part before other operations, so the cut depth does not need to be very large. A facing pass can also be used to finish the top of a part *after* other operations to improve surface finish and deburr holes left from milling and drilling. The common design feature of face mills is that they have a larger diameter in order to cut a wide path and reduce the total number of passes needed. The individual design can vary quite a bit, so we will now look at some common types.

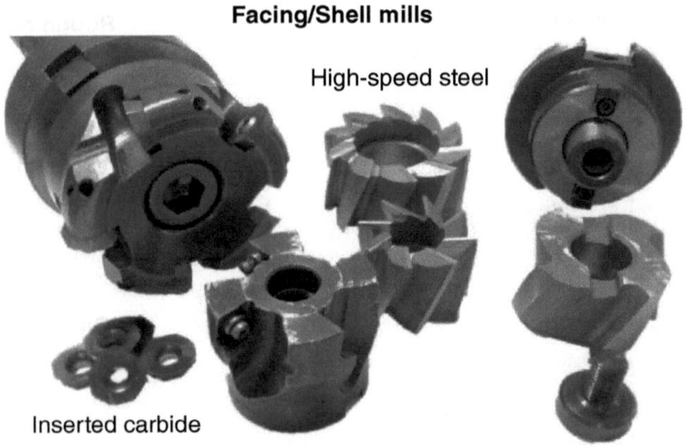

Figure 17.18 Shell mills are attached to an arbor with the appropriate dimensions for the mill. Shell mills come in a variety of sizes and can be high-speed steel or inserted carbide.

17.3.2.1 Shell Mills

Shell mills use a removable cutting head that mounts to a tool holder. Shell mills can be high-speed steel or inserted carbide, and the number of cutting edges varies, typically with more cutting edges being present on high-speed steel versions. The arbor and the shell mills may be purchased separately, and the combination determines the machine it can be used with. Cutting with these tools requires greater machine horsepower since the cut radius is large and multiple teeth may be engaged simultaneously. Common shell mills are shown in Fig. 17.18.

17.3.2.2 Fly Cutters

Fly cutters often use a single cutting point that is mounted to cover a large radius when rotated. The cutting tool can be very similar to, or exactly the same as, a single point cutting tool for the lathe. Fly cutters are very simple and inexpensive. The tool is typically high-speed steel, which can be easily re-ground, or brazed carbide (Fig. 17.19). Even though the tool covers a significant area in a single pass, the feed rate across the part should be modest since a single cutting edge is doing all of the work.

17.3.3 Slitting and Slotting Saws

Slitting saws are flat tools used to cut off features and make small grooves in a part. **Slotting saws** are used to create thin slots in a workpiece of specified width determined by the blade. Slotting saws are able to cut on the sides and are designed with tooth set and/or with side clearance so they do not bind while cutting. Both are closely related to circular saws (at least more so than they are to end mills). They follow the basic considerations discussed in Secs. 14.2.1 and 14.2.2.3, but they are uniquely applied in milling situations, most often with metallic workpieces. The cut rate will be much lower than in other circular saw processes, but they are the most efficient means to cut thin, deep slots. Since they must rotate with the vertical spindle, they are used to cut into the side of a part unless used on a horizontal mill. Slitting saws are shown in Fig. 17.20.

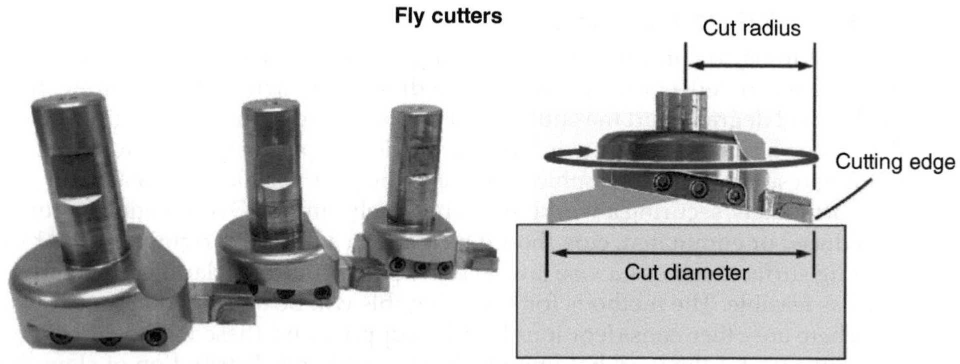

FIGURE 17.19 Fly cutters come in a variety of sizes and designs. They are primarily used to face parts, and the maximum cut width (or diameter) is determined by the size of cutter and how far the cutting tool extends out from the tool holder.

17.4 Workholding

Although there are many important decisions about cutting tools, milling processes are equally dependent on proper workholding. Workholding in milling operations must be rigid enough to withstand the forces generated by the milling processes and allow access to the surfaces that are to be milled. The workpiece may start with any geometry (flat, round, contoured, etc.) and be of any size within the machine's capability. This leaves a great number of possibilities, and we will consider some common situations.

FIGURE 17.20 Slitting (or *slotting*) saws are used to create thin recesses in parts that would be difficult to create otherwise. Their tooth design follows similar considerations to circular saws (see Sec. 14.2.1). The three uppermost tools are slotting saws, as evidenced by the edge relief, and three blades at the bottom of the image are slitting saws as they are flat on the sides.

17.4.1 Workpiece Support

The first thing to consider in workholding is how the workpiece is supported. If the workpiece can deflect during cutting, the dimensions will not be accurate, the surface quality will degrade and the cutting tool will wear more quickly. Quality and efficiency depend on a rigid part. In a worst-case scenario, the part may be ejected from the workholding resulting in catastrophic damage to the part, tool, and/or machine.

Fortunately, securing a workpiece is relatively simple. Overhanging features should be reduced or eliminated, cuts should be made in a direction rigidly backed by a workholding surface, such as a vise jaw, and the part should be clamped in as many directions as feasible. The methods for achieving this will be discussed in more detail later.

There are other considerations besides support, and these include access (you cannot mill a part if the workholding blocks the part) and distribution of clamping force. Improper and redundant support has the ability to warp the part, which will result in an uneven geometry after the part is unclamped.

17.4.2 Workpiece Positioning

Workpieces should be rigidly held, but they must also be accurately positioned. For instance, if supporting an overhanging workpiece using a device such as a **machinist's jack**, care must be taken not to apply too much force from underneath, or the workpiece may be lifted up. Also, if supports are not all of equal height, the part will be angled vertically. For this reason, machinist's parallels, or simply parallels, are often used. These concepts are shown in Fig. 17.21.

Parallels are precision ground spacers that have dimensions precisely matched so that a part can be elevated and supported by them while remaining parallel to the table. Parallels often come in sets with numerous pairs of varying heights. These are often used to elevate a workpiece above the vise's jaws, but they can be used in other circumstances as well. For instance, when drilling the top surface does not need to be higher than the vise jaws, face milling requires it.

Likewise, positioning a part parallel to the travel of the table is important. Although a workpiece can be intentionally tilted to create an angled feature, it is more common

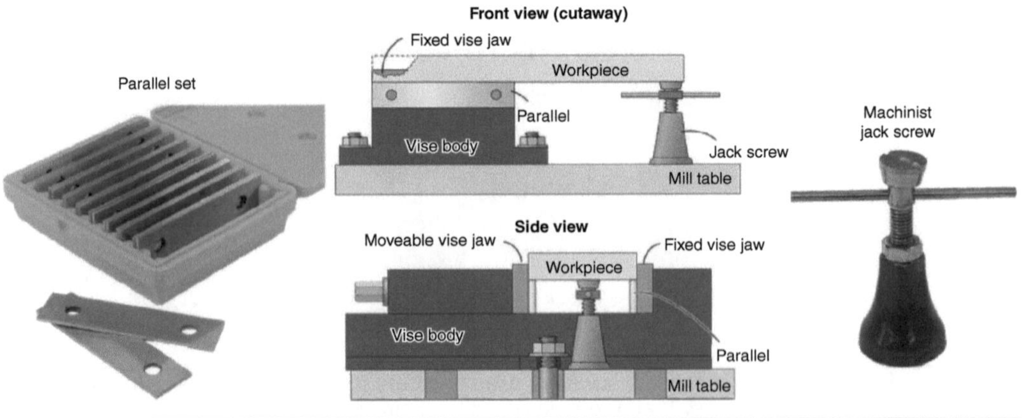

FIGURE 17.21 Workpiece positioning and support can be achieved through various accessories. Parallels are used to elevate a workpiece and jack screws are used to support overhanging sections.

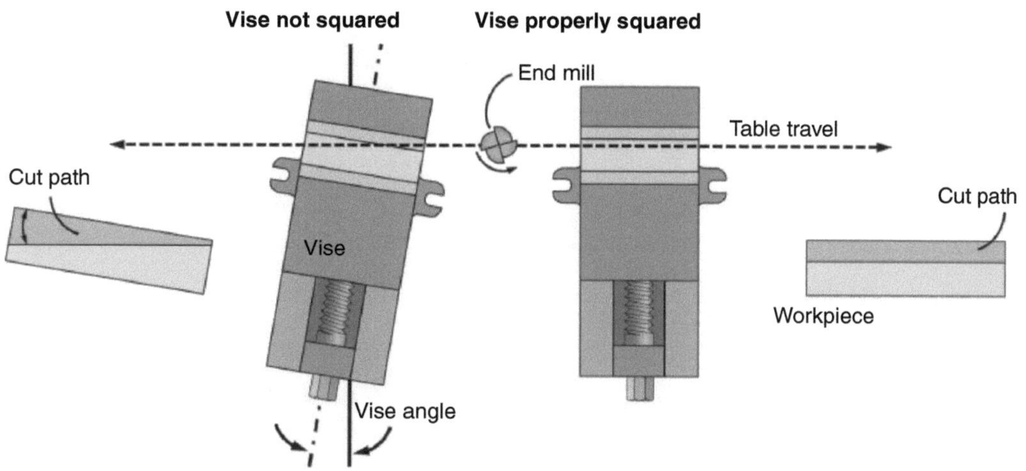

FIGURE 17.22 Proper alignment of workpieces often depends on the proper alignment of workholding. In this case, the consequences of a misaligned vise are shown. Though the angle is exaggerated, even small angles of tilt can result in a part being out of tolerance.

that a square corner be created and that is accomplished by carefully aligning the part. Although this can be done using a reference surface on the part, it is common to align the workholding using a surface that is known to precisely flat.

The most common application where this alignment process is practiced is in the setup of vises (Fig. 17.22). Although the design and use of vises is left for Sec. 17.4.3, we can see how their setup will affect part geometry. A vise should have a very flat jaw and have a bottom surface parallel to the table. The table is the ultimate reference point, so the accuracy of the vise is critical.

To ensure the vise is properly aligned with table travel, a dial indicator, mounted in the mill spindle, is run across the fixed jaw of the vise, which is more accurate than the moveable jaw. If the indicator moves, the jaw is not parallel ("squared"). The vise is adjusted until the dial indicator reads a constant number across the entire width of the jaw.*

17.4.3 Vises

Vises are workholding tools that clamp a workpiece between surfaces, known as the **jaws**. The jaws can be flat or shaped, and they are most commonly present as a pair, one of which does not move (the fixed jaw) and the other which does move (the moveable jaw). This is much like your jaw. Your top teeth do not move and you bottom teeth must do all of the movement.

The size of a vise (capacity) is determined by the maximum space between the two jaws, but the width is also an important variable. Most milling vises are 6 in vises,

*The process for squaring the vise can be quick, but if not done right, it may take considerable time. The most beneficial practice is to snug the vise to the table and continuously tap it with a hammer to straighten it out as you move the table. This gives you instant feedback so you can quickly stop the indicator movement.

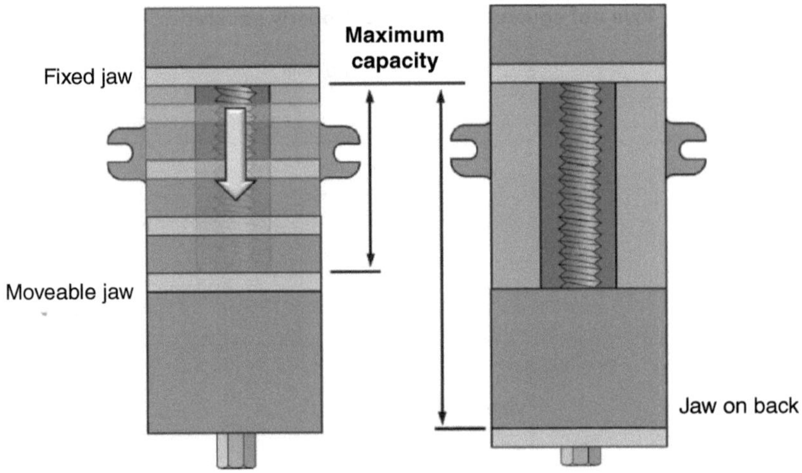

FIGURE 17.23 The capacity for a vise is dependent on the maximum space between holding surfaces. The jaw spacing can be increased on some vises by mounting a jaw on the rear surface, but only a small portion of the jaw is able to hold a part in this configuration.

meaning their jaws can be spaced by a maximum of 6 in, and they are also 6 in wide. Some vises can have the jaw mounted on the back of the moveable side to add more space, but this greatly reduces the depth at which a part can be clamped. In line with the workholding considerations, the greatest amount of surface should be clamped to enhance part rigidity (Fig. 17.23).

A vise mounted at an angle to the table travel is a bad thing if you want a square feature, but there are times when a specific angle is needed, and the ability to set the vise at a precise angle is beneficial. This can be accomplished by mounting the vise on a swivel base, and many vises can be purchased with this base as an accessory, or it can be purchased later on. Both types of vises mount to the mill table in a similar fashion, using bolts and the T-slots, which are simply T-shaped slots in which retaining nuts can be placed. These slots are important for other workholding as well (Fig. 17.24).

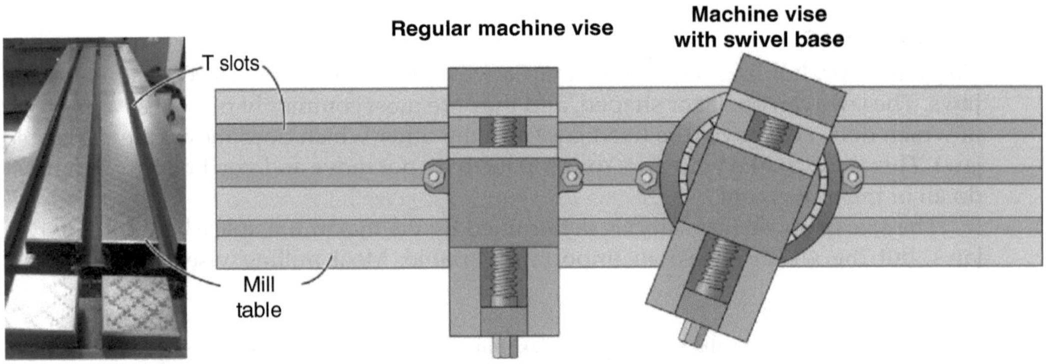

FIGURE 17.24 A vise can be precisely tilted to a desired angle using a swivel base. The concept is shown here. The different vises are shown in Fig. 17.25.

Vises are general-purpose tools, and because of that, they appear in many forms. The design will vary depending on the intended application, and even within specific applications, the design can vary. Some vises are high precision while others are heavy duty without great precision. Some are hydraulically operated and some are screw operated. Some swivel and tilt in multiple axes, but many are only adjustable in regard to the clamping surfaces.

For instance, a drill press vise does not need to be as rigid as a machine vise since it is primarily supporting load during holemaking operations (see Chap. 15). Sine vises are precision, tilting vises that are used to make accurate angles. Some compound vises can tilt in multiple axes to create compound angles. Some vises have flat sides and can be turned 90° to hold workpieces in that configuration. A *small* sampling of the types of vises you may find in a typical manufacturing environment are shown in Fig. 17.25.

17.4.4 Clamp Sets

Clamp set are an alternative to dedicated workholding devices, and the most common type is the step clamp set. **Step clamp sets** are general-purpose workholding tools that have a variety of bolts, nuts, and blocks that can be adjusted to suit the need of the workpiece. These sets typically contain a range of sizes to accommodate a ever changing situations. One of these clamp sets and the way it is used to hold a workpiece are shown in Fig. 17.26.

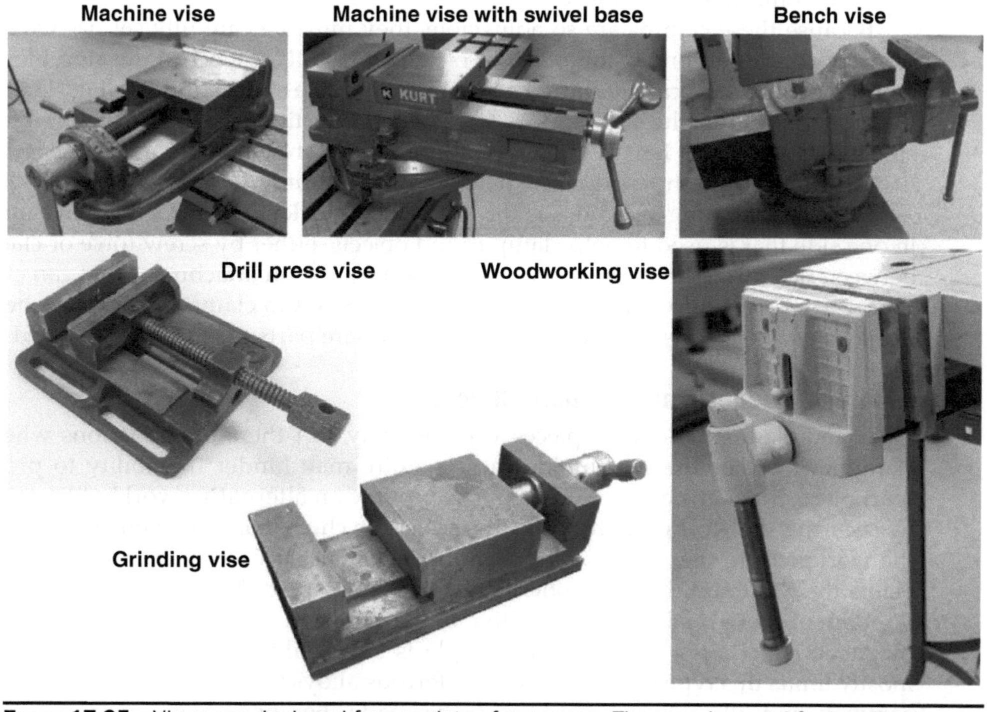

FIGURE 17.25 Vises are designed for a variety of purposes. They can be used for handwork, grinding, milling, drilling, etc. Machine vises are the most common in milling, but others are shown here as well.

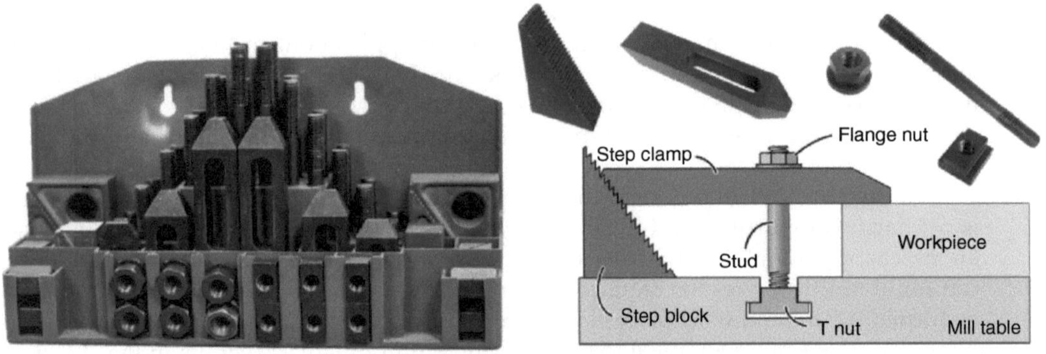

FIGURE 17.26 Step clamps allow adjustability in how workpieces are fastened to the mill table. They are often sold in sets with parts that may be combined in a variety of ways to accommodate many workpiece geometries.

They are called step clamps because the primary positioning features resemble steps and the height is adjusted by interlocking the step clamp with the step block. This must be done carefully to ensure the clamp is parallel to the mill table. An angle between the two will result in less efficient clamping and, in the worst case, an unstable setup that may come apart under force.

Because these clamps are so adjustable, they must be carefully set up. One of the primary considerations is the position of the stud with respect to the step block and workpiece. The stud can be shifted along the slot in the step clamp (see Fig. 17.26), and the amount of force the stud applies to each is proportional to its distance. Whichever side it is closer to will have a higher force applied. As shown in Fig. 17.27, the larger the distance, L_2, is from the workpiece the lower the clamping force becomes.

Another clamping accessory is the V-block. **V-blocks** are blocks that have an angle in one side that is used to help clamp round objects, either by screw force or clamping between flat surfaces. For instance, a V-block may use an attachment that can clamp a round workpiece into it, or two V-blocks may be used to clamp a workpiece between the jaws of a vise. This is shown in Fig. 17.28. Square parts may also be clamped at 45°.

17.4.5 Magnetic and Vacuum Chucks

Vises and clamps hold workpieces very securely, but there are situations where the clamping force or the positioning these require may hinder the ability to process a workpiece. Magnetic and vacuum chucks provide an alternative workholding method by securing the workpiece from below. Magnetic chucks use magnetic force to attract the workpiece to the surface, and vacuum chucks use a vacuum pump to remove air below the work and the atmospheric pressure on the piece holds it in place.

Both of these methods are sensitive to the type and geometry of the workpiece. A magnetic chuck only works if the sample is susceptible to the magnetic field. This mostly limits this type of workholding to ferrous alloys (see Sec. 4.2), but they are very common and allow many products to be made. For nonferrous metals and alloys (Sec. 4.3), or nonmagnetic ferrous alloys like austenitic stainless steel, the work can be held in a magnetic vise that is held to the magnetic chuck, but this largely negates the

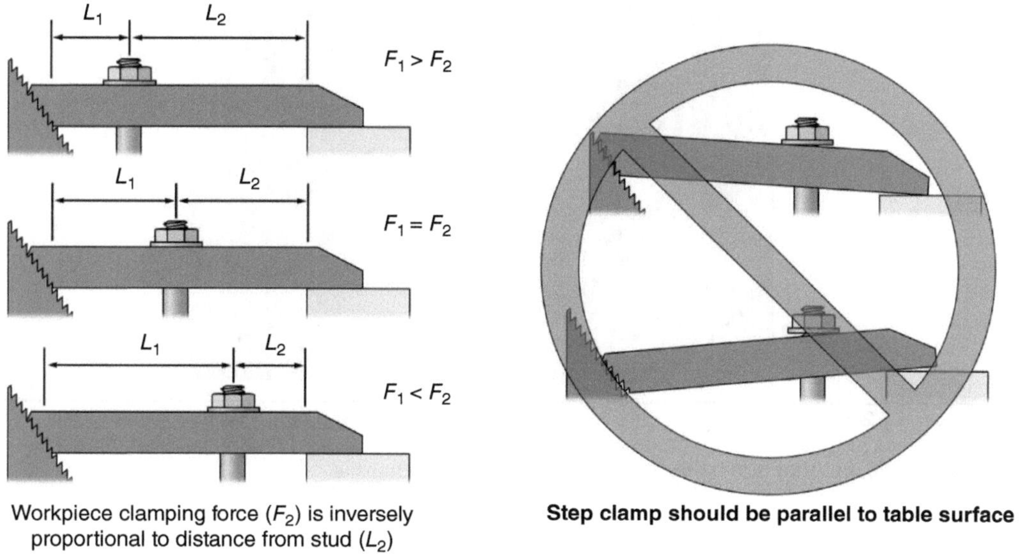

FIGURE 17.27 The way step clamp sets are used is variable and important. The closer the stud is to the workpiece, the greater the force that it will apply there. Additionally, the step clamp should be parallel to the table surface to ensure solid holding.

purpose of the chuck. The magnetic pull is activated by turning a handle or other device. A magnetic chuck is shown in Fig. 17.29.

Vacuum chucks use a rubber gasket that is run though grooves in the chuck surface. This gasket is placed so that it is within the workpiece perimeter. The maximum area should be allowed as holding force is directly proportional to the area the vacuum is applied to. If the gasket does not seal well, either from workpiece variations or a damaged gasket, holding power will be lost. Also, the surface of the workpiece cannot be cut through or vacuum will be lost. This limits the feature type possible with this workholding.

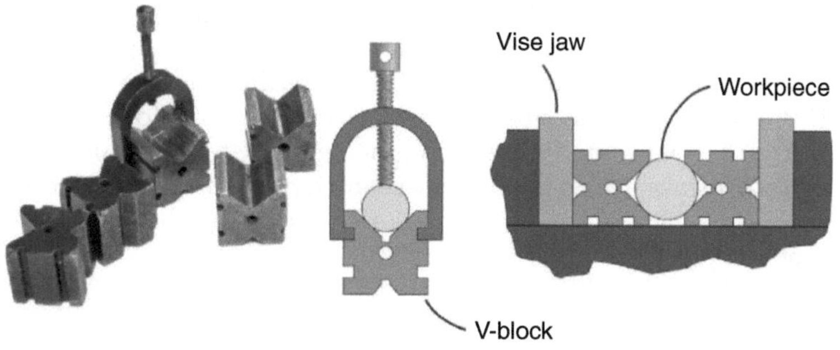

FIGURE 17.28 V-blocks are often used to secure round objects, and some common ways they are applied are shown here.

FIGURE 17.29 Magnetic chucks are alternative workholding methods that allow for different workpiece geometries to be processed. These are very common in grinding applications (see Sec. 14.6), but they can be used in milling applications as well.

Magnetic chucks are often used in surface grinding (Sec. 14.6.2.1), but they can also be used in milling applications. Vacuum chucks can be used in light-duty milling operations, but they are often limited to polymer or wood processing, which generates lower forces. Both methods work well for thin sheet goods since force is evenly spread out and there is a large holding area. These thin parts are also likely to buckle if held tightly in a vise.

17.4.6 Rotary Tables and Indexing (Dividing) Heads

Although the mill is primarily used to create straight features, round profiles can be created as well. To do this, a rotary table is used. A **rotary table** is a device to which a workpiece is mounted and that can be bolted to the mill table and independently rotated. It can be rotated to create circles or round profiles, or it can be rotated to a desired angle between cuts to create straight angles, very much like a swivel base for a vise (see Fig. 17.24). Bolt hole circles (and any other feature) can also be placed at intermittent angles this way, though mills equipped with digital readout can be precisely and quickly positioned using the hole coordinates as well.

To more accurately position a workpiece rotationally, an indexing head is used. An **indexing head**, or **dividing head**, is a rotary workholding devise that can be precisely positioned by means of a plate with holes of exact distance.* The rotation rate is geared to a handle, and a specified number of turns will achieve a known angle of rotation (Fig. 17.30). This is used to create features of exact angle to one another (e.g., two holes, 45° apart) and is particularly useful in spacing many features such as a bolt hole pattern or gear teeth, especially on the outside diameter of round workpieces. The ability to tilt the head from a vertical to a horizontal orientation also allows holes and other features to be made at precise angles to the surface.

*The process of determining the number of rotations is based on the gear ratio of input and output shafts as well as the number of holes in the dividing plate. You can find more about this in metalworking handbooks.

Figure 17.30 Circular profiles and specific angular orientations can be created using rotary tables and indexing heads.

17.5 Milling Processes

Milling covers a diverse range of processes, and many of them have been discussed in the context of other topics so far in this chapter. Also, general holemaking processes can be performed on the mill, but those are described in Chap. 15. What we will detail here are some finer points about milling operations and specific tools to help complete the job.

17.5.1 Basic Milling

The most basic techniques for material removal during milling include face, peripheral, and shoulder milling. These can all be accomplished with an end mill as shown in Fig. 17.1. Additionally, slots and holes can be created by varying the depth, position, and movement of the end mill. One tool does a great deal of work, but, as discussed in Sec. 17.3, there are important considerations concerning that tool.

One of the most important decisions during milling is the use of climb or conventional milling, as described in Sec. 13.5.2 and shown in Fig. 13.16. The proper determination of speeds and feeds (Sec. 13.5.1) is also critical. Assuming you are familiar with those settings, then the major choices are the cut depth and step over (which were also discussed in Sec. 13.5.3).

As a general rule, the deeper the cut is, the shallower the step over should be. This reduces the overall force, which depends on both. It is often best practice to use a smaller cut depth and large step over as this reduces the amount of vibration (chatter) for the same feed rate. Often, a much higher feed rate can be used with small cut depths, and this can be faster than taking one or two deep cuts.

The way a tool is used depends on the operation and the machine. In CNC machining (Sec. 22.3), it is common to use high feeds and smaller cut depths, but this may not be as practical for short operations on a manual mill. You have to balance speed with accuracy and fatigue (moving that table back and forth can be tiring!).

Especially with small diameter end mills, deep cuts (greater than the diameter) can cause unwanted tool deflection. That can be reduced by slowing down the feed rate. If the feed rate is slowed, then the speed should also be reduced to prevent burning the tool up. If you change one parameter, do not forget to adjust others.

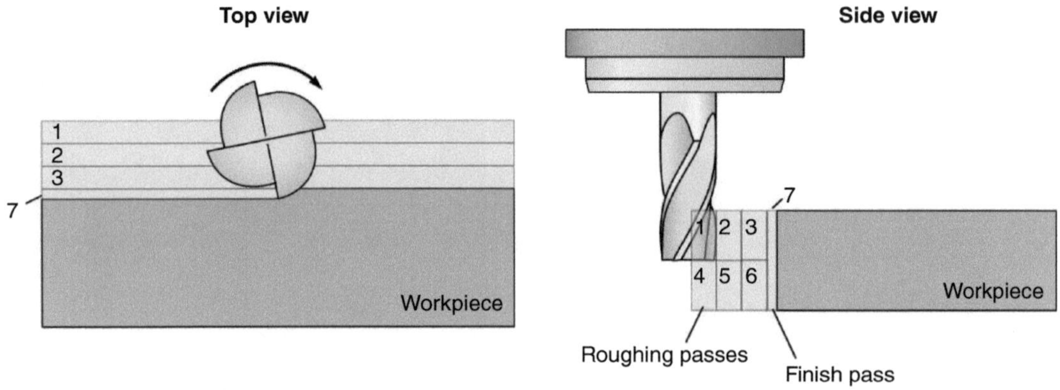

Figure 17.31 Machining is done through a series of passes. The cut quality generally decreases as the amount of material removed increases. A light finish pass is used to leave a good surface and accurate dimension.

When it comes to making slots, it is a good technique to use an end mill that is smaller than the desired feature size. Cutting a ¼-in slot with a ¼-in end mill is risky since it is likely that the slot will be slightly oversize. Even if it is within tolerance, the surface finish may not be acceptable. Cutting the slot in two (or more) passes will allow you to dial in the exact dimensions, and to do so with good quality.

In general, taking a finish pass at full depth is a good idea. It removes machining marks from previous passes and leaves a uniform finish (Fig. 17.31). At least 0.005 to 0.010 in should be left for a finish pass to give the end mill something to remove, and high feed may be necessary due to chip thinning (see Fig. 13.18 and associated discussion). Also, even if the machining beforehand was done with conventional milling, the finish pass should use climb milling for the best result.

17.5.2 Pockets

A special type of feature in milling is known as a pocket. A pocket is a recessed area that is enclosed along its perimeter. Because a pocket is recessed and enclosed, the end mill must be lowered into the recess. This can be done by several methods, and the most common are plunge, ramp, and helix milling.

Plunge milling is the vertical movement of the end mill into the part until the desired depth is reached. Ramp milling is a combined movement vertically and horizontally that is terminated when the end mill reaches the desired depth. The angle of ramp can be determined by the user. Helix (or helical) milling is when the end mill is moved in a spiral progression down into the part. This is similar in shape to a spring. All three methods are shown in Fig. 17.32.

End mills, even center-cutting end mills, are not really efficient at plunging vertically. The ramp and helix milling methods help balance the vertical plunge with a longitudinal movement. Non-center-cutting end mills can be lowered into a workpiece as along as the angle is fairly shallow. The maximum ramp angle is determined by the design and should be specified by the manufacturer.

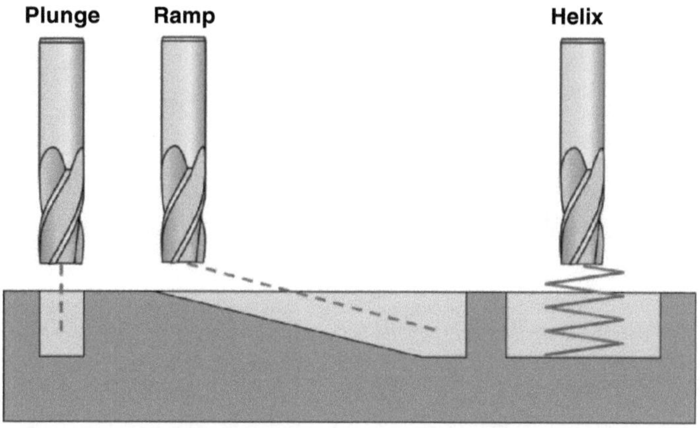

Figure 17.32 The way an end mill enters into a workpiece can be varied. Most manual operations will use a plunge entrance, but computer-controlled milling often uses ramp and helix methods.

Obviously, moving the table in a helix is something that would require three arms and a great deal of skill, neither of which are practical expectations. Even ramping is not common manually, and these more advanced entry methods are only common in computer-controlled equipment (see Sec. 22.3.1).

The good news is that once the end mill reaches the desired depth, milling can proceed as normal. The way the milling progresses will determine the cut characteristics, and when done by hand, fairly simple, straight line movements are the most likely method. Modern computer software can make very complicated patterns possible which maximize cutting efficiency and tool life. Some basic patterns are shown in Fig. 17.33. Note that a raster pattern requires a combination of climb and conventional milling, but a spiral pattern can be accomplished with climb milling only.

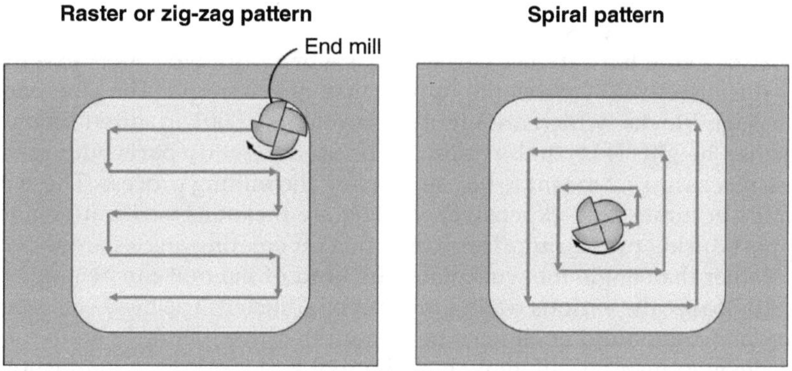

Figure 17.33 Pocket milling requires removing an area larger than the end mill. How the end mill is moved will determine cutting characteristics.

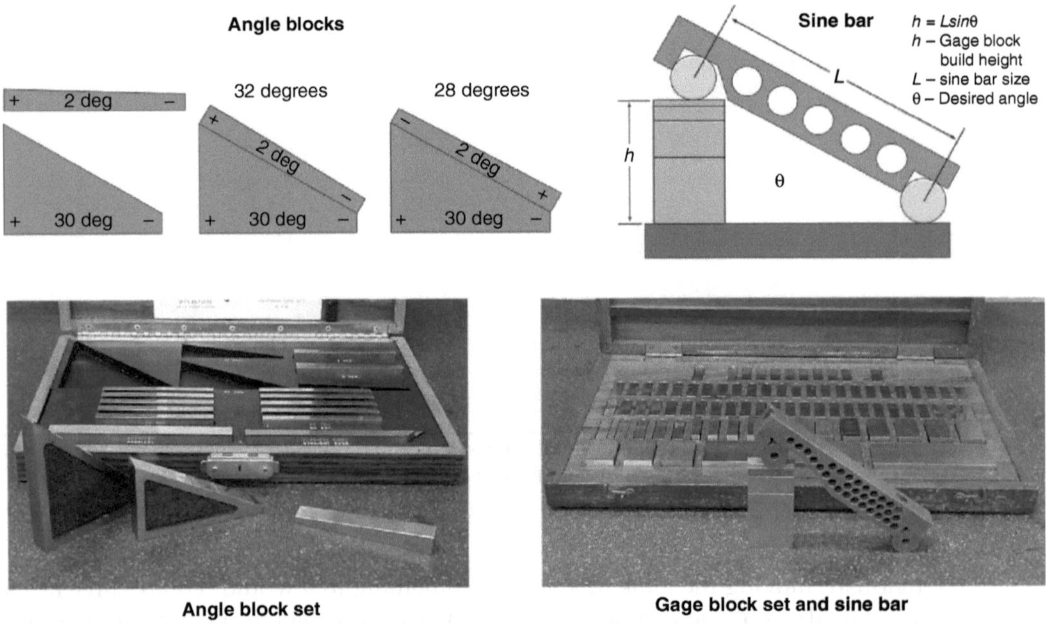

Figure 17.34 Precision angles can be set and checked using angle blocks or a sine bar with gage blocks. These are precision measuring instruments and must be handled with care to maintain accuracy.

17.5.3 Milling Angles

Angles can be created in a variety of ways using a mill. By tilting the vise on a swivel base (see Fig. 17.24), angles can be made in the X-Y plane, and likewise a part can be positioned and clamped at an angle. Very small-angled surfaces can be created using a chamfer mill (Fig. 17.17). To create varying angles over greater distances, other methods are required.

One method to create angled features is to angle the workpiece. This may be accomplished by using special vises that can rotate in multiple axes. In this way, any desired angle or angles can be accurately created. Another method is to use **angle blocks**, which are precisely angled spacers that can be used alone or in combination to achieve specific angles.

Also, a **sine bar** can be used, which is a bar with two rollers exactly spaced apart, and this effectively creates the hypotenuse of a triangle. The sine bar can be paired with **gage blocks**, which are highly accurate (± 0.0001 in, often much less) blocks of specified height. This combination can be used to set up parts with great accuracy, but these blocks are *not* meant to be used during the milling process. These tools are exceptionally accurate (and expensive), so they are meant as verification and setup, not as support during processing. These methods for creating angles are shown in Fig. 17.34.

Rather than angle the workpiece, the head of the mill can be angled. Refer back to Fig. 17.7 to see the various ways a mill can be adjusted. The head can rotate in two directions, and compound angles can be created this way. Just keep in mind that the forces that develop may be different since the tool and workpiece interactions will change. Moving the table up now also changes the x- and/or y-axis relationship to your part geometry when holding tolerances in multiple directions. An example of tilting a mill head is shown in Fig. 17.35.

FIGURE 17.35 One way to create an angled surface is to tilt the mill head as shown here.

17.6 Why It Matters

We have determined that milling covers a range of equipment and processes, even beyond simple categories such as horizontal and vertical milling. For instance, even though it is outside the scope of our present discussion, common woodworking processes such as planing, jointing, and routing are technically milling processes. Milling machines are general tools, not dedicated to a specific purpose, so there are many products that can be made using them.

Milling machines excel at creating square products, but through a combination of attachments and accessories, they can also be used to create round and angle features. The adjustability of milling equipment allows for a wide variety of features to be created with a single tool, and the accuracy of the machine allows for high-precision work.

Tooling comes in a variety of designs. Some of the design features are related to material removal, but some design choices also influence the features that can be created. For instance, a ball end mill is the only suitable choice for creating a round groove, but either a corner radius end mill or a ball end mill may be used to create a corner radius. The flute number, helix angle, and tool material all influence the cut characteristics, and understanding the basic effects of these choices can save a lot of trouble and money with an inferior setup.

Part of the setup that determines quality is the workholding. Although machine vises are the most common, the wide range of options allows for workpieces of varying sizes and geometries to be secured to the mill table. The orientation of the workpiece can be varied, and this is for better or for worse. Where angles are not desired, care must be taken to ensure that the workholding aligns the workpiece to the mill table travel. Otherwise, the angle desired can also be verified through precision setup fixtures and measuring equipment.

Through a few combinations of tooling and technique, infinite possibilities for part geometry exist (Fig. 17.36). While certain cutting tools are best suited for particular jobs, even a single tool can be applied in several ways. Some of these processes include facing, plunging, pocketing, counterboring, peripheral milling, etc. Because there are many options, the operator (you) must be judicious in correctly choosing and executing the setup and production of parts during milling. You must know the basics before that is possible.

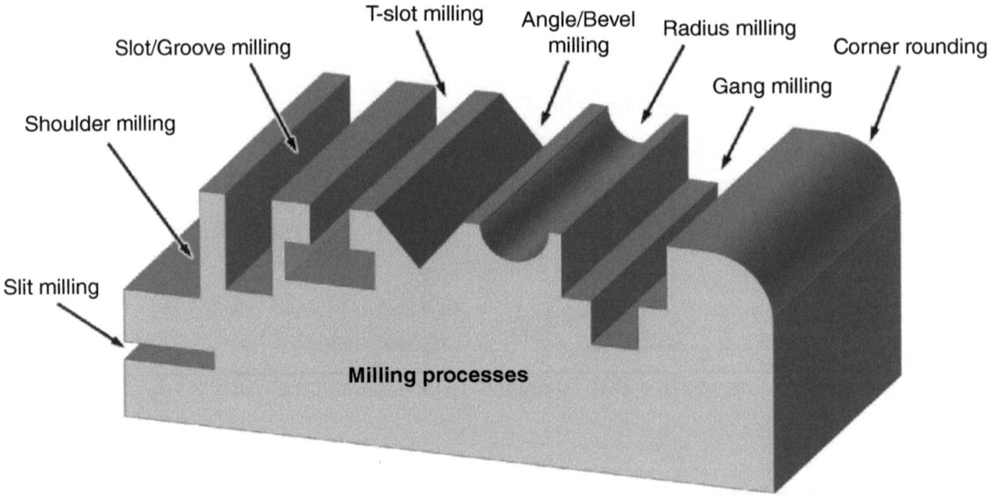

FIGURE 17.36 Milling is capable of creating many features, some even in several different ways. This diversity allows for many products to be created on the mill with relatively little tooling. A small sampling of the features created using milling is given here.

17.7 The How and Why of Milling

17.7.1 Chapter Review

1. Why is it important to define milling when discussing manufacturing?
2. How do face and peripheral milling differ?
3. Why is plunge milling not common for holemaking?
4. Draw and label the six coordinate axes associated with milling.
5. What is the primary difference between vertical and horizontal mills? What are some ways in which they are similar?
6. What is the benefit of a bed-type mill over a knee-type mill?
7. What is the quill, and what are the functions of the controls related to it?
8. When should the quill be lowered? When is it better not to lower it?
9. What is meant by tramming the head of mill? Why is it important?
10. How do horizontal mill cutting tools differ from those used in vertical milling?
11. Why are horizontal mills more rigid than vertical mills?
12. Describe gang milling. What tool factors determine the part profile geometry created using gang milling?
13. What is an end mill? How can it be used to remove material?
14. Describe the importance of the number of flutes when choosing an end mill.
15. How does the helix angle affect cutting characteristics?

16. How can you determine if an end mill is center cutting? Why is this important?
17. Describe the difference between square, corner radius, and ball end mills.
18. How does a corner rounding end mill differ from a corner radius end mill?
19. What step must precede the use of T-slot and dovetail end mills?
20. How does a roughing end mill make heavy cutting more efficient?
21. What are face mills used for? Why are they well-suited for the task?
22. What is an advantage with using a fly cutter over a face mill? A disadvantage?
23. What is the difference between slitting saws and slotting saws?
24. Why is workholding rigidity so important to part quality?
25. What is the purpose of parallels?
26. If applying considerable cutting force, is it better to do so toward the fixed jaw or moveable jaw of the vise?
27. Why must the vise be checked as square before cutting?
28. How can the capacity of some vises be extended beyond their typical capacity? How does this affect the way the workpiece is held?
29. What is the purpose of a swivel base for a vise?
30. How does a machine vise differ from a drill press vise?
31. Describe how step clamp sets allow a variety of workpieces to be secured to the mill table.
32. Why is the stud placed closer to the workpiece than the step block?
33. What are some uses of V-blocks?
34. Name two advantages and disadvantages of magnetic and vacuum chucks.
35. What are rotary tables and indexing heads used for? How do indexing heads provide greater precision?
36. Why is a lighter cut taken as the final pass in milling operations?
37. What are the benefits of ramp and helix entry into a pocket? Why are these rarely used when manual milling?
38. What is the use for angle blocks? How are they simpler than a sine bar and gage blocks?
39. When would the mill head be tilted instead of the workpiece?

17.7.2 Concept Connections and Extensions

1. What are the functional differences between vertical and horizontal mills? Which one do you consider more versatile and why?
2. Why are there so many different types of end mills and other mill cutting tools? Describe some of the features.
3. How do the number and characteristics of flutes control the cutting action of an end mill?

4. What are some workholding methods and things that can go wrong with the setup?
5. Develop an analogy to explain climb milling and contrast it to conventional milling. Use common examples that an average person will understand. Refer to Fig. 13.16 and associated discussion.
6. How do clamp sets vary in effectiveness with the way they are set up? Relate this to the concepts covered in Chap. 7.
7. How are some of the techniques discussed in Sec. 17.5 practical only with CNC machining (Sec. 22.3.1)?

CHAPTER 18

Welding and Cohesive Assembly

It is often more efficient to create products from assemblies of parts, rather than a single-piece structure. Even though some production methods can create complex shapes, such as casting and machining, these methods have other limitations that make them impractical. For example, consider an automobile chassis. It is a large assembly of many components. These components are combined through permanent connections that result in a structure similar to being one piece. Because of the size and the amount of open space, there is no better method to create this product than as a permanent assembly.

18.1 How It Works

Cohesive assembly results in a permanent bond between two components and involves atomic or molecular mixing of the materials. The most common ways to achieve this bond include melting or chemically dissolving the two components at their interface and then allowing them to solidify as a single unit.

Most thermal cohesive bonding processes are simply referred to as welding. **Welding** is a category of processes that cohesively bond materials through the application of intense heat and/or pressure, which results in the localized fusing of the two materials. This general process has been around for more than a century,* and it now has many embodiments, and new techniques are still being developed. Cohesion can also be achieved through extremely high force and friction, which mixes the materials in a still solid state, but this still represents a limited portion of the manufacturing industry (these "cold methods" will be discussed more in Sec. 18.3 for metals and Sec. 18.4.3 for polymers).

The primary distinction in cohesive assembly is the microstructural unity between the two components after they are assembled. In metals this results in the atomic mixing of the two materials. In polymers it results in molecular mixing of the polymer chains. Where heat is used, the melted area is known as the **weld pool** (or **weld puddle**), and once this area solidifies, the bond cannot be disassembled without breaking, cutting, or remelting the material.

*Welding by thermal methods, which is the basis for modern methods, was first developed in the late 19th century.

Material compatibility in cohesive bonding

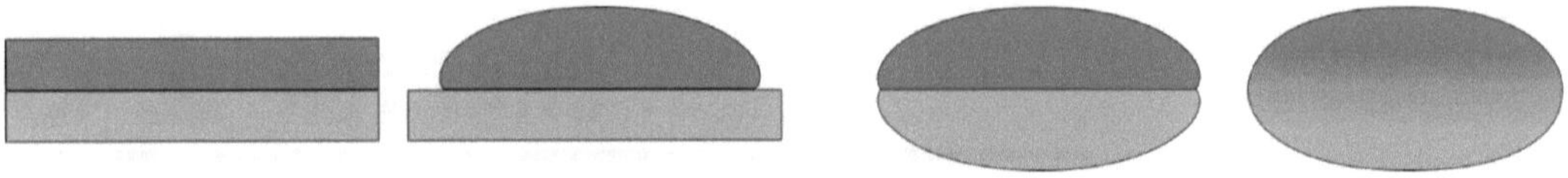

Figure 18.1 The compatibility of materials determines the ability to cohesively bond them by heat or chemical dissolution. Various incompatibilities are shown here. Note that if two materials are both dissolved by a single chemical, they are likely chemically compatible with each other.

18.1.1 Material Choice

Because cohesive bonding results in the mixing of two materials, the materials are very important. Oil and water do not mix well, and they will quickly separate naturally. This can happen with cohesive bonding as well. When the two materials are combined in liquid or molten form, they can easily mix, but they can also easily separate (Fig. 18.1). The outcomes are determined by their chemical compatibility.

The general concept behind this was discussed in Secs. 3.2.4.2 and 3.2.4.3 for metals in regard to alloying and phase diagrams. Metals are primarily joined by localized melting, as we will see in welding methods (Sec. 18.2). The creation of a cohesive bond in metals is related to the ability of the two materials to mix readily at the *same temperature*. For that reason, the two parts should have a compatible chemistry and a similar melting temperature.

Because we are bound by the laws of chemistry and physics, we cannot create a strong bond between steel and aluminum because their melting points are too different. Even though the melting points of steel and copper are closer together, we cannot create a good bond because they will not mix well. There are some possibilities for welding dissimilar metals in cold methods, and those will be discussed in Sec. 18.3. The vast majority of cohesive bonding in metallic materials combines parts of the same or very similar materials.

Polymers have similar considerations. For instance, if the melting temperatures are too far apart, one material will be very soft while the other is not yet molten. A cohesive bond will not form, though an *adhesive* bond might (see Chap. 19). A unique consideration for polymers is that thermosets will not melt at all (see Sec. 5.3.2.5), so they cannot be combined with thermoplastics by simply heating the two.

Both thermosets and thermoplastic may be dissolved chemically. The same problem arises that if two polymers are of incompatible chemistry, they will not mix. Also, as discussed more in Sec. 18.4.4, different polymers may not both dissolve using the same solvent.

18.1.2 Structural Effects of Welding

A welded area takes two materials with independent microstructures and combines them together. The structure and composition will often vary somewhat in the welded area and within close proximity to it. The intense heat of thermal welding erases any prior structure where the materials melt, but it can also change the *solid* structure near the welded area (Fig. 18.2). The way the heat is generated, the amount of heat needed and the effects the process has on the properties of the assembly are dependent on the method and the materials being welded.

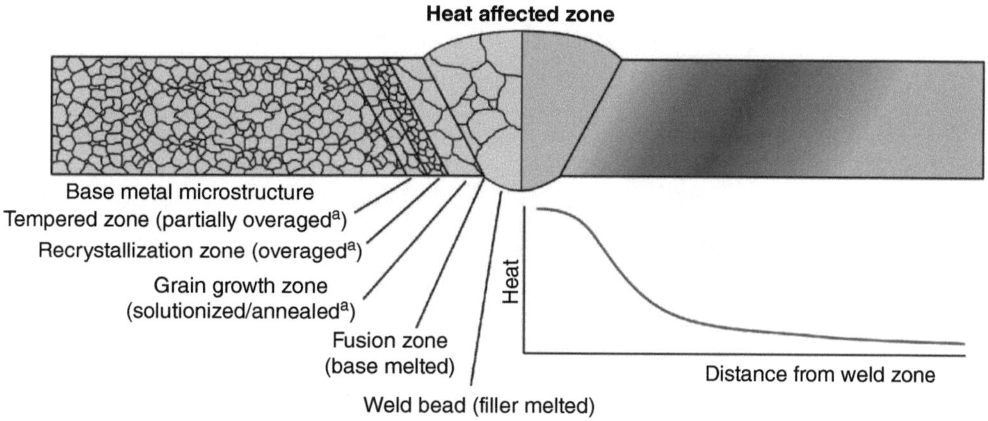

[a]These changes are specific to precipitation hardened metals, such as many aluminum alloys.

Figure 18.2 The heat required to fuse metal components changes the microstructure near the weld. Depending on the composition and condition of the base metal, this may cause a variety of changes, typically resulting in softening.

Discussing the way welding affects structure is pretty simple, but it requires a substantial background to really understand. For your benefit, you should be comfortable with the structure of metals (Chap. 3) and polymers (Sec. 5.3) and the general effects of temperature (Sec. 8.2). The short story is that heat can change chemistry, geometry, internal stress, and atomic and molecular structure. You do not need to know all of these to create a good weld, but knowing these can help you prevent creating bad and potentially dangerous welds.

The primary consideration in heat welding is heat! While not surprising, the way it can change a part or assembly may be. The amount and distribution of heat will determine the extent of the effects, both good and bad. Welding methods that spread out heat more will have a greater area *around* the weld pool that will be affected by the welding operation. This area is a sort of collateral damage, since it is not directly involved in the weld but it is directly affected by it. The best case scenario would be a weld that contains the heat entirely in the molten area. Of course, this is typically not the case.

Based on the thermal conductivities of the materials being welded and the way the heat is applied, this may be quite impractical. Certain materials are better at holding heat near the weld zone, but others spread it out quite readily (see Table 8.2 for a list of thermal conductivities). For instance, stainless steel has a lower thermal conductivity than aluminum, and therefore it results in a smaller heated area around the weld. The welding method also determines how concentrated the heat is that is being applied, and various methods will be discussed a little later. The heated zone is determined by a combination of the material being welded and the method used.

The area around a weld that receives significant heat is known as the **heat-affected zone** (HAZ). This zone is susceptible to basic thermal softening through annealing or even reversal of prior hardening such as quenching or precipitation hardening. The whole assembly should be thermally conditioned after welding if this is a concern. Typically, the weld area will be of lower strength and higher ductility than the rest of the structure, but in assemblies that not thermally conditioned, the difference is not significant enough to cause concern.

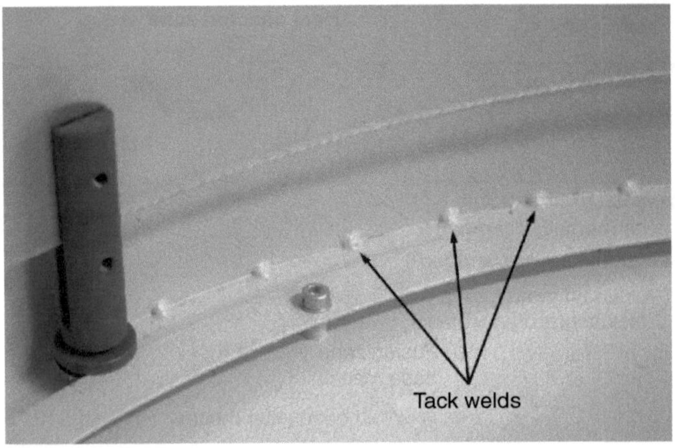

FIGURE 18.3 Welded assemblies need not be welded along the full length. Stitch welds and tack welds placed intermittently along a joint are sufficient for light-duty applications, such as this machine guard.

Heat can have geometric effects as well. Uneven heating may result in residual stress after the weld is complete and the assembly may distort. Even during welding, the two pieces may move due to the heat and cause the weld joint to grow larger. For this reason, a weld joint is **tack welded** in place, which means that small welds are placed at the ends or incrementally along a long joint to hold it in position when the main weld is being applied. The cause of thermal distortion is described more in Sec. 8.2.4. A series of tack welds can also be used to hold parts together that are not structurally critical, such as a sheet metal safety guard, as shown in Fig. 18.3. Longer segments are known as **stitch welds** and they serve the same purpose to reduce weld time, material, and energy while providing the needed strength.

A final note on the results of welding is related to chemical reactions. The high heat involved in thermal welding processes enhances diffusion rates and allows for chemical bonding to take place. A common example of the adverse effects of welding is found in stainless steel. Stainless steel contains a high level of chromium, which makes it stainless (see Sec. 4.2.6). When the stainless steel contains too much carbon, the chromium will bond to the carbon and cease to provide a corrosive protection. This results in chromium carbide forming on grain boundaries and results in **grain boundary sensitization**, which accelerates corrosion and can catastrophically weaken the structure. Low-carbon grades of stainless steel, designated with an "L" after their number (e.g., 316L), are used in welded assemblies to prevent this.

18.1.3 Basic Weld Joints and Methods

Parts can be welded together in just about any configuration, but there are some common joints and naming conventions worth learning. The weld joint is determined by the orientation of two surfaces and the position of the weld in relation to those surfaces. Welds can be performed on straight or contoured sections, but the two surfaces must be in close proximity. Although additional filler material can be added, for greatest strength the two components should be in contact to ensure the weld completely fuses them, or for thick components a small gap is left (~1/8 in).

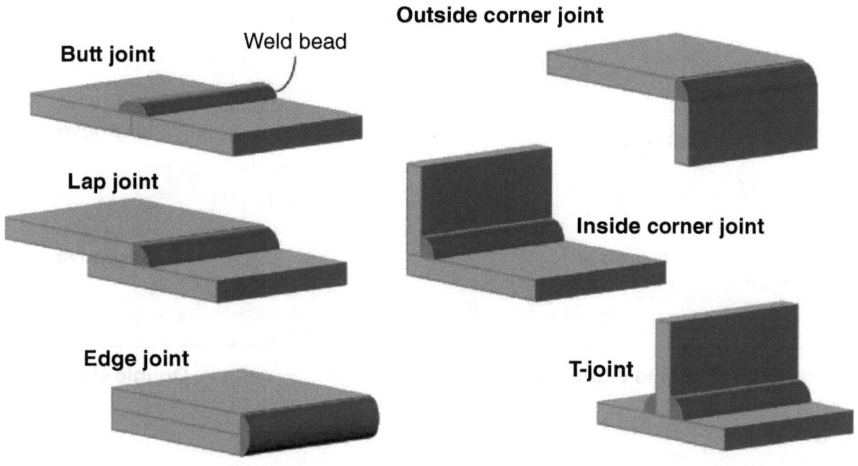

Figure 18.4 Many products and structures can be assembled using basic joints. They may appear as shown here, or involve variations.

The joints we will define are the butt joint, edge joint, lap joint, T-joint, and corner joint. A **butt joint** positions the edges of two components parallel to one another, and it results in the joint being the same thickness as the component thickness. This is often used to combine thin parts on edge, but it can be used with any thickness. A **lap joint** overlaps two pieces such that the welded connection consists of one edge and the face of the other part. When two faces are in contact and they are welded flush to one another, it is an **edge joint**. When the edges of two parts are set at an angle to one another, it is a **corner joint**. This can be created on the inside or outside of the corner. When two parts intersect in a "T" shape, is called a **T-joint**. These joints are shown in Fig. 18.4, but variations are commonly used. For instance, bicycle frames are most often made from round tubing at various angles, none of which clearly meet the criteria for the flat examples in Fig. 18.4.

A single type of joint can be welded differently. Consider a butt joint in two thin pieces (sheet metal, for instance). A weld bead does not need to be very large (including depth) to completely connect the two parts using a **surface weld** (or *bead weld*). Consider the same joint in much thicker plates. To reach entirely though the joint, the weld pool would have to extend very deep into the material. To help reach deeper sections, a groove is added to the joint in what is then termed a **groove weld**. The groove can be made by beveling one or both pieces to be welded on one or both sides to create single and double V-groove welds. If a concave edge is used, it is called a J-groove, and two J-grooves adjacent to one another form a U-groove. Examples of these variations are given in Fig. 18.5.

The two parts are often placed with a small gap between them, and the edges may be beveled to a point (a **feather edge**) or a small flat section is present (the **root face**). The gap ensures the filler rod can reach to bottom of the weld for complete cohesion, so the gap is larger when using larger fill rod. A feather edge is more likely to melt through since there is less material, but when a backing plate is used, it is preferred over a root face to help all three combine.

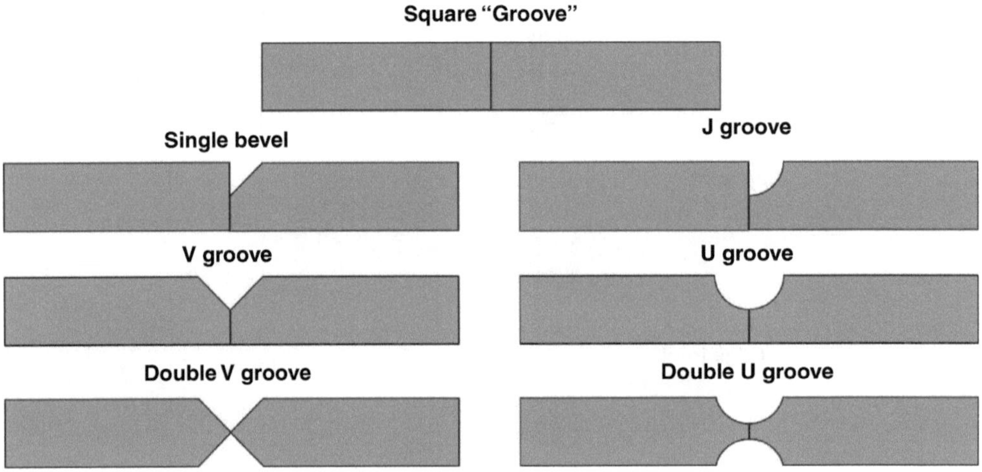

FIGURE 18.5 Two components can be prepared in a variety of ways before welding. Adding grooves allows for deeper, and more complete weld penetration, and the assortment shown is only a selection of the many variations possible.

Although the weld bead can be adjusted in size based on the fill material rate, there is a limit on how large a single weld bead can be. Even for very powerful welding equipment, it is common practice to weld using multiple beads to fill a large groove (e.g., 2 in thickness). This can be done from a single side or from both sides. When filling a large V-groove, for instance, the first pass, known as the **root pass**, connects the two pieces. The next pass, traditionally called the **hot pass**, is used to remove contamination (slag, dissolved gas, etc.) from the root pass by using greater heat. Even if this is not the intention, the second pass is still often called the hot pass. **Filler passes** are used to build up the material in joint. The final pass is called the **cover pass**. These welds are shown in Fig. 18.6. Note that each "pass" may consist of several beads. As described in Sec. 8.2.4, uneven heating can lead to distortion. By approaching the weld from both sides the stress can be balanced out more effectively. Allowing time for the weld to cool between passes can help as well but requires more time. Water quenching is not advised as it may increase the stress in the material. An example of the process for welding thick components is given in Fig. 18.6, and the order of welding operations is numbered.

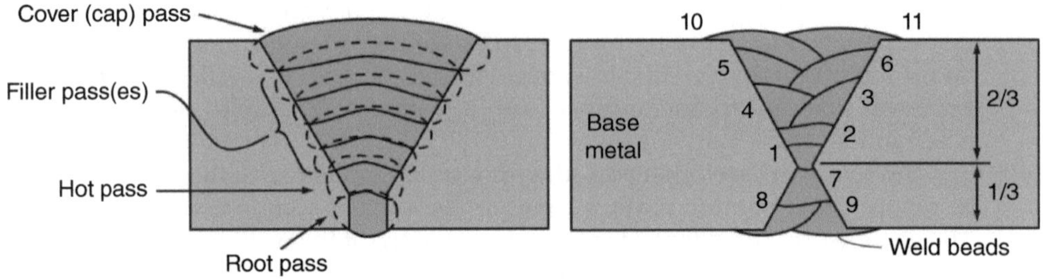

FIGURE 18.6 Welding thick components should be done using multiple welding passes. The procedure varies based on thickness, but an example of the order of operations is given here.

In addition to the order, the amperage is also varied, and typically increases as the welds progress.

To ensure proper fusion and reduce distortion and internal stress, a joint may be preheated before welding. This reduces the shrinkage in the weld zone as the material cools (thermal expansion/contraction is discussed in Sec. 8.2.3), and it also controls the weld microstructure, which may differ under rapid cooling, especially in carbon steel (see Sec. 9.3) where hardenability is high. The amount of heat is determined by the amount of material being heated (including any backer plate or support structure) and the ambient conditions (cold parts must be heated longer). After welding has commenced, the weld area is sufficiently heated by the process, as long as weld beads are applied without excessive delay. The welding process can also build excessive heat, especially in thin parts, so a proper balance should be maintained. Heating after welding can also be used to stress relieve the joint. The temperature is modest and cooling should occur slowly. Most often preheating and stress relieving are performed with a torch and require some experience or the aid of temperature indicators to achieve the proper heat.

The same considerations of weld order and heat should be applied to joints involving two components that meet at an angle. This is the case in corner joints, T-Joints, and even lap joints for thicker materials. The weld bead should fill the space between two surfaces, and to do so the weld must be balanced across the two with what is called a **fillet weld** (see corner and lap joints in Fig. 18.4). In cases where the edges of an assembly may not be appropriate for welding, a hole can be created in one part and the two are then overlapped. This hole is then filled with weld bead to connect the two using a **plug weld**. When a slot is used, it is called a **slot weld**. These welds are shown in Fig. 18.7. In thin materials, it is often much more efficient to spot weld or seam weld the assembly as described in Sec. 18.2.6.1, but this is not always possible. In certain cases, the joint may not be accessible or the weld bead would interfere with operation (where close tolerances and/or fitment of adjoining parts are needed), and in this case welding from the back side is preferable.

Regardless of the orientation of pieces and the method used to weld parts, there must be some pattern for adding the filler. If you want to fill a joint along a bathtub, you would probably apply caulk in a straight line, directly in the middle of the joint. That works for filling a gap, and in many cases works for welding components (a straight bead is called a **stringer**). Welding must achieve proper fusion between parts, which

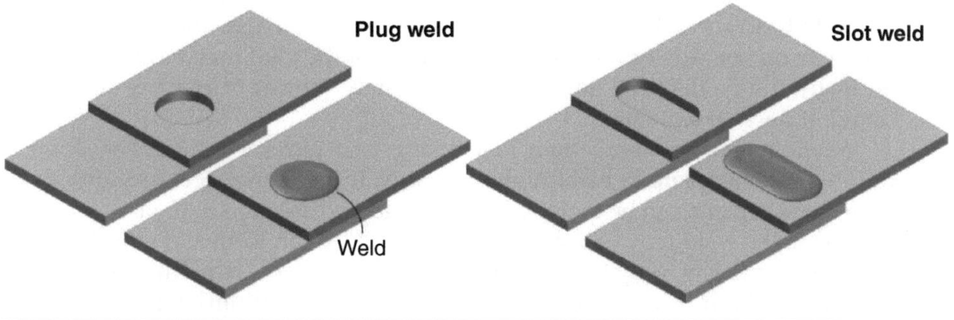

Figure 18.7 Plug and slot welds are used to combine overlapping materials without welding along the intersection of both parts.

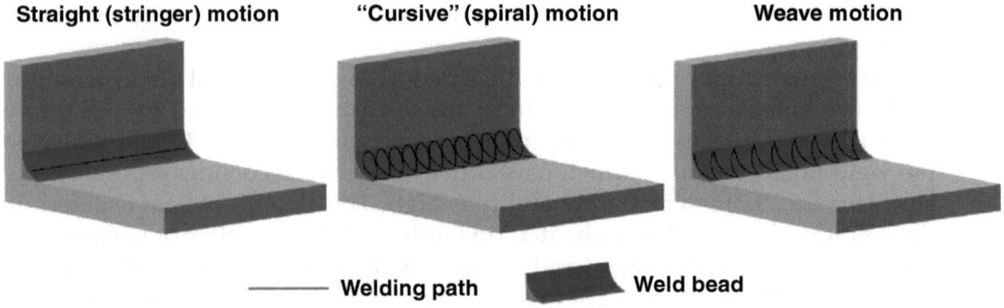

FIGURE 18.8 Different motions during welding result in different patterns. Appearance is important in some applications, but strength is important in all.

often necessitates other methods. These methods vary the welding motion to cover a broader area and ensure proper cohesion. The motion may be any repeated variation, but often it is a straight line, one-dimensional oscillation, two-dimension loop (like a cursive "e"), or weave. Examples are given in Fig. 18.8, but these are only a small selection of the possibilities.

18.1.4 Weld Quality

One of the most obvious indicators of weld quality is appearance. Although some very ugly welds can hold well and very pretty ones not hold well at all, welds of uniform appearance typically indicate a consistent penetration into the base material. That penetration is very important, because if the weld does not sufficiently combine the underlying, base material, the weld will be weak, no matter what it looks like.

The reason appearance can be telling is that the way in which the weld bead sits on the part will provide some indication of the success. For instance, if there was not enough heat or the weld progressed too quickly, the weld bead will stand off the surface. If the heat is too high, motion too slow or not enough filler was used, the weld will appear as depression in the surface. The effects of heat, wire speed, and cover gas are shown in Fig. 18.9. Note that what is considered "high" heat and "low" speed will vary for the type and thickness of the material being welded.

Also, if there are specks of metal around the weld, called **spatter**, it is likely the weld progression was too fast or filler was not added fast enough [especially in metal inert gas (MIG) welding]. If the weld is porous looking, there was likely insufficient cover gas. The weld bead should be evenly spaced across the two parts, and if the weld appears to be mostly on one side or the other, the strength of the joint may be compromised. There are many things that can go wrong. Fortunately, getting it right is not terribly hard, though it does take practice.

Welding requires more than just fusing two parts together. The bond must be uniform and complete to be of highest quality. If the workpieces are dirty or oxidized, these contaminants can be included in the weld and weaken it. If gases are dissolved in weld pool, the weld may become porous and oxides can form and embrittle the weld. If heat is not sufficient, only the surface of the joint will be fused. The weld may appear to very "pretty," but if it does not hold, it is worse than an ugly weld that does. Beauty should be full depth when judging a weld (Fig. 18.10). All of these quality factors determine the performance conditions during use.

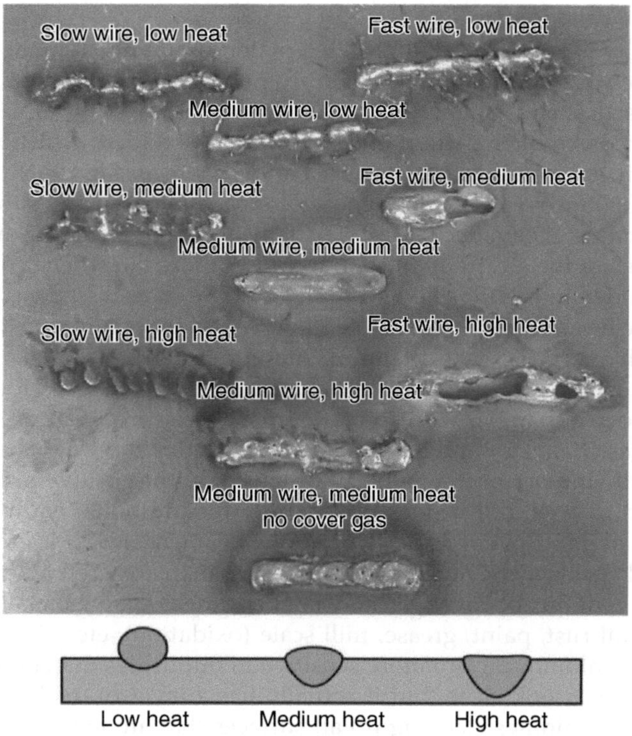

FIGURE 18.9 There are a variety of factors that determine weld quality and appearance. Some of the effects of those settings in MIG welding are shown here, but the concepts are universal.

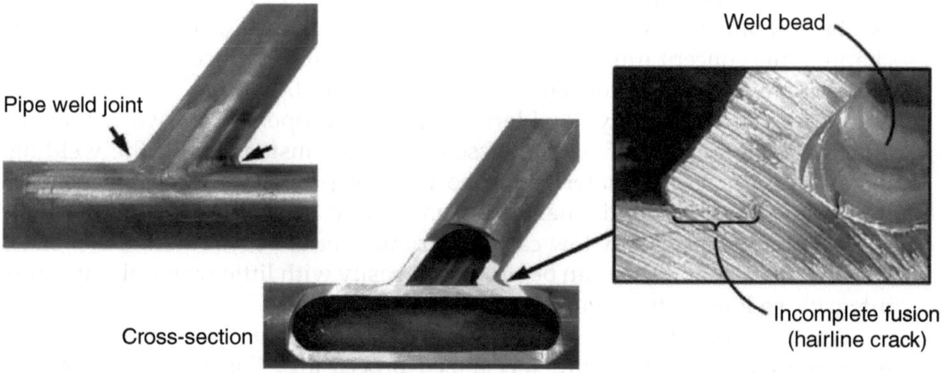

FIGURE 18.10 The purity and complete penetration of a weld are critical to achieving full strength. The weld should not be a weak point in the assembly. Even "good looking" welds are unacceptable if they are not structurally sound.

18.1.4.1 Weld Purity

Weld purity is very important. Molten materials are much more reactive with gases in the air such as oxygen and hydrogen. To prevent these reactions, a protective environment is used, such as an inert **cover gas**, which displaces the ambient atmosphere while the material is molten. Alternatively, **flux** can be added, which reacts with unwanted elements in the molten material and separates them as slag which can be easily removed.

The flux or cover gas can be chosen for each material and situation, and some basic understanding is beneficial. Flux is usually added to a surface that will be gas welded or as a part of the electrode in stick welding (more later). The cover gas is determined by the welder and changed by swapping out the compressed gas cylinder. Maintaining a clean weld can require careful attention. This is especially true for metals that strongly bond with oxygen, such as titanium. Titanium must be welded in an oxygen-free environment, sometimes even requiring shielding gas to the back side of a weld or welding to be done in a chamber with an inert atmosphere such as argon.

The gas chemistry and flow rate will determine the weld characteristics. Flow rate is especially important in windy conditions or with sensitive materials. Flux cannot be controlled as well, but it does not require the additional equipment associated with compressed gas. Cover gas determines the heat transfer and protective properties, but the "ideal" gas may be much more expensive than an acceptable gas.

The use of flux or cover gas may not be sufficient if the metal surfaces are not clean. Any and all rust, paint, grease, mill scale (oxidation), etc., should be removed. Some metals, like aluminum, are more sensitive to impurities than others. To make it worse, the metal may look clean, but it actually has a layer of oxide that must be removed. Spend a few minutes cleaning up all surfaces that are to be welded. Anything on the surfaces of a welded joint is likely to be incorporated into the weld when the metal is melted and reduce the quality. A wire brush is commonly used to remove oxide and prepare the surface, but using a single wire brush on multiple types of materials can contaminate the welds, too!

Of course, not all materials are created equal, and some have higher weldability than others. **Weldability** (or *joinability*) is a measure of the ease with which materials can be combined by welding. As will become clear throughout the remainder of the chapter, this concept must be put into context with the process, as not all methods of welding are equally appropriate for a single material.

In general, weldability considers the material properties, the ways it can be welded, and the design safety of welded assemblies. For instance, will the welding process negatively impact the material composition or properties? Stainless steel was mentioned above as a potential hazard when welded because of chromium depletion for corrosion resistance. Only low carbon varieties should be welded. Other steels are not sensitive to this, and they can be processed easily with little worry about compromised integrity, so long as the weld is good.

Another consideration is whether the weld can be performed by a variety of techniques, or does it require specialized processing? For example, since titanium is *very* sensitive to atmospheric gases, a "clean" welding process, such as TIG/GTAW (see Sec. 18.2.5), is preferred. Also, the amount of heat the area adjacent to the weld receives (refer back to Fig. 18.2) can be reduced by choosing a method such as electron beam welding, but the method is impractical for many applications.

Finally, how safe are welded assemblies? Does the welded structure have to be designed within close tolerances to be robust? Is there room for error without comprising strength? Both material and process considerations play directly into the safety aspects, and the use will determine the room for error. For instance, if a weld cracks on a bicycle frame, the rider may be injured, but if a weld on a natural gas pipeline operating at 1500 psi fails, the potential for a catastrophic event is much higher.

For this reason, quality control is much more stringent in critical structural applications. Weld quality can be checked afterward by methods such as a simple visual inspection, but more advanced methods such as radiographic, magnetic particle, liquid penetrant, and ultrasonic inspection can be used to reveal defects. Each method has different strengths and weaknesses, and each requires a trained inspector to interpret the results. The weld can weaken over time, so regular inspection schedules are needed for large structures, such as bridges and pipelines. In the case of pipelines, they are often buried, which makes inspection impractical by methods applied before they are covered. In these cases, inspection can be achieved by sending a remote inspection vehicle through the pipeline, and it collects quality data as it passes along the pipe. These inspection vehicles are called "pigs," and they may use magnetic field, ultrasonic frequency, or any other suitable inspection method.

18.2 Metal Welding Processes

Metals are the most common materials to be welded. This is due to a combination of high strength and modest melting temperature. Large structures like sky scrapers and cruise ships can be assembled on-site using welding. Welding can be done using a variety of equipment and techniques, so knowing which one to use, or what a specific technique is capable of, is useful. As a metalworking discipline, welding is one the most diverse and nuanced. There are numerous standards for welding processes [see American Welding Society (AWS) standards, for example], and certifications are available for specific welding methods and techniques (e.g., overhead, vertical, etc.). Some of these standards and certifications are given in Table 18.1. To pass these tests, the weld

Welding Qualifications		
Joint Position/Type	**Plate Position/Setup**	**Pipe Position/Setup**
1	Horizontal plates/joint	Horizontal, rolled
2	Vertical plates/horizontal joint	Fixed vertical
3	Vertical plates/vertical joint	—
4	Horizontal overhead joint	—
5	—	Fixed horizontal
6	—	Fixed at 45 degrees
G	Grooved (beveled) butt joint	
F	Fillet joint	
R	Restricted position	

Note: Each setup can be performed with varying welding methods (e.g., MIG/GMAW 3F plate or Stick/SMAW 6GR pipe).

Table 18.1 Common Welding Certification Positions and Techniques is inspected (visually, x-ray, etc.) and/or mechanically tested to ensure proper strength is achieved. Specific applications of welding, such as structural steel, railroad, aerospace, etc., are also specifically designated. Not all welding processes use a single method. TIG/GTAW welding may be used for the root and hot pass and followed by stick/SMAW welding for the rest of the weld. This is preferred in applications where the root pass should be "clean" (no slag, spatter, etc.) that would contaminate or damage a system such as in piping.

18.2.1 Oxy-Fuel (Gas) Welding

Oxy-fuel welding (also known **gas welding**) uses a high-temperature flame to melt the metal base material (Fig. 18.11). This flame is generated using a mixture of oxygen and a flammable gas, called the fuel. The ratio of gas to fuel and the chemistry of the fuel determine the amount of heat generated. The most common fuels include acetylene, MAPP (methylacetylene propadiene), propylene, hydrogen, natural gas, propane, and others. Acetylene burns the hottest of these gases at about 6300°F (3482°C), but it also has some drawbacks.

Acetylene is stored as a liquid (it is dissolved in acetone), and it should always be positioned vertically. Above 15 psi it becomes unstable and can spontaneously combust. The rate at which the acetylene is used should not exceed 1/7th the total tank volume per hour. Other fuels, like MAPP or natural gas, do not have these issues, but they burn at lower temperature. Based on application, convenience, and usage practices, they can be quite competitive.

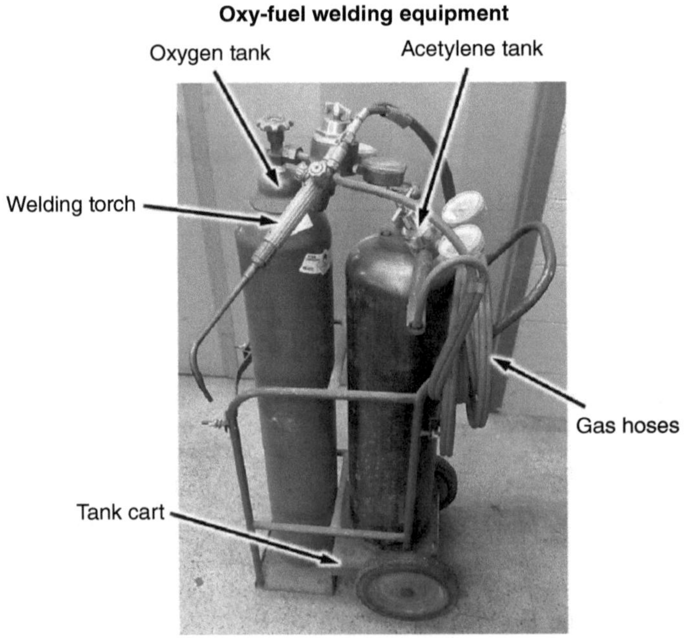

FIGURE 18.11 Basic oxy-fuel setup. The portability of oxy-fuel is a key advantage.

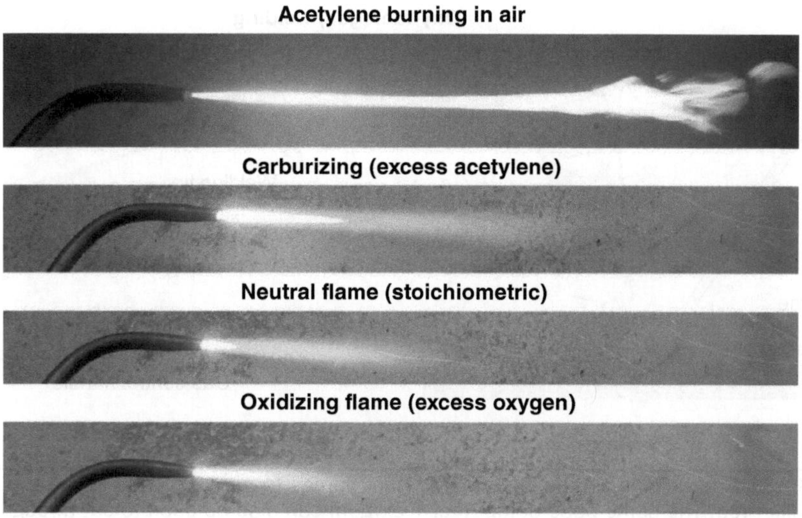

Figure 18.12 The ratio of oxygen and fuel changes the welding characteristics of the flame. Too much acetylene can add carbon to the weld (carburizing), but too much oxygen can oxidize it. Too much oxygen also results in a loud hissing sound, and the oxygen should be reduced until the sound goes away and a small central cone is present near the tip of the torch. This is known as a neutral flame.

18.2.1.1 Flame Properties

Because the heat source is a flame, the characteristics of that flame are important. The ratio of oxygen to fuel will determine the heat and chemistry of the flame (Fig. 18.12). The oxygen and the fuel react to create the flame, but the reaction is not always perfect. In a stoichiometric reaction, all of the chemical inputs (e.g., oxygen and acetylene) are converted to other chemicals (e.g., carbon dioxide and water).

When the mixture is nearly "perfect," it produces a **neutral flame**, which contains no excess oxygen or carbon, and is the most common flame for welding. When using common hydrocarbon fuels (e.g., acetylene, MAPP, etc.), excess fuel will result in excess carbon in the flame. This **carburizing flame** can change the base metal composition by adding unwanted carbon to the weld area. Similar issues can arise from an **oxidizing flame,** which introduces excess oxygen that will react with the molten metal and compromise the weld. The torch begins hissing loudly when excessive oxygen is flowing. With more oxygen the flame will go out with a loud pop.

18.2.1.2 Oxy-Fuel Welding Process

The process of welding using oxy-fuel starts by lighting the torch and adjusting the gas mix to create the appropriate flame (usually a neutral flame). First acetylene is started by opening the valve on the welding handle, and this acetylene is ignited. Copious amounts of soot and smoke develop without oxygen, and the acetylene flow is increased until the smoke and soot are minimized. An aggressive, but not very hot, flame is created. Oxygen is then added by opening the valve on the handle and is increased until the appropriate mix is achieved.

This creates a very hot flame that is then applied to weld area. The hottest part of the flame, and therefore the portion used to heat the metal, is the inner cone. The torch

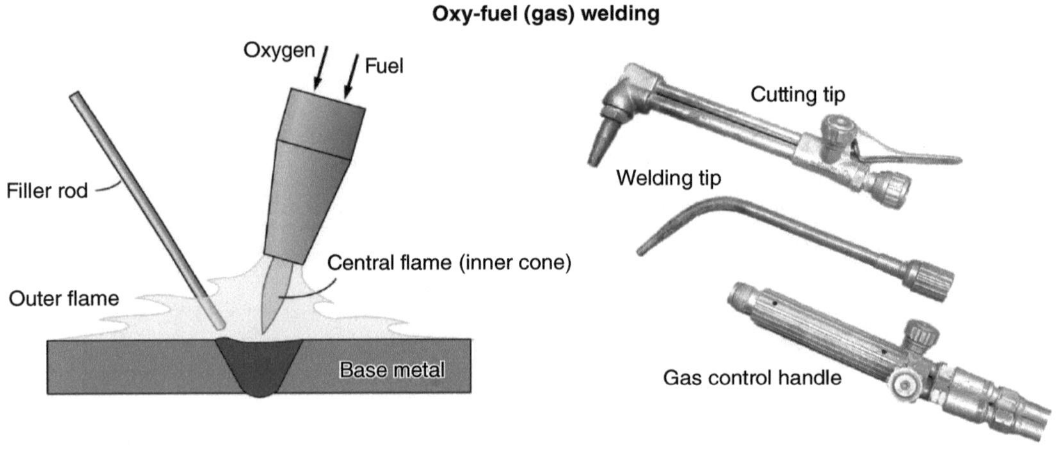

Figure 18.13 Gas welding uses a high temperature flames to melt the base metal and requires a separate filler rod. The same base setup (fuel + oxygen) can also be used to cut through metal.

must be held very close to the metal surfaces, and the outer flame area will spread out across the surface, heating the surrounding area, but not enough to melt it. To add material to the weld area, a separate filler rod is melted simultaneously. This general process and the torches for welding and cutting are shown in Fig. 18.13.

The outer flame is important because it can preheat the surrounding metal so that the inner cone can more quickly melt the weld area. It is also important because it creates a large HAZ as described in Sec. 18.1.2. Depending on the technique chosen, the weld pool can be made larger or smaller. Pointing the flame, "downstream," in the direction yet to be welded, is known as **forehand welding**. Pointing the flame "upstream," where welding is already complete, is known as **backhand welding**.* These techniques are shown in Fig. 18.14.

Because the outer flame will preheat or postheat the weld area, it affects the weld size and material response. Forehand welding creates a larger weld pool, and backhand welding reduces the cooling rate of the weld and can help relieve stress. The material being welded will determine the appropriate method, as some metals melt more easily, and some are more sensitive to cooling rate. Arc welding methods are less affected by the use of forehand or backhand techniques since the heat is much more concentrated.

18.2.1.3 Oxy-Fuel and Plasma Cutting

In addition to welding, oxy-fuel can be used to cut through metal. This is done using a different type of torch that can deliver a blast of additional oxygen. The purpose of this oxygen is to burn away the metal. The process starts with a preheating flame similar in nature to the welding flame, but distributed through multiple, smaller orifices in the torch tip. Once the metal has reached its **ignition temperature**, or the temperature at

*Forehand and backhand welding are sometimes called **push welding** and **pull welding**, especially in MIG welding.

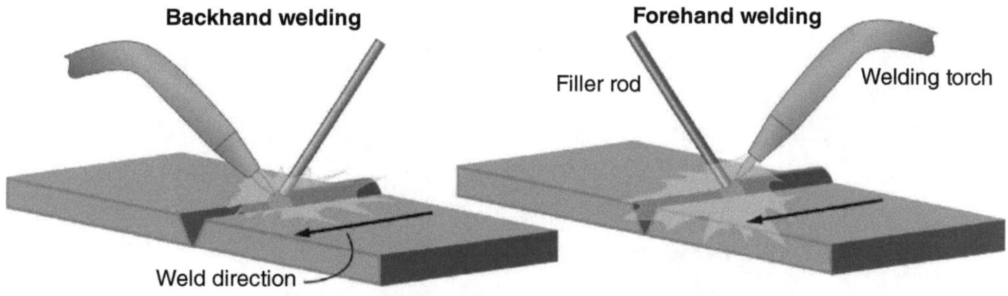

FIGURE 18.14 The direction of welding will affect the weld characteristics. The two primary welding techniques are forehand and backhand methods as shown here.

which it will burn in the presence of oxygen, the additional high-pressure oxygen lever is depressed and blows the metal away.

In steel the ignition temperature is reached when the part turns "cherry red" [>1300°F (~700°C)]. As the metal burns, it releases more heat, which helps keep the next section of material above the ignition temperature until the cut is complete. If the torch is moved too quickly, the metal will fall below the ignition temperature and the torch will begin **gouging** the surface, which means that it will create a divot or trench in the part. There are cases where this is desirable, but specialized setups are preferred to do this with greater quality.

Oxy-fuel cutting can be used to make complex cut profiles in very thick materials, or multiple parts can be stacked and cut together. The cut quality is acceptable for some industrial applications, but consumer products typically require secondary processes for greater quality. The torch settings, rate of cutting, and the operator skill all determine the quality of the final product. An alternative method, known as **plasma cutting**, uses electrical arc and gas flow to create hot plasma with temperatures in excess of 20,000°F (~11,100°C). Because the temperature is much higher, the method is preferred in production cutting work where thicker materials can be cut faster, but the equipment is not as portable as it requires a high-power source of electricity. Plasma cutting has similar setup considerations as arc welding, discussed in Sec. 18.2.2. These cutting processes are shown in Fig. 18.15.

18.2.2 Arc Welding Considerations

The source of heat is critical to weld quality, and arc welding provides distinct advantages over gas welding. First, an electrical arc has a higher temperature, easily exceeding 7000°F (3871°C), depending on the setup. Second, the arc can be exactly controlled through electrical settings and the heated zone is much smaller than what a gas flame produces. Finally, the purity of the weld can be more tightly controlled since flame chemistry is not a factor, and a separate cover gas can be used.

The benefits and disadvantages under arc welding are varied, as the utility of the process has spawned many variations. The method chosen will depend on the versatility, portability, capability, and cost. We will look at some of the key methods, but welding, particularly arc welding, is an industry that could (and does!) have entire books written about it alone. These sections should provide you with a basic understanding, but welding, at least good welding, also requires experience.

FIGURE 18.15 When a cutting tip is used, additional oxygen raises the heat enough to melt the metal. An alternative method, plasma cutting, produces a much higher temperature and is capable of cutting faster and through thicker materials.

18.2.2.1 Basic Electrical Considerations

Before we get into individual processes, we should take a moment to consider what common features these methods have. For instance, to create an arc, and therefore a weld, there are some key elements that are needed. These elements consist of

1. A source of electricity
2. An electrode to transfer the arc
3. A workpiece to be welded
4. A ground to complete the circuit for the flow of electricity

If there is any failing in these features, the weld will not be successful. The welding machine, or welder, provides the means to generate the large amount of electricity needed. This can be done by plugging it into an appropriate outlet (most welders require more than a typical wall outlet provides), or by an integrated generator that is driven by a gas or diesel-powered engine if no power is available on the job site. The basic requirements for arc welding are shown in Fig. 18.16.

18.2.2.2 Arc Polarity

Polarity can be a confusing topic in welding, but it is important to the weld characteristics. The flow of electricity can happen in one of two ways, and we will consider them from the perspective of the electrode. In **direct current electrode negative (DCEN)** polarity, also called **straight polarity**, the electrode is negatively charged and electrons flow from the electrode to the workpiece. In **direct current electrode positive (DCEP)**, also called **reverse polarity**, the electrode is positively charged and electrons flow from the workpiece to the electrode. There is also a third option that uses both polarities. **Alternating current (AC) welding** switches between DCEN and DCEP, and the rate and pattern of the oscillation can be varied. So although there are three common settings (i.e., DCEN, DCEP, and AC), there are only two options for polarity.

The reason polarity is important is because whenever voltage is applied, it controls the flow of electrons. As described throughout Chap. 2, there is a balance of electrical

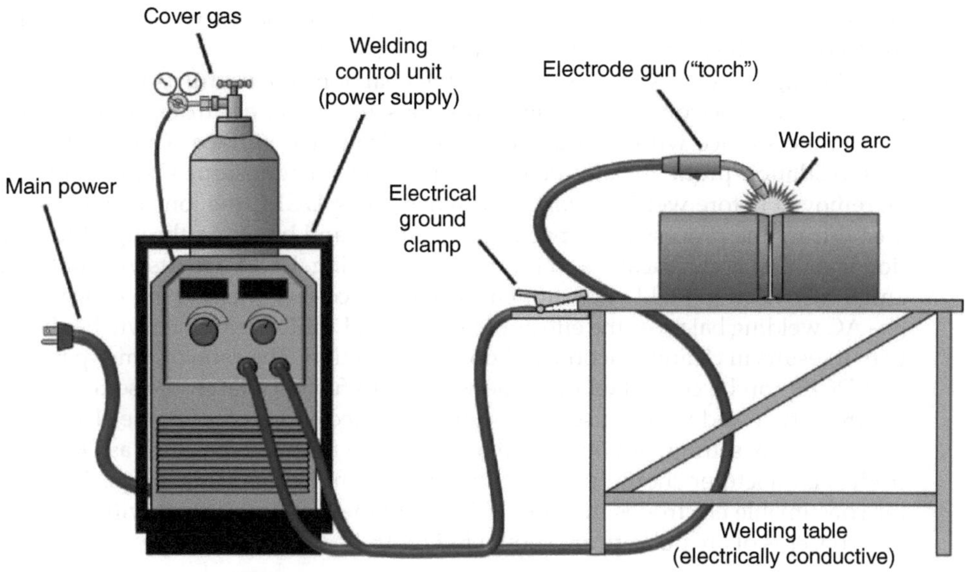

Figure 18.16 The basic requirements for arc welding are shown. Welding cover gas is only needed in certain methods where flux is not used, and a welding table is not necessary if the ground clamp can be connected directly to the workpiece. The primary consideration is the flow of electricity.

charge in materials, otherwise ions are formed. In an arc, the extreme temperature and electrical potential causes gas atoms/molecules to become ionized. When an electron is removed, the positively charged cation will be affected by the applied voltage. The ions will accelerate toward the negative charge and the electrons toward the positive charge. This results in the ions and electrons going to opposite sides of the arc. Which one goes to the electrode and which one goes to the workpiece is determined by the arc polarity.

Electrons and ions are very different in their properties. Electrons have negligible mass and their "flow" is more efficient than ions. Therefore, the positively charged side receives more energy (~2/3 of the total), and this results in higher temperature there. Here is where most of the confusion is based. For nonconsumable electrodes, which should not melt (e.g., in TIG/GTAW, Sec. 18.2.5), DCEN (cooler electrode) is preferable and more energy can be applied to the workpiece to create a deeper weld. For consumable electrodes, such as rod (Stick/SMAW, Sec. 18.2.3) or wire (MIG/GMAW, Sec. 18.2.4), DCEP is often preferred because the electrode should be thoroughly melted to ensure proper transfer of the metal across the arc and achieve a deeper weld penetration. Confused? Just remember, the heat should go where you want most of the melting (i.e., the positive side). So, using a play on words, it is *negative* (DCEN) to melt a nonconsumable electrode, but it is *positive* (DCEP) to melt a consumable one.

Hopefully it makes sense, because now we will bend the rules. The reason is that on very thin materials, smaller weld penetration may be preferable, so you might use the opposite polarity (in addition to a lower overall electrical current). To prevent damage to a non-consumable electrode, a larger diameter electrode is used to accommodate the additional heat. Each welding electrode and polarity setting should be individually considered! Manufacturers specify the proper polarity for most electrodes. Actually changing polarity can be as simple as flipping a switch on the welder, or you may have to disconnect the electrical cables, which is also fairly easy.

Although electrons transfer most of the heat, ions are useful as well. Ions are heavy compared to electrons, and when those ions reach the electrode or workpiece, they impact the surface with enough energy to "clean" the surface. This can be important with oxidation-prone metals such as aluminum or magnesium. Oxides should always be removed before welding starts, but when using DCEP, the ions help clean the weld area during the process by impacting the surface and breaking through the thin oxidation layer that is electrically and thermally insulating. DCEP results in a shallower weld and is also used to weld very thin parts where excessive melting is a concern.

AC welding balances the effects of DCEN and DCEP. Because it switches between the two, it results in cleaning action and good penetration. The ratio of time spent on DCEN and DCEP can be controlled on some welders to further dial in the settings. Again, the electrode type and welding settings should be selected together. Some electrodes can be used with any setting, but others are only suitable for one type, such as DCEP. The electrode manufacturer should provide compatibility notes, and this is especially important for consumable electrodes that are designed to melt in conjunction with the base metal. The concepts of arc polarity are shown in Fig. 18.17.

18.2.3 Stick Welding (SMAW)

Stick welding, also known as **shielded metal arc welding (SMAW)**, uses an electrical arc to simultaneously melt the base metal and the electrode. The electrode is the filler rod, so it is consumed during welding. To protect the weld from unwanted contamination, the metal electrode is coated with a material that vaporizes and envelopes the weld pool. This process and related equipment are shown in Fig. 18.18.

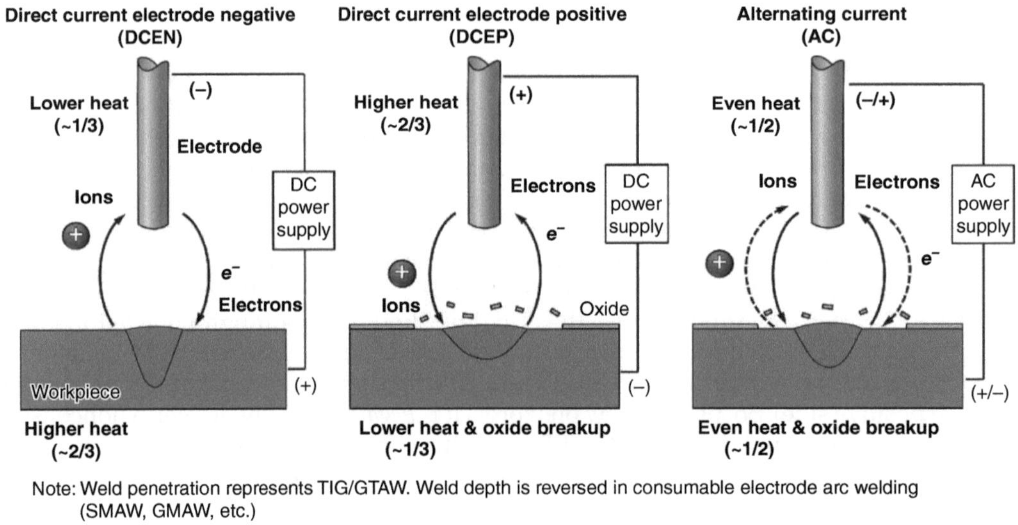

FIGURE 18.17 Electrode polarity dictates the direction of electrical flow, which determines the distribution of heat. DCEN results in deeper weld penetration for nonconsumable electrodes, but DCEP creates a deeper weld in consumable electrode processes and also breaks up surface oxidation. AC welding balances the two and creates a medium weld depth with surface breakup.

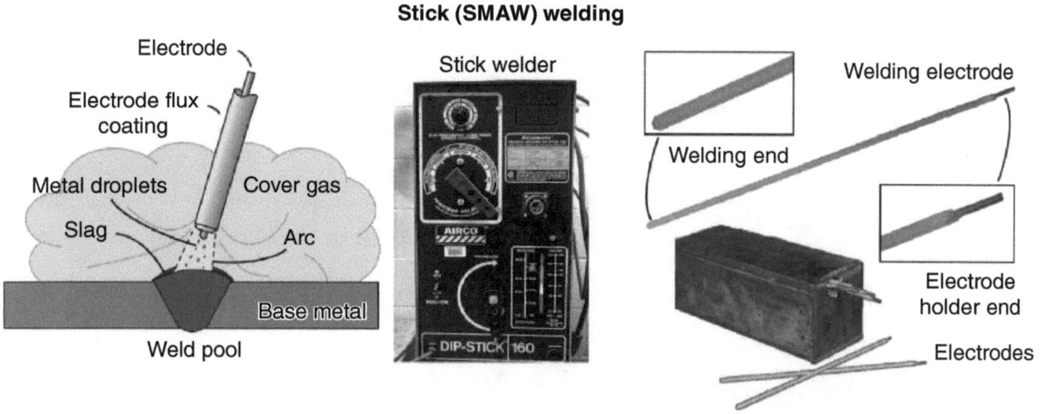

FIGURE 18.18 The basic process and common equipment for stick/SMAW welding are shown here.

The electrode coating, also called the flux coating, varies depending on the application and manufacturer specification, but it typically consists of a variety of ingredients including cellulose, metal carbonates, titanium dioxide, silicates, alloying elements, processing agents, etc. These various ingredients create a reducing environment, react with contamination to form slag, help restore or maintain alloy content, adjust fluidity, etc.

Because the electrode is consumable, the electrode (or stick) must be maintained near the surface by progressively lowering it as it is consumed. The proper gap must be maintained (usually ¼ in or less) to maintain the arc yet prevent short circuiting the weld by touching the surface. If the electrode is too far away, spattering will increase. To allow the flux to do its job, only pull welding is used. If the electrode coating chips off during handling or working, that section should not be used to weld or the rod in the damaged portion can be used on a piece of scrap until a good section is available.

Small weld beads can be made by simply moving in a straight line. To make the weld wider, a side-to-side "weaving" motion may be applied. The quality and appearance of the weld are highly dependent on the skill with which they are made. It is beneficial to ensure that the weld can be completed without restriction of movement before starting. Because sticks have a finite length, starting a second stick at the end of the prior bead is a good skill to practice.

18.2.4 MIG Welding (GMAW) and Flux Core (FCAW)

Metal inert gas (MIG) welding, also known as **gas metal arc welding (GMAW)**, is a method that uses a continuous spool of wire as a consumable electrode, and the wire is automatically fed to maintain the weld bead. This results in the closest welding gets to a "point-and-shoot" process. Unlike stick welding, it can be performed without interruption over very long distances. This is because the wire can be on a large spool, the size of which dictates the maximum weld length (which is almost always, *much* longer than needed for a single weld). Also, MIG welding can be done in a push- or pull-type direction, the better of the two is a matter of preference (and plenty of debate). The general process and equipment for MIG welding is shown in Fig. 18.19.

The weld is protected from contamination by the flow of a cover gas. The gas chemistry can be changed by changing the compressed gas tank, but common gases include

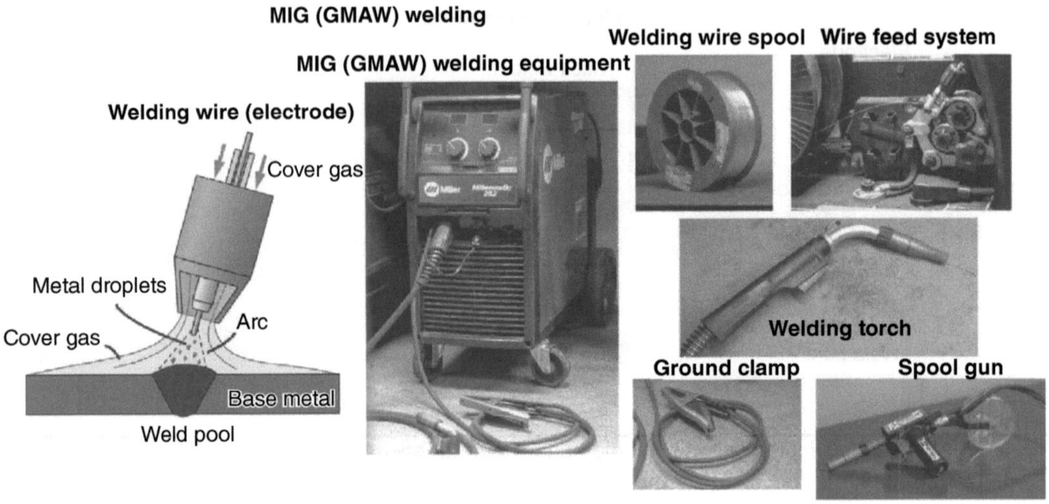

FIGURE 18.19 The basic process and common equipment for MIG/GMAW welding are shown here.

argon, helium, carbon dioxide, and mixes of these. MIG welding is often performed with a 75% argon, 25% carbon dioxide mix. Because carbon dioxide is not technically an inert gas, you may see this referred to as MAG, or metal *active* gas, welding.

Because the wire must be fed through a long passage leading to the torch (or gun), there is a tendency for softer filler materials to get jammed. These materials, such as aluminum, can be fed using an alternate spool on the torch. Because the torch must be maneuvered, the spool size is much smaller to reduce weight. Alternatively, rather than pushing the wire through, as is common, a pull-type feeding mechanism can be used.

The main controls on the MIG welder are the wire speed and amperage (heat). For thicker material both can be increased, and the manufacturer often provides a table of appropriate values for varying material thickness. If travel speed (by the operator) is too fast, the electrode will lose consistent contact and start sputtering excessively. Slow welding will result in a much thicker weld bead and may overheat and melt a hole in the base material, especially in sheet metal. Once the settings are dialed in, the process is very user-friendly.

Because MIG welding can be performed continuously, it lends itself well to automation. Robotic arms can have the necessary equipment attached and be programmed to weld complex contours with excellent repeatability. Also, since MIG welding is a cleaner process, no slag needs to be removed, which further simplifies the processing.

A variation on the MIG welding process replaces the cover gas with flux, similar to the way stick welding is protected. The flux can be brittle and flake off when on the exterior, and this prevents it from being put on a spool and fed through roller as done in MIG. To prevent these issues, the flux is contained *within* the wire, and so the process is called **flux core arc welding** (FCAW) (Fig. 18.20).

The same MIG equipment can be used some minor modifications. For instance, the rollers that feed the wire may be exchanged for more aggressive knurled version and the polarity will be reversed (polarity was discussed in Sec. 18.2.2.2). Otherwise, flux core welding proceeds in the same way. Similar to stick welding, the weld must be "pulled" in order for the flux to protect the weld without interfering with the weld pool.

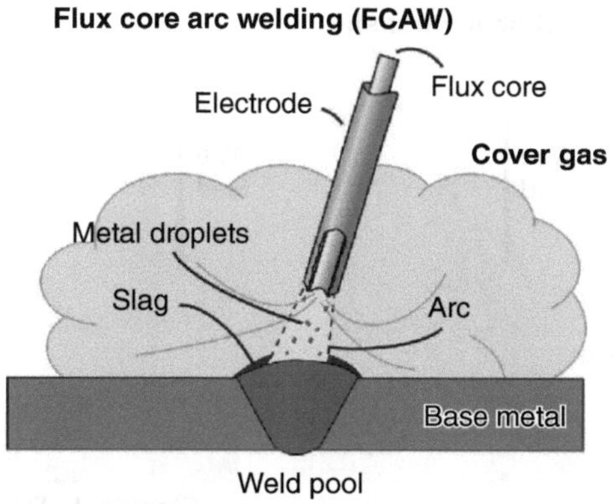

FIGURE 18.20 Flux core arc welding (FCAW) shares many similarities to MIG welding. The wire structure is the primary difference, and FCAW does not require a cover gas.

18.2.4.1 Metal Transfer and Shielding Gas Selection

The way metal is transferred from a consumable electrode to the joint (weld puddle) is worth mentioning, as it affects the weld properties. There are numerous variations on how it happens, but the primary categories we will consider are spray transfer, globular transfer, short-circuit transfer, and pulsed transfer. **Spray transfer** uses a large electrical current to melt the wire and send a steady stream of small metal droplet to the weld pool. It results in good penetration and low spatter, but it can burn through thin materials. It is best suited for horizontal welding positions, but can be performed in other orientations. Argon-rich welding mixes (e.g., 80% Ar) are necessary. **Globular transfer** sends large drops of molten metal to the weld pool and can result in deep penetration with pure CO_2 shielding gas (cheaper than Ar). The large drops are not as controlled as in spray transfer, so spatter can be excessive and requires more wire and clean-up time after welding. This transfer type is not typically desired, though it may result from improper settings or shielding gas when aiming for spray transfer.

Short-circuit transfer melts the wire after it has come in contact with the workpiece by the high current in the small wire. This short-circuit and melt process happens tens to hundreds of times a second, creating a sort of "sizzling" sound. The method can be used with high CO_2 gas concentrations and uses lower voltage, thereby reducing the heat input. This makes it suitable for thin components. The final method, **pulsed transfer**, uses short intervals of higher current to control the transfer of metal droplets. Ideally, one droplet will be transferred with each pulse, and because high current is not continuously applied, the total heat input and likelihood of burn through are reduced. It uses gas mixtures similar to spray transfer, has very low spatter, and is suitable for welding in all positions (not just horizontal). The various metal transfer mechanisms are shown in Fig. 18.21.

18.2.5 TIG Welding (GTAW)

Tungsten inert gas (TIG), also known as **gas tungsten arc welding (GTAW)**, is a process that uses nonconsumable tungsten electrode, and filler rod is added separately. In many ways, TIG welding is similar to gas welding, but the amount of control over heat and the purity of the welds are far superior. Tungsten is the most versatile welding method and

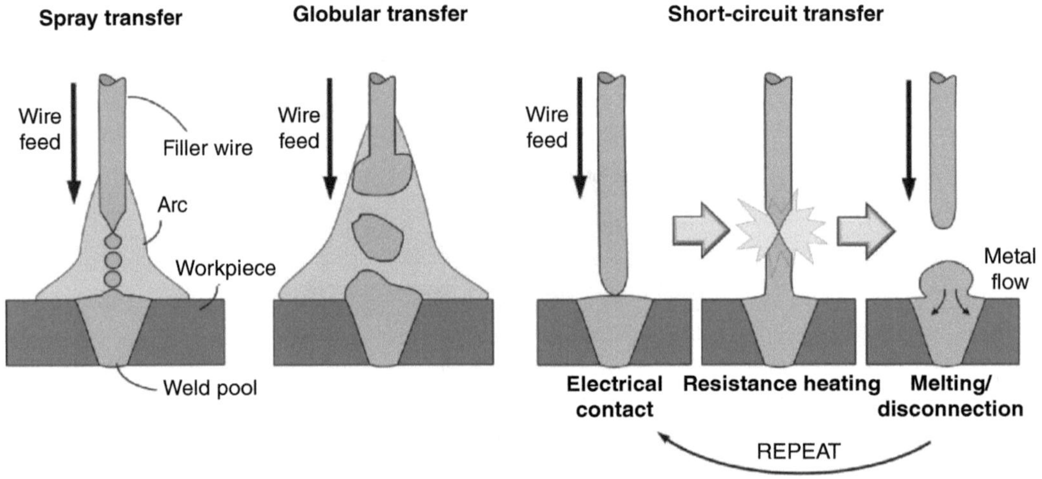

Figure 18.21 The metal transfer process in MIG/GMAW welding can occur by a variety of mechanisms, as shown here.

is used in general-purpose applications as well as with the most demanding metals and structures. The general setup and common equipment for TIG welding are shown in Fig. 18.22.

Tungsten is used for the electrode since it has such a high melting point and will not melt at the high arc temperatures (see Table 4.12). Although pure tungsten is used, it is also often mixed with thoria (thoriated), ceria (ceriated), lanthanum (lanthanated), or other rare earth elements. The compatibility of a particular electrode must be determined in conjunction with the mode of welding and the material being welded.

The electrode tip shape is important, and for DC welding a sharp tip is desired (Fig. 18.23). The electrode should be ground using a special tungsten grinding machine or on an abrasive wheel (diamond is best) dedicated to the purpose, since other materials

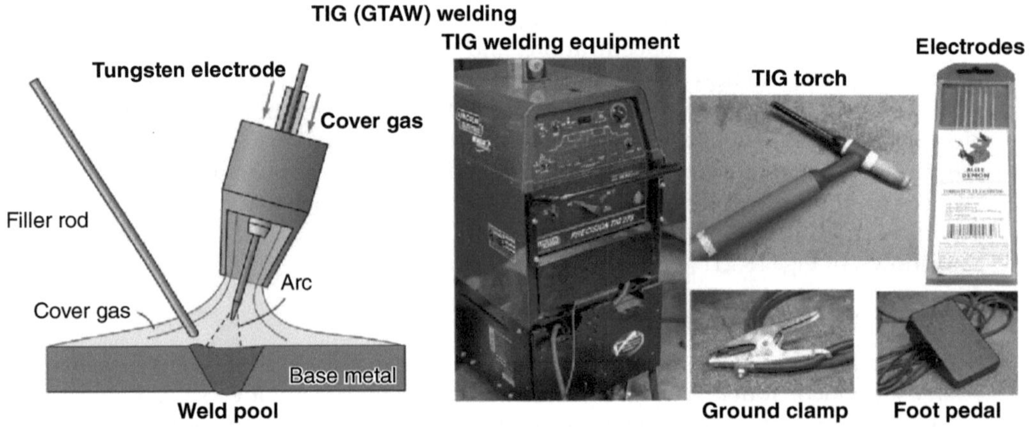

Figure 18.22 The general process and common equipment used in TIG/GTAW welding is shown.

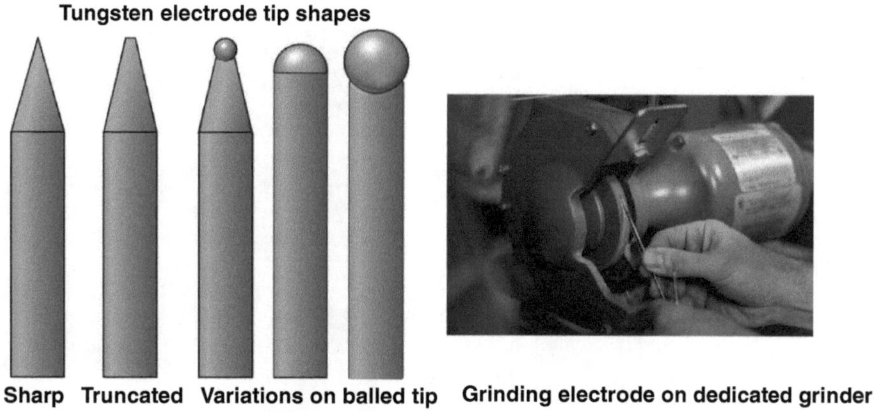

FIGURE 18.23 The tip shape and condition is very important to the arc quality during TIG welding. Variations on the tip are shown, each being desirable for specific materials and welding conditions.

can contaminate the electrode and ruin arc stability and electrode life.* The sharpened tip should be no more than 2.5 times the diameter, and grinding should be done parallel to the length of the rod with a fine grit abrasive to assist formation and control.

A sharp point helps concentrate the arc and eases starting the arc. A small flat is put the end to create a truncated cone for high amperage applications as this prevents transferring tungsten to the weld pool or unwanted balling of the tip. The electrode tip is intentionally ball-shaped in certain applications, such as when doing AC welding of aluminum. This is done by sharpening the electrode to a sharp point then striking an arc using an electrode positive polarity to melt the tip *slightly*.

The reason the electrode is subject to such scrutiny is that TIG welding can be performed in a variety of configurations. It can be done in DCEN, DCEP, or using AC. The importance of these settings was discussed in Sec. 18.2.2.2, and the ability to set the current allows for precisely setting the weld characteristics (roughly one amp per 0.001 in thickness in steel).

TIG welding is commonly performed by first striking an arc using a foot pedal, which is very much like an accelerator. The further down you push the pedal, the more current, and therefore heat, you get. The torch is held in one hand and the filler rod is held in the other. The torch must be kept close to the weld surface, but if the electrode touches the surface, it will create an electrical short and stick. This will contaminate the electrode and it should be resharpened. TIG welding requires that you pay attention to both hands and your foot simultaneously, so it does require a little finesse. You can also weld at constant current (heat) by "scratch" starting the arc. This entails bring the electrode into contact with the workpiece and quickly retracting to begin the weld. This does not require a foot pedal, but it also requires some skill and is only preferable where welding position will make it difficult or impossible to operate the pedal.

*Chemical sharpeners can also be used by heating the tungsten then dipping it into the sharpening compound. This is usually a last resort rather than a preferred method since it does not give as much control over tip geometry.

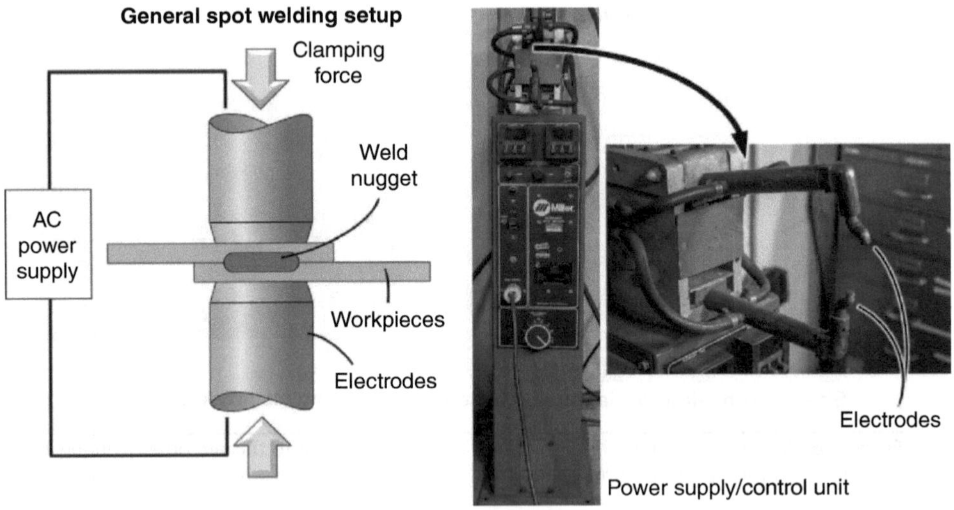

Figure 18.24 Spot welding uses concentrated electrical current to fuse components together. The concept and example equipment are shown here.

Once the weld pool is established, the amount of heat will determine the travel rate, and the filler rod must be added incrementally as the pool is advanced. There are a great many variations on TIG welding, and it goes to a depth we simply do not have space for. Welding, and TIG especially, is almost as much an art form as a science, and there are many opinions about the best way to do it. So long as the joint is sound, the rest is just preference.*

One final capability of TIG that we will consider is the use of the arc without filler rod. Since TIG allows you to create a very intense and well-controlled weld pool, this can be used to fuse two parts together without adding filler rod in a process known as **fuse welding** or **autogenous welding**. This should be used with some restraint as the joint ends up thinner because the material filling it is "borrowed" from the two pieces. Also, the two pieces must be in *very* close contact or the weld pool will not cross between them.

18.2.6 Resistance Welding

Resistance welding uses high electrical current flowing *through* the parts to be assembled, which causes them to heat up and bond. This is different than other welding methods discussed so far because there is no heat generated by an arc or flame, there is no torch, etc. Resistance welding works by applying a large amount of current between two electrodes. As the current flows, anything between the electrodes is heated up due to its electrical resistivity. This process is shown schematically in Fig. 18.24.

The electrodes are often made from copper so that they do not heat up as much as the workpieces (low electrical resistivity), and they can also be water cooled. The major variables in resistance welding are the amount of current (amperage) and the weld time. Most resistance welding is used on thin parts, especially in sheet metal assemblies.

*This assumes you are doing the work safely and cost effectively, of course!

Resistance welding can create strong joints, can be automated through robotic operation, and requires no consumables such as filler rod or wire. Because it relies on two electrodes to clamp onto the parts, holding the work pieces must be considered as well as the workpiece size, which may be limited by the electrode layout, but these considerations vary with technique, and we will look at two of the most common next.

18.2.6.1 Spot Welding

Spot welding creates a small fusion zone between two electrodes clamped onto the parts. The weld size and penetration is dependent on machine and material properties, but it is usually closely correlated to the electrode size and shape. Common electrode shapes include flat, domed, truncated, offset, etc. Because melting the materials requires a large electric current, it is only efficient in thin materials (less than 1/8th in). The process can be completed in less than a second and the surrounding area is not significantly heated, thereby reducing concerns associated with the HAZ (Sec. 18.1.2).

The welded section, called the **weld nugget**, is completely fused and quite strong. Because of this, a series of spot welds placed intermittently along a joint is typically sufficient to create a suitable assembly for applications that are typically supported by other components. For instance, spot welding is used extensively in automotive assembly to build the body from sheet metal and attachments at thin sections. This process is more easily automated than other welding methods.

A related method, **projection welding**, uses the same principles as spot welding but the contact area is reduced by a projection of some size and shape to control the welding process. The pressure applied is sufficient to flatten the components so that the final assembly is comparable to spot welding. Multiple projections can be used to create more than one weld in a single operation, and the principles of the method can be applied to any assembly of components with a limited contact area where the welding occurs. Examples include the welding of threaded inserts, brackets, and overlapping components such as wire mesh fencing, dishwasher racks (also vinyl coated), and wire containers, such as shopping carts. Representative projection welding processes are shown in Fig. 18.25.

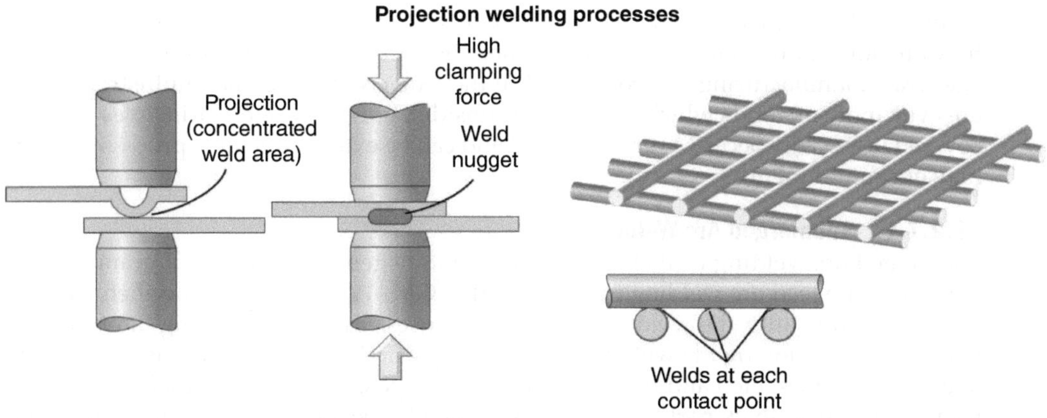

Figure 18.25 Projection welding can be used to create assemblies by controlling the weld zone with raised areas or by welding over many contact points.

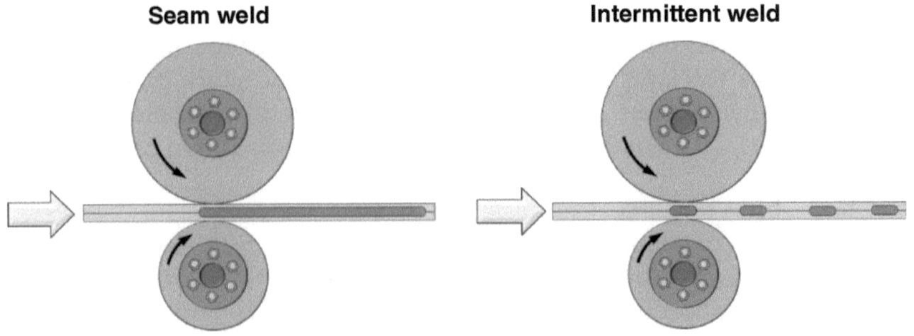

Figure 18.26 Seam welding can be done continuously or intermittently to achieve varying weld intervals. The equipment uses rollers as electrodes to achieve this.

18.2.6.2 Seam Welding

Although a series of spot welds can be created by moving the spot welder or assembly from point to point, stronger joints can be made more efficiently by maintaining continuous contact with the workpieces. This can be accomplished using **seam** welding, where the workpieces are fed through wheels that double as the electrodes and can apply electrical current continuously or intermittently (Fig. 18.26). In a **continuous seam weld**, the electrodes are powered throughout the operation and create an uninterrupted weld zone. In **intermittent seam welding** (also called **roll spot welding**), the part is advanced a set amount and a spot weld is created. This is faster than regular spot welding as the wheels are already in contact and advance the work. A continuous seam weld will be stronger than an intermittent one. The difference is similar to a zipper and buttons. The zipper (continuous) leaves no gap of disconnected material as buttons (intermittent) will. The continuous weld requires more weld time and higher energy input, and should only be used where necessary.

18.2.7 Other Welding Methods

There is a great variety of welding methods in addition to the ones already discussed. Although the following methods are summarized rather briefly, it does not indicate they are not commercially important. Rather, they are less likely to be encountered in a "general" manufacturing environment, but they are critical to certain industries. There are even more methods which we will not consider, but many of the principles we have covered (especially Sec. 18.1) are universal and can be extended to new processes as you may encounter them.

18.2.7.1 Submerged Arc Welding

Submerged arc welding (SAW) covers a joint with a generous coating of granular flux, applies a wire feed arc welding head (like MIG/GMAW, Sec. 18.2.4), and then recovers the unused flux after the weld solidifies. The process can be used along large components by either moving the welding head or the part. The weld is clean (no spatter, limited smoke, fumes, etc.) and has a high material deposition rate. This makes it suitable for welding large, thick parts such as heavy-walled pipe, bridge spans, and other structural members. Unused flux is recovered, often by means of a vacuum system, and it is reused. The primary limitation with SAW is that the flux is placed on the joint using

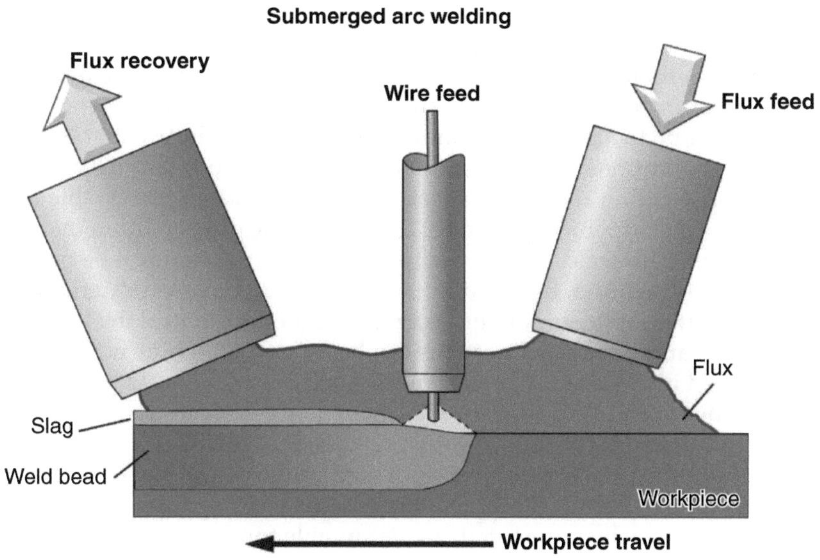

Figure 18.27 Submerged arc welding (SAW) is a common technique for joining large sections efficiently.

gravity, so the weld orientation must be horizontal and relatively flat. Pipe can be welded by rotating the joint under the welding head. The general process is shown in Fig. 18.27.

18.2.7.2 Electron and Laser Beam Welding

Most of the welding processes we have looked at so far result in a relatively large HAZ. As the weld depth increases, the lateral dimensions also increase. The more concentrated the heat source, the smaller the HAZ. Because of that the HAZ for gas welding is larger than arc welding, but there are applications where a deep, narrow weld is required, and that requires a method that can impart a large amount of heat in a very concentrated area to melt the metal very quickly (Fig. 18.28).

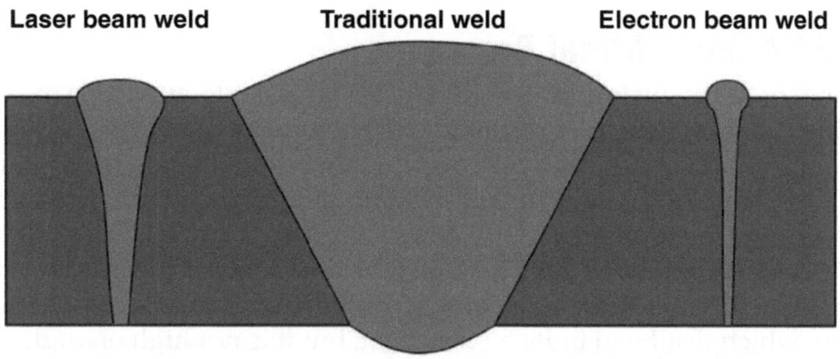

Figure 18.28 Electron and laser beam welding create deep welds with very small heat-affected zones when compared to traditional welding. The cost of equipment is higher and they are used for specialized, high-value products or those where very small welds on thin material are needed. Actual profile shapes depend on the applied beam power and spot size.

Electron beam (e-beam) welding uses a concentrated flow of high energy electrons to melt a metal in a very small area quickly. The electrons are created using very high voltage under high vacuum so air molecules do not interfere with their travel. Therefore, higher vacuum results in greater transferred energy and a deeper the weld. These electrons can be focused and accurately aimed using electromagnetic lenses for high-precision applications. Even from that brief description, you may get the impression that e-beam welding requires a specialized setup and is not suitable for most general applications, such as bicycle frames, and the applications require the accuracy and/or a high aspect ratio weld profile. Because the electron beam can be precisely controlled for energy and position, the weld characteristics can be precisely engineered. The beam can be oscillated, split, and otherwise manipulated to control the weld profile and cosmetic appearance, and weld patterns can be created with high accuracy. Most often the beam is fixed in place and the workpiece is moved as the weld progresses.

Laser beam welding uses a very high energy laser to quickly melt the metal and create a fusion zone. The laser energy and spot size can be controlled to distribute heat, with higher energy and a smaller spot size resulting in deeper welds, though not as pronounced as e-beam welding. Unlike e-beam welding, the process is effectively performed in air, is easily automated, and is suitable for high-production applications. The laser source can be positioned feet away from the workpiece, allowing the weld to be placed in otherwise difficult-to-reach locations. Alternatively, laser energy can be used in conjunction with arc welding (particularly MIG/GMAW) to increase weld deposition rates. Laser energy can also be used to cut, etch, and perform various operations by modifying the energy of the beam.

Neither laser nor e-beam welding require shielding gas, they can produce deep, narrow welds, and they are typically autogenous. Either can be used to weld small spots or create a continuous seam and they produce clean, high-quality welds without requiring highly skilled operators. The setup of the processes is critical, but where multiple parts are being made, the process is primarily controlled by the equipment. Both processes have safety considerations regarding the beam. E-beam welding produces x-rays as a byproduct of the energy transfer from the electrons to the material, and in laser welding the beam may be reflected off the surface and can cause bodily harm, especially eye damage. Arc welding has comparable concerns, but in e-beam and laser welding the equipment setup generally addresses primary hazards.

18.3 Cold Welding Metal Processes

Welding requires fusion of two distinct components such that no distinct interface remains. That is difficult to achieve without melting or dissolving the components. Since most metals have relatively high melting points (at least compared to polymers), it can be difficult to achieve fusion without high temperature. The ductility of metals is critical to making it possible.

Cold welding includes processes that fuse materials together without exceeding the melting point. Fusion is usually accomplished through considerable pressure and friction, which results in quite a bit of heat, but it is not high enough to be considered a traditional, liquid-state welding process. This type of welding process requires specialized equipment and is most often more expensive to produce parts. For this reason, it is limited to applications with specific requirements that cannot be achieved by traditional welding processes.

18.3.1 Friction and Friction Stir Welding

Friction welding uses high pressure and abrasive motion between parts that are welded at the site of the friction. Often one part of the assembly is rotated or continuously oscillated against a stationary component to create frictional heating. The amount of force can be varied during processing to control the heat input, but the process culminates with the two parts having a high temperature at their interface and being forced against one another while they cool. You can try this with two pieces of chocolate, applying pressure and friction will allow the two pieces to bond, even though they have not fully melted. Since metals are often ductile, they too can be "smeared" together with sufficient heat and pressure. The easiest way this can be achieved is by rotating one component against another, fixed component. By applying varying pressure during the process, the two can be bonded together, and this can be applied to solid or hollow structures.

The process must be carefully designed so that rotation stops once the weld begins to solidify. Otherwise, the weld can be damaged. The amount of interference (overlap) of the two components during welding will result in excess metal at the interface (i.e., *flash*) that can be left as is or removed later. A distinct advantage of this process over melting the components is the ability to join dissimilar metals. Because of the plastic deformation (discussed in Sec. 7.4.3) at the interface, impurities are pushed out of the joint, and this method generally results in high-strength welds.

Friction stir welding (FSW) uses a rotating, nonconsumable tool to physically stir material together at a joint (Fig. 18.29). It has similarities with friction welding and milling operations (Chap. 17). The rotating tool (called a *pin* or a *probe*) is most often carbide (see Sec. 6.2.2), so it is strong at high temperature and wear resistant. The shape of the tool is important because it determines the stirring characteristics. High force is required to create the weld, so extremely rigid, often expensive equipment is required. FSW produces a high-quality joint, can weld dissimilar metals, and has limited heat effects on the surrounding material. Because of the force generated, the workpieces must be rigidly clamped and the setup time is significantly higher and versatility and simplicity are lower than traditional welding. Due to the setup requirements, the most common weld arrangements involve butt joints.

18.4 Polymer Welding Processes

There are two primary methods for welding polymers: heat and dissolution. Due to their low melting points and chemical reactivity, most polymers can be welded. The joint types correspond to those in metals (see Fig. 18.4), and much of the preparation is related as well (bevels, cleanliness, etc.). The concepts of melting point differences are also important, and material chemistry is important too, especially for dissolution methods. Because polymers can be so easily molded (Sec. 11.3) and formed (Sec. 12.3), welding operations are often avoided. That does not mean they are uncommon, however, and we will look at several of the most utilized methods.

18.4.1 Gas Welding Polymers

Gas welding polymers is much different than gas welding metals. The required temperature is much lower, so the source of heat is different. **Polymer gas welding** uses hot air to produce a molten polymer base material into which a filler rod is laid. Polymers

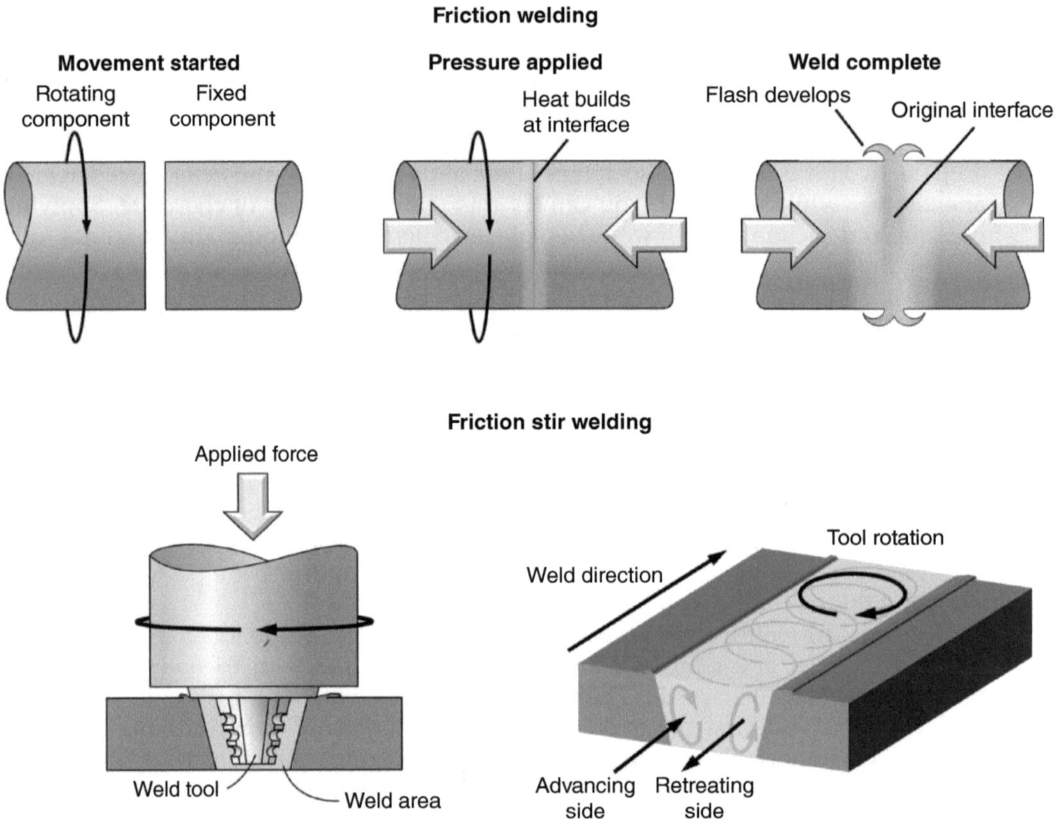

Figure 18.29 Friction welding uses the heat generated by friction between two components to soften the material at the interface and bond the metals. Friction stir welding uses a separate, nonconsumable tool to mix materials at the joint interface.

do not melt and become fluid like metals. Rather than creating a molten weld pool, the base material is heated until soft or semi-molten, then the filler material (also heated by hot air flow) is laid along joint in conjunction with the advancing heat. The heat gun has adjustable output and has a small, interchangeable nozzle to change the distribution of heat. The type of polymer will determine the appropriate welding temperature, but thermosets are not welded with heat, though they can be welded otherwise. Most often the parts to be welded are of the same polymer, though some dissimilar polymers may be welded this way. The equipment for gas welding polymers is relatively inexpensive, and the method is often used for repair work or small-scale fabrication. A gas welding tool and example filler rods are shown in Fig. 18.30.

18.4.2 Heat and Pressure Tool Welding

Heat and pressure welding includes various methods that use the direct contact of a heated tool with the plastic parts to melt and bond them together directly, or the components are heated first and then pressed together immediately after. Unlike

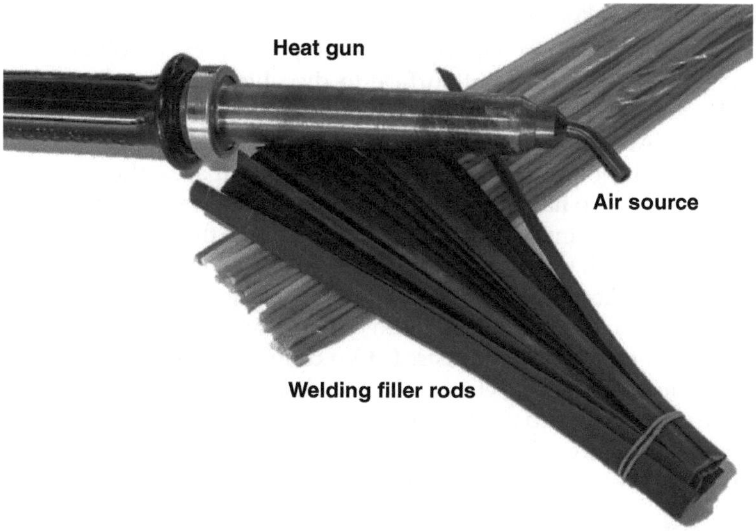

FIGURE 18.30 Plastics are often welded using a specialized heat gun connected to compressed air, as shown here.

hot gas welding, this method can use tools of varying design to directly melt and bond components together without the need for any filler material. The tool can resemble a soldering iron, with interchangeable tips to achieve varying melting characteristics. The heat causes the polymer in the immediate vicinity of the tip to melt, and as long as components are in intimate contact, when they cool they will create a cohesive bond.

There are many variations on this general method. For instance, two components may be held in contact with a hot surface and then put in contact with one another before cooling. The hot tool may be a wire, plate, or any other shape suitable for the desired joint. Two pieces of plastic may be pressed between hot platens to stick them together, such as vacuum storage systems for food. Many packaging applications use heat applied at the edges to seal the package closed. This can be done in a production setting using heated rollers (or heat then rollers) to continuously feed the components through (look back to Sec. 18.2.6.2, for a similar concept). Any method where the polymer is directly heated by thermal transfer before welding could fit under this category.

18.4.3 Friction Welding

Another method that uses heat to weld is friction welding. This corresponds directly to friction welding in metals (Sec. 18.3.1) where pressure and motion are used to create heat at the interface. That heat is sufficient to cause the polymer to melt and the components to bond. The motion can be linear vibration or rotation. Round parts (plastic wheels, travel mug liners, etc.) are often bonded using spin welding, which requires fast rotation and high pressure to combine the components. Excessive heat and pressure will result in excessive flash, which is undesirable and may result in the parts being unacceptable.

18.4.4 Solvent Welding

One of the most common methods for joining polymers is **solvent welding**, in which a solvent is applied to the joint surface to dissolve and fuse the polymer components. Often this solvent vaporizes relatively quickly, so the two components must be assembled quickly. After they are in contact, they bond at the molecular level, and as the solvent escapes, the joint solidifies. The larger the joint area, the longer it will take for the solvent to escape, so handling should be minimized until full hardness is reached.

A common polymer solvent is acetone. This is the active ingredient in nail polish remover, and it will damage most plastics that is comes in contact with (including furniture finishes). Acetone will not dissolve all plastics (like olefins, such as polyethylene or polypropylene), and there are often better choices, though their availability and cost may be problematic for hobbyists. If two different polymers are to be combined, it is important to check their chemical compatibility, as provided in Table 18.2. More extensive polymer compatibility charts are available from chemical suppliers, material suppliers, and many laboratory supply companies. Polymers that are *not* compatible with a solvent are desirable for solvent welding.

18.4.5 Ultrasonic and Radio-Frequency Welding Methods

The use of high-frequency vibration or energy waves can be used to melt and fuse polymers. **Ultrasonic welding** uses high-frequency (tens of kHz) vibration to heat and weld two components. One of the components often has small protrusions (spikes) on it to concentrate the heat and localize the melting. The two components are forced together during the operation using a fixed anvil, and the vibration is induced by the ultrasonic *horn* or *sonotrode*, and this creates a strong bond. The method is typically quick, only taking a second or less, and it can be automated to create assemblies of a wide range of polymers (even dissimilar ones). It can also be used to activate certain adhesives. It is used extensively to create packaging, especially where the contents would be damaged by heat (e.g., electronics, food, medical, etc.).

Radio-frequency (RF) welding uses high-frequency (tens of MHz) generators to melt specific areas of polymer components to be joined. The radio waves cause the polymer molecules with dipoles (see discussion in Sec. 2.3.2) to oscillate, and this oscillation results in internal heating that melts the material. The polymer must be susceptible to the high-frequency waves so that the energy is absorbed rather than transmitted. Polymers that are not suitable for this method include polyethylene, polystyrene, and fluoropolymers.

18.5 Ceramic Welding Processes

Ceramics are very difficult to weld because they typically have very high melting temperatures and are chemically inert and brittle. These characteristics make melting, dissolution, and frictional welding impractical. For these reasons, ceramics are often used as separate components or secured using fasteners or adhesives (see Chap. 19). The nearest ceramic welding process sprays metallic powder and oxygen onto a very hot ceramic base material. As the metal oxidizes, it releases additional heat and bonds to the substrate forming an oxide build-up. This is used to repair industrial furnaces where the ceramic lining is degraded or damaged and has the benefit of not requiring the equipment to be cooled as would be necessary to exchange the refractory lining. The most likely ceramic to be welded is glass, and that is typically for decorative products and some laboratory glassware.

Polymer/Solvent Compatibility									
	Solvent								
Polymer/Plastic	Acetone	Benzene	Ethyl Alcohol	Isopropyl Alcohol	Methyl Ethyl Ketone	Methylene Chloride	Tetrahydro-furan	Toluene	Xylene
Acrylonitrile Butadiene Styrene (ABS)	D	D	B	C	D	D	D	D	D
Buna N (Nitrile)	D	D	A	C	D	D	D	D	D
Epoxy	B	A	A	A	B	A	A	A	A
Natural Rubber	D	D	A	A	D	D	D	D	D
Polyethylene (PE)	C	D	B	A	D	D	D	D	D
Nylon	A	A	A	A	A	D	A	D	D
Polypropylene (PP)	B	D	A	A	A	D	C	D	D
Polystyrene (PS)	D	D	B	A	D	D	D	D	D
Polytetrafluoroethylene (PTFE)	A	A	A	A	A	A	A	A	A
Polyvinyl chloride (PVC)	D	D	A	A	D	D	D	D	D

Note: Some common examples of polymers and solvents are given here. A, no effect; B, minor effect; C, moderate effect; and D, significant effect. These interactions represent general findings for broad categories of polymers. Some variation will occur within each category and in different environments (e.g., elevated temperature).

TABLE 18.2 Polymer and Solvent Compatibility

18.6 Why It Matters

Welding is critical to creating permanent and durable assemblies. Imagine if everything had to be bolted or screwed together. Fasteners have important roles, but they are susceptible to loosening stretching and breaking in ways that welds are not. Furthermore, welding provides a unique method for repairing components that is not practical by adding fasteners. Large structures such as pipelines, oil rigs, skyscrapers, etc. are only feasible and reliable over long service lives by utilizing welding.

As you know by now, welding is a diverse set of techniques. Some, such as TIG welding are capable of a great deal of finesse and flexibility needed in critical assemblies and sensitive materials. Others, like stick welding, are practical and functional for assembling less sensitive but equally important projects (e.g., pipelines). More specialized methods, such as e-beam welding can achieve precise control over the weld and be applied in thick or very thin parts.

Without plastic welding, packaging as we know it would not exist. Think of any food product that comes in a bag. It is most likely welded by heat and pressure, or by some high-frequency process. Many other packages are also welded closed. The next time you are struggling to open something, chances are you can thank polymer welding for it.

So whether you are riding the train (those tracks are welded), shaving (those little blades are welded on), or out riding your bike (the frame is welded), there are countless items you encounter in your day-to-day routine that involve welding and the various processes for doing it.

18.7 The How and Why of Welding and Cohesive Assembly

18.7.1 Chapter Review

1. Describe what a cohesive assembly is.
2. Name two characteristics that affect the proper cohesion of two materials.
3. Describe how the application of heat affects the area around the weld. What role does the material being welded play in the size of this area?
4. Name the various regions of the HAZ and what changes occur in each.
5. What are tack welds used for?
6. How can the composition of the material being welded affect the final product? Use stainless steel as an example.
7. How does the thickness of the parts being welded affect the process?
8. Why is a small gap left between parts for some welded joints?
9. Describe the difference in weld passes (first to last) when filling a large joint.
10. What steps can be used to reduce thermal distortion during welding?
11. Why would a plug or slot weld be used?
12. What is the benefit of cursive or weave motion when welding?
13. How can weld appearance indicate quality? What is the limit to this?

14. What does a weld bead with high porosity indicate about the welding setup?
15. What role do flux and cover gas play?
16. Why should areas to be welded be cleaned first? How is this typically done?
17. What does weldability indicate?
18. What is a 6G weld? 3G?
19. What are benefits and disadvantages of oxy-fuel welding?
20. Why is acetylene popular for oxy-fuel welding? What are its drawbacks?
21. Describe the differences between a carburizing, oxidizing, and neutral flame in oxy-fuel welding.
22. What audible indication is there that an oxy-fuel flame has too much oxygen?
23. Can a carburizing flame ever be a good thing?
24. What part of the flame should be used to create the weld pool?
25. Describe the difference in forehand and backhand welding and the benefits of each.
26. How do oxy-fuel welding and cutting differ?
27. What is the primary benefit of plasma cutting over oxy-fuel?
28. What are the key components to arc welding?
29. How does the polarity of the welder affect the welding process?
30. What benefit does DCEP welding have for the surface condition of the part?
31. Why is DCEP a good thing for MIG welding and a bad thing for TIG?
32. Can DCEP ever be the right choice for TIG? What electrode considerations should be made?
33. What is the benefit of AC welding over DCEP and DCEN?
34. How does SMAW shield the weld pool and reduce impurities?
35. What is an SMAW electrode coated with, and what are the functions?
36. Why is SMAW unsuitable for automation?
37. How is the GMAW electrode supplied to the weld?
38. What trouble can arise when using soft metal wire, like aluminum, for GMAW? How is it addressed?
39. What is used to protect the weld pool in GMAW? How is it different for FCAW?
40. How is FCAW similar to SMAW? How is it different?
41. Describe the benefits of spray transfer in arc welding.
42. When looking at a weld, how might you know if globular transfer was occurring?
43. When is short-circuit transfer useful?
44. How is GTAW similar to oxy-fuel welding? How is it different?
45. Why are GTAW electrodes made from tungsten?

46. How does electrode tip shape affect performance in GTAW?
47. What role does the foot pedal have in GTAW welding?
48. What is autogenous welding?
49. How is resistance welding different than arc welding?
50. How can spot welding be used to create long joints?
51. How is projection welding used to create assemblies of many components?
52. What benefit does seam welding have over a series of spot welds?
53. Why is submerged arc welding popular for very large components?
54. What are the benefits and disadvantages of e-beam and laser welds?
55. What is meant by "cold" welding?
56. What is the role of friction in friction welding?
57. What limitations are there for creating assemblies by friction welding? What are some benefits?
58. How is friction stir welding different from friction welding?
59. What type of joint is most common for friction stir welding? Why?
60. What is the tool made of in friction stir welding? Why is that important?
61. What methods are used to weld polymers?
62. Why is welding less common in polymers than metals?
63. How is gas welding different for metals and polymers?
64. What can be used to create welds by heat and pressure? What some common applications?
65. What is the most fundamental consideration in solvent welding?
66. What is one of the most common solvents for polymers, and in what household item is it located?
67. If a polymer is compatible with a solvent, what does that indicate about its use for welding?
68. What challenge arises when welding polymers like PTFE?
69. How do ultrasonic and radio-frequency methods melt the polymers being welded?
70. Why are ceramics not often welded?

18.7.2 Concept Connections

1. MIG wire for welding steel is often coated with copper. Use the concepts discussed in Secs. 3.2.4.2 and 8.5.2 to justify this combination.
2. Describe the similarities between friction stir welding and vertical milling (Sec. 17.2.1). How does the tooling differ? Now compare rotational friction welding with lathe work.

3. Many lightweight bicycle frames are made from aluminum. What welding method would you suggest to join the parts? Why? Describe your answer based on a bicycle frame you own or an image of one.
4. If you have to harden parts in a welded assembly, would you harden them before or after welding? Thermal conditioning is described in Sec. 9.3. Use terminology from that chapter and this one to justify your answer.
5. Let's say you are a bolt salesperson. How would you convince a customer to use your products instead of welding? Threaded fasteners are discussed in Sec. 19.2.2.
6. Describe the similarities between MIG welding and fused-filament deposition (Sec. 21.2.1). How are they different? Could you operate MIG the same way? What considerations would need to be made?
7. Consider the sustainability of products that are welded and bolted. Which requires more embodied energy? How might it affect shipping space required and, therefore, the fuel usage?

CHAPTER 19

Mechanical and Adhesive Assembly

The ancient Greek scientist and philosopher, Aristotle, coined the phrase, "The whole is greater than a sum of its parts." Every day we use countless objects that would have little value if we only had the individual parts that make it up. A vehicle would be a pile of body panels, tires, connecting rods, axles, a steering wheel, and so on. A computer would simply be a screen, circuits, drives, and keys. Without combining many complex and often dissimilar components, even the most precisely created parts are mostly useless. If you have ever bought furniture that must be assembled, you know that just having the right parts is not enough. The parts must be brought together and secured in the proper arrangement. The way the chaos is converted to a finished product is through a wide array of fasteners and adhesives, carefully selected to achieve the proper function. From the largest structures to the smallest devices, the choices and methods in creating assemblies are universally important.

19.1 How It Works

Cohesive assemblies were discussed in some detail in Chap. 18. The primary difference between those methods and the ones discussed in this chapter is the nature of the bond. In cohesive assembly, as discussed there, the two components are fused together such that no distinct interface between the components remains. Two parts become one at the atomic or molecular level. Now we will consider ways of combining parts where the two components remain distinct after they are joined. There are two distinct categories that satisfy this description: mechanical assembly and adhesion.

19.1.1 Mechanical Assembly

Mechanical assembly involves the joining of two components using a separate member to secure the assembly together. This may create a permanent or reversible assembly, and the means of securing the assembly can be a single component or a more complex system. For example, two boards can be assembled by a single screw. Alternatively, a bracket, comprised of a metal brace and several screws, may be used. The components may be dissembled by removing the fixture holding them together, either by design or otherwise. A cell phone case is designed to semipermanently enclose the phone, but it is also designed to be reversible so it can be changed at will. While it is on, it should not loosen, move, or come apart. When it is time to come off, it should be simple, not require special tools, and be quick.

When building a deck, the boards are not intended to be exchanged often, so they should be secured in way that permanence is emphasized. There may come a day when those boards will be removed, but the primary goal is to keep them in place over a long period of time. Some mechanical assemblies must be destroyed to break the assembly apart. One example is a rivet, which must drilled out to remove, and nails can be removed, but they are often bent or otherwise damaged and should not be reused. We will discuss each of these fasteners in this chapter, and there are many others that operate differently. Indeed, some mechanical assemblies do not require any fasteners.

19.1.1.1 Considerations in Mechanical Assembly

Although fasteners and other methods for mechanically assembling components can be used fairly effectively without much thought, there are important considerations to maximize their effectiveness. The method that is simplest, fastest, and that achieves the desired result is optimal. There is no reward for overbuilding an assembly, but there are real consequences for underbuilding it. If in doubt, error on the side of safety.

There are many potential considerations for mechanical fastening, but we will focus on a few common ones. These are related to strength, longevity, and disassembly. When assembling plastic parts, it is not beneficial to use high-strength steel bolts, just as it is dangerous to use low-strength bolts in critical structural assemblies. Strength is not the only consideration, and the compatibility of materials is also important. As discussed in Sec. 8.5.2, corrosion can be inhibited or accelerated by design choices. If the wrong combination of metals is used, the bolt itself may rapidly corrode and weaken. The fastener should be the same material or the more passive (cathodic) member of the assembly due to its small volume.

The simplest fastener or joint should also be used. If the assembly is meant to be permanent, a more expensive removable fastener is a poor choice. It does not make sense to add capability that is not required. If the assembly should be removable, the number of fasteners should be minimized. In cases where a tight seal is necessary, such as an oil pan, it may make sense to have many bolts distributed around the perimeter. In this case, strength is not the primary consideration, so smaller bolts can be used. Also, keep in mind that some assemblies are not meant to come apart easily, but they may need to someday. Components on an engine, like a water pump, are not designed for easy removal, but if one goes bad, there must be a way to remove it.

You may have heard of "torque specs" before (torque is discussed in Sec. 7.3.3). These are specifications for how tight a bolt should be to properly hold the assembly. Over- or under-tightening the fasteners can damage the assembly or insufficiently hold it. Threaded fasteners with fine threads are stronger than coarse threads and each rotation results in a smaller advancement of the bolt. This reduces the ratio of the torque to the stress applied to the bolt. Also, if the thread contact is clean and lubricated, the torque will be smaller for a given bolt stress as it reduces the frictional resistance. The strength of the bolt and the amount of torque should be known to ensure the proper clamping force and to avoid breaking the bolt. A basic equation to determine the force, F(lb), exerted by a bolt of diameter, D (in), at torque, T(ft-lb), with a known coefficient of friction, k (unitless), is

$$F = \frac{12T}{kD} \tag{19.1}$$

Equation (19.1) indicates that torque and force are linearly proportional. If torque doubles, the force will double. It is also determined by the bolt diameter, and a bolt half the

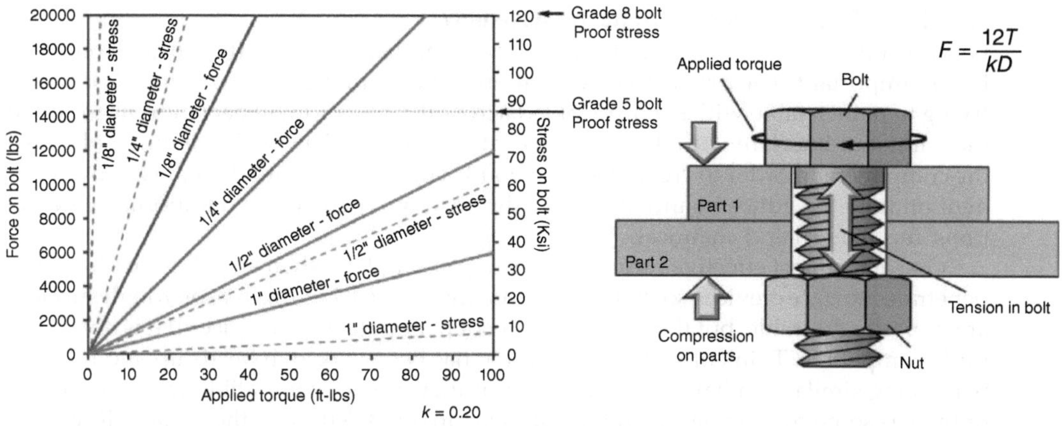

Figure 19.1 The proper size of mechanical fasteners is determined by the required force to hold components together and the expected loading during use. Here the relationship between tightening (torque) and applied force and stress is compared. Note that stress rises faster than force as diameter is reduced. Smaller fasteners are more likely to break for the same holding force as larger ones. Values assume $k = 0.20$. Actual value depends on materials in contact and presence of lubricant.

diameter will produce twice as much force. It might sound tempting to use smaller bolts then, but in reality, the stress on a smaller bolt increases even faster and the bolt will fail before achieving the same clamping force (consider reviewing Sec. 7.3.1 about the stress/force relationships). Because of this, the working stress of the bolt should be considered, which includes the clamping force and any additional force during use, so that a safe margin is maintained between the expected stress and failure stress (Fig. 19.1). Even for a single size of bolt, varying strength is available. The different grades of bolts are described more in Sec. 19.2.2.4.

19.1.2 Adhesive Assembly

Adhesion processes use a separate material (the adhesive) to bond together distinct components at their interface without the need for physical overlap or interference. This includes all material types (i.e., metals, polymers, ceramics, and composites). The adhesive layer can be liquid, semisolid, or solid, such as glue, paste, and tape. Their primary operation is dependent on creating a strong attraction between surfaces, but the exact mechanisms are dependent on the specific adhesive. If you make a trip to the local hardware or craft store, you will be met with hundreds of choices for joining anything from tile to fabric. Each formulation is created to withstand the requirements of the application (water resistance, flexibility, temperature variations, etc.). Adhesives are designed to achieve attractive forces by van der Waals forces (see Sec. 2.3.2.2) or stronger bonds (known as chemisorption*), or the adhesive is designed to penetrate the surface of the components, harden and lock them together. Multiple effects can be operating simultaneously, and the science behind adhesive bonding is still being actively researched.

*Chemisorption occurs when a substance (the adsorbate) forms bonds with a surface. If the substance and surface do not form new bonds, it is physisorption.

19.1.2.1 Considerations in Adhesive Assembly

Because adhesion relies on combining two parts at their surfaces, the surface condition is very important. The primary consideration is the cleanliness of the surfaces. Imagine trying to paint a table with sand sprinkled across it. The sand will become embedded in the paint and interfere with its ability to coat the surface. The adhesive must fully coat the component in order to create a strong bond. The strength of that bond is also dependent on how the adhesive interacts with the surface. There are two important interactions: mechanical and chemical.

One of the most effective means for creating strength is when an adhesive physically penetrates surface crevices so that it locks into the surface. For this reason rough surfaces are typically desirable, but the adhesive will not penetrate these crevices if it is not chemically compatible. Think of water on a car that has been waxed. It beads up and rolls off. Something similar can happen to adhesives if they are not compatible with the surface or there is some contamination. If the adhesive does not "stick" to the surface, it will not hold parts together. The adhesive should "wet" the surface (spread out on it).

Adhesives are difficult to use with some materials, like PTFE (see Sec. 6.3.1.7), because they do not bond well. In other cases, a primer, or a layer that acts to improve the bonding strength, can be used. Most adhesives are thinned with a solvent so it will flow more readily into the surface features, and after it has evaporated the adhesive hardens. On large, overlapping surfaces it will take longer for the solvent to evaporate. Some adhesives, especially for polymers, will use a solvent that also dissolves the component surface to create a sort of welded assembly, though it may not be completely cohesive (polymer welding is described in Sec. 18.4). The concepts of wetting and mechanical interlock for adhesives are shown in Fig. 19.2. The surface should be clean and slightly rough (where possible) to enhance the holding power.

The design of the joint is also important with adhesive assembly. When loaded in varying directions, the adhesive material will carry a different proportion of the load (see loading configurations in Sec. 7.2.1). Under tension, the adhesive is the focal point of the stress and the strength is determined by its properties. This includes whether it is brittle or not and the maximum stress it can withstand before breaking. When you peel a sticker off of a product, the peeling action focuses the stress on a small portion of the joint and this reduces the total force necessary to separate the joint. A standardized peel

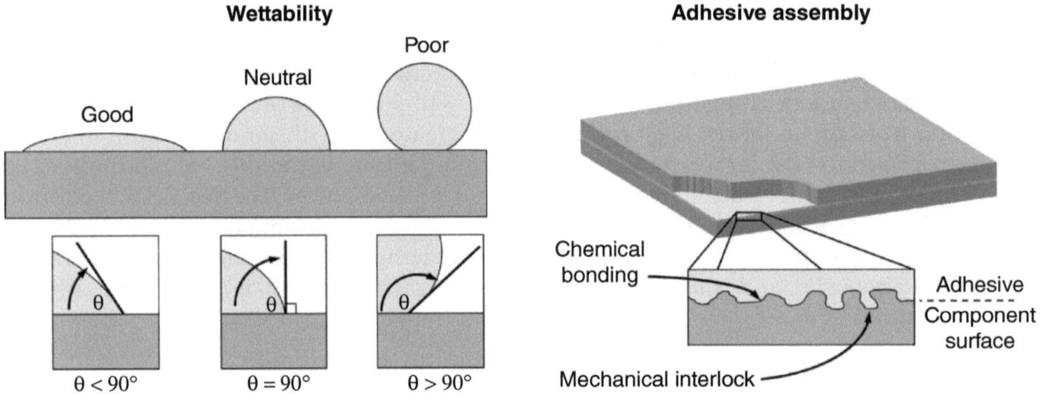

FIGURE 19.2 For an adhesive to work well, it should result in a chemical and mechanical bond.

test is a common way of testing the strength of adhesives. Shear force and impact strength are also important as they can preferentially damage the adhesive. Compressive force is most desirable, because the force is carried by the components and forces the joint closed so that the adhesive does not carry the load alone. Joints that increase the bonded surface area and shield the adhesive from unwanted stresses are often used.

19.2 Mechanical Assembly Methods

Mechanical fasteners hold components together by applying force and/or physical interference to keep parts together. There are thousands of variations and millions of possible combinations to achieve this end. Fortunately, there are typical ways this is done that we will focus on and try to distill the concepts to the most common. As discussed in the prior section, there are considerations regarding the materials, design, application, and cost that should inform any choice.

19.2.1 Permanent Physical Interlock

Physical interlock assembly uses interference between two components to lock them together. This can be done permanently using the material of the components or a third component. Although it may be possible to reverse the connection, it cannot be done without damaging one or more of the components involved.

19.2.1.1 Sheet Metal Seams

Physical interlock is often applied in sheet metal where the metal is folded over to form a seam. The seam can be used to eliminate a sharp edge and add strength, or two sheets can be folder together to create an assembly. Examples of some common sheet metal seams are shown in Fig. 19.3. Because of the tight bend radii, a thin, ductile metal

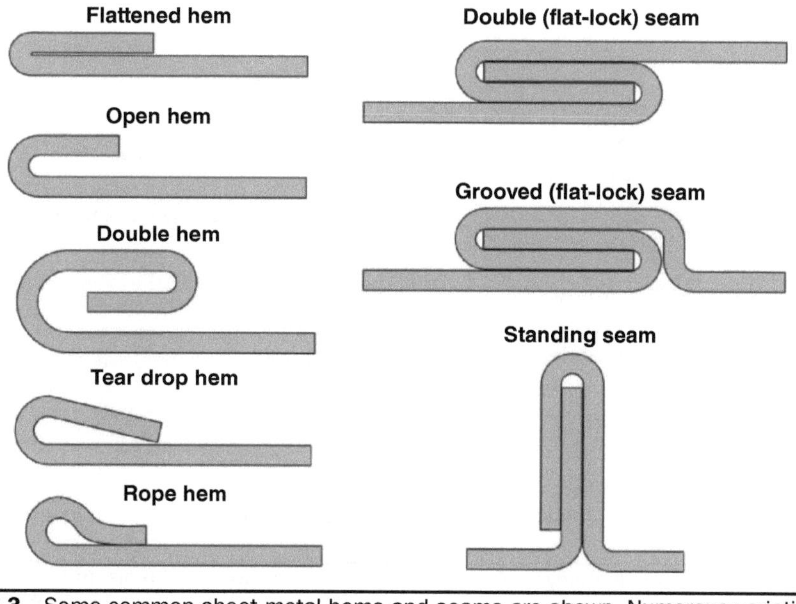

FIGURE 19.3 Some common sheet metal hems and seams are shown. Numerous variations involving these and other bends are used to assemble sheet metal products.

is required. You can even replicate this challenge with junk mail. Single sheet paper is easily folded, but the glossy card stock mailers require more force and will tear at the surface when bent sharply. When the metal is not highly ductile, an open seam or tear drop hem will be preferred. For how these bends are made, refer to Sec. 12.2.4.1.

19.2.1.2 Rivets

Rivets are fastening devices characterized by having two ends of greater diameter than a central shaft, or *stem*, with the purpose of holding components together between them. The rivet starts with one end, or *head*, and a stem. The rivet is inserted in a preexisting hole and the other end is deformed to create a secondary head. After the two are complete, the assembly cannot be disassembled without damaging the rivet or the components being held together. The stem of a blind rivet is notched so it intentionally breaks within the head and leaves no sharp surface. This type of riveting process, known as "pop" riveting, is shown in Fig. 19.4. In larger rivets, as would be used in bridge structures, solid rivets are deformed by impact to "mushroom" out the stem on the opposite side. These can also be "semi-tubular," where the end to be deformed is hollow (not the whole length), which makes it easier to deform the end out.

The rivet head shape can be hemispherical, conical, flat, or other variations, and the shaft can be solid or partially hollow. They can come in a range of sizes for varying applications. For instance, a bridge or large building will require much larger rivets of stronger material than is required for a sheet metal assembly. In some cases, rivets can only be accessed from one side of a joint. In this case, a blind rivet is used, which has a central shaft with an enlarged end that deforms the back side of the rivet. This is most common in small rivets, and design should reduce the necessity for this where possible. Some examples of rivet types and applications are given in Fig. 19.5.

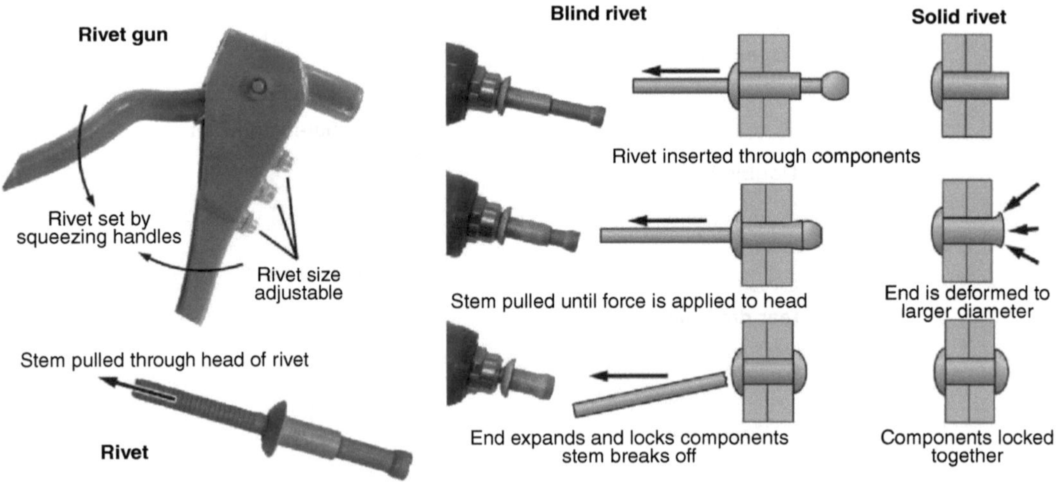

FIGURE 19.4 Rivets come in many forms, but the most popular for small projects are blind rivets which can be installed with an inexpensive rivet gun. Larger rivets are installed using impact to deform one end, rather than pulling a central stem through the rivet sleeve.

Figure 19.5 Rivets are used in a variety of applications from ships and bridges to small consumer products. An assortment of small rivets is shown, as is the box of a truck.

19.2.2 Threaded Fasteners (Screws and Bolts)

The most common reversible assembly method is to use threaded fasteners. These are included on almost every product containing multiple parts. From eyeglasses to dishwashers, the assembly uses threaded fasteners to combine parts that are structurally and aesthetically important. The concepts of threads are presented in Sec. 15.5.1 (thread designations and terminology) and Sec. 16.5.6 (multiple lead threads and various thread forms). Refer to those sections if unfamiliar with the general concepts. Most threaded fasteners are created by rolling the threads, which gives them better strength than cut threads (see rolling operations in Sec. 12.2.3). In this chapter, we will focus on the design and application differences. There are two primary categories of threaded fasteners that are commonly used and confused. These are screws and bolts. We will start by considering how they are defined and what makes them different.

19.2.2.1 Features of Screws and Bolts

Have you ever "screwed" in a bolt? Even basic terminology can make the differences between screws and bolts seem minor, but they are very important. Considering the central features of the two. The greatest difference between a screw and a bolt is that a screw is "driven," or caused to rotate, from a feature within the head of the screw, and a bolt is driven by an external feature of the head, typically a hexagonal profile. This is not a flawless definition though, as some fasteners may be tightened using either. Using an alternative definition, a bolt would pass through a hole in the components and require a nut to be tightened, and any threaded fastener that threads into a component directly is a screw. Seem confusing? In practice, the terms are often used interchangeably, and a bolt may have features more commonly associated with a screw and vice versa. For instance, if you do an internet search for an image of a lag bolt, you get the same results as when you search for a lag screw. We will consider specific examples of screws and bolts in the following sections, but be aware that you may hear or come across varying terminology.

There are features common to both screws and bolts. The primary considerations are the form of the threads and the head features. The threads are the most critical, as they determine how far the screw or bolt advances with each rotation, the strength of the assembly, the need for mating threads in the components or a nut to create the assembly. Threads used with wood will differ from those for concrete, plastics, etc. The screw or bolt head will determine the tool used to tighten it, and some geometries are more efficient at transferring torque, easier to align, etc. A deeper feature that engages

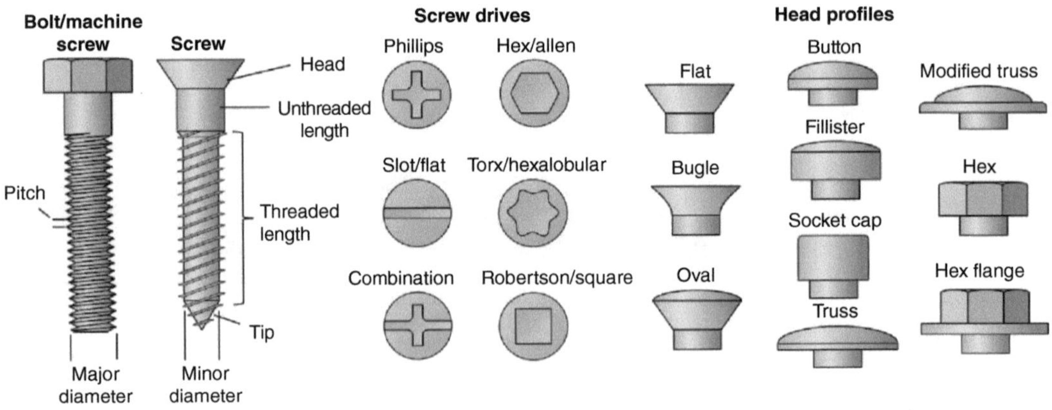

FIGURE 19.6 Threaded fasteners come in many varieties. Some of the key differences are related to the thread form, method of driving them, and the head profile. Some common forms are given here, but there are many more.

with the drive geometry will prevent "cam out," where the tool slips in the head. The profile of the head determines how much, if any, of the head protrudes after tightening, how the force is distributed, and the holding characteristics. A wider head spreads force over a greater area and accommodates holes that are somewhat oversized or off-center in bolted assemblies. Examples of common threaded fastener features are given in Fig. 19.6. There many other variations available and thousands of possible thread, head, and profile combinations.

19.2.2.2 Standard Screws

There are two basic types of screws. The first does not require mating threads and the second does. A standard screw, often simply termed a *screw*, is common for assembling wood, plastic, and sheet metal by forming an internal thread as it is inserted. These screws most often have a tapered minor diameter that makes starting the screw in the hole easier. The thread pitch varies based on application, but the pitch is much greater (coarser) than bolts of the same diameter. The tips of screws often come to a sharp point (called a *gimlet point* when threaded), but they may also be flat or tapered with a flat end (a truncated cone or semi-conical). Most screws benefit from a preexisting hole (a pilot hole), but some have a modified tip that acts as a drill to remove material ahead of the thread and/or that helps cut threads into the material, rather than deforming the material to create them.

The larger the screw diameter, the more critical a pilot hole becomes. The pilot hole locates the screw, but even more importantly it reduces the force needed to drive the screw into position and reduces the residual stress on the components being assembled. This is particularly important with materials which are susceptible to breakage, such as wood and plastics, but it depends on the specific material. For example, softwoods should have a pilot hole no smaller than 70% of the screw's minor diameter, but hardwoods should have a hole no smaller than 90% of the screw's minor diameter. The softer wood can deform more easily, and the additional compressed material increases the strength of the assembly. If the pilot hole is too large, the screw will rip out more easily. If it is too small, the screw may break while being inserted (see Fig. 15.21 for a similar example).

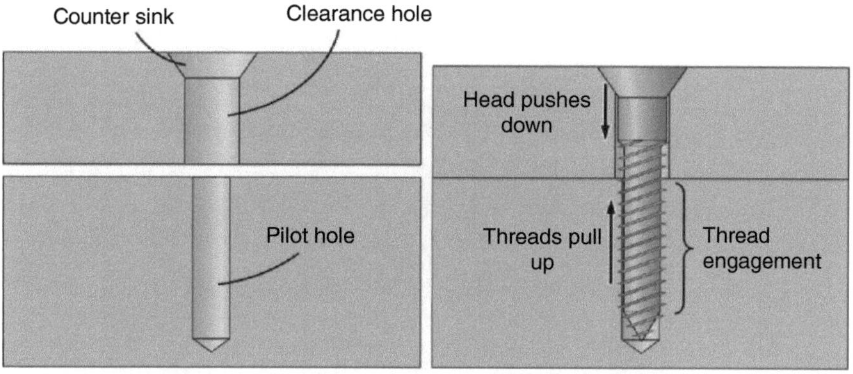

FIGURE 19.7 Screws are very simple to install, but the be done properly, some hole preparation is required.

To ensure proper compressive holding is developed, one of the two components should provide clearance, and the other should interlock with the screw threads. This combination results in the engaged component being drawn toward the screw head, and the second component is compressed between. If the screw is equally driven through both components, the two will not be drawn together. They will only be held in place (no compression) by the threads. Although you may be able to "get away" without using a clearance hole and pilot hole in your assembly, it is not advised. The basic features to consider are shown in Fig. 19.7.

Since screws can serve a diverse range of applications, the design of the screw will vary. Some typical screws include wood screws, masonry screws, sheet metal screws, drywall screws, etc. Each will have thread features designed to work with the material and conditions in which they are used. For instance, drywall screws should be driven just below the surface of the drywall so they disappear after the wall is finished. They also should not tear the surface paper on the drywall, so their head is larger in proportion to the threads than other screws. The head is also bugle-shaped, so it is not good for applications such as wood, where a countersink of a different geometry is used. It is important to use the right screws for the application! Some common screws are shown in Fig. 19.8.

19.2.2.3 Machine Screws

The second type of screw is called a machine screw. It has standardized threads intended to match corresponding threads in a nut or component. A machine screw is not designed to form threads in a component. Machine screws follow standardized conventions that are described in more detail in Sec. 15.5.1. Machine screws come in a wide array of sizes with most being available with coarse or fine threads. For instance, a ¼-in-diameter machine screw can have 20 threads per inch (tpi) or 28 tpi. The finer the thread is, the stronger the screw will be. This is because the threads do not extend as deeply into the shaft of the screw and more material will support the load. Finer threads also result in less forward motion per turn. This can be bad if you want the screw to advance quickly, but it is useful for finer adjustment, and the screw is less likely to come loose. The disadvantages are that the threads are more susceptible to seizing up from dirt or other contamination and galling (deformation leading to seizing). What thread is considered "fine" is determined by thread convention, but the smaller the diameter the smaller the

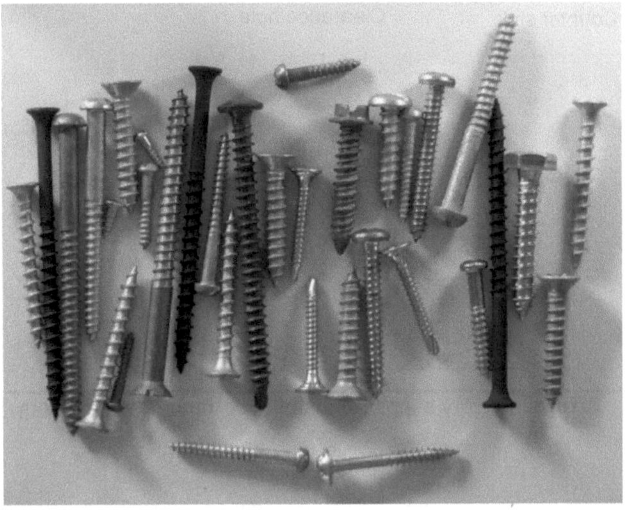

Figure 19.8 Screws can be used to assemble parts quickly and permanently. The thread characteristics, material they are made from and head style determine the appropriate applications.

thread pitch becomes. 28 tpi is a fine thread on a ¼-in screw, but 32 tpi is considered *coarse* on a #8 screw (0.164 in diameter).

Like standard screws, machine screws come in a variety of shapes and sizes for specific applications or general use. One of the most common machine screws is the socket head cap screw (SHCS). These screws use a hex (allen) drive to tighten and offer a good, general-purpose solution for many needs. Many of head profiles shown in Fig. 19.6 can also have hex drives, and machine screws can use the other drive geometries as well (phillips, slot, etc.). Long machine screws will tend to have an unthreaded portion near the head, assuming that only the lower threads will actually be engaged. Fully threaded and partially threaded options exist for most screws around 1 in long or greater, though the minimum length increases with diameter. An assortment of machine screws is shown in Fig. 19.9.

19.2.2.4 Bolts

Bolts have many similarities to machine screws, and as mentioned earlier in the chapter, the terms are used interchangeably. We will consider the primary difference to be that bolts are driven by an external drive (typically a socket). Both rely on a mating thread, and bolts are more commonly associated with a corresponding nut, but it is not necessary. Both come with same considerations of thread pitch, thread length, and other features. One of the common distinctions with bolts is the concept of bolt "grade," and we will focus on that aspect.

Bolts of the same size can have quite different properties. By far, the most common bolt material is steel, but even among steels varying strengths are possible (refer back to Secs. 4.2.2 to 4.2.6 for a small sampling of this diversity). The reason this is important is because not all applications require the same strength. A bolt in a classroom desk will be under different stresses and strains than a bolt holding together components on a forklift. Even though larger bolts could be used, it often makes more sense to use stronger bolts. For this reason, various bolt grades are available.

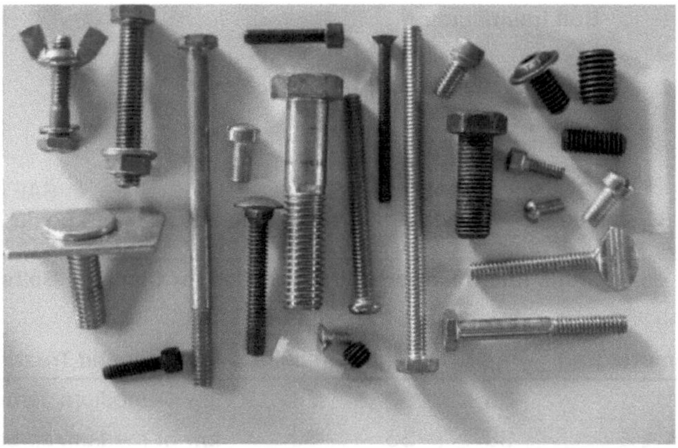

FIGURE 19.9 Machine screws are available in a wide variety of sizes and styles. They require a matching thread to create an assembly. This may be a nut or internal thread on a part. They can be used to create reversible assemblies, but they are widely used without the expectation that the assembly will be separated.

Bolts are designated by SAE, ASTM, or ISO grade identification. The more common reference is to the SAE grade,* which includes popular grade 2, grade 5, and grade 8 bolts. The strength and specifications for some common bolt types are given in Fig. 19.10. Cost increases with the strength of the bolts, and the cheapest bolt that satisfies the need of the application should be used. Misuse of the product should be considered as well to increase the safety under unexpected circumstances. A nut of equal or greater strength should be used with the chosen bolt.

19.2.2.5 Screw and Bolt Finishes

The appearance of screws and bolts, such the color and surface texture, can vary widely. Specifically, a single size of bolt may appear dull gray, highly reflective, or black. Screws vary even more, with some being colored (e.g., green or blue) and also having a variety of gray finishes. The difference in appearance is due to the material or surface treatment of the fastener. A common example is with wood screws. These screws may be designed for interior or exterior applications (the same goes for hinges, brackets, etc. that are used with screws). Screws that will be outside in the elements should have higher corrosion resistance (corrosion is described in Sec. 8.5).

To achieve this, steel screws may be pre-oxidized (such as "black oxide" screws) or coated with zinc, either by plating or galvanizing (i.e., diffusion bonded). A plated screw is suited for interior use, but not exterior use, especially with pressure-treated lumber, which contains chemicals that will rapidly corrode the screw. This is because the plating can chip off and the protection will be lost, but the galvanized coating is more robust and protective. Coating may also be done with ceramics or polymers to reduce corrosion, thread friction, etc. Even better than using a coating, stainless steel screws are intrinsically corrosion resistant and can be used instead, but their higher cost must be justified. Other appearances, such as "nickel" or "bronze" may be used for

*Bolts are covered under SAE J429 (Mechanical and Material Requirements for Externally Threaded Fasteners), and nuts are covered under SAE J995 (Mechanical and Material Requirements for Steel Nuts).

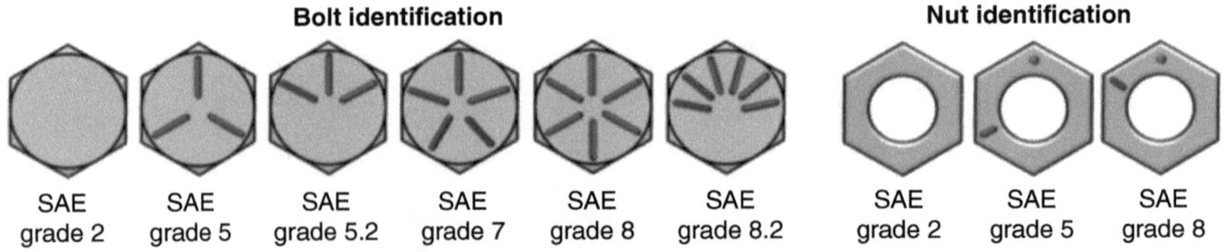

Grade Identification and Properties for Common Bolts					
		Minimum Strength (ksi)			
SAE Grade	**Diameter (in.)**	**Proof**	**Yield**	**Tensile**	**Material and Treatment**
2	¼–¾	55	57	74	low or medium carbon steel
2	7/8–1½	33	36	60	low or medium carbon steel
5	¼–1	85	92	120	medium carbon steel, quenched and tempered
5.2	¼–1	85	92	120	low carbon martensitic steel[a], quenched and tempered
7	¼–1½	105	115	133	medium carbon alloy steel, quenched and tempered
8	¼–1½	120	130	150	medium carbon alloy steel, quenched and tempered
8.2	¼–1	120	130	150	low carbon martensitic steel[a], quenched and tempered

[a]Low carbon martensitic steel is an alloy with high hardenability and toughness. Martensitic transformation is described throughout Sec. 9.3.6, especially Sec. 9.3.6.2.

Figure 19.10 Common SAE bolt grades are described here, but other grades of bolts are available with different markings. Grade 2 bolts have no markings. Marking on nuts may vary somewhat. Proof stress is determined at the onset of plastic deformation and is often used interchangeably with yield stress, which is similar.

decorative and not functional purposes. A variety of screw finishes are shown in Fig. 19.8. Methods such as plating, oxidizing, and coating are described in Sec. 20.2.

19.2.2.6 Nuts

Nuts are made to match the thread specification of a bolt or machine screw, but they can come in a variety of designs (Fig. 19.11). The majority of nuts have a hexagonal profile, but some have a square profile. The width across the flats is standardized for the thread, but the thickness of the nut and the vertical profile can change. For instance, low profile nuts are thinner than standard nuts, acorn nuts (or cap nuts) have a rounded top that covers the end of the screw, flanged nuts have a wider contact surface than standard nuts, wing nuts have tabs that can be used to tighten them by hand, castle nuts have slots in the top of them in which a pin or wire can be inserted to hold them from loosening, and locking nuts have a nylon insert that fits tightly against the threads and helps prevent the nut from loosening. There are also nuts that have a serrated bottom to lock it in place and others have built-in washers (washers are discussed in Sec. 19.2.4).

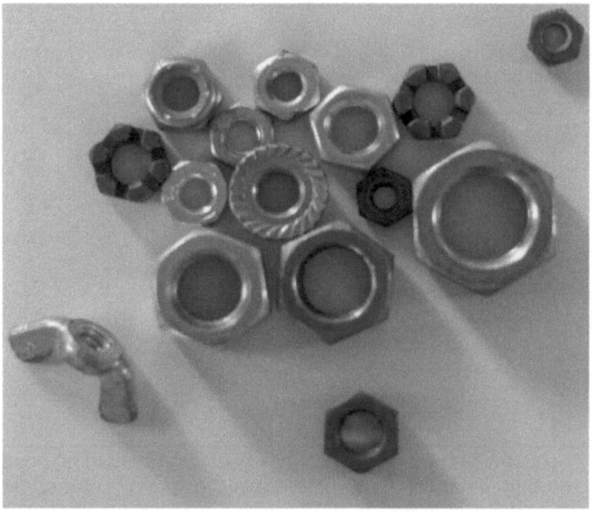

FIGURE 19.11 Nuts are available in many sizes, designs, and finishes. The primary considerations in selection include strength, corrosion resistance, and holding characteristics, such as self-locking features.

19.2.3 A Note on Nails

Nails are widely used for assembly, but they do not hold as well as threaded fasteners. Many of their characteristics are related to the discussion for screws and bolts (variation in diameter, head size and effects, material, coating, etc.), so they will not be detailed here. Nails are a permanent choice for connecting components, and they are cheaper and easier to use than screws. Where suitable, the use of nails is preferred to reduce cost, the need for predrilling, and for faster assembly. The use of nail guns can allow dozens of nails to be installed per minute, a rate that cannot be achieved with screws. Nails can be used for structural applications and even in finishing where very small nail heads are used to be inconspicuous. The size of a nail is often termed as "penny." For instance, a 10-penny nail. This terminology is associated with a specific diameter and length, and the unit of the penny is "d." A 10-penny nail is noted as 10d.* Some common nails and the penny designations are shown in Fig. 19.12.

19.2.4 Washers

Washers are typically rings used to distribute the clamping force of bolts and screws, but in some cases they will have other shapes and functions (Fig. 19.13). The diameter (inner and outer) and thickness scale with the bolt size, but specific designs have a much larger outer diameter to help spread out force so as not to damage the parts being assembled. Other washers act to prevent the fastener from loosening over time by adding some spring tension. This is often accomplished by using a split washer, star washer, or another type of washer that is bent so that when a fastener is tightened against it,

*It is suggested that the unit "d" was originally related to the Roman coin, the denarius, but it was later changed to the "penny" in English. The quantity was associated with the cost per weight of nails.

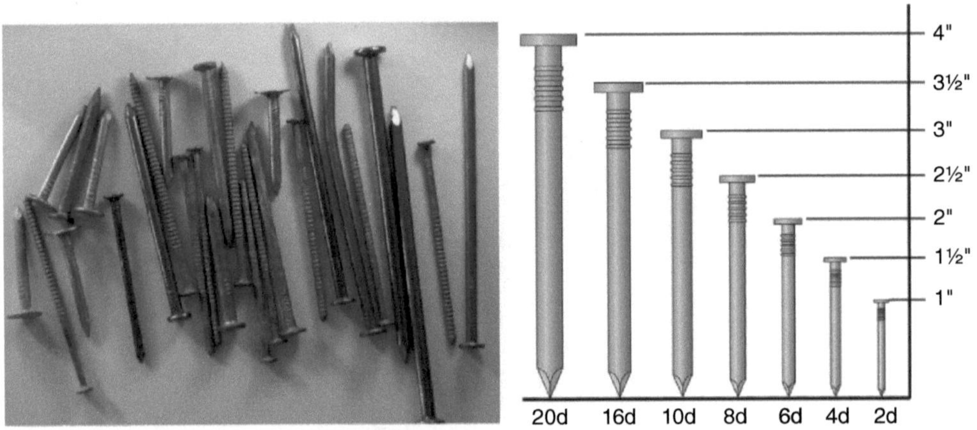

FIGURE 19.12 Nails are fast to install, but they have limited holding power compared to other methods. The common designation for nail size is the "penny," and some common sizes are shown.

it will be compressed. These may also bite into the material to prevent rotation. Multiple washers can be paired as well. Washers come in a variety of materials, including a range of metals and polymers (often nylon).

19.2.5 Other Fasteners

There are many other ways to hold assemblies together. Some rigidly hold components together, other only keep the two from moving too far apart, but do not apply force. These include pins, clips, retaining rings and variations on these. Some of these are easily removed, and some can be removed, but they are not removed regularly. Pins can slide through components easily or fit tightly between them. Where two parts are held together, a tapered pin can be used that is hammered or pressed into place. Alternatively, a pin called a roll pin or split pin has a slit along the side that fits tightly in place but is more easily removed. Pins can also be held in placed by inserting a clip into a groove in

FIGURE 19.13 Washers come with varying inside and outside diameters to accommodate various fasteners. Some also serve to lock the fastener in place and avoid loosening due to vibration.

Figure 19.14 There are many variations of clips and pins to hold assemblies together. Each has specific benefits for ease of assembly, holding power or for the specific type of components.

the pin using a snap ring, or a second pin, called a snap pin, can be added to prevent the main pin from sliding out. Cotter pins are inserted through a hole and bent over to prevent movement. There are dozens of additional fasteners that could be described, but all operate on some premise of restricting motion by snapping, pressing, or bending them to lock the components together (Fig. 19.14).

19.3 Adhesive Assembly Methods

Adhesive assembly uses the interaction between surfaces to create a strong connection between parts. The strength of the connection can be *very* strong. If you have done woodworking, you know that gluing joints is very common, and you may have even heard that the joint is stronger than the wood itself. This can be true, as the wood around the glued area will fracture first under high stress. "Glue" is a popular term, but it is not very precise. Adhesives can have a wide array of chemistries, properties, and methods of application. Holding two surfaces together is not their only role either. In many cases, the adhesive also seals a joint such that fluids are contained (or prevented from entering), dampening vibration and other functions in addition to holding parts together. We will look at common methods and materials for creating adhesive assemblies.

19.3.1 Soldering/Brazing of Metals

Metals can be used to create an adhesive joint. That might sound odd, especially if you have already read through Chap. 18 on welding. Metals are used extensively for cohesive bonds, but they are regularly used for adhesive ones as well. The greatest difference between the two is that the *adhesive assembly of metals does not melt the base material.* A filler material with a lower melting point becomes liquid while the components being assembled are still solid. There are two primary categories of adhesive assembly in metals: soldering and brazing. Their principal difference is temperature.

19.3.1.1 Soldering

Soldering is a method of assembling components where a metallic filler material (typically a wire) is melted and applied to the surfaces at a temperature below 840°F (450°C) or below. There are not many metals that melt at such a low temperature (some are described in Sec. 4.4.2), and most solder is based on alloys that have a deep eutectic point. This means the mixture of metals will melt at a much lower temperature when combined than when separated (see Sec. 3.2.4.3 for more on phase diagrams).

Historically, solders were based on lead (Pb) alloyed with tin (Sn). At their eutectic composition (61.9% Sn) the alloy melts at 183°C, which is 50°C lower than Sn (T_M = 213°C) and 144°C lower than Pb (T_M = 327°C). Due to the toxicity of Pb, there have been a variety of Pb-free solders developed. Most of these are based on Sn, silver (Ag), copper (Cu), antimony (Sb), bismuth (Bi), zinc (Zn), and other low melting point metals. Common Pb-free solders include Sn-Sb (95%/5%) and Sn-Ag-Cu (96.5%/3%/0.5%).

The melting point is important, but it is not everything. To apply solder the base metals should be heated and the solder should be melted by the hot components. Soldering is widely applied in electronics and electrical applications. In this case, the wires or junctions would be heated, often using a restively heated soldering "iron." If the base material is too cold, the solder will not flow into spaces between parts. If you have soldered, you may know that a drop of solder will instantly freeze when it hits the table. This will happen similarly if the parts are cold. The solder hardens, but it does not create a good bond.

The condition of the surfaces is also important, and just like in welding, a flux material can and should be used to ensure a strong bond between the components. A separate flux paste can be applied to the joint or a flux core wire can be used. In all cases it is very important that the components be closely spaced. Solder is not intended to fill gaps. The composition of the solder will also determine its performance (strength, electrical conductivity, corrosion, etc.). There are many options available, including solders containing Pb, so the choice should be carefully weighed to balance cost, performance, and safety. The general process for soldering (and brazing) is shown in Fig. 19.15.

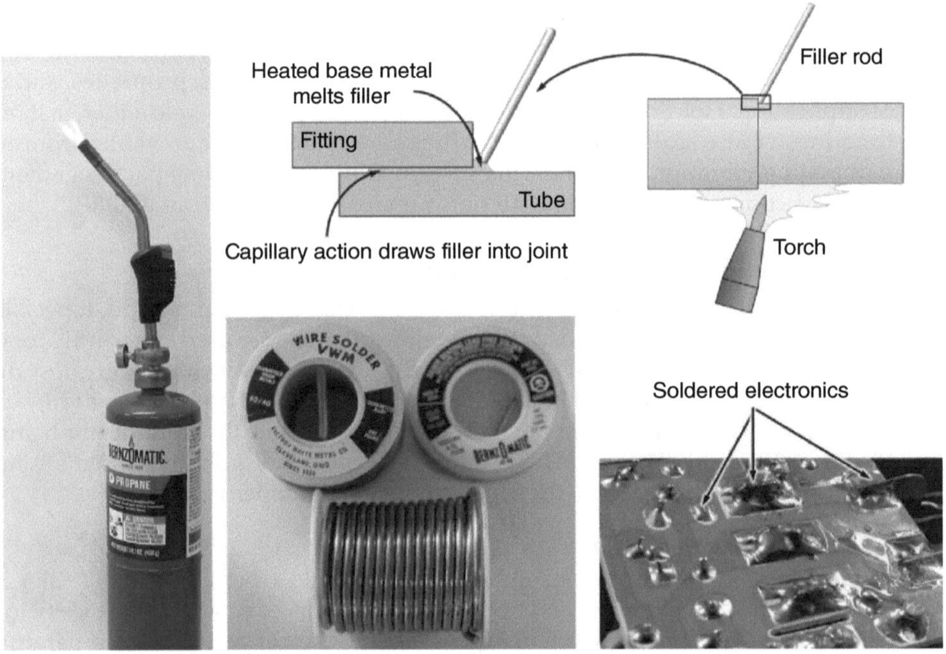

Figure 19.15 Soldering and brazing are closely related, but brazing requires a higher temperature to melt the filler. Both require heating a base material and melting the filler rod using the hot components, not direct heat.

19.3.1.2 Brazing

Brazing is a method of assembling components where a metallic filler material (typically rod or wire) is melted and applied to the surfaces at or above 840°F (450°C). You may notice this definition is nearly identical to soldering. That is because the processes and considerations are very similar. Brazing is done at a higher temperature, and brazing is sometimes even called "hard" soldering (soldering as described in the previous section is then called "soft" soldering). Because of the higher temperature, brazing is often done using a torch fueled by acetylene, propane, or other flammable gas to heat the components faster and to a higher temperature. Small handheld torches are common for work such as plumbing and other small jobs. In a production environment, a larger gas tank and separate torch may be used.

The brazing alloys are typically alloys of silver, copper, tin, and often cadmium. The alloy will determine the melting temperature and the flow characteristics, and some brazing alloys are thicker when melted and more suitable for filling gaps. Indeed, a method where the filler material does not penetrate a gap at all is termed "braze welding." Brazed joints can have high strength and are used in structural applications with a range of metals such as brass, stainless steel, copper, nickel, and others. Brazing can also be used to combine dissimilar metals, and those that would be difficult or impossible to weld. For instance, carbide cutting inserts may be brazed onto a steel body (refer to Sec. 13.4.2.1). Cast iron can also be repaired using brazing, as it is tricky to weld.

Brazing can also be done in a furnace. In that case the brazing filler material is placed at the joint and the whole assembly is heated until the brazing melts and fills the joint. This has the particular advantage that many parts can be repeatable processed. Like soldering, brazing may use a flux to reduce contamination at the joint. The higher temperature of brazing may also benefit from a protective atmosphere, just like some welding processes (see Sec. 18.1.4.1). Brazing is shown in Fig. 19.15.

19.3.2 Organic and Synthetic Adhesives

Organic and synthetic adhesives create an assembly by chemical and/or mechanical attraction between parts and are often based on polymers. **Organic adhesives** are often based on naturally derived materials, but by definition they are principally comprised of carbon-based substances. **Synthetic adhesives** are created through precisely controlled chemical processes, and they too may be carbon-based, but they are not limited to it. Both types of adhesives can be applied to a carrier material, as in duct tape, or they may be in liquid, gel, or paste form and applied directly to the surfaces and form an assembly during contact. Before we get too far down the road, we should take a moment to consider the vast array of products that fit this description. You are probably familiar with many of them.

19.3.2.1 Pressure-Sensitive Adhesives

No doubt you have used different types of "tapes," such as electrical, painter's, masking, or packing tapes. These all belong to a category referred to as **pressure-sensitive adhesives** (or **adhesive tapes**), which create a strong attraction by being pressed against one or more surfaces (Fig. 19.16). The strength of the tape itself is determined by the carrier material, or *backing*, to which the adhesive is applied. You can rip through duct tape relatively easily by hand. Packing tape cannot be easily torn and some packing tape is reinforced with polymer strands to *prevent* tearing. As always, the application determines the appropriate design.

Figure 19.16 Pressure-sensitive adhesives, or tape, can have a range of properties. Some are meant to weakly adhere, tear easily prevent electrical conduction, etc.

For instance, electrical tape should be electrically insulating and adheres to polymers well (i.e., wire insulation). It should be conformable and withstand mild abrasion without tearing. Packing tape should adhere to cardboard and its physical width should allow covering large joints. If you want a simple example of the trouble with improper design, try closing a cardboard box with electrical tape or wrapping a wire with packing tape. The grade of adhesive and the properties of the backing material *both* need to satisfy the intended use. Another example is with sticky notes. These have a weak adhesive that does not leave behind residue, but it is strong enough to hold the weight of the note itself. Sometimes a strong adhesive is not beneficial.

19.3.2.2 Types and Properties of Adhesives

We have talked about adhesives in terms of their holding power, but there is much more to it that just that. Some adhesives are activated by water or heat, and others harden by solvent evaporation or the reaction of two ingredients. Because there is such diversity, they are used in very different ways. We can start by defining some terms.

You have probably heard of "super" glue or rubber cement. These terms have some history behind them. For instance, **glue** has been traditionally defined as being derived from natural sources. These were often vegetable starch or animal protein products. **Cements** are often identified with rubber-based adhesives. While some traditional materials are still available, the terms are often used interchangeably, and "adhesive" is a more general term covering the others as well.

Most modern adhesives are based on synthetic polymers as they can be more tightly controlled and provide better properties such as bond strength, inherent (polymer) strength, water resistance, temperature resistance, chemical resistance, etc. For that reason, it may be beneficial to review the concepts in Secs. 5.3 and 6.2.4. Specifically, adhesives are often defined as being thermoplastics or thermosets.

Thermoplastic adhesives, sometimes called thermoplastic resins, include acrylic and vinyl adhesives. Cyanoacrylate, or "super" glue, bonds components by polymerization that is driven by moisture. That means it will harden more slowly in dry air than in humid air. Also, it does not "instantly" bond surfaces where there is a gap, because the surface of the adhesive will harden and slow the curing process by creating a barrier to further moisture. Test it. Put a drop of super glue on a flat surface and see how long it takes to fully harden. If you (or preferably a "friend") push your finger onto that drop, it spreads out and instantly adheres it to the surface. Vinyl adhesives can have a range of properties, but one of the most common types is "white" glue, also called *Elmer's glue*. It uses water as a solvent so it dries more slowly than some other adhesives, but it also cleans up more readily.

Because thermoplastics soften with heat, some are melted and then applied. This includes some traditional glues that come as flakes that are melted and spread on surfaces, but "hot" glues are much more common. Hot glues, or more accurately, **hot-melt adhesives**, are often based on thermoplastic polymers with low glass transition temperatures that enhance the flowability, or ease of flow, of the material. These adhesives do not require solvents or chemical reaction and are entirely controlled through heating. This means there is no solvent given off during use, but premature cooling of the adhesive, especially when used over large surfaces, can be a problem. The adhesive has moderate flexibility that can be beneficial when combining flexible parts, but the joint will lose strength at modest temperature and is not suitable for those environments where elevated temperature will be encountered. These adhesives also rely strongly on mechanical interlock at the surfaces to hold the assembly together (Fig. 19.17).

Thermoset adhesives (or resins) include common examples such as formaldehydes and epoxies. Formaldehyde-based polymers, such as urea-, phenol-, and resorcinol-formaldehydes, are used extensively in engineered wood products such as particle board and plywood (see Sec. 6.4.1.6). The reasonable cost, good strength, and moisture resistance are key benefits. Epoxy is also a common adhesive polymer that is versatile and quite strong. It is often purchased as a two-part compound and mixed directly before application, though some can be hardened with heat. An application that is not a "traditional" assembly method is use as the matrix material for composites such as carbon fiber or fiberglass. See Sec. 6.3.2.3 for more on epoxy and Sec. 6.4.1 for more about polymer-matrix composites. Thermosetting resins are also used to bond the grit in abrasives as well (see Sec. 14.6.1).

Figure 19.17 Adhesives create a strong, permanent bond between surfaces, and they are specifically formulated for many tasks depending on the materials being bonded and the environment it will be used in. Some, like hot melt glue on the left, require heat to be applied, but others, like the construction adhesive on the right, are applied at room temperature.

Figure 19.18 Metallic components exposed to modest temperature can be assembled by specialized adhesives.

19.3.2.3 Elastomeric Adhesives

Some applications require a flexible adhesive, especially for sealing joints. Elastomers are described in Sec. 6.3.3, and their properties are used to create assemblies and to fill gaps between surfaces. One the most common is silicone. These adhesives, or caulks, are used extensively to seal joints in bathroom and kitchen environments or on exterior joints to prevent water damage. There are many grades of silicone caulk, such as RTV* sealant for automotive gaskets. Other elastomeric adhesives, such as polyurethane roofing caulk, can satisfy a broad range of applications. These are easy to apply, but they are often not intended for structural applications. Silicone is described in more detail in Sec. 6.3.3.5.

19.3.2.4 Modified Adhesives

Adhesives are primarily used to bond parts together, but there are applications where their properties do no match the need. For instance, most adhesive materials do not conduct electricity well. To increase the conductivity fine particles or flakes of a conductive material can be added. Examples include silver, nickel, gold, or graphite. Another modification is to the color of the adhesive. This is achieved by adding dyes or pigments to help match the color of the parts being bonded. The adhesive can also be reinforced using steel powder or other materials. A popular metal adhesive, JB-Weld, uses an epoxy and steel mixture that is resistant to elevated temperature, changes in temperature and vibration (Fig. 19.18). Other variations include putties that can be applied in thicker amounts and without dripping.

19.4 Why It Matters

The saying, "familiarity breeds contempt," may be more accurate with fasteners and adhesives than any other techniques and materials in this book. They are common. We use them every day, but often the most convenient one is not the best

*RTV stands for room temperature vulcanizing.

one. Often the easiest installation method is not the strongest. As long as it works, does it matter?

The best answer is, "yes." The honest answer is, "maybe." If you are building a bird house, it is not going to be critical that each hole is properly piloted, clearance drilled, and countersunk. The fasteners do not need to be stainless steel to last for decades. If using adhesive, neglecting surface preparation and a specific outdoor formula for wood is probably not going to cause short-term failure. If you handle it before the glue has properly set, it will probably not fall apart. We bend the rules all the time and it works out OK (read the directions on the label of an adhesive, for instance, and see how well you are doing).

"OK" and "probably" are unacceptable in manufacturing. Products should be engineered to achieve the best result at the lowest cost. Reliability and performance are critical, and unlike home projects, the right approach is well worth the time. It makes sense to research options, assess various assembly designs (parts and the assembly method) and to do the work necessary to prepare the components for assembly. There are many options, and only some of the most common ones were addressed here.

If there is any critical take-away, it is "know your options." This not only informs you to possible alternatives, it may also prevent you from using an inferior assembly method that requires more effort. Using the right fastener or adhesive *and* doing it the right way can do more than improve the quality of your products. It can also save you time and money! Imagine using half the amount of the right adhesive and getting higher strength, for example. Think of using smaller bolts that are stronger. Where possible, eliminate the need for assembly methods such as adhesives and fasteners altogether. Where needed, make the job easier and more effective.

19.5 The How and Why of Mechanical and Adhesive Assembly

19.5.1 Chapter Review

1. How does assembly add value to products?
2. Contrast cohesive, mechanical, and adhesive assembly. Focus on how they are used and the permanence of them.
3. How might mechanical fasteners be underestimated in terms of their simplicity? What affect might this have on the assembled product?
4. What benefits are there to using fine-threaded fasteners? What disadvantage might there be in dirty environments?
5. Use Eq. (19.1) to determine the force applied by a bolt (steel-on-steel (lubricated), $k = 0.2$) for the following:
   a. Torque of 200 in·lb, diameter of 0.25 in
   b. Torque of 20 ft·lb, diameter of 0.375 in
   c. Torque of 15 in·lb, diameter of 0.125 in
   d. Torque of 35 ft·lb, diameter of 0.625 in
6. Use Eq. (19.1) to determine the force applied by a bolt (aluminum-on-aluminum (dry), $k = 1.1$) for the parameters in question 5.

7. Use Eq. (19.1) to determine the torque needed to apply the following forces to a bolt (steel-on-steel (dry), $k = 0.7$) for the following:
   a. Force of 20 lb, diameter of 0.375 in
   b. Force of 150 lb, diameter of 0.5 in
   c. Force of 12 lb, diameter of 0.125 in
   d. Force of 500 lb, diameter of 0.75 in
8. What is the maximum torque before failure when using a ¼-in grade 5 bolt? How does it differ for grade 8?
9. Describe the difference in adhesives that act by van der Waals forces and those that include chemisorption.
10. What interactions occur between an adhesive and the surfaces it combines to hold it together? What role does chemical compatibility play?
11. What is a physical interlock? Is this a strong connection? Where is it often used?
12. How do rivets hold components together?
13. What consideration must be made when using solid rivets that is not necessary for blind rivets?
14. What they benefits of solid rivets?
15. Why can the differences between screws and bolts be confusing?
16. Write your own definition that describes the difference between screws and bolts.
17. Why are there so many different drive shapes for screw heads?
18. Why are there so many different head profiles for screws?
19. What are the benefits of a deep drive recess and certain shapes of screw drive profiles?
20. What role does a pilot hole play in screwed assemblies?
21. Is a clearance hole really necessary when joining two components? Share any experience you may have doing it either way.
22. Why is a pilot hole smaller in soft wood than hard wood?
23. What is a possible drawback of using the wrong type of screw (e.g., using a wood screw in drywall and vice versa)?
24. How is a machine screw different from a standard screw, such as a wood screw?
25. What is meant by a "fine" thread? How can the same thread pitch be considered fine and coarse?
26. What is the purpose of a partially threaded screw?
27. How are machine screws and bolts different? How are they similar?
28. What does the "grade" of a bolt indicate?
29. What is meant by "proof stress?"
30. What roles does the finish of screw play?

31. One type of bolt or machine screw can be used with many different nuts of the same thread. What should you consider when choosing a nut?
32. Look at Fig. 19.11. Toward the center there is nut with a serrate bottom shown. What is the use of this design?
33. How does the size of a nail correspond to its "penny" value?
34. What are the benefits of nails over screws? The disadvantages?
35. What is the primary function of a washer?
36. How can a washer prevent the nut from loosening?
37. Why are there so many different types of fasteners?
38. How are soldering and brazing different from welding?
39. What is difference between soldering and brazing?
40. Why are lead-free solders becoming more common?
41. Briefly describe the process of soldering. What is the order of operations?
42. Carbide cutting tools are brazed onto steel bodies. Why cannot soldering be used?
43. What is the difference between synthetic and organic adhesives? Which one is more versatile?
44. When is a weak pressure-sensitive adhesive a good thing?
45. What pitfalls might there be to using tape that is not intended for the application?
46. What type of material are most adhesives made from?
47. Why does superglue take so long to harden in open air, but it hardens "instantly" when pressed between surfaces?
48. What is a benefit of hot melt glue? Disadvantages?
49. Where are elastomeric adhesives used?
50. How can an adhesive be used to "fix" metal components?

19.5.2 Concept Connections and Extensions

1. Consider the differences in zinc-plated screws and galvanized screws. What are the benefits and disadvantages of the two? Use concepts in Secs. 8.5.2.3 and 20.2.2.
2. Describe the decision-making process for choosing bolt size and grade. Is better to have a larger or a stronger bolt?
3. Draw a general stress-strain curve and label proof stress. Where on the curve should a bolt operate most of its lifetime? See Sec. 7.4.1 for more on stress-strain curves.
4. Low carbon martensitic steel is used in grade 8.2 bolts. How can low carbon steel be hardened through martensitic transformation? Martensitic transformation is described throughout Sec. 9.3.6, especially Sec. 9.3.6.2.
5. Why is the exact mixture of elements important in soldering wire? Relate this to the phase diagram in Fig. 3.16.

6. Describe how thermoplastics and thermosets can both be used as adhesives. The two are described in Sec. 5.3.2.5.
7. Are composites adhesive assemblies? Give your thoughts and justifications.
8. Abrasive wheels rely on adhesive bonds between particles. Use the discussion in Sec. 14.6.1 to describe the properties that adhesive should have.
9. Consider electrical tape and how elastic it is. After you stretch it, it slowly returns to its original shape (or close to it). What mechanical property is this related to? How might it benefit the holding ability of the tape? Refer to Sec. 7.5.3.
10. Consider hot melt adhesives. How does this adhesive category compare to welding (Chap. 18)? How does it compare to soldering and brazing (Sec. 19.3.1)?
11. Let us say you are a welding equipment salesperson. How would you convince a customer to use your products instead of threaded fasteners? Welding is discussed in Chap. 18.
12. Why would heating a bolted assembly help loosen it? Review Sec. 8.2.3, if needed.

CHAPTER 20
Finishing

You may have heard the term, "finished product," but there are multiple ways that can be understood. Something is finished when the work is complete, and that extends across disciplines. Think about the last thing you finished. Maybe it was a homework assignment, dinner, or some task. Once you finish that "operation," no further effort is needed. Therefore, finishing operations in manufacturing are typically the last steps before a component is integrated into an assembly or a product is sent to market (not including packaging, shipping, etc.). This requires that all of the details of size, operation, surface quality, and any other factors that determine the proper condition of the component or product are complete. The finishing operation or operations do not need to be the absolute end of the processing, but any later modification should not compromise the "finish" of the part.

20.1 Introduction

In manufacturing, **finishing** describes operations that produce a functional and/or aesthetic quality that is not altered in later processing and is desirable to the consumer. The functional aspect may be incorporated with the aesthetics, such that both are achieved with a single operation. Most of the finishing operations we will discuss modify a surface to improve corrosion resistance (Sec. 8.5), improve friction properties (reducing and resisting wear), to change the surface chemistry, or incorporate other physical or chemical changes to improve the performance of a product. The term "finish" can be used as a noun and a verb. For instance, we could say that, "Kevin applied clear *finish* to the table top." Alternatively, "Kevin *finished* the table top with clear coat." Be aware that either form is common, and they are used interchangeably by many people. The verb form is more informative in most cases since the type of finish is also specified. For the purpose of this discussion, "finish" and "finishing" will denote manufacturing operations, not products.

What constitutes finishing varies for different materials and applications. Wood might be stained, metal might be plated, plastics might have decorative elements included during earlier stages (e.g., during molding, such as in Sec. 11.3) even though they determine the final appearance as well. Each category of material will be discussed separately, but many techniques can be applied to any category of material and similar results can be achieved through different processes. The main outcomes of these processes include one or more of the followings:

- Decreasing corrosion
- Reducing/increasing friction
- Reducing wear
- Altering the chemical compatibility (e.g., hydrophobic/hydrophilic)
- Improving electrical conductivity
- Modifying aesthetics

20.2 Metals

Metals are primarily finished to prevent corrosion. Most metals are highly reactive to environmental factors, such as moisture and oxygen. This requires common metals, such as steel, to be protected from rusting. Other metals that are relatively stable and do not degrade readily may be treated to improve their wear resistance and/or appearance.

20.2.1 Temporary Coatings

Some coatings may be temporary, and they are only used during processing, shipping, or other periods before the product is ready for use. For example, many end mills are made of high speed steel (Sec. 13.4.1). These tools are susceptible to corrosion, so they are often shipped with a light coating of oil or cosmoline (a waxy coating). The coating may also serve to preserve sharp edges and protect people handling them from danger (Fig. 20.1). This type of coating can be readily removed with a solvent such that it does not affect performance. This is suitable if a manufacturer is shipping components from a distance or the parts will be stored for an extended period. These items are not expected to be exposed to significant moisture or other chemical attack. For example, if you go to an auto parts store, you do not want to buy rusty brake rotors, but a more durable coating, such as paint, would be inappropriate. These temporary coatings are

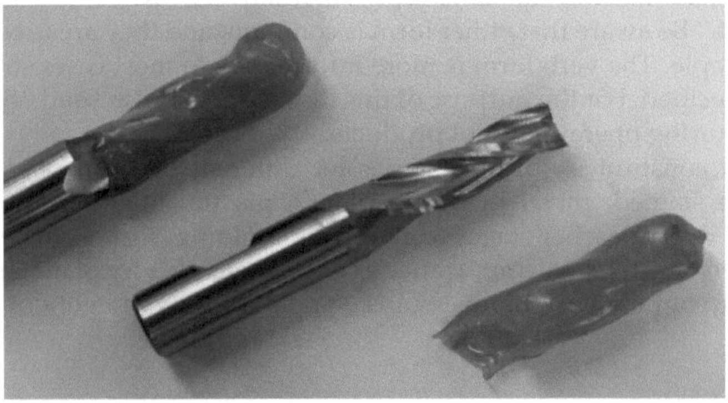

Figure 20.1 Temporary coatings are used to prevent corrosion and damage to a product or component until it is put into service, such as the polymer layer that protects the cutting edges of this end mill.

frequently used before a product is purchased or as an intermediate step in processing. In some cases, a comparable coating may be applied during use by the consumer. There are products available for protecting bare metal surfaces that may be applied during storage or intermittently during the lifetime of the product. The primary protective mechanism is the displacement of water.

20.2.2 Plating

Plating operations apply a very thin layer of metal to another material. The metal layer is most often deposited on a substrate (or component) in a liquid solution containing metallic ions. By applying an electrical charge, the metal ions are directed to the component and the thickness of the plating is determined by a combination of processing factors, including the electrical potential and the amount of time. The process for plating is known as electroplating (sometimes electrogalvanizing) and is a good way to apply an even coating over complex surfaces. The mechanisms of the process are reminiscent of the galvanic corrosion process described in Sec. 8.5.2.2 and shown in Fig. 8.21. One metal electrode disintegrates into the solution and is deposited on the electrically charged parts.

An alternative process, called **electroless plating**, deposits the metallic ions from solution by a chemical reaction with a reducing agent. Nickel is most commonly used, and the process requires a source of nickel ions such as nickel chloride ($NiCl_2$), nickel acetate [$Ni(CH_3CO_2)_2 \cdot 2\ H_2O$], or other compound, such as hypophosphorous acid [$Ni(H_2PO_2)_2$]. A reducing agent, such as sodium hypophosphite ($NaH_2PO_2.H_2O$), is added to the solution such that the metal ions are selectively reduced at the surface of the component, and that results in deposition. The process is autocatalytic, and therefore the deposition is driven by the deposits and can be regulated through the addition of other additives. The primary benefit of electroless plating over electroplating is that the deposition is more uniform and works well on complex geometries with fine features. It also does not require the electricity of electroplating, but it does have chemical disposal considerations.

Plating is used in a variety of aesthetic and functional applications. For instance, zinc coatings on steel reduce corrosion. Chrome-plated steel produces a shiny, rust-resistant surface. Gold plating produces the appearance of gold without the cost and can also increase electrical conductivity. Nickel plating on brass pipe fittings can reduce corrosion and changes the appearance from a golden color to a silver one. Examples of plated products are given in Fig. 20.2.

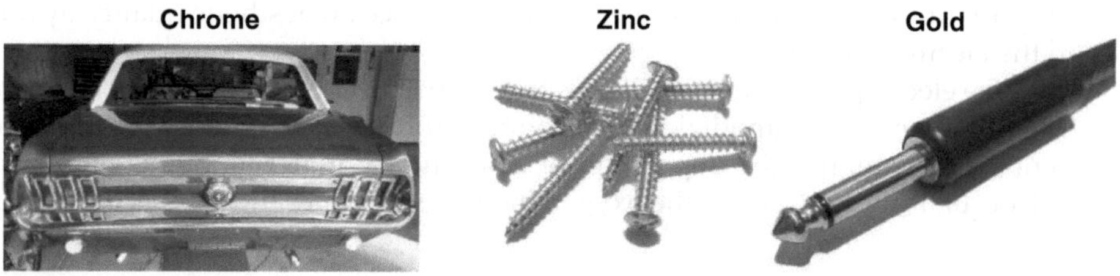

FIGURE 20.2 Metal plating is used for aesthetic and functional purposes. It improves appearance (chrome) and reduces corrosion (chrome and zinc) as well as improving electrical contact (gold).

20.2.3 Galvanizing

The purpose and effects of galvanizing are described in Sec. 8.5.2.3. Briefly, a sacrificial metal coating is applied to another metal. The metal coating is more anodic, and therefore it will corrode preferentially when the two are combined. This is most often used in steel structures and hardware coated with zinc, though the base metal and coating can be any paring. Unlike plating, galvanizing creates a diffusion bond between the coating and the base metal. This means that there is atomic-level mixing of the two (more on diffusion in Sec. 8.4), and this creates a stronger and more durable coating.

In applications such as screws, a zinc-plated surface may chip or peel on the screw threads, and this is a problem in harsh environments, especially when used in pressure-treated lumber, which can accelerate corrosion of steel. A hot-dip galvanized screw provides a stronger coating that remains intact during installation and is better suited to more corrosive environments. The hot-dipping process involves submerging a steel component into a molten zinc bath (it can be very large, such as light pole). A galvanized structure can often be identified by a large visible grain structure appearing as sections of varying shades of gray (see Fig. 8.24, for example). The size of these grains may be controlled as part of the processing to achieve a certain appearance or "spangle,"* and when very small, simply appears uniformly gray.

The zinc coating will increase the part dimensions, so critical geometries should be designed to accommodate this. Small components, especially machine screws (Sec. 19.2.2.3), can be problematic since the zinc coating will either fill in fine threads or have to make up the majority of the thread form, thereby weakening the structure. In these cases, stainless steel would be preferred. In highly corrosive environments, galvanizing will not last long, so a more corrosion resistant alloy would be preferred as well.

20.2.4 Anodizing and Other Surface Reactions

The addition of oxygen to the surface of metals results in metal oxides that are typically harder, more wear resistance, and environmentally stable than the base material. The use of these surface treatments may also have other benefits such as improving the retention of lubricants and aesthetic qualities. The primary consideration is the base metal, because these processes rely on the component material to form the coating.

20.2.4.1 Anodizing

Anodizing is a treatment that electrolytically oxidizes a metal to create a protective layer. This process can be applied to many metals, but aluminum is by far the most common followed by titanium. The most important factor is that the oxide film produced does not spall off (see Sec. 8.5.2.5 for these conditions). This allows a durable coating to be created that can serve to reduce corrosion, decrease wear, reduce galling (important for threaded components), and produce an aesthetic quality by adding dye to the electrolytic bath.

Like electroplating, the anodizing process uses a chemical bath and electrical circuit. The difference is that anodizing relies on the buildup of an oxide layer from the component's structure, not the deposition of a secondary material. Some geometric change will occur as the addition of the oxygen increases volume, and this becomes increasingly

---

*The only other place you have probably heard this term is in "star-spangled banner." The word refers to a small glittering object.

important with thicker deposits. Most anodized layers are relatively thin (<0.004 in, most often <0.001 in), so it is not a critical parameter in many applications. The oxide layer becomes increasingly porous as the thickness increases. This can be beneficial as it can retain lubricant, but it requires sealing to maximize corrosion resistance. Thick layers can also be susceptible to fissures during processing and use, especially from heating and cooling due the difference in thermal expansion (Sec. 8.2.3). Anodizing is described as a surface conditioning process in Sec. 9.3.8.2, and an anodized component is shown in Fig. 9.15.

20.2.4.2 Bluing/Blackening and Black Oxide Coating

Bluing operations, also referred to as blackening and black oxide treatments,* are used to create a uniform layer of magnetite (iron oxide in the form Fe_3O_4). Like anodizing, the process relies on the base metal being suited to the treatment. Because bluing produces iron oxide, the metal must be ferrous, and it is often applied to plain carbon steels, which do not have high corrosion resistance. Iron forms multiple oxide chemistries when it rusts. Magnetite is a bluish-black form that is more stable than the familiar red-orange ferric oxide (hematite) associated with deteriorating steel structures.

The bluing process is most widely associated with firearms. The exposed steel structure will rapidly rust if not treated, and bluing provides some protection, but it is important to combine the porous magnetite layer with a water displacement chemical like oil, which is held in the thin oxide film. The oxide layer can be produced by "cold" chemical treatment (often with selenium dioxide) or by "hot" processes that use elevated temperature and a chemical bath, such as potassium nitrate or sodium hydroxide, or gaseous treatment (fume bluing) where moisture and the fumes of an acid (e.g., nitric or hydrochloric) are present. The common outcome for each configuration is the formation of a durable oxide layer. Black oxide coating is a thicker version of the same basic process (still <0.0005 in thick). This is used in hardware such as nuts and bolts to reduce friction and corrosion (Fig. 20.3).

Figure 20.3 Black oxide treatment creates a durable finish that resists corrosion in steel parts such as firearms, nuts and bolts, and cutting tools. The cost is relatively low, but depending on the method used to apply the coating, it can be susceptible to damage.

*Black oxide treatment can also be applied to copper, its alloys, zinc, and others, but for our discussion ferrous metals will be the focus. The goal and outcomes are very similar for the various metals.

20.2.4.3 Electropolishing

Electropolishing is a method of smoothing the surface of metals by applying an electrical potential in a corrosive electrolyte (typically an acid) to remove sharp corners or surface irregularities. Rough surfaces have a combination of "peaks" and "troughs," and the electropolishing process removes material from the peaks faster than valleys due to a higher current density (concentration of electrical flow). Over time, those peaks are reduced until a smooth surface is created, which will be less susceptible to the acid. This self-regulating behavior, known as anodic leveling, allows for rough surfaces to be smoothed without excessive removal of material. It is often used in stainless steel as it leaves a very smooth, uniform surface that is easier to sterilize and free of stress induced by mechanical polishing. The basic setup for electropolishing is shown in Fig. 20.4, but it is closely related to multicomponent corrosion, shown in Fig. 8.21, and the sharp peaks undergo dissolution of metal atoms as shown in Fig. 8.20. The discussion on these figures will also be useful in understanding the mechanisms of electropolishing.

20.2.5 Coatings

A coating is an applied layer of a different material that changes the appearance or function of the product. Unlike a plating operation, which deposits a metal, a coating may be a different class of material. For example, metal components can be finished by adding a polymer coating. Paint is perhaps the most common example of a coating and can be applied to any base material. Another example is the rubber coating of a tool handle, such as an adjustable wrench, for added grip and comfort. This prevents loss of grip when the surface is wet or oily. The primary difference with a coating and other surface treatments is that it remains distinct in composition and is not atomically integrated with the base material through heat, chemical reaction, or other means. There will be a clear interface between the base material and the coating. With an oxidation process, for example, there will also be an interface, but it will tend to be more diffuse. Plating will

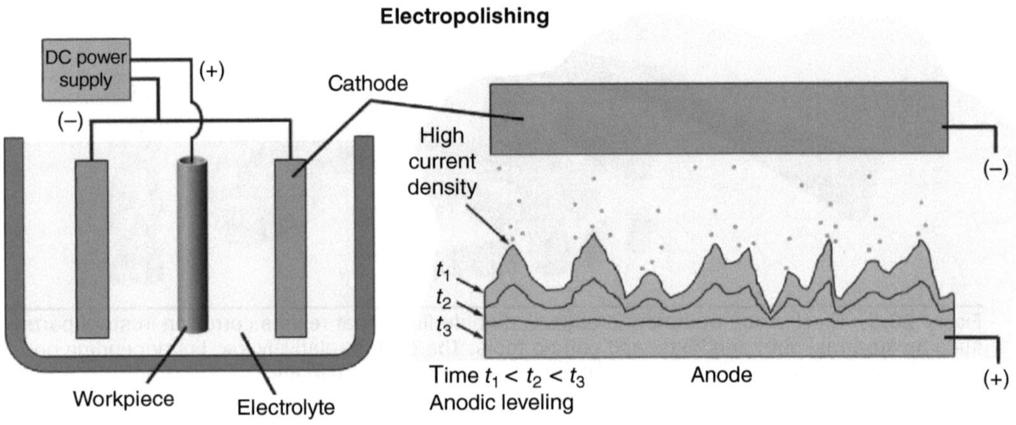

Figure 20.4 Electropolishing smooths a surface by dissolving the metal faster at peaks due to a higher current density. Over time the surface self-levels until very smooth. This process is known as anodic leveling.

retain an interface, but with proper application it should be more discrete. Plating, however, can be considered a coating as well.

20.2.5.1 Paint

Paint is the most common coating. There is a multitude of paints that are formulated for specific uses, with an array of appearances, and that have varying performance. They have applications ranging from interior walls to underwater features to engine components. The aesthetics and performance are determined by the application, and new examples are being formulated constantly. For example, paints that look like stone, that look like metal, that handle high temperature, that bond to plastic, that prevent rust in metals, and that do just about anything are available. It would require a second book to describe all the considerations that go into these products and the application of them. Even a glance at just the spray paint aisle (or more formally, *aerosol* paint) should emphasize the diversity in paint (see Fig. 20.5). And that is only the tip of the iceberg. The key factors for successfully applying most paints are

- Surface preparation (clean, no debris or loose material, slightly rough for adhesion)
- Chemical compatibility (bonds strongly to base material, primer used if necessary)
- Built-up thickness (multiple, thinner layers, sufficient drying time between coats)
- Top coat where needed (clear coat to improve durability, change appearance)

Application may be done by methods such as immersion, spraying, or brushing a liquid. The viscosity varies greatly, with thin paints being sprayed more easily than thick ones (wall paint). The most common method for drying is the evaporation of a solvent, but some also require baking at elevated temperature to fully harden the paint. The ingredients of paint also vary. Because of the diversity it is difficult to generalize, but most will contain a combination of the pigment or dye for color, binders, and a liquid dispersant (solvent). One of the critical differences in coatings is whether a dye or pigment is used. In brief, a **dye** is a colorant that is fully dissolved and **pigments** are suspended in the liquid. Where pigments are used, an additive to keep the pigment evenly distributed is also used. This is often called a dispersant. In slip casting, the same principle was achieved using a deflocculant (see Sec. 11.4.2). Additional ingredients

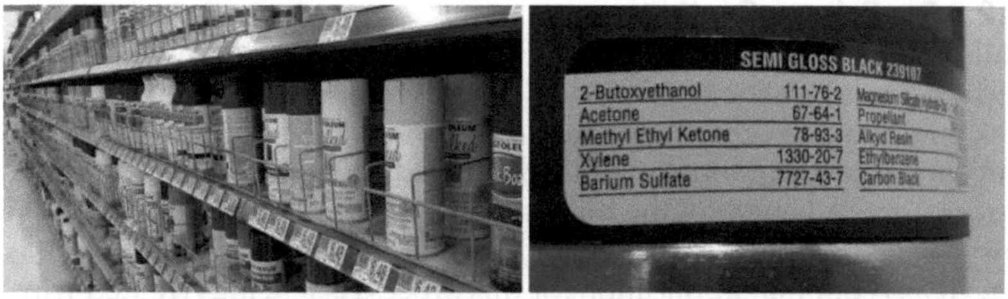

FIGURE 20.5 Paint comes in thousands of formulations and can be applied with a range of methods. Even one type, aerosol paint, comes in various colors and textures with performance characteristics such as corrosion protection, chalkboard finish, and high-temperature durability.

may include extenders to thicken the paint at low cost, plasticizers (see Sec. 10.2.3.1), ceramic particles, metallic flake, and other decorative elements. Water-based paints may also contain a humectant to retain moisture during use. Just like the appearance, performance characteristics, such as hardness/scratch resistance, flexibility, thermal stability, chemical and moisture compatibility, and friction are all determined by the ingredients.

20.2.5.2 Powder Coating

Powder coating is a dry finishing process where fine particles of a polymer resin, often mixed with other ingredients, are applied to a surface using electrostatic charge and then cured in a separate step using heat. The powder contains the pigment and additives to control the curing and flow of the heated resin to create a uniform, durable coating. The process to apply the powder relies on an electrical potential between electrodes within the powder coating gun. As the particles pass between the electrodes they become charged and attracted to the grounded part. The particles and part may be combined using a variety of techniques, but the simplest is to spray a mist of fine powder toward the part. As the two approach one another, the powder is attracted to the surface and sticks to it, much like a balloon sticks to your hair when you rub them together. The attraction is sufficient to create a thorough coating, even on complex surfaces with fine features that are difficult to reach otherwise.

After the application of the powder, the component is heated to cure the powder, which flows together and coats the surface. The temperature depends on the type of coating [thermoplastic/thermoset and type of polymer (refer back to Sec. 5.3.2.5)], with epoxy and polyester being common (Secs. 6.3.2.3 and 6.3.2.2, respectively). Curing is often done around 400°F (204°C). Although the process requires special spray guns and some basic electrical setup, it can be done by a hobbyist with enough ambition, and low cost kits are available. For small-scale and one-off work, it does require more equipment and time than painting. It is much more efficient in production environments as unused sprayed powder can be recycled.

Powder coating produces a tougher finish than most paints, and the technique is used wherever a durable finish is needed (Fig. 20.6). Examples include machinery, wheels, and other components for vehicles, as well as household appliances. In addition to its durability, it requires no hazardous solvents, can be done efficiently in large batches, and does not run or drip. The process time, particularly drying/curing, is less than most paints.

20.2.5.3 Polymer Coatings

While paint and many powder coating materials are primarily polymers, some coatings produce a solid plastic finish that encases the underlying metal. These coatings are very thick compared to others and provide substantial protection for the underlying metal. They are often combined with galvanized steel to further improve the corrosion performance. A benefit of the thick polymer coating [often polyvinyl chloride (PVC), or simply called "vinyl"] is that a product can be colored, though black is most common (Fig. 20.7). A popular application for this process is fencing (welded wire and chain link). Imagine trying to paint chain link fence! The PVC coating is extruded and higher quality grades are also bonded to the wire to ensure it does not peel off. Different classes of coatings will vary in the way the PVC is applied to the steel and how thick the coating is.

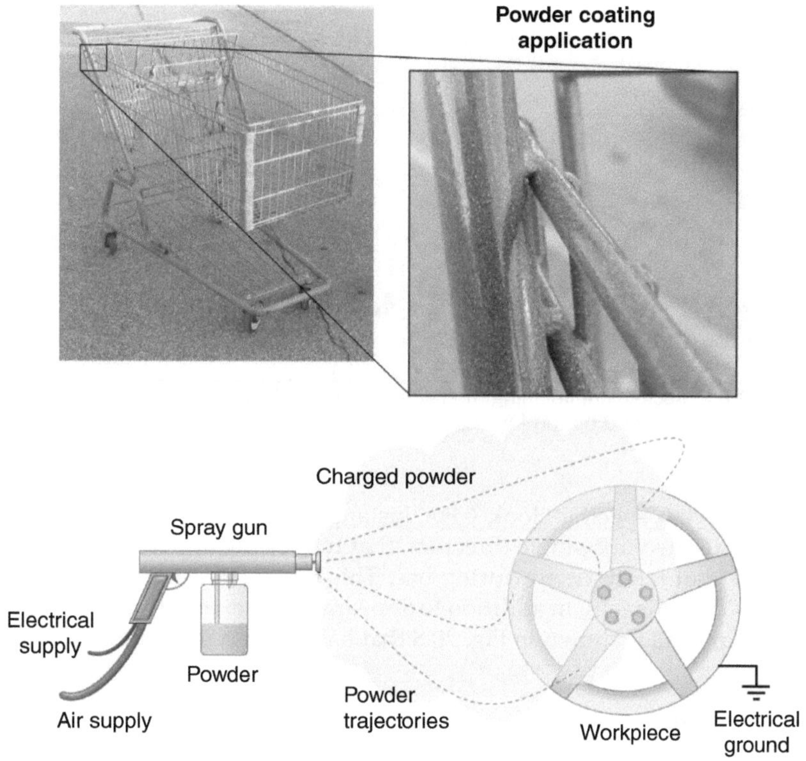

Figure 20.6 Powder coating produces a durable finish on complex surfaces by attracting the coating material electrostatically.

20.2.6 Mechanical Treatments

Metals can be treated mechanically to improve their mechanical properties, as discussed in Sec. 9.2, but they are also processed for aesthetic and functional purposes. The most basic distinction is in surface roughness. Some metals are highly polished to have a mirror-like appearance, others are textured to increase friction, produce a matte finish,

Figure 20.7 Thick polymer coatings can be used to protect metal from corrosion. The layer is much thicker than paint or powder coat and also allows aesthetics to be improved.

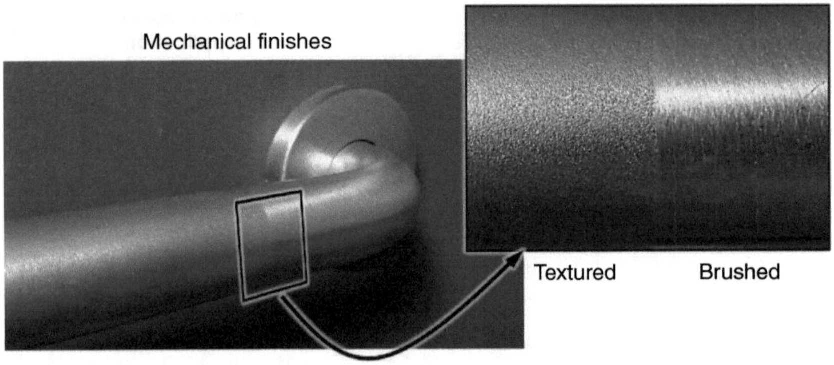

Figure 20.8 Mechanical finishing modifies the material's existing surface for function and aesthetics. It is a good option for materials which are intrinsically corrosion resistant and durable.

or to achieve a specific "look." For instance, a "brushed" surface has fine scratches, typically aligned in a single direction, that reduce the appearance of smudges and any scratches that might occur during use. This type of finish is popular in stainless steel household appliances. In addition to appearance, a textured surface is useful to increase grip. A handrail is shown in Fig. 20.8 that has both brushed and textured surfaces.

20.3 Polymers

Polymers are unique in their ability to incorporate decorative elements into their structure. Whereas the color of a metal is based on the metallic elements (shades of gold or gray), most polymers are naturally varying degrees of white, off-white, translucent, transparent, or other hues that are easily colored. This allows decorative elements to be applied in the molding or casting process, and postproduction finishing operations can be eliminated or reduced.

20.3.1 Coloration

Coloration is the process by which color is introduced throughout the bulk of a plastic component (Fig. 20.9). This has the distinct advantage that the color cannot be

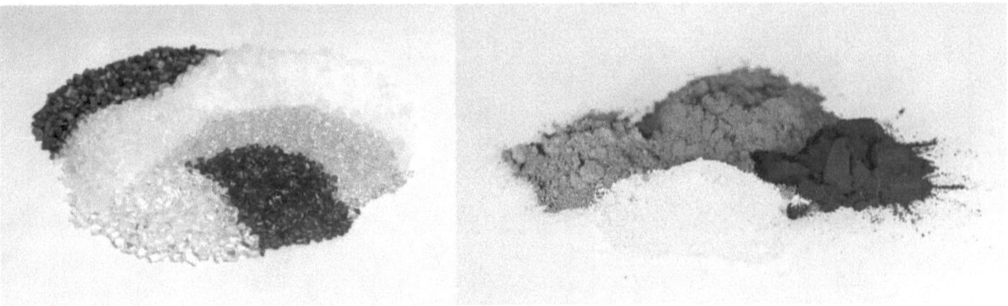

Figure 20.9 Polymers are colored before molding and extrusion by using pellets that are colored (left) or adding pigment powders (right) to the pellets and mixing during processing.

scratched off, but it may fade due to environmental degradation (see Sec. 8.5.3). The color can be introduced to the plastic component before the molding operation or during. Specifically, a "masterbatch" of concentrated pigment pellets may be mixed with natural (uncolored) pellets, or uniform, precolored pellets can be used. The pellets may be mixed before being introduced to the process, or the mixture can be controlled from multiple hoppers at the machine. The colorant can be added directly (not as a pellet) and mixed during the molding operation. In addition to pigments, other materials, such as metallic flake, can be added to produce other visual effects.

20.3.2 Lamination

Lamination includes processes that combine two or more distinct layers. The layers may have differing chemistry, color, or other functional purposes. The layers can be combined by methods such as extrusion, calendering, or pressing. The common result achieves distinct layers that are adhered together, and this may be applied to very thin or thick layers. Consider a foil-lined chip bag, as well as many other food packaging applications. Thin layers are laminated for decorative and functional purposes (i.e., prevent spoilage). Look at Fig. 5.35 and the associated discussion. Many of the things we use every day and do not think twice about have a significant amount of engineering that goes into them.

Another common example is "laminate" flooring. This type of flooring uses multiple layers of polymers for decorative (often "wood look") and functional purposes to create a wear-resistant surface. This decorative layer is applied to a thick core, which is often a composite wood product. A cheaper flooring option is linoleum, which is a relatively thin, durable polymer laminate that is often textured in addition to having various color patterns. Again, each layer contributes to the performance and appearance of the product. Even the cover of this book is a polymer laminate, though the base layer is paper.* The decorative and water resistant polymer lamination increases the durability, ensuring it will be a close companion for years to come.†

20.3.3 Printing

Despite the range of appearance effects possible with polymers, applying detailed logos, pictures, and other graphics is still important. Stickers, or *decals*, for instance, are printed on polymer sheet with an adhesive backing. The print is applied to the polymer, and the decal can be applied to almost any surface. Printing can be done directly on a product, such as a travel mug. This allows for the print to be applied with the product material as the background (only lettering or a logo), or the print can include a background to effectively change the appearance. For example, in Fig. 20.10, the travel mug appears red with print on it, but removal of the bottom rubber sleeve reveals the mug is actually white.

There are many methods for printing on plastics, but we will not go through each of them. One of the most widely used (including the mug in Fig. 20.10) is screen printing with UV-curable ink. This can be rapidly applied in a production line as the ink can be cured in seconds allowing subsequent handling. This may be done using multiple

*Some will rightly note that paper, a wood product, is composed of macromolecules like polymers are. We will consider them to be different.

†And ensuring it survives unexpected use as a coaster, door stop, or whatever else you might do to it!

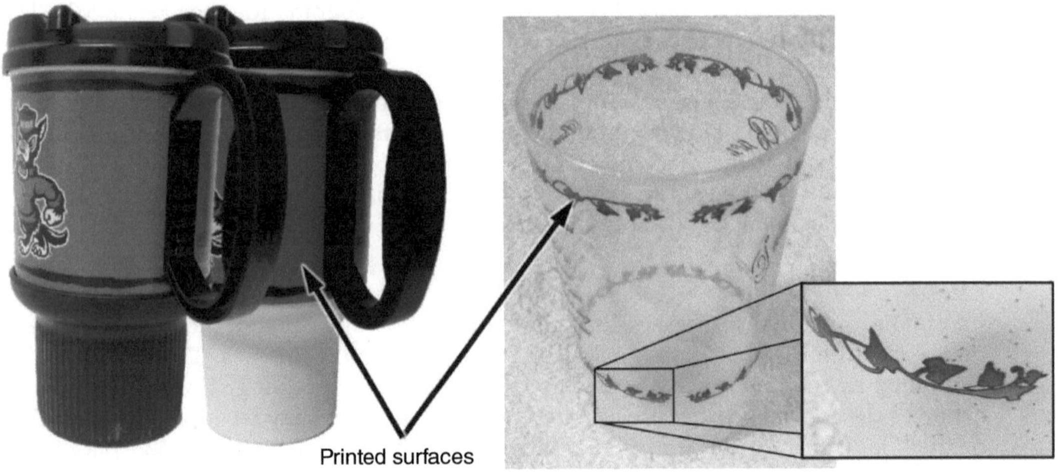

Figure 20.10 Polymers are often decorated by printing. As shown, printing can contribute the majority of the color, or it can be applied with the polymer as the background.

colors in a single printing process and the underlying plastic can be one of the colors in the image, especially where a full background print is used.

Other printing methods include **letterpress printing**, which uses raised features with ink on them that is transferred to the surface (flexible versions for irregular surfaces are known as letterflex or flexographic printing depending on the ink characteristics). **Offset printing** transfers ink to a roller that then transfers the image to the surface. This may include more than one color, which is applied in one step. This is known as **offset lithography** when the ink is applied to select locations on the roller by chemical compatibility rather than a raised pattern. For instance, nonpolar ink can be applied to specific areas of the roller by treating those areas with a nonpolar base material (e.g., grease) and the rest of the roller with a polar material (i.e., water).

20.3.4 Metallization

Metals have a unique reflective appearance, but they are not as easy to mold into complex shapes as polymers are, and they are considerably denser. If the structural properties of metal are not necessary but the aesthetics are, plastics can be treated to look like metals. The process of metallization coats a surface with metal. That surface may be any material, but it has become especially popular with polymers. The ease of manufacturing polymers, in addition to their light weight, makes them excellent candidates for many nonstructural components. For instance, the interior of most cars are entirely plastics of varying types. Many of the trim pieces, knobs, switches, etc. that appear to be metal are really plastic that has been metallized. On the exterior of vehicles, most trim (grille, badges, door handles, etc.) are also metallized plastic. This gives the desired appearance with lower weight, lower cost, and greater fuel efficiency. Other factors such as heat transfer, electrical conductivity, reflectivity, and wear resistance also make metallization a good choice for certain applications.

The processes to create this surface include electroplating and electroless deposition, which are described for metals in Sec. 20.2.2. Polymers lack the electrical conductivity of

Figure 20.11 Metallization is used to give plastics the appearance of metallic materials with less weight and easier processing. The metallic coating can be susceptible to peeling under heavy use over time, such the door handle in the center.

metals, so they must be treated to be electrically conductive for electroplating. The creation of the conductive surface may require several etching and activation steps in various baths of acids and metal salts. An alternative method uses metal heated in a vacuum chamber until it vaporizes and coats everything in the chamber, including polymer components. This coating is very thin (about 0.0001–0.0005 in) and uniformity varies depending on the distance from the source of the metal vapor and complexity of surface. To increase uniformity the parts are rotated during deposition. A similar process, physical vapor deposition (PVD), used the impact of gas ions on a metal target to create metal vapor. This process used to creates many useful coatings, some of which are for cutting tools, as described in Sec. 13.4.4. As with other coatings, metallization is susceptible to damage over time as evidenced in the center image in Fig. 20.11. Other examples of metallized products are also shown.

20.4 Ceramics

Ceramics are largely immune to corrosion and very wear resistant. This makes some of the critical applications in metals and polymers unnecessary. The primary functions of finishing operations with ceramics are to eliminate porosity and change the aesthetics of the product.

20.4.1 Coating

Like all materials, ceramics can be painted to change their appearance. This is not common practice since ceramics are not prone to corrosion and can be decorated more durably with other coating methods. We will consider glazing (described next) and methods to coat other materials with ceramics.

20.4.1.1 Glazing

Glazing is the process where a coating of compounds is applied to the surface of a ceramic and heated to melt and fuse the coating into an impervious, glassy layer. The chemistry of the glaze determines color, and various metallic compounds (oxides and carbonates) are used to achieve these colors. The required temperature is also dependent on the chemistry, as certain glazes will melt (or *fire*) at different temperatures (a look back at Sec. 11.4.3 may help here). The ingredients of glazes will include a glass-forming compound (often SiO_2) and a flux, which helps reduce the melting temperature of the SiO_2. The flux is often a metal oxide with a low melting point, such as lead oxide

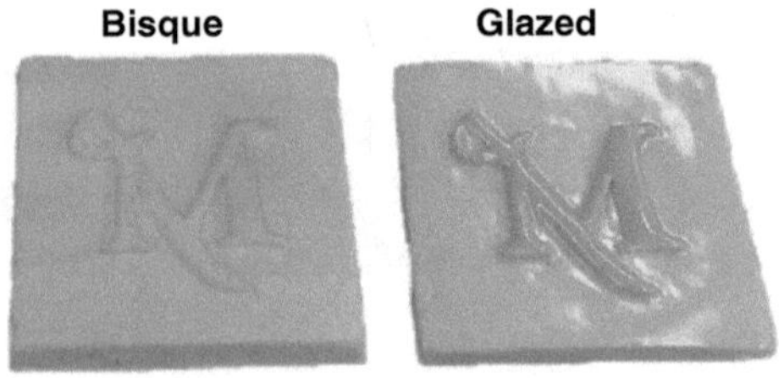

Figure 20.12 Glazing has functional and decorative uses. In addition to allowing specific colors to be applied, it also creates a durable coating that is impervious to water.

(PbO, problematic due to toxicity), sodium oxide (Na_2O), potassium oxide (K_2O), and others. Additional ingredients include stiffeners to reduce the tendency for the glaze to run when it melts, and additives to control the opacity and gloss of the fired glaze. An example of bisque and glazed tiles is given in Fig. 20.12.

20.4.1.2 Thermal and Cold Spray Coating

Thermal spray coating uses intense heat to melt a material and gas pressure to propel the molten droplets toward a surface where they rapidly solidify. This continual process quickly builds up a coating that can be very thick (tens or hundreds of thousandths of an inch) and can be applied to any substrate able to withstand the temperature [e.g., 300°F (148°C)]. The melting process can be driven by flame, arc, plasma, detonation, or other means of melting the coating material. The high temperature [as high as 29,000°F (16,000°C) for plasma] is capable of melting high-melting point metals and ceramics, as well as other materials with lower temperature. The molten droplets impact the surface at high velocity and flatten out to a "pancake" shape. The thermal spray process is shown in Fig. 20.13.

Thermal spray coatings can be used to repair worn components that are expensive to replace, to increase wear and corrosion resistance, to increase electrical conductivity, and it can even be used to produce free-standing components. Because the bond is mechanical in nature, surface prep is critical. Typical preparation includes cleaning/degreasing and roughening, often by sand blasting or comparable technique (see Sec. 14.6.3.1). The microstructure (described for metals in Sec. 3.1, conceptually similar for other materials) of the coating is unique due to the spreading of the droplets upon impact. This creates an anisotropic structure such that the layers (lamellae) are oriented parallel to the surface. This can benefit the mechanical properties (see Fig. 9.1 and associated discussion), as long as the principal forces are in the same direction, much like the grain of wood makes it stronger along the grain than across it.

Cold spray coating is similar in concept to thermal spray, but instead of melting the coating material, it propels particles at supersonic velocity to bond the coating to the base material. "Cold" is a relative term, as the gas stream, typically nitrogen or helium, may be heated to hundreds of degrees [e.g., 200–900°F (~100–500°C)]. The primary difference is that the material being sprayed is not melted, and therefore it retains the original characteristics of the feedstock powder. It creates a stronger bond in some cases as the impact causes physical mixing of the coating and base material. This is most useful in metals, as their ductility allows significant deformation of the surface upon impact. Advantages include lower porosity and a simpler setup as there is no need for flame, arc, or other equipment to melt the powder.

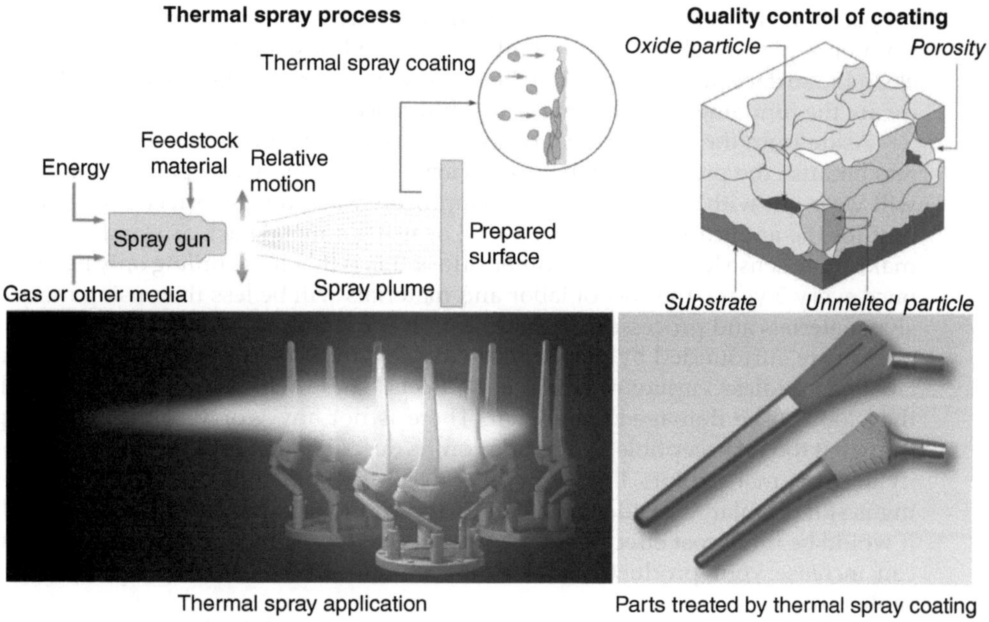

FIGURE 20.13 Thermal spray coating allows materials of very different properties to be combined. The specific application of coating orthopedic implants is shown here. (*Reproduced by permission of Oerlikon Metco.*)

20.4.2 Mechanochemical Polishing

Polishing processes are used to create exact dimensions and flat surfaces in some ceramic applications. **Mechanochemical polishing** combines abrasive action with chemical attack to remove material and produce a smooth surface. Although the concept can be applied broadly, this is most extensively used in silicon-processing to create wafers for producing integrated circuits. In this case, extremely flat, uniform surface are required for the fine features of the circuits. The process is termed chemical mechanical planarization (CMP) and may achieve angstrom-scale flatness (atomically flat or very near). The combination of abrasive and chemical action removes material and ensures enhanced smoothness with less surface damage, similar to electropolishing in metals (Sec. 20.2.4.3).

20.5 Why It Matters

We are accustomed to finishing operations, like painting. We simply grab a brush or spray can and apply the paint. The amount of thought that has gone into the product often goes unnoticed. We can get better results by considering it a little more. It mattered to the many companies that compete for your business in this space, and it matters for achieving the best result, not just an acceptable result.

What we are finishing is the primary consideration. Some methods of finishing, like decals, are pretty much universal. Other techniques, like plating, need a specific chemistry and process controls to generate the right result. The basic process requires surface preparation and considering the compatibility between the materials. If you use wood stain on a plastic cabinet, you might be surprised how bad it looks. If you neglect to clean the surface properly, you will be in the same boat. It is the "details," which in reality are the fundamentals, that make the difference.

Aside from the foundational concepts of whether a process is appropriate, it matters because the process may be insufficient. You can perfectly paint a handrail, but it will wear out with use. It is not that the process was done incorrectly; it is just not the best for the application. Powder coating or using a metal such as stainless steel might make more sense long term. The other side of this is that if the railing only needs painted once every 5 years, the cost of labor and materials will be less than using more expensive materials and processes. The application is important!

We are surrounded by plastics these days. Part of their popularity is the ability to create an endless variety of colors and effects. These will not wear off either, though they too can be damaged somewhat. There is not any worry of the color chipping, though it too is susceptible to degradation, as described in Sec. 8.5.3. Even with the ability to create many colors, the use of other decorative finishing operations such as printing is still popular. Consider whether an appearance can be created during molding or if would be more cost effective to use printing or decals. By knowing your options, you can increase your productivity, reduce cost, and meet customer expectations at the same time.

20.6 The How and Why of Finishing

20.6.1 Chapter Review

1. What purpose(s) does finishing serve?
2. How is the term *finish* used as a noun and a verb?
3. What is the most common purpose for finishing metals?
4. When is a temporary coating used? Give an example.
5. Name the decorative and functional values of plating.
6. How does electroless plating work?
7. How must nonconductive surfaces be treated before electroless plating?
8. How is galvanizing different from plating?
9. What geometric considerations should be made when plating, galvanizing, or applying other coatings?
10. Anodizing creates a layer of what on the surface of the component?
11. What are the functional benefits of anodized surfaces? What about aesthetics?
12. Can a thick anodized coating be a bad thing?
13. Bluing is a form of oxidizing ferrous metals. How is it different black oxide coating?
14. In what applications is a black oxide coating popular? Why is it used?
15. Describe how electropolishing achieve a "self-leveling" effect?

16. What applications is electropolishing used for?
17. In what form are polymer coatings most universally applied?
18. How do coatings differ from surface treatments such as anodizing and galvanizing?
19. Consider the factors for successfully applying paint. Which are most likely to be ignored or prone to error?
20. What are key ingredients in paint? Why is it rare for the ingredients to be listed with any detail?
21. What is the difference between a pigment and a dye?
22. How do most paints "dry?"
23. Paint and powder coating both create polymer coatings. How is powder coating superior?
24. Why would you not always use powder coating?
25. Why can powder coating cover complex geometries so well?
26. What is the purpose of heating the powder coated components?
27. How are "polymer coatings," often *vinyl* coatings, different from paint or powder coating?
28. What are mechanical finishes used for? What is their benefit over other finishes?
29. Why are polymers important for laminates? Name some common applications.
30. Why is printing so important to polymers in particular?
31. Why would metallization be used instead of using a metal component?
32. How is metal coated onto a polymer?
33. What functional purpose might metallization have?
34. What about their properties make it less common to finish ceramics?
35. What is the primary function of glazing?
36. How can ceramics be coated onto other materials?
37. What are thermal spray coatings used for?
38. Why can the substrate be of a material with low melting point but still be coated with a high-melting point material using thermal spray?
39. Why is surface prep important for thermal spray coating? Be specific.
40. How is cold spray different than thermal spray?
41. What is the microstructural advantage of cold sprayed coatings?

20.6.2 Concept Connections and Extensions

1. Consider a nonstick coating on a frying pan. What factors listed in Sec. 20.1 would be included in this finishing operation? Compare the performance of Teflon™ and Thermolon™ coatings. Research each of these and use PTFE (Sec. 6.3.3.4) and SiO_2 (Sec. 6.2.1.3) as the models for these materials, respectively.

2. What is the importance of diffusion in creating durable coatings? What drives the diffusion process and how does this affect the way the coatings can be applied? Refer to Sec. 8.4 for additional information.
3. Consider how a mirror is made. What is the primary material? What processes in this chapter might be used to create the reflective appearance?
4. How are plating operations similar to multicomponent corrosion (see Sec. 8.5.2.2)? What is the primary difference?

CHAPTER 21

Additive Manufacturing

There are few technologies that fundamentally change the way manufacturing is done. The industrial revolution, for instance, ushered in extensive use of mechanized equipment and mass production, which was aided by the development of steam power. The development of computing, and eventually the reduction in scale of the required circuitry, allowed machines to be programmed to do multiple tasks in a flexible way. This was enabled by developments in silicon processing to create microscale and nanoscale transistors. Modern manufacturing relies on incremental developments to achieve scale and efficiency, but these incremental advancements can yield "disruptive" changes. Additive manufacturing is the result of years of development in automation, control, and other support technologies, but it presents a fundamental change in approach to how products are made and what is possible.

21.1 How It Works

Additive manufacturing (AM) creates a three-dimensional structure by adding material in a layer-by-layer manner until the structure is complete. This definition is quite broad, and that is a benefit of AM. Structures can be created from nearly any material (polymers, metals, ceramics, and even composites), and the size of the structures can range from being smaller than an insect to the scale of a car or small building. The fundamental limitations to the product are the size of the machine (the part is created within a machine) and the resolution of the machine (the smallest accurate movement and material allowance). The term "additive" indicates this method is different from "subtractive" manufacturing, such as machining. Other methods "add" material (like welding and casting), but AM is principally concerned with building the bulk of the structure by selectively adding material in free space, not with a mold.

AM is also known as 3D printing or rapid prototyping. These terms reflect the original method and use of the technology. The term printing arose because the method can be performed by depositing a fine layer of material at precise locations. This is similar to printing a page with ink on it, but in AM each layer has a certain thickness, and the printing is done on top of the previous layer. This is typically accomplished by lowering the part (or raising the printing head) for the next layer to be deposited. Imagine running a single page through the printer hundreds or thousands of times, with each pass having a slightly different pattern. In effect, that is what happens in AM. A 3D object is broken into slices and each "slice," or layer, is printed separately, building height as it goes (Fig. 21.1).

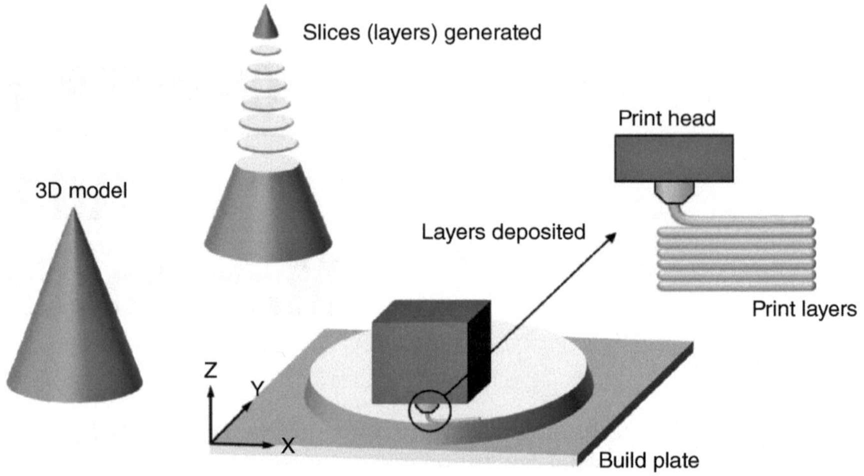

FIGURE 21.1 General process for creating structures using additive manufacturing. Note that this representation is based on extrusion of the build material.

The way printing (i.e., the unification of the layers) occurs depends on the method, and we will consider some of the most common. New technologies are constantly being developed, and there are a variety of websites devoted to the topic.*

21.1.1 Applications

AM opens up a unique possibility to create exceptionally complex shapes with features that would be otherwise impossible to make. It also allows shapes that are relatively simple to create by other production methods to be produced without dedicated tooling such as molds or dies (the kind of stuff needed in Chaps. 11 and 12). This allows prototypes to be made without investing the time and money necessary to design and build this tooling. This is not without limitations and considerations, however, and we will look at the most important capabilities and challenges associated with AM.

First, we will consider what AM does that can already be done. It makes 3D shapes. That is not impressive, and in most production situations, AM is not a good choice. For instance, you can print a screw using AM, but you cannot make thousands an hour like thread rolling operations can. You can make odd shapes with fine detail, but investment casting can too. AM is most useful in making small quantities of parts or prototypes *before* investing in dedicated tooling. For instance, a flange that must fit in a tight spot can be printed and tested to ensure that it will work as intended, but the actual production will be done by forming, casting, or machining. The prototype ensures that a problem is not encountered later in the production process after thousands of dollars have already been invested in the particular design.

AM is also good for making a few custom parts. Most AM methods are still pretty slow. They have gotten *much* better, but when compared to traditional manufacturing, the vast majority of parts can be produced faster using other techniques. The most successful

*For example: 3dprint.com; 3dprinting.com; additivemanufacturing.com, etc.

processes in AM are those that are too complex, require specialized tooling for traditional processes, and that are not being produced in mass quantities. Custom dental implants or other one-off biomedical devices are examples of where AM is the best option. Getting repair parts to military personnel could also be faster if it can be printed in a nearby location rather than shipping from a central warehouse that may be on the other side of the globe. Most of the examples you will find online about 3D printing involve decorative geometric wireframe shapes and novelty items. These are interesting to the hobbyist, but commercially, there has to be financial incentive to invest in the equipment and use it. In select advanced manufacturing outfits, it makes sense. Critical aerospace, defense, and medical applications are among the most developed presently.

21.1.2 Design Considerations

There is a general, "anything goes" attitude about AM, and that is partly true. What can be created by 3D printing is only limited by the equipment or method. Can you print a fully functioning automobile? No. There is more to it. You *can* create functional assemblies, but there are some basic considerations when printing.

The most basic considerations are the print conditions. Most machines allow modifications to the resolution, orientation, and scale of the virtual design. The resolution is based on the equipment, but in noncritical prints the process can be made faster by changing the layer thickness, print speed, and part parameters such as the shell thickness and infill percentage. The **shell thickness** is thickness of the exterior layer of the printed object. Many objects are printed to be mostly hollow with only a thin layer on the outside, which is solid. This brings us to **infill**, which is the amount of material within the printed shell (typically expressed in percentage), and it may exist in varying designs. For instance, the infill is often hexagonal, but it may be square or any other shape the user defines. Most prints will use 10% to 25% infill with 2 to 5 shell layers. If the part is supposed to be structural, 100% infill can be used, which creates a solid part but dramatically increases the print time and material requirement. Infill density is demonstrated in Fig. 21.2. One of the important advantages of AM is that reduced material usage is possible.

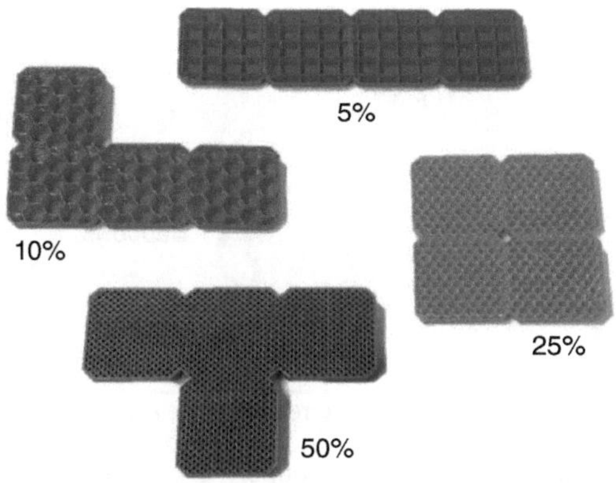

FIGURE 21.2 Infill density can be used to save material while retaining strength. Higher density will increase strength, but requires more print time and material (examples prepared by Nicholas Bozelli and Andrew Miller, Millersville University).

Another common requirement is support. Whenever overhanging features are to be created, it may be necessary to add a **support** (very much like scaffolding), which is printed from the build plate to the feature and prevents that area from sagging or the print from tipping due to an imbalance. Most software will automatically add supports, but users can also define them. Additionally, a special layer, known as a **raft**, is added to anchor the part to the plate using the same material just with a specific design to allow it to be removed from the part later.

These additional structural features must be removed before the part is complete, and in machines that accommodate multiple build materials, the supports may be printed from a material that is dissolved by water or other solvent that the build material is not susceptible to. This allows the supports to be dissolved without damaging the main part. There may be some hand cleaning, scraping, sanding, or other work to fully remove these, but it is much less than if the supports are made from the same materials as the part. A final step with many polymer components is acetone vapor smoothing. The acetone dissolves the surface of the polymer allowing it flow slightly and smooth out. The part is typically held above a bath of heated acetone, but the process can be performed at room temperature over a longer time. The typical steps and features AM are represented in Fig. 21.3. These considerations are primarily based on fused filament deposition, the most popular polymer AM method, but as we will discuss next, there are many variations, and their considerations are different.

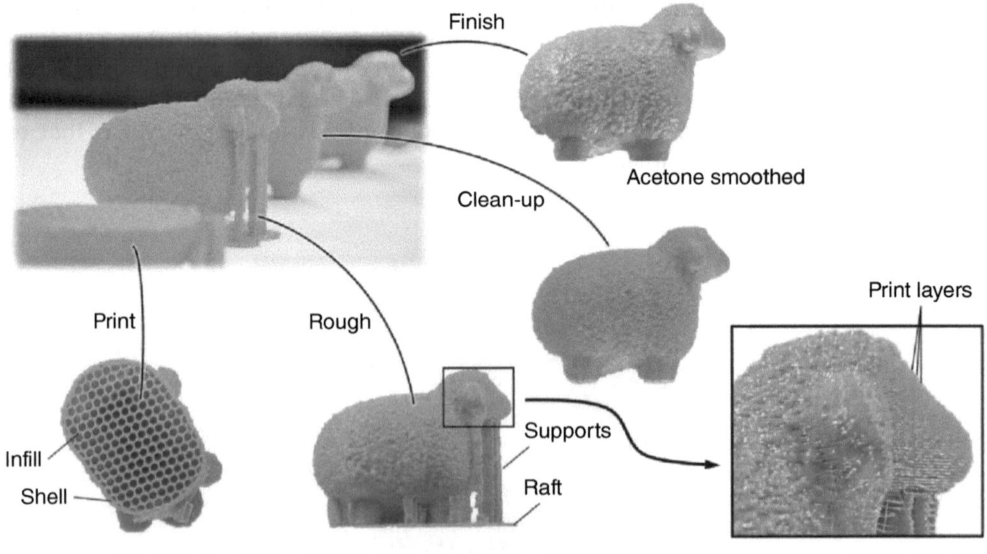

Figure 21.3 Additive manufacturing requires printing a solid object from a precursor material that is solidified/bonded to create a free-standing component. The printing process is done layer-by-layer and often requires supplemental support. The supporting material is removed and the part is finished by sintering, resin impregnation, or other treatment.

21.2 Types of Additive Manufacturing

AM is best known for its versatility in creating complex objects. Part of its versatility is also that it can be accomplished by any means that results in a 3D structure being built by layers in a geometrically controlled pattern. Two primary designations may be used: direct AM and indirect AM. **Direct AM** creates a complete (full-strength) part in the machine. **Indirect AM** requires posttreatment to infill, sinter, or otherwise impart full strength. There are several categories of AM, and under each there may be multiple versions using similar techniques. These various methods have diverse cost, maintenance, durability, and capability considerations that will be highlighted for each.

21.2.1 Fused-Filament (Extrusion) Deposition

The most popular method for creating parts by AM is **fused-filament deposition** (FFD), which uses the extrusion of a molten polymer filament (wire) to build layers that quickly solidify and bond together. Most people who have seen 3D printers have seen this method. The terms fused deposition modeling (FDM) and fused-filament fabrication (FFF) are also used, with FDM being a patented method and FFF being an alternative, unconstrained term. The filament is often PLA (polylactic acid, see Sec. 6.3.1.5) or ABS (acrylonitrile butadiene styrene, see Sec. 6.3.1.6). The polymer should have a relatively low melting temperature and wide glass transition temperature range to ensure it flows well but holds its shape and does not decompose (see Sec. 5.3 for more on polymer structure and Sec. 8.2.6.1 for temperature effects).

FFD is widely used for educational and hobby activities, mostly to print predesigned 3D parts available through open-source file sharing websites. This allows the files to be downloaded, scaled, and printed without the need for any advanced software. These same printers will accept any design in the proper format (.stl file type). They range in size, but most are not much larger than a microwave, and they are priced for individual buyers not making profit on their work. With the growing popularity of 3D printers, the cost has come down significantly. A hobbyist can purchase a build-it-yourself version for a few hundred dollars, and preassembled, higher quality versions are available for a few thousand dollars. More commercially viable options may be $20,000 or more, but the build quality, resolution, and reliability are greatly improved. Anyone who has done much printing on low-cost models has experienced the constant maintenance and adjustment that is necessary for heavily used hobby models. For some, this is simply part of the experience.

This method relies on several key components:

- The filament
- The feed system
- The heated extrusion nozzle
- A build plate (sometimes heated)
- The positioning system (motors and hardware)

The filament is typically about 1/16 in. in diameter but extrudes to a smaller diameter and allows for finer resolution. The filament is fed into the heated nozzle by rollers

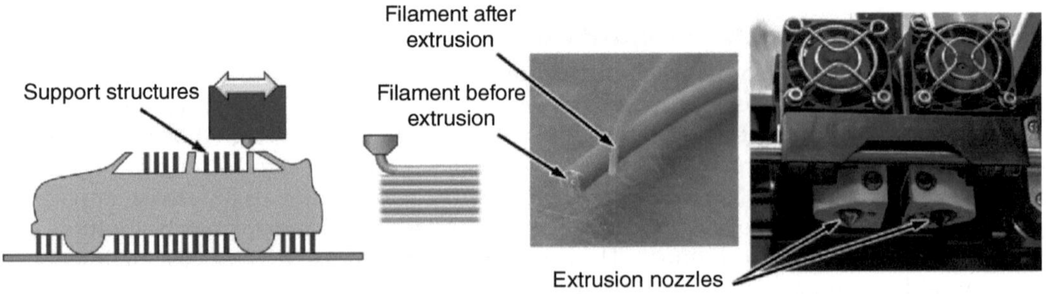

FIGURE 21.4 Fused filament printing creates solid structures by extruding and fusing together layers of plastic. The filament is fed into a heated extrusion nozzle with a small opening to achieve good resolution and low cost.

and the rate of motion of the nozzle can be used to regulate the amount of material being deposited. The part being created is attached to a build plate that supports the structure during printing. Some materials require a heated build plate to achieve a good bond between the part and the plate. For instance, ABS is typically heated to around 230 to 250°F (110 to 120°C), but PLA is often printed without heat. The forces encountered in these systems are very small, but some adhesion is needed to prevent shifting of the material. The nozzle is typically moved in the *x*- and *y*-axes, and build plate is moved in the *z*-axis (Fig. 21.4). The accuracy and resolution of the machine are largely determined by the accuracy of the positioning motors and the rigidity of their guides, but the print material and machine settings are also important.

The filament determines the look and qualities of the part being made. The color is the most noticeable difference, but the material will affect the strength and weatherability. Although all polymers are susceptible to degradation (see Sec. 8.5.3), PLA, which is the most widely used FFD material, degrades rather quickly when exposed to the elements. This limits its use in exterior applications, though its modest mechanical properties are also a major limitation. Filament with varying coloration, flexibility, luster, sparkle, and even glow-in-the dark varieties are available. The extrusion temperature and rate vary for these different materials, so an appropriate setting for one may not work well for another. Multiple filament types can be incorporated into a single component by using multiple extruders in one machine. The machine uses the appropriate extruder based on the material or color needed in a specific section of the build. Examples of equipment for FFD are given in Fig. 21.5. The example prints in Fig. 21.2 and Fig. 21.3 were created using this method.

21.2.2 Stereolithography (Photopolymerization)

Stereolithography (SLA, for stereolithography apparatus) uses light to polymerize a liquid resin that makes the exposed area solid, and the first methods for creating 3D objects by AM used this method. Wherever the liquid resin is exposed to the light, it becomes solid. New resin is then placed between the light source and the previous layer, and the process is repeated to deposit the next layer. The polymerization happens rapidly, so the process can be fast. Unlike filament extrusion methods discussed in the previous section, there is no requirement for heating the build plate or extrusion nozzle and the resolution can be much better as the liquid resin can be cured with greater

FIGURE 21.5 Common FFD equipment and components are shown here. The basic requirements are filament, an extruder, and position control in three axes. This method is the lowest cost and most widely used due to its popularity with hobbyists.

dimensional accuracy. Disadvantages of the process include higher equipment and materials costs and the limitation of printing one material only, but these are being addressed in newer equipment.

There are two general approaches to SLA: control the location of light on a large area of resin or control the deposition of resin under constant light. Both achieve a similar end, but the requirements vary greatly.

21.2.2.1 Mask and Projection Stereolithography

In stereolithography, the light source can be moved throughout the build area to expose specific locations to the light and create layers. This is similar to printing on paper where ink is only deposited at certain locations. Instead of moving a print head with a light source, the light itself can be directed to specific locations. In this *projection SLA* method, the light can be turned on and off and moved across the area using a mirror or mirrors to control the location. Unlike FFD, physical travel time is not necessary. The light can be quickly projected in a specific pattern using a UV laser and mirror system to trace the layer pattern. This is something like sketching an image without moving your hand. You can redirect the pencil tip to create a pattern by tilting its orientation. This can be done in under a second with a laser, and eliminates moving components for positioning (other than the mirrors).

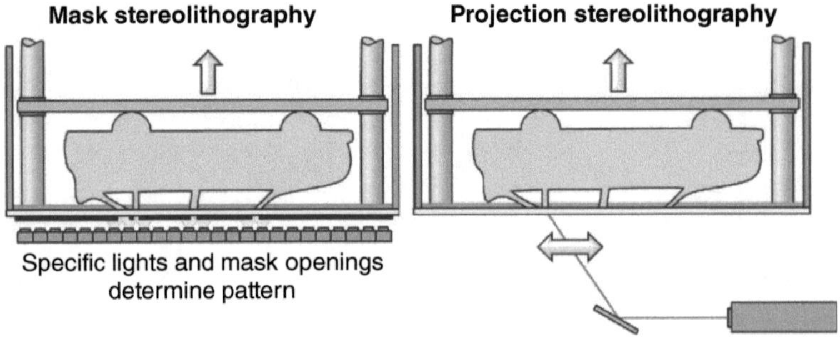

FIGURE 21.6 Stereolithography can be performed using a point source of light or a broad beam blocked by a mask. Both methods may be performed by exposing the surface of a tank of resin to light, or as shown here, in an upside down configuration.

Alternatively, a mask that controls the light over the entire build area may be used (known as mask SLA or MSLA). This method is similar in a way to making shadow puppets, except instead of blocking light to make a dark pattern, you block everything except the pattern where light is allowed through. Wherever light is blocked the resin remains liquid and wherever it is exposed, it solidifies. Each layer requires a different mask to change the cross section of the part. This would require an extremely large set of physical masks (similar to projection reels from the movies with individual frames) or, more reasonably, a mask that can be adjusted for each layer. Newer technology uses an array of light sources and a digital mask that can quickly block or allow light in specific patterns across a large area. This has the particular advantage of allowing the entire build platform to be exposed simultaneously to produce large parts, multiple copies of the same part or an assortment of parts. Example setups for mask and projection SLA are shown in Fig. 21.6.

To create a 3D object the solid part must be lowered or raised (depending on configuration) to allow another thin layer of resin to be exposed. In "right-side up" SLA a tank of resin is used and light hits the top surface. After each layer is formed, the object drops so that new resin covers the previous layer (often aided by a wiper that spreads the resin). The process is then repeated until the part is complete. In upside down SLA the light source is below the object being printed and it requires only a thin layer of resin to be present, not a full tank. Light shines through the transparent plate to solidify the material. This adds an additional consideration of "peel" force required to separate the solid material from the transparent plate as it is pulled upward. Newer approaches, which create a "dead" zone where the solidified material never touches the plate, are now available and allow faster and continuous (not layer by layer) parts (Fig. 21.7).

21.2.2.2 PolyJet

PolyJet AM is a photopolymerization method that creates 3D objects by depositing and quickly curing a resin using a print head that may contain one or many nozzles (Fig. 21.8). The primary benefit of this technique is that multiple colors and material types can be printed together. The resolution is better than SLA, and the print speed can be higher for small parts, but it is typically lower when making parts larger than cube dimensions of 5 in (~127 mm) per side.

This is because the print head motion becomes a limiting factor for large areas, which SLA can expose simultaneously. PolyJet is mostly used for small prototype work

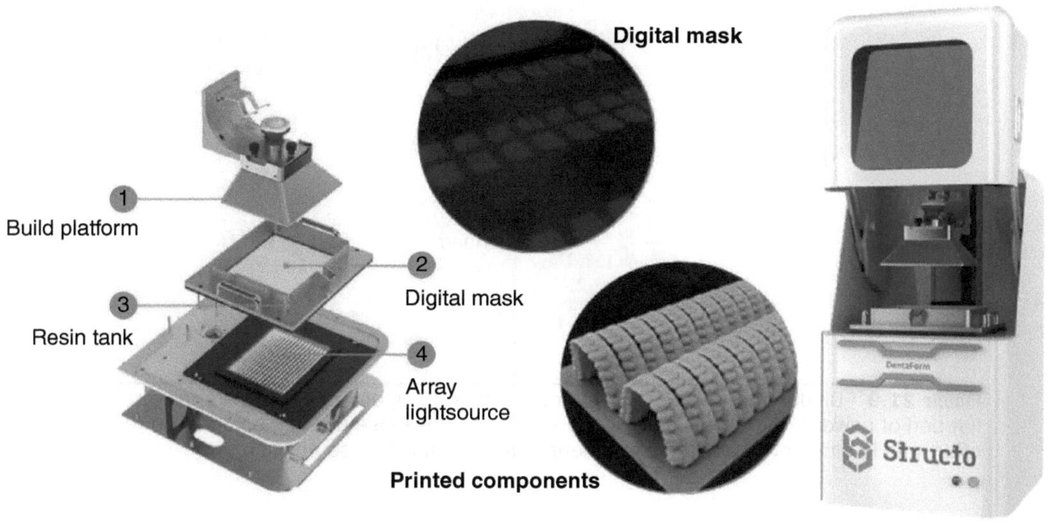

FIGURE 21.7 This mask stereolithography (MSLA) equipment simultaneously exposes areas across the build plate to generate multiple components quickly, and the mask is adjusted for each slice during the build. The parts are pulled upward during printing in an upside-down orientation (see also Fig. 21.6). This reduces the resin requirement during printing. (*Courtesy of Structo, Pte. Ltd.*)

and allows materials with different mechanical properties to be combined that approximate the final product. For instance, a flexible hinge can be printed to connect rigid components. The resolution is excellent (<0.001 in) but machine and materials costs are typically quite high (tens to hundreds of thousands of dollars depending on size and capability).

21.2.3 Powder Bed Printing

The methods in the previous sections have a common theme of creating solid structures from liquid (or molten) polymers. These methods are unsuitable for metals, ceramics, and even many polymers. An alternative approach uses powdered material and fuses it together using heat or a binder (glue) that is applied at precise locations across the surface. These methods are known as **powder bed printing**, as they start with a layer of

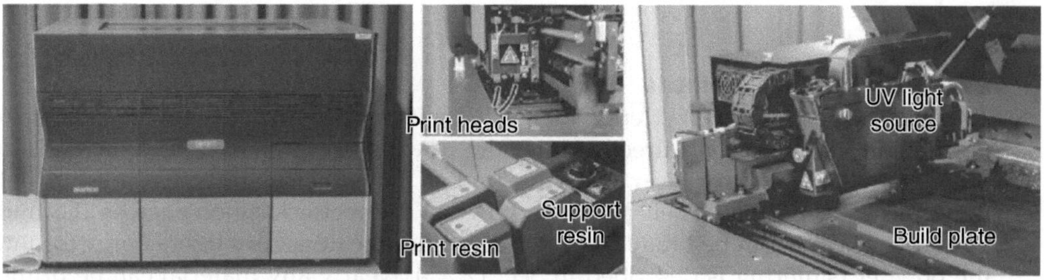

FIGURE 21.8 PolyJet AM allows high resolution and high throughput for small builds. A variety of materials can be printed and the resolution is very good.

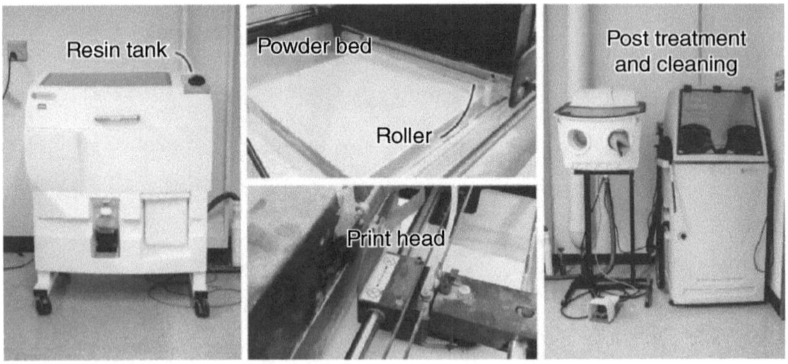

Figure 21.9 Binder methods fuse powders together using an adhesive material sprayed onto a flat bed of powder. The part is completely surrounded by loose powder, so clean-up requires some effort and is aided by dedicated equipment to recover the excess powder.

powder that is solidified in select areas, lowered and a new layer of powder is applied and the process is repeated.

21.2.3.1 Binder Methods

Binder methods for AM (sometimes simply called Binder Jet) apply an adhesive to combine powder particles in selected locations (Fig. 21.9). The material applied to the powder may cure from reaction with the atmosphere, but heat, ultraviolet, or microwave irradiation are often used to speed up and control the solidification process. The liquid binder is sprayed onto a thin layer of powder, cured and the build plate is lowered by a small increment. Powder is spread over the surface and the next layer is completed until the part is fully formed.

Once finished, the part is removed from the powder, cleaned, and fully cured (if needed). Some build materials require infiltration to achieve full handling strength, and liquid resins such as cyanoacrylate are used (see Sec. 6.3.1.8). If the part is a metal, a final sintering process is needed. If the material is a polymer, heat may be used to fuse the powder, or the binder itself may be sufficient for the application. Ceramics rely on the binder to impart the necessary strength until drying and firing and glazing can be applied as normal (the heat treatment of ceramics is discussed in Sec. 11.4.3). The powder size will limit the resolution, so fine powder (<0.001 in across) is typically used.

21.2.3.2 Sintering Methods

To avoid the posttreatment strengthen needed with some powder types after binder jet production, methods that directly sinter the powder are preferred. The methods are often known as **selective laser sintering**, and as the name suggests, these use a high-power laser to partially melt and fuse powder in precise locations across the build platform. The powder bed may be heated to reduce the necessary laser energy to sinter the powder (sintering is described in Sec. 12.2.5). Nearly any material can be printed using this method, and the ability to mix powders (e.g., alloy metals) is an advantage, but multiple materials cannot be printed in distinct sections like PolyJet or FFD.

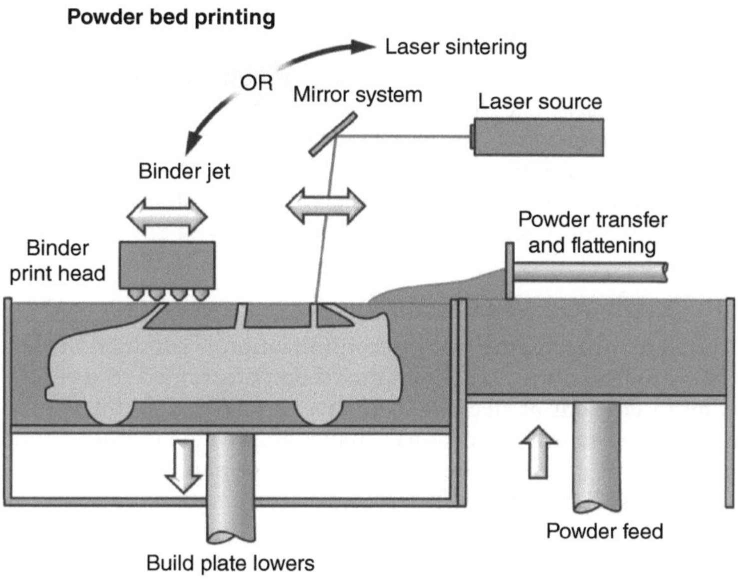

FIGURE 21.10 Powder bed printing builds a component within a volume of loose powder that increases in size with each printed layer. Binder and laser sintering methods are shown, but they are used separately.

One of the more interesting advantages is the ability to create parts within parts, as the powder supports the solidified layers so secondary components can be created that "float" within the open space of another component. Many parts can be nested together in all three dimensions during a single run, and this makes the technique more practical for prototyping and limited production runs. Depending on the materials and the final properties required, postsintering heat treatment to reduce porosity may still be needed. The general concepts of powder bed printing are shown in Fig. 21.10.

21.2.4 Lamination

Laminated object manufacturing (LOM) is a method that applies a series of thin layers, and it cuts away the unnecessary material from each layer. Like the laminates described in Sec. 6.4.3, the structure is determined by the properties of each layer, though in LOM it is typical to use a single material repeated over the many layers. The thickness of the layers will determine the resolution, which is less than most other AM methods. To build the part, a single layer is rolled/pressed into position, often with heat to adhere it in place, and the sheet is then cut using a knife or laser to define the part boundaries. The supporting material is cut into smaller segments to make it easier to remove the part later. Unlike powder bed methods, the excess material is not directly reusable. Also, the part is typically only practical if printed to full density as removal of the sheets from a complex interior cavity is impossible. Advantages include being relatively inexpensive and capable of quickly producing large parts. This method often uses a reel-to-reel setup and can be used with metallic foil, paper, plastic, and wood sheets, so long as the material is flexible and strong enough.

21.3 Special Considerations

The disruptive nature of AM is not based on the technology it uses or the materials it applies. AM is *enabled* by developments in technology, but it is distinguished by the changes it makes to what can be produced and how it can be produced. Especially important is the rising interest in at-home 3D printing and the improving affordability of the equipment. This has been suggested to mark a shift in the way goods can be produced and brings up interesting ethical questions.

21.3.1 Distributed Manufacturing

Distributed manufacturing is the decentralization of production such that parts can be made at many locations globally, rather than constrained to a central facility. Production is more efficient at dedicated, highly specialized facilities, but the transport of products can present an additional challenge. While central facilities are still highly favored and more practical, the ability to produce parts at various sites or even at home using a low-cost 3D printer is changing the conversation.

The emergence of the so-called AM **cottage industry**, which is simply producing goods in an individual's home, could change the way parts are made and sold *if* the parts can be produced with identical specifications, or alternatively, if the ability to customize the product for the specific area or individual is beneficial. One way this could work in practice is that instead of buying a product, that product could be downloaded and printed by the customer. The customer would have the ability to specify or modify the digital file such that the design was specifically suited to their desires.

The limitations of the most affordable hobby 3D printers would reduce the efficiency and product complexity, so the concept does not present a viable alternative to traditional production and distribution, *yet*. If the quality and capability increase for a competitive price, which they are likely to, more substantial products will become possible. Remember, there was a time when computers were only practical for governments, universities, and major corporations. Alternatively, distributed manufacturing can be practical for large companies now, which can shift production to various sites around the globe with the needed capability. This requires distributed resources, which is costly if those resources are not fully utilized on regular basis. The key difference is that 3D printers can create almost any type of part (i.e., versatility), whereas specialized equipment is designed to create a specific kind of part or closely related variants.

21.3.2 Intellectual Property and Legal Issues

The concern with at-home manufacturing is that there is no control on the production of items protected under patent law or other legal constraints. For instance, 3D scanners are also becoming more affordable and can be used to make an accurate digital model of nearly any object. It records the exterior geometry and converts it into a file that can be transferred to a 3D printer and recreated. While many of these scanners (especially low-cost versions) require some skill to clean up the data collected, they raise the possibility of making a duplicate of a product without the company who designed it receiving any compensation.

This subverts **intellectual property (IP)** rights, which are the legal protection of products covered by trademarks, patents, copyrights, and other designations of ownership over designs and ideas. Because the companies behind products have invested time and money to produce them, they rightly seek compensation for the use of those products. Normally that is accounted for in the sticker price. If you make a copy at home, much like illegally downloaded music or movies, the individuals who put the effort in are not compensated.

This is not new, but the scale at which it can happen is. People have always been able to make look alike products using traditional manufacturing methods at home. A simple example is a wood cutout painted to look like a famous cartoon character. The person did not create the character. They copy someone else's work to avoid paying for the available merchandise or to produce a product not available but closely related. Indeed, the idea of self-replication of 3D printers by printing a new one is popular. The diversity of materials required (metals, polymers, wiring, circuitry, etc.) makes this impractical, though many components can be created and only the support systems are needed, which will reduce the cost. With most consumer models of scanners and printers, the products are limited to decorative and minimally functional items.

This becomes much more concerning when the products made at home are illegal or side-step current regulations. This has been most thoroughly discussed in regard to firearms. Though most printers cannot produce a firearm that is reliable long-term, the ability to successfully discharge a few rounds is possible. Also, it is possible to create critical components used to assemble standard, off-the-shelf parts, such as the barrel, firing, pin, and other components involved in the firing of munition. These assembly items are not subject to the full heat and pressure of the firing action, which makes them more reliable, but they allow the creation of firearms that are not registered or traceable. There are laws restricting the ownership of homemade firearms in most countries.

21.3.3 Making the "Impossible"

Equipment for AM is getting more reliable, quicker, and cheaper. That has elevated it from a novelty to a serious and growing facet of the manufacturing landscape. Although the method has become more practical in many regards, it is still the subject of frequent developments that change the way we consider manufacturing.

21.3.3.1 Geometrically Difficult

AM allows geometries that cannot be created by other methods. Parts with extremely complex geometries that would be impossible or impractical to make by methods such as casting or machining are simple to create. The creation of parts within parts and functional assemblies in a single run are also possible. These structures can be made by printing a dissolvable support between the parts that is removed after printing is complete and allows the part to move (Fig. 21.11). Parts with complex interior geometries and fine cavities are possible. Designs may also be simpler to make using AM than traditional methods. Although the component is technically possible, the design is not practical except by 3D printing.

21.3.3.2 Living and Nontraditional Materials

An area that traditional manufacturing cannot compete is in the production of living "products." Ongoing developments are enabling individual cells to be printed and supported to create tissue structure for applications such as regenerative medicine. One of the goals of such work is to be able to print organs for transplant candidates, especially using their own cells so that rejection of the implant would not occur. As you might imagine, printing living cells is not simple, and one of the challenges is positioning

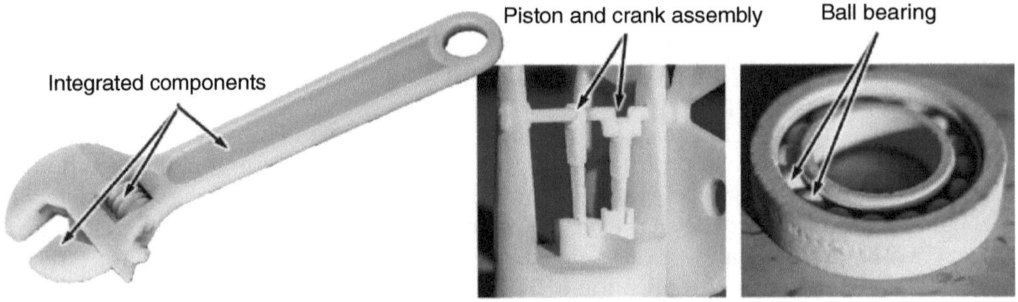

FIGURE 21.11 3D printing can combine dissolvable support material with the build material such that assemblies can be created. This allows components to be integrated with one another in ways that difficult or impossible to achieve by traditional methods.

them and holding the shape of the object. Methods cannot damage the cells, such as heat or UV light would, so a binder method is typical. The binder must be a biocompatible material, and this scaffolding is often a natural polymer, but work continues in the use of self-contained droplets and other techniques to improve the survivability of cells.

The broadening of the use of AM has also begun impacting the construction industry. This is sometimes called large-scale AM or freeform construction, and it is still limited in its use, but it demonstrates the concepts of the AM can be extended to create large structures. For instance, concrete can be printed similar to other extrusion methods with a fiber-reinforced mixture that is paste-like. The resolution and layer thickness essentially scale with the technique, such that layers may be an inch or more thick and several inches wide, depending on the extrusion nozzle. The resolution does not need to be as high as smaller components and the greater volume of material reduces build time. The printer structure can be assembled on site such that small buildings can be built in place, or the buildings can be printed and moved to the site (Fig. 21.12).

The strength of the underlying material is a concern for larger structures, as concrete takes time to fully strengthen and as the mass increases from additional layers, the bottom may not hold the weight in tall buildings. Also, the use of traditional rebar and other reinforcing agents cannot be included as they are in traditional construction, so

FIGURE 21.12 Additive manufacturing is not limited to small parts. The techniques are transferrable to larger structures, such as the concrete printing pictured here. (*Courtesy of Andrey Rudenko, Total Kustom, LLC.*)

the strength and durability of the structure will be affected. The use of modified concrete compositions may help compensate for this, but for small buildings this is not a central concern. For larger structures, building codes are not adapted for this type of construction and standards will likely need to be determined in the future.

21.4 Why It Matters

Manufacturing is an ever-changing discipline. Some of the methods we employ extensively every day have not really changed much for the last 100 years. Others did not exist 20 years ago. Being aware of the implications of new technology is important, because some day that technology will not be new anymore; it will be routine. The ability to utilize methods like AM can provide a competitive advantage and cost savings if understood and properly applied.

Knowing the basic methods of AM matters because the results will differ. If you want to print full color parts, SLS methods are not a good option. If you want high resolution, FFD is not your best bet. What you need from a system should be more important than what it *can* do. For instance, if you do not need to print in the highest resolution with multiple build materials simultaneously, PolyJet can be a bit of a burden. The resin expires, the machine requires regular cleaning and maintenance, and the build cost is relatively high. They are great machines, *if* you plan to use them constantly and for the full gamut of their capability.

The marketplace is changing, and the types of products people expect today have a fairly consistent theme: custom. People want choice and uniqueness in many of the products they buy. You do not have to go back very far to realize purple chain saws, and camouflage tuxedos are fairly recent developments. Customers want the function of the product with a personal touch. Three-dimensional printing gives that kind of versatility and freedom of design. Although at-home printing it is largely limited to enthusiasts now, the potential to expand custom manufacturing to individuals is worth considering.

Perhaps the most significant reason to take AM seriously is that it allows on-demand production of complex parts, some of which are impossible to make otherwise. It probably will not fundamentally change the way simple, high-volume products (e.g., screws), are made, but it will change (and has changed) the way the products are developed and how low production, high-value parts are produced.

21.5 The How and Why of Additive Manufacturing

21.5.1 Chapter Review

1. How is additive manufacturing fundamentally different than other manufacturing approaches described in this book?
2. Described the process differences in additive and subtractive manufacturing.
3. Additive manufacturing is often viewed as being able to create anything. What is the most basic limitation to what can be created?
4. Why are casting and welding not considered additive manufacturing?
5. Additive manufacturing is valuable for parts that can be made other ways. Why?

6. What industries are most likely to use additive manufacturing to produce parts? Why not more?
7. Why would you change the print parameters when creating parts?
8. What is the role of the infill and the shell?
9. Why are overhanging features problematic? How are they addressed?
10. Why would multiple build materials be used in one print?
11. What finishing steps are applied after printing? How might this vary with each process?
12. What kind of raw material does FFD use?
13. Why is FFD so popular among hobbyists?
14. What filament materials are most common for FFD printing?
15. Why is the build plate sometimes heated in FFD? Which print material is it commonly heated for?
16. What are the limitations of PLA in applications? Why is still so popular?
17. How is the print material solidified in stereolithography?
18. Why can a layer be created so much faster with projection SLA than with FFD? Why can MSLA be even faster?
19. How can MSLA expose the entire build area at once?
20. How does upside down SLA save resin compared to right-side up?
21. What is the function of a "dead zone" in upside down SLA?
22. How does PolyJet differ from project SLA?
23. What benefits does PolyJet have over other SLA methods?
24. Why are support structures not required in powder bed printing?
25. How is a solid part made in binder methods of powder bed printing?
26. Why is powder size so important in powder bed printing?
27. What advantage in material types does powder bed printing have?
28. Describe what processing may have to occur after a part is printed by powder bed methods.
29. How are SLS and projection SLA similar?
30. What clean up procedure is needed for powder bed methods?
31. What kind of products is LOM used for?
32. Why are internal cavities not possible with LOM?
33. What is the benefit of producing goods at a centralized facility? How may additive manufacturing change that?
34. How can additive manufacturing infringe on intellectual property rights?
35. Is it practical to recreate commercial products at home using 3D printing?
36. What is unique about the creation of assembled products using 3D printing?
37. How are moveable assemblies "released" after printing? Why is it necessary?

38. How are living structures possible with additive manufacturing? How is this unique compared to traditional manufacturing?
39. What role might additive manufacturing have in the future of construction? What are the limitations?

21.5.2 Concept Connections and Extensions

1. What are the similarities between traditional profile extrusion (Sec. 12.3.4.1) and FFD printing?
2. What are the differences between powder metallurgy and SLS (Sec. 12.2.5) of metal powders? How might this affect the performance of SLS components relative to traditional powder metallurgy parts?
3. Do you feel that printing a product at home is a violation of intellectual property? What factors might influence your decision one way or another?
4. What are your impressions of printing living tissue? What benefits can you imagine as a result? How might this be controversial?
5. Compare the mask process for photolithography (Sec. 22.4.2.1) and that of mask stereolithography. What is the difference in the result? How do you think this is achieved?
6. What process(es) might you envision being used to color parts in 3D printing? Specify the printing method and the finishing method (see Chap. 20).
7. PLA is a popular print material. Use the discussion in Secs. 6.3.1.5 and 8.5.3 to determine its benefits and disadvantages.
8. Casting and 3D printing can both create complex geometries with internal features. In what circumstances would you recommend 3D printing rather casting and molding (Chap. 11)?

CHAPTER 22

Advanced Manufacturing

Manufacturing is much different from the 19th-century representation it often receives. It is no longer a dark, dirty, and dangerous profession. In most cases, manufacturing is "advanced" in the sense that it is tightly controlled, efficient, and incorporates many modern methods and materials. This chapter focuses primarily on so-called "nontraditional" manufacturing, computer-controlled manufacturing, and nanoscale materials and production methods.

22.1 How It Works

"Advanced" is a difficult word to define. At one point, steam engines were advanced. That indicates the definition does not have a fixed reference. In fact, we will consider **advanced manufacturing** to include manufacturing processes and materials that utilize state-of-the-art scientific understanding and technology to improve the quality and/or efficiency of processes or applications. Advanced manufacturing is a constant pursuit of a better way. One prominent example is additive manufacturing, discussed in Chap. 21, but there are many other aspects that are rapidly developing as well. In some cases, these methods are specialized and limited in their use. Some may gain ground as the technology matures. Others may fade away as alternative processes are developed. One thing is sure; advancements will continue.

22.2 Nontraditional Manufacturing

Nontraditional manufacturing uses methods of shaping materials that do not conform to the broadest classifications of manufacturing, such as casting, welding, machining, forming, etc. These methods may use electrical, chemical, laser, or water energy to remove material, and each has unique advantages and disadvantages. We will look at some of the most common methods that are used in production or support roles. All require computer feedback to monitor and control the cutting process, and the fundamentals of this are described in Sec. 22.3.1. There are other nontraditional techniques, such as electrochemical machining, that we will not consider here.

22.2.1 Electrical Discharge Machining (EDM)

Electrical discharge machining (EDM) uses rapid pulses of electricity to create a spark between an electrode and the workpiece to vaporize a small amount of material. The individual sparks do not remove much material, but over the millions of sparks occuring across the surface, bulk removal occurs. If you have witnessed an electrical arc on wiring or circuitry, you see that the result can be very damaging and leaves a charred, sometimes cratered surface (ever run into this when using jumper cables). When controlled, this process becomes valuable for machining. The use of electrical arc can be used for various manufacturing processes, such as welding (Sec. 18.2.2), but EDM is unique in that it uses electrical arcs to precisely remove material to create components or modify existing components.

22.2.1.1 Basic EDM Process and Considerations

The use of electrical arc has important consequences on the way the process is applied. The process works by applying a voltage to the tool electrode (anode) and the workpiece (cathode) that is sufficient to create and arc between the two. Electrode pairings are also important to corrosion (Sec. 8.5.2.2) and welding characteristics (Sec. 18.2.2.2), and some of the basic considerations are also discussed there. The electrode is often made of graphite or copper (often paired with tungsten to improve durability), and the workpiece must be conductive. Tool steels and other hard metals are often cut by this method as they are difficult to work by other techniques.

A computer controls the "on-time" and "off-time" of the arc so that the material is being vaporized during part of the cycle (on-time), and the vaporized material can be removed during the rest of the cycle (off-time). This burn-and-flush process is very important. Imagine the last time you were shocked by static when grabbing a doorknob or other metallic surface. You are not shocked until you are very close to the surface. The same applies to EDM. The arc will not occur unless the electrode is very close to the workpiece, but the machining by-products (tiny metal droplets) will literally short-circuit the process if they are not removed.

To control the arc and clear the **spark gap** (space between electrodes), a **dielectric fluid** is used, which is typically high-purity water or oil that has specific characteristics to facilitate the ionization and flush away the material. It should be noted the "cutting tool" (the electrode) never touches the workpiece. Instead, the EDM process starts by building a voltage between the electrode and workpiece until the fluid ionizes and a burst of electrical current transfers to the workpiece. The surface of the workpiece melts and the vapor bubble created by the arc collapses and the process starts again. The on- and off-times during this cycle are only one to a couple of hundred microseconds in duration. It is important to maintain (filter and condition) and replace the dielectric as its properties change over time, often resulting in slower cut rates. The basic process is shown in Fig. 22.1.

These considerations are universal to EDM processes, but the technique can be applied in a variety of ways. There are three primary categories of EDM we will discuss:

1. Sinker/ram EDM
2. Wire EDM
3. Small hole EDM

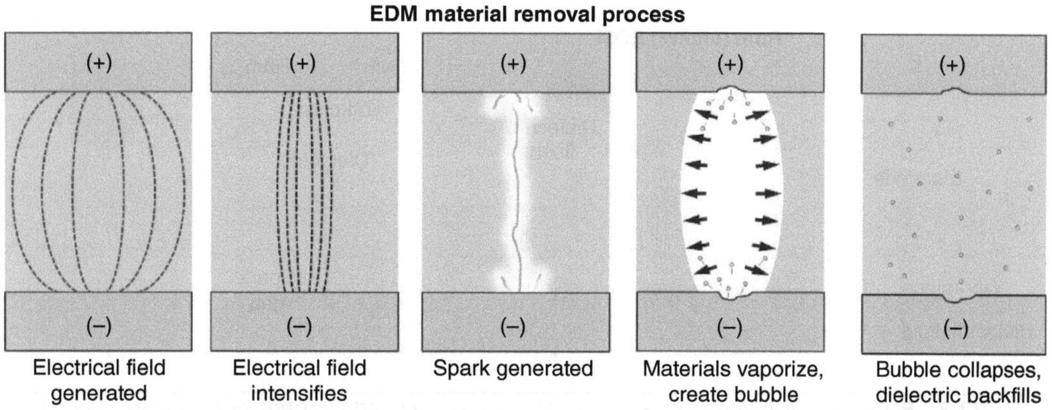

FIGURE 22.1 Electrical discharge machining removes material by creating an electrical spark that vaporizes a small amount of material. The dielectric fluid acts to control the spark characteristics and flush the removed materials from the area.

22.2.1.2 Sinker/Ram EDM

Sinker or **ram EDM** uses a three-dimensionally featured electrode to transfer that geometry to a surface (Fig. 22.2). The electrode is slowly lowered into the part, removing material as it goes. The workpiece is often in a bath of dielectric oil, and the oil is circulated through the cut area using nozzles. The electrode may also be cycled vertically to aid in flushing the cut area. This is especially important when a large cut area is used.

A negative of the geometry to be created the workpiece is first machined into the electrode (Fig. 22.3). The electrode then transfers the shape into a cavity. This process is commonly used to create molds for processes such as injection molding, die casting, and some forming operations. Depending on the amount of detail and depth of cut, the electrode can wear away to a degree that is unacceptable. A rough shape may be created

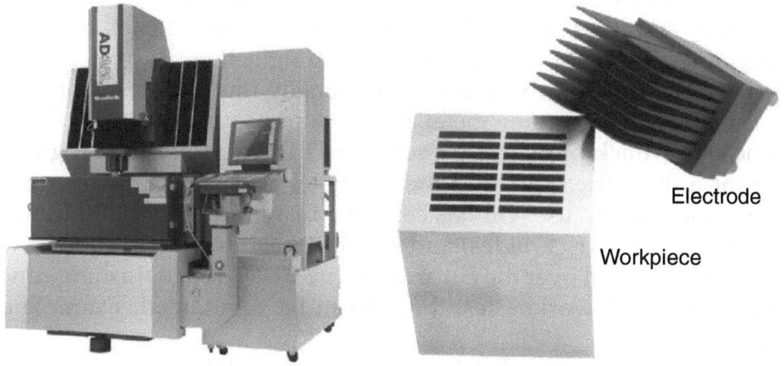

FIGURE 22.2 Ram/sinker EDM uses a solid electrode to transfer the geometry to the workpiece. The electrode is often copper or graphite and can create complex geometries in the workpiece. (*Courtesy of Sodick, Inc.*)

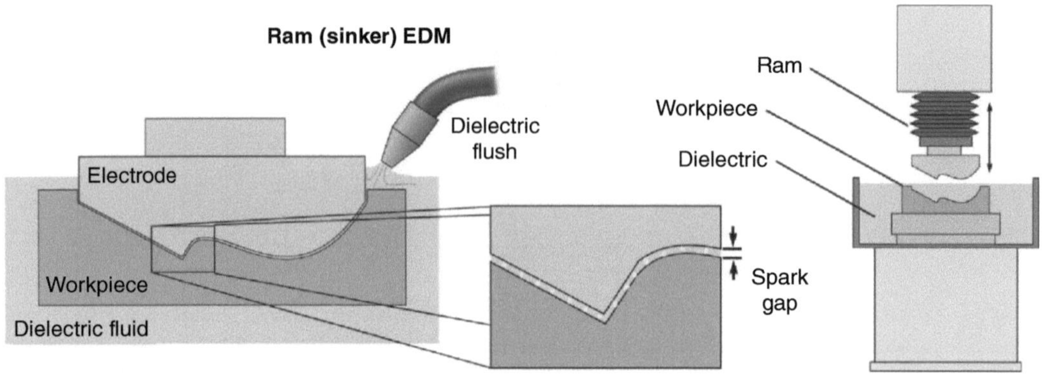

FIGURE 22.3 Ram EDM, also known as sinker EDM, uses an electrode to transfer the desired geometry to the part.

using one electrode and a finishing electrode may be used at the end. Sharp corners and fine features will wear away faster than flat areas, and this should be considered and eliminated where possible. The electrode may also be rotated in conjunction with the plunge motion such that helical features can be created, such as threads or other internal geometries.

22.2.1.3 Wire EDM

Wire EDM uses a thin wire-electrode to remove material in a thin region as it progresses through the part. This is very much like a bandsaw, as the wire moves past the part during cutting. There is one key difference: EDM wire is not reused. Like other EDM electrodes, the wire degrades as the workpiece is machined. To ensure a consistent cut, fresh wire is continually supplied to the cut area. The used wire is collected and recycled but not reused. The wire is very thin (e.g., 0.001–0.013 in) and made of brass or molybdenum, and may be coated to improve performance (e.g., zinc-coated brass). The process is subjected to a kerf width so that the cut is always slightly larger than wire, just as with saws (Sec. 14.2.1). The basic setup for wire EDM is shown in Fig. 22.4.

Because wire EDM cuts along a thin straight line, it is often used to create two-dimensional profiles. It can create exceptionally fine detail as there is little to no force during cutting to distort the metal. Multiple parts can be cut simultaneously by stacking sheets, or thick parts can be cut. Wire EDM usually uses high-purity, deionized water for a dielectric fluid, and the machine has subsystems for purifying the water continuously during use. Oil has certain advantages, especially in materials prone to absorbing water (like carbides) and in micromachining applications. Often the dielectric is simply flushed over the cut area in wire EDM. When flushing conditions are poor, such as very thick workpieces, the cut can be submerged. Example of equipment is shown in Fig. 22.5.

The wire does not need to be perfectly vertical. Instead, one or both of the heads can be moved relative to one another to create complex taper angles. This allows tapers to be cut as well as different shapes to be cut on the top and bottom of the part (e.g., a circle on top and square on bottom). When a profile is to be cut though the center of a part, a hole for the wire to start in is required. The wire has to feed through

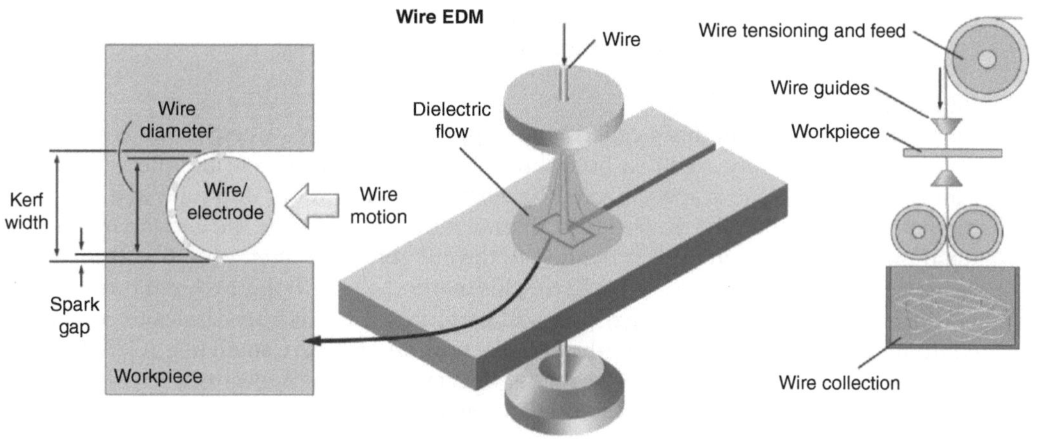

Figure 22.4 Wire EDM uses a thin wire as an electrode. Fresh wire is continually fed through the cut area to maintain a consistent cut quality.

this hole. Aligning and feeding such a small wire is not trivial, but modern wire EDM machines have the ability to self-feed the wire [typically called automatic wire threading (AWT)]. When the guides are close, this can be done by simply feeding the wire as usual. At farther distances, a high-pressure jet of dielectric will direct the wire to the receiving guide.

Improper cutting conditions can lead to wire breakage, but a machine that can self-feed the wire can restart the cut and continue the work. This allows machines to work unattended overnight and on weekends. This is particularly useful as EDM processes are slow compared to some traditional techniques, taking hours for modestly sized parts. Modern machines have improved the cut rates due to advanced electrical and motion controls. EDM is not often a production technique unless the material type or

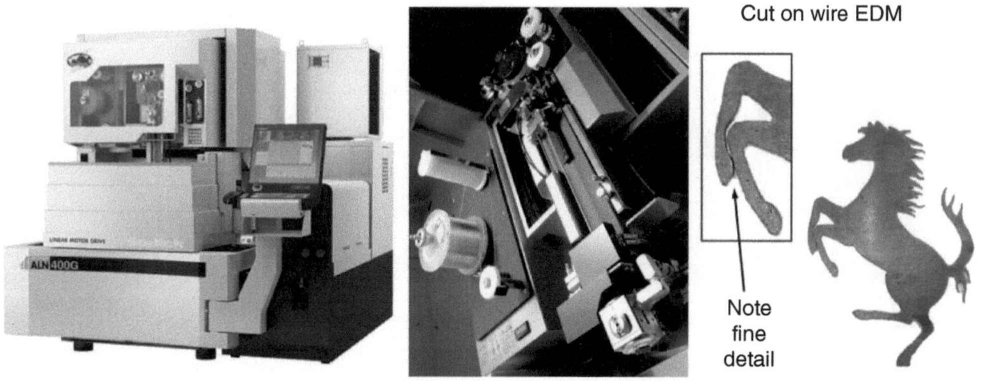

Figure 22.5 Wire EDM is a versatile technique for creating complex contours (right image). An example of the equipment is shown on the left, and the wire tensioning and guide system is shown in the center. (*Equipment images courtesy of Sodick, Inc.*)

geometry is such that it cannot be effectively created by other means. Instead it is often used on tooling and other low-volume or custom parts.

22.2.1.4 Small Hole EDM

Small hole EDM (sometimes called hole popping EDM) uses an electrode with a small diameter to effectively "drill" a hole through a workpiece. The technique is similar to ram EDM, but the use is specialized. The electrode is a cylinder or a tube with an actual diameter smaller than the desired final diameter to accommodate the kerf (spark gap). To flush the area, dielectric fluid is flushed to the cut zone and often plumbed through the hollow center of the electrode (Fig. 22.6). This method allows holes to be cut through very hard materials without breakage, very soft materials without burrs, holes to be made in contoured and angled surfaces without deflection, and for very small (e.g., 0.0001 in) and very deep (e.g., 24 in) holes to be made. Specially designed electrodes with multiple internal channels and electrode rotation are used to improve cut uniformity and produce flat-bottom holes. This is impractical by traditional drilling operations (Chap. 15). Because a pressurized flush is used, the removal rate can be much higher than ram EDM and depths of 2 in/min are achievable. The electrode is supported near the surface by a ceramic-sleeved guide tube of appropriate diameter. This increases accuracy, but very deep holes are still prone to experience drift (going off-axis).

Small hole EDM is particularly useful for removing broken tools from parts. For instance, a hardened tap (Sec. 15.5.2) that is broken off in an expensive component is difficult to remove, but the part is too valuable to scrap. Small hole EDM can burn the tap out without damaging the part and the hole can be fixed.

22.2.2 Water Jet Machining

Water jet machining uses a narrow, high-velocity stream of water to erode materials and create a profile. Maybe you have come across videos of people cutting through watermelons (and other things) using a pressure washer. Water jet takes this to the extreme. This method is capable of working on any material, regardless of electrical conductivity, hardness, and other physical and mechanical properties. For example, a water jet

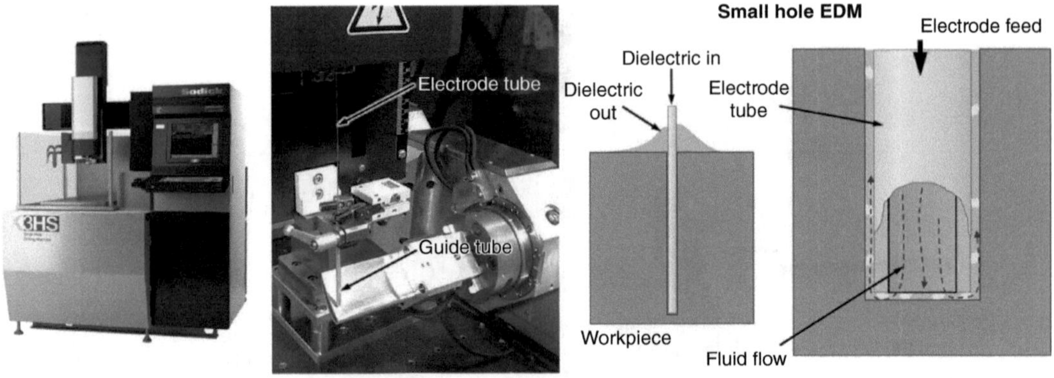

Figure 22.6 Small hole EDM uses specialized hollow electrodes to create holes with a small diameter and large depth. By flushing dielectric fluid through the electrode, the removal of debris is very efficient. (*Photographs courtesy of Sodick, Inc.*)

machine can cut carpet, rubber, wood, metal, glass, etc. Because it *can* cut it, does not mean the process will be efficient. The thinner and softer the material, the faster the cut will be. The process is most often used to produce two-dimensional profiles, and multiple sheets of material can be stacked to produce copies in one setup. Thick materials can be cut (e.g., 10 in of steel, 2 ft of polymer foam), but the cut time increases and accuracy suffers somewhat. Nevertheless, water jet is versatile and capable for many projects.

The general process (Fig. 22.7) involves pressurizing water to extreme levels (e.g., 100,000 psi), and pumping it through a very fine orifice (e.g., 0.010 in). Due to this high pressure, the jet is traveling at supersonic speeds upon exiting the nozzle (e.g., 1000–2000 mph). The pumps and water velocity make the process quite loud (95 dB, or 75 dB submerged) and hearing protection is required. The workpiece is held on a series of thin supports so water can travel through the work and be recycled.

To make the process more effective, abrasive particles may be added to the water. This abrasive water jet machining process removes material more efficiently in general and is critical to cutting hard materials such as ceramics. To achieve the mixture, abrasive is added to the water flow near the nozzle. In order to withstand the extreme pressure and abrasion, the orifice that restricts the water to a fine stream is made of a very hard, wear-resistant material [typically ruby (i.e., alumina, see Sec. 6.2.1.1), diamond or a proprietary ceramic]. The abrasive is also a hard ceramic, such as garnet [or silicates (see Sec. 6.2.1.3)]. This mineral is particularly useful as it fractures into sharp fragments so that even as particles become smaller, they do not get dull.

22.2.3 Laser Machining

Laser machining (or laser cutting) uses high-power laser energy to melt and/or vaporize the workpiece material. The technique is most often applied to cut through sheet and plate materials, but notching and cutting other features in tubing and formed sheet metal can be done with certain equipment. Laser machining is used to engrave surfaces and cut through material to form complex profiles. The equipment ranges from table top hobbyist versions to massive gantry systems. The price scales with size

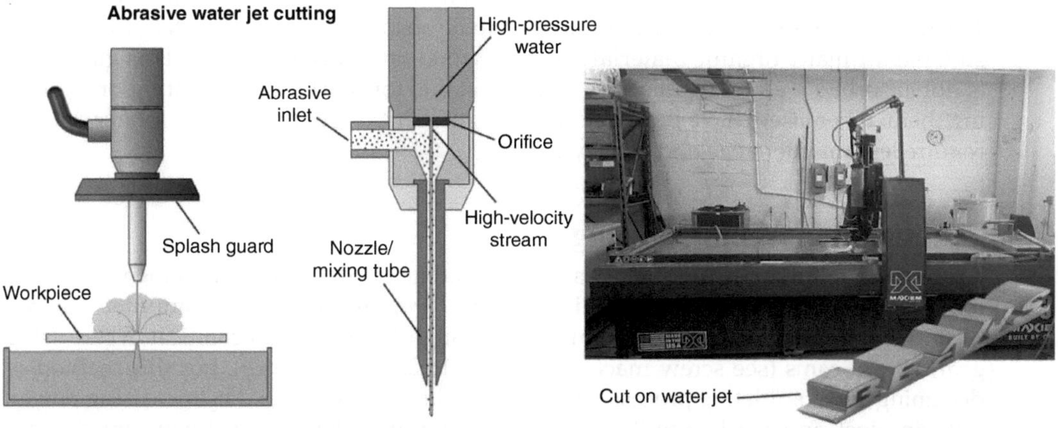

Figure 22.7 Abrasive water jet cutting combines a high-velocity stream of water with small abrasive particles to cut through materials. (*Photograph courtesy of R.H. Bolick & Company, Inc.*)

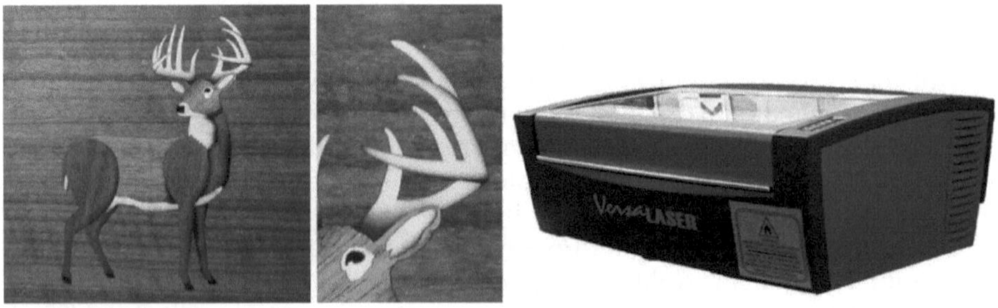

FIGURE 22.8 Lasers can be used to cut and etch materials precisely. The power of the laser determines the type and thickness of material it can remove. The fine features and control of laser cutting are demonstrated in this marquetry example. (*Marquetry example prepared by Ryan Haugh, Millersville University*).

and capability, from a few thousand to a few million dollars. The type of work ranges from small, finely detailed parts to inches thick steel geometries feet in size. The former is demonstrated in Fig. 22.8, where precisely sized and cut wood veneer is inlaid into a wooden surface. The fitment is critical to the appearance, as is the orientation of grain and other effects, such as shading.

The most critical component to the cutting ability is the laser itself. A small unit capable of engraving and cutting thin wood and polymer sheets may have a 30 to 40 W laser, while an industrial unit capable of cutting inches thick steel will be thousands of watts. To compare, a typical laser pointer is limited to 5 mW, so an industrial laser cutter may have one million times more power. The more energy the laser can put into the material, the faster it can cut through thick sections. The laser power can be dialed back for thinner materials, and to etch/mark wood, plastics, and metals. For instance, end mills, socket sets, and other metal components can be labeled with their size and details in a way that will not wear off but does not compromise the surface.

The laser can be of several types. Common examples are CO_2, YAG (yttrium aluminum garnet), and fiber lasers. The primary difference is the wavelength of light generated. For instance, YAG has a wavelength that is 10 times smaller than CO_2 (CO_2 is 10.6 μm, YAG is 1.06 μm), and this allows it to be absorbed better by metals, but it is not suitable for many organic materials such as wood and polymers, as well as glass. CO_2 is not absorbed well by metals and is only suitable for marking, not cutting or engraving them. Fiber lasers have the same wavelength as YAG, but they have a longer life and require less maintenance and energy to run.

22.3 Automated Manufacturing

Automated manufacturing includes processes that are designed, implemented, monitored, and controlled with the aid of digital programming and computer control. Historically processes could be completed through the use of mechanical cycles using gearing and cams (see screw machines in Sec. 16.2.3, for instance), but this method of designing and completing processes is becoming increasingly rare. The increasing power, reducing cost, and nearly infinite versatility of computers have made them a much more effective choice. This topic could be (and is!) the subject of a textbook of equal volume, so this section will only touch on some major themes to whet your appetite for future study.

22.3.1 Computer-Controlled Equipment

We will not focus on the plethora of equipment that can be automated through computer control. You can pretty much add a computer to assist in the movement or measurement of any process described in Chaps. 11 to 20. Additive manufacturing (Chap. 21) and the nontraditional machining techniques described in the previous sections are fully dependent on it. A brief look at the benefits and general setup of common tools is worth looking at, but our primary focus will be on how the machines are controlled, not the specific incarnations of the technology. Most automated machine tools are operated by **computer numerical control (CNC)**, which is machine control through computer-programmed sequences of commands that result in machine movement and that activate functions. In this way, a lathe, mill, plasma cutter, etc. is called a CNC lathe, CNC mill, CNC plasma cutter, and so on. Not all machines use the term CNC. For instance, water jet cutters are controlled via computer, but there is not a manual alternative, so the abbreviation is dropped.

We will look at two of the most common CNC machines: mill and lathe. These machines are generally the same as the manual versions, but they have specific benefits of automatically changing tools, applying coolant to allow faster cutting speeds, and they are capable of creating complex parts quickly. Some CNC lathes are "flat bed" style, like manual engine lathes (see Fig. 16.4), but most are slant bed lathes that have the tooling mounted behind the spindle, and the machine guides (the ways) are mounted at an angle (see Fig. 22.9).

22.3.1.1 Basic Considerations

All of these methods and concepts boil down to a need to communicate the ideas of a human to a machine, and the need for that machine to faithfully execute its instructions repeatedly. The primary benefits of automated processes are repeatability, efficiency, and the extension of human capital. A machine can make the same exact movements to produce the same exact product or step in a manufacturing process all day long every day without fatigue. That ensures that quality and productivity are consistent, and one person can monitor a handful of machines working simultaneously. This control scheme is the basis for automated machining, but the principles extend to many facets of production. Machines are very good workers and even better listeners. They will do what they are told, whether right or wrong. A skilled programmer is needed to give them the correct message.

For instance, if the programmer giving instructions to the machines tells it to move too fast or forgets to activate the spindle, it will result in a broken tool, possible machine damage, or best-case scenario, a waste of time. It is critical that the programmer communicates his or her ideas exactly in a way the machine understands and can execute. Newer machines not only execute the commands, but they also monitor and interpret the progress of the operation. For instance, by monitoring spindle load in a milling operation, the machine can continuously modify the cutting conditions to maintain the greatest efficiency both for tool life and part production (this concept is described in Sec. 13.5.3). The automatic and continuous monitoring and adjustment of processing conditions is known as **adaptive control**.

All automated processing requires a communication of geometry (three-dimensional shape) and production settings (feeds, spindle speed, laser power, deposition rate, etc.). This information is transferred to the machine as coordinates in space (usually x-, y-, and z-axes) and machine control to activate and regulate

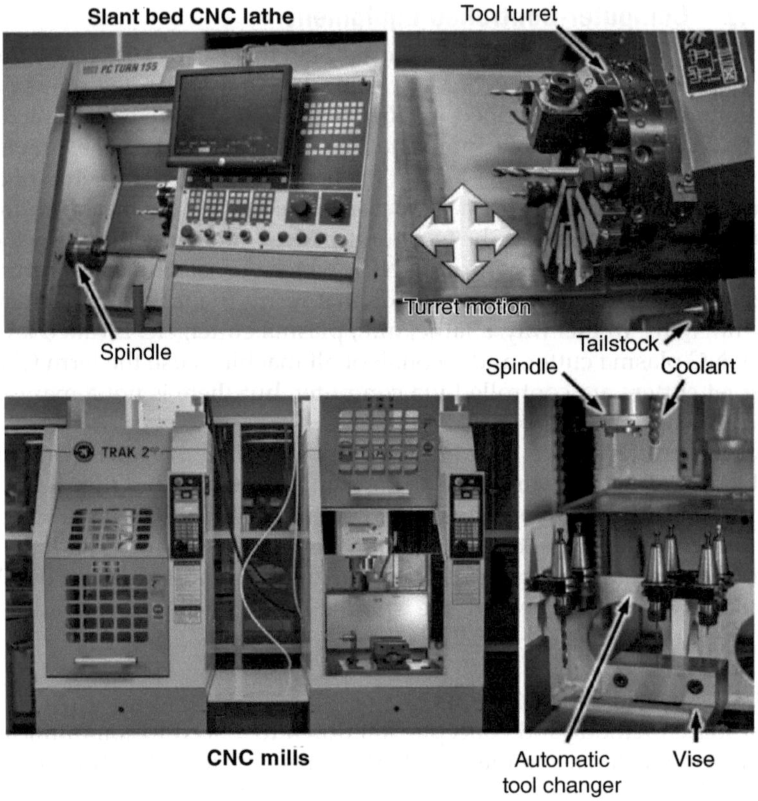

FIGURE 22.9 CNC technology increases the versatility and efficiency of turning and milling. Examples of that equipment are shown here, but many other features and equipment types are available. Other CNC mills are shown in Fig. 17.3.

machine functions and process controls. Since machines do not speak English, Chinese, Swahili, or any other human language, programmers must be bilingual; translating their instructions to the machine language. Programming languages in automation can vary, with some control systems (especially robotics) using Python, C++, or Java. Most machine tools (lathes, mills, etc.) use a language known as G-code, which we will consider next.

22.3.1.2 Introduction to G-Code Programming

G-Code is a semi-standardized machine programming language that controls machine position, motion, and functions through the use of short commands consisting of letters and numbers. This has a structure like a paragraph, which is made up of sentences, words, and letters. In G-code, the machine program is made up of lines, also called **blocks**, and those blocks are made up of **words**, and each word is made up of **characters**. Just like there are rules for sentence structure and grammar, there are rules of how these words and characters can be used (see Table 22.1). Before we look at the rules, the term "semi-standardized" in the definition of G-code should be noted. G-code is very similar between machine manufacturers, and there are G-code standards (yes, *plural*).

Code Designation	Example(s)
Character	G, 4, X, N
Word	G90, M08, F10
Block	N10 G01 X2.0 Y3.75
Program	
N1 M05 G40 G70 N2 M09 N3 T3 M06 N4 G00 G90 X0.7588 Y0.7151 S2000 M03 N5 G43 H3 Z0.2 N6 G00 X0.7588 Y0.7151 Z0.2 N7 Z0.05 N8 G01 Z –0.01 F7.5 S2000 N9 X0.7702 Y0.7269 F15.0 N10 X0.759 Y0.7376 ...	

Table 22.1 Terminology for G-Code

Like British and American English are largely similar but with variations, G-code for each machine may differ. The machine manual is your best resource for determining the proper use of G-code.

As you might have guessed, the letter "G" is important in this code. Known as **preparatory codes**, G words often control the machine motion and cycle type. For instance, you can tell a machine to move as fast as possible in a straight line between two points (G00) or to move at a specified rate (G01), usually expressed in inches per minute or inches per revolution (lathe). You can also tell it to move in a circular arc going clockwise (G02) or counterclockwise (G03). You can set up a drilling cycle (e.g., G81) or a turning cycle on lathe (e.g., G71). These cycles will also have additional words that inform the machine how deep to drill, how much to turn with each pass, whether to leave material for a finishing operations, etc.

There are other functions that support the manufacturing process, especially in machining operations. These are often controlled by miscellaneous functions, called **M-codes,*** and they are responsible for initiating tool changes (M06), turning on coolant (M08), starting the spindle (M03/M04 for CW/CCW), and many other functions. Even more must be specified to further tell the machine what tool it is to use (T), what the length (H) or diameter (D) of the tool is, what spindle speed to use (S), what feed rate to use (F), etc. Hopefully you get the point. Everything can and must be specifically detailed so the machine performs the work accurately. Physically setting up the machine and tooling is just as important, because the machine must know exactly where the part is and where the tool is to produce an accurate result. It is not trivial, but with some practice it is not too difficult. For you reference, G- and M-codes are provided in appendix C.

*Note that M-codes are *part* of G-code, not a separate type of language.

Because each point in space needs to be specified and every detail of motion needs to be included, even relatively simple parts may require hundreds of lines of code. Many require thousands or tens of thousands of blocks. This makes coding by hand (typing each line individually) a cumbersome process that is prone to error. Only simple programs and small modifications to existing programs are done this way. Most code is generated by computer software, which we will discuss next, but one thing is for sure, if you are able to program manually, you will be much better at troubleshooting code generated by a computer.

22.3.1.3 Computer-Aided Manufacturing

Computer-aided manufacturing (CAM) typically refers to software-driven design and programming of processes. A closely related term is **computer-integrated manufacturing** (CIM), which describes the software and hardware sides together and the communication between equipment in a process. The hardware portion was briefly discussed above, so here we will focus on software programs. Software is critical to modern manufacturing because it allows a user-friendly interface between the designer, programmer, and the machine/machine operator (sometimes all the same person).

Increasingly, part designs are not recorded on paper, but they are enabled by dedicated software programs and stored digitally. This allows someone to design a part and send that part to be manufactured with less chance of misinterpretation than when looking at a blueprint on paper. For instance, dimensions may be left off a blueprint (part drawing), but if you have a digital model of the part, you simply determine the dimension in the software. There is potential for issues when converting a drawing between different programs, but those experienced in these software packages should be able to easily avoid these problems or at least catch them. This does eliminate the need for hard-copy part drawings, but it can save significant time, especially if a model has to be redrawn in software using the paper version. Sending the file directly reduces misinterpretation and time reproducing work that was already done. Some industries with tight control on designs may prefer not to release their design files, but they are also more likely to do the work in-house.

In the past, creating three-dimensional models was performed in one software program, and the machining parameters were created in another. This requires one program interpreting data from another, which is not always perfect. Now many programs are capable of both the design and machining. Some other terms that come up are computer-aided design (CAD) and computer-aided engineering (CAE). These terms encompass the ability to create and test components and assemblies virtually within the software. Complex assemblies and individual components can be mechanically, thermally, and otherwise tested without ever making a physical part. The powerful analysis packages available can accurately reproduce the expected conditions of use and alert the designer to potential problems with fit, material choice, or other design parameters.

Once the design is acceptable, the machining parameters (tool type, settings, etc.) can be applied. To determine these factors, collectively called **toolpaths**, the size of the starting stock, the type of machine available tools and workholding—all of the physical requirements—must be incorporated into the machining code. Fortunately, the toolpaths can be simulated in three dimensions and include the part, workholding, tool holders, and all other details. Not all of that is needed for each part, but most programs are capable of including at least parts of it. This allows the programmer to visualize the machining process and identify problem areas. Once the simulation provides an acceptable result,

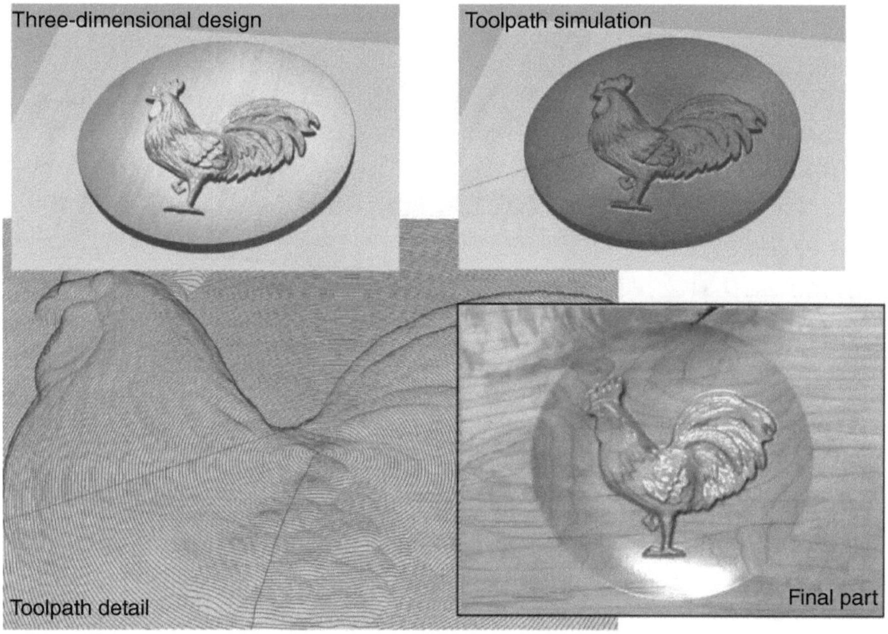

FIGURE 22.10 The production process includes designing the part, setting up toolpaths for machining, and finishes with the removal of material. This basic process is the same for simple parts, as shown here, as well as complex assemblies.

the G-code is automatically produced by choosing the control type for that specific machine (remember, G-code varies some between machine manufacturers). This general process is shown in Fig. 22.10.

22.3.2 Multifunction and Hybrid Processing

Although machining processes such as turning and milling were considered separately, there are many parts that require both operations. This can be done using separate machines, but that also requires the investment, floor space, workholding, etc., that are required for two separate machines. This also requires the time to load/unload and move parts from one machine to the other. Newer developments in control software and hardware have led to the development of "multitasking" tools that incorporate milling and lathe capabilities. Some may have a milling head that can swivel and extend in multiple directions as well as two lathe spindles and auxiliary cutting capabilities in one machine. This allows for the front portion of a turned component to be created and then transferred to a secondary spindle where the back half is completed. This makes programming much more complex, but the time and cost per part is reduced. The machine cost is quite high, as it incorporates the capability of two or more machines, so the investment only makes sense if many parts will be created using it and no other method makes sense.

In Chap. 21, additive manufacturing was introduced as an alternative process to subtractive manufacturing (namely, machining). The unique capability to create complex designs from the "ground up" was only tempered by the time and expense of the method.

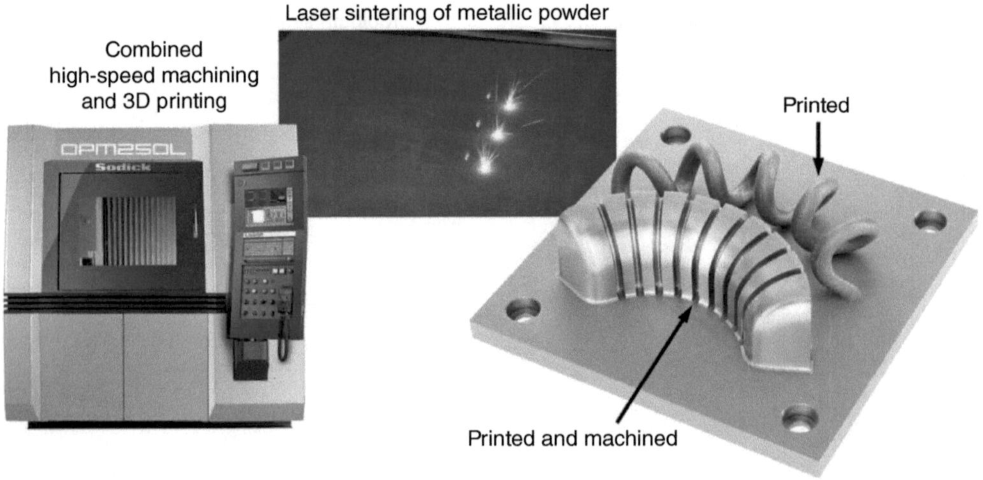

FIGURE 22.11 Hybrid machining allows material to be built up through additive manufacturing and removed by traditional machining. (*Courtesy of Sodick, Inc.*)

These two approaches need not be in competition with each other. Rather, they can be used together. For instance, selective sintering of a powder bed can be used to fuse and then machine the part. The machining can be performed after each layer is deposited, which allows for finished (smooth!) surfaces to be created on complex internal features such as cooling passages. An example of this technology is given in Fig. 22.11.

Hybrid machining is also performed using methods such as laser deposition welding. This method has some features in common with powder bed selective laser sintering (Sec. 21.2.3.2), but instead of working from a fixed layer of powder, the powder is shot into the area where the laser and underlying materials intersect. When the powder reaches this zone it is immediately welded to the surface of the part, functionally similar to thermal spray (Sec. 20.4.1.2). This allows features to built-up on the surface of an already machined part, and then it is machined to size as well. In that way, one machine can deposit and remove material as desired. This also allows hard materials to be deposited at critical locations on a part for enhanced wear resistance without the cost or difficulty of making the entire part from the same material. This also allows repairs and production of parts at a rate of 10 times faster than comparable powder bed methods, and there is no need for supports as the part can be rotated so that deposition is always supported by underlying layers.

22.3.3 Robotics and Integrated Processing

The computer control of machine tools that perform processes is critical in manufacturing, but there is an increasing prevalence of automation to move, measure, and test products and components as well. While a thorough treatment of robotics and automation would be valuable, it is beyond the scope of our work. There are many automated systems that track products up until they reach your house (some afterward too!), but we will consider the systems that are the most directly involved in the production of components.

The term robotics covers a huge range of possibilities. We will not concern ourselves with Roombas and R2-D2 technology. We will consider industrial robots, which include mechanized systems for lifting and positioning parts, and tools to perform operations in the manufacturing of products. These often have a fixed base or one that moves along a single line rather than moving freely through an area. They can be mounted on walls or overhead in order to perform the necessary tasks.

They perform repetitive actions such as placing parts in and removing them from a machine tool before and after a process is completed. They can lift heavy loads with high precision repeatedly, and they can also maneuver equipment, such as welders, paint sprayers, or parts for assembly with excellent precision and repeatability. They are not sensitive to sound or air quality, the way human workers are, so they are particularly useful in hazardous environments. They do not get distracted by their phone, either, so they are always "focused" on the work. The way parts are transported to these robots often involves conveyors and sensor systems, which employ a good deal of automation as well.

Work cells can be constructed where a robot has a central position within several pieces of equipment and as each phase of production is completed, it moves the part to the next machine. This allows a number of material blanks to be provided at the beginning and a finished part is output at the end. Additionally, human workers can prepare various workspaces that are serviced by a single robot, such as in welding. After parts are setup for welding, a button is pushed and the robot will come to perform the work when it is done with other pending duties. Multiple stations can be run by one person and a robot this way. There is a good deal of planning and engineering to set up this type of operation, but it is very efficient afterward. Larger work cells may involve several robots.

These robots are typically either SCARA (Selective Compliant Assembly Robot Arm) or more capable articulated robots, which can rotate around multiple joints in varying directions to access difficult areas and maneuver complex parts into position. Rotation around these various joints is denoted by specific letters for each joint or "J" numbers, such as J1, J2, J3, etc.

The features of these robots include their payload, their maximum speed and accuracy, the type and ability to mount multiple end effectors (the "business" end of the robot for grabbing, twisting, or otherwise performing the work), etc. Additionally, vision systems allow the robot to find parts that are not perfectly arranged, or for it to identify the correct parts out of an assortment or parts. Industrial robots can be programmed using code alone, but some also allow the operator to "teach" the robot the path by physically moving it as it records the movements. This allows complex maneuvering to be programmed more efficiently. Some basic SCARA and articulated robots are shown in Fig. 22.12.

22.4 Nanotechnology in Manufacturing

Nanotechnology involves the study and development of materials and devices with nanoscale features (10^{-9} m, ~4×10^{-8} in). It is about *really* small stuff. You may have heard "nano" applied to music players or other small products. It has become a buzzword for selling technology or anything small scale, but not everything that is called nano represents nanotechnology. In its most basic form, nanotechnology manipulates individual atoms or molecules to build useful structures. The top end of the scale (about 100 nm)

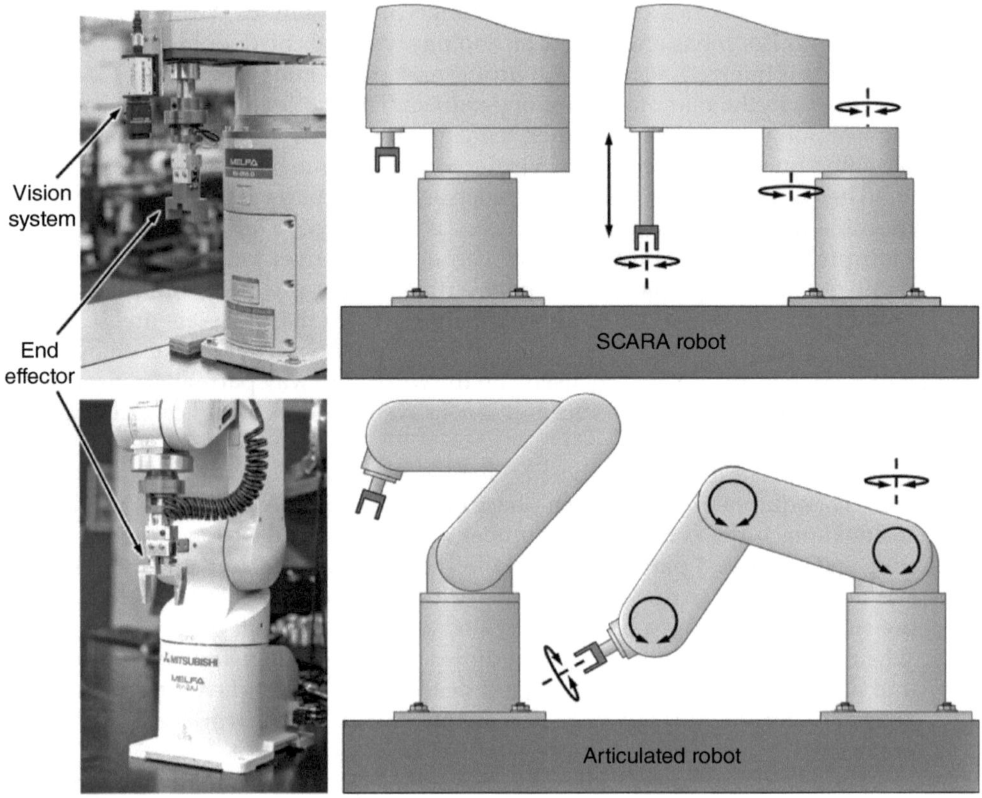

Figure 22.12 Robots come in a wide variety of styles and capabilities. The ones shown here are relatively small compared to some industrial robots, but the designs are very similar.

is about 1/1000th the diameter of a hair. Although the idea of nanoscale design is not new, the understanding of materials at this scale and ability to precisely control production of devices have only recently become practical in a variety of applications. The production of nanomaterials and nanoscale devices is not trivial, and despite the massive amount of research being conducted, there are still significant hurdles associated with cost and scale of production. Even with the challenges, the potential benefits are sufficient to maintain the effort. We will look at the reasons that nanoscience is so interesting for advanced materials and technology and the current applications.

22.4.1 Nanoscale Materials

Nanoscale materials are unique. The properties of a material actually change when very small. For instance, metals change color when they are turned into nanoscale particles. Gold might appear red and silver may look green. While this might be useful as a nonfading pigment, there are much more useful changes that occur. Electrical, mechanical, and chemical properties also change. Catalysts become more effective, nanoparticles can be tagged with medicine to target tumors, and small devices can be created that are integral to our modern technology, such as smartphones.

One of the key characteristics of nanomaterials is their high surface-area-to-volume ratio (SA:V), but this principle operates at any scale. For instance, if you put powdered sugar into a hot drink, it will dissolve immediately. If you drop in a sugar cube, it takes time to dissolve. A greater surface area means there is more area for reactions to occur. When the size of particles decreases for the same volume, reactions can happen faster. This can make batteries more powerful and quicker to recharge, and it can cause materials to behave differently.

The small size can also be beneficial to shrink technology. Computers were first run using vacuum tubes, which act like on/off switches for binary code. You have probably never seen one, but some were not much smaller than a soda can, and they require a lot of electricity and produce a lot of heat. This made computers very big, filling large rooms, and limited the practical computing power. The invention of the transistor greatly reduced the size and heat load, but these were still individual components added to a circuit. The technology that allows us to put computers in everything from dog bowls to washing machines is the integrated circuit. This involves silicon processing (discussed below) to create nanoscale transistors within a piece of silicon so that millions of these devices can fit into the space of a dime. This is largely what has driven the size reduction and increase in power in computers and electronic devices.

Studying materials at the nanoscale has also led to discoveries about new forms of materials and that has resulted in methods to create those materials. For instance, carbon may be in the form of graphite or diamond in our common experience, but it has other allotropes like Buckminster fullerene (or *buckyballs*), which are small cages of carbon atoms with a structure something like a soccer ball, or they may be nanotubes, which are rolled up sheets of graphite that have exceptional strength and electrical conductivity, and there are more. The idea is that these tiny structures have extraordinary potential to create lighter, stronger, and more efficient products, but producing them in large quantities with consistent quality is challenging. We will consider some techniques to do so next.

22.4.2 Nanoscale Manufacturing

Because of the small scale of nanomaterials and devices, the production methods are significantly different. In many cases, the creation of features requires atomic-level control where traditional tools are incapable. This can be addressed in two ways: top-down and bottom-up processes. We will look at the principal uses and differences between them.

22.4.2.1 Top-Down Techniques (Silicon Processing)

Top-down nanomanufacturing involves the iterative removal and addition of material to create features in an existing substrate. This can be thought of as subtractive manufacturing, where the bulk of a material exists before the component does. It is very different in many regards to traditional processing, because it removes and adds materials in precise patterns to build an assembly. Imagine making an automobile by starting with a block of steel, removing most of it, then adding rubber, plastic, glass, etc. to create the final product. This is an imperfect way to describe the processing, but it does provide a concept much different than we have described so far.

Silicon processing, or the manufacture of devices based on silicon, is the primary use of top-down fabrication. Silicon is used because it is a semiconductor than can be modified to conduct electricity in different ways by adding elements to give it excess

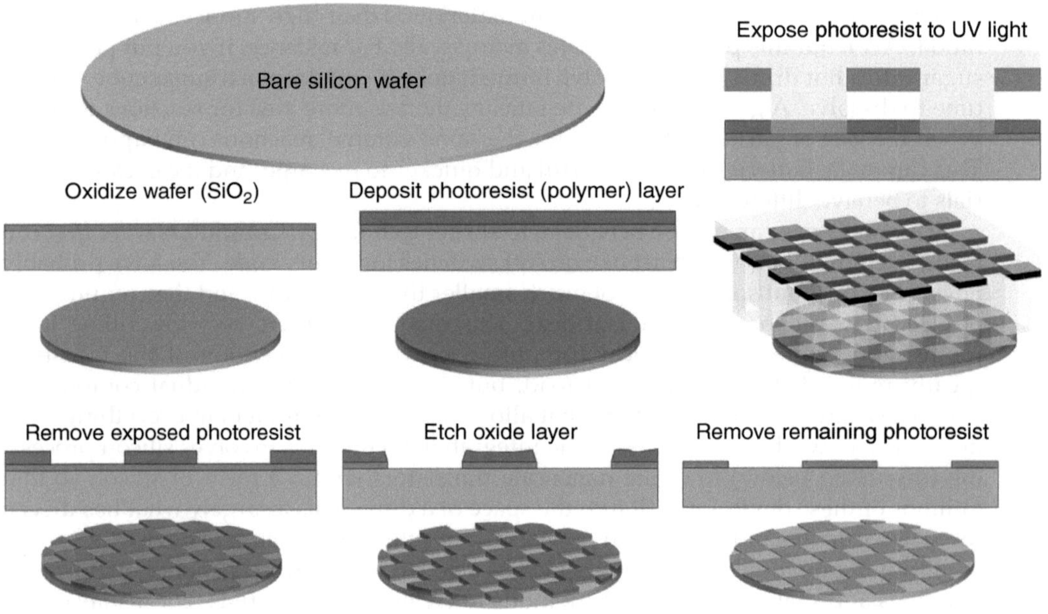

FIGURE 22.13 The process for creating nanostructures used in electronics often follows a top-down scheme, such as the etching process shown here. The process involves the precise patterning to add and remove layers.

electrons (*n*-type) or depleting them (*p*-type, also termed as excess "holes," or missing electrons). These *p-n* junctions allow electricity to be controlled in an on/off manner. To change the conductivity of silicon, it must be "doped" by implanting atoms at high velocity; basically shooting them at the silicon wafer. Boron is a common *p*-type element, and phosphorous is a common *n*-type element. All of this is done in a carefully controlled pattern with nanometer resolution.

The creation of integrated circuits for electronic devices has driven the advancements of technologies and is the most developed industry in nanotechnology. To create an integrated circuit, a pure silicon wafer (a thin "slice" of silicon) undergoes a series of etching and deposition steps to create circuits (Fig. 22.13). Circuitry is built-up in layers and consists of millions of transistors. Manufacturers can now fit 100 million transistors per square millimeter and the number will only continue to increase.*

At each stage, the circuitry must be exactly placed to make a complete chip. A method called **photolithography**, where light is used to create a pattern, is used extensively. This is very much like mask stereolithography discussed in Sec. 21.2.2 for additive manufacturing. A photopolymer (called photoresist) is exposed to light to either make it insoluble or soluble to a solvent that washes away all of the photoresist except what is needed. The light is often of very small wavelength (deep UV) to increase the resolution

*There is a limit to the size of transistors where they will not function. They are already pushing the limits of production technology and will become increasingly difficult to produce at a smaller scale, hence the interest in alternative methods such as quantum computing.

and decrease the critical feature size that can be created. The masks are often created using an electron beam that etches away the pattern at an even greater resolution.

The photoresist acts to control where etching and deposition occur. **Etching** includes processes where a layer of material is removed, typically in a defined pattern. Using a combination of physical and chemical attack, the silicon and other layers can be etched using plasma within an evacuated chamber. The ionized gas impacts the surface of the wafer and chemicals, in the plasma such a fluorine, chemically etch the material. This process is known as reactive ion etching (RIE) and can be enhanced by including a magnetic field to control the plasma (inductively coupled plasma RIE or ICP-RIE). This type of etching is known as "dry" etching, but "wet" etching using a chemical bath or spray can also be used. Hydrofluoric acid, a mixture of sulfuric acid and hydrogen peroxide known as Piranha, and other acids are used to remove silicon, organic residues, and metals during these processes. The wafer is repeatedly rinsed with exceptionally pure deionized water between steps, as trace elements that can contaminate the circuitry exist in regular water.

In addition to etching, the deposition of materials is critical to building circuitry. **Deposition** involves the addition of thin layers of metals or ceramics (insulators) to create features or cover existing features. To deposit metals the wafer may be placed in a vacuum chamber and a metal or metals are heated until they vaporize, causing the metal vapor to spread throughout the chamber and onto the wafer. The metal may be vaporized using resistive heat, electron beam, or plasma bombardment. The thickness of the coatings can be controlled to within nanometers using these techniques. Some of the related equipment for these processes is shown in Fig. 22.14.

Many of these processes are conducted in a vacuum chamber to allow the uninterrupted action of the process (plasma requires low pressure and atoms cannot move through air effectively) and to avoid contamination. Any air in the chamber may be incorporated into the structure and cause the failure of the whole device. When the circuitry is only tens of nanometers across, any defects are catastrophic. Just the action of electrons moving through the circuits can cause the metal to drift (called **electromigration**) and results in a short circuit. Because of the sensitivity to contamination, silicon processing is performed in cleanrooms where the air is purified to be around 1,000,000 times cleaner than the air in a typical room (Fig. 22.15). Not all cleanrooms are equivalently

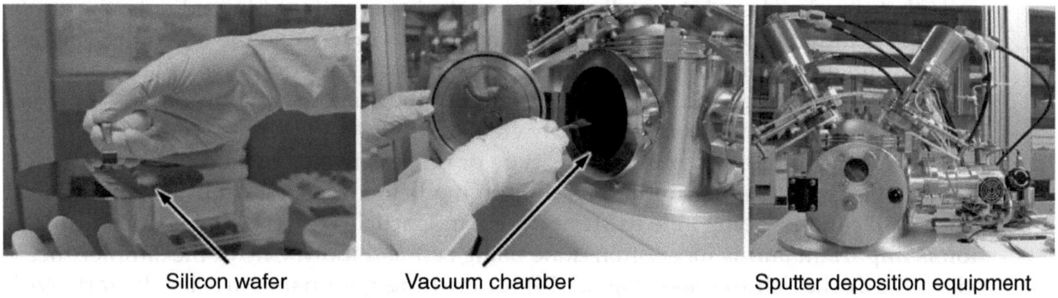

FIGURE 22.14 Top-down nanomanufacturing is extensively applied in silicon processing. As shown here, a silicon wafer is transferred between tools to add and remove layers of materials in specific patterns. *(Courtesy of Robert Ehrmann, Center for Nanotechnology Education and Utilization, Pennsylvania State University.)*

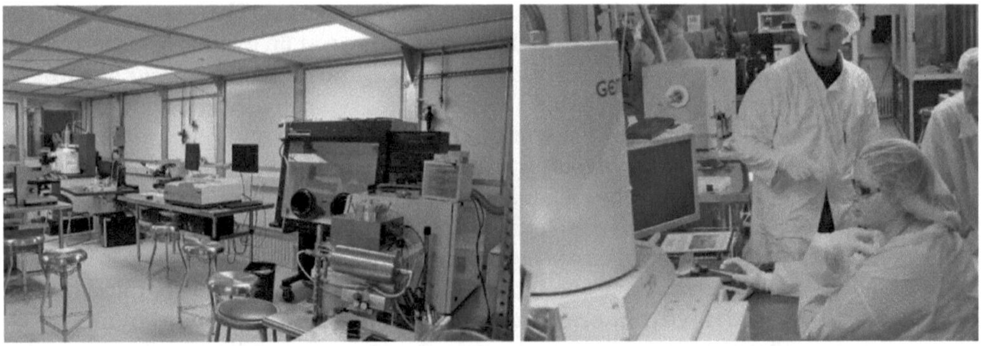

General clean room Light-sensitive area

FIGURE 22.15 A clean room environment is necessary to prevent contamination of nanoscale devices. The higher the purity of the clean room, the more stringent the filtration and usage protocols. In addition to debris unwanted exposure to light can damage the devices, and specialized lighting is used in certain areas, like those for photolithography.

"clean" and they are referred to by their "class," with a class 10 being cleaner than a class 100 but dirtier than a class 1. Personnel working in a cleanroom must wear protective suits so as not to contaminate the workspace, not for their own safety. Even specialized paper must be used to reduce contamination.

22.4.2.2 Bottom-Up Techniques (Chemical Synthesis)

The top-down fabrication methods just discussed are extremely versatile and the developments over the last few decades have allowed extremely small devices to be created efficiently enough to reduce the price for more powerful computing resources. An alternative way to create nanoscale features is to use the principles of physics and chemistry to guide the process. A simple example is Italian dressing. The oil and water separate over time, oil on top and water on bottom, due to their differences in polarity and density. It happens this way every time without fail.

Bottom-up nanofabrication uses raw materials that interact in such a way that they self-assemble into the desired shape, size, or distribution desired. The key to this is that the raw ingredients are added together and a product "emerges." Of course it is not quite that simple. There is typically precise control over ratios of ingredients, the application of heating/cooling, time and other inputs that determine the outcome. The biggest advantage of this is that there is no requirement for equipment with nanoscale control. Basic chemical and physical laws govern the outcome, and through the proper inputs nanoscale features can be reliably created.

The most common example of this type of process is the creation of nanoparticles. Nanoparticles are finding numerous uses in medicine, sensors, catalysis, energy, and other applications. Being able to "tune" the nanoparticles to serve the needed application is important and is most often done using self-assembly. One of the simpler methods is the Turkevich method (Fig. 22.16) for creating gold nanoparticles. It starts with gold salt (chloroauric acid, $HAuCl_4$) that is dissolved in water. Sodium citrate is then added to the heated solution and it causes the gold atoms to cluster into small (10–20 nm) particles. Normally the gold would form larger particles, but the citrate coats the gold and stops it from combining with other gold atoms. The amount of citrate

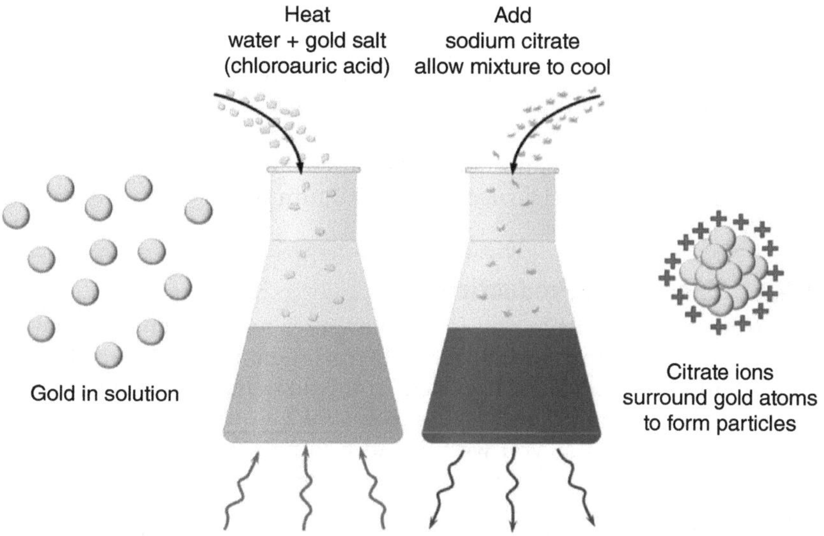

FIGURE 22.16 The Turkevich method for creating gold nanoparticles is a simple technique. It requires dissolving gold in water and then reducing the gold and capping it as nanoparticles using citrate. The size of the particles is determined by the ratio of gold to citrate.

added controls the final size, with more citrate resulting in smaller particles. Other methods achieve a similar end: small clusters of gold atoms capped with "ligands" (molecules) that prevent further growth of the particles, and those ligands can be used to add other molecules to the nanoparticles. In a similar way, surfaces and other structures can be modified to form nanoscale structures using self-assembly.

22.5 Sustainable Manufacturing

Manufacturing is material and energy intensive. There is no way around it. Converting raw materials into useful products requires the input of natural resources, but environmental concerns are becoming more central to manufacturing decisions. This is partly due to a shared sense of stewardship for the planet, but it is also being driven by government regulation and increasing energy prices. Just as consumer-level purchasing decisions involve considerations of the source, lifetime, and recyclability of the product, companies must consider what long-term effects their manufacturing practices and products will have. This makes manufacturing more costly and more difficult. It so happens that some of the lowest cost and most effective practices are bad for the environment. It will always be cheaper to dump your waste out back than to have someone to come pick it up and transport it for proper disposal; at least in the short term. The long-term consequences may end up costing more in the end, but new solutions to reduce the energy input and use more environmentally friendly materials are allowing for cost to remain low while simultaneously reducing negative impacts.

So what is sustainable? We will consider sustainable manufacturing to include methods and products that can be practiced/produced indefinitely from financial and environmental perspectives. Both are important. If it is not financially sustainable, the company goes out of business. If it is not environmentally sustainable, the local impact can be health degradation of people, animals, and plants. Collectively, unsustainable manufacturing has a global impact; potentially resulting in loss of quality of life for populations in distant locations. The topic of sustainability and global responsibility is something we will not venture into. Our concern is what decisions are involved in manufacturing processes and material choices. Rarely are these decisions entirely perfectly aligned with financial success or maximum sustainability. We end up somewhere between.

22.5.1 Energy and Production

Converting materials from one form to another requires energy. Many processes are designed to reverse the second law of thermodynamics (see Sec. 8.2.1); specifically to reduce randomness and create order. Unfortunately, most of the useful products we have are not in a naturally occurring form. Take a quick look around you. Do you see anything that appears as you would find it in nature? Even wooden furniture, cotton towels and granite countertops have undergone numerous processing steps to reach their present form. They are "natural" materials, but they are not in a truly natural form. Even houseplants have been transported, are probably in a synthetic soil, and may be fertilized with processed ingredients. All of this requires energy. Whether it is heat, electricity, fuel, or any other form, the process of creating our societies is energy intensive.

Energy is required, but it can be sustainable. Solar, hydroelectric, wind, nuclear, and other "sustainable"* energy sources are gaining market share. This allow for the same manufacturing practices to be applied but the overall sustainability to increase. It is much more efficient when the manufacturing practices also increase in efficiency, however, and newer equipment is designed to perform the same operations with less energy. Again, this is financially sensible as electricity is part of the manufacturing cost.

It is not just incremental improvements in equipment such as machine tools, furnaces, and support systems, but new methods for extracting metals from ore, new cutting tools, biodegradable lubricants, and many other advancements are fundamentally changing the way materials are converted into products. Increased use of recycled materials (and new ways to recover materials for recycling) and the reduced use of materials for packaging and structural applications are enabled by lighter, stronger materials, and new engineering approaches.

Lighter materials in the transportation industry and more efficient operation are also reducing the energy required during the use of products. For instance, "high-efficiency" washing machines can reduce the water and energy over the lifetime of the product and greatly outweigh the energy consumed in producing the product during manufacturing (Fig. 22.17). This use phase energy requirement dominates some products, such as ovens, hairdryers, electric tea kettles, computers, and other electronic devices. In these cases, the engineering design of the product controls the total efficiency (i.e., energy consumption during use), and the manufacturing stage is less impactful *in comparison*. If you are in the business of producing electric tea kettles, it is still within your control to reduce the energy input during manufacture.

---

*Note that sustainable and renewable are not the same, and one energy source may be one or both. It is generally considered sustainable if the rate of use is low or the available resources are high.

Figure 22.17 Sustainability is often associated with energy consumption. Products like "high-efficiency" appliances can save significant electricity and water over their extended use phase, but other products may be more dependent on the energy required to produce and dispose of them.

22.5.2 Embodied Energy of Materials

The idea of energy input to products has important consequences. It will not require the same energy to make a gear out of metal or plastic. The amount of energy required to produce a material, component, or finished product as a result of each stage of production and transportation is the **embodied energy**. Considering the difference between metal and plastic, we can compare some of the key differences (Table 22.2).

Both materials follow a similar progression of raw material to finished product, but the energy at each stage varies. The extraction and refining stages are energy intensive for both, but the amount of energy required to refine titanium is much greater than that of steel, so it is important to consider specific cases in practice. Metals tend to require more energy to extrude, mold, and machine (stronger, high melting temperatures) and they are denser which requires more energy to move. The component may last longer, and the plastic version will have to be replaced during the lifetime of the product, and therefore its embodied energy will be doubled.

Metal	Plastic	Both
Mining	Extraction of oil	Transportation
Refining ore	Refining of oil	Cleaning
Production of stock	Polymerization	Inspection
Machining/extrusion	Extruding/molding	Finishing

TABLE 22.2 When Considering the Material to Use for Product, the Amount of Energy Needed to Make It Useful Should Be Included. Materials Requiring Minimal Processing and Transport Are Best

It is difficult to account for all of the factors contributing to embodied energy, especially in complex products, such as a vehicle. Knowing the amount of energy required to make and transport each component requires a great deal of bookkeeping, but there are methods of doing it. The details are beyond the scope of our discussion, but it is important to consider that there are many factors that play into it. Something else to consider is how far back in history will you consider directly relevant to the current calculation. Each product is made using machines that are made from products that are made by machines, and so on. A single product can have an infinite embodied energy if we include enough steps. For instance, a car may be delivered to the dealership on a tractor trailer. Will you then include the embodied energy of that tractor trailer? The cargo ship that delivered the tires?

If we go back to the discussion of natural materials also being processed, we can now see the benefit. They require minimal processing compared to materials such as polymers. The embodied energy of natural materials is the lowest, and high-melting point metals such as Ti and W are among the highest (Fig. 22.18). Polymers tend to be on the lower end of nonnatural engineering materials, and ceramics and metals have wide ranges, but ceramics are lower overall. This does not mean their performance is similar. A smaller volume of metal may serve the purpose better than a large amount of a polymer and actually have a lower embodied energy in practice.

So how can we improve the sustainability of manufacturing? Make it more efficient, and keep it local. Due to the lower labor costs, many raw materials are shipped to

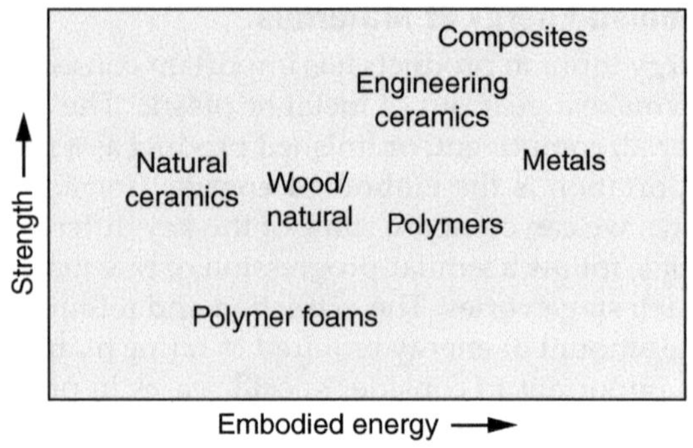

FIGURE 22.18 The energy required to produce materials and products in known as the embodied energy. Plotting this against the strength, or other pertinent characteristics, can aid sustainable design. Notice that most high-strength materials also have a high embodied energy.

distant locations, like China, to be turned into products and then shipped back to be sold. Manufacturing near the source of sale will greatly reduce transportation energy requirements. The idea of placing manufacturing facilities near the source or intentionally sourcing local resources can be helpful, but it is only practical to a certain extent. Resources are not evenly distributed throughout the globe, and most governments do not want to be dependent on products coming from one place. There are practical limitations. Considering alternative materials with less energy of processing and more even distribution globally can also help.

22.5.3 Material Life Cycle

Sustainability is not only determined by the processes that go into creating a product, it also includes how the product is handled at the end of its lifetime. For example, an egg carton may be used for a week or two then thrown away. The use phase of the product is brief, but their disposal varies greatly. For instance, a quick glance at the egg section in the grocery store reveals paper, foam, and clear plastic cartons (Fig. 22.19). Which one is best?

That is a more complex question than it appears. Are we considering price? Thermal insulating properties and protection of the eggs from impact? Recyclability? Biodegradability? Obviously there are many potential responses and reasons for each. A prominent egg producer lists reasons to use all three, and they do. They highlight the benefits of all three. So why not use one type? Obviously the reasons are not straightforward. Table 22.3 gives some of the prominent differences between the choices.

So which one is better for the environment? You will find a common theme with egg producers and grocers that state that it takes "three times more material in waste paper" to make the pulp paper carton. Also it is noted that the polystyrene producer has won awards for sustainability. It should be expected that less material goes into the foam. Foam is mostly air! More material does not necessarily mean more embodied energy, and awards do not make a process sustainable, they can simply indicate it is better than it used to be.

There are issues with pulp paper, too. Just because it can biodegrade, does not mean it will in modern landfills. Unless you compost your carton, it may have no real benefit in that regard. There is a lot of noise surrounding sustainability and anyone can claim it. Backing it up with transparent data is different. Unfortunately, each side of the debate is free to misrepresent the truth, so the consumer should be as informed as possible when making decisions around purchase. We have not even touched on what it means to raise the chickens producing the eggs sustainably and ethically. That aspect

Carton Material	Performance Benefits	Environmental Considerations
Polystyrene foam	Excellent impact and thermal protection	Recyclable in select locations
Polyethylene terephthalate (PET) plastic	Transparent, minimal acceptable impact and thermal protection	Made from recycled content, widely recyclable
Pulp paper	Modest impact and thermal protection	Made from recycled content recyclable and biodegradable

Table 22.3 The Choices Between Materials in Products and Packaging Have Multiple Aspects to Consider. The Consumer, Producer, and Seller May Have Different Perspectives on These Considerations

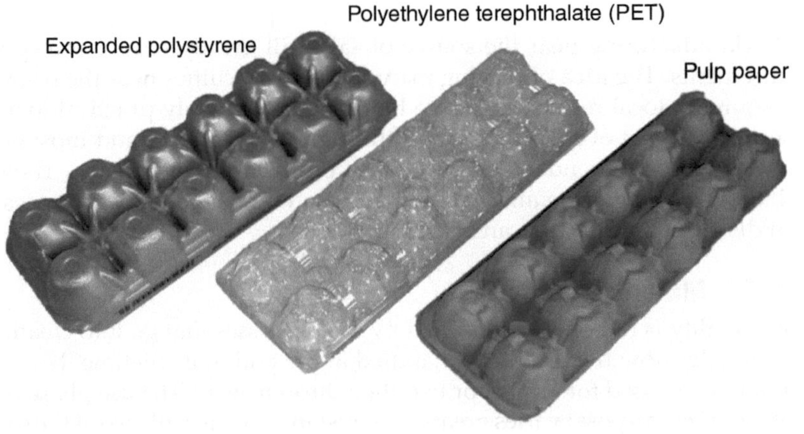

FIGURE 22.19 Some products are available in multiple materials, like these egg cartons. The best choice for the environment is not always as simple as it seems, and the performance of each material must also be considered.

of production is outside our range of consideration, but it is very much a part of the broader conversation of sustainability. If we all had our own chickens, there would be no packaging and transport energy, but it takes resources to feed, house and otherwise raise chickens. There is no one-size-fits-all answer to the problem.

The simplest actions that anyone can do, even without specific interest in sustainability, are follow disposal laws and recycle where possible. Do not forget the three "Rs"*: reduce, reuse, recycle. Buy products with less packaging or buy larger quantities, which requires less packaging per volume. Look for items that are made from recycled materials. Just buy less stuff, period, and give it away when you are done if it still has value to someone. If you throw it away it does no one any good.

22.6 Why It Matters

We have covered a lot of ground in this chapter. The reason these "advanced" concepts matter is because they enable new possibilities and set the tone for future progress. Without advancement, we would still be using steam engines, or horses, or water wheels, and so on. Newer developments like nontraditional manufacturing techniques have allowed greater throughput and capability. If you are not familiar with the options available, you might miss out on a better solution.

Perhaps the most important advancement to manufacturing is automation. As unfortunate as it is, people can be liabilities in production. They are prone to get tired or distracted. They need to eat. They are not perfect. Machines are capable of doing the same exact job repeatedly without fail until they simply wear out. People are still the best option for designing and planning production (though artificial intelligence has made some progress in this area), so there should be a working relationship between programming a machine to do it and a person solving problems and innovating.

*Some would say these are reading, writing, and arithmetic, which are also important.

The advancements we make in technology can outpace our understanding of it. Nanomaterials and devices are a prime example. There are some questions as to the safety of nanoscale materials because of their unique properties. Nonetheless they are the most researched of any category, and they are already appearing in commercial products. Knowing more about what "nano" really means is one part of you being an informed consumer, but it may also directly impact your production decisions. Are workers going to be exposed to these materials? Are you working with a commercial product containing them? What special considerations are needed in these cases? These all point to the importance of being familiar with the topics. As someone who works to develop nanoscale materials and processes, I wish I could say more. It is sufficient to say that great potential exists with these technologies, but creating safe products efficiently is a significant challenge. That is why they are so interesting.

The same "grand challenge" theme exists with sustainability. The concepts are becoming more familiar in everyday discussion, but there are special considerations in manufacturing. Manufacturing converts raw materials into useful goods. That requires a great deal of energy and resources and produces significant waste. Being efficient in manufacturing is good practice regardless. Energy costs money, so efficient methods mean greater profit. Reducing waste means lower disposal costs. Manufacturing is the greatest contributor to embodied energy for most products, so it is a natural target for improvement. By considering materials and products from cradle to grave (or cradle to cradle in a truly sustainable scheme), better choices for materials and processes can be made. If anything, I hope this book helps you make those choices.

22.7 The How and Why of Advanced Manufacturing

22.7.1 Chapter Review

1. In your own words, describe what is meant by "advanced" manufacturing. How does it change with time?
2. How do we define the tool in nontraditional machining? How is this different than traditional methods?
3. Why are nontraditional machining methods capable of producing such fine features and complex profiles?
4. What removes material in EDM?
5. Why is it bad for the electrode to touch the workpiece in EDM?
6. Describe what happens during the on-time and off-time in EDM.
7. In your own words, summarize what happens during a single spark process in EDM.
8. What are the three types of EDM machine. How are they different?
9. In sinker EDM, describe the considerations that necessary with the electrode geometry.
10. How is particulate flushed from the cut area?
11. What traditional cutting tool is wire EDM similar to? How is it different?
12. What determines the kerf width of a wire EDM?

13. How are tapers cut using wire EDM?
14. How does a high-pressure water jet help guide the wire when rethreading?
15. What is likely to happen when a wire EDM is not programmed with the correct settings?
16. What is meant by "small" in small hole EDM?
17. How can small hole EDM be used for repair work?
18. Why does a small hole EDM guide tube have a ceramic liner?
19. What is the benefit of adding abrasive to water jet? Is it always needed?
20. What is the role of the orifice in water jet? Why is made out of materials like ruby or diamond?
21. Why cannot tempered glass be cut using water jet? Refer to Sec. 10.3.2 if needed.
22. What determines the cut capacity (thickness) of a laser system?
23. Describe what materials can be cut with laser, water jet, and EDM, and compare their capabilities.
24. How is laser machining used to process the surface and profile of parts?
25. Why is the laser wavelength important to cut characteristics? How can a laser that can cut metal not be suitable to cut wood?
26. What does it mean to "automate" manufacturing?
27. What beneficial features do CNC mills and lathes have that increase productivity?
28. Why would a CNC machine be used to make a single part, rather than manual machining?
29. How is a slant bed different from a flat-bed lathe (see Fig. 16.4)?
30. How are machines better able to produce many parts quickly and more accurately than even highly skilled workers?
31. How can machines be inferior to a skilled worker?
32. What is the purpose of adaptive control?
33. Give your own definition of G-code.
34. Describe how the structure of G-code is similar to written language?
35. What kind of machine parameters do preparatory codes control?
36. What kind of machine parameters do miscellaneous functions (M-codes) control?
37. How are the tasks of design and manufacturing addressed in software? What are the benefits and disadvantages?
38. What is meant by the term "toolpaths" in regard to CAM software?
39. What are the benefits of hybrid machining centers? Why are they not more common?
40. How can additive manufacturing and machining be combined into a single machine. Name two ways.

APPENDIX B

Mechanical Properties of Common Materials

Mechanical Properties of Common Metals (Customary Units)					
Metal	**Modulus of Elasticity (E), psi × 10^6**	**Modulus of Rigidity (G), psi × 10^6**	**Tensile Strength, psi × 10^3**	**Yield Strength, psi × 10^3**	**Total Elongation, %**
Cast steel	28.5	11.3	60–125	30–90	33–14
Cold-rolled steel	29.5	11.5	84	76	18
Stainless steel 18-8	27.6	10.6	85–95	30–35	60–55
All other steels	28.6–30.0	11.0–11.9	50–65	30–40	40–30
Cast iron	13.5–21.0	5.2–8.2	18–60	8–40	0
Copper	15.6	5.8	32	5	58
Brass	15.9	6	40–120	8–80	60–3
Aluminum alloys	9.9–10.3	3.7–3.9	30–60	10–50	33–15
Monel metal	25	9.5	79	30	48
Magnesium alloys	6.3	2.5	21–45	11–30	17–0.5
Titanium	15–16	6.5	130	120	10
Molybdenum	48–52	17.4	100	75	30

Mechanical Properties of Common Metals (SI Units)					
Metal	**Modulus of Elasticity (E), GPa**	**Modulus of Rigidity (G), GPa**	**Tensile Strength, MPa**	**Yield Strength, MPa**	**Total Elongation, %**
Cast steel	196	78	413–861	861–207	33–14
Cold-rolled steel	203	79	579	524	18
Stainless steel 18-8	190	73	586–655	655–207	60–55
All other steels	197–207	76–82	345–448	448–207	40–30
Cast iron	93–145	36–56	110–413	413–55	0
Copper	107	40	220	34	58
Brass	110	41	276–827	827–55	60–3
Aluminum alloys	68–71	25–27	207–413	413–69	33–15
Monel metal	172	65	544	207	48
Magnesium alloys	43	17	145–310	310–76	17–0.5
Titanium	103–110	45	896	827	10
Molybdenum	331–358	120	689	517	30

Data adapted from Tables 3.1.1 and 3.1.3 in Avallone, E. A., I. T. Baumeister, and Ali Sadegh. *Marks' Standard Handbook for Mechanical Engineers*. 10th ed. New York: McGraw-Hill, 2006.

Fracture Toughness of Common Materials

Material	Toughness (K_{IC}) MPa√m	Toughness (K_{IC}) ksi√in
AISI 1144 Steel	66	60
AISI 4130 Steel	110	100
300-M Steel (650°C temper)	152	138
300-M Steel (300°C temper)	65	59
18-Ni Maraging Steel (vacuum melted)	176	160
2014-T651 Aluminum	24	22
2024-T351 Aluminum	34	31
7075-T651 Aluminum	29	26
Titanium (Ti-6Al-4V)	66	60
Acrylonitrile-Butadiene-Styrene (ABS)	3	2.7
Acrylic	1.8	1.6
Epoxy	0.6	0.5
Polycarbonate (PC)	2.2	2.0
Polyethylenterephthalat (PE)	5	4.5
Polyester	0.6	0.5
Polystyrene (PS)	1.2	1.1
Polyvinylchloride (PVC) Rigid	2.4	2.2
Polyvinylchloride (PVC) Plasticized	3.4	3.1
Soda-Lime Glass	0.8	0.7
Magnesia (MgO)	2.9	2.6
Alumina	4	3.6
Zirconia Toughened Alumina (15% ZrO_2)	10	9.1
Silicon Carbide	3.7	3.4
Silicon Nitride	5.6	5.1
Dolomitic Limestone	1.3	1.2
Westerly Granite	0.9	0.8
Concrete	0.2	0.2

Material Adapted from Table 18.2 in Young, Warren Clarence, and Richard Gordon Budynas. *Roark's Formulas for Stress and Strain*. Vol. 7. New York: McGraw-Hill, 2002.

APPENDIX C

Common G-Code in Machine Programming

Common G-Code (Positioning, Machine Functions, and Cycles)			
Mill Codes		Mill Codes	
Code	**Description**	**Code**	**Description**
G00	rapid positioning	M00	compulsory program stop
G01	linear interpolation (movement)	M01	optional program stop
G02	circular interpolation (CW)	M02	program end
G03	circular interpolation (CCW)	M03	spindle on (CW)
G17	select XY plane	M04	spindle on (CCW)
G18	select XZ plane	M05	spindle off
G19	select YZ plane	M06	tool change
G20	programming units of inches	M08	coolant on
G21	programming units of millimeters	M09	coolant off
G28	return to machine reference (home)	M30	program end and rewind
G40	cutter radius compensation off		
G41	cutter radius compensation left	FXX	feedrate (in/min, mm/min)
G42	cutter radius compensation right	NXX	block number
G43	negative tool length offset	SXXXX	spindle speed
G44	positive tool length offset	HXX	tool length offset number
G54	work offset coordinate	DXX	tool diameter offset
G80	cancel canned (drill) cycle	RXX	radius value for circular interpolation
G81	simple drilling cycle	IXX, JXX	vector values for circular interpolation
G82	simple drilling cycle with dwell	TXX	tool number
G83	peck drilling cycle		
G84	tapping cycle		
G90	absolute coordinate reference		
G91	incremental coordinate reference		

Common G-Code (Positioning, Machine Functions, and Cycles)			
Lathe Codes		**Lathe Codes**	
Code	**Description**	**Code**	**Description**
G00	rapid positioning	M00	compulsory program stop
G01	linear interpolation (movement)	M01	optional program stop
G02	circular interpolation (CW)	M02	program end
G03	circular interpolation (CCW)	M03	spindle on (CW)
G17	select XY plane	M04	spindle on (CCW)
G18	select XZ plane	M05	spindle off
G19	select YZ plane	M06	tool change
G20	programming units of inches	M08	coolant on
G21	programming units of millimeters	M09	coolant off
G28	return to machine reference (home)	M30	program end and rewind
G40	nose radius compensation off		
G41	nose radius compensation left	X, Y, Z	absolute coordinate reference
G42	nose radius compensation right	U, V, W	incremental coordinate reference
G54	work offset coordinate	FXX	feedrate (in/min, mm/min)
G70	finishing cycle	NXX	block number
G71	turning cycle	SXXXX	spindle speed
G72	facing cycle	RXX	radius value for circular interpolation
G73	pattern repeating cycle	IXX, JXX	vector values for circular interpolation
G96	constant surface speed (rpm)	TXXXX	tool number and wear offset
G97	direct rpm		

APPENDIX D

Speed and Feed Values for Machining Operations

Common Machine Settings for Milling§					
Using High-Speed Steel Cutting Tools			Using Carbide Cutting Tools		
Material	Cutting Speed (ft/min)	Cutting Speed (m/min)	Material	Cutting Speed (ft/min)	Cutting Speed (m/min)
Aluminum alloys	300–450	7.6–11.4	Aluminum alloys	1200–1800	30.5–45.7
Brass	200–300	5.1–7.6	Brass	800–1500	20.3–38.1
Bronze	100–180	2.5–4.6	Bronze	800–1000	20.3–25.4
Magnesium	600–1000	15.2–25.4	Magnesium	1000+	25.4+
Steel (low carbon)	100–120	2.5–3.0	Steel (low carbon)	600–720	15.2–18.3
Steel (medium carbon)	75–90	1.9–2.3	Steel (medium carbon)	300–450	7.6–13.7
Steel (high carbon)	60–80	1.5–2.0	Steel (high carbon)	360–480	9.1–12.2
Alloy steels*	40–60	1.0–1.5	Alloy steels*	400–600	10.2–15.2
Stainless steel	60–80	1.5–2.0	Stainless steel	600–800	15.2–20.3
Titanium	80–100	2.0–2.5	Titanium	50–250	1.3–6.4
Plastics	350–500	8.9–12.7	Plastics	500–600	12.7–15.2

(Continued)

Common Machine Settings for Turning§					
Using High-Speed Steel Cutting Tools			Using Carbide Cutting Tools		
Material	**Cutting Speed (ft/min)**	**Cutting Speed (m/min)**	**Material**	**Cutting Speed (ft/min)**	**Cutting Speed (m/min)**
Aluminum alloys	150–225	3.8–5.7	Aluminum alloys	300–450	7.6–11.4
Brass/bronze	275–350	7.0–8.9	Brass/bronze	600–700	15.2–17.8
Magnesium	350–500	8.9–12.7	Magnesium	800–1250	20.3–31.8
Steel (low carbon)	225–300	5.7–7.6	Steel (low carbon)	550–700	14.0–17.8
Steel (medium carbon)	200–275	5.1–7.0	Steel (medium carbon)	450–600	11.4–15.2
Steel (high carbon)	175–250	4.4–6.4	Steel (high carbon)	400–500	10.2–12.7
Low alloy steels* ()	150–200	3.8–5.1	Low alloy steels* ()	325–450	8.3–11.4
Stainless steel	80–150	2.0–3.8	Stainless steel	300–400	7.6–10.2
Titanium	100–160	2.5–4.1	Titanium	375–600	9.5–15.2
Plastics	350–600	8.9–15.2	Plastics	600–1800	15.2–45.7

Customary units†: $N_s = \frac{12V}{\pi D}$; $\text{rpm} = \frac{12\,(\text{cutting speed})}{\pi\,(\text{diameter})}$; SI units‡: $N_s = \frac{1000V}{\pi D}$

§Broad guidelines only. Machine settings rely on various factors, including the cut depth, feed rate, machine rigidity, the grade of cutting tool, tool coatings, etc., as well as the condition of the material, such as cold work, hardening, annealing, and other treatments. Use caution when choosing these settings. Consult tool manufacturer for best practices. Values adapted from: Cormier, Denis, and Ronald A. Walsh. *McGraw-Hill Machining and Metalworking Handbook*. McGraw-Hill Professional Publishing, 2005.

*Includes alloys with nickel, chrome and molybdenum (e.g., AISI-SAE 4340).

†Units of feet per minute (cutting speed) and inches (diameter).

‡Units of meters per minute (cutting speed) and millimeters (diameter).

Suggested Feed Rates for HSS End Mills§			
Material	**inch/tooth**	**Material**	**inch/tooth**
Magnesium and alloys	0.011	Cast steel	0.006
Aluminum and alloys	0.011	Low-carbon steel	0.005
Free-cutting brasses and bronzes	0.011	Medium-carbon steel	0.005
Medium brasses and bronzes	0.007	Alloy steel (soft)	0.004
Hard brasses and bronzes	0.005	Alloy steel (medium)	0.003
Copper	0.006	Alloy steel (hard)	0.002
Cast iron (soft)	0.008	Stainless steel	0.003
Cast iron (medium)	0.007	Monel	0.004
Cast iron (hard)	0.006	Titanium and alloys	0.004
Malleable iron	0.006	Plastics	0.007

Feed rate†: $f_m = f_t n N_s$; $\frac{\text{feed}}{\text{min}} = \left(\frac{\text{feed}}{\text{tooth}}\right)$(number of teeth)(rpm)

§Broad guidelines only. Machine settings rely on various factors, including the cut depth, rpm, machine rigidity, the grade of cutting tool, tool coatings, etc., as well as the condition of the material, such as cold work, hardening, annealing, and other treatments. Use caution when choosing these settings. Data adapted from Figure 7.69 in Cormier, Denis, and Ronald A. Walsh. *McGraw-Hill Machining and Metalworking Handbook*. McGraw-Hill Professional Publishing, 2005.

†Units determined by inputs (e.g., inches per minute, meters per minute, etc.).

Glossary

adaptive control the automatic and continuous monitoring and adjustment of processing conditions

addition polymerization reaction where monomers combine to form a larger chain without any waste products

additive manufacturing (AM) creates a three-dimensional structure by adding material in a layer-by-layer manner until the structure is complete

adhesion joining of components using a third material to bond the other together through a combination of chemical and physical interactions, results in distinct interfaces between parts

adhesive tapes see pressure-sensitive adhesive

adjustable blade reamers accommodate larger adjustments in diameter than expansion reamers to achieve the desired geometry

advanced manufacturing manufacturing processes and materials that utilize state-of-the-art scientific understanding and technology to improve the quality and/or efficiency of processes or applications

age hardening process following solutionizing where atoms come out of solution to form very fine precipitates that impart strength to the alloy, rate and result are dependent on temperature

aging see age hardening

air bending bending operation where the tooling provides ample space between surfaces, such that the travel of the tooling determines the bend angle, not the geometry of it

aircraft extension drills extra-long drills for reaching difficult areas

allotropy the ability for an element or material to assume various crystal structures, often as a result of temperature or pressure change

alloy a metallic material composed of two or more elements

alloy steels see low alloy steels

alpha and near-alpha alloys titanium alloys characterized by the hexagonal close packed (alpha) structure

alpha-beta alloys titanium alloys containing a combination of hexagonal close packed (alpha) and body-centered cubic (beta) structures

alternating copolymer copolymer where every other monomer in the backbone is of a different type (i.e., ABABABAB pattern)

alternating current (AC) welding variable welding polarity where the electrode rapidly switches between being positively and negatively changed

alternating tooth bevel (ATB) saw tooth design including narrower teeth for reducing tear out in fibrous materials and reducing cutting force

alumina aluminum oxide, a ceramic with many applications

aluminum alloys metallic mixtures where aluminum is the primary element

aluminum titanium nitride (AlTiN) hard compound of titanium, aluminum, and nitrogen where aluminum is present in a higher quantity than titanium, often used as a tool coating

amorphous without long-range order

amorphous alloys metallic materials with no long-range crystallographic (crystal lattice) order

amorphous carbon carbon with no long-range crystal structure

amorphous polymer a polymer with little to no alignment of polymer chains

angle blocks precision ground blocks of varying angles used to setup, measure or support workpieces depending the type of block

angle of twist measure of the angular displacement of member under torsion

anion an ion with less protons than electrons and a negative charge

anisotropic material a material that behaves differently in different directions

annealing thermal conditioning process using high temperature and slow cooling, results in lower internal stress, and in metals, a soft, ductile structure

anodizing creation of surface oxidation on a component to improve mechanical, corrosion, or aesthetic properties

anodizing process where a metal is electrolytically oxidized to create a protective and/or decorative layer

aramid type of polyamide polymer with aromatic rings on each side of the amide group in the molecular backbone, also called Kevlar

aromatic ring a group of atoms connected in the form of a circle, typified by the structure of benzene

articulated robot robot with multiple rotary joints allowing complex movement with many degrees of freedom of motion to fully access a work volume dictated by the size of the robot

artificial aging age hardening at elevated temperature to accelerate the precipitation process

aspect ratio the ratio of the depth of a feature to its width or cross-section

atactic side groups of polymer molecule oriented randomly along its length

atomic bonds the interaction of electrons between two or more elements such that the atoms are attracted to one another

atomic number the number of protons in an atom of element

austenite face-centered cubic phase of ferrous alloys, also known as gamma (γ) iron

austenitic stainless steels stainless steel with face-centered cubic crystal structure, primarily attributed to a high nickel content

austenizing temperature the temperature above which a ferrous alloy converts from body-centered cubic structure (ferrite) to face-centered cubic (austenite)

autogenous welding weld where the base materials are melted and fused together without any filler material being added

automated manufacturing processes which are designed, implemented, monitored, and controlled with the aid of digital programming and computer control

backboring process where a recessed diameter is created at the far end of a hole, often used when the workpiece cannot be reoriented to provide easier access

backhand welding welding processes where the heat source is pointed in the opposite direction from the welding progression, particularly important in oxy-fuel welding

backward extrusions see indirect extrusions

bainite plate-like phase consisting of ferrite and cementite in which the ferrite phase has a high dislocation density, making it stronger than typical ferrite

ball end mill end mill with straight sides and semicircular bottom profile, used to create corresponding radii in a workpiece or improve surface finish in three-dimensional profiles

basic diameter diameter of a threaded component measured from the midpoint of the threads

bending force applied transversely that distorts a member from its original axial geometry

beryllium copper alloy of copper with beryllium as the primary alloying element

beta alloys titanium alloys characterized by the body-centered cubic (beta) structure

bevel an angle cut where the cut is parallel to the long axis of the workpiece (extends along the length)

bimetal blade saw blade where the teeth are tool steel and attached to a carbon steel body

binder additive manufacturing applies an adhesive to combine powder particles in select locations, hardened through reaction with atmosphere, heat, or irradiation

bisque firing a thermal process used impart strength to a ceramic component and create a porous condition suitable for later glazing processes

black oxide oxidation treatment applied to ferrous metals to preferentially create a layer of magnetite [Fe (II, III) oxide]

blank holder a device which holds the metal to be drawn tightly enough to reduce wrinkling but without overly constraining its motion

blanking punching/stamping operation where the desired part is cut from the interior of a larger workpiece

blanking dies tooling used in forging which imparts a general shape to a part in preparation for later forming operations

blast media the characteristics of the abrasive used in sand blasting

bleeder cloth used on top of the peel ply to absorb excess resin in composite layup

blind risers riser that is contained within the mold assembly (not visible from the outside)

block copolymers copolymer with multiple units of each monomer combined in distinct sections

blocks (G-code) a full line of G-code used to program CNC equipment, made from one or more blocks

blow molding a polymer molding processes where gas pressure is used to expand a polymer to fill the mold

blowing agent see foaming agent

blushing the tendency for a polymer to change in appearance to be opaque and/or lighter in color (or white) in an area of high deformation, such as a bend

body-centered cubic type of crystal structure where a single atom resides in the center of the cubic unit cell

boiling point elevation the raising of the boiling point by mixing ingredients

bolt generic term for fastening devices using an externally threaded section and corresponding nut to hold parts together

boring process of removing material from the inside diameter of a rotating workpiece

boron carbide lightweight ceramic made from boron and carbon

bottoming bending operation where the tooling completely compresses the material between surfaces

bottoming tap tap with little to no length of partial thread cutting depth, difficult to start by hand, but fully threads blind holes

bottom-up nanofabrication processes for creating nanoscale materials using self-assembly by naturally occurring chemical or physical forces without the need for specific step-by-step input

boule a single crystal cylinder of high-purity silicon

brass alloy alloy of copper with zinc as the primary alloying element

brazing a method of assembling components where a metallic filler material (typically rod or wire) is melted and applied to the surfaces at or above 840°F (450°C)

bronze alloy alloy of copper with elements other than zinc, nickel, and beryllium as the primary alloying addition, often designated by the primary alloying element (e.g., aluminum bronze)

built-up edge (BUE) the accumulation and adherence of material at the cutting edge during machining

bulk metallic glasses see amorphous alloys

bulk molding compound thermoset polymer mixed with a high proportion of fillers to create a dough-like consistency used in molding

bulk molding compound (BMC) polymer molding resin mixed with extensive fillers and reinforcement, with longer chopped fibers than sheet molding compounds

bull nose end mill see corner radius end mill

bump-type knurling tool knurling tool that has form rollers forced against one side of the workpiece

burnishing the mechanical working of a surface to increase strength and surface quality

burrs sharp, protruding areas of material left from cutting or other processes

butt joint joint where two flat surfaces are held parallel to one another

by-products chemicals that result from a process but do not contribute to the final product

capped steel similar to rimmed steel, but with reduced segregation of elements

carbide inserts interchangeable carbide cutting tools of a set geometry used with a corresponding tool holder

carbon backbone carbon atoms forming the central structure of a polymer

carbon fiber-reinforced polymers (CFRP) composite composed of carbon fibers in a polymer matrix

carbon-carbon composite ceramic matrix composite with a carbon matrix and carbon reinforcement used in high-temperature applications, often created by pyrolyzing carbon-reinforced polymer composites

carbon-ceramic a composite with carbon as the matrix and a ceramic reinforcement phase, often silicon carbide

carbonitriding process of adding a combination of carbon and nitrogen to a component by diffusion in solid or gas phases

carbonization reduction of a material to carbon only, often by heating in the absence of oxygen

carborundum see silicon carbide

carburizing process of adding carbon to a component by diffusion in solid or gas phases

carburizing flame a flame containing unreacted fuel (extra carbon)

carriage (lathe) the portion of the lathe that allows movement of the cutting tool in the longitudinal direction (along the length of a workpiece)

case hardening process used to create a hardened outer layer on a component by changing the composition or selectively conditioning that surface

casting the production of parts by melting the material and using a mold to create features

catalyst of polymerization, a chemical that initiates and/or accelerates the chemical reaction

cathodic protection technique where a component or structure is made the cathode of an electrochemical cell to protect it from corrosion

cation an ion with more protons than electrons and a positive charge

Celsius temperature scale in which water freezes at 0°C and boils at 100°C

cement (ceramic) ceramic primarily composed of limestone with other minerals and clay, depending on the type

cementite iron carbide phase in ferrous alloys, imparts strength and hardness

cement (adhesive) rubber-based adhesives

center drill see combination drill and countersink

center-cutting end mill an end mill with cutting edges on the bottom that extend to the center

centerless grinding cylindrical grinding operation where the workpiece is not supported on the ends, but rather is supported and advance by a secondary, regulating wheel

centrifugal casting casting process where a mold is rotated around its axis to force material to edges and results in hollow part

centrifuge casting casting process where a mold is rotated around an axis outside of the mold

ceramic cutting tools cutting tools made from engineering ceramics (e.g., aluminum oxide) capable of withstanding high temperatures

ceramic matrix composite (CMC) a composite where a ceramic is used to combine or encapsulate the dispersed phase

cermet term for cutting tools made from ceramic particles bound together by a metallic matrix (i.e., ceramic + metal), often excludes tungsten carbides

cesium chloride structure unit cell typified by the structure of CsCl, where two elements create a body-centered cubic-type structure

chain-growth polymerization see addition polymerization

chamfer end mill conical end mill which produces chamfered edges on workpiece

chamfers angled surface created on workpiece corners for functional and/or aesthetic purposes

characters (G-code) a single unit in a word of G-code, a letter, number, or symbol, used to program CNC equipment

charge the initial amount of material used for a single component or molding operation

chemical blowing agents chemical additives that produce vapor as a result of a chemical reaction and create pores

chemical degradation breakdown of a material by direction action from contact with a chemical

chip a discrete piece of material removed during machining

chip load the amount of material removed by each cutting edge of a cutting tool

chip thickness the thickness of the material being cut/sheared from the surface during machining

chip thinning phenomenon where the chip load becomes drastically smaller than predicted by the feed rate because of a small radial engagement

chro-moly steels steels with primary alloying additions of chromium and molybdenum

chuck key device for tightening a keyed chuck

cis-isoprene polyisoprene where side groups are on one side of the polymer molecules making the polymer more flexible

clearance angles see relief angles

climb milling removing material with a mill by moving the workpiece with the rotation of the tool (i.e., tool rotation pulls the part into the cut, the peripheral cutting edge moves in the opposite direction of tool motion)

closed cell foams a porous material which has pores that are not connected to each other and do not create a continuous pathway through the material

closed-die forging see impression die forging

cluster mill a rolling stand with multiple rollers which act to support one another against the pressure of the rolling process

coarsening an increase in grain size and other microstructural features

coefficient of thermal expansion a material's change in dimensions proportional to the change in temperature

cohesive assembly assembly method which results in atomic/molecular mixing of the components, leave no distinct interface between parts

coining process where a pattern is created by indenting a part under high pressure

co-injection molding injection molding process where two or more separate materials are injected into a single mold

cold extrusion extrusion performed below a metal's recrystallization temperature

cold rolling process where rolling is performed at low or modest temperature (below the material's recrystallization temperature)

cold spray coating propels particles at supersonic velocity to bond the coating to the base material

cold welding welding process that occurs by severe pressure and deformation below the material's melting point

cold work the application of deformation at a temperature below a material's recrystallization temperature such that defect accumulation results in higher strength

cold-chamber die casting die casting process where the molten metal is transferred to the injection hardware for each shot

coloration processes which result in the distribution of color throughout a polymer product

combination blades circular saw blades with tooth design that balances efficiency between rip cuts and cross cutscombination die; see compound die

combination drill and countersink short, rigid drill used to create a tapered hole (60° angle) in which a center is located to support a workpiece in a lathe, also used to create a starting hole for other drills

commodity plastics common, extensively used polymers with standard properties

compacted graphite iron cast iron with a combination of vermicular and nodular graphite inclusions with properties between those of gray iron and ductile iron

composite lumber see wood-plastic composites

composites a material comprising two distinct phases, typically of different structures or materials (e.g., polymer + ceramic)

compound die a tooling design where multiple processes can be performed with one cycle, either by a single station or with multiple stations in a larger die

compound rest (lathe) allows tool motion on the lathe to be set at any angle relative to spindle axis

compression force applied in a manner that shortens a member axially

computer numerical control (CNC) machine control through computer-programmed sequences of commands that result in machine movement and activate functions

computer-aided manufacturing software-driven design and programming of processes

computer-integrated manufacturing integration of software (computer-aided design and programming) and hardware in manufacturing

concentration cell corrosion corrosion process where the concentration of corrosive products varies over a part, often accelerating corrosion in crevices or other small recesses where the local chemistry changes with time

concentration gradient a change in composition over distance, such that the concentration is higher in one position than another

concentricity having two or more diameters with the same center point

condensation polymerization reaction where monomers undergo direct reaction with one another and produce by-products in the process

conditioning the treatment of a material to modify its mechanical properties

cone number a visual system for indicating temperature in ceramic firing where a standard pyrometric cone begins to soften and slump

continuous fibers a dispersed fiber phase in a composite, where the fibers extend from one boundary of the component to another without interruption

continuous manufacturing technique by which large quantities are produced without interruption of processing

continuous saw blades saw blades that are in a continuous loop and allow for constant cutting

continuous seam weld seam weld where the fused region is uninterrupted along the full length of the joint

contour cut cut including non-straight segments, may have any fixed radius or one that varies along the cut line

conventional milling removing material with a mill by moving the workpiece against the rotation of the tool (i.e., tool rotation pushes the part away, peripheral cutting edge moves in the same direction as tool motion)

coolants cutting fluids that act to lubricate and remove heat from the cutting area

coordination number the number of atoms a single atom is in contact with based on the crystal lattice

cope the top half of a sand casting mold

copolymer polymer composed of more than one type of monomer

corner joint joint where two parts are held perpendicular to one another such their edges align and are welded together

corner radius end mill end mill with a cylindrical profile, but which has the bottom corners rounded to increase strength and/or create a radius on an interior corner of the workpiece

corner rounding end mill end mill with a concave radius that produces a corresponding radius on the outside corner of a workpiece

cottage industry production of goods in an individual's home

Coulomb force the electrostatic force that causes attraction between objects of opposite electrical charge and repulsion of objects with the same charge

counterboring creation of a concentric flat-bottom hole of larger diameter at the surface of a workpiece, often sized to accommodate the head of a bolt so that is flush with or recessed into the surface

countersinking modification to a hole to create a conical recess to aid in locating a mating part or capable of receiving matching fasteners

covalent bonds a primary bond where electrons are shared between two atoms

cover gas gas flow used to displace oxygen from an area being welded to improve purity

cover pass final welding pass(es) at the surface of a joint requiring multiple passes

creep the tendency for a material to undergo plastic strain at a stress level below its typical yield stress over long periods of time, often exacerbated by temperature

critical quench rate the minimum rate required to result in hardening, varies by alloy

critical resolved shear stress (CRSS) stress required to initiate plastic deformation, determined by the slip system of a metal

cross cuts saw cuts perpendicular to wood grain

cross-linking the formation of strong, primary bonds between polymer molecules

cross-slide (lathe) the portion of the lathe that allows movement of the cutting tool in the transverse direction (along the diameter of a round workpiece)

cryogenic concerning temperatures below –238°F (–150°C)

crystal lattice the three-dimensional pattern and position of atoms within a crystal

crystal structure see crystal lattice

crystallinity the degree of order of a structure, determined by the presence and extent of a repeated pattern, or in polymers, the alignment of polymer chains

cubic boron nitride (CBN) very hard phase of boron nitride used in cutting tools

cubic zirconia zirconia with a cubic crystal structure, often used in jewelry as a low-cost diamond simulant

cupronickel alloy of copper where nickel is the primary alloying element

curling creation of a cylindrical feature by bending

cut knurling knurling process where the surface is cut to create the desired pattern

cutting speed the proper surface speed for a specific combination of tool and workpiece materials

cutting tool a tool that removes material during processing, often by physical interaction

cyanoacrylates fast-acting polymer adhesive based on polymethylmethacrylate

cylindrical grinding process where a part is abrasively machined while rotated to create cylindrical surfaces

dead center a device for holding the free end of a workpiece in the lathe that does not rotate with the part and requires a lubricant

dead risers riser through which metal flows after reaching the mold cavity

dealloying preferential removal of a more active element from an alloy

decarburized steel steel with carbon removed by the introduction of a reacting agent

deep drawing forming operation where metal is stretched into a cavity that has a depth greater than the diameter or smallest cross-section distance

deep holeg a hole that is three or more times deeper than its diameter

deep hole drilling drilling operations where multiple smaller cut depths are used and the drill is fully removed from the hole remove waste material

defect density the number of defects per unit volume of material

defects of atoms, disorder when compared to the ideal crystal lattice

definite cutting edge layout cutting tools with specifically placed cutting edges

deflocculant an additive that assists in keeping particles suspended in a liquid

delocalized describing electrons not fixed with relation to any one atom or pair of atoms

dense carbon-carbon composites fully dense carbon-carbon composite with high strength, created by multiple cycles of polymer infiltration around a carbon reinforcement followed by pyrolysis

deposition processes that add material to a surface, specifically by atomic-scale processes in nanomanufacturing

depth of cut (D.O.C.) the distance between the surface of the workpiece and the position of the cutting tool during a machining operation

diamond crystalline allotrope of carbon with extremely high hardness

diamond cubic structure unit cell structure of two interpenetrating face-centered cubic cell structures, exemplified by carbon in the form of diamond

diamond cutting tools cutting tools made from diamond, or coating with diamond for cutting very abrasive materials

diamond-like carbon a special type of amorphous carbon that is very hard and used as a wear-resistant coating

die casting metal casting technique where molten metal is forced into a mold at high pressure

die swell distortion of an extruded polymer profile due to elastic recovery of the extrudate and uneven solidification

dielectric fluid liquid used in electrical discharge machining that has specific characteristics to facilitate ionization and flush away the material removed to maintain a stable cut, typically high-purity water or oil

diffusion coefficient parameter describing the rate of diffusion of one material through another and dependent on the pairing

diffusivity the ability of something to be diffused through another

digital readout positioning measurement system used on mills, lathes, and other equipment that displays in digital format and directly measures motion so that backlash is inconsequential

dipole an object with separate positive and negative charges, particularly of atoms or molecules

direct additive manufacturing creates a completed (full-strength) part in the machine in one step

direct current electrode negative (DCEN) polarity of a welding process where the electrode has a negative charge and attracts positive ions (lower heat at the electrode)

direct current electrode positive (DCEP) polarity of a welding process where the electrode has a positive charge and attracts electrons (higher heat at the electrode)

direct extrusions material is passed through an extrusion die using force in the same direction as the extrusion (also known as forward extrusions)

directional an atomic bond that is between only two atoms

discontinuous fibers a dispersed fiber phase in a composite where the fibers are shorter than the part dimensions and separated axially by matrix material

dispersed phase the phase of material that is distributed within the matrix phase of a composite for structural or functional enhancement

dissolution the chemical breakdown of a material by dissolving

distributed manufacturing the decentralization of production such that parts can be made at many locations globally, rather than constrained to a central facility

dividing head see indexing head

double bond an atomic bond where two electrons are involved

dovetail end mills end mill with a larger diameter at the base than at the top, used to create matching joints for assemblies

draft angular allowance on the side of a mold to more easily remove a part or pattern

drag the bottom half of a sand mold

drape forming use of vacuum pressure to shape a sheet of plastic around a mold with positive geometry (extends upward)

draw filing advancing a file over a surface in a direction perpendicular to its longitudinal axis (i.e., "sideways")

draw ratio the ratio between depth of draw and the smallest opening width after drawing

drawing out forging processes that decrease the cross section and increase the length of part

dressing the process of removing the outer layer to expose new grit on a grinding wheel

drill chuck a general-purpose device for centering and holding round objects, often drills

drill drift device for separating self-holding tapers

drill mill (end mill) a multipurpose cylindrical end mill with conical bottom, used to drill holes and mill workpiece profiles

drop forging forging by impact from large tooling either released or propelled down upon the component

dubbing process to flatten to the cutting edge of a drill for cutting materials such as brass and plastics

ductile (nodular) iron cast iron characterized by graphite in the form of nearly spherical "nodules"

ductile-to-brittle transition temperature temperature below which a metal rapidly becomes brittle

duplex stainless steels type of stainless steel comprising both body-centered and face-centered cubic phases

dye colorant that is fully dissolved in a liquid

edge joint joint where the faces of two components are held together and welded along their edges

edging forging operations where material in directed inward by the tooling, increasing the part cross section

elastic modulus a material property describing the amount of axial stress required to cause a given amount of strain

elasticity see elastic modulus

electrical discharge machining uses rapid pulses of electricity to create a spark between an electrode and the workpiece to vaporize a small amount of material

electroless plating plating process that deposits the metallic ions from solution by a chemical reaction with a reducing agent

electromigration the deleterious movement of atoms under high electric current that causes them move and create short circuits, particularly troublesome in nanoscale electronics

electron beam (e-beam) welding welding process using a high-energy electron beam to melt the material

electron shells the distinct layers of electrons that surround the nucleus of an atom

electronegativity the tendency for an atom to attract the electrons in a bond where a higher value indicates greater attraction

electropolishing electrochemical process which removes material preferentially from sharp edges and results in the creation of a smooth surface

embodied energy the amount of energy required to produce a material, component, or finished product as a result of each stage of production and transportation

embossing similar in operation to shallow drawing, a workpiece is formed into a shallow cavity, often for decorative purposes though it may add rigidity

end mills cutting tools designed to remove material with cutting edges on their periphery and end, can be used to cut laterally or vertically

endurance limit stress level below which a component can operate indefinitely without failing by fatigue

engineering ceramics ceramic materials that are carefully controlled for composition and/or structure, typically not sourced from nature

engineering plastics polymers with advanced properties used in select applications

entropy tendency of a system to transition to thermodynamic equilibrium, often by increasing the disorder to create a uniform mixture or temperature

epoxies polymers with epoxide groups on the ends of molecules, most often associated with two-part hardening adhesives

equilibrium a balanced state of lowest energy that is stable and will not change unless acted on

erosion corrosion corrosion process accelerated by the abrasive action of moving fluid and/or particulate

etching processes that remove material by chemical attack, in some cases also incorporating physical attrition (i.e., plasma etching)

eutectic composition the composition at which a phase change (e.g., melting point) in a mixture is minimized

evaporative pattern casting casting process where a molten metal directly burns out a pattern that acts as a placeholder

expanded polystyrene polystyrene processed to create a foamed structured

expansion reamers accommodate small adjustments in diameter to achieve the desired geometry

expendable mold casting casting processes which use a mold that is destroyed to remove the part (may be recyclable)

extenders polymer additives used to fill volume so as to reduce cost

extrudate the material/component produced during extrusion

extrusion processes which force a material through a shaped orifice to dictate the part cross section

extrusion blow molding blow molding process where a hollow tube of material is extruded into an open mold before the mold is closed and the polymer is expanded

face mills large end mills designed specifically to remove material from the top surface of a workpiece

face-centered cubic type of crystal structure where atoms reside in the center of each face of the cubic unit cell

facing process of removing material from the end of a part in the lathe

Fahrenheit temperature scale in which water freezes at 32°F and boils at 212°F

fatigue the weakening or failure of a component over many cycles of stress below its yield stress

feather edge condition where the edge of a component has been beveled to a sharp point

feed rate rate at which a cutting tool progresses through a workpiece and removes materials

ferrite body-centered cubic phase of ferrous alloys that is mostly iron and relatively soft and ductile, also known as alpha (α) iron

ferritic stainless steels stainless steel with high chromium content only, less costly than other grades

fiberglass see glass-reinforced plastic

Fick's first law quantitative description of diffusion where the rate of change of concentration is proportional to the position

filing processes that use a file with numerous, small teeth distributed across a surface to remove material

filler passes series of welding passes on a joint requiring multiple passes, used to fill the region between the parts

fillet weld weld bead created at the intersection of two perpendicular or otherwise intersecting surfaces

film polymer stock with thickness of 0.1 in or less

finish cuts light cuts used to produce the final dimension and good surface finish

finishing general term for operations that produce a functional and/or aesthetic quality that is not altered in later processing and which is desirable to the consumer

finishing dies tooling used in forging that creates the final features and dimensions in a part

first law of thermodynamics states that the total energy of an isolated system is constant

fixed mandrel device for creating hollow section in extrusion by causing material to flow around it

flame retardants polymer additives that decrease the flammability of a plastic, often by evolving gas such that a flame cannot be sustained

flanging any bending operation that is not performed along a straight line

flash leakage of material between mold pieces such that a thin edge protrudes at parting lines

flask a complete sand mold, consisting of a cope and drag

flexible foam porous polymer that is highly elastic, used extensively for seat cushions

floating tap holder specialized holder used with powered tapping and automated equipment that accommodates imperfections in the ratio of linear and rotational rates

flux (welding) chemical additive that reacts with contamination to increase purity in the weld and allow removal of the unwanted material

flux (diffusion) rate of flow of a substance through an area

flux core arc welding welding process that uses a consumable wire electrode with flux contained in the interior to protect the weld from contamination

foaming agent an element or compound added to create porosity within a material

forehand welding welding processes where the heat source is pointed in the same direction as the welding progression, particularly important in oxy-fuel welding

forging forming of a material using compressive force, where part features are dictated by the tooling geometry

formability property of a material that allows it to be reshaped without breaking

forming processes where physical deformation is used to shape a part in a volume-conservative manner

forward extrusions see direct extrusions

fracture toughness the energy required to extend a crack in a given material

free radicals reactive chemical species that break bonds in other molecules

freezing point depression the lowering of the freezing point by mixing ingredients

fretting corrosion accelerated corrosion caused by the abrasive action of particulate

friable the measure of how readily an abrasive particle will fracture under pressure

friction stir welding (FSW) use of a rotating tool applied in a joint to physically mix materials below their melting point

friction welding cohesively joining two component using friction to heat and mix the materials

frost zone area beyond the die in a blow mold where the polymer solidifies

fullering forging operations where material is directed outward from the tooling, decreasing the part cross section

full-mold process sand casting process where a template is packed directly in sand and burned out during pouring

functional fillers polymer additives used to fill volume and improve the properties

fuse welding see autogenous weld

fused-filament deposition type of additive manufacturing process where a polymer filament is heated and extruded from a nozzle to build layers

gage blocks precision blocks of known size, used for setup and measurement

gallium nitride semiconductor compound used in light-emitting diodes (LEDs) and other applications

galvanic corrosion corrosion process where different metals are in contact with each other and an electrolyte, results in the more active (anodic) metal corroding more rapidly and protecting the more passive (cathodic) metal

galvanizing process where steel is coated with zinc, most often by dipping the steel component in molten zinc, improves corrosion resistance

gang milling milling process where multiple tools of varying geometry are combined to create a more complex profile without changing setups, most common in horizontal milling

gas metal arc welding (GMAW) welding process that uses a consumable wire electrode and shielding gas to protect the weld

gas tungsten arc welding (GTAW) welding process that uses a nonconsumable tungsten electrode to melt the base material and filler rod

gas welding see oxy-fuel welding

gates features in a casting mold that control the flow of the material into the cavity

G-code a semi-standardized machine programming language that controls machine position, motion, and functions through the use of short commands consisting of letters and numbers

gel coat the top layer of matrix phase material used to protect the underlying reinforcement and enhance appearance

geometric die type of thread-cutting die that is closed before the threading operation and snaps opens at the end, can be reset to quickly thread components without having to reverse the machine or die

glass transition temperature temperature below which a polymer becomes brittle

glass-reinforced plastic a composite composed of glass fibers in a polymer matrix

glazing (finish) a decorative and impervious coating created by applying glaze and firing (heating) the component

glazing (abrasives) the condition where the particles become dull and less efficient at cutting

globular transfer arc welding filler transfer where the electrode is transferred to workpiece as large, irregular droplets

glue generic term for adhesives, originally referring to naturally derived products

gouging process where heat is applied to produce a trench in a workpiece

grade strength of the bonds between abrasive particles

graft copolymer copolymer with a single monomer backbone and secondary monomer chains branching off of that backbone

grain boundary sensitization phenomenon where the grain boundaries of a material are chemically changed and their resistance to corrosion is diminished

graphene a single layer of carbon atoms, or one atomic sheet of graphite

graphite carbon with regular order in sheets of atoms stacked on one another, used in pencils, lubricants and electrical conductors

graphitization use of very high temperature to convert disordered carbon to graphite

G-ratio see grinding ratio

gray iron cast iron characterized by graphite in the form of flakes

green compact an unsintered powder compact in powder metallurgy

green sand special casting sand with binders and additives to perform well when moist

green state see green compact

greenware an unfired ceramic component that contains a high moisture content

grinding ratio the ratio of material removal rate of the workpiece with respect to that of the wheel

grit number designation for grit size that uses the minimum mesh size (number of openings per square inch) through which the abrasive particles will pass, such that a large number corresponds to a small size

grit size size of abrasive particles

groove weld a weld created within a beveled region to allow deeper penetration and enhance strength of the joint

gum tape used to seal a vacuum bag for resin infiltration of a matrix material

gun drills straight, often single flute drills with integrated coolant passages used for drilling very deep holes

gutta-percha see trans-isoprene

hammer forging forging by repetitive impacts

hand reamers reamers that can be used by hand with appropriate handle

hand saws saw blade positioned and powder by hand

handedness description of the direction a screw must be turned to advance or retract it, when turned clockwise, right-handed bolts tighten and left-handed bolts loosen

hardenability the ease of achieving full hardness in an alloy through quenching where a lower minimum cooling rate indicates a higher hardenability

hardener a chemical additive that causes a polymer to harden by polymerization or cross-linking

hardness a material's resistance to indentation and abrasion

head (mill) the top-most portion of the mill, houses the motor spindle and other controls

headstock (lathe) the portion of the lathe where the spindle, gearing, and motor are contained

heat-affected zone extended volume of material near the welded joint where thermal effects from the welding process result in structural changes

heat and pressure welding processes using the direct contact of a heated tool with the plastic parts to melt and bond them together or which heats the components and then presses them together immediately after

heat treatment see thermal conditioning

helix angle measure of how steeply the cutting edge spirals around the outside diameter of an end mill

hemming bending operation where the edges of a part are folded over

heteropolymer see copolymer

Hevea rubber see cis-isoprene

hexagonal close packed (hcp) type of crystal structure where atoms form a hexagonal base and top of a unit cell with three coplanar atoms residing between

high-carbon steel alloy of iron and carbon with more than 0.6% carbon and up to 1% carbonhigh-density polyethylene (HDPE) polyethylene with no branching so that molecules pack together tightly

high-speed machining (HSM) subset of automated machining that uses very high rpm and feed rates to increase productivity

holemaking operations processes that create holes in a workpiece or modify existing holes

homologous temperature the temperature of a material divided by its melting temperature using an absolute temperature scale

homopolymer polymer composed of one type of monomer

honeycomb sandwich structure a sandwich structure with a low-density core with hexagonal cells reminiscent of a honeycomb

honing processes where a fine surface finish is achieved, often applied to the interior of cylindrical objects

hook tooth saw tooth with positive rake

Hooke's law physical principle that states that force and extension/compression of a spring are linearly proportional

hot extrusion extrusion performed above a metal's recrystallization temperature

hot hardness the hardness of material at elevated temperature

hot pass second welding pass on a joint requiring multiple passes

hot rolling rolling performed above a metal's recrystallization temperature

hot work forming processes performed above a metal's recrystallization temperature. defect accumulation does not occur and large amounts of strain can be accommodated

hot-chamber die casting die casting process where the injection hardware is constantly submerged in molten metal

hot-melt adhesives adhesives based on thermoplastic polymers to allow melting at modest temperature and applied between surfaces before cooling and solidifying

HSK taper specialized taper of high accuracy for high-speed spindles

hybrid composite a composite with two or more dispersed phases

hydraulic cement cement that hardens by reaction with moisture

hydraulic press equipment that uses fluid power to drive tooling and generate high forces

hydrogen damage embrittlement and other defects caused by the inclusion of hydrogen within a material

ignition temperature temperature at which a metal will begin to burn, important for thermal cutting processes

imperfections see defects

impression die forging forging operations where features are created in all three-dimensional by constraining material deformation

impurity the presence of a secondary, often unwanted, element

independent chuck workholding devices used on the lathe that have jaws that open and close separately, each having its own adjustment

indexable insert carbide insert that can be rotated to provide multiple cutting edges

indexing head a milling accessory that is mounted to a machine table and tilts and rotates to allows circular features or patterns to be created using precise relationships between the handle motion and the device gearing (aided by plates with holes of known spacing)

indirect additive manufacturing part created requires posttreatment to infill, sinter, or otherwise impart full strength

indirect extrusions material is passed through an extrusion die using force in the opposite direction of the extrusion (also known as backward extrusions)

infill the amount of material within a 3D printed component, typically expressed in percentage, and which may exist in three-dimensional geometries

ingot standard workpiece for forging operation with specified size and geometry, typically of very simple shape

initiator see catalyst

injection blow molding blow molding process where a parison is expanded by gas pressure to fill the mold

injection molding polymer molding process where molten plastic is forced into a mold at high pressure

inoculant an elemental addition to cast iron that enhances nucleation and results in the formation of graphite spheres rather than flakes

integrated hinge a hinge that is of the same material and continuous with the parts it connects

intellectual property (IP) legal protection of products covered by trademarks, patents, copyrights, and other designations of ownership over designs and ideas

interfacial bonding the type and quality of connection between phases at their boundary (e.g., between a fiber and the matrix phase)

intergranular corrosion preferential corrosion on grain boundaries caused by compositional differences

intermittent seam welding seam weld where the fused region is applied in discrete segments along the joint

interpolation method of estimating a value between other, known values

interstitial alloy secondary atoms exist between the primary atoms in normal crystal lattice positions

interstitial diffusion movement of atoms through a crystal lattice where diffusing atoms exchange positions between the regular lattice positions

investment casting expendable mold casting process that requires a sacrificial template to produce a ceramic shell for a mold

ionic bonds a primary bond where electrons are exchanged, such that one atom in the pair receives one or more electrons from the other

ions atoms with an imbalance between electrons and protons such that a positive or negative charge results

isotactic all side groups positioned on one side of a polymer molecule

isotopes atoms with an unequal number of protons and neutrons

isotropic material a material that behaves the same in all directions

jaws the clamping surfaces of a workholding device (e.g, vise or chuck)

jig saws a type of reciprocating saw where the tool rests on the work surface and the blade is perpendicular to the cutting direction, used for creating contoured cuts, common in woodworking

jobber length drill general-purpose drill length

Kelvin absolute temperature scale associated with the Celsius scale, water freezes at 273K and boils at 373K

kerf the width of cut from a saw blade or other tool (e.g., wire EDM), typically larger than tool body

killed steel steel deoxidized while in a molten state to reduce porosity upon solidification

kilns furnace used to thermally process ceramics

knurling process which creates a geometric pattern in the surface of workpiece, often used to enhance grip on a section a workpiece

Kolsky bar test see Split-Hopkinson bar test

laminate see laminate composite

laminate composite a category of composites where layers of varying structure and/or composition are combined

laminated object manufacturing applies a series of thin layers and cuts away the unnecessary material from each layer

lap joint joint created where two plate or sheet materials are overlapped and welded along the edge of one part and the face of the other

lapping process where extremely fine abrasive is used to create a mirror-like finish on a workpiece

laser beam welding welding process using a high-power laser beam to melt the material

laser cutting see laser machining

laser machining cutting processes using a high-power laser to melt the material

latex naturally occurring rubber compound

lathe dog device for holding a workpiece between centers in the lathe such that the workpiece is supported against the cutting force by transferring the force to a faceplate

lattice diffusion movement of atoms through a crystalline structure

lattice friction see Peierls stress

leaching removal of additives from a polymer over time by chemical attack, heat, or other diffusive processes

lead low melting point element of high density, atomic number 82

lead number number of individual thread sets on a screw adjacent to one another

Leidenfrost effect the phenomenon where rapid boiling forms a vapor barrier that insulates the surface, problematic in quenching operations

letterpress printing uses raised features with have ink on them which is transferred to the surface

line defect deviation from ideal position involving a row of atomic positions

linear low-density polyethylene (LLDPE) polyethylene with minor branching that prevents molecules from packing together tightly but not as much as low-density polyethylene

live center a device for holding the free end of a workpiece in the lathe that rotates with the part

live risers riser through which metal flows before reaching the mold cavity

living hinge see integrated hinge

load frame see universal tester

longitudinal cross section cross section corresponding to the longest axis of a workpiece

lost wax casting see investment casting

lost-foam process sand casting process similar to investment casting where polymer foam is coated in a ceramic shell and the foam is burned out during pouring, the ceramic shell is supported by sand

low-alloy steels steel with elements added in quantities between 1% and 4%

low-carbon steel alloy of iron and carbon with less than 0.3% carbon

low-density polyethylene (LDPE) polyethylene with branching that prevents molecules from packing together tightly

low-volume manufacturing production of one or a few parts, often for custom work or small quantity orders

machine screw a screw with standardized thread geometry used with a corresponding thread in a nut or component

machinist jack screw-driven device used to add support to overhanging workpieces, commonly applied to milling operations

magnesium alloys metallic mixtures where magnesium is the primary element

major diameter the diameter of a threaded component

malleable iron cast iron created by reheating and cooling white iron, which produces compact spheres of graphite, similar to ductile iron

martensite body-centered tetragonal phase of iron and carbon formed by rapid quenching and produces the highest hardness in steels

martensitic stainless steels stainless steel with high chromium and carbon content such that it can be hardened by thermal conditioning

masonry drill bit specialized drill with carbide tip for drilling through concrete, brick or other ceramics

match-plate pattern two halves of a pattern attached to a single, central plate

material removal rate (MRR) the volume of material removed as a function of time

matrix phase the phase of material which combines or encapsulates the dispersed phase in a composite

matter general term describing all material existing in physical space

mechanical assembly the joining of two components using physical interference or a separate member to secure the assembly together

mechanical fasteners devices for assembling components using force or physical interference

mechanical forming applying mechanical force to shape a polymer

mechanical tester see universal tester

mechanochemical polishing combines abrasive action with chemical attack to remove material and produce a smooth surface

medium-carbon steel alloy of iron and carbon with up to 0.6% carbon, and more than 0.3% carbon

medium-density fiberboard (MDF) composite engineered wood product made from wood fibers compressed and adhered together

medium-volume manufacturing modest part quantities often processed in large batches

melt index or melt flow index, a method that measures a polymer's flow rate through a small orifice at a given temperature

melt strength the mechanical stability of a heated polymer melt such that it will not break apart under tension

mer see monomer

metal inert gas (MIG) welding arc welding process that uses a continuous wire electrode and shielding gas

metal matrix composite (MMC) a composite where a metal is used to combine or encapsulate the dispersed phase

metal straighteners devices used to straighten curved stock (often that arrives coiled up) by applying compensating bends in opposite directions as it progresses through a set of rollers

metallic bonds a primary bond between metallic elements that results in a delocalized electron structure

metallic glasses see amorphous alloys

metallization process of coating nonmetallic materials with a layer of metal for aesthetic or functional purposes

metalloids materials with characteristics between those of metals and nonmetals, often referred to as semiconductors

metals materials composed of metallic elements or mixtures of elements characterized by metallic bonds

microstructure the microscopic features of material

milling processes machining processes using a rotating cutting tool to remove material from a workpiece using a combination of lateral and vertical motion

miscellaneous functions (M-codes) codes that control machine settings and features such as spindle rotation, coolant flow, tool changes, etc.

misruns the failure of material to reach all mold features before cooling, results in a partially formed component

miter an angle cut where the cut is perpendicular to the short axis of the workpiece (extends across the width)

miter gauge accessory used with circular saws and vertical band saws to hold the workpiece at an angle during cutting

miter saw sawing equipment purpose-built for cutting angles where a circular blade is mounted to an arm with a base that can be rotated to varying angles and which has linear travel parallel to the blade for cutting wide workpieces

mixed ceramics materials created from two or more ceramics to improve toughness

modulus of resilience the amount of energy required to deform a material up to is yield point, measured as the area under the elastic potion of the stress strain curve

modulus of rigidity a material property describing the amount of rotational force required to cause a given amount of strain

mold a template with three-dimensional geometry used to dictate features in a material

molding processes that use a three-dimensional cavity to create part features

molecular mass the collective mass of all atoms included in a molecule

molecular weight see molecular mass

molybdenum refractory metallic element, atomic number 42

moment arm the perpendicular distance between a force (or component of force) and the point of rotation or attachment along a member

monomer a single chemical unit in a polymer, composed of multiple elements in a group

morphology the form of a substance, such as appearance and geometric features

moving mandrel extrusion process for creating hollow extrusions where a central mandrel moves with the extruded metal

mulling the process of mixing casting sand to smooth out the consistency and distribute additives

nanocomposites composite material incorporating nanometer-sized phase(s)

nanofabrication general term for processes that create materials and devices with nanoscale features, typically applied to silicon processing

nanotechnology science and application of materials and devices with critical dimensions and features of 100 nm or less in at least one dimension

natural aging age hardening at room temperature, occurs over very long periods

negative draft mold sides that diverge as when moving away from the parting line (angle out, away from each other), locks a part into the mold

neoprene elastomer with good oil resistance, also known as chloroprene

network polymer an extensive collection of cross-linked polymer molecules

neutral axis the position within the cross section of a member that experiences no extension or compression under bending, location of zero stress

neutral flame a flame with a stoichiometric balance such that no unused fuel or oxygen remains

neutral rake angle rake angle where the cutting face is perpendicular to the cutting surface

niobium refractory metallic element, atomic number 41

nitriding process of adding nitrogen to a component by diffusion

nitrile butadiene rubber version of polybutadiene developed to overcome shortcomings when in contact with oils

noble metals characterized by a resistance to corrosion

nonferrous metals metals and alloys that do not have iron as the primary element

non-hydraulic cement cement that hardens by reaction with carbon dioxide

nonmetals materials composed of nonmetallic elements or mixtures of elements characterized by covalent or ionic bonds

nontraditional manufacturing methods of shaping materials that do not conform to the broadest classifications of manufacturing, such as casting, welding, machining, forming, etc., may use electrical, chemical, laser, or water energy to remove material

normalizing thermal conditioning process using high temperature and air cooling, results in a structure with modest strength and ductility

novolac resins phenolic polymer hardened by addition of a curing agent

number average molecular mass method of calculating the average molecular mass using total molecular mass divided by the number of molecules in the sample

nut an internally threaded component designed to be used with a bolt and compress components together between them

nylon see polyamides

offset boring head a device with adjustable tool position used in the mill to create holes with arbitrary diameter using a single-point cutting tool

offset lithography offset printing process where ink is applied to select locations on the roller by chemical compatibility rather than a raised pattern

offset printing transfers ink to a roller, which then transfers the image to the surface

olefins chemicals composed of only carbon and hydrogen, also known as hydrocarbons

open cell foams a porous material that has pores that are connected to each other and the surface to create a continuous pathway

open risers riser that is open to the exterior of the mold assembly (visible from the outside)

open-die forging forging operation where metal flow is not restricted laterally

orbital saws sawing equipment which moves a straight blade in an elliptical motion to aid the clearance of chips between cuts

organic adhesives adhesives based on naturally derived substances, or general term for adhesives with carbon-based chemistry

oriented strand board (OSB) composite engineered wood product made from chips of wood at varying angles to give it more uniform properties

overaged late stage of age hardening where the precipitates become excessively large and the strength begins to drop

oxidation a chemical bond being formed that results in the loss of an electron from the base material, often referring to bonding with oxygen but not limited to it

oxidizing flame a flame containing unreacted oxygen

oxy-fuel welding welding processes using oxygen combined with a fuel gas to drive combustion and produce a flame for heating the material

packing factor the occupied volume within a unit cell when representing atoms as hard spheres

parallels precision ground spacers with parallel surfaces used to elevate and space workpieces, particularly in the mill

parison a preform used in blow molding that provides the proper amount and distribution of material to fill the mold as well as being able to anchor the part during the gas injection process

partially stabilized zirconia zirconia that has been treated to retain the normally unstable tetragonal crystal structure at room temperature

particle board low-cost composite engineered wood product made from small chips of wood

particle reinforced composites composites with a dispersed strengthening phase that is non-fibrous and composed of discrete objects with comparable size in three dimensions

passive property of a metal which resists corrosion through oxidative processes, or that forms a protective oxide film preventing further corrosion

pearlite a phase consisting of a combination of ferrite and cementite oriented in fine layers (lamellae)

peck drilling drilling operations where multiple smaller cut depths are used and short retract motions between each cut act to "break" the chip and help remove waste material

peel ply a layer of plastic used in the processing of plastic that releases from the matrix phase after curing and is used to prevent the bleeder cloth from adhering to the composite

peening impacting a surface to impart a compressive strain state that improves resistance to wear and fatigue

Peierls stress the force that must be overcome to cause dislocation motion, and therefore atoms to slip and produce permanent deformation in a material

percent elongation measurement of the change in length of a material by deformation without breaking, indicative of ductility

percent ionic character the tendency of an atomic bond to behave as an ionic bond, and may vary from 0% (covalent only) to 100% (ionic only), based on difference in electronegativity

percent reduction in area measurement of the change in area of a material by deformation without breaking, indicative of ductility

permeability ability to transfer liquid or gas through a material, often polymers

phase boundaries on a phase diagram, lines that separate distinct phases and indicate a phase change when crossed

phase diagrams composition and temperature dependent maps of structure and atomic solubility for mixtures of elements

phase segregation the separation of elements that do not mix completely, results in distinct regions rich in one element or another

phases a state of material with unique structure (e.g., solid, liquid, gas, crystalline, fibrous, etc.) and composition (e.g., pure, alloyed, etc.)

phenolics phenol formaldehyde resins with high hardness

photolithography process used in nanomanufacturing where areas of a polymer layer are exposed to light to create patterns for latter processes

photopolymer a resin that polymerizes and solidifies after exposure to light, often ultraviolet

physical blowing agents chemical additives that vaporize at elevated temperature to create pores

physical interlock the assembly of two components by intersection of those components

piercing (forging) process where a hole is created by displacing material

piercing (stamping) punching/stamping operation where one or more features are cut from the interior of a larger workpiece

piezoelectric effect phenomenon where electrical charge is generated in a material by mechanical stress and vice versa

pigments colorant that is suspended in a liquid, not dissolved

pilot drill a drill of smaller diameter used to create a locating hole to which another, larger drill is then applied

pilot hole a hole smaller diameter used to locate and reduce the work needed by a larger drill

pitting corrosion type of corrosion created by a loss of a passive protection in a small area, such that the specific area corrodes rapidly and creates a pit in the surface

plain carbon steel an alloy with primary ingredients of iron and carbon, other alloying additions are minimal

planar defect deviation from ideal position involving a plane of atomic positions

plasma cutting thermal cutting process using gas heated to create an extremely hot plasma

plasticized PVC PVC with additives to reduce the glass transition temperature and increase flexibility

plasticizers polymer additives that decrease the glass transition temperature of a plastic and thereby increase the flexibility at a given temperature

plating processes that apply a thin layer of metal to component, most often by electrochemical reactions

plug a physical feature that assists vacuum or drape forming by forcing the polymer sheet to more closely match the mold features before negative or positive pressure is applied

plug tap tap with intermediate length of cutting surfaces of partial thread depth, balances ease of starting with ability to thread blind holes to an acceptable depth

plug weld weld bead created using a hole in one part to provide a connection to a surface below

plywood composite engineered wood product made from layers of wood with alternating orientation

point defect deviation from ideal crystal lattice involving a single atomic position

polar moment of inertia the ability of a section to resist torsion as a function of its cross-sectional shape and size

polyacrylonitrile polymer used for fibrous filters, textiles, and as a precursor to carbon fibers

polyamides polymers with an amide group in the backbone, also called nylon

polybutadiene polymer similar to polyisoprene but without the methyl side group and lower strength

polybutylene terephthalate polymer with structure similar to polyester terephthalate, but with two ethylene groups (a butylene group) in the backbone

polycarbonate amorphous polymer with very high toughness, used in safety glasses and other impact resistant products

polycrystalline composed of multiple crystal orientation, or grains

polycrystalline diamond diamond material made from many small, single-crystal diamond particles

polyesters polymer with an ester group in the polymer backbone

polyethylene commodity polymer with only carbon and hydrogen atoms

polyethylene terephthalate common polyester material used in beverage bottles and many other container applications

polyisoprenes polymer structure of natural and synthetic rubbers

PolyJet additive manufacturing process that uses photopolymerization to create three-dimensional objects by depositing and quickly curing a resin using a print head that may contain one or many nozzles

polylactic acid thermoset polyester used in 3D printing and biological applications

polymer a large molecule (macromolecule) composed of simpler subunits repeated over the length of the molecule

polymer additive a chemical added to a plastic that is not part of the polymer molecules

polymer gas welding polymer welding process using heated gas, typically air, to melt the material

polymer matrix composite (PMC) a composite where a polymer is used to combine or encapsulate the dispersed phase

polymerization chemical processes by which monomers are bonded together in the formation of a polymer molecule

polymethylmethacrylate (PMMA) amorphous polymer often used for applications where clarity is needed

polymorphism see allotropy

polyoperative cutting processes cutting operations that use numerous cutting surfaces to remove material, often tens or hundreds

polypropylene commodity polymer with only carbon and hydrogen atoms and a methyl side group

polysiloxanes non-carbon-based polymer with a silicon-oxygen molecular backbone

polystyrene commodity plastic with a phenyl side group (molecular ring)

polytetrafluoroethylene chemical-resistant, low friction polymer used for nonstick coatings and other applications

polyurethanes versatile category of polymer consisting of a urethane group in the molecules and which may be made to be highly flexible or rigid

polyvinyl chloride commodity polymer with one hydrogen atom in the monomer exchanged for a chlorine atom

porous carbon-carbon composites carbon-carbon composite with voids throughout to improve the thermal insulation properties

Portland cement see cement (ceramic)

positive draft mold sides that converge as the move away from the parting line (angle in, toward each other), allows part to be extracted from the mold

post-forming see sizing (extrusions)

powder bed printing uses powdered material and fuses it together using heat or a binder (glue) that is applied at precise locations across the surface

powder coating a dry finishing process where fine particles of a polymer resin, often mixed with other ingredients, are applied to a surface using electrostatic charge and then cured in a separate step using heat

powder metallurgy processes where metal powder is pressed into shapes and sintered to create components

precious metals see noble metals

precipitation hardening see age hardening

precipitation-hardening stainless steels stainless steel that can be strengthened through precipitation hardening

precision the ability to produce the same result repeatedly over time

preparatory codes (G-code) codes beginning with the letter "G" that control the machine motion and cycle type

press brake equipment designed to hold tooling used in the bending and forming of materials

press forging forging by the slow, continuous application of force

pressure forming same concept as vacuum and drape forming, but uses positive pressure to form a polymer sheet into or around a mold

pressure knurling knurling process where the surface is deformed to create the desired pattern

pressure-sensitive adhesives adhesive that creates a strong attraction by being pressed against one or more surfaces and is often supported on another material or backing

primary bonds strong bonds that provide the principal attraction between atoms

profile extrusion see extrusion

profile rolling rolling processes where a complex cross section is created, often by using multiple stages of rollers to shape the material

progressive die stamping tooling with multiple stations such that a strip of material is progressively punched, bent and otherwise processed over multiple strokes of the press brake

projection welding resistance welding where the points of contact between two or more components are fused together

proportional limit the point at which the stress and strain are no longer linearly proportional

pull saws hand saws that cut on a pull stoke, toward the user

pull-type broach broach that cuts while force is applied from the end of minimum cut cross section (pulled through the hole)

pulsed transfer specially controlled arc welding filler transfer where the electrode is transferred to workpiece using rapid pulses of electrical energy instead of a constant current

punching operation process where a geometric feature is removed from the cross section of a workpiece by application of a shaped tool that shears the feature(s) out

pure aluminum elemental aluminum with no significant alloying addition

pure copper elemental copper with no significant alloying addition

pure magnesium elemental magnesium with no significant alloying addition

pure titanium elemental titanium with no significant alloying addition

push saws saws that cut on a forward stroke, away from the user

push-type broach broach the cuts while force is applied from the end of maximum cut cross section

pyrometric cone a device for indicating temperature visually by softening at a corresponding, known temperature

quartz highly pure form of silica

quench medium see quenchant

quenchant the substance that removes heat from the component being quenched

quill (mill) feature on a mill that allows the tool to be quickly lowered by hand, often used for drilling operations

radial arm saw sawing equipment that uses a circular blade guided along an overhanging support arm, commonly used for cross cuts in woodworking, but is capable of other operations

radiation damage defects introduced in a material by radiative sources, such as light and gamma rays

radio-frequency (RF) welding use of very high frequency (MHz) vibration to heat and weld components together

raft base layer used in additive manufacturing to anchor the printed part during the build

rake angle angle of the cutting face relative to the workpiece, determines cutting force and chip flow

ram (EDM) see sinker EDM

ram (mill) adjustable horizontal component of a mill that supports the head

ramming the process of tightly packing sand around a pattern

random copolymer copolymer with alternating monomer blocks that occur with no repeated pattern

random cutting edge layout cutting tools without specific placement or orientation of cutting edges, common in abrasives

Rankine absolute temperature scale associated with the Fahrenheit scale, water freezes at 491°R and boils at 671°R

reaction injection molding injection molding process where two or more chemicals are injected into the mold and react with one another to fill the mold

reamers specialized cutting tools which remove a small amount of material from the interior of a hole in order to create a precise diameter with good surface finish

rebar reinforcement bar made of steel, used to increase the crack resistance of concrete and cement under tension

reciprocal saw see reciprocating saws

reciprocating saws sawing equipment that moves a straight blade in a cyclic linear motion

recirculating ball screw type of screw that uses ball bearings in semicircular grooves to allow for a tighter fit, less backlash, and lower friction

recrystallization the conversion of elongated, highly strained grains into smaller strain-free grains with increasing temperature

recrystallization temperature the temperature where recrystallization occurs and above which a metal does not accumulate stress and strain in its microstructure during forming operations

refractory metals metals with characteristically high melting points, typically 2000°C (~3600°F) or above

regulating wheel wheel positioned opposite of the grinding wheel in centerless grinding that supports, regulates the rotational motion of, and advances the workpiece during abrasive machining

reinforced reaction injection molding reaction injection molding process where a secondary material/component is placed in the mold before injection and enclosed after

reinforcement phase the phase of material in a composite used to provide strength or enhance other mechanical properties

relief angles geometric features on a cutting tool that prevent the tool flank from rubbing against the workpiece

relief cuts intermittent saw cuts made in a waste section of material to decrease blade binding in areas of tight curvature

resistance welding welding two piece together by passing a large electrical current through them such that they melt and fuse at that point

resole resins phenolic polymer hardened by heating

retained austenite austenite phase remaining after the formation of martensite with less than 100% conversion

reverse polarity see direct current electrode positive (DCEP)

rhenium refractory metallic element, atomic number 75

riddling the process of sifting casting sand through a mesh screen to achieve a fine consistency

rigid foam porous polymer which is not highly flexible

rigid PVC PVC with a high glass transition temperature and relatively low flexibility

rigid tapping tapping operation that requires precise coordination between rotation and linear progression to match the thread pitch

rimmed steel steel that has little to no deoxidation, results in an outer "rim" of highly pure steel with alloying elements segregated to the center

ring forging forging operation specifically designed to create rings by incrementally rotating and forging the material

ring rolling forging operation specifically designed to create rings by continuously rotating and forming the material

rip cuts saw cuts parallel to wood grain

riser features in a mold that act as a reservoir for additional material to be provided to the cavity during solidification of the material

rivet fastening devices characterized by having two ends of greater diameter than a central shaft, or stem, with the purpose of holding components together between them

robot machine that can be programmed to move and perform tasks with minimal human interaction

roll bending process using opposing rollers to cause a component to bend as it progresses through them

roll forging forging process where shaped rolls are used to create impressions along the length of a part

roll spot welding version of intermittent seam welding where the components are advanced through rollers that also act to weld the parts together

rolling forming processes where compressive force is applied by rotating cylinders that also advance the material forward

rolling stands equipment used to support and control the rolls in a rolling process

root diameter minimum diameter of a threaded component, measured from the bottom of the threads

root face flat region remaining on the edge of a beveled part at the base of the region to be welded

root pass first welding pass on a joint requiring multiple passes

rotary broaching broaching that use rotational motion in addition to linear progress through the hole, used to create helical features

rotary swaging process where a multi-section die is repeatedly opened and closed while simultaneously rotating around the outside of a part

rotary table a milling accessory that is mounted to a machine table and rotates to allows circular features or patterns to be created more efficiently

rotational molding polymer molding process where the polymer is heated to become molten and the mold is rotated in multiple directions to cost the interior of the mold

roughing cuts heavy cuts resulting in quick material removal with lower accuracy and rougher surface finish

roughing end mill end mill with scalloped cutting edges that reduces contact force for heavy cutting, also results in poor surface quality compared to regular end mills

rubber toughening addition of an elastomeric compound to a polymer to increase toughness

rule of mixtures rule of thumb whereby the properties of a mixture or composite are proportional to the weighted average of the individual components

runners features in a mold along which material is directed to transfer it from the sprue(s) to the mold cavity

sand blasting process where abrasive grit is propelled toward a workpiece surface using compressed air

sand casting casting technique where sand is used to create the mold

sand core an expendable sand template used to create internal features during sand casting

sanding general term for abrasive machining process used to smooth a surface or remove material using abrasives supported on a flexible backing

sandwich panel see sandwich structure

sandwich structure a composite composed of face sheets with different properties than the core, often has low density and high stiffness

SCARA robot selective compliance assembly robot arm (SCARA) with four axes of motion, primarily in one plane, typical in assembly operations

Schmid factor see critical resolved shear stress (CRSS)

scission breaking of atomic bonds by reactive chemicals

scissor-type knurling tool knurling tool that has form rollers forced against opposing sides of the workpiece to reduce deflection

screw generic term for fastening devices using an externally threaded section to hold parts together

screw machine length drill short drill length used in confined spaces, such as screw machines

screw machines a mechanically actuated lathe with the ability to time and control multiple tools simultaneously for increased productivity

scroll saws reciprocating saw that uses a very fine blade and is capable of creating detailed contours, common in woodworking

seam weld the use of resistance welding to connect two components along the length of a joint

second law of thermodynamics states that the entropy of a system will increase as thermodynamic equilibrium is approached

selective laser sintering uses a high-power laser to partially melt and fuse powder in precise locations across the build platform

self-driving center device for holding a workpiece between centers in the lathe such that the workpiece is supported against the cutting force using pressure, a center that digs into the workpiece

self-holding taper a low-angle taper that holds the components together tightly by friction

self-interstitial an extra atom exists between the atoms in normal crystal lattice positions

self-opening die see geometric die

self-releasing taper a high-angle taper that acts to center the components but does not hold them together by friction

semi-centrifugal casting casting process where a mold is rotated around central axis and results in a solid part

semi-killed steel partially deoxidized steel with higher yield than killed steel

sequence of abrasives the proper progression of abrasive grit to achieve the smoothest surface with the least time by sequentially stepping down in grit size (or up in grit number)

shallow drawing forming operation where metal is stretched into a cavity that has a depth less than the diameter or smallest cross-section distance

shape rolling see profile rolling

shear force caused by opposing applied loads that result in stress and deformation parallel to those loads

shear angle the angle at which a material deforms by shear during a cutting operation ahead of the cutting edge

shear modulus see modulus of rigidity

shear strain strain caused by shear stress, results in angular displacement rather than axial

shear stress stress on a cross section such that adjacent planes will move in opposing directions if sufficient force is applied

shearing a process by which a material is separated using closely offset forces in opposite directions

sheet polymer stock with thickness greater than 0.1 in

sheet metal metal stock that has a thickness much less than its area, typically below 1/8 in

sheet molding compounds (SMC) polymer molding resin mixed with extensive fillers and reinforcement, with shorter chopped fibers than molding compounds and used in thinner sections

shell mills large end mills that can be attached to a shaft, allows exchange of the cutting tool with a single tool holder

shell reamers an interchangeable reamer where multiple cutters can be used with a single holder

shell thickness thickness of the exterior layer of a 3D printed object

shielded metal arc welding (SMAW) welding process with a consumable electrode of fixed length that is coated in flux to protect the weld from contamination

short shots see misruns

short-circuit transfer arc welding filler material repeatedly contacts the workpiece and melts by resistive heating

shot particulate, short segments of wire, or other media used to peen the surface or part

shrinkage reduction in volume as a molded component solidifies and/or cools

SiAlON variation on silicon nitride ceramic cutting tool mixed with other element (i.e., silicon-aluminum-oxynitride)

side group molecule extending from the central backbone of the polymer or other molecule

side risers riser connected to a runner, located adjacent to the mold cavity

silica silicon dioxide, used for glass and other applications

silicates form of silicon dioxide where other elements, such as hydrogen, sodium, etc., make up a significant portion of the material and notably change the properties

silicon carbide common ceramic used in abrasives

silicon nitride a lightweight ceramic often used in small structural components and silicon processing as a protective insulating layer

silicone see polysiloxanes

sine bar a general-purpose setup device having a known distance between two points and used in conjunction with gage blocks to precisely setup angles

single bond an atomic bond where one electron is involved

single crystal diamond diamond with a single orientation of its crystal lattice, used in highly specialized tooling applications

sinker EDM uses a three-dimensionally featured electrode to transfer that geometry to a surface by progressively lowering it

sintered metals see powder metallurgy

sintering process where particles are fused together by diffusion at elevated temperature but below their melting point

sizing (of fibers) process and coating to protect fibers and improve bonding with the matrix

sizing (extrusion) the modification of an already extruded section to precisely control geometry

skilled labor workers who are trained in and have particular knowledge related to their tasks, and who can work independently.

slip a mixture of clays and other materials suspended in water to allow those materials to be poured into molds for ceramic components

slip casting casting process for ceramics where slip is poured into a mold, allowed to harden and form a shell, and the remaining liquid is poured out forming a hollow part

slip directions the direction of motion of a two-dimensional sheet of atoms during deformation

slip planes a two-dimensional sheet of atoms with a specific orientation along which slip may occur

slip rolls see roll bending

slip system the combination of slip plane and slip direction that describes the deformation motion within a crystal lattice

slitting saws circular saw blades designed to be used on an arbor with a milling machine, primarily used for separating parts or for very thin grooves

slot weld weld bead created using a slot in one part to provide a connecting surface to a surface below

slotting saws circular saw blades designed to be used on an arbor with a milling machine, primarily used for creating grooves

small hole EDM uses an electrode with a small diameter to effectively "drill" a hole through a workpiece

smelting high-temperature process by which metal ore is refined to produce pure metal

S-N curve plot of the stress versus number cycles that shows the maximum number of cycles a material can withstand at a given stress level

sodium chloride structure unit cell typified by the structure of NaCl, where two elements create a face-centered cubic-type structure

soft jaws jaws that are made from a material that can be machined to accommodate a specific workpiece

soldering a method of assembling components where a metallic filler material (typically a wire) is melted and applied to the surfaces at a temperature below 840°F (450°C) or below

solid pattern a single-piece pattern used to create a mold cavity

solid solution strengthening the increase in strength caused by alloying where atomic size differences inhibit dislocation motion

solutionizing process by which elements in an alloy are distributed evenly in solution at elevated temperature and quenched to prevent their separation at room temperature

solutionizing temperature the temperature above which the elements in an alloy are fully soluble in one another or exhibit significantly increased solubility

solvent welding welding processes using the dissolution of materials from two components to allow mixing, typically by applying a solvent to the surfaces to be joined

spark gap space between electrodes in electrical discharge machining

spatter the ejection of small droplets of material from a weld resulting in specks of solidified metal around the weld

specific stiffness the elastic modulus of a material divided by its density

specific strength the strength of a material divided by its density (strength-to-weight ratio)

spider mandrel see fixed mandrel

spiral flute taps tap with flutes that aid in the removal of chips

spiral point tap tap with end geometry that pushed chips forward

split collet see spring collet

split pattern a pattern for creating a mold cavity that can be disassembled for ease of use

split point drills drill tip geometry where the land of the drill has been modified to allow cutting to extend to the center and reduce the need for predrilling

Split-Hopkinson bar test high strain rate test where an incident bar impacts a sample against a transmitted bar to measure its response

spot drill short, rigid drill used to create a hole for a later drill to follow

spot facing creation of a concentric flat-bottom hole of larger diameter at the surface of a workpiece with minimal depth so as to create a flat surface

spot welding resistance welding applied to a small area

spray transfer arc welding filler transfer where the electrode is transferred to workpiece as a constant stream of small droplets

spring collet workholding device used in lathe and other applications where the body of the holder is compressed inward to hold a part by forcing it against a matching taper

sprues mold features where liquid/molten material is introduced into a mold

square end mills end mill with a cylindrical profile

stabilizers polymer additives used to reduce degradation, such as that caused by UV-exposure, heat or other sources

stainless steels steel with a chromium concentration of 10.5% or more, characterized by corrosion resistance

stamping general term describing operations including bending and punching

standard taper a commonly recognized taper angle matching tooling or equipment of the same taper

statics study of mechanics that describes the forces on a structure in equilibrium with the assumptions that the structure is perfectly rigid and motionless

steady-state diffusion diffusion of a substance occurring at a constant rate

step clamp sets adjustable workholding system with interchangeable parts allowing diverse workpiece geometries to be clamped to the machine table or other suitable surface

step-growth polymerization see condensation polymerization

stereolithography additive manufacturing processes that uses light to polymerize a liquid resin, which makes the exposed area solid

stick welding see shielded metal arc welding (SMAW)

stiffness see elastic modulus

stitch welds use of short, intermittent weld beads to hold parts together, saves time, energy and material for low-stress applications

stoichiometric (materials) composition of a real compound is exactly represented by the chemical formula (e.g., Al_2O_3)

stoichiometric (reaction) of a chemical reaction, having an ideal balance of reactants such that there is no deficit or excess in the products formed

straddle-type knurling tool see scissor-type knurling tool

straight cut cut that is perpendicular or parallel to the workpiece surfaces (no angle other than 0 or 90°)

straight flute taps general-purpose tap design with flutes that are parallel to the length of the tap

straight polarity see direct current electrode negative (DCEN)

straight saw blade saw blade of fixed length and limited stroke length

straight shank drill shank that is cylindrical

strain the change in length, width or other dimension, divided by the original dimension

strain hardening physical phenomenon by which a material becomes stronger and harder after deformation

strain rate sensitivity the tendency of a material to behave in a more brittle fashion when stress is applied at a high rate

strength of materials study of mechanics that describes the forces on a structure in equilibrium and determines the states of stress and strain (deformation) within the members

stress the force per area on a section

stress corrosion cracking formation and propagation of cracks in highly stressed regions by accelerated corrosion

stress relaxation a drop in stress on a component, such as a bolt, over a long period of time as it stretches

stringer weld bead created using a straight line motion

stroke length the maximum length of cutting motion in a single pass of saw blade

structure spacing between abrasive particles

substitutional alloy secondary atoms replace the primary atoms in normal crystal lattice positions

subtractive manufacturing manufacturing processes which remove material to create a desired geometry

subzero cooling process where a component is cooled below room temperature to fully convert retained austenite and improve stability of quenched components

superplastic ability for a material to undergo extensive permanent deformation without breaking, often by a factor of 2 (i.e., 200%) or more

support (additive manufacturing) additional material that acts as a scaffold to prevent sagging or failure of overhanging areas during the printing process and removed after completion

surface broaching process where features are cut into the surface of a workpiece using a broach

surface grinding process where a part is abrasively machined on one or more flat surfaces

surface speed the velocity at which the cutting edge and workpiece pass one another, often as the result of rotation of one or the other

surface weld a weld created on the surface of two components where no recessed region exists, suitable for very thin parts

swelling detrimental increase in size due to the absorption of liquids or gases

syndiotactic side groups of a polymer molecule located in alternate positions with a regular, repeated pattern

synthetic adhesives adhesives created through precisely controlled chemical processes

synthetic polyisoprenes polyisoprene not derived from natural sources and with better control of properties

table (mill) the work surface of a milling machine where parts and workholding are mounted during milling

table saw sawing equipment with a circular blade mounted to a flat surface such that workpieces are passed through the protruding blade, common in woodworking

tack welds small points of weld used to temporarily hold parts together or assemble parts which minimal structural demands

tacticity the order or arrangement of side groups in polymers

tailstock a moveable device on the lathe, oriented coaxial to the spindle, that is used to support long workpieces and can be used to drill holes in the end of a workpiece

tantalum refractory metallic element, atomic number 73

tap drill a drill of proper size to accommodate the threads of a standard tap size

tap handle attachment used to drive a tap by hand

taper an angled surface, typically describing a circular component with increasing/decreasing diameter

taper sleeve an adapter that allows for tooling and equipment with different tapers to be used together

taper tap tap with an extended length of cutting surfaces of partial thread depth, allows easier starting for threading, especially when tapping by hand

tapered end mill end mill with gradual taper, used to create molds with specified draft or other need for low taper angles

tapping creation of threads on the inside of a hole using a solid-body cutting tool

temper process of reheating a quench-hardened component to reduce hardness and increase toughness

temper designation indication of the thermal condition procedure used to process the given alloy

tempered glass glass that has been treated to impart a compression stress at the surface which suppresses crack formation

tension force applied in a manner that extends a member axially

texture the alignment of microstructural features and resulting anisotropy

thermal conditioning the application of heating and cooling cycles to modify the mechanical properties of a material

thermal conductivity the ability for a material to pass heat through its structure

thermal degradation the breakdown of material at high temperature by a change in composition or structure

thermal spray coating uses intense heat to melt a material and gas pressure to propel the molten droplets toward a surface where they rapidly solidify

thermochromism the phenomenon where a material changes color with temperature

thermodynamics discipline concerning the amount and transfer of energy in systems (i.e., how heat and other forms of energy cause changes)

thermoforming polymer forming processes that use heat to improve their formability

thermoplastic polymers polymer that decreases in strength at elevated temperature to allow molding and forming

thermoset polyesters polyester polymers with double bonds in the molecules that allow cross-linking to occur

thermoset polymers polymer that increases in strength at elevated temperature due to increased cross-linking

third law of thermodynamics states that the entropy of a system will approach a constant, minimum value as the temperature approaches absolute zero

thread angle included angle of a single thread (e.g., 60°)

thread cutting die a tool for cutting threads on the exterior of component using cutting edges positioned around the circumference that have the desired thread geometry

thread pitch the distance between adjacent threads at corresponding points, often the peak or trough

time-temperature-transformation (TTT) diagram thermal conditioning plot indicating the phase(s) formed at varying rates of cooling

tin common alloying element and low melting point element, atomic number 50

tin pest disintegration of tin during phase transformation at low temperature, especially in pure tin

titanium alloys metallic mixtures where titanium is the primary element

titanium aluminum nitride (TiAlN) hard compound of titanium, aluminum and nitrogen where titanium is present in a higher quantity than aluminum, often used as a tool coating

titanium carbide hard ceramic often used in cutting tools

titanium carbo-nitride (TiCN) compound of titanium, carbon and nitrogen, harder and more wear resistant than titanium nitride, often used as a tool coating

titanium dioxide ceramic commonly used as a whitening additive among other uses

titanium nitride (TiN) hard compound of titanium and nitrogen, often used as a tool coating

T-joint joint where two parts are held perpendicular to one another such the edge of one is held against the face of another

tool coating a thin coating applied to a cutting tool to reduce friction, wear, or improve other properties

tool steel steel with high carbon and/or alloy content, primarily used for its strength and hardness

toolpaths the calculated tool trajectories for removing material in an automated machining process

tooth set differential tooth offset on a saw blade where each successive tooth has a varying angle or direction of offset

top risers riser connected directly to the mold cavity and located above

top-down nanomanufacturing processes for creating nanoscale devices by iterative deposition and removal of material

torque applied rotational force

torsion stress in a cross section caused by rotational force (i.e., torque)

torsional shear stress rotational stress on a cross section such that adjacent planes will move in opposing direction if sufficient force is applied

toughness a material's resistance to fracture, measured by the area under the stress strain diagram

tow size the number of fibers in a single strand of woven reinforcement fabric, typically expressed in thousands of fibers (i.e., 4k is 4000 fibers per strand)

trace elements small quantities of unwanted impurities with a maximum concentration but no minimum

tramming process of ensuring that the spindle is perpendicular to the table on the mill by adjusting the tilt in two axes

transformation toughening the ability for a material to undergo structural transformation that increases its resistance to crack propagation, exemplified by zirconia

trans-isoprene polyisoprene where side groups are on varying sides of the polymer molecules making the polymer more rigid

transverse cross section cross section corresponding to short axis, or across the width

trimming finishing operation to remove flash from the edges of parts

triple bond an atomic bond where three electrons are involved

triple-chip grind (TCG) saw tooth design including wider teeth with greater strength and resistance to chipping

truing the process of flattening or reshaping a grinding wheel to restore the original geometry

T-slot end mills end mill with a short cutting section of greater diameter than the shank, used to create an inverted "T" profile common for creating features in equipment used for mounting workholding or other workpieces

tungsten refractory metallic element, atomic number 74

tungsten carbide hard ceramic often used in cutting tools

tungsten inert gas (TIG) see gas tungsten arc welding (GTAW)

turning processes which rotate a workpiece and remove material from the diameter to create a circular cross section

turret lathes type of manual lathe with the ability to mount multiple tools and quickly change between them for increased productivity

ultimate tensile strength (UTS) the point of highest stress on an engineering stress-strain curve, transition point from uniform to non-uniform plastic deformation

ultrahigh molecular weight polyethylene (UHMWPE) polyethylene with exceptionally long, unbranched molecules

ultrasonic welding use of high frequency (kHz) vibration to heat and weld components together

uniaxial pressing compaction of parts using force applied along one axis, with the cross section being contained by a die

unified numbering system (UNS) standardized system for designating metals and alloys

universal chucks workholding devices used on the lathe that have jaws which open and close simultaneously

universal tester a machine capable of performing a diverse range of tests using a moving cross-head to apply force upward or downward, depending on the type of test

upsetting forging processes that increase the cross section and decrease the length of part

vacancy a normal crystal lattice position without an atom present

vacancy diffusion movement of atoms through a crystal lattice where diffusing atoms and vacancies exchange positions corresponding to the regular lattice positions

vacuum forming use of vacuum pressure to shape a sheet of plastic into a mold cavity (negative geometry)

valance shell the outermost electron layer of an atom

van der Waals force attractive force caused by the collective action of fluctuating atomic or molecular dipoles

variable geometry cutting tool where tooth geometry and layout are varied, common in saw blades

V-blocks workholding accessory with a V-shaped groove in one or more side that assists in hold round parts along their diameters

vermicular worm-like, often in reference to graphite inclusions in compacted graphite iron

viscoelasticity the delayed strain response to an applied load by energy absorption

vises general-purpose workholding devices that apply pressure from opposing surfaces (jaws), most often flat and parallel

vitrification process where ingredients within a ceramic are fused together permanently at high temperature

vulcanization process used to cross-link natural rubber to increase the strength by heating in the presence of sulfur

walking drill the flexing and unwanted movement of drill across a surface when no provision has been made for the drill to start, more common as drill size increases

washers devices used to spread out force applied by screws and bolts, and that may also act to prevent loosening of those fasteners and other functional purposes

water jet machining uses a narrow, high-velocity stream of water to erode materials and create a profile

weathering the collective damaging effects of exposure to the outdoor environment

webbing wrinkles or folds in a polymer sheet that are created during forming, especially at sharp corners or large differences in height

weight average molecular mass method of calculating the average molecular mass using the mass fraction of each molecule divided by the total mass of molecules of that size

weld nugget the volume of material fused together in a spot weld

weld pool the molten region during welding

weld puddle see weld pool

weldability measure of the ease with which welded assemblies can be created using a particular metal or alloy

welding processes that use heat, pressure and/or chemical dissolution to result in cohesive assembly of parts

whisker-reinforced ceramics ceramic matrix composite with micron to nanoscale ceramic fiber reinforcement that increase crack resistance, often used in cutting tools

white iron cast iron with high proportion of cementite and is very hard and brittle, typically a precursor to malleable iron

wire EDM uses a thin wire electrode to remove material as it progresses through the part

wood-plastic composites composite engineered wood product with a high percentage of thermoplastic polymer, often produced to resemble lumber

words (G-code) a single command in a line of G-code, typically a letter and numbers, used to program CNC equipment

work envelope the size limit of a machine such that parts the exceed those dimensions cannot be fully processed

work hardening see strain hardening

wrought iron a form of iron with no significant alloying additions, most suitable for forming operations

yield point see yield stress

yield stress the amount of stress on a material such that the response changes from elastic to plastic deformation

Young's modulus see elastic modulus

zero draft mold sides that are perpendicular to the parting line and parallel to one another

zero rake angle see neutral rake angle

zeroth law of thermodynamics states that when two, separate objects are at equilibrium with a third object, they are at equilibrium with each other

zinc common alloying element and low melting point element, atomic number 30

zinc oxide ceramic used as a pigment and processing additive

zincblende unit cell structure of two interpenetrating face-centered cubic cell structures of different elements, exemplified by beta-type ZnS

zirconia zirconium oxide, a versatile ceramic with varying crystal structures and properties

Index

C

D

G

H

I

M

S

U

V

W

Common Trigonometric Functions

Right Triangles

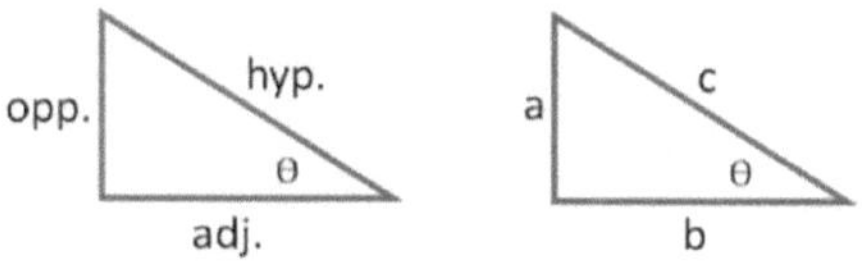

$$\sin\theta = \frac{opposite}{hypotenuse} = \frac{a}{c}$$

$$cos\,\theta = \frac{adjacent}{hypotenuse} = \frac{b}{c}$$

$$tan\,\theta = \frac{opposite}{adjacent} = \frac{a}{b}$$

$$a^2 + b^2 = c^2$$

$$tan\,\theta = \frac{\sin\theta}{\cos\theta}$$

Oblique Triangles

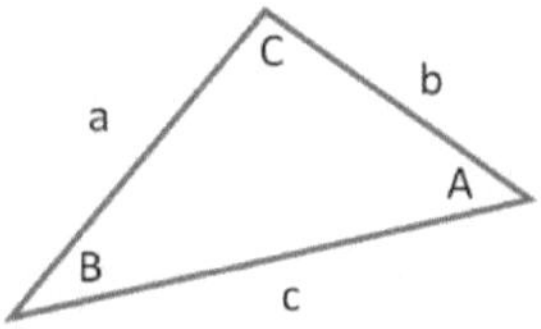

Law of Sines

$$\frac{\sin A}{a} = \frac{\sin B}{b} = \frac{\sin C}{c}$$

Law of Cosines

$$a^2 = b^2 + c^2\text{-}2bc\cos A$$
$$b^2 = a^2 + c^2\text{-}2ac\cos B$$
$$c^2 = a^2 + b^2\text{-}2ab\cos C$$

Radian to Degrees: $radians \times \frac{180}{\pi}$

Degrees to Radians: $degrees \times \frac{\pi}{180}$

Properties of Common Section Geometries

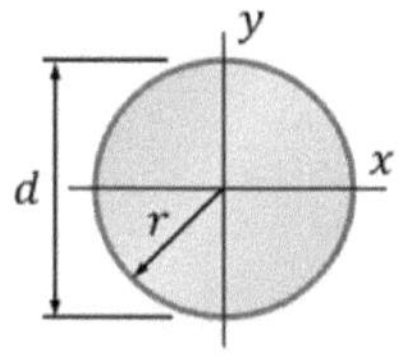

$$A = \pi r^2 = \frac{\pi}{4}d^2$$
$$I_x = I_y = \frac{\pi}{4}r^4$$
$$J = \frac{\pi}{2}(r_1^4 - r_2^4)$$

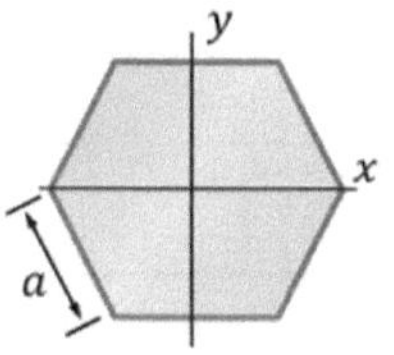

$$A = \frac{3\sqrt{3}}{2}a^2$$
$$I_x = I_y = \frac{5\sqrt{3}}{16}a^4$$
$$J = \frac{5\sqrt{3}}{8}a^4$$

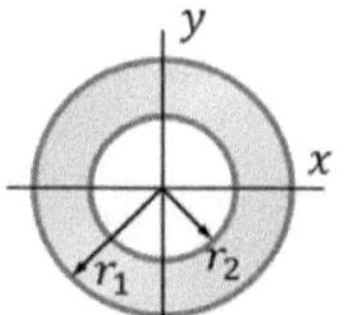

$$A = \pi(r_1^2 - r_2^2)$$
$$I_x = I_y = \frac{\pi}{4}(r_1^4 - r_2^4)$$
$$J = \frac{\pi}{2}(r_1^4 - r_2^4)$$

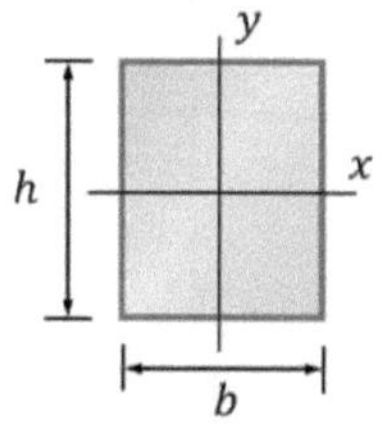

$$A = bh$$
$$I_x = \frac{bh^3}{12} \quad I_y = \frac{b^3h}{12}$$
$$J = \frac{bh}{12}(b^2 + h^2)$$

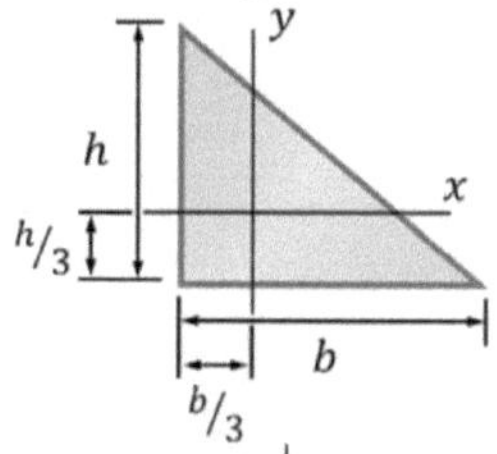

$$A = \frac{bh}{2}$$
$$I_x = \frac{bh^3}{36} \quad I_y = \frac{b^3h}{36}$$
$$J = \frac{bh}{36}(b^2 + h^2)$$

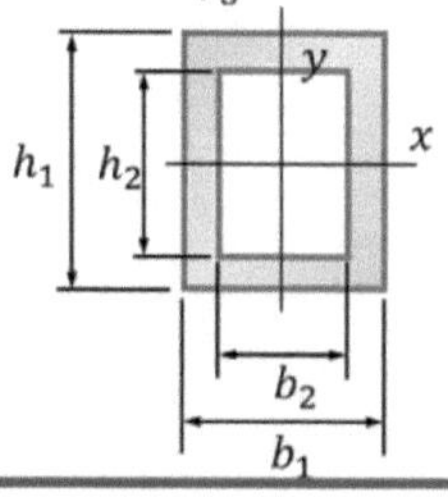

$$A = b_1h_1 - b_2h_2$$
$$I_x = \frac{bh_1^3}{12} - \frac{bh_2^3}{12}$$
$$I_y = \frac{b_1^3h}{12} - \frac{b_2^3h}{12}$$
$$J = I_x + I_y$$

Properties of Common Solids

Cuboid

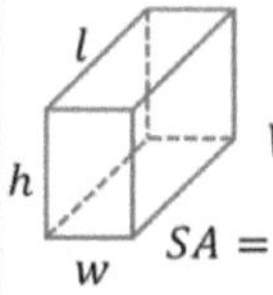

$$V = lwh$$
$$SA = 2hw + 2lw + 2hl$$

Right Angle Cone

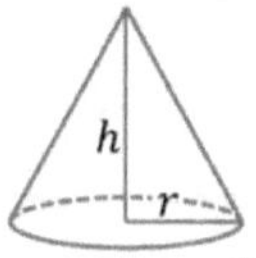

$$V = \pi r^2\frac{h}{3}$$
$$SA = \pi r\left(r + \sqrt{h^2 + r^2}\right)$$

Hexagonal Prism

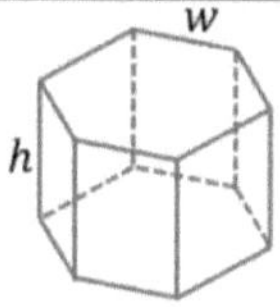

$$V = \frac{3\sqrt{3}}{2}w^2h$$
$$SA = 6wh + 3\sqrt{3}w^2$$

Cylinder

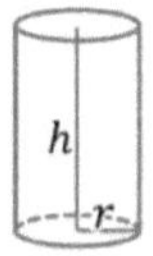

$$V = 2\pi r^2h$$
$$SA = 6ah + \frac{\sqrt{3}}{2}a^2$$

Right Angle Conical Frustum

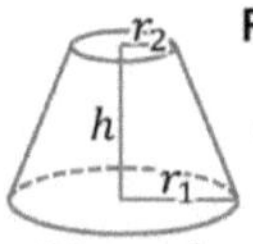

$$V = \frac{\pi h}{3}(r_1^2 + r_1r_2 + r_2^2)$$
$$SA = \pi(r_1 + r_2)\sqrt{(r_1 - r_2)^2 + h^2} + \pi r_1^2 + \pi r_2^2$$

Sphere

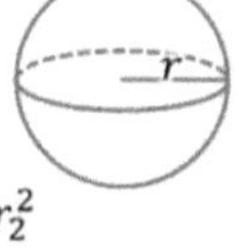

$$V = \frac{4}{3}\pi r^3$$
$$SA = 4\pi r^2$$

Drill Size Designations and Specifications

Drill Designation	Decimal Size	Tap Drill for Thead	Clearance Drill for Thread
80	.0135"		
79	.0145"		
1/64	.0156"		
78	.0160"		
77	.0180"		
76	.0200"		
75	.0210"		
74	.0225"		
73	.0240"		
72	.0250"		
71	.0260"		
70	.0280"		
69	.0292"		
68	.0310"		
1/32	.0313"		
67	.0320"		
66	.0330"		
65	.0350"		
64	.0360"		
63	.0370"		
62	.0380"		
61	.0390"		
60	.0400"		
59	.0410"		
58	.0420"		
57	.0430"		
56	.0465"		
3/64	.0469"	0-80	
55	.0520"		
54	.0550"		
53	.0595"	1-64, 72	
1/16	.0625"		
52	.0635"		0-80
51	.0670"		
50	.0700"	2-56, 64	
49	.0730"		
48	.0760"		1-64, 72
5/64	.0781"		
47	.0785"	3-48	
46	.0810"		
45	.0820"	3-56	
44	.0860"		
43	.0890"	4-40	2-56, 64
42	.0935"	4-48	
3/32	.0938"		
41	.0960"		
40	.0980"		
39	.0995"		
38	.1015"	5-40	
37	.1040"	5-44	3-48, 56
36	.1065"	6-32	
7/64	.1094"		
35	.1100"		
34	.1110"		
33	.1130"	6-40	
32	.1160"		4-40, 48
31	.1200"		
1/8	.1250"		
30	.1285"		5-40, 44
29	.1360"	8-32, 36	
28	.1405"		
9/64	.1406"		
27	.1440"		6-32, 40
26	.1470"		
25	.1495"	10-24	
24	.1520"		
23	.1540"		
5/32	.1562"		
22	.1570"		
21	.1590"	10-32	

Drill Designation	Decimal Size	Tap Drill for Thead	Clearance Drill for Thread
20	.1610"		
19	.1660"		
18	.1695"		8-32, 36
11/64	.1719"		
17	.1730"		
16	.1770"	12-24	
15	.1800"		
14	.1820"	12-28	
13	.1850"		
3/16	.1875"		
12	.1890"		
11	.1910"		
10	.1935"		
9	.1960"		10-24, 32
8	.1990"		
7	.2010"	1/4-20	
13/64	.2031"		
6	.2040"		
5	.2055"		
4	.2090"		
3	.2130"	1/4-28	
7/32	.2188"		
2	.2210"		12-24, 28
1	.2280"		
A	.2340"		
15/64	.2344"		
B	.2380"		
C	.2420"		
D	.2460"		
E	.2500"		
1/4	.2500"		
F	.2570"	5/16-18	1/4-20, 28
G	.2610"		
17/64	.2656"		
H	.2660"		
I	.2720"	5/16-24	
J	.2770"		
K	.2810"		
9/32	.2812"		
L	.2900"		
M	.2950"		
19/64	.2969"		
N	.3020"		
5/16	.3125"	3/8-16	
O	.3160"		
P	.3230"		5/16-18, 24
21/64	.3281"		
Q	.3320"	3/8-24	
R	.3390"		
11/32	.3438"		
S	.3480"		
T	.3580"		
23/64	.3594"		
U	.3680"	7/16-14	
3/8	.3750"		
V	.3770"		
W	.3860"		3/8-16, 24
25/64	.3906"	7/16-20	
X	.3970"		
Y	.4040"		
13/32	.4062"		
Z	.4130		
27/64	.4219"	1/2-13	
7/16	.4375"		
29/64	.4531"	1/2-20	7/16-14, 20
15/32	.4688"		
31/64	.4844"	9/16-12	
1/2	.5000"		
33/64	.5156"	9/16-18	1/2-13, 20
17/32	.5313"	5/8-11	

Drill Designation	Decimal Size	Tap Drill for Thead	Clearance Drill for Thread
35/64	.5469"		
9/16	.5625"		
37/64	.5781"	5/8-18	9/16-12, 1
19/32	.5938"		
39/64	.6094"		
5/8	.6250"		
41/64	.6406"		5/8-11, 1
21/32	.6562"	3/4-10	
43/64	.6719"		
11/16	.6875"	3/4-16	
45/64	.7031"		
23/32	.7188"		
47/64	.7344"		
3/4	.7500"		
49/64	.7656"	7/8-9	3/4-10, 1
25/32	.7812"		
51/64	.7969"		
13/16	.8125"	7/8-14	
53/64	.8281"		
27/32	.8438"		
55/64	.8594"		
7/8	.8750"	1-8	
57/64	.8906"		7/8-9, 1
29/32	.9062"		
59/64	.9219"	1-12	
15/16	.9375"	1-14	
61/64	.9531"		
31/32	.9688"		
63/64	.9844"	1 1/8-7	
1	1.000"		

Pipe Thread Designations (NPT)

Pipe Thread	Tap Drill (in)	Pipe Thread	Tap Drill (in)
1/8-27	11/32	1/2-14	23/32
1/4-18	7/16	3/4-14	59/64
3/8-18	37/64	1-11 1/2	1 5/32

Metric Thread Designations

Metric Thread	Tap Drill (mm)	Approx. Tap Drill (in)	Clearance Drill (mm)
M1.6 x 0.35	1.25	0.0492	1.8
M1.8 x 0.35	1.45	0.0571	2
M2 x 0.4	1.6	0.0630	2.4
M2.2 x 0.45	1.75	0.0689	2.8
M2.5 x 0.45	2.05	0.0807	2.9
M3 x 0.5	2.5	0.0984	3.4
M3.5 x 0.6	2.9	0.1142	3.9
M4 x 0.7	3.3	0.1299	4.5
M4.5 x 0.75	3.7	0.1457	5
M5 x 0.8	4.2	0.1654	5.5
M6 x 1	5	0.1969	6.6
M8 x 1.25	6.7	0.2638	9
M8 x 1	7	0.2756	9
M10 x 1.5	8.5	0.3346	11
M10 x 1.25	8.7	0.3425	11
M12 x 1.75	10.2	0.4016	13.5
M12 x 1.25	10.8	0.4252	13.5
M14 x 2	12	0.4724	15.5
M14 x 1.5	12.5	0.4921	15.5
M16 x 2	14	0.5512	17.5
M16 x 1.5	14.5	0.5709	17.5
M18 x 2.5	15.5	0.6102	20
M18 x 1.5	16.5	0.6496	20
M20 x 2.5	17.5	0.6890	22
M20 x 1.5	18.5	0.7283	22
M22 x 2.5	19.5	0.7677	24
M22 x 1.5	20.5	0.8071	24
M24 x 3	21	0.8268	26
M24 x 2	22	0.8661	26

NONMETAL	METALLOID
METAL	VERY RARE/SYNTHETIC*

Key		
Density (g/cc) →	7.87 26 Fe Iron 1535 2795	← Atomic Number
Melting Temp. (°C) →		← Melting Temp. (°F)

1 H rogen -434								
3 i ium 357	1.85 4 Be Beryllium 1278 2332							
11 a ium 208	1.74 12 Mg Magnesium 649 1200							
19 K ssium 147	1.54 20 Ca Calcium 839 1542	2.99 21 Sc Scandium 1539 2802	4.51 22 Ti Titanium 1660 3020	6.09 23 V Vanadium 1890 3434	7.14 24 Cr Chromium 1857 3375	7.44 25 Mn Manganese 1244 2271	7.87 26 Fe Iron 1535 2795	8.89 27 Co Cobalt 1495 2723
37 b dium 102	2.63 38 Sr Strontium 769 1416	4.47 39 Y Yttrium 1523 2773	6.51 40 Zr Zirconium 1852 3366	8.58 41 Nb Niobium 2468 4474	10.28 42 Mo Molybdenum 2617 4743	43 Tc Technetium	12.45 44 Ru Ruthenium 2310 4190	12.41 45 Rh Rhodium 1966 3571
55 s ium 83	3.65 56 Ba Barium 725 1337		13.31 72 Hf Halfnium 2150 3902	16.68 73 Ta Tantalum 2996 5425	19.26 74 W Tungsten 3407 6165	21.03 75 Re Rhenium 3180 5756	22.61 76 Os Osmium 3045 5513	22.65 77 Ir Iridium 2410 4370
87 r cium	5.50 88 Ra Radium 700 1292		104 Rf Rutherfordium	105 Db Dubnium	106 Sg Seaborgium	107 Bh Bohrium	108 Hs Hassium	109 Mt Meitnerium

LANTHANIDE SERIES

6.16 57 La Lanthanum 920 1688	6.77 58 Ce Cerium 798 1468	6.48 59 Pr Praseodymium 931 1708	7.00 60 Nd Neodymium 1010 1850	61 Pm Promethium	7.54 62 Sm Samarium 1072 1962

ACTINIDE SERIES

10.07 89 Ac Actinium 1047 1917	11.72 90 Th Thorium 1750 3182	15.37 91 Pa Protactinium 1554 2829	18.97 92 U Uranium 1132 2070	20.48 93 Np Neptunium 640 1184	19.74 94 Pu Plutonium 641 1186

								(G) H Heli -272
			2.46 5 B Boron 2300 4172	3.51 6 C Carbon 3550 6422	(G) 7 N Nitrogen -210 -346	(G) 8 O Oxygen -218 -361	(G) 9 F Fluorine -220 -363	(G) N Nec -249
			2.70 13 Al Aluminum 660.5 °C	2.33 14 Si Silicon 1410 2570	1.82 15 P Phosphorous 44 111	2.06 16 S Sulfur 113 235	(G) 17 Cl Chlorine -35 -30	(G) A Arg -189
8.91 28 Ni Nickel 1453 2647	8.92 29 Cu Copper 1084 1982	7.14 30 Zn Zinc 420 787	5.91 31 Ga Gallium 30 86	5.32 32 Ge Germanium 937 1719	5.72 33 As Arsenic 613 1135	4.82 34 Se Selenium 217 423	(L) 35 Br Bromine -7 19	(G) K Kryp -157
12.02 46 Pd Palladium 1552 2826	10.49 47 Ag Silver 962 1763	8.64 48 Cd Cadmium 321 610	7.31 49 In Indium 156 313	7.29 50 Sn Tin 232 450	6.69 51 Sb Antimony 631 1167	6.25 52 Te Tellurium 450 841	(L) 53 I Iodine 114 236	(G) X Xen -112
21.45 78 Pt Platinum 1772 3222	19.32 79 Au Gold 1064 1948	(L) 80 Hg Mercury -39 -38	11.85 81 Tl Thallium 304 578	11.34 82 Pb Lead 328 622	9.80 83 Bi Bismuth 271 521	9.20 84 Po Polonium 254 489	85 At Astatine	(G) R Rad -71
110 Ds Darmstadium	111 Rg Roentgenium	112 Cn Copernicium	113 Nh Nihonium	114 Fl Flerovium	115 Mc Moscovium	116 Lv Livermorium	117 Ts Tennessine	O Ogan

5.25 63 Eu Europium 822 1512	7.89 64 Gd Gadolinium 1311 2392	8.25 65 Tb Terbium 1360 2480	8.56 66 Dy Dysprosium 1409 2568	8.78 67 Ho Holmium 1470 2678	9.05 68 Er Erbium 1522 2772	9.32 69 Tm Thulium 1545 2813	6.97 70 Yb Ytterbium 824 1515	9.84 L Lute 1656

95 Am Americium	96 Cm Curium	97 Bk Berkelium	98 Cf Californium	99 Es Einsteinium	100 Fm Fermium	101 Md Mendelevium	102 No Nobelium	L Lawre

Printed by Libri Plureos GmbH in Hamburg,
Germany

9 781260 122312